Potato Breeding: Theory and Practice

John E. Bradshaw

Potato Breeding: Theory and Practice

Springer

John E. Bradshaw
Edinburgh, UK

ISBN 978-3-030-64413-0 ISBN 978-3-030-64414-7 (eBook)
https://doi.org/10.1007/978-3-030-64414-7

This Springer imprint is published by the registered company Springer Nature Switzerland AG
The registered company address is: Gewerbestrasse 11, 6330 Cham, Switzerland

Preface

The potato (*Solanum tuberosum*) is the world's fourth most important food crop after maize, rice and wheat with 377 million tonnes fresh-weight (FW) of tubers produced in 2016 from 19.2 million hectares of land, in 163 countries, giving a global average yield of 19.6 t ha^{-1} (http://faostat.fao.org). About 62% of production (234 million tonnes) was in Asia (191), Africa (25) and Latin America (18) as a result of steady increases in recent years, particularly in China and India. Indeed, China (99 million tonnes) is now the number one potato producer in the world and India (44) is second, with the Russian Federation (31) third, Ukraine (22) fourth and the USA (20) fifth. As a major food crop, the potato has an important role to play in the United Nations "2030 Agenda for Sustainable Development" which started on 1 January 2016 (http://faostat.fao.org). The agenda includes 17 goals, the second of which is to end hunger, achieve food security and improved nutrition, and promote sustainable agriculture. By 2030, the aim of the agenda is to "ensure access by all people, in particular the poor and people in vulnerable situations, including infants, to safe, nutritious and sufficient food all year round". By then, the world population is expected to reach 8.5 billion and continue to increase to 9.7 billion in 2050. For potatoes, the need is to increase production and improve nutritional value to alleviate micronutrient deficiencies ('hidden hunger') during a period of climate change. A key aspect of this endeavour will be the breeding of new cultivars for a wide range of target environments and consumers.

Potato breeding during the twentieth century involved planned artificial hybridizations followed by multi-stage, multitrait selection over as many as eight clonal generations.

From the 1960s, programmes typically started each year with as many as 100,000 seedlings from 200 to 300 crosses (conventional breeding). The two main weaknesses of such breeding are the number of clonal generations required to select a new cultivar and the inability of intense early-generation selection to affect most economically important traits, which are quantitative in nature. Nevertheless, continued progress worldwide in adapting potatoes to new environments, farming practices and uses (markets) was made by cycles of such hybridization and selection, usually among the developing elite germplasm. Sometimes new traits were required, such as resistances to emerging pest and disease problems, and sometimes broader genetic bases were sought to deal with perceived plateaus in progress for traits such as yield. Thus, the twentieth century saw the use of potato landraces and wild relatives in introgression breeding and in base broadening. However, a striking feature of breeding in the twentieth century was the longevity of use of cultivars that became widely grown (e.g. Maris Piper in the UK). It can therefore be argued that new cultivars did not contribute as much as might have been expected to the improvement of potato crops worldwide.

Ways to make potato breeding faster, more efficient and more effective have become available since the 1990s (e.g. progeny testing, estimated breeding values from pedigree information, diagnostic molecular markers for marker-assisted introgression and selection, SNP arrays and genotyping by sequencing for genomic selection, *Agrobacterium*-mediated genetic transformation, and site-directed DNA sequence modifications including gene editing). They either have been, or are being, integrated into conventional breeding programmes of the kind described and discussed over 25 years ago by Bradshaw and Mackay (1994). However, it will be argued in this book, that we need to rethink the way that potato breeding is done in order to make the best use of advances in technology and genetic knowledge following the publication of the potato genome sequence in Nature on 14 July 2011 (Potato Genome Sequencing Consortium 2011). The challenge for a new generation of potato breeders and biotechnologists is to design breeding programmes that integrate marker-assisted selection of specific alleles, genomic selection of unspecified alleles and phenotypic selection, having decided when a gene editing or transgenic approach is more appropriate, given consumer acceptability of the latter. It therefore seems timely to re-examine options for breeding vegetatively propagated tetraploid cultivars of potato and to compare them with breeding potatoes for TPS (true potato seed) propagation, including diploid F_1 hybrid breeding. It will be important to consider long-term crop improvement (increasing the frequencies of desirable combinations of alleles over sexual generations) as well as short-term cultivar production (combining alleles in a single genotype). Indeed, the former is the driver of progress in the latter. It will also be important to consider the effective use of all of the germplasm available to potato breeders, namely modern cultivars, landraces and wild relatives; and the need to go beyond the barriers of sexual hybridization through genetic transformation.

The aim of the book is to provide more information and greater detail on the theory and practice of potato breeding than I have found possible in review articles (e.g. Bradshaw 2009; Bradshaw 2017) and book chapters (e.g. Bradshaw and Mackay 1994; Bradshaw 2007a, 2007b; Bradshaw and Bonierbale 2010). The book updates the information provided by Bradshaw and Bonierbale (2010), incorporates the theory from Bradshaw and Mackay (1994) and my book on plant breeding (Bradshaw 2016), and extends the theory to cover genomic selection and diploid F_1 hybrid breeding. It is divided into three parts and comprises nine chapters in what I regard as a logical order for considering scientific potato breeding. Part I deals with the history of potato improvement from domestication to present-day cultivars and potato genetics: in other words, the necessary background information and knowledge for designing a potato breeding programme in the twenty-first century. It includes an explanation of the major revision of the taxonomy of wild tuber-bearing *Solanum* species and the modifications to the classification of cultivated potatoes, both of which breeders need to appreciate. Part II deals with deciding breeding objectives and translating them into selection criteria in the order in which I think they need to be incorporated into the programme design: first, increasing potato yields as consideration of this determines the environments in which the breeding programme is conducted; second, improving potato quality as this needs to be done by assessing the limited number of tubers from the yield trials, at harvest, after storage and after cooking; and third, improving resistance to the most important of the many diseases and pests of potatoes, in laboratory, glasshouse and field tests with the limited number of seed tubers not required for the yield trials. Part III deals with the actual breeding methods and germplasm available for achieving the desired objectives: first, the use of landraces and wild relatives of potato in introgression breeding, base broadening and population improvement; second, breeding clonally propagated cultivars as the way to deliver potato improvement to farmers' fields; third, breeding potato cultivars that can be propagated through true potato seed (TPS) as an alternative way to deliver potato improvement to farmers' fields; and fourth, gene editing and genetic transformation as ways of making further improvements to already successful and widely grown cultivars.

In writing the book, I have drawn on my experiences over a 20-year period as a potato breeder and geneticist at the former Scottish Crop Research Institute in Dundee (now the James Hutton Institute). Throughout this time, I benefited from discussions with colleagues and the wider potato breeding community, particularly members of EAPR (The European Association for Potato Research) and EUCARPIA (The European Association for Research on Plant Breeding). I also benefited from participation in teaching, working groups and editorial work. In writing another book for Springer, I have once again received much help and encouragement from Kenneth Teng, and also appreciate the help from Rahul Sharma and Raja Dharmaraj of Springer during the production stage. The book was completed during our coronavirus lockdown from 24 March to 28 May 2020, during which time the support of my partner Shiona Mackie was much appreciated. I hope that the book will of value to a new generation of potato breeders as they face the challenge of feeding a growing world population during a period of climate change. Globally, a large number of breeding programmes will be required to produce the necessary range of adapted cultivars for the wide range of agro-ecological zones and regions within zones in which potatoes are grown, and the different potato characteristics needed for the various end uses of potato crops, namely staple food, vegetable for cash and crop for processing. For any given programme, the key to success will be to focus on the limited number of objectives that will have most impact on potato improvement, and to choose the right germplasm and breeding method to achieve those objectives. I wish the new generation of potato breeders every success in their endeavours.

Edinburgh, Scotland
20 July 2020

John E. Bradshaw

References

Bradshaw JE (2007a) Potato-breeding strategy. In: Vreugdenhil D (ed) Potato biology and biotechnology advances and perspectives. Elsevier, Oxford, pp 157–177
Bradshaw JE (2007b) Breeding potato as a major staple crop. In: Kang MS, Priyadarshan PM (eds) Breeding major food staples. Blackwell Publishing, Oxford, pp 277–332
Bradshaw JE (2009) A genetic perspective on yield plateau in potato. Potato J 36:79–94
Bradshaw JE (2016) Plant breeding: past, present and future. Springer, Switzerland
Bradshaw JE (2017) Review and analysis of limitations in ways to improve conventional potato breeding. Potato Res 60:171–193
Bradshaw JE, Mackay GR (1994) Breeding strategies for clonally propagated potatoes. In: Bradshaw JE, Mackay GR (eds) Potato genetics. CAB International, Wallingford, pp 467–497
Bradshaw JE, Bonierbale M (2010) Potatoes. In: Bradshaw JE (ed) Root and tuber crops, handbook of plant breeding, vol 7. Springer, New York, pp 1–52
Potato Genome Sequencing Consortium (2011) Genome sequence and analysis of the tuber crop potato. Nature 475:189–195

Contents

Part II Breeding Objectives

Part III Breeding Methods

Part I
Introduction

Chapter 1
Domestication to Twenty-First-Century Potato Cultivars

1.1 Introduction

1.1.1 Global Importance of the Potato

In 2016, the potato (*Solanum tuberosum*) was the world's fourth most important food crop after maize, wheat and rice (Table 1.1), with 377 million tonnes fresh-weight (FW) of tubers produced from 19.2 million hectares of land, in 163 countries (http://faostat.fao.org). Although the increases in production and yield of potatoes from 1966 to 2016 were not as great as those for the cereals, in 2016 the average yield of potatoes on a dry-weight basis compared favourably with the three cereals. Furthermore, 62% of potato production (234 million tonnes) was now in Asia (191), Africa (25) and Latin America (18) as a result of steady increases in recent years, particularly in China and India. Indeed, China (99 million tonnes) is now the number one potato producer in the world and India (44) is second, with the Russian Federation (31) third, Ukraine (22) fourth and the USA (20) fifth. Perhaps then it was not surprising that the United Nations named 2008 as the International Year of the Potato in recognition of its contribution as a major food staple to their Millennium Development Goals of providing food security and eradicating poverty. The potato also has an important role to play in the United Nations '2030 Agenda for Sustainable Development' which started on 1 January 2016 (http://faostat.fao.org). The agenda includes 17 goals, the second of which is to end hunger, achieve food security and improved nutrition, and promote sustainable agriculture. By 2030, the aim of the agenda is to 'ensure access by all people, in particular the poor and people in vulnerable situations, including infants, to safe, nutritious and sufficient food all year round'. By then, the world population is expected to reach 8.5 billion and continue to increase to 9.7 billion in 2050.

Where potatoes are a staple food, the need is to increase production and improve nutritional value to alleviate micronutrient deficiencies ('hidden hunger') during a period of climate change. In countries where food security has been achieved and

J. E. Bradshaw, *Potato Breeding: Theory and Practice*,
https://doi.org/10.1007/978-3-030-64414-7_1

Table 1.1 Crop production in 1966 and 2016 (http://faostat.fao.org), on a fresh-weight basis except for 2016 yields in parentheses which have been corrected for moisture content assuming cereals are 15% water and potatoes are 80% water

Crop	Production million tonnes		Area million hectares		Yield t/ha	
	1966	2016	1966	2016	1966	2016
Maize	245.6	1060.1	111.2	188.0	2.21	5.64 (4.79)
Wheat	303.8	749.5	215.8	220.1	1.41	3.41 (2.90)
Rice	261.2	741.0	125.7	159.8	2.08	4.64 (3.94)
Potato	282.0	376.8	21.5	19.2	13.1	19.6 (3.92)

the potato is grown as a vegetable, the need is to increase potato usage in an economically and environmentally sustainable way. The same is true for potatoes destined for processing. Finally, the potato is even being considered for human life support in space (Wheeler 2009). In all of these endeavours, a key aspect will be breeding new cultivars for a wide range of target environments and consumers.

1.1.2 Potato Breeding

The ultimate success of a new breeding programme depends on starting with the right objectives and germplasm. Choose the wrong objectives and nobody will want to grow your new cultivars which will have taken you much time, effort and resources over a period of some 12 years. Choose the wrong germplasm and you will fail to achieve your objectives despite all of your hard work. In this chapter, we are going to review the genetic improvement of potatoes from domestication to the present day. We will then be able to ask and answer the questions that will enable us to decide breeding objectives and starting germplasm. Only then can we get down to the details of available breeding methods.

1.2 Wild Relatives as Food

Wild tuber-bearing *Solanum* species grow in a wide range of habitats in 16 countries from the south-west of the USA (38°N), through Central America and the Andes, to Argentina and adjacent Chile (41°S) (Spooner and Hijmans 2001). The tuber is a subterranean swollen stem which evolved to survive from season to season as a dormant storage organ, and in which the form of energy storage is almost entirely starch. Today, we know that our cultivated potatoes also contain significant amounts of protein, minerals, vitamins, micronutrients and phytonutrients which include antioxidants (Singh and Kaur 2016). We also know that the journey from gathering wild tubers to cultivation and domestication started early in the human colonization of the Americas.

The earliest evidence of wild potato use comes from the extremes of their distribution in North America and Chile. Wild potato remains were found in a late Pleistocene settlement in south-central Chile dated to around 12,500 years before present (Ugent et al. 1987; Moseley 2001). Then, more recently, Louderback and Pavlik (2017) extracted well-preserved starch granules from ground stone tools found at North Creek Shelter, southern Utah and dated to between 10,900 and 10,100 years before present. The granules were identified as those of *S. jamesii*, which is known to be highly nutritious, having twice the protein, zinc and manganese content of cultivated *S. tuberosum* and three times the calcium and iron content. Thus, a summer-active and highly productive herbaceous perennial would have provided a reliable, year-round source of carbohydrate and minerals that significantly improved dietary quality.

1.3 Domestication and Glycoalkaloids

The taxonomy of the wild tuber-bearing *Solanum* species is complicated and has undergone major revisions as explained by Spooner (2016). Spooner et al. (2014) recognized 107 species, all classified as *Solanum* section *Petota* (tuber-bearing species) and partitioned into three nuclear clades. The three nuclear clades were similar to the four clades based on plastid DNA restriction site data but with plastid clades 1 and 2 merged, that is, nuclear clades are 1 + 2, 3 and 4. Li et al. (2018; figure 4) have provided the most recent and comprehensive taxonomy based on a phylogenetic analysis of 201 accessions of *Solanum* section *Petota* species, comprising 146 wild accessions and 21 diploid cultivated accessions from clade 4, 14 wild accessions from clade 1 + 2, 18 wild accessions from clade 3 and 2 non-tuber-bearing outgroup species (*S. etuberosum* and *S. palustre*). The phylogenetic trees were generated from 66,666 high-quality SNPs (single-nucleotide polymorphisms), identified from whole-genome sequencing, using singular value decomposition (SVD) quartets (two other methods gave similar results). All analytical methods strongly corroborated the partitioning of section *Petota* into outgroup, clade 1 + 2, clade 3 and clade 4 (by far the largest). They separated members of clade 4 into subclasses of cultivated, wild north (Peru) and wild south (Argentina, Bolivia and Chile). They corroborated the northern members of the *S. brevicaule* complex as the progenitors of cultivated potato (Spooner et al. 2005a). They failed to separate cultivated *S. tuberosum* subspecies *phureja* and *stenotomum* into clades, supporting placing these names into synonymy based on prior microsatellite data (see below). They support much of the recent synonymy of the wild species in the *S. brevicaule* complex and elsewhere in clade 4. Finally, they place *Solanum verrucosum*, the sole A-genome diploid species from Mexico, firmly in the southern South American subclade of clade 4. The reader is referred to the original paper for all of the detail (Li et al. 2018). However, some genebanks still use the classification of Hawkes (1990) which recognized 219 wild tuber-bearing species, arranged into 19 series of subsection *Potatoe* of section *Petota* of subgenus *Potatoe* of genus *Solanum*. Hence, breeders need to be aware of this older classification.

As just mentioned, it was Spooner et al. (2005a) who provided molecular taxonomic evidence for a single domestication in the highlands of southern Peru, from the northern group of members of the *S. brevicaule* complex of diploid species typified by *S. bukasovii* (now *S. candolleanum*: Spooner et al. 2014). However, it has proved difficult to date the earliest cultivation of potatoes. Hawkes (1990) concluded that the potato is an ancient domesticate based on preserved food plant remains found at various excavated sites on the coast of Peru and one site in the high Chilca canyon, south of Lima (Engel 1970). The Oxford Radiocarbon Accelerator dated the tuber remains found by Engel to about 7000 years before present. Rumold and Aldenderfer (2016) provided direct microbotanical evidence for the early use of potatoes at Jiskairumoko, an early village site in the western Titicaca Basin of the high south-central Andes (3890 m above sea level) dating to the Late Archaic to Early Formative periods (3400–1600 BCE, i.e. 5420–3620 years before present). Of 141 starch micro remains recovered from 14 ground stone tools, 50 were identified as consistent with cultivated or domesticated potatoes, based on reference to published materials and a study of wild and cultivated potato starch morphology. However, the authors could not say with certainty that the 50 micro remains were from cultivated/domesticated potatoes and not from wild potato species. Nevertheless, the authors were able to cite references to the evidence for small-scale farming being in place by 1600 BCE. In other words, domestication can probably be considered complete by this date. Moseley (2001) speculated on the process of domestication before the spread of potato cultivation throughout South America. Early Andean cultures were likely to be have been 'vertical' ones, moving up and down the mountains with the changing seasons. It is likely that the herding of animals and the domestication of llama and alpaca preceded the settled cultivation of crops. Seasonal visits to favoured sites for food plants would mean that domestication would proceed gradually, with favoured types being re-planted to ensure harvests at subsequent re-visits.

Rumold and Aldenderfer (2016) also speculated that grinding could have removed toxic steroidal glycoalkaloids from potato tubers, thus making them safe to eat. However, it is usually assumed that lower levels of glycoalkaloids were achieved through selection for less bitter-tasting tubers during domestication (Simmonds 1995). Today tubers with concentrations of steroidal glycoalkaloids above 20 mg 100 g^{-1} FW are considered unsafe for human consumption, resulting in symptoms typically associated with food poisoning (Friedman and Levin 2016). Interestingly, Johns and Alonso (1990) found that some genebank accessions of *S. bukasovii* (now *S. candolleanum*: Spooner et al. 2014), now considered the most likely progenitor of cultivated potatoes, had tuber glycoalkaloid levels (9–29 mg 100 g^{-1} FW) which were consistently close to the levels found in many clones of cultivated diploid potatoes (*S. tuberosum* Stenotomum Group). They concluded that exploitation and domestication of this species would have required little or no selection for lower glycoalkaloid content, unlike their samples of other former candidates for domestication, namely *S. canasense*, *S. leptophyes* and *S. sparsipilum*, with levels of 42–164 mg 100 g^{-1} FW. Johns and Alonso (1990) also found that *S. jamesii* had high levels of glycoalkaloids (115–128 mg 100 g^{-1} FW), so perhaps grinding was used for detoxification of potatoes at North Creek Shelter (see previous section). Whatever the method of detoxification, it seems fair to credit the early Andean farmers with making the potato an edible crop. In fact, Hardigan et al. (2017) have provided evidence from a *Solanum* section *Petota* diversity panel for farmer selection at two loci in the steroidal glycoalkaloids pathway, the squalene synthase locus (*SQS*) and the *GLYCOALKALOID METABOLISM* 9 locus (*GAME9*).

Finally, Hawkes (1990) concluded that Andean farmers either baked their fresh potatoes in the embers of a fire or cooked them in an earth oven on hot stones, because after domestication the potato was grown for at least four millennia prior to the development of ceramics. We now know that the potato needs to be cooked because of the indigestibility of its ungelatinized starch (Burton 1989). Today such cooking is frequently by baking, boiling, steaming, roasting, deep-fat frying or microwave cooking, although in the Andes a broad diversity of additional preparation methods is still employed.

1.4 Cultivated Potatoes of South America

1.4.1 Classification of Cultivated Potatoes

The result of domestication was a diploid cultigen *S. tuberosum* Stenotomum Group ($2n = 2x = 24$) from which all other cultivated potatoes were derived (Fig. 1.1). The widely accepted classification of cultivated potatoes is that of Dodds (1962) as modified by Spooner et al. (2007) using molecular data, although other schemes can be found in the literature. Dodds (1962) classified cultivated potatoes into five informal groups within one species (*S. tuberosum*) in which Andigena (tetraploid), Chaucha (triploid), Phureja (diploid) and Tuberosum (tetraploid) groups were derived from Stenotomum (diploid). Chaucha is the triploid hybrid of Stenotomum and Andigena and like Stenotomum is confined to the central Andes of Peru and Bolivia. Phureja was selected from Stenotomum by Andean farmers for lack of tuber dormancy and faster tuber development so that they could grow up to three crops a year in the lower, warmer, eastern valleys of the Andes. Phureja potatoes were therefore able to spread into northern Ecuador, Colombia and Venezuela and are the second most widely cultivated type in South America after Andigena (Hawkes 1990). Interestingly, Ghislain et al. (2006) found that 32 out of 102 accessions of Phureja in the CIP (International Potato Centre) collection of landraces were triploid or tetraploid, not diploid, in agreement with Hawkes (1990) that not all Phureja potatoes are diploid. This provides justification for not using ploidy as a species criterion. Goniocalyx (diploid) has been recognized as a northern subgroup of Stenotomum, noted for tubers with a bright yellow flesh (Hawkes 1990). Andigena potatoes are grown throughout the upland Andes of South America, presumably because farmers found the tetraploid superior to the diploids for yield and other traits. Tuberosum potatoes were selected from Andigena types for tuber production in long days in coastal Chile and are referred to as Chilean Tuberosum. They are a genetically

S. brevicaule complex of diploid species e.g.
S. bukasovii (now *S. candolleanum*) (2*x*)

(wild) ↓ (domestication in southern Peru)

S. acaule (4*x*) × *S. tuberosum* Stenotomum Gp (2*x*) × *S. megistacrolobum* (2*x*)

(wild) ↓ (cultivated) | | ↓ (wild)

S. juzepczukii (3*x*) | ↓ *S. ajanhuiri* (2*x*)

↓ ↓ Phureja Gp (mostly 2*x*)

unreduced gamete(s)

× ← Andigena Gp (4*x*) × Stenotomum Gp (2*x*)

↓ Wild species? →↓

Andigena Gp (4*x*)

↓ Wild species? →↓ ↓

S. curtilobum (5*x*) Tuberosum Gp (4*x*) (Chilean) Chaucha Gp (3*x*)

S. tuberosum Andigenum Gp (Andigena, Chaucha, Phureja, Stenotomum)

S. tuberosum Chilotanum Gp (Chilean Tuberosum)

Fig. 1.1 Origin of cultivated groups (Gp = Group) of *S. tuberosum* (Dodds 1962; Spooner et al. 2007) and cultivated species with bitter taste and frost tolerance (*S. ajanhuiri*, *S. juzepczukii* and *S. curtilobum*) (2, 3, 4 and 5*x* = diploid, triploid, tetraploid and pentaploid) (modified from Bradshaw 2019 with permission)

distinct group of potatoes with a different cytoplasm to Andigena potatoes (Raker and Spooner 2002; Hosaka 2004).

Spooner et al. (2007) also regarded Andigena, Chaucha, Phureja, Stenotomum and Tuberosum as a single species *S. tuberosum*, but now divided into just two cultivar groups. These are the Andigenum Group of upland Andean landraces containing diploids, triploids and tetraploids, and the Chilotanum Group of lowland tetraploid Chilean landraces. Spooner et al. (2007) also recognized the three frost-tolerant species cultivated in the Andes as separate hybrid species derived from crosses between domesticates and wild relatives (Fig. 1.1).

1.4.2 *Frost-Tolerant Species and chuño Production*

Three frost-tolerant species are cultivated in the Andes: *S. ajanhuiri* (diploid), *S. juzepczukii* (triploid) and *S. curtilobum* (pentaploid) (Fig. 1.1). These 'bitter potatoes' are grown at high altitudes (up to 4500 m for *S. juzepczukii*) in the central Andes of Peru and Bolivia (Hawkes 1990). *S. ajanhuiri* is the hybrid of *S. tuberosum* Group Stenotomum with the wild frost-resistant diploid species *S.*

megistacrolobum, *S. juzepczukii* is the hybrid of *S. tuberosum* Group Stenotomum with the wild frost-resistant tetraploid species *S. acaule* and *S. curtilobum* is the hybrid between an unreduced gamete of triploid *S. juzepczukii* and a normal gamete of *S. tuberosum* Group Andigena. In the high, cold mountains, an ancient method of potato preservation was chuño production where potatoes were in effect freeze-dried using the very low night temperatures and intense sunlight of the day (Hawkes 1990). The frozen potatoes were trampled by foot to eliminate any remaining water, and then washed and dried. The washing removed toxic glycoalkaloids. The outcome was a dehydrated product that could be stored over several years and used when fresh potatoes were scarce. Chuño production has continued in use to the present day, yielding a product highly prized by the peoples of the Andes.

1.4.3 Origin of Tetraploid Potatoes

First, we need to consider how *S. tuberosum* Group Andigena arose from Group Stenotomum. Sukhotu and Hosaka (2006) concluded from chloroplast and nuclear DNA markers that Group Andigena arose from Group Stenotomum through sexual polyploidization from unreduced gametes many times at many places in the fields of Group Stenotomum. This would explain the chromosome behaviour and tetrasomic inheritance of tetraploid *S. tuberosum*, the details of which have been revealed using a high-density linkage map in a biparental mapping population (Bourke et al. 2015). In other words, tetraploid *S. tuberosum* can be regarded as the autotetraploid of diploid Group Stenotomum. The tetraploids were subsequently modified through occasional selection by Andean farmers of natural hybrids with neighbouring wild species to give present-day Group Andigena. Scurrah et al. (2008) demonstrated that closely related species growing around farmers' fields can hybridize with Group Andigena and that some hybrid progeny would be selected by present-day Andean farmers. Furthermore, Hardigan et al. (2017) have provided data from a *Solanum* section *Petota* diversity panel that suggests that wild *Solanum* species assisted the spread of cultivated potatoes by transmitting alleles for tolerance of new ecological factors, enabling colonization of non-native habitats as the cultivated potatoes migrated south following domestication and polyploidization.

Second, we need to consider how *S. tuberosum* Group Tuberosum arose from Group Andigena. We need to appreciate that there are six distinct types of cytoplasm in potato, namely M, P, A, W, T and D as determined by Hosaka and Sanetomo (2012). For now, it is sufficient to say that P (Phureja) and A (Andigena) are subcategories of M and T (Tuberosum) and D (wild species *S. demissum*) are subcategories of W. Hosaka (2004) suggested that Chilean Tuberosum (T) cytoplasm is derived from the southern wild species *S. tarijense*. Hence, Group Tuberosum is not simply Group Andigena potatoes that were selected to tuber in long days. Furthermore, Spooner et al. (2007) showed that the T cytoplasm is found at low frequency in Andean landraces, including some diploids, indicating that the T cytoplasm moved northwards as well as becoming predominant in Chilean germplasm. Nevertheless, as mentioned earlier, it is clear that the long-day adapted landraces of coastal Chile are genetically distinct from the short-day adapted ones of the Andes (Raker and Spooner 2002). Although uncertainty remains about the exact origin of Chilean Tuberosum, Hardigan et al. (2017) have provided a likely scenario based on Bolivian and Argentinian species alleles in Chilean-derived Tuberosum, namely that Andigena tetraploids interbred with wild species en route to their eventual destination in southern Chile where long-day adaptation was required for tuberization.

There is also uncertainty as to why tetraploid potatoes were superior to their diploid ancestors, and hence were selected in preference to them to become the major cultivated types in South America. One possible clue comes from the work of Stupar et al. (2007). They developed a synthetic autopolyploid series in potato (primarily Group Phureja) that included one monoploid (1*x*) clone, two diploid (2*x*) clones and one tetraploid (4*x*) clone, in order to explore phenotypic and transcriptomic (about 9000 genes) changes associated with autopolyploidization. Interestingly, the diploid plants were the most vigorous and generated the greatest biomass with the monoploid inferior to both the diploids and the tetraploid. However, the diploid and tetraploid plants had similar gene expression patterns. Therefore, the eventual superiority of tetraploid potatoes may have come from their increased potential for heterozygosity rather than polyploidy per se, and this has implications for future breeding.

1.4.4 Reproductive Biology of Potatoes

Potatoes, like their ancestral wild species, reproduce by sexual means through botanical seed and also by producing tubers. They flower and set true seed in berries after natural pollination by insects capable of buzz pollination (e.g. some bee species), which releases pollen from their poricidal anthers (Scurrah et al. 2008). Outcrossing is enforced in cultivated and most wild diploid species by a single *S*-locus, multiallelic, gametophytic self-incompatibility system (Dodds 1965). While self-incompatibility does not operate in tetraploid *S. tuberosum*, 40% (range 21–74%) natural cross-pollination was estimated to occur in Group Andigena in the Andes (Brown 1993) and 20% (range 14–30%) in an artificially constructed Andigena population (Glendinning 1976). This level of cross-pollination is sufficient to maintain genetic variation and heterozygosity in a population as shown by the early population geneticists. The stable equilibrium for the simple situation of just two alleles, *A* and *a*, at a locus in a tetraploid population which is partly self-pollinated was derived by Haldane (1930). The algebra is quite straightforward but very tedious and hence will not be repeated here. One starts with the five possible genotypes and their frequencies at equilibrium and then derives the five genotype frequencies in the next generation, based on a constant probability s that any plant will be self-fertilized and a probability $1\text{-}s$ ($= \lambda$ in Haldane's paper) that it will cross with some plant chosen at random from the population. Chromosomal segregation is assumed (i.e. no double reduction). As there is no change in frequency from one generation to the next at equilibrium, the relevant equations can be solved to express the five genotype frequencies in terms of s (λ in Haldane's paper) and the ratio of the frequency of A (p) to a (q) ($p/q = u$ in Haldane's paper). The results for $s = 0.2$, 0.4, 0.6 and 0.8 are given in Table 1.2 for $p = q = ½$. It can be seen that with 40% ($s = 0.6$) cross-pollination the *AAaa* genotype is most frequent whereas with 20% ($s = 0.8$) cross-pollination *AAAA* and *aaaa* are the most frequent genotypes, followed by *AAaa*. If an equilibrium population is inbred further by self-pollination of all individuals each generation, inbreeding depression occurs for traits where non-additive genetic effects are in the same positive direction over loci; for example, for a single locus affecting yield: if *AAAA* = *AAAa* = *AAaa* = *Aaaa* = 20 t/ha and *aaaa* = 10 t/ha, the population yield will decline from 19.375 to 18.186 to 17.1755 to 15 t/ha as s increases from 0 to 0.6 to 0.8 to 1.

Sexual reproduction therefore creates an abundance of diversity by recombining the variants of genes that arose by mutation, and as we have just seen, potatoes are therefore highly heterozygous individuals that display inbreeding depression on selfing. The genetically unique seedlings that grow from true seeds produce tubers that can be replanted as seed tubers and hence distinct clones can be established and maintained by asexual (vegetative) reproduction. Thus, the reproductive biology of potatoes was ideal for creating and maintaining variation. As a consequence, farmers were able to select and vegetatively propagate what became the cultivated landraces of potatoes in South America. We can envisage a potentially complicated genetic structure for such landraces. Firstly, they could make up metapopulations of landraces which are grown by communities of farmers. Secondly, populations of clonal crops consist of two components, established clones and sexually produced volunteers. Farmers could spare volunteer plants when weeding and examine those that survive to harvest. They could then incorporate some of them into their stock of clonal propagules (seed tubers), possibly assigning them to the landraces they most resemble. Each landrace could therefore be a diverse assemblage of multiple clones sharing phenotypic characteristics rather than a single clone. Hence, landraces should be viewed as evolving entities, in contrast to modern cultivars which are expected to be maintained true to type (the definitive stock), and where seed

Table 1.2 Genotype frequencies at equilibrium for a single locus with alleles A and a at frequencies $p = q = ½$ under mixed selfing and random mating with tetrasomic inheritance, ignoring the phenomenon of double reduction, where s is the proportion of selfing (from Bradshaw 2016 with permission)

Genotype	$s = 0$	$s = 0.2$	$s = 0.4$	$s = 0.6$	$s = 0.8$	$s = 1$
AAAA	$p^4 = 0.0625$	0.0874	0.1239	0.1814	0.28245	$p = 0.5$
AAAa	$4p^3q = 0.25$	0.23865	0.21865	0.1836	0.1202	0
AAaa	$6p^2q^2 = 0.375$	0.3479	0.3149	0.2700	0.1947	0
Aaaa	$4pq^3 = 0.25$	0.23865	0.21865	0.1836	0.1202	0
aaaa	$q^4 = 0.0625$	0.0874	0.1239	0.1814	0.28245	$q = 0.5$

production systems to achieve this are a very important final stage in the breeding and multiplication of new cultivars.

1.4.5 Landraces of Potatoes

Following its creation in Lima, Peru in 1971, The International Potato Centre (CIP) assembled a collection of potato landraces native to Latin America (https://cipotato.org/genebankcip/process/potato/potato-cultivated/). It began with a donation from the Peruvian National Potato Programme of approximately 1800 potato accessions comprising traditional cultivars (landraces). The collection was placed in the CIP Huancayo Experimental Station of Santa Ana-INIA Huancayo. During its 45 years of existence, the number of cultivated potato accessions peaked at 17,326. However, after the staff at CIP had eliminated mixtures, atypical plants, and those with virus symptoms, and had also identified and eliminated duplicates, they were left with the current collection of 4727 accessions. These include 4421 traditional landrace cultivars from 17 countries (mainly from the Andean region) and improved cultivars. The landrace collection comprises 289 Stenotomum +102 Goniocalyx, 206 Phureja, 14 *S. ajanhuiri*, 235 hybrids (diploid and tetraploid), 36 *S. juzepczukii*, 121 Chaucha, 3233 Andigena, 179 Chilean Tuberosum and 6 *S. curtilobum*. The improved cultivars comprise ones released mainly by Latin-American countries and genetic stocks used to identify resistance to diseases. The entire clonal collection is now conserved in vitro and distributed internationally as tissue-cultured materials. This global collection is maintained in trust and is distributed with the Standard Material Transfer Agreement (SMTA) under the terms of the International Treaty for Plant Genetic Resources for Food and Agriculture (ITPGRFA). All accessions in the collection are maintained and available for use in research, breeding and training by humanity, now and into the future (https://cipotato.org/genebankcip/process/potato/potato-cultivated/).

The collection shows that South American farmers retained a much wider variety of tuber shapes and skin and flesh colours than is seen in wild species (Glendinning 1983; Simmonds 1995) and also greater than we see today in modern cultivars. Interestingly, potato tuber diversity was one of many examples given by Darwin, in *The Variation of Animals and Plants under Domestication* (Darwin 1868), of the principle that the valuable and selected parts of all cultivated plants show the greatest amount of modification. The wild relatives of cultivated potatoes have small tubers, round or oval in shape, with white flesh, and skins that are white or have a faint bluish-purple flush. In contrast, the tuber shapes of landraces were either regular, from compressed through elliptic to long ones, or unusual irregular ones such as coiled and concertina-shaped. Skin colour was pink, red, blue or purple, due to anthocyanins, or white, and the distribution of pigments could result in pigmented eyes or eyebrows, and splashed, scattered, spectacled, or stippled tubers. Flesh colour was yellow or orange, due to certain carotenoids, or white, and anthocyanin pigmentation could also occur (Ortiz and Huamán 1994). Interestingly, molecular genetics research by De Jong et al. (2003) and Zhang et al. (2009) provided data that were consistent with the hypothesis of Dodds and Long (1955) that the allele for red skin was selected just once during the domestication of the potato.

South American farmers must have selected their potatoes for appropriate maturity and dormancy, higher yields and harvest index, and resistance to abiotic and biotic stresses. The results of Victorio et al. (1986) on harvest index are of particular interest. They grew 10 potato genotypes to maturity at two contrasting sites in Peru; one in La Molina at 230 m above sea level, in the warm (16.3 °C to 23.1 °C) coastal desert region where irrigated potatoes are grown in a season of length 120 days; the other in Huancayo in cooler (6.2 °C to 20.9 °C) conditions at 3273 m above sea level, which is typical of Andean potato farming with a season of 150 days and higher tuber production (yields) at a higher dry matter content (percentage). Good growth was obtained at both sites for the 9 cultivated genotypes whereas the wild species *S. acaule* grew slowly and only produced a small number of miniature (<0.5 cm diameter) tubers. The cultivated genotypes comprised six Tuberosum × Andigena hybrids, one Chaucha, one Stenotomum and one *S. curtilobum*. The latter grew well but differently to the other eight genotypes. The harvest index of the 9 cultivated genotypes ranged from 0.73 to 0.85 at Huancayo but dropped to 0.33 to 0.75 in La Molina. The only consistent differences between the two sites were shorter plants at Huancayo with an increased tuber dry matter percentage (20 to 28% compared with 14 to 21%). These results raise the question of whether or not worldwide there is still scope to select potatoes for a higher harvest index and better adaptation to the environments in which they are now being grown. Interestingly in a recent review, Haverkort and Struik (2015) quote 0.75 as the typical harvest index of potato compared with 0.4–0.6 for cereals.

It is perhaps fitting to end this section with an acknowledgement of the achievements of the Incas in the Andes during the fifteenth century (Moseley 2001). They developed a sophisticated agriculture appropriate for the high altitude, cold and dry climate around their Peruvian capital of Cuzco. They were able to grow crops of maize, potatoes and other food plants during summer on terraces built of dressed stone which trapped the heat of the sun and aided irrigation; then store the produce for use during winter. The Incas also had a suite of locally domesticated plants and animals. Perhaps there are still lessons for us today when thinking about what farming systems are appropriate for different parts of the world.

1.5 Introduction of Potatoes to Europe

The Spanish conquerors of the Incas were the first Europeans to see potatoes being cultivated in South America, with the first recorded sighting by the expedition of Jiménez de Quesada in what is now Colombia in 1537 (Hawkes 1990). Thirty years later (1567) we have the first record of cultivated potatoes outside of South America with their export from Gran Canaria in the Canary Islands to Antwerp in Belgium (Hawkes and Francisco-Ortega 1993). This was 6 years before they were first recorded in Spain in 1573 in the market archives of the Hospital de La Sangre in Seville (Hawkes and Francisco-Ortega 1992). Hence, potatoes were probably first introduced from South America into the Canary Islands around 1562, and from there to mainland Europe (Hawkes and Francisco-Ortega 1993).

The early introductions of potatoes to Europe included the one shown in the first water-colour painting of a potato dated 1588 (late maturing, red-skinned tubers of irregular shape with deep eyes) and the one shown in the first printed illustration of 1597 (not as late maturing, white-skinned tubers of irregular shape with deep eyes) (Hawkes 1990). The painting was sent by Philippe de Sivry, Prefect of Mons in Belgium, to the herbalist Clusius in Vienna in 1589 and the illustration was by the Englishman, John Gerard, in his Herball of 1597 (Salaman 1926). The cooking methods mentioned by Gerard were baking in embers and boiling in water which subsequently became the main method because it is quicker, as little as 15 min in boiling water at 100 °C compared with 75 min baking at 180 °C. Nevertheless, hot baked potatoes became popular again on the streets of London in Victorian times, in the 1850s (Reader 2008), and are still popular today.

It was often assumed that these early introductions came as ships' stores from Colombia and were of Columbian, or possibly Peruvian, origin and hence were primarily tetraploid Group Andigena potatoes. Then as the growing of potatoes spread north-eastwards across Europe, selection took place for tubering in the long summer days of northern Europe. However, extant Canary Island potatoes comprise both Andean- and Chilean-type landraces so Rios et al. (2007) suggested that there were multiple early introductions of both types. Furthermore, they suggested that the early European potato was selected from the Chilean introductions because they were better adapted to European conditions. Potato introductions from South America were reviewed by Glendinning (1983), but one cannot say with certainty how many there were and what their contribution was to the subsequent spread of the potato in and from Europe, as reviewed by Hawkes (1990). It therefore seemed safest to assume that the early introductions of cultivated potatoes to Europe came from both the Andes and coastal Chile (Hosaka et al. 1994; Spooner et al. 2005b; Rios et al. 2007). However, Ames and Spooner (2008) then analysed DNA from 49 herbarium specimens and confirmed the presence in Europe of Andean potatoes from around 1700 and Chilean potatoes from 1811. Incidentally, Charles Darwin recorded in his journal of the voyage of HMS Beagle around the world, eating 'wild potato tubers' on 7 January 1835 near the northern end of the Chonos Archipelago (45°S) in Chile (Bettany 1889). Ames and Spooner (2008) concluded that the original introductions of potatoes did come from the Andes but that the Chilean potato became predominant in Europe long before the late blight epidemics of 1845 onwards. Hence, the late blight epidemics were not the stimulus for the introduction of Chilean potatoes. Using a *Solanum* section *Petota* diversity panel, Hardigan et al. (2017) found that significant allelic diversity (comparable to that in wild species) existed in the nineteenth century Chilean Tuberosum founders of North American breeding programmes, but their small number ensured a group of individually heterozygous but closely related descendants. Interestingly, molecular analyses of old Japanese cultivars were consistent with them being derived from Group Andigena through early European potatoes (Hosaka et al. 1994) while recent molecular analyses have clustered all Indian potato cultivars, including putatively remnant Andean populations, with Chilean Tuberosum (Spooner et al. 2005b). We now know from the work of Kloosterman et al. (2013) that maturity in potato is controlled by the *StCDF1* locus on chromosome 5, and selection of genetic variant *StCDF1.2* at this locus results in early maturity and the ability to tuber in long days. Hardigan et al. (2017) explored haplotype diversity at the *StCDF1* locus and found introgression of truncated alleles encoding shortened StCDF1 proteins from wild species, particularly *S. microdontum*, in long-day-adapted cultivars. The research of Gutaker et al. (2019) has confirmed and extended many of these recent findings. They sequenced 88 individuals that included 29 historical herbarium specimens spanning the years 1660–1896, obtained from various European museums including specimens collected by Darwin during the voyage of the Beagle. This set included three Chilean historical samples and 26 European historical samples. The other individuals were 43 potatoes constituting a subset of South American modern samples representing landrace diversity and 16 European modern cultivar samples. The reader is referred to the original paper for all of the results and discussion. A brief summary of the conclusions is as follows. European potatoes being grown during the period 1650–1750 were closely related to Andean landraces, and were presumably derived from Andean tetraploids. Neither group of potatoes contained alleles of *StCDF1* (*2*, *3* and *4*) for adaptation (tubering) to long-days. Allele *1* (unadapted to long-days) is disrupted by a putative transposon in allele *3* and by different seven base-pair insertions in alleles *2* and *4*, resulting in frameshift mutations and a gain of function because of the way tuberization is controlled. Then from 1750 to 1850 Chilean Tuberosum was introduced into Europe and an admixture of Chilean and Andean genotypes occurred. European potatoes became progressively more similar to historical Chilean individuals while becoming increasingly distant from Andean ones. This trend was disrupted between 1846 and 1891, presumably as a result of the late blight epidemics. The three long-day allelic variants of *StCDF1* were detected in Europe from 1810 but differences between European and Chilean variants suggest that they may have arisen independently. Strictly speaking, modern European potato cultivars behave as facultative short-day plants since they tuberize in long days, but tuber differentiation is still accelerated following transfer of the plants to shorter day lengths. Finally, the researchers were able to detect the use of wild species by European breeders during the twentieth century.

1.6 Transition to Major Food Crop Worldwide

After their introduction to Europe, potatoes initially remained a botanic curiosity, being grown and studied in physic gardens for interest and medicinal purposes. Their potential as a food crop was first seen in Ireland at the end of the seventeenth century and throughout the eighteenth century (Burton 1989). A suitable climate and soil combined with societal and economic factors resulted in their increase in importance as a food crop, and as a consequence, the population of Ireland increased in size from two million in 1700 to 8.5 million in 1845 (Reader 2008). However, over-reliance on the potato meant that the late blight epidemics of 1845, 1846, 1848 and 1849 resulted in famine in Ireland with profound societal consequences both at the time and to the present day (Zadoks 2008; Nally 2011). Ironically, it was food shortages that proved to be the stimulus to potato cultivation throughout Europe during the eighteenth century because military and economic strength depended upon adequately fed manpower (Burton 1989; Reader 2008). Frederick the Great of Prussia, for example, in 1744 enforced by decree the cultivation of potatoes in what is now northern Germany and Poland, and in France, Parmentier did much to encourage potato cultivation after famine in 1770. According to Burton (1989), the first reference to dried potato was by Parmentier in 1781 in the form of biscuits made from boiled potatoes for use by sailors. He also discussed methods of drying cooked potatoes in forms to be reconstituted by the addition of boiling water. *Pommes frites* became a culinary item in France and Belgium around 1800 and were introduced into England in the 1860s where they were soon sold with fish, thus establishing a 'fish and chips' tradition which for many years was the most sustaining meal of many poor families. Although the potato was brought to Russia from Holland by Peter the Great at the end of the seventeenth century, again it was a royal decree by Catherine the Great in 1765 that led to its spread from around St Petersburg to all parts of the country (Hawkes 1992). Thus the eighteenth century saw potatoes accepted as a food throughout Europe while the nineteenth century saw their ascendancy as a major food crop (Burton 1989).

During the seventeenth and eighteenth centuries, many European countries developed widespread political and commercial interests in the rest of the world and the European (British, Dutch, French, Portuguese and Spanish) colonists and missionaries took with them their common crops including potatoes (Burton 1989). Thus, potatoes were introduced into Africa, China, India, North America and Oceania. Potato production expanded worldwide during the nineteenth century, but it was the expansion in China and India during the second half of the twentieth century that led to these countries becoming the two most important producers in the world. The increases in production from 1966 to 2016 (Table 1.3) in China and India were through increases in the area of potatoes planted (three-fold and four-fold, respectively) and increases in yield per hectare (two-fold and 2½-fold, respectively). The same period saw a decline in potato production in Europe from 224 to 118 million tonnes (Table 1.3). It can also be seen in Table 1.3 that in 2016, there was still a large difference in average yields between Africa (13.9 t/ha) and North America (32.4), whereas the average difference between Asia (18.7 t/ha) and Europe (21.4) was not so great (Table 1.3). Average yield variation between individual countries was even larger, from 0.8 (Central African Republic) to 49.0 t/ha (New Zealand), with a global average of 19.6 t/ha in 2016 (http://faostat.fao.org). Increases in production in Asia are expected to continue. Scott et al. (2019) expect demand for potatoes in India to increase to between 58 and 74 million tonnes per year by 2030.

Table 1.3 Potato production in 1966 and 2016 (http://faostat.fao.org)

Region	Production million tonnes		Area million hectares		Yield t/ha	
	1966	2016	1966	2016	1966	2016
Africa	2.29	24.50	0.27	1.77	8.35	13.86
N America	16.43	24.32	0.72	0.75	22.75	32.41
C America	0.41	2.54	0.05	0.10	8.08	26.74
S America	7.05	15.50	0.97	0.91	7.27	17.00
Asia	30.30	190.52	3.16	10.19	9.58	18.70
China	18.02	99.12	2.00	5.82	9.00	17.05
India	4.08	43.77	0.48	2.13	8.51	20.55
Europe	224.41	117.56	16.24	5.48	13.82	21.45
Oceania[a]	0.93	1.66	0.05	0.04	17.72	42.28
World	281.96	376.83	21.49	19.25	13.12	19.58

[a]Australia and New Zealand

1.6.1 Changes in Farming Methods

It is important to appreciate that the potato was introduced into a Europe, and then a world, in which farming methods were beginning to change and hence new farming systems were evolving. In Britain, for example, in the eighteenth century, land enclosure saw traditional open-field strip farming replaced by modern scientific agriculture. The result was new crop rotations, including break crops and nitrogen-fixing forage crops, increases in livestock production, and the use of inventions such as the 4-coulter plough, seed drill and horse-hoe (Riddet 1925). Innovation continued in the nineteenth century, but it was the twentieth century that saw changes in potato production in Europe and North America that eventually resulted in increases in yield: for example, in Britain from 16 t/ha in 1920 to 22 t/ha in 1960, and then almost doubling to 40 t/ha in 2000, with an average of 46.5 t/ha for years 2013–2016 (British Potato Council and AHDB Statistics). Modern chemical fertilizers were already available by 1900; clean seed tubers were guaranteed from the 1920s through cultivar registration and seed certification; modern chemical herbicides, insecticides and fungicides became available from the 1950s; mechanization and larger fields have been trends since the 1960s; and extensive irrigation has been used since the 1980s (Bradshaw 2009a). Furthermore, post-harvest losses have been reduced by the use of controlled environment stores since the 1960s. Interestingly, Scott et al. (2019) have estimated that cold storage facilities in India will need to double by 2030 to accommodate the projected demand for potatoes. But did new cultivars contribute to any of these improvements in potato production and storage? That is a key question that we will need to address along with the issue of whether or not industrial agriculture is the way forward in other parts of the world; for example, for smallholder farming in Asia, Africa and Latin America? There was, however, another development during the twentieth century that proved to be one of the key drivers of modern potato breeding.

1.6.2 Development of Potato Processing

The twentieth century saw the development of potato processing with processors requiring cultivars that had specific characteristics to meet the demand for quality products (Kirkman 2007). Industrial production of crisps (chips) started in the 1920s and French fries in the 1950s, both initially in the USA. Since then potato processing has grown into a global industry which is still expanding. In North America and some European countries, as much as 60% of the crop is processed. Furthermore, processors are building factories in countries where the potato is primarily grown as a staple food, and this is a trend that is likely to continue. Today the major processed food products are potato crisps (chips), French fries and other frozen products, followed by dehydrated products, chilled-peeled potatoes and canned potatoes (Kirkman 2007).

The first potato starch plant was established in the USA in the 1830s (Treadway 1959). Since then the industry has developed in North America and Europe (Burton 1989), and more recently in China and elsewhere, and high dry matter starch cultivars have been bred. As a result of effective chemical and physical modification, potato starch is used in a wide variety of industrial products, including food ingredients, sizing agents for paper and textiles and starch-based plastics (Li et al. 2006).

1.6.3 Potato Production in the Twenty-First Century

By 2001, potatoes were being grown in 149 countries from latitudes 65°N to 50°S, and at altitudes from sea level to 4000 m (Hijmans 2001). Potatoes can in fact be grown wherever it is neither too hot (ideally average daily temperatures below 21 °C) nor too cold (above 5 °C), and there is adequate water from rain or irrigation (Govindakrishan and Haverkort 2006). The effect of temperature on the balance between photosynthesis and respiration means that maximum light use efficiency occurs at 18 °C (Haverkort and Struik 2015). In practice, potatoes are grown as a summer crop in the tropical highlands of Bolivia, Peru and Mexico, and in the lowland temperate regions of the world (North America, western and eastern Europe, northern China and Australia and New Zealand); all the year round in parts of China and Brazil and in the equatorial highlands of South America (e.g. Ecuador and Colombia) and East Africa (e.g. Kenya and Uganda); as a winter crop in the lowland subtropics (e.g. northern India and southern China); and as spring and autumn crops in the Mediterranean (e.g. North Africa). In a country like Japan that goes from 24°N to 46°N, the potato is grown as a summer crop in the north and a winter crop in the south, with double cropping in spring and autumn in some regions. In a slightly different classification into six potato production systems, Haverkort and Struik (2015) put the emphasis on the need for irrigation, namely for summer crops in the north-west of the USA, autumn crops in the Mediterranean, and winter crops where rice is grown in the rainy summer. In other production systems, potato crops may benefit from some irrigation.

As potatoes cannot be grown all of the year round in most parts of the world, it is normal to have to store both seed tubers for planting the next crop and ware tubers for consumption. Hence, post-harvest infrastructure in terms of road transport and cold storage facilities is also an important aspect of successful potato production and has implications for breeding objectives.

Potential yields depend on the length of growing season, day-lengths and temperatures, but models of potato crop growth can be used for more sophisticated calculations of potential yields, as reviewed by Haverkort and Struik (2015). The growing season can be as short as 75 days in the lowland subtropics, where 90–120 days is the norm, and as long as 180 days in the high Andes. In the lowland temperate regions where planting is done in spring and harvesting in autumn, crop duration is typically 120–150 days, and yields are potentially high. Beukema and van der Waag (1990) estimated that potential yields in their chosen tropical, subtropical and temperate countries/regions could vary from 40 to 140 t/ha depending upon the length of growing season and temperatures. However, average fresh-weight yields in these countries in 2016 were still much lower than the potential yields (about 50% of potential on average; range 24–65%). More recent figures have been quoted by Haverkort and Struik (2015) for average over potential yields: 72% (65 compared with 90 t/ha) in most productive areas of the Netherlands; 35% in less productive areas of Chile; and 60% (31 compared with 55 t/ha) in Northern Maine, USA. Haverkort and Struik (2015) also make the point that somewhere around 60% may be close to an economic optimum in terms of the costs of inputs. Nevertheless, there are examples in the literature where high potential yields have been realized in practice. MacKerron (1985), in Scotland, reported achieved yields for four crops that were 102, 101, 94 and 85% of their estimated potential yields. The achieved yields of 80 to 64 t/ha compared with UK average yields of around 40 t/ha. Kunkel and Campbell (1987) reported an even higher maximum measured yield of 124 t/ha for cultivar Russet Burbank in the Columbia Basin, Washington, USA, compared with predicted yields of 140 and 160 t/ha based on intercepted radiation and length of growing season, respectively (i.e. the former is more accurate). Haverkort and Struik (2015) did in fact conclude that for the potato there may not be much scope to increase yield potential, provided cultivars matched their growth cycle from planting until crop maturity with the length of the available growing season. Therefore, increased yields will have to come from closing the gaps between actual yields and potential yields. The reasons for the gaps will be an important issue when deciding breeding objectives and methods. But what will be the effects of climate change?

1.6.4 Climate Change

In 2008, Haverkort and Verhagen (2008) reviewed the likely consequences of climate change on potato production based on an Intergovernmental Panel on Climate Change report (IPCC 2007). Climate change will bring a rise in temperature, an increase in carbon dioxide concentration in the air, and an altered precipitation pattern. In parts of the world such as the Mediterranean and Sahelian Africa, potato yields will decrease as the suitable heat-free period of the year for production becomes shorter. In these areas breeding for heat tolerance will be important because of the adverse effect of high temperatures on tuberization. Furthermore, with a higher evaporative demand, water will be used less efficiently (but see later in this section) and breeding for drought tolerance will also be important. In contrast, potato yields in temperate climates may increase due to higher carbon dioxide concentrations in the air and a longer growing season. In northern Europe, there will be a decreasing number of days with frost, more rain in winter and less in summer, with more erratic but heavier rainstorms. Water availability for irrigation will be important to maintain yields and quality.

In their literature review and assessment of 11 global crops in 2010, Jaggard et al. (2010) concluded that the increase in CO_2 concentrations from 370 ppm in 2010 to 550 ppm in 2050 would increase potato yields by 36% whereas the increase in O_3 concentration from 50 to 60 ppb would decrease them by 7.5%, so leading to a net increase of 28.5%.

More recently, Haverkort et al. (2013) explored the effects of increases in CO_2 concentration on potato yields in contrasting South African agro-ecosystems between 1960 and 2050. Over this 90-year period CO_2 concentration increases from 315 to a predicted 550 ppm and temperatures are projected to rise significantly by about 1.9 °C (the commonly quoted figure is 2 °C above pre-industrial levels). They concluded that for all four agro-ecosystems, the possible negative effects of rising temperatures and reduced availability of water are more than compensated for by the positive effect of increased CO_2 levels on photosynthesis and hence on water use efficiency and crop productivity. Furthermore, potato growers can advance planting dates to avoid heat stress in late spring and summer and make other beneficial adjustments to the growing season if suitable cultivars are available (Franke et al. 2013).

Finally, in their most recent review, Haverkort and Struik (2015) concluded that higher temperatures and elevated CO_2 levels compensate each other in areas where potato threatens to become marginal with higher temperatures reducing the length of the available growing season. Where the growing season is limited by the frost-free period, climate change will lead to higher yields due to a lengthened growing season and the effect of CO_2 on photosynthesis.

Climate change is also likely to bring new challenges in reducing losses from pests and diseases. Higher temperatures and longer growing seasons are likely to alter the geographic ranges of pests and diseases and the pressure they exert on potato production (Haverkort and Verhagen 2008). As a consequence, breeders may need to alter some of their priorities.

1.6.5 Improving the Nutritional Value of Potatoes

One change in breeding priorities during the twenty-first century has been increased emphasis on improving the nutritional value of potatoes. In 2008, the United Nations (UN) 'International Year of the Potato', simply improving potato production fitted the UN Millennium Development Goals of providing food security and eradicating poverty. Move on 8 years to 1 January 2016 and we have the start of the UN '2030 Agenda for Sustainable Development' (http://faostat.fao.org) which includes 17 goals, the second of which is to end hunger, achieve food security and improved nutrition, and promote sustainable agriculture. The background to the agenda is as follows. In 2015, out of a world population of 7.3 billion, around 800 million people suffered from hunger and more than two billion from one or more micronutrient deficiencies, known as 'hidden hunger', especially in women, infants and children. Ironically about 1.9 billion people were overweight, of whom 600 million were obese. Furthermore, the world population is expected to reach 8.5 billion by 2030 and 9.7 billion in 2050. By 2030 the aim of the agenda is to 'ensure access by all people, in particular the poor and people in vulnerable situations, including infants, to safe, nutritious and sufficient food all year round'. For potatoes, the need is to improve nutritional value as well as to increase production.

As a staple food and vegetable, the potato is a major supplier of carbohydrate (starch) in the human diet. The potato tuber also provides significant amounts of protein, with a good amino acid balance, minerals, vitamins, micronutrients and phytonutrients which include antioxidants. It is high in dietary fibre, especially when eaten unpeeled with its skin. Fresh potatoes are virtually free of fat and cholesterol and have a water content of about 80%. Details of chemical composition can be found in the book *Advances in Potato Chemistry and Technology second Edition* (Singh and Kaur 2016), along with analytical techniques. In a later chapter, we shall need to consider realistic targets for improving nutritional value. The complication is that potatoes are stored at various temperatures for varying lengths of time and then cooked or processed in various ways before being eaten; factors that can affect chemical composition, often but not always in a deleterious way. Furthermore, following food digestion, the nutrients must be bioavailable and hence taken up by the human body and utilized; and there are still issues over the bioavailability of some nutrients. Only after all of these considerations can the chemical composition of potato tubers be related to recommended daily allowances (RDA) and intakes (RDI) of nutrients and sensible targets set.

1.7 Potato Breeding in the Nineteenth Century

It is now time to return to potato improvement and to consider developments in potato breeding. Until the nineteenth century, potato improvement continued to exploit the genetic variation that arose from natural pollination of potato flowers by insects capable of buzz pollination. The genetically unique seedlings that can be grown from the true botanical seeds produce tubers that can be replanted as seed tubers. Hence, growers (selectors) could establish distinct clones and maintain them by asexual (vegetative) reproduction. Named potato cultivars produced in this way appeared in Europe and North America from the 1730s (Reader 2008). The use of artificial hybridization was advocated as early as 1807 by Knight in England (Knight 1807), but William Paterson in Dundee, Scotland was typical in propagating new cultivars from botanical seed in naturally occurring berries produced from open-pollination. He started this propagation work in 1853 and his famous cultivar Victoria was made available to growers from 1860 (Paterson 1871). James Clark in England was probably the first breeder to produce successful cultivars from deliberate cross-pollinations, for example, Magnum Bonum (Victoria × Early Rose) in 1876. He had a good commercial relationship with the Sutton family who multiplied, marketed and distributed all of his later and successful cultivars such as Magnum Bonum (Wilson 1993). Suttons (as it is now called) was established as a seed company in England in 1806 by John Sutton, and from 1837 expanded as John Sutton & Son, concentrating on bulbs, flowers and vegetables, grass seed and seed potatoes. Turning to crops in general, Suttons was just one of a steady stream of seed companies that were established in Europe and North America between the middle of the eighteenth century (e.g. Vilmorin in France in 1743 for cereals) and the end of the nineteenth century (e.g. Burpee in the USA in 1876 for vegetables). Trade in seed led to the naming of cultivars and by the late nineteenth century, an enormous number were being promoted in seed catalogues, having been produced by farmers and growers, hobby breeders, seedsmen and seed companies. There were also exchanges of germplasm between Europe and North America. Deliberate hybridization in plant breeding was common practice by the end of the nineteenth century, having begun at its beginning. The First International Conference on Hybridization and Cross-Breeding was organized by the Royal Horticultural Society in London in 1899, by when plant breeding can be considered established both as a discipline and a profession, distinct from earlier crop improvement by farmers (Kingsbury 2009; Bradshaw 2016). However, in potatoes, the raising of seedlings from seed of self-set berries remained a common practice which continued into the twentieth century. North America's most popular potato cultivar, Russet Burbank, released in 1914, was descended from Rough Purple Chili through three generations of open-pollination (Ortiz 2001). Rough Purple Chili proved to be an important Chilean Tuberosum introduction into the USA in 1851 (Goodrich 1863) because its descendants were widely used as female parents in crosses with European Tuberosum at the end of the nineteenth century (Ortiz 2001). Modern potato breeding started later in India and China in the 1930s, but with rapid expansion since 1948 and 1978, respectively (Gaur and Pandey 2000; Jin et al. 2004).

1.8 Potato Breeding and Genetics in the Twentieth Century

The twentieth century saw the development of crop improvement by scientific breeding methods. The foundations were laid by Mendel's paper of 1865 (Mendel 1865), on the laws and mechanism of inheritance, and Darwin's book of 1876 (Darwin 1876), on the effects of cross and self-fertilization in 57 species of flowering plants from 52 genera and 30 families. Mendel's work had finally been rediscovered and appreciated in 1900. An English translation of Mendel's paper was published in 1901 by the Royal Horticultural Society of London in volume 26 of their journal and promoted by William Bateson in his book *Mendel's Principles of Heredity*, published in 1902 (Bateson and Mendel 1902). At the same time, cytologists were on the verge of confirming that chromosomes were the physical carriers of the genetic information. Breeders could now appreciate that the genetic make-up of the gametes produced from meioses in anthers and ovules, and the ways in which they are combined by the mating system, determines the genetic variation available to them for the selection of new cultivars. Advances in genetics and cytology (cytogenetics) during the first half of the twentieth century therefore informed the development of new plant breeding methods, including the use of genes from the wild relatives of cultivated plants. However, the principal cultivated potato (*Solanum tuberosum*) proved a difficult species for genetic research because of its complex inheritance patterns, and for cytological investigations because of its small and relatively numerous chromosomes. Indeed, it was not until the late 1930s and early 1940s that geneticists recognized that it was a tetraploid ($2n = 4x = 48$) that displays tetrasomic inheritance (Lunden 1937; Cadman 1942). As a consequence, potato breeding in the early twentieth century remained empirical and unsophisticated, as can be seen in two papers on experiments in crossing potatoes by John H. Wilson of St Andrews University, Scotland (Wilson 1907, 1916). He bred two cultivars, Kate Kennedy and Bejant, from his first cross in 1901 with British Queen as female parent and pollen producing Myatt's Kidney as male parent, a cross that produced just 133 seeds. He describes a number of crosses in his two papers and comments on the numbers of offspring with different flower colours and different tuber colours, both in the skin and flesh. Nevertheless, in his first paper he concludes: 'It remains to be shown whether the potato conforms to Mendel's law to such a degree as to render its application of practical value to the breeder'. Despite initial lack of genetic knowledge, as the twentieth century progressed, a number of important developments in potato breeding took place.

1.8.1 Importance of UPOV and CIP

Initially, the research work was done primarily in universities and newly founded government-funded research institutes. Breeding was also undertaken by these organizations as well as commercial companies, the latter benefiting from the research. However, the real stimulus for commercial breeding came in 1961 with the adoption in Paris of the International Convention for the Protection of New Varieties of Plants (UPOV, from the French name) which led to Plant Breeders' Rights and Royalty Income, usually lasting 30 years, in most industrialized countries. UPOV headquarters are in Geneva (Switzerland) and by December 2017, 75 countries were members (www.upov.int). Protected cultivars can be used as parents for further breeding, something that is considered very important for future progress. Hence, the use of patents to protect biotech crops (see later) and prevent this from happening is a cause for concern. A further important development was the foundation in 1971 of the International Potato Centre (CIP) in Lima, Peru. Since

1971, CIP has been supported by the Consultative Group on International Agricultural Research (CGIAR), whose aim is the eradication of human hunger and poverty through research and whose funding is heavily but not exclusively dependent on the World Bank. In 2010, CGIAR chose Agropolis International in Montpellier, France as the headquarters for its consortium of 15 International Research Centres and took up residence in 2011. CIP, like the other research centres, makes improved germplasm available to individual countries for release as cultivars or for further breeding through National Agricultural Research Systems (NARS). Peru was chosen as home to CIP because it is a centre of biodiversity for tuber-bearing *Solanum* species. Hence, germplasm collecting and genebanks seems a good starting place for a brief summary of potato breeding in the twentieth century.

1.8.2 Germplasm Collecting and Genebanks

There have been numerous collecting expeditions, from those pioneered by the Russians in the 1920s (Hawkes 1990) to the more recent ones of the 1990s (Spooner and Hijmans 2001). Reference books have been written on the potatoes of Argentina, Brazil, Paraguay and Uruguay (Hawkes and Hjerting 1969), Bolivia (Hawkes and Hjerting 1989; Ochoa 1990), Peru (Ochoa 2004) and North and Central America (Spooner et al. 2004). The collecting expeditions led to the establishment of a number of potato germplasm collections (genebanks) worldwide. The world collection is held at CIP in Lima, Peru. The other main germplasm collections are the Commonwealth Potato Collection (CPC, Dundee, Scotland), the Dutch-German Potato Collection (CGN, Wageningen, The Netherlands), the Groß Lusewitz Potato Collection (GLKS, IPK, Groß Lusewitz, Germany), the Potato Collection of the Vavilov Institute (VIR, St Petersburg, Russia), the US Potato Genebank (NRSP-6, Sturgeon Bay, USA) and Potato Collections in Argentina, Bolivia, Chile, Colombia and Peru. The genebanks usually comprise clonal collections of potato landraces native to Latin America and collections of wild potato species conserved as botanical seed. Accessions held in the genebanks can be found on their websites along with information about their resources and the services they provide. For example, CIP also has an herbarium collection and a cryopreservation programme (https://cipotato.org/genebankcip/process/potato/potato-cultivated/).

Today, we know that the wild species form a polyploid series from diploid ($2n = 2x = 24$) to hexaploid ($2n = 6x = 72$) (Hawkes 1990). Nearly all of the diploid species are outbreeders, whereas the tetraploids and hexaploids are mostly self-compatible allopolyploids that display disomic inheritance (Hawkes 1990). Genomes were classified into five groups A, B, C, D and P by Matsubayashi (1991), with a sixth group E recognized in closely related non-tuber-bearing species. Spooner et al. (2004) summarized the putative genome compositions of the polyploid species, with additional information from molecular phylogenetics added by Gavrilenko (2011), but further research is required to resolve their origins. Of particular interest to potato breeders is the crossability of wild and cultivated species. This has been determined through artificial pollinations done across many years (Jansky 2006). The results can be explained primarily but not exclusively in terms of endosperm balance number (EBN), which can be regarded as the effective rather than the actual ploidy of the species (Johnston et al. 1980). Five groups of species can thus be recognized based on ploidy and EBN: diploid EBN = 1, diploid EBN = 2, tetraploid EBN = 2, tetraploid EBN = 4 and hexaploid EBN = 4 (Hawkes and Jackson 1992). Hybridizations between species with the same EBN number are usually successful. In a later chapter, we shall see that breeders can achieve sexual hybridizations between cultivated potatoes (*S. tuberosum*) and their wild relatives largely but not entirely by manipulation of ploidy with due regard to EBN (Ortiz 1998, 2001; Jansky 2006; Ortiz and Mihovilovich 2020).

1.8.3 Introgression of Genes from Wild and Cultivated Species

The potatoes which were introduced from South America into Europe and from there to the rest of the world lacked adequate resistances to the diseases and pests which became problems as the crop increased in global importance. Introgression (backcross) breeding started in 1908 in Germany and 1909 in Great Britain with the recognition of *S. demissum* as a source of genes for resistance to late blight, caused by the oomycete pathogen *Phytophthora infestans* (Muller and Black 1951). Although *S. demissum*, like the principal cultivated potato, has an EBN of 4, early hybridizations and introgressions were by trial and error. By the end of the 1980s, just six wild species had been used extensively in the breeding of successful cultivars in Europe, together with cultivated *S. tuberosum* Groups Andigena and Phureja (Ross 1986). *S. demissum* and *S. stoloniferum* had provided resistances to late blight and viruses, *S. chacoense* and *S. acaule* resistances to viruses, and *S. vernei* and *S. spegazzinii* resistances to potato cyst nematodes. Likewise, just three species, *S. demissum, S. chacoense* and *S. acaule*, had been used extensively in North America (Plaisted and Hoopes 1989). Hence, at the beginning of the 1990s, wild potato relatives were still a largely untapped source of potentially desirable genes which had proved difficult to successfully utilize in potato breeding. Nevertheless, by the 1990s a lot more was known about how to achieve sexual hybridizations between cultivated potatoes and their wild relatives (Ortiz 1998, 2001; Jansky 2006). Furthermore, since the 1990s, molecular marker-assisted introgression has offered the possibility of faster progress than can be achieved by traditional backcrossing (Hermsen 1994; Barone et al. 2001; Barone 2004; Iovene et al. 2004). We shall look at the theory and practice of introgression breeding in a later chapter. Interestingly, Ortiz and Mihovilovich (2020) point out that many clones bred worldwide show a genetic bottleneck in cytoplasmic diversity due to the continuous use of cytoplasmic-based male sterility 'lineages' derived from *S. demissum* or *S. stoloniferum*.

An insight into introgression breeding history has been provided by Vos et al. (2015). They combined in a 20 k Infinium SNP array, a non-redundant subset of 15,138 previously identified SNPs and 4454 SNPs originating from a project known as SolCAP (14,530 SNPs proved usable). They deliberately included a large proportion of SNPs with a low minor allele frequency (MAF). They genotyped a total of 569 potato genotypes (537 tetraploids and 32 diploids), which included ones from different countries and market sectors, complemented with historically important cultivars and progenitors. SNPs polymorphic in any one of the 48 cultivars released before 1945 were defined as 'pre-1945' genetic variants, and usually continued to be polymorphic after 1945. This genetic variation most likely represents the material that was brought to Europe from the Americas between the 16th and the nineteenth century. SNPs monomorphic before 1945 but polymorphic in one or more of the 489 genotypes released after 1945 were assumed to be the result of introgression breeding and were defined as 'new' or 'post-1945' variants. The year 1945 was chosen because cultivar Craigs Bounty (released in 1946) was the oldest cultivar with wild species in its pedigree. The researchers found that 24.1% of SNPs were post-1945 and that their average MAF was 1.4% compared with 18.0% for pre-1945 SNPs. Introgression segments differed greatly in size from very small (genotype Y-66-13-636) to complete chromosomes (VTN 62–33-3), with 95 out of the 97 SNPs first observed in Craigs Bounty forming three introgression segments on chromosomes 5, 10 and 12. Just ten genotypes (VTN 62–33-3, Lenape, Mara, Urgenta, VE 71–105, AM 78–3704, Maris Piper, Craigs Bounty, Ulster Glade and VE 66–295) were responsible for the introduction of 50% of post-1945 SNPs. Some of the introgression segments (haplotypes) introduced soon after 1945 had reached frequencies of up to 19% (which means absent in 43% of cultivars and present as single copy in 40% and as two copies in 14%) compared with 4% (which means absent in 85% of cultivars and present as single copy in 14%) for recently introduced ones. However, 50% of all post-1945 SNPs remained below an allele frequency of 1% and 549 SNPs were no longer polymorphic in cultivars released after 2005. Interestingly, when SNP allele frequencies in the 48 pre-1945 cultivars were examined in 81 post-2005 cultivars and a subpopulation of 59 starch cultivars, the majority (6441 or 42%) had hardly changed in allele frequency during a century of potato breeding, and 96% of the genetic variants present in those ancestral cultivars remained polymorphic in modern cultivars. In other words, there had been little erosion of genetic variation.

As potatoes are heterozygous outbreeders, use of the same recurrent parent during introgression would result in a self of the recurrent parent and hence inbreeding depression. In practice, this was avoided by using different Tuberosum parents for each backcross. The results of introgression were therefore entirely new cultivars. Hence, potato breeding could still be viewed as essentially cycles of inter-crossing clones with complementary traits followed by selection of new clones with desirable combinations of traits.

1.8.4 Diploid Breeding

Some cultivated potatoes are diploids, and for them breeding is naturally done at the diploid level with the production of diploid cultivars. But can the breeding of tetraploid potatoes be done at the diploid level with a return to the tetraploid level for cultivar production? It would certainly overcome some of the problems and complexities of working with a tetraploid both for breeding and genetic studies (Hougas and Peloquin 1958; Ortiz and Peloquin 1994). Ivanovskajya (1939) accidently found diploid (dihaploid) potato plants among the progeny of a cross between tetraploid Tuberosum and diploid Phureja. However, it was Hougas et al. (1958) who initiated potato dihaploid production on a large scale by a systematic utilization of $4x \times 2x$ crosses between $4x$ Group Tuberosum cultivars and $2x$ Group Phureja 'pollinators' that induced dihaploid production by a form of parthenogenesis. Strictly speaking, dihaploids are haploids of $4x$ Group Tuberosum. Subsequently, the frequency of dihaploids was increased by the use of elite pollinators (IVP35, IVP48 and IVP101, with the latter being the best), which combined superior haploid-induction frequency with homozygosity for the seed marker 'embryo-spot' whose absence in the offspring indicated putative dihaploids (Hougas et al. 1964; Ross 1986). Ortiz and Mihovilovich (2020) have provided evidence for PL-4, produced at CIP from the cross IVP35 × IVP101, being an even better inducer than IVP101. It was soon found that dihaploids could be crossed with many wild tuber-bearing species which were also diploid. However, the dihaploids were usually male sterile, and most dihaploids, diploid species and their hybrids displayed gametophytic self-incompatibility. Therefore, true breeding lines which could readily be selfed and crossed and which displayed disomic inheritance could not be easily produced and inheritance studies remained difficult. De Jong and Rowe (1971) found some self-fertile hybrids in crosses between Group Phureja or Group Stenotomum and dihaploids of Group Tuberosum and were able to study the effects of inbreeding in cultivated diploid potatoes. In most families, the inbreeding depression for yield could be explained by the expected loss of heterozygosity in a diploid species. Later, a dominant self-incompatibility inhibitor was found in *S. chacoense* by Hanneman (1985) and the gene (*Sli*) mapped to a distal end of chromosome 12, thus showing that it is independent of the *S* locus on chromosome 1 (Hosaka and Hanneman 1998). However, it was Lindhout et al. (2011, 2018) and Stockem et al. (2020) who demonstrated the feasibility of diploid F_1 hybrid breeding for true potato seed propagation through the use of the *Sli* gene. They produced inbred progenies that combined a high level of homozygosity with good plant performance and good tuber quality and yield. These were then crossed to produce diploid F_1 hybrids which were uniform and showed good tuber quality. Hence, the end product would be a diploid cultivar. Jansky et al. (2016) have argued that it is worth considering the possibility of converting the potato from an asexually propagated tetraploid crop into an inbred seed-propagated diploid one to take full advantage of the modern genetics and genomics tools available to improve gain from selection. This would be an extremely ambitious long-term programme of research and development. We shall look at the theory and practice of this recent breeding method for potatoes in a later chapter. For now, we need to return to the second half of the twentieth century and the use of dihaploids in introgression and base broadening where the end product is a tetraploid cultivar.

1.8.5 Introgression and Base Broadening

Peloquin and his co-workers (Hermundstad and Peloquin 1987; Jansky et al. 1990) developed a novel breeding strategy using dihaploids to introgress specific characteristics and to broaden the genetic base of potato in a way similar to that envisaged by Chase (1963) in his analytic breeding scheme. The hybrids between dihaploids of tetraploid Tuberosum and diploid wild species with an EBN of 2 will often form tubers in long-day growing conditions (Jansky et al. 2004; Jansky 2009). When they also produce $2n$ gametes by FDR (first division restitution), much of the genetic diversity of the wild species can be efficiently transferred to the tetraploid offspring in $4x$ x $2x$ crosses and results in about 25% of the wild species genes in the final product (Tai 1994; Jansky 2009). Tai (1994) concluded, however, that dihaploid-wild species hybrids need to be improved before they are used in $4x$ x $2x$ crosses, for example, through population improvement by recurrent selection (Rousselle-Bourgeois and Rousselle 1992). In contrast, diploid hybrids with *S. raphanifolium* with resistance to cold-sweetening after storage for 3 months at 2 °C, have been used by Hamernik et al. (2009) to produce commercially acceptable tetraploid offspring without extensive backcrossing to cultivated germplasm. These should be valuable parents for producing new cultivars suitable for processing.

1.8.6 Base Broadening

During the second half of the twentieth century, potato breeders started to use population improvement schemes on germplasm which they considered unadapted compared with modern cultivars. Their aim was to produce adapted parents from a wider genetic base for feeding into the breeding of finished cultivars. The first programmes recognized the value of the cultivated species of South America for broadening the genetic base of European and North American programmes. They demonstrated that through simple mass selection under northern latitude, long-day summer conditions, Group Andigena will adapt and produce 'Neotuberosum' parents suitable for direct incorporation into European and North American potato breeding programmes (Simmonds 1969; Glendinning 1975; Rasco et al. 1980; Munoz and Plaisted 1981; Tarn and Tai 1983; Maris 1989). Neotuberosum germplasm provided adapted sources of resistance to important pathogens including *Globodera rostochiensis* (golden potato cyst nematode) and *Potato virus Y* (PVY) (Plaisted 1987). Likewise, Carroll (1982) and Haynes (Haynes and Lu 2005) produced populations of Group Phureja/Group Stenotomum adapted to long-day conditions in Europe and North America, respectively. Direct hybridization of Carroll's improved diploid population with tetraploid potato cultivars via unreduced pollen grains ($4x$ x $2x$ crosses) resulted in tetraploid hybrids, some of which were superior to standard tetraploid cultivars in both total and marketable yield (Carroll and De Maine 1989). Furthermore, diploid cultivars, such as Mayan Gold, have now been produced from this population, but they are targeted at niche markets because their yield is only two-thirds that of Tuberosum potatoes. Despite these breeding efforts, relatively few clones of Neotuberosum (long-day Group Andigena) and long-day Group Phureja/Group Stenotomum have been used to any extent in the breeding of modern cultivars. Part of the reason for this is that while adaptation to tuberization in long days was quickly achieved, other problems remained. Neotuberosum clones lacked the regularity of tuber shape of intensively selected Group Tuberosum clones and long-day-adapted Group Phureja clones lacked tuber dormancy.

CIP on its establishment recognized the need to make broad-based germplasm and candidate cultivars from the world collection available to national programmes in developing countries, particularly germplasm with durable resistance to late blight. Probably the best-known broad-based germplasm from CIP's breeders are the *R*-gene (race-specific) free B populations with quantitative resistance to late blight. B1 was a population of short-day-adapted Andigena potatoes, B2 was derived from crosses of the Andigena clones with long-day-adapted Tuberosum clones lacking *R*-genes, and B3 was a population which additionally contained resistances from wild species but from which *R*-genes were meant to have been removed by progeny testing. The aim was to help with the development of improved cultivars, for a wide range of environments, which possessed high and stable levels of resistance to late blight in combination with resistances to viruses and suitable tuber type, and culinary and processing quality (Trognitz et al. 2001; Landeo 2002; Lindqvist-Kreuze et al. 2014). The population improvement involved recurrent selection for quantitative resistance to late blight and other desirable traits under high endemic disease pressure in the Andean highlands together with selection in those geographical areas where the new cultivars were to be grown, for example, in short days in East Africa (Mulema et al. 2004, 2008). Evaluation under long days in Argentina (Trognitz et al. 2001) and in South Korea was included to broaden the range of adaptation of the largely short-day adapted population. Clones with good general combining ability were identified for use as parents in the local breeding programmes of the National Agricultural Research Systems in developing countries. Landeo (2002) reported good progress in selecting for resistance in Population B, large additive genetic variances and high heritability estimates for resistance, and stability of resistance across diverse environments and pathogen populations in tropical environments.

Parallel efforts at CIP developed a lowland tropic virus-resistant population denoted LTVR (Bonierbale et al. 2003; Bradshaw and Bonierbale 2010; Khan et al. 2015). This population combines the heat tolerance and early bulking ability of *S. tuberosum* germplasm bred under the summer conditions of the Northern Hemisphere on the one hand, with virus resistance from native *S. tuberosum* Group

Andigena and Neotuberosum on the other. Recurrent selection with evaluations in a range of environments on the Peruvian coast led to the development of multiplex progenitors with extreme resistance to PVY and PVX, and the selection of virus-resistant clones better adapted to the warm conditions of the lowland tropics in each selection cycle. More recently, advanced germplasm from European and other long-day breeding programmes was incorporated to improve resistance to *Potato leafroll virus* (PLRV), while enhancing earliness and market traits. Furthermore, Bonierbale et al. (2020) have provided a detailed case study of CIP's 'Agile Potato' project which seeks to develop potato cultivars for fresh and processing use in the short crop production windows of Asia, currently left fallow in the cereal-based systems of the subtropical lowland and temperate regions.

In the Central Potato Research Institute of India, there has been interest in Tuberosum (female) × short-day Andigena hybrids in breeding for the sub-tropical plains where the potato crop is grown under short days (Gopal et al. 2000; Kumar and Kang 2006). A number of Indian potato cultivars have already been developed from Tuberosum × Andigena crosses, including Kufri Pukhraj, Kufri Giriraj and Kufri Chipsona II (Birhman et al. 2008). In a later chapter, we will look at the details of some population improvement schemes.

1.8.7 Lack of Diversity in Cytoplasms of Potato

Mihovilovich et al. (2015) have provided interesting results on the lack of diversity in potato cytoplasms. They start by reminding readers that classical European and American (Group Tuberosum) cultivars have exclusively T-type chloroplast DNA which has been shown to be in complete association with β-type mitochondrial DNA, a combination (T/β) that is sensitive to nuclear genes that condition male sterility. The prevalence of the T/β cytoplasm in common potato gene pools is a consequence of a forced use of these genotypes as female parents because of pollen sterility and the more attractive agronomical attributes of their derivatives in comparison with those from the (A) cytoplasm of Andean landraces. However, danger of cytoplasmic uniformity may also occur from the widespread use of cultivars with other cytoplasmic genomes associated with male sterility; for example, cultivars carrying the Ry_{sto} gene for resistance to *Potato virus Y* exhibit complete male sterility caused by association of W-type chloroplast DNA with the characteristic γ-type mitochondrial DNA derived from *S. stoloniferum* (W/γ cytoplasm). In contrast, F_1 and backcrossed progenies carrying *S. demissum* cytoplasm (W/α) produce abundant and normal-looking pollen, but this pollen is non-functional on *S. tuberosum*, again meaning *S. demissum*-derived blight-resistant clones have been used as female parents. In contrast, Hosaka and Sanetomo (2012) showed that cytoplasm types derived from cultivated diploid landraces (Phureja) did not contribute to male or female sterility and hence could help to alleviate the problems of sterility in potato breeding programmes. These authors developed a potato cytoplasm classification method which distinguished potato cytoplasm into six distinct types (M, P, A, W, T and D) where the P- and A-type cytoplasms and the T- and D-type cytoplasms were relatively distinct types within the M- and W-type cytoplasms, respectively, each of which has diverse cytoplasmic variations. The A-type cytoplasm is most prevalent in tetraploid Andigena potatoes, the T-type in tetraploid Tuberosum potatoes, the P-type in diploid Phureja potatoes (Phureja) and the D-type in potatoes derived from *S. demissum*.

Mihovilovich et al. (2015) analysed the cytoplasmic diversity of CIP potato breeding germplasm. Cytoplasm types (I assume D includes/is W/α and W includes/is W/γ) were assigned to 978 genotypes consisting of 265 foreign accessions used as input germplasm, 642 breeding lines developed by CIP (including B3 and LTVR germplasm mentioned above) and 71 cultivars released from CIP material. The researchers found 440 T (45.0%), 368 D (37.6%), 110 W (11.2%), 53 A (5.4%), 5 M (0.5%) and 2 P (0.2%) types. Not surprisingly, there was a high percentage of D cytoplasm in the late blight resistant B3 population (81.5%). Furthermore, the frequency of T-type cytoplasm decreased significantly from input germplasm to the breeding gene pool created by CIP and its released (output) germplasm. The opposite was true for D-type cytoplasm. Less represented cytoplasm types, W and A, increased their frequencies in the breeding gene pool and released germplasm, except for W-type that was barely represented in the released germplasm. The presence of M- and P-type cytoplasms was negligible in all groups. Interestingly, out of the 77 male parents with cytoplasm type information, 48 had T-type cytoplasm and were used to develop 302 genotypes. The researchers also revealed that there were unexplained inconsistencies (presumed errors) in cytoplasm type between genotypes and their maternal parents, as were found for 68 (17%) out of 404 genotypes for which they had maternal cytoplasm type information. The researchers concluded that the CIP breeding programme, as many others worldwide, had experienced a genetic bottleneck in terms of cytoplasmic diversity due to the unintended and continuous use of cytoplasmic-based male-sterile maternal lineages derived from *S. demissum* and *S. stoloniferum* in parental line and cultivar development. Nonetheless, the finding of male-fertile T-type breeding lines must have alleviated the problem to a certain extent, thus enabling CIP breeding to progress. Furthermore, CIP has probably been distributing male-fertile maternal breeding lines within T-type cytoplasm due to the presence of a fertility-restoring gene *Rt* previously described by Iwanaga et al. (1991). The reader is referred to the paper by Mihovilovich et al. (2015) for more information on the CIP germplasm.

1.8.8 True Potato Seed

We saw early in this chapter that potatoes reproduce sexually through botanical seed and asexually (vegetatively) by producing tubers. Hence, farmers and breeders can grow genetically unique plants from the true botanical seeds, let each plant produce genetically identical tubers (a clone), and then decide which clones to replant as seed tubers the next season. They can thus establish distinct clones and maintain them by asexual (vegetative) reproduction as distinct landraces and cultivars, and this is the normal method of potato propagation. However, there are circumstances where propagation by true potato seed (TPS) is an attractive proposition. When CIP was founded in 1971, TPS propagation appeared most attractive in the torrid zones of the lowland tropics and subtropics. Here, the difficulty of establishing a TPS crop, later maturation and less uniformity could be outweighed by three advantages (Golmirzaie et al. 1994). Seed costs are reduced due to the much smaller amounts of planting material required. Planting time is flexible because the farmer does not have to consider the physiological age and condition of seed tubers. Finally, tubers are free from tuber-borne diseases with the possible exception of the few caused by true seed-borne viruses. In practice, TPS potatoes have been established in Bangladesh, China, Egypt, India, Indonesia, Nicaragua, Peru, Philippines, southern Italy and Vietnam (Almekinders et al. 1996; Ortiz 1997; Simmonds 1997). They have proved extremely useful for food security, particularly for small-scale farmers (Chujoy and Cabello 2009). When Chilver et al. (2005) reviewed on-farm profitability and prospects for TPS, they concluded that widespread geographic adoption was unlikely in the immediate future, but that investment in a small but sustained TPS breeding effort could be justified in both China and India.

Breeding cultivars for TPS started at CIP in 1972 with the aim of high yields and acceptable uniformity. None of the breeding methods available at the time could deliver genetic uniformity and hence all of them involved selection for acceptable uniformity. Methods and progress were reviewed by Golmirzaie et al. (1994) and Ortiz (1997). Tetraploid cultivars were produced, either from $4x$ x $4x$ crosses in which heterosis was exploited between Group Tuberosum and Group Andigena (Simmonds 1997), or from $4x$ x $2x$ crosses in which the $2x$ parent produced a high frequency of $2n$ gametes by FDR so that 83% of its heterozygosity was transmitted to the offspring (Golmirzaie et al. 1994; Clulow et al. 1995), or from $2x$ x $2x$ crosses in which both parents produced $2n$ gametes, again with a very high frequency of $2n$ pollen produced by FDR (otherwise the offspring will contain more diploids than tetraploids) (Ortiz and Peloquin 1991). Simmonds (1997) argued, however, that diploid TPS cultivars should not be ruled out for the future, and De Maine (1996) produced TPS families of long-day-adapted Phureja that appeared as uniform for tuber size and shape as selected clones. More recently, as mentioned above, Lindhout et al. (2011, 2018) and Stockem et al. (2020) have demonstrated the feasibility of producing diploid inbred lines, through the introduction of a dominant gene (*Sli*) from *S. chacoense* that inhibits gametophytic self-incompatibility, and true F_1 hybrids for maximum heterosis and uniformity. We will look at the details of breeding and multiplication of TPS cultivars in a later chapter, along with the technology to produce hybrid TPS, including F_1 hybrids.

1.8.9 Genetically Modified Potatoes

Potato transformation using *Agrobacterium*-mediated systems was developed during the 1980s, first with *A. rhizogenes* (Ooms et al. 1986) and then more successfully with *A. tumefaciens* (Stiekema et al. 1988). The technique can be used to insert one or more agriculturally important genes into the potato genome and hence allows targeted improvements of successful and widely grown cultivars. The resulting crops are called genetically modified (GM), biotech or transgenic crops. Commercialization requires the demonstration that GM crops and products are safe to grow and eat or feed to animals. Regulations governing GM crops, food and feed are also often designed to address any other public concerns and to allow farmer and consumer choice (labelling and traceability) over whether or not to grow GM crops and eat GM products. However, the regulations vary with country. In a later chapter, we shall explore the details of genetic transformation and explain the differences between cisgenesis, intragenesis and transgenesis. We shall see that the choice of promoter (controls gene expression) is important for gene expression and come across examples where success depended on tuber expression of the inserted gene rather than constitutive expression throughout the plant. We shall also see that selecting the best transgenic potato clones for commercialization requires a lot of work; namely seeking the best constructs (for transformation) and promoters and eliminating any undesirable transformants or somaclonal variants that arise from the transformation per se and in vitro regeneration, respectively. Finally, we shall also examine how different genes can be stacked in the same cultivar using co-transformation, multi-gene cassettes or re-transformation with further genes.

Major improvements to already successful potato cultivars have been demonstrated using various types of *Agrobacterium*-mediated genetic transformation. Cisgenesis has been used to transfer R-genes, cloned from cross-compatible wild potato species, into late blight (*Phytophthora infestans*) susceptible cultivars (Jo et al. 2014; Haverkort et al. 2016). Gene silencing by RNA interference has successfully targeted reducing sugars and cold sweetening (Rommens et al. 2006), asparagine and acrylamide production (Chawla et al. 2012), enzymic browning and blackspot bruising (Chi et al. 2014) and late blight resistance (Sun et al. 2016). Finally, transgenesis has been used to introduce genes (and sometimes traits) not found in cross-compatible wild relatives; for example, resistance to Colorado potato beetle and viruses PLRV and PVY (Duncan et al. 2002), resistance to potato tuber moth (Douche and Grafius 2005), improved protein content and amino acid balance (Chakraborty et al. 2000) and improved nutritional value through inulin production (Hellwege et al. 2000), altered carotenoid content (Ducreux et al. 2005; Morris et al. 2006; Chitchumroonchokchai et al. 2017) and enhanced accumulation of vitamin B6 (Bagri et al. 2018). This list of traits and references is representative rather than complete, but certain types of traits are currently missing, namely yield, photosynthetic efficiency and efficiencies of nutrient and water use. We shall look at some examples in more detail in a later chapter.

However, the period 1995–2013 saw a reluctance to accept the first genetically modified (GM) potatoes in Europe and North America, and it remains to be seen if the same will be true of the new GM potatoes now available for human consumption in the USA. Monsanto were the first to commercialize GM potatoes for processing, namely Newleaf™ cultivars in North America from 1995. An account can be found in the review by Davies (2002). Despite extensive evaluation, leading processing and fast-food outlet companies in North America were reluctant to purchase GM potatoes, because of consumer concerns over GM potato products, and Monsanto stopped marketing GM potatoes in 2001. There was even more resistance to GM potatoes in the European Union. As a consequence, BASF Plant Science failed to commercialize Amflora, a pure amylopectin starch potato, and Fortuna, a late blight (*P. infestans*) resistant potato with two resistance genes from the wild tuber-bearing Mexican species *Solanum bulbocastanum.* On 16 January 2012, BASF announced that it was stopping all of its GM crop developments targeted for the European market due to lack of acceptance of GM technology in Europe, and in 2013, the company relocated from Germany to the USA. A current attempt to commercialize GM potatoes is that of the J.R. Simplot Company in the USA. The first of their Innate™ potatoes were deregulated by the US Department of Agriculture (USDA) on 7 November 2014 and voluntarily reviewed by FDA (US Food and Drug Administration) who concluded on 20 March 2015 that they were as safe and nutritious as their conventional counterparts. By August 2017, the second generation of Innate™ potatoes had received approval in Canada as well as the USA. They combine processing quality (reduced bruising and blackspot, reduced asparagine which contributes to reduced acrylamide, and lower reducing sugars for cold storage as well as reduced acrylamide) with resistance to late blight (www.simplot.com).

In this book, I am going to assume that GM potatoes will be accepted and hence contribute to potato improvement. A whole chapter will be devoted to the subject.

1.8.10 Conventional Potato Breeding

Potato breeding in the twentieth century involved planned artificial hybridizations followed by multi-stage, multitrait selection over as many as eight clonal generations. In so far as potato breeders made progress in potato improvement it was through cycles of such hybridization and selection, mainly among their developing elite germplasm. We have seen the use of potato landraces and wild relatives in introgression breeding and in base broadening. We have also seen the possibility of further targeted improvements of successful and widely grown cultivars by genetic transformation. Nevertheless, the driving force for potato improvement was the cycles of hybridization and selection. New cultivars were in a sense the by-product of this process, albeit the means by which improvements were transferred to farmers' fields. There has certainly been no shortage of new cultivars. The 2014 World Catalogue of Potato Varieties (Pieterse and Judd 2014) listed more than 4800 cultivars from some 100 countries covering all potato growing regions in the world, although a much smaller number have been widely grown. Up until the 1960s breeding programmes were relatively small by modern standards and the time to breed a new cultivar was relatively short, often just 7–10 years from crossing to registration or National Listing (Bradshaw 2009b). Then, from the 1960s, breeders felt the need to increase the sizes of their programmes, both to assess more traits and to assess potential cultivars in more environments. Simmonds (1969) was certainly of the view that greatly increased numbers of seedlings were required in potato breeding because success is proportional to number and raising 100,000 seedlings from 200 to 300 crosses became the norm (Mackay 2005; Bradshaw 2009b; Jansky

2009; Bradshaw and Bonierbale 2010; Lindhout et al. 2011; Eriksson et al. 2016). The two main weaknesses of such breeding are the number of clonal generations required to select a new cultivar and the inability of intense early-generation selection to affect most economically important traits, which are quantitative in nature. Before moving on to consider ways of making conventional breeding faster, more efficient and more effective, it is worth pausing to examine the effect of selection by breeders on allele frequencies at loci across the potato genome because they should be the basis of progress in potato improvement. It is first worth reminding ourselves of some simple population genetics for quantitative traits under tetrasomic inheritance.

In Table 1.4, we have an additive genetic model (genotypic values 10 down to 6) and a single locus in equilibrium. As the frequency of the desirable allele *A* increases from zero to one (fixation), the population mean steadily increases from 6 to 10 and the genetic variance (all additive) peaks when the allele frequency is 0.5. Hence, if selection for the trait is effective, we would expect to see steady improvements in the trait over the cycles of crossing and selection and steady increases in the frequency of the desirable allele. In practice, many loci are expected to be segregating, but if they are additive in effect the same argument applies to them all.

However, for a trait like yield, we know that there is non-additive as well as additive genetic variation. Hence, in Table 1.5, we have a non-additive genetic model (genotypic values either 10 for *A*--- or 6 for *aaaa*) for a single locus in equilibrium. This time there is a rapid increase in the population mean, the total genetic variance (V_G) and the additive genetic variance (V_A) peak at a low frequency of the desirable allele *A* (around 0.1), and there is a steady decline in the proportion of additive genetic variance. Hence, selection for the trait will be effective when the desirable alleles are at a low frequency. We shall return to these issues in more detail in later chapters, particularly the proportion of the genetic variance that is additive as this

Table 1.4 Changes in population mean and genetic variance ($V_G = V_A$ all additive) as allele frequency *A* increases for a simple additive genetic model

A	*AAAA* = 10	*AAAa* = 9	*AAaa* = 8	*Aaaa* = 7	*aaaa* = 6	Mean	$V_A = V_G$
Frequency	p^4	$4p^3q$	$6p^2q^2$	$4pq^3$	q^4		
0.0	0.0	0.0	0.0	0.0	1.0	6.0	0.00
0.1	0.0001	0.0036	0.0486	0.2916	0.6561	6.4	0.36
0.2	0.0016	0.0256	0.1536	0.4096	0.4096	6.8	0.64
0.3	0.0081	0.0756	0.2646	0.4116	0.2401	7.2	0.84
0.4	0.0256	0.1536	0.3456	0.3456	0.1296	7.6	0.96
0.5	0.0625	0.25	0.375	0.25	0.0625	8.0	1.00
0.6	0.1296	0.3456	0.3456	0.1536	0.0256	8.4	0.96
0.7	0.2401	0.4116	0.2646	0.0756	0.0081	8.8	0.84
0.8	0.4096	0.4096	0.1536	0.0256	0.0016	9.2	0.64
0.9	0.6561	0.2916	0.0486	0.0036	0.0001	9.6	0.36
1.0	1.0	0.0	0.0	0.0	0.0	10.0	0.00

Table 1.5 Changes in population mean and total (V_G) and additive genetic variance (V_A) as allele frequency *A* increases for a simple non-additive genetic model

A	A--- = 10	*aaaa* = 6	Mean	V_G	V_A	V_A/V_G
Frequency	$1 - q^4$	q^4				
0.0	0.0	1.0	6.0000	0.0000	0.0000	–
0.01	0.0394	0.9606	6.1576	0.6056	0.5965	0.985
0.05	0.1855	0.8145	6.7420	2.4174	2.2347	0.924
0.1	0.3439	0.6561	7.3756	3.6101	3.0611	0.848
0.2	0.5904	0.4096	8.3616	3.8692	2.6844	0.694
0.3	0.7599	0.2401	9.0396	2.9192	1.5812	0.542
0.4	0.8704	0.1296	9.4816	1.8049	0.7166	0.397
0.5	0.9375	0.0625	9.7500	0.9375	0.2500	0.267
0.6	0.9744	0.0256	9.8976	0.3991	0.0629	0.158
0.7	0.9919	0.0081	9.9676	0.1286	0.0098	0.076
0.8	0.9984	0.0016	9.9936	0.0256	0.00066	0.026
0.9	0.9999	0.0001	9.9996	0.0016	0.000016	0.010
1.0	1.0	0.0	10.0000	0.0000	0.0000	–

determines the response to selection over cycles of crossing and selecting. But what have breeders actually done in practice?

Hirsch et al. (2013) made a genome-wide (3763 SNPs) assessment of historical and contemporary cultivars in North America. They concluded that 150 years of potato breeding had resulted in selection of alleles in biosynthetic pathways that are integral to phenotypic traits such as pigmentation (yellow market class) or carbohydrate amount and composition (processing market class), but not in substantial changes in heterozygosity and allele dosage.

Hardigan et al. (2017) used a panel of 67 genotypes to capture genome variation in North American cultivars, South American landraces and diploid wild species. They identified 2622 genes putatively under selection in either North American cultivars (compared with wild species) or Andean landraces (compared with wild species), but with only 14–16% shared by North American cultivars and Andean landraces. They concluded that a limited gene set drove domestication and early improvement of cultivated potato and that adaptation of upland (*S. tuberosum* Andigena Group) and lowland (*S. tuberosum* Chilotanum Group or Tuberosum Group) populations targeted distinct loci. They uncovered signatures of selection in genes controlling carbohydrate metabolism, glycoalkaloid biosynthesis, the shikimate pathway, the cell cycle and circadian rhythm. They put forward hypotheses for why selection by ancient farmers and modern breeders might have affected these pathways. They also found that the reduced sexual fertility that accompanied the shift to asexual reproduction in cultivars was reflected in signatures of selection in genes regulating pollen development and gametogenesis.

The paper referred to earlier by Li et al. (2018) also provides insight into selection during domestication. They reported a comprehensive assessment of wild and cultivated potato species based on genomic analyses of 201 accessions of *Solanum* section *Petota*. They sequenced the genomes of the 201 accessions and identified 6,487,006 high-quality single-nucleotide polymorphisms (SNPs) from 167 accessions in clade 4 of *Solanum* section *Petota*, comprising 146 wild and 21 cultivated diploid potato accessions with a broad geographic distribution (16.4% of SNPs were located in coding sequences of which 541,208 were synonymous and 515,492 were non-synonymous SNPs). Large-effect mutations were present in 4025 of the potato genes (10%). Genome-wide genetic variation analysis showed that the diversity of wild potatoes is higher than that of cultivated potatoes, with much higher genetic diversity in the agronomically important disease resistance genes in wild potatoes. Furthermore, by exploiting information about known quantitative trait loci (QTL), they identified 609 genes under selection (selective sweeps identified regions with significantly lower levels of polymorphism in cultivated potato compared with wild potato). These included those correlated with the loss of bitterness in tubers (tuber glycoalkaloid content on chromosome 1) and those involved in tuberization (*GA20ox-1*, which is involved in GA biosynthesis, and *CONSTANS*, which mediates photoperiodic responses), two major domesticated traits of potato, as well as late blight resistance (on chromosomes 3 and 11), tuber length (on 6), tuber width (on 5) and tuber number (on 10). However, no starch metabolism-related genes were identified in the selection sweeps. In addition, they analysed the genome of *S. candolleanum* (the progenitor of cultivated potatoes) and identified 529 genes lost in cultivated potato, including many involved in signal transduction pathways, as well as seven for resistance to late blight. Collectively, they considered the molecular markers generated in their study to provide a valuable resource for the identification of agronomically important genes useful in potato breeding. The reader is referred to their original paper for the details (Li et al. 2018).

1.8.11 Potato Breeding in China and India

As China and India are now the world's leading potato producers, it is worth finishing this section on potato breeding in the twentieth century with these countries. Readers interested in cultivar development and promotion in Africa and their contribution to improving potato production in Africa are referred to the review by Nyongesa and Nganga (2018).

As already mentioned, modern potato breeding started in India and China in the 1930s, but with rapid expansion since 1948 and 1978, respectively (Gaur and Pandey 2000; Jin et al. 2004). The successful expansion of potato production in India was the result of adapting the crop to the different agroecological conditions in India as explained by Kawar et al. (2018). Initially, potatoes were grown as a summer crop in the high northern hills (1600 masl to over 2400 masl) as the climate is similar to European conditions. Subsequently, production spread to the lower plains (< 300 masl), which now account for more than 80% of the potato acreage, and the big success story of Indian potato breeding is the production of cultivars suitable for the short-day, winter conditions of the subtropical plains. Today, the potato growing areas of India can be classified as high (>1600 m), mid (800–1600 m) and low (300–800 m) hills, the north-western, central and eastern plains (<300 m) and the low-altitude (300–600 m) and high-altitude (600–1200) plateau. Each area has its own requirements for new cultivars, but not surprisingly, there is a heavy emphasis in India on breeding early-maturing, short-duration cultivars suited for cultivation in the plains. These give the greatest flexibility in managing crop sequences such as rice-potato-wheat. As an example, Kawar et al. (2018) described the successful breeding of an early-maturing, moderately late-blight resistant cultivar, Kufri Khyati, for the Indian plains which was released in 2008 from a cross made in 1992. Since the year 2000, there has been increasing demand for processed potato products with several Indian and multinational companies establishing factories for chips (crisps), French fries and flakes. They have improved agronomic practices and introduced suitable processing cultivars. The ICAR-Central Potato Research Institute in India has also been breeding processing cultivars of which the most recent is Kufri FryoM, a medium maturity cultivar suitable for French fries which was released in 2018 for the Northern and Central plains (Luthra et al. 2020). Today, Gujarat is not only the fourth largest producer of potatoes in India, but also a leader in potato processing, having the processing plants of three large companies, and is now an exporter of French fries to other Asian countries. However, as only 6% of the potatoes produced in India are processed, there is still much scope for expansion (The Times of India, 28 and 29 January 2020).

Jin (2018) has provided an account of potato production and breeding in China, a few key points of which are as follows. Cultivated tetraploid potatoes are thought to have been introduced from Europe and America by various routes at different times, possibly starting at the end of the sixteenth century. However, the earliest written records of potatoes in China are found in a chronicle written in 1682. By the beginning of the nineteenth century, there were large areas devoted to potato cultivation in the provinces of Yunnan, Guizhou, Shanxi, Shaanxi and Gansu, whereas cultivation in the north-eastern provinces of Liaoning, Jilin and Heilongjiang developed more slowly, possibly as late as the early twentieth century. As remarked earlier, it was the rapid expansion in potato production over the last 40 years that resulted in China becoming the largest potato producer in the world from 1993. Since the 1970s, most of the cultivars grown in China have been bred by the national breeding programmes (today involving over 80 agencies), and the number of new cultivars has increased significantly since the 1990s. Their genetic background has also increased and includes cultivars from the former Soviet Union, Eastern Europe, North America and improved germplasm from CIP, the International Potato Centre in Lima, Peru. Interestingly, between 1930 and 2010 four major changes in the cultivars being grown in China can be recognized. Virus-free seed potato production also started in China in the 1970s but has only been widely used in the last 10 years and still only accounts for about 35% of production. In other words, most farmers still save some of their crop for planting the next season. Today, potatoes are the fourth most important food crop in China and are produced throughout the country in four agroecological zones: the northern single-cropping zone (48% of total area), the central plain double-cropping zone (10%), the southern winter-cropping zone (5%) and the south-western mixed-cropping zone (37%). Some 61% of production goes for table use, 16% for processing, including starch production, and 12% for seed tuber production, which is done in the north at high latitudes and in the south-west at high altitudes. Breeding objectives are those appropriate for these end uses and the agroecological zones.

1.9 Potato Breeding and Genetics in the Twenty-First Century

Ways to make potato breeding faster, more efficient and more effective have become available since the 1990s; for example, progeny testing, estimated breeding values from pedigree information, diagnostic molecular markers for marker-assisted introgression and selection, SNP arrays and genotyping by sequencing for genomic selection, *Agrobacterium*-mediated genetic transformation and site-directed DNA sequence modifications including gene editing. They either have been, or are being, integrated into conventional breeding programmes of the kind described and discussed over 25 years ago by Bradshaw and Mackay (1994). However, is this the best way forward or do we need to rethink the way that potato breeding is done in order to make the best use of continuing advances in semi-automated and automated phenotyping and genotyping technology and genetic knowledge?

Genes underlying qualitative traits can now be mapped directly onto dense molecular-marker maps as individuals can be classified into distinct categories for both trait and marker. Quantitative trait loci (QTLs) can also be mapped indirectly through associations between trait scores and molecular markers. This can now be done in tetraploid as well as diploid potatoes for a biparental F_1 population (Hackett et al. 2014) and for a diverse set of genotypes in a collection of interest to the breeder (D'hoop et al. 2014; Rosyara et al. 2015; Sharma et al. 2018). However, these methods often lack the statistical power to confidently detect QTLs of small effect, particularly in populations of the size currently being used; for example, a QTL explaining only 5% of the variation in a population of size 200 (Hackett et al. 2014). Since 1995 miniaturized microarrays have been used for gene expression profiling of thousands of genes in parallel, and since 2004, 'Next-Generation Sequencing (NGS)' methods have been used for the rapid sequencing of genomes and hence for SNP (single-nucleotide polymorphism) discovery. Arguably the biggest advance in genetic knowledge came with the publication in *Nature* on 14 July 2011 of 86% of the sequence of the 844 Mb-genome of the potato, with a prediction of 39,031 protein-coding genes (Potato Genome Sequencing Consortium 2011). Potato breeders need to use this increased genetic knowledge and combine it with the technological advances to aid the discovery of desirable genes (alleles) for achieving their goals. Gene discovery has also been advanced through the increasing power of comparative genomics. Grube et al. (2000), for example, aligned the chromosomes of potato, tomato (*Solanum lycopersicum*) and pepper (*Capsicum annuum*), having corrected for chromosome rearrangements, and found that phenotypic resistance to diseases and pests mapped to corresponding genome regions in the three species more often than by chance. Homologous regions across the three species were often associated with resistance to very different pathogens (e.g. a bacterium, a virus and a nematode). Mazourek et al. (2009) were able to go further and confirm that the *Bs2* gene in pepper (*Capsicum chacoense*) for resistance to the bacterium *Xanthomonas campestris* pv. *vesicatoria* and the *Rx* and *Gpa2* genes in potato for resistance to Potato Virus X and the nematode *Globodera pallida*, respectively, are orthologous (descended from a common ancestor). In other words, homologous *R*-genes reside in common genome locations across species.

Genome resequencing studies in plant and animal species have revealed that they contain core sets of genes required for growth and development, as well as dispensable genes that are missing in individuals, leading to the concept of a pan-genome for the species. The potato is no exception. Hardigan et al. (2016) compared a panel of 12 monoploid (including three doubled monoploids) clones with the doubled monoploid Group Phureja clone DM1-3 516 R44 (DM) used to sequence the potato genome and to generate the potato reference genome. The 12 monoploids were generated via anther culture using germplasm primarily composed of *S. tuberosum* Phureja Group landraces with limited introgression of Stenotomum Group, Tuberosum Group and *Solanum chacoense*. The researchers identified thousands of deleted genes in the monoploid panel. Despite an abundance of missing genes, each homozygous clone (except one) was able to flower and tuberize, suggesting they possessed the core gene set required for development and reproduction. Dispensable genes were defined as those affected by deletion in at least one flowering and tuberizing clone, with the copy number variation (CNV) spanning at least half an exon within the gene. Of 8888 (22.6%) genes overlapping deletions, among these clones, 7183 were classified as dispensable. An additional 1429 non-deleted genes were predicted to contain SNPs encoding premature stop codons, indicating that at least 8612 (21.9%) genes in DM may be dispensable. The researchers defined the core potato gene set of 30,401 genes (77.4%) as all annotated DM genes not impacted by deletions or premature stop codons in the study panel. This opens up a whole new area of research for potato genetics and understanding the potato genome, as does my next and final example of current research.

It is a paper by Pham et al. (2017) that gives an indication of the genetic variation that exists among cultivated potatoes. They focused on genomic and transcriptomic variation within six commercial, elite tetraploid cultivars of *S. tuberosum*. Using a combination of whole-genome DNA and RNA sequencing, they determined the genomic diversity of four round white chip-processing cultivars from North America (Atlantic, Kalkaska, Missaukee and Snowden), a round white fresh market cultivar (Superior) and a fresh market russet cultivar (Russet Norkotah), relative to the reference potato genome of DM. Variation in the form of SNPs, indels (insertions/deletions) and CNV was determined on a genome-wide scale in each cultivar. The results, but not the researchers' discussion of them (see original paper) were as follows.

Approximately 8.4 million SNPs, with one SNP per 46 nucleotides in the ~390 Mb of the genome assayed, and 320,000 to 424,000 indels were identified in each cultivar. Furthermore, copy number variation (CNV) was extensive with between 16,098 and 18,921 genes in each cultivar exhibiting duplication or deletion. The most abundant SNPs (66–71%) were bi-allelic sites with at least one reference (DM) allele (*A*) and at least one alternate (cultivar) allele (*B*). In all six cultivars, the combined number of simplex (*AAAB*) and triplex (*ABBB*) sites was substantially higher than the number of duplex sites (*AABB*). The majority of indels shared one allele with DM with the second allele being novel to the cultivar. Intergenic regions represented approximately 77% of the sampled genome and contained ~43% of the SNPs and ~ 40% of the indels among the six cultivars. Among SNPs in the genic regions of all cultivars, 417,413 (52.3%) were predicted to cause missense mutations, 370,420 (46.46%) silent mutations and 9504 (1.2%) nonsense mutations. For indels, 4929 (0.25%) in total were predicted to cause frame-shift mutations. The collective allelic variation in tetraploid potato suggests that single base substitutions of a single allele (simplex, triplex SNPs), which result in multiple alleles at each locus, are the dominant type of intra-genome variation in cultivated potato. Moving on to gene expression, and using a cut off of four significant SNPs per gene, between 2180 and 3502 genes showed preferential allele expression (PAE) in leaves, and between 3367 and 5270 showed PAE in tubers. The dosage of allele expression at significant sites tended to be skewed towards the reference (DM) allele. Between 20.1% and 22.2% of genes were estimated to have copy number deletions in each cultivar, and between 5.8% and 8.2% of genes were estimated to contain duplications. In total, 4351 genes shared deletions among all six cultivars in the study, and 848 genes shared duplications. Genes with evidence of PAE were significantly enriched in the conserved angiosperm gene set whereas copy number variants were significantly under-enriched in core plant genes, suggesting that there may be fitness costs associated with copy number variation in genes highly conserved among the angiosperms. In total, 528 genes had correlations between copy number and gene expression. Copy number variable genes tended to be evolutionarily recent, lowly expressed and enriched in genes that show increased expression in response to biotic and abiotic stress treatments suggestive of a role in adaptation. The researchers concluded that in addition to allelic variation of coding sequence, the heterogenous nature of the tetraploid potato genome contributes to a highly dynamic transcriptome which is affected by allele preferential and copy number-dependent expression effects.

While this new genetic and genome research is being done, the challenge for a new generation of potato breeders and biotechnologists is to design breeding programmes that integrate marker-assisted selection of specific alleles, genomic selection of unspecified alleles and phenotypic selection, having decided when a gene editing or transgenic approach is more appropriate, given consumer acceptability of the latter. It will be important for them to consider long-term crop improvement (increasing the frequencies of desirable combinations of alleles over sexual generations) as well as short-term cultivar production (combining alleles in a single genotype).

1.10 Conclusions

In this chapter, we have reviewed the genetic improvement of potatoes from domestication to the present day. We can now ask sensible questions about breeding objectives and starting germplasm for future potato breeding.

1. *Target Environment*: We have seen that the potato is a global crop which is grown in a wide range of environments. Hence, the first question that we need to answer is what is the target environment (agroecological zone) for our breeding programme in terms of length of growing season, daylengths, temperatures and water availability? Are existing cultivars well or badly adapted to these conditions? Can the growing season be extended through selection for heat or frost or drought tolerance? What are the factors that are limiting yields under current conditions? Are there disease and pest problems? Are there problems with the fertility of the soil? What will be the impact of climate change?
2. *Target End Use*: We have seen that the potato is grown as a staple food for local consumption, as a vegetable for cash and as a crop for processing into food or starch. Hence, the second question we need to answer is who is the target end user of the crop? We can then add quality criteria to that of yield in the target environment, including nutritional value.
3. *True Potato Seed or Seed Tuber Crop*: We have seen that potato crops can be grown from true potato seed (TPS) as well as from seed tubers. Hence, the third question we need to answer is which is the best way of growing the crop in our target environment and for our target end use? The production of diploid F_1 hybrid cultivars for TPS propagation raises the issue of their yields compared with those of tetraploid cultivars for clonal propagation and whether or not we will see a progressive move to diploid cultivars.
4. *Germplasm*: We have seen that modern cultivars were bred from landraces of cultivated potatoes which occupied a wide range of environments in South America. These modern cultivars themselves occupy a wide range of environments throughout the world. We have also seen that cultivated potatoes have over 100 wild relatives that are also tuber-bearing. There has been both natural introgression of genes from wild relatives in South America and also deliberate introgression by modern breeders. Hence, our fourth question is which is the most appropriate germplasm for achieving our objectives as quickly as possible and by which breeding method?
5. *Major Genes and QTLs or Polygenes*: We now have increasing knowledge about the genetics of economically important traits and about desirable alleles for achieving our objectives. Hence, our fifth question is can we make use of major genes and QTL (quantitative trait locus) alleles of large effect in order to achieve our objectives, or do we need to treat the trait as a polygenic one where we need to assemble as many alleles of small effect as possible? Furthermore, can we make faster progress by using molecular marker-assisted selection for the former and genomic selection for the latter?

6. *Genetically Modified Potatoes*: We have seen that genetic transformation has been available for potato improvement since the 1990s. Hence, our sixth and last question is can we make major improvements to already successful potato cultivars more easily by genetic modification and/or gene editing?

The rest of this book is an attempt to answer these questions and to explain the theory and practice of the most appropriate breeding method for a given situation.

References

Almekinders CJM, Chilver AS, Renia HM (1996) Current status of the TPS technology in the world. Potato Res 39:289–303

Ames M, Spooner DA (2008) DNA from herbarium specimens settles a controversy about origins of the European potato. Am J Bot 95:252–257

Bagri DS, Upadhyaya DC, Kumar A, Upadhyaya CP (2018) Overexpression of *PDX-II* gene in potato (*Solanum tuberosum* L.) leads to the enhanced accumulation of vitamin B6 in tuber tissues and tolerance to abiotic stresses. Plant Sci 272:267–275

Barone A (2004) Molecular marker-assisted selection for potato breeding. Am J Potato Res 81:111–117

Barone A, Sebastiano A, Carputo D, Frusciante L (2001) Molecular marker-assisted introgression of the wild *Solanum commersonii* genome into the cultivated *S. tuberosum* gene pool. Theor Appl Genet 102:900–907

Bateson W, Mendel G (1902) Mendel's principles of heredity. Digitally printed version 2009. Cambridge University Press, Cambridge, 212p

Bettany GT (1889) Darwin's journal during the voyage of HMS Beagle round the world, 5th edn. Ward, Lock and Co., London

Beukema HP, van der Waag DE (1990) Introduction to potato production. Pudoc, Wageningen, pp 23–24

Birhman RK, Gopal J, Kaushik SK, Kang GS, Kumar R, Joseph TA, Luthra SK (2008) Inventory of potato germplasm (Group Andigena) collection, Technical Bulletin No. 46. CPRI, Shimla

Bonierbale MW, Amoros W, Landeo J (2003) Improved resistance and quality in potatoes for the tropics. Acta Hortic 619:15–22

Bonierbale MW, Amoros WR, Salas E, de Jong W (2020) Potato breeding. In: Campos H, Ortiz O (eds) The potato crop: its agricultural, nutritional and social contribution to humankind. Springer, ebook, Cham, pp 163–218

Bourke PM, Voorrips RE, Visser RGF, Maliepaard C (2015) The double-reduction landscape in tetraploid potato as revealed by a high-density linkage map. Genetics 201:853–863. https://doi.org/10.1534/genetics.115.181008

Bradshaw JE (2009a) A genetic perspective on yield plateau in potato. Potato J 36:79–94

Bradshaw JE (2009b) Potato breeding at the Scottish Plant Breeding Station and the Scottish Crop Research Institute: 1920-2008. Potato Res 52:141–172

Bradshaw JE (2016) Plant breeding: past, present and future. Springer, Cham

Bradshaw JE (2019) Improving the nutritional value of potatoes by conventional breeding and genetic modification. In: Qureshi AMI, Dar ZA, Wani SH (eds) Quality breeding in field crops. Springer Nature, Cham, pp 41–84

Bradshaw JE, Bonierbale M (2010) Potatoes. In: Bradshaw JE (ed) Root and tuber crops, handbook of plant breeding 7. Springer, New York, pp 1–52

Bradshaw JE, Mackay GR (1994) Breeding strategies for clonally propagated potatoes. In: Bradshaw JE, Mackay GR (eds) Potato genetics. CAB International, Wallingford, pp 467–497

Brown CR (1993) Outcrossing rate in cultivated autotetraploid potato. Am Potato J 70:725–734

Burton WG (1989) The potato, 3rd edn. Longman Scientific & Technical, Harlow

Cadman CH (1942) Autotetraploid inheritance in the potato: some new evidence. J Genet 44:33–52

Carroll CP (1982) A mass-selection method for the acclimatization and improvement of edible diploid potatoes in the United Kingdom. J Agric Sci Camb 99:631–640

Carroll CP, De Maine MJ (1989) The agronomic value of tetraploid F1 hybrids between potatoes of group Tuberosum and group Phureja/Stenotomum. Potato Res 32:447–456

Chakraborty S, Chakraborty N, Datta A (2000) Increased nutritive value of transgenic potato by expressing a non-allergenic seed albumin gene from *Amaranthus hypochondriacus*. Proc Natl Acad Sci U S A 97:3724–3929

Chase SS (1963) Analytic breeding in *Solanum tuberosum* L. – a scheme utilizing parthenotes and other diploid stocks. Can J Genet Cytol 5:359–363

Chawla R, Shakya R, Rommens CM (2012) Tuber-specific silencing of asparagine synthetase-1 reduces the acrylamide-forming potential of potatoes grown in the field without affecting tuber shape and yield. Plant Biotechnol J 10:913–924

Chi M, Bhagwat B, Lane WD, Tang G, Su Y, Sun R, Oomah BD, Wiersma PA, Xiang Y (2014) Reduced polyphenol oxidase gene expression and enzymatic browning in potato (*Solanum tuberosum* L.) with artificial microRNAs. BMC Plant Biol 14:62. https://doi.org/10.1186/1471-2229-14-62

Chilver A, Walker TS, Khatana V, Fano R, Suherman R, Rizk A (2005) On-farm profitability and prospects for true potato seed (TPS). In: Razdan MK, Mattoo AK (eds) Genetic improvement of Solanaceous crops, volume I: potato. Science Publishers Inc, Enfield, pp 39–63

Chitchumroonchokchai C, Diretto G, Parisi B, Giuliano G, Failla ML (2017) Potential of golden potatoes to improve vitamin A and vitamin E status in developing countries. PLoS One 12(11):e0187102. https://doi.org/10.1371/journal.pone.0187102

Chujoy E, Cabello R (2009) The canon of potato science: 29. True potato seed (TPS). Potato Res 50:323–325

Clulow SA, McNicoll J, Bradshaw JE (1995) Producing commercially attractive, uniform true potato seed progenies: the influence of breeding scheme and parental genotype. Theor Appl Genet 90:519–525

D'hoop BB, Keizer PL, Paulo MJ, Visser RG, van Eeuwijk FA, van Eck HJ (2014) Identification of agronomically important QTL in tetraploid potato cultivars using a marker-trait association analysis. Theor Appl Genet 127:731–748. https://doi.org/10.1007/s00122-013-2254-y

Darwin C (1868) The variation of animals and plants under domestication, volume 1. Digitally printed 2010. Cambridge University Press, Cambridge, 411p

Darwin C (1876) The effects of cross and self fertilisation in the vegetable kingdom. Digitally printed 2009. Cambridge University Press, Cambridge, 482p

Davies HV (2002) Commercial developments with transgenic potato. In: Valpuesta V (ed) Fruit and vegetable biotechnology. Woodhead Publishing Limited, Cambridge, pp 222–249

De Jong H, Rowe PR (1971) Inbreeding in cultivated diploid potato. Potato Res 14:74–83

De Jong WS, De Jong DM, De Jong H, Kalazich J, Bodis M (2003) An allele of dihydroflavonol 4-reductase associated with the ability to produce red anthocyanin pigments in potato (*Solanum tuberosum* L.). Theor Appl Genet 107:1375–1383

De Maine MJ (1996) An assessment of true potato seed families of *Solanum phureja*. Potato Res 39:323–332

Dodds KS (1962) Classification of cultivated potatoes. In: Correll DS (ed) The potato and its wild relatives. Texas Research Foundation, Renner, pp 517–539

Dodds KS (1965) The history and relationships of cultivated potatoes. In: Hutchinson JB (ed) Essays in crop plant evolution. Cambridge University Press, Cambridge, pp 123–141

Dodds KS, Long DH (1955) The inheritance of colour in diploid potatoes. 1. Types of anthocyanidins and their genetic loci. J Genet 53:136–149

Douches DS, Grafius EJ (2005) Transformation for insect resistance. In: Razdan MK, Mattoo AK (eds) Genetic improvement of Solanaceous crops, volume I: potato. Science Publishers Inc, Enfield, pp 235–266

Ducreux LJM, Morris WL, Hedley PE, Shepherd T, Davies HV, Millam S, Taylor MA (2005) Metabolic engineering of high carotenoid potato tubers containing enhanced levels of β-carotene and lutein. J Exp Bot 56:81–89

Duncan DR, Hammond D, Zalewski DJ, Cudnohufsky J, Kaniewski W, Thornton M, Bookout JT, Lavrik P, Rogan GJ, Feldman-Riebe J (2002) Field performance of transgenic potato, with resistance to Colorado potato beetle and viruses. HortScience 37:275–276

Engel F (1970) Exploration of the Chilca Canyon, Peru. Curr Anthropol 11:55–58

Eriksson D, Carlson-Nilsson U, Ortíz R, Andreasson E (2016) Overview and breeding strategies of table potato production in Sweden and the Fennoscandian region. Potato Res 59:279–294

Franke AC, Haverkort AJ, Steyn JM (2013) Climate change and potato production in contrasting South African agro-ecosystems 2. Assessing risks and opportunities of adaptation strategies. Potato Res 56:51–66

Friedman M, Levin CE (2016) Glycoalkaloids and calystegine alkaloids in potatoes. In: Singh J, Kaur L (eds) Advances in potato chemistry and technology, 2nd edn. Elsevier, Amsterdam, pp 167–194

Gaur PC, Pandey SK (2000) Potato improvement in sub-tropics. In: Khurana SMP, Shekhawat GS, Singh BP, Pandey SK (eds) Potato, global research & development-volume 1. Indian Potato Association, Shimla, pp 52–63

Gavrilenko T (2011) Application of molecular cytogenetics in fundamental and applied research of potato. In: Bradeen JM, Kole C (eds) Genetics, genomics and breeding of potato. Science Publishers, Enfield, pp 184–206

Ghislain M, Andrade D, Rodriguez F, Hijmans RJ, Spooner DM (2006) Genetic analysis of the cultivated potato *Solanum tuberosum* L. Phureja Group using RAPDs and nuclear SSRs. Theor Appl Genet 113:1515–1527

Glendinning DR (1975) Neo-Tuberosum: new potato breeding material. 2. A comparison of Neo-Tuberosum with unselected Andigena and with Tuberosum. Potato Res 18:343–350

Glendinning DR (1976) Neo-Tuberosum: new potato breeding material. 4. The breeding system of Neo-Tuberosum, and the structure and composition of the Neo-Tuberosum gene pool. Potato Res 19:27–36

Glendinning DR (1983) Potato introductions and breeding up to the early 20th century. New Phytol 94:479–505

Golmirzaie AM, Malagamba P, Pallais N (1994) Breeding potatoes based on true seed propagation. In: Bradshaw JE, Mackay GR (eds) Potato genetics. CAB International, Wallingford, pp 499–513

Goodrich CE (1863) The origination and test culture of seedling potatoes. Trans NYS Agric Soc 23:89–134

Gopal J, Chahal GS, Minocha JL (2000) Progeny mean, heterosis and heterobeltiosis in *Solanum tuberosum* × *tuberosum* and *S. tuberosum* × *andigena* families under a short day sub-tropic environment. Potato Res 43:61–70

Govindakrishan PM, Haverkort AJ (2006) Ecophysiology and agronomic management. In: Gopal J, Khurana SMP (eds) Handbook of potato production, improvement, and postharvest management. Food Products Press, New York, pp 179–229

Grube RC, Radwanski ER, Jahn M (2000) Comparative genetics of disease resistance within the Solanaceae. Genetics 155:873–887

Gutaker RM, Weiß CL, Ellis D, Anglin NL, Knapp S, Fernández-Alonso JL, Prat S, Burbano HA (2019) The origins and adaptation of European potatoes reconstructed from historical genomes. Nat Ecol Evol 3:1093–1101

Hackett CA, Bradshaw JE, Bryan GJ (2014) QTL mapping in autotetraploids using SNP dosage information. Theor Appl Genet 127:1885–1904

Haldane JBS (1930) Theoretical genetics of autotetraploids. J Genet 22:359–372

Hamernik AJ, Hanneman RE, Jansky SH (2009) Introgression of wild species germplasm with extreme resistance to cold sweetening into the cultivated potato. Crop Sci 49:529–542

Hanneman RE Jr (1985) Self fertility in *Solanum chacoense*. Am Potato J 62:428–429

Hardigan MA, Crisovan E, Hamilton JP, Kim J, Laimbeer P, Leisner CP, Manrique-Carpintero NC, Newton L, Pham GM, Vaillancourt B, Yang X, Zeng Z, Douches DS, Jiang J, Veilleux RE, Buell CR (2016) Genome reduction uncovers a large dispensable genome and adaptive role for copy number variation in asexually propagated *Solanum tuberosum*. Plant Cell 28:388–405. https://doi.org/10.1105/tpc.15.00538

Hardigan MA, Laimbeer FPE, Newton L, Crisovan E, Hamilton JP, Vaillancourt B, Wiegert-Rininger K, Wood JC, Douches DS, Farré EM, Veilleux RE, Buell CR (2017) Genome diversity of tuber-bearing *Solanum* uncovers complex evolutionary history and targets of domestication in the cultivated potato. PNAS 114(46):E9999–E10008. https://doi.org/10.1073/pnas.1714380114

Haverkort AJ, Struik PC (2015) Yield levels of potato crops: recent achievements and future prospects. Field Crop Res 182:76–85

Haverkort AJ, Verhagen A (2008) Climate change and its repercussions for the potato supply chain. Potato Res 51:223–237

Haverkort AJ, Franke AC, Engelbrecht FA, Steyn JM (2013) Climate change and potato production in contrasting South African agro-ecosystems 1. Effects on land and water use efficiencies. Potato Res 56:31–50

Haverkort AJ, Boonekamp PM, Hutten R, Jacobsen E, Lotz LAP, Kessel GJT, Vossen JH, Visser RGF (2016) Durable late blight resistance in potato through dynamic varieties obtained by cisgenesis: scientific and societal advances in the DuRPh project. Potato Res 59:35–66

Hawkes JG (1990) The potato: evolution, biodiversity & genetic resources. Belhaven Press, London

Hawkes JG (1992) History of the potato. In: Harris PM (ed) The potato crop, 2nd edn. Chapman & Hall, London, pp 1–12

Hawkes JG, Francisco-Ortega J (1992) The potato in Spain during the late 16th century. Econ Bot 46:86–97

Hawkes JG, Francisco-Ortega J (1993) The early history of the potato in Europe. Euphytica 70:1–7

Hawkes JG, Hjerting JP (1969) The potatoes of Argentina, Brazil, Paraguay and Uruguay. A biosystematic study. Oxford University Press, Oxford

Hawkes JG, Hjerting JP (1989) The potatoes of Bolivia. Their breeding value and evolutionary relationships. Oxford University Press, Oxford

Hawkes JG, Jackson MT (1992) Taxonomic and evolutionary implications of the endosperm balance number hypothesis in potatoes. Theor Appl Genet 84:180–185

Haynes KG, Lu W (2005) Improvement at the diploid species level. In: Razdan MK, Mattoo AK (eds) Genetic improvement of Solanaceous crops, volume I: potato. Science Publishers Inc, Enfield, pp 101–114

Hellwege EM, Czapla S, Jahnke A, Willmitzer L, Heyer AG (2000) Transgenic potato (*Solanum tuberosum*) tubers synthesize the full spectrum of inulin molecules naturally occurring in globe artichoke (*Cynara scolymus*) roots. PNAS 97(15):8699–8704. https://doi.org/10.1073/pnas.150043797

Hermsen JGT (1994) Introgression of genes from wild species, including molecular and cellular approaches. In: Bradshaw JE, Mackay GR (eds) Potato genetics. CAB International, Wallingford, pp 515–538

Hermundstad SA, Peloquin SJ (1987) Breeding at the 2*x* level and sexual polyploidization. In: Jellis GJ, Richardson DE (eds) The production of new potato varieties. Cambridge University Press, Cambridge, pp 197–210

Hijmans RJ (2001) Global distribution of the potato crop. Am J Potato Res 78:403–412

Hirsch CN, Hirsch CD, Felcher K, Coombs J, Zarka D, Van Deynze A, De Jong W, Veilleux RE, Jansky S, Bethke P, Douches DS, Buell CR (2013) Retrospective view of North American potato (*Solanum tuberosum* L.) breeding in the 20th and 21st centuries. G3 3:1003–1013

Hosaka K (2004) Evolutionary pathway of T-type chloroplast DNA in potato. Am J Potato Res 81:153–158

Hosaka K, Hanneman RE Jr (1998) Genetics of self-compatibility in a self-incompatible wild diploid potato species *Solanum chacoense*. 2. Localization of an S locus inhibitor (*Sli*) gene on the potato genome using DNA markers. Euphytica 103:265–271

Hosaka K, Sanetomo R (2012) Development of a rapid identification method for potato cytoplasm and its use for evaluating Japanese collections. Theor Appl Genet 125:1237–1251

Hosaka K, Mori M, Ogawa K (1994) Genetic relationships of Japanese potato cultivars assessed by RAPD analysis. Am Potato J 71:535–546

Hougas RW, Peloquin SJ (1958) The potential of potato haploids in breeding and genetic research. Am Potato J 35:701–707

Hougas RW, Peloquin SJ, Ross RW (1958) Haploids of the common potato. J Hered 49:103–107

Hougas RW, Peloquin SJ, Gabert AC (1964) Effect of seed-parent and pollinator on the frequency of haploids in *Solanum tuberosum*. Crop Sci 4:593–595

Iovene M, Barone A, Frusciante L, Monti L, Carputo D (2004) Selection for aneuploid potato hybrids combining a low wild genome content and resistance traits from *Solanum commersonii*. Theor Appl Genet 109:1139–1146

IPCC (2007) Climate change synthesis report, 2007. http://www.ipcc.ch/pdf/assessment-report/ar4/syr/ar4_syr.pdf

Ivanovskajya EV (1939) A haploid plant of *Solanum tuberosum*. C R Acad Sci URSS 24:517–520

Iwanaga M, Ortiz R, Cipar MS, Peloquin SJ (1991) A restorer gene for genetic cytoplasmic male sterility in cultivated potatoes. Am Potato J 68:19–28

Jaggard KW, Qi A, Ober ES (2010) Possible changes to arable crop yields by 2050. Philos Trans R Soc B 365:2835–2851

Jansky S (2006) Overcoming hybridization barriers in potato. Plant Breed 125:1–12

Jansky S (2009) Breeding, genetics and cultivar development. In: Singh J, Kaur L (eds) Advances in potato chemistry and technology. Academic, Burlington, pp 27–62

Jansky SH, Yerk GL, Peloquin SJ (1990) The use of potato haploids to put 2*x* wild species germplasm into a usable form. Plant Breed 104:290–294

Jansky SH, Davis GL, Peloquin SJ (2004) A genetic model for tuberization in potato haploid-wild species hybrids grown under long-day conditions. Am J Potato Res 81:335–339

Jansky SH, Charkowski AO, Douches DS, Gusmini G, Richael C, Bethke PC, Spooner DM, Novy RG, De Jong H, De Jong WS, Bamberg JB, Thompson AL, Bizimungu B, Holm DG, Brown CR, Haynes KG, Sathuvalli VR, Veilleux RE, Miller J, Bradeen JM, Jiang J (2016) Reinventing potato as a diploid inbred line–based crop. Crop Sci 56:1412–1422. https://doi.org/10.2135/cropsci2015.12.0740

Jin L (2018) Potato production and breeding in China. In: Wang-Pruski G (ed) Achieving sustainable cultivation of potatoes volume 1: breeding improved varieties. Burleigh Dodds, Cambridge, pp 259–270

Jin L, Qu DY, Xie KY, Bian CS, Duan SG (2004) Potato germplasm, breeding studies in China. In: Proceedings of the fifth world potato congress. Kunming, China, pp 175–178

Jo K-R, Kim C-J, Kim S-J, Kim T-Y, Bergervoet M, Jongsma MA, Visser RGF, Jacobsen E, Vossen JH (2014) Development of late blight resistant potatoes by cisgene stacking. BMC Biotechnol 14, 50. http://www.biomedcentral.com/1472-6750/14/50

Johns T, Alonso JG (1990) Glycoalkaloid change during the domestication of the potato, *Solanum* Section *Petota*. Euphytica 50:203–210

Johnston SA, der Nijs TPM, Peloquin SJ, Hanneman RE Jr (1980) The significance of genic balance to endosperm development in interspecific crosses. Theor Appl Genet 57:5–9

Kawar PG, Kardile HB, Raja S, Dutt S, Kumar R, Manivel P, Bhardwaj V, Singh BP, Govindakrishnan PM, Chakrabarti SK (2018) Developing early-maturing and stress-resistant potato varieties. In: Wang-Pruski G (ed) Achieving sustainable cultivation of potatoes. Burleigh Dodds, Cambridge, pp 143–167

Khan MA, Munive S, Bonierbale M (2015) Early generation in vitro assay to identify potato populations and clones tolerant to heat. Plant Cell Tissue Organ Cult 121:45–52. https://doi.org/10.1007/s11240-014-0677-z

Kingsbury N (2009) Hybrid: the history & science of plant breeding. The University of Chicago Press, Chicago/London, 493p

Kirkman MA (2007) Global markets for processed potato products. In: Vreugdenhil D (ed) Potato biology and biotechnology advances and perspectives. Elsevier, Oxford, pp 27–44

Kloosterman B, Abelenda JA, Carretero Gomez MDM, Oortwijn M, de Boer JM, Kowitwanich K, Horvath BM, van Eck HJ, Smaczniak C, Prat S, Visser RGF, Bachem CWB (2013) Naturally occurring allele diversity allows potato cultivation in northern latitudes. Nature 495(7440):246–250

Knight TA (1807) On raising of new and early varieties of the potato (*Solanum tuberosum*). Trans Hortic Soc Lond 1:57–59

Kumar R, Kang GS (2006) Usefulness of Andigena (*Solanum tuberosum* ssp. *andigena*) genotypes as parents in breeding early bulking potato cultivars. Euphytica 150:107–115

Kunkel R, Campbell GS (1987) Maximum potential potato yield in the Columbia Basin, USA: model and measured values. Am Potato J 64:355–366. https://doi.org/10.1007/BF02853597

Landeo JA (2002) Durable resistance: quantitative/qualitative resistance. In: Lizarraga C (ed) Proceedings of the global initiative on late blight conference, 11–13 July 2002, Hamburg, Germany, pp 29–36

Li X-Q, Scanlon MG, Liu Q, Coleman WK (2006) Processing and value addition. In: Gopal J, Khurana SMP (eds) Handbook of potato production, improvement, and postharvest management. Food Products Press, New York, pp 523–555

Li Y, Colleoni C, Zhang J, Liang Q, Hu Y, Ruess H, Simon R, Liu Y, Liu H, Yu G, Schmitt E, Ponitzki C, Liu G, Huang H, Zhan F, Chen L, Spooner D, Huang B (2018) Genomic analyses yield markers for identifying agronomically important genes in potato. Mol Plant 11:473–484

Lindhout P, Meijer D, Schotte T, Hutten RCB, Visser RGF, van Eck HJ (2011) Towards F1 hybrid seed potato breeding. Potato Res 54:301–312

Lindhout P, de Vries M, ter Maat M, Ying S, Viquez-Zamora M, van Heusden S (2018) Hybrid potato breeding for improved varieties. In: Wang-Pruski G (ed) Achieving sustainable cultivation of potatoes volume 1: breeding improved varieties. Burleigh Dodds, Cambridge, pp 99–122

Lindqvist-Kreuze H, Gastelo M, Perez W, Forbes GA, de Koeyer D, Bonierbale M (2014) Phenotypic stability and genome-wide association study of late blight resistance in potato genotypes adapted to the tropical highlands. Phytopathology 104(6):624–633

Louderback LA, Pavlik BM (2017) Starch granule evidence for the earliest potato use in North America. Proc Natl Acad Sci U S A 114:7606–7610. https://doi.org/10.1073/pnas.1705540114

Lunden AP (1937) Arvelighetsundersokelser i potet. (Inheritance studies in the potato.). Meldinger fra Norges Landbrukshøishkole 17:1–156

Luthra SK, Gupta VK, Tiwari JK, Kumar V, Bhardwaj V, Sood S, Dalamu, Kaur RP, Kumar R, Vanishree G, Kumar D, Mhatre P, Chakrabarti SK (2020) Potato breeding in India, CPRI Technical Bulletin No 74 (revised). ICAR-Central Potato Research Institute, Shimla

Mackay GR (2005) Propagation by traditional breeding methods. In: Razdan MK, Mattoo AK (eds) Genetic improvement of Solanaceous crops volume I: potato. Science Publishers, Enfield, pp 65–81

MacKerron DKL (1985) A simple model of potato growth and yield: Part II. Validation and external sensitivity. Agric For Meteorol 34:285–300

Maris B (1989) Analysis of an incomplete diallel cross among three ssp. *tuberosum* varieties and seven long-day adapted ssp. *andigena* clones of the potato (*Solanum tuberosum* L.). Euphytica 41:163–182

Matsubayashi M (1991) Phylogenetic relationships in the potato and its related species. In: Tsuchiya T, Gupta PK (eds) Chromosome engineering in plants: genetics, breeding and evolution, Part B. Elsevier, Amsterdam, pp 93–118

Mazourek M, Cirulli E, Collier SM, Landry LG, Kang B-C, Quirin EA, Bradeen JM, Moffett P, Jahn M (2009) The fractionated orthology of *Bs2* and *Rx/Gpa2* and the comparative genomics model of disease resistance in the Solanaceae. Genetics 182:1351–1364

Mendel G (1865) Experiments in plant hybridisation. English translation with introduction by R.A. Fisher, 1965, Oliver & Boyd, Edinburgh. 95p

Mihovilovich E, Sanetomo R, Hosaka K, Ordoñez B, Aponte M, Bonierbale M (2015) Cytoplasmic diversity in potato breeding: case study from the International Potato Center. Mol Breed 35:137. https://doi.org/10.1007/s11032-015-0326-1

Morris WL, Ducreux JM, Fraser PD, Millam S, Taylor M (2006) Engineering ketocarotenoid biosynthesis in potato tubers. Metab Eng 8:253–263

Moseley ME (2001) The Incas and their ancestors: the archaeology of Peru, 2nd edn. Thames and Hudson, London

Mulema JMK, Olanya OM, Adipala E, Wagoire W (2004) Stability of late blight resistance in population B potato clones. Potato Res 47:11–24

Mulema JMK, Adipala E, Olanya OM, Wagoires W (2008) Yield stability analysis of late blight resistant potato selections. Exp Agric 44:145–155

Muller KO, Black W (1951) Potato breeding for resistance to blight and virus diseases during the last hundred years. Z Pflanzenzuchtung 31:305–318

Munoz FJ, Plaisted RL (1981) Yield and combining abilities in Andigena potatoes after six cycles of recurrent phenotypic selection for adaptation to long day conditions. Am Potato J 58:469–479

Nally DP (2011) Human encumbrances. University of Notre Dame Press, Notre Dame, 348p

Nyongesa M, Nganga N (2018) Improving potato cultivation to promote food self-sufficiency in Africa. In: Wang-Pruski G (ed) Achieving sustainable cultivation of potatoes. Burleigh Dodds, Cambridge, pp 271–284

Ochoa CM (1990) The potatoes of South America: Bolivia. Cambridge University Press, Cambridge

Ochoa CM (2004) The potatoes of South America: Peru. 1: the wild species. International Potato Center, Lima

Ooms G, Bossen ME, Burrell MM, Karp A (1986) Genetic manipulation in potato with *Agrobacterium rhizogenes*. Potato Res 29:367–379

Ortiz R (1997) Breeding for potato production from true seed. Plant Breed Abstr 67:1355–1360

Ortiz R (1998) Potato breeding via ploidy manipulations. In: Janick J (ed) Plant breed revs volume 16. Wiley, New York, pp 15–86

Ortiz R (2001) The state of the use of potato genetic diversity. In: Cooper HD, Spillane C, Hodgkin T (eds) Broadening the genetic base of crop production. CAB International, Wallingford, pp 181–200

Ortiz R, Huaman Z (1994) Inheritance of morphological and tuber characteristics. In: Bradshaw JE, Mackay GR (eds) Potato genetics. CAB International, Wallingford, pp 263–283

Ortiz R, Mihovilovich E (2020) Genetics and cytogenetics of the potato. In: Campos H, Ortiz O (eds) The potato crop: its agricultural, nutritional and social contribution to humankind. Springer, ebook, pp 219–247

Ortiz R, Peloquin SJ (1991) A new method of producing inexpensive 4*x* hybrid true potato seed. Euphytica 57:103–108

Ortiz R, Peloquin SJ (1994) Use of 24-chromosome potatoes (diploids and dihaploids) for genetical analysis. In: Bradshaw JE, Mackay GR (eds) Potato genetics. CAB International, Wallingford, pp 133–154

Paterson W (1871) On propagating new varieties of potatoes. Trans High Agric Soc 1871:135–139

Pham GM, Newton L, Wiegert-Rininger K, Vaillancourt B, Douches DS, Buell CR (2017) Extensive genome heterogeneity leads to preferential allele expression and copy number-dependent expression in cultivated potato. Plant J 92(4):624–637. https://doi.org/10.1111/tpj.13706

Pieterse L, Judd J (2014) World catalogue of potato varieties, 6th edn. Agrimedia GmbH, Clenze

Plaisted RL (1987) Advances and limitations in the utilization of Neotuberosum in potato breeding. In: Jellis GJ, Richardson DE (eds) The production of new potato varieties. Cambridge University Press, Cambridge, pp 186–196

Plaisted RL, Hoopes RW (1989) The past record and future prospects for the use of exotic potato germplasm. Am Potato J 66:603–627

Potato Genome Sequencing Consortium (2011) Genome sequence and analysis of the tuber crop potato. Nature 475:189–195

Raker CM, Spooner DM (2002) Chilean tetraploid cultivated potato, *Solanum tuberosum*, is distinct from the Andean populations: microsatellite data. Crop Sci 42:1451–1458

Rasco ET, Plaisted RL, Ewing EE (1980) Photoperiod response and earliness of *S. tuberosum* ssp. *andigena* after six cycles of recurrent selection for adaptation to long days. Am Potato J 57:435–448

Reader J (2008) Propitious esculent. William Heinemann, London

Riddet WM (1925) The turnip crop. In: Paterson WGR (ed) Farm crops, volume II. Root crops. Gresham Publishing Co. Ltd., London, pp 69–118

Rios D, Ghislain M, Rodriguez F, Spooner DM (2007) What is the origin of the European potato? Evidence from Canary Island landraces. Crop Sci 47:1271–1280

Rommens CM, Ye J, Richael C, Swords K (2006) Improving potato storage and processing characteristics through all-native DNA transformation. J Agric Food Chem 54:9882–9887

Ross H (1986) Potato breeding – problems and perspectives. In: Advances in plant breeding, vol 13. Paul Parey, Berlin/Hamburg

Rosyara UR, De Jong WS, Douches DS, Endelman JB (2015) Software for genome-wide association studies in autopolyploids and its application to potato. Plant Genome 9. https://doi.org/10.3835/plantgenome2015.08.0073

Rousselle-Bourgeois F, Rousselle P (1992) Creation et selection de populations diploide de pomme de terre (*Solanum tuberosum* L.). Agronomie 12:59–67

Rumold CU, Aldenderfer MS (2016) Late Archaic–Early formative period microbotanical evidence for potato at Jiskairumoko in the Titicaca Basin of southern Peru. PNAS 113(48):13672–13677. https://doi.org/10.1073/pnas.1604265113

Salaman RN (1926) Potato varieties. Cambridge University Press, Cambridge

Scott GJ, Petsakos A, Suarez V (2019) Not by bread alone: estimating potato demand in India in 2030. Potato Res 62(3):281–304

Scurrah M, Celis-Gamboa C, Chumbiauca S, Salas A, Visser RGF (2008) Hybridization between wild and cultivated potato species in the Peruvian Andes and biosafety implications for deployment of GM potatoes. Euphytica 164:881–892

Sharma SK, MacKenzie K, McLean K, Dale F, Daniels S, Bryan GJ (2018) Linkage disequilibrium and evaluation of genome-wide association mapping models in tetraploid potato. G3 8:3185–3202. https://doi.org/10.1534/g3.118.20037

Simmonds NW (1969) Prospects of potato improvement. In: 1968–69. Scottish Plant Breeding Station, Pentlandfield, pp 18–38

Simmonds NW (1995) Potatoes. In: Smartt J, Simmonds NW (eds) Evolution of crop plants, 2nd edn. Longman Scientific & Technical, Harlow, pp 466–471

Simmonds NW (1997) A review of potato propagation by means of seed, as distinct from clonal propagation by tubers. Potato Res 40:191–214

Singh J, Kaur L (2016) Advances in potato chemistry and technology, 2nd edn. Elsevier, Amsterdam

Spooner DM (2016) Species delimitations in plants: lessons learned from potato taxonomy by a practicing taxonomist. J Syst Evol 54:191–203

Spooner DM, Hijmans RJ (2001) Potato systematics and germplasm collecting, 1989-2000. Am J Potato Res 78:237–268

Spooner DM, van den Berg RG, Rodriguez A, Lara-Cabrera SI (2004) Wild potatoes (Solanum section Petota; Solanaceae) of North and Central America. Syst Bot Monogr 68:1–209

Spooner DM, McLean K, Ramsay G, Waugh R, Bryan GJ (2005a) A single domestication for potato based on multilocus amplified fragment length polymorphism genotyping. Proc Natl Acad Sci U S A 102:14694–14699

Spooner DM, Nunez J, Rodriguez F, Naik PS, Ghislain M (2005b) Nuclear and chloroplast DNA reassessment of the origin of Indian potato varieties and its implications for the origin of the early European potato. Theor Appl Genet 110:1020–1026

Spooner DM, Nunez J, Trujillo G, del Herrera MR, Guzman F, Ghislain M (2007) Extensive simple sequence repeat genotyping of potato landraces supports a major reevaluation of their gene pool structure and classification. Proc Natl Acad Sci U S A 104:19398–19404

Spooner DM, Ghislain M, Simon R, Jansky SH, Gavrilenko T (2014) Systematics, diversity, genetics, and evolution of wild and cultivated potatoes. Bot Rev 80:283–383

Stiekema WJ, Heidekamp F, Louwerse JD, Verhoeven HA, Dijkhuis P (1988) Introduction of foreign genes into potato cultivars Bintje and Desiree using an *Agrobacterium tumefaciens* binary vector. Plant Cell Rep 7:47–50

Stockem J, De Vries M, Van Nieuwenhuizen E, Lindhout P, Struik P (2020) Contribution and stability of yield components of diploid hybrid potato. Potato Res 63(3):345–366

Stupar RM, Bhaskar PB, Yandell BS, Rensink WA, Hart AL, Ouyang S, Veilleux RE, Busse JS, Erhardt RJ, Buell CR, Jiang J (2007) Phenotypic and transcriptomic changes associated with potato autopolyploidization. Genetics 176:2055–2067

Sukhotu T, Hosaka K (2006) Origin and evolution of Andigena potatoes revealed by chloroplast and nuclear DNA markers. Genome 49:636–647

Sun K, Wolters A-MA, Loonen AEHM, Huibers RP, van der Vlugt R, Goverse A, Jacobsen E, Visser RGF, Bai Y (2016) Down-regulation of Arabidopsis DND1 orthologs in potato and tomato leads to broadspectrum resistance to late blight and powdery mildew. Transgenic Res 25:123–138. https://doi.org/10.1007/s11248-015-9921-5

Tai GCC (1994) Use of 2*n* gametes. In: Bradshaw JE, Mackay GR (eds) Potato genetics. CAB International, Wallingford, pp 109–132

Tarn TR, Tai TCC (1983) Tuberosum × Tuberosum and Tuberosum × Andigena potato hybrids: comparisons of families and parents, and breeding strategies for Andigena potatoes in long-day temperate environments. Theor Appl Genet 66:87–91

Treadway RH (1959) Potato starch. In: Talburt WF, Smith O (eds) Potato processing. The Avi Publishing Company, Inc., Westport, pp 374–389

Trognitz BR, Bonierbale M, Landeo JA, Forbes G, Bradshaw JE, Mackay GR, Waugh R, Huarte MA, Colon L (2001) Improving potato resistance to disease under the global initiative on late blight. In: Cooper HD, Spillane C, Hodgkin T (eds) Broadening the genetic base of crop production. CAB International, Wallingford, pp 385–398

Ugent D, Dillehay T, Ramirez C (1987) Potato remains from a late Pleistocene settlement in South Central Chile. Econ Bot 4:17–27

Victorio RG, Moreno U, Black CC Jr (1986) Growth, partitioning and harvest index of tuber-bearing Solanum genotypes grown in two contrasting Peruvian environments. Plant Physiol 82:103–108

Vos PG, Uitdewilligen JGAML, Voorrips RE, Visser RGF, van Eck HJ (2015) Development and analysis of a 20K SNP array for potato (*Solanum tuberosum*): an insight into the breeding history. Theor Appl Genet 128:2387–2401

Wheeler RM (2009) Potatoes for human life support in space. In: Singh J, Kaur L (eds) Advances in potato chemistry and technology. Academic, Burlington, pp 465–495

Wilson A (1993) The story of the potato through illustrated varieties. Balding and Mansell, Cambridgeshire, 120p

Wilson JH (1907) Experiments in crossing potatoes. Trans High Agric Soc 1907:74–92

Wilson JH (1916) Further experiments in crossing potatoes. Trans High Agric Soc 1916:33–55

Zadoks JC (2008) The potato murrain on the European continent and the revolutions of 1848. Potato Res 51:5–45

Zhang Y, Cheng S, De Jong D, Griffiths H, Halitschke R, De Jong W (2009) The potato *R* locus codes for dihydroflavonol 4-reductase. Theor Appl Genet 119:931–937

Chapter 2
Potato Genetics: Avoiding or Embracing Tetrasomic Inheritance

2.1 Introduction

Some knowledge of potato genetics is required for understanding the theoretical basis of breeding methods. This chapter therefore covers those aspects of potato genetics that are not readily available in textbooks of genetics because the principal cultivated potato (*Solanum tuberosum*) is a tetraploid ($2n = 4x = 48$) that displays tetrasomic inheritance (Lunden 1937; Cadman 1942). Furthermore, as some breeding is also done at the diploid level with a return to the tetraploid level through unreduced ($2n$) gametes, we need to consider meiosis in diploids which produce $2n$ gametes, as well as in autotetraploids, before dealing with tetrasomic inheritance for qualitative and quantitative traits. First, however, it is useful to summarize from Chap. 1 the germplasm available to potato breeders and to deal with crossability, dihaploids, the gametophytic incompatibility system and self-fertility; or put another way, methods of avoiding rather than embracing tetrasomic inheritance.

2.2 Potato Germplasm and Crossability

Together, the cultivated species of potatoes and their wild relatives form a polyploid series from diploid ($2x$) to hexaploid ($6x$) (Hawkes 1990). Nearly all of the diploid species are outbreeders, whereas the tetraploids and hexaploids are mostly self-compatible allopolyploids that display disomic inheritance (Hawkes 1990), the exception being the principal cultivated potato which can be regarded as an autotetraploid with tetrasomic inheritance. The taxonomy of cultivated potatoes and their wild relatives is complicated and has undergone major revisions as explained by Spooner (2016).

The widely accepted classification of cultivated potatoes is that of Dodds (1962), as modified by Spooner et al. (2007), although other schemes can be found in the

J. E. Bradshaw, *Potato Breeding: Theory and Practice*,
https://doi.org/10.1007/978-3-030-64414-7_2

literature. Spooner et al. (2007) classified Andigena (4*x*), Chaucha (3*x*), Phureja (2*x*), Stenotomum (2*x*) and Tuberosum (4*x*) landraces of potatoes as a single species *S. tuberosum*, divided into two cultivar groups. These are the Andigenum Group (Andigena, Chaucha, Phureja and Stenotomum) of upland Andean landraces, containing diploids, triploids and tetraploids, and the Chilotanum Group (Tuberosum) of lowland tetraploid Chilean landraces. Spooner et al. (2007) also recognized the three frost-tolerant species (*S. ajanhuiri* (2*x*), *S. juzepczukii* (3x) and *S. curtilobum* (5*x*)) cultivated in the Andes as separate hybrid species derived from crosses between domesticates and wild relatives.

There are two main classifications of wild tuber-bearing *Solanum* species. The newer one of Spooner et al. (2014) recognizes 107 species, all classified as *Solanum* section *Petota* (tuber-bearing species) and partitioned into three nuclear clades. The older one of Hawkes (1990) recognizes 219 species, arranged into 19 series of subsection *Potatoe* of section *Petota* of subgenus *Potatoe* of genus *Solanum.* Breeders need to be aware of both classifications, but are more concerned about the crossability of species than their taxonomy.

The crossability of species has been determined through artificial pollinations done over many years (Jansky 2006). The results can be explained primarily but not exclusively in terms of endosperm balance number (EBN), which can be regarded as the effective ploidy of a species (Johnston et al. 1980). In crosses between species with the same EBN the hybrids have a normal endosperm for nourishing the hybrid embryo, whereas in crosses between species with different EBNs the endosperm degenerates. The endosperm develops following the fusion of a sperm nucleus from the male parent with two polar nuclei from the female parent to give a triple fusion nucleus and it is the genetic composition of this nucleus that is important. Under the EBN hypothesis, the endosperm is normal when the three nuclei have the same EBN and hence a 2 maternal to 1 paternal ratio of endosperm balance factors. Despite attempts to understand the genetic and biological basis of the EBN concept, EBNs are determined experimentally relative to species assigned an arbitrary value of 1 (Hermsen 1994). Five groups of species can thus be recognized based on ploidy and EBN: diploid EBN = 1, diploid EBN = 2, tetraploid EBN = 2, tetraploid EBN = 4 and hexaploid EBN = 4 (Hawkes and Jackson 1992). Diploid *S. tuberosum* has EBN = 2 and tetraploid *S. tuberosum* has EBN = 4. Hybridizations between species with the same EBN number are usually successful. In a later chapter, we shall see how breeders can achieve sexual hybridizations between cultivated potatoes (*S. tuberosum*) and their wild relatives with different EBNs by manipulation of ploidy with due regard to EBN (Ortiz 1998, 2001; Jansky 2006).

2.3 Dihaploids

An important development in potato breeding and genetics was the production of relatively large numbers of dihaploids from tetraploid *S. tuberosum* and the discovery that they could be crossed with many wild species which were also diploid.

Ivanovskajya (1939) accidently found diploid (dihaploid) potato plants among the progeny of a cross between tetraploid Tuberosum and diploid Phureja. However, it was Hougas and Peloquin (Hougas and Peloquin 1958; Hougas et al. 1958) who initiated potato dihaploid production on a large scale by a systematic utilization of 4*x* × 2*x* crosses between 4*x* Group Tuberosum cultivars and 2*x* Group Phureja 'pollinators' that induced dihaploid production by a form of parthenogenesis. Strictly speaking, dihaploids are haploids of 4*x* Group Tuberosum. Subsequently the frequency of dihaploids was increased by the use of elite pollinators which combined superior haploid-induction frequency with homozygosity for the seed marker 'embryo-spot' whose absence in the offspring indicated putative dihaploids (Hougas et al. 1964). Dihaploids need to be confirmed by counting chromosomes in mitotically dividing root cells or by counting chloroplasts in stomatal guard cell pairs or through flow cytometric analysis, as reviewed by Ortiz and Mihovilovich (2020). Well-known elite pollinators are IVP35, IVP48 and IVP101, with the latter being the best (Ross 1986). Ortiz and Mihovilovich (2020) have provided evidence for PL-4, produced at CIP from the cross IVP35 × IVP101, being an even better inducer than IVP101.

It was soon found that dihaploids (EBN = 2) could be crossed with many wild tuber-bearing species, which were also diploid (EBN = 2), as well as with diploid landraces of *S. tuberosum* (EBN = 2). However, the dihaploids were usually male sterile and most dihaploids, diploid landraces and wild species, and their hybrids displayed gametophytic self-incompatibility. Therefore, true breeding lines which could readily be selfed and crossed and which displayed disomic inheritance could not be easily produced and inheritance studies remained difficult. Nevertheless, crosses involving diploid *S. tuberosum* and crossable wild species proved useful for mapping purposes from 1988 onwards as a range of molecular markers became available. Genes underlying qualitative traits could now be mapped directly onto the dense molecular-marker maps as individuals could be classified into distinct categories for both trait and marker (Mann et al. 2011). Quantitative trait loci (QTLs) could also be mapped indirectly through associations between trait scores and molecular markers. The aim for potato breeders was and is the discovery of diagnostic molecular markers that can be used for marker-assisted selection of desired major genes (alleles) and QTL alleles of large effect. Some traits such as tuber shape can be treated as a qualitative trait (round and long) or a quantitative trait (length/width ratio), but the underlying major gene or QTL should map to the same locus.

2.4 Linkage Maps Including Ultra-High-Density Map and Genome Sequences

The first potato map was developed using just 65 F_1 progeny clones from a diploid interspecific cross [*S. tuberosum* Gp Phureja × (*S. tuberosum* Gp Tuberosum × *S. chacoense*)] and restriction fragment length polymorphism (RFLP) markers previously mapped in tomato (Bonierbale et al. 1988). It comprised 135 markers (RFLP + 11 isozyme + 1 morphological marker) which were assigned to 12 linkage groups and had a total map length of 606 cM. The second potato map, in contrast, was developed using 67 progeny clones from an intraspecific backcross (BC_1) involving two diploid *S. tuberosum* clones with wild species in their pedigrees [(H8 1.691/1 × H82.309/5) × H82.309/5] (Gebhardt et al. 1989). It comprised 141 loci (potato RFLPs) and had a total map length of 690 cM. It was extended to 304 loci and a total map length of 1034 cM by Gebhardt et al. (1991). Additional mapping work by Gebhardt et al. (1991) and Tanksley et al. (1992) allowed the linkage maps of potato and tomato to be aligned and the numbering of chromosomes (linkage groups) to be agreed. Conservation of nucleotide sequences was found, but also differences in both linkage group and total map length. The work revealed that potato and tomato differ by five major paracentric inversions.

Mann et al. (2011) reviewed the progress that had been made with linkage maps by 2011 and summarized 16 significant maps for potato and its wild relatives in terms of mapping population type (F_1 or BC_1), parents, number of progeny (49–246), marker type and number, and map length (403–1170 cM). The marker types used have been restriction fragment length polymorphism (RFLP), random amplified polymorphic DNA (RAPD), cleaved amplified polymorphic sequence (CAPS), sequence characterized amplified region (SCAR), amplified fragment length polymorphism (AFLP), simple sequence repeat (SSR), Diversity Array Technology (DArT) and more recently and extensively a wide range of single-nucleotide polymorphism (SNP) genotyping platforms such as Kompetitive Allele Specific PCR (KASP, Semagn et al. 2014). An explanation of how these polymorphisms (except CAPS and KASP) arise at the DNA level can be found in Chap. 3 of my book *Plant Breeding: Past, Present and Future* (Bradshaw 2016), which explains why RFLPs and SSRs are single locus, multi-allelic, codominant markers whereas RAPDs and AFLPs are multilocus, single allelic, dominant markers. An account of Mendelian genetics and linkage maps can be found in Chap. 4 and QTL analysis in Chap. 6 of *Plant Breeding: Past, Present and Future* (Bradshaw 2016). However, for a full-sib family of an outbreeding species such as the diploid potato species we have been considering, the reader is referred to Chap. 8 of *Genetic Mapping in Experimental Populations* by Van Ooijen and Jansen (2013). A few key points and issues are as follows.

With outbreeding species up to four alleles may segregate and the number can vary across loci from two to four (with inbreeding species the limit is two alleles). Furthermore, segregation is already present in the first generation after the cross (after crossing homozygous lines it is the second and later generations that reveal segregation). Van Ooijen and Jansen (2013) explain segregation type and linkage phase. If A_1 and A_2 are the alleles of the first parent and A_3 and A_4 are the alleles in the second parent, we can write:

Grandparents	**GP_1**	A_1-	×	**GP_2** A_2-		**GP_3**	A_3-	×	**GP_4** A_4-
Parents		**P_1**	↓					↓	
			A_1A_2		×		**P_2**	A_3A_4	
Segregating					↓				
population		**F_1**	A_1A_3	A_1A_4	A_2A_3	A_2A_4			

In practice, the genotypes of the grandparents may be unknown. Depending on the number of different alleles, including the so-called null allele *0* which does not produce an observable signal (e.g. band on gel, so that Q_1Q_1 cannot be distinguished from Q_10), seven distinct segregation types can be recognized.

Let us consider two loci *A* and *B*, each with four alleles, separated on the same linkage group (chromosome) by a recombination frequency (proportion) of *r*, and let us concentrate on the female parent. The two alleles at the two loci can be in coupling (A_1–B_1) or repulsion (A_1–B_2) linkage.

Coupling (A_1–B_1/A_2–B_2) gives recombinant gametes A_1–B_2 ($\frac{1}{2}r$) and A_2–B_1 ($\frac{1}{2}r$) and non-recombinant gametes A_1–B_1 ($\frac{1}{2}$(1-*r*)) and A_2–B_2 ($\frac{1}{2}$(1-*r*)) with the frequencies shown in brackets.

Repulsion (A_1–B_2/A_2–B_1) gives recombinant gametes A_1–B_1 ($\frac{1}{2}r$) and A_2-B_2 ($\frac{1}{2}r$) and non-recombinant gametes A_1–B_2 ($\frac{1}{2}$(1-*r*)) and A_2–B_1 ($\frac{1}{2}$(1-*r*)) with the frequencies shown in brackets.

There will be fewer recombinants than non-recombinants so that the linkage phase in the female parent will be 'obvious' from the numbers in the segregating population and likewise for the male parent (there will be a linkage map for the female parent and another for the male parent), and we can estimate *r* from the proportion of recombinants. If we knew that A_1 and B_1 had both come from grandparent $\mathbf{GP_1}$ we would also know that they were linked in coupling in parent $\mathbf{P_1}$, but the genotypes of the grandparents are likely to be unknown.

With four different alleles at a locus, we have segregation type $A_1A_2 \times A_3A_4$ and complete information about the two parental meioses. However, now consider segregation type $A_1A_1 \times A_1A_2$ at locus *A* and segregation type $B_1B_2 \times B_3B_4$ at locus *B*. In the female parent we have: A_1–B_1/A_1–B_2 and hence locus *A* segregation type provides no information about the female meiosis (no discrimination between recombinants and non-recombinants at *A* and *B* loci), whereas in the male parent we have either A_1–B_3/A_2–B_4 (A_1–B_3 in coupling) or A_1–B_4/A_2–B_3 (A_1–B_3 in repulsion) and hence complete information about the male meiosis.

If finally, we consider segregation type $A_1A_2 \times A_1A_2$ at locus *A* and segregation type $B_1B_2 \times B_3B_4$ at locus *B*, we find that locus *A* segregation type provides partial information about the meiosis in both the female and male parent. If, for example, the linkages in the cross are A_1–B_1/A_2–$B_2 \times A_1$–B_3/A_2–B_4, the segregating population is as shown in Table 2.1. Half of the genotypes/phenotypes are informative about linkage (shown in bold: they discriminate between recombinants and non-recombinants).

It can be shown that out of the seven segregation types, only type $A_1A_2 \times A_3A_4$ with four different alleles provides complete information about meiosis in both parents. By comparing pairwise recombination frequency estimates under all possible linkage phases it is usually possible to distinguish the linkage phases in both parents (the likelihood of one phase is usually much larger than that of the other). The two most common segregation types in practice are $A_1A_2 \times A_1A_1$ and $A_1A_1 \times A_1A_2$ and their use for forming the linkage maps of the female and male parent, respectively, is called a two-way pseudo-testcross strategy. Van Ooijen and Jansen (2013) also explain the estimation of the pairwise recombination frequency in the full-sib

Table 2.1 Offspring from cross $A_1–B_1/A_2–B_2 \times A_1–B_3/A_2–B_4$ (Rec = recombinant): genotypes/phenotypes in bold are informative about linkage (i.e. discriminate between recombinant and non-recombinant)

	Female	Non-Rec		Rec	
Male	**Gametes**	$A_1–B_3$	$A_2–B_4$	$A_1–B_4$	$A_2–B_3$
Non-rec	$A_1–B_1$	$\mathbf{A_1–B_1}$ $\mathbf{A_1–B_3}$	$A_1–B_1$ $A_2–B_4$	$\mathbf{A_1–B_1}$ $\mathbf{A_1–B_4}$	$A_1–B_1$ $A_2–B_3$
	$A_2–B_2$	$A_2–B_2$ $A_1–B_3$	$\mathbf{A_2–B_2}$ $\mathbf{A_2–B_4}$	$A_2–B_2$ $A_1–B_4$	$\mathbf{A_2–B_2}$ $\mathbf{A_2–B_3}$
Rec	$A_1–B_2$	$\mathbf{A_1–B_2}$ $\mathbf{A_1–B_3}$	$A_1–B_2$ $A_2–B_4$	$\mathbf{A_1–B_2}$ $\mathbf{A_1–B_4}$	$A_1–B_2$ $A_2–B_3$
	$A_2–B_1$	$A_2–B_1$ $A_1–B_3$	$\mathbf{A_2–B_1}$ $\mathbf{A_2–B_4}$	$A_2–B_1$ $A_1–B_4$	$\mathbf{A_2–B_1}$ $\mathbf{A_2–B_3}$

family including linkage phase determination. Joint or separate (female and male) parental recombination frequencies can be estimated. The reader is referred to their book for the details. Computer programs such as JoinMap 5 (Van Ooijen 2018, https://www.kyazma.nl/index.php/JoinMap) and MapQTL 6 (Van Ooijen 2009, https://www.kyazma.nl/index.php/MapQTL) provide the software for constructing a genetic linkage map and doing a QTL analysis, respectively. Hence constructing linkage maps and doing QTL analyses in diploids can be considered routine, in so far as any experimental work is routine!

The linkage maps included an ultra-high-density (UHD) genetic map of 10,365 AFLP markers in a cross between two diploid heterozygous *S. tuberosum* clones (SH83-92-488 and RH89-039-16) which produced 136 F_1 offspring clones of which 130 were usable (van Os et al. 2006). The markers were grouped into bins such that each bin was separated by a single recombination event from the adjacent bin. This required the development and use of new computer software to cope with the large number of markers, namely RECORD that could produce accurate marker orders in a relatively short time (van Os et al. 2005a) and SMOOTH that could remove genotyping errors (van Os et al. 2005b). Together the two programs made it possible to construct a reliable and robust framework map that accelerated gene isolation by map-based cloning and provided a genome-wide physical map through the anchoring of BAC (bacterial artificial chromosomes) contigs. However, of particular interest for potato breeding were the results on the distribution of recombination events per chromatid. Analysis of the distribution of recombination events per chromatid showed that out of 3120 chromatids, 1379 (44%), 1505 (48%), 228 (7.3%) and 7 (0.22%) showed zero, one, two and three recombination events, respectively. Evidently, the vast majority (93%) of the meiotic bivalents in the parents experienced only one chiasma, resulting in two recombinant and two non-recombinant chromatids. Furthermore, no significant differences in recombination frequencies were observed between the female and male meioses. These results mean that large linkage blocks (haplotypes) are transmitted from parents to offspring.

RH89-039-16 (RH) was chosen by the Dutch-led global consortium to sequence the potato genome. A potato genomic bacterial artificial chromosome (BAC) library of 78,000 clones was fingerprinted and aligned into about 7000 physical map contigs. The BAC-ends were sequenced and approximately 30,000 BACs were anchored to the ultra-high-density genetic map of potato, composed of 10,000 unique AFLP™ markers. A BAC by BAC sequencing strategy was then adopted but overall progress was slow. The heterozygosity of RH limited the progress of physical mapping and was expected to complicate assembly of the genome sequence. Hence whole-genome shotgun sequencing of a doubled monoploid of Group Phureja DM1–3 516 R44 (DM) was undertaken as it was expected to eliminate the complexity in assembly. The draft genome sequence from DM was used to integrate sequence data from RH. These two genotypes represent a sample of potato genomic diversity; DM with its fingerling (elongated) tubers was derived from a primitive South American cultivar whereas RH more closely resembles commercially cultivated tetraploid potato. On 14 July 2011, in *Nature*, the consortium published 86% of the sequence of the 844 Mb-genome of the potato, with a prediction of 39,031 protein-coding genes (Potato Genome Sequencing Consortium 2011). This has proved an aid to gene discovery and the development of molecular breeding methods and will continue to be an important resource for potato breeding and genetics. However, as the 'draft' DM sequence still needed improvement, in 2019, Solynta and Plant Breeding, Wageningen University and Research decided jointly to sequence one of the more homozygous selections of the Solynta diploid hybrid breeding programme. This unique genotype, called Solyntus, is the first potato (*Solanum tuberosum*) genotype that is highly homozygous, relatively vigorous and self-compatible. Solyntus was sequenced using Oxford Nanopore reads supplemented with Illumina reads to generate an improved reference genome for the cultivated potato. It comprises 185 contigs (contig N50 = 7.55 Mb) covering 716 Mb organized in pseudo-molecules using RaGoo and DMv4.03 (https://www.plantbreeding.wur.nl/Solyntus/).

Clone DM was also used by Hardigan et al. (2016) to explore the pan-genome of potato. Genome resequencing studies in a number of plant and animal species had already revealed that they contain core sets of genes required for growth and development, as well as dispensable genes that are missing in individuals. Hardigan et al. (2016) compared a panel of 12 monoploid (including three doubled monoploids) clones with the doubled monoploid Group Phureja clone DM. The 12 monoploids were generated via anther culture using germplasm primarily composed of *S. tuberosum* Group Phureja landraces with limited introgression of Group Stenotomum, Group Tuberosum and *Solanum chacoense*. The researchers identified thousands of deleted genes in the monoploid panel. Despite an abundance of missing genes, each homozygous clone (except one) was able to flower and tuberize, suggesting they possessed the core gene set required for development and reproduction. Dispensable genes were defined as those affected by deletion in at least one flowering and tuberizing clone, with the copy number variation (CNV) compared with DM spanning at least half an exon within the gene. Of 8888 (22.6%) genes overlapping deletions among the clones, 7183 were classified as dispensable. An additional 1429 non-deleted genes were predicted to contain SNPs encoding premature stop codons, indicating at least 8612 (21.9%) genes in DM may be dispensable. The researchers defined a core potato gene set (pan-genome) of 30,401 genes (77.4%) as all annotated DM genes not impacted by deletion or premature stop in the study panel. This opens up a whole new area of research for potato genetics and understanding the potato genome in terms of the pan-genome of the cultivated potato (genes present in all cultivated potatoes), covering traditional landraces (diploids to pentaploids) and modern potato cultivars of the Andigenum and Chilotanum gene pools. The reader is referred to Hardigan et al. (2016) for more detail than can be given here although a few more findings are as follows. The number of SNPs relative to DM ranged from 800,333 to 4,764,182. Of these, 2.4 to 4.4% were located in coding regions and 70.1 to 75.7% were intergenic, with 0.67 to 0.84 ratios of synonymous to non-synonymous changes in coding SNPs. Copy number variants (CNVs) were defined as duplications or deletions compared with DM. Several thousand CNVs were called in each monoploid ranging from 500 bp to 575 kb (3.0-kb median in the panel), with total CNV calls per individual varying from 2978 to 10,532. A total of 92,464 CNVs were identified in the panel collectively affecting 30.2% of non-gap sequence in the DM v4.04 reference genome. Ratios of duplication and deletion were highly conserved, with duplications comprising 29.2 to 33.2% of total CNVs per clone. Structural variation was most common in intergenic sequence and on a genome-scale was often more prevalent in pericentromeric regions with lower frequency observed in the gene-dense euchromatic arms of chromosomes, particularly in regions with high rates of recombination. In total, 11,656 potato genes (29.7%) overlapped CNV calls, with 9001 genes (22.9%) affected in at least half their annotated gene model. Presence/absence variation (PAV) represents a form of CNV in which genes lack copies in the reference genome but are present in non-reference individuals. In total, 1256 putative PAVs lacking high-quality read coverage from DM were classified as true PAVs.

2.5 Gametophytic Incompatibility System

We will now return to the reproductive biology of potatoes. By the mid-1960s, it was assumed from the available evidence that outcrossing occurs in cultivated diploid landraces and most wild diploid species of potato, and is enforced by a single multiallelic *S*-locus gametophytic self-incompatibility system (Dodds 1965). In such a system, pollen is rejected when there is a match between the single *S*-haplotype in the haploid pollen and either of the two *S*-haplotypes in the diploid pistil which are co-dominantly expressed. Cipar et al. (1964) obtained hybrids between clones of cultivated diploid ($2n = 2x = 24$) Phureja potatoes and a dihaploid ($2n = 2x = 24$) of a tetraploid ($2n = 4x = 48$) Andigena clone. The tetraploid Andigena was self-fertile because the gametophytic self-incompatibility system breaks down in tetraploids, at least when two different alleles are present in the pollen (Lewis 1943, 1954). Cipar et al. (1964) found that their hybrids were vigorous and had excellent flowering ability and male fertility, unlike the dihaploids. They therefore used the hybrids along with Phureja and Stenotomum clones to analyse the inheritance of incompatibility from the success or failure of crosses. The cut-stem technique of pollinating in a glasshouse is described in the paper along with the genotypes expected from hypothetical crosses (i.e. one, two or four classes depending on the *S* genotypes of the parents). Diallel crosses within the full-sib families were used for cross-compatibility and self-incompatibility analysis (Table 2.2). The researchers concluded that a gametophytic system of incompatibility was operative in Phureja and Stenotomum and that their self-incompatible parental clones had either one ($S_iS_j \times S_iS_k$) or no ($S_iS_j \times S_kS_l$) *S* allele in common. They also concluded that their (dihaploid) Andigena was an *S* allele homozygote (S_iS_i) with an *S* allele not shared with the heterozygous (S_jS_k) Phureja clone. There were no reciprocal differences in male fertility in the hybrids between the Phureja and (dihaploid) Andigena to complicate this interpretation. Thus, self-incompatibility became recognized as the norm in diploid potatoes.

One of the cDNA clones used by Gebhardt et al. (1991) in their linkage mapping coded for one allele (S_{r1}) of the *S*-locus in potato (Kaufmann et al. 1991) so that the self-incompatibility locus was mapped to chromosome 1, proximal to the reference marker TG24. Kaufmann et al. (1991) had described the allelic diversity and gene structure of the *S* locus as revealed by the isolation and characterization of genomic and cDNA clones encoding *S*-associated major pistil proteins from three alleles (S_1, S_{r1}, S_2). The most prominent example in the paper of Gebhardt et al. (1991) of distorted segregation ratios was caused by the self-incompatibility locus (*S*).

Progress by 2011 in understanding the genetics of compatibility and incompatibility in S-RNase-based systems can be seen in the review of McClure et al. (2011). By then it was known that the polymorphic *S*-locus for self-incompatibility in the Solanaceae, Rosaceae and Plantaginaceae encodes at least two genes. S-RNases (ribonucleases) determine the specificity of pollen rejection in the pistil and *S*-locus F-box proteins fulfil this function in pollen. These S-RNases are thought to function as *S*-specific cytotoxins as well as recognition proteins. Thus, incompatibility results from the cytotoxic activity of S-RNase, while compatible pollen tubes evade S-RNase cytotoxicity. Mutations affecting pollen compatibility were considered especially important for defining models of compatibility and incompatibility. McClure et al. (2011) thought that the *Sli* gene for self-compatibility in potato had potentially important implications for understanding the mechanism of

Table 2.2 Gametophytic incompatibility system: diallel crosses within full-sib families

Table 2.2 A $S_1S_2 \times S_1S_3$ gives two classes S_1S_3 and S_2S_3

	S_1S_3	S_2S_3
S_1S_3	Incompatible	Two classes
S_2S_3	Two classes	Incompatible

Table 2.2 B $S_1S_2 \times S_3S_4$ gives four classes S_1S_3, S_1S_4, S_2S_3 and S_2S_4

	S_1S_3	S_1S_4	S_2S_3	S_2S_4
S_1S_3	Incompatible	Two classes	Two classes	Four classes
S_1S_4	Two classes	Incompatible	Four classes	Two classes
S_2S_3	Two classes	Four classes	Incompatible	Two classes
S_2S_4	Four classes	Two classes	Two classes	Incompatible

self-incompatibility. However, we shall concentrate on its implications for practical breeding.

More recently, Ye et al. (2018) created self-compatible diploid potatoes by knocking out the self-incompatibility gene *S-RNase*, using the CRISPR–Cas9 system, in the diploid self-incompatible *S. tuberosum* Group Phureja S15–65 clone. The single guide RNA was designed to target the relatively conserved domain of the S-RNase protein and could target S_{p3} and S_{p4} alleles simultaneously. They ultimately obtained five mutant diploid clones, each containing mutations in at least one *S-RNase* allele. As *S-RNase* is a co-dominant gene, the mutation of either allele would enable pollen containing the same *S*-haplotype to extend into the style, thereby resulting in self-compatibility. Although the rate of fruit setting varied among the lines, the number of seeds per fruit was sufficient for further propagation, ranging from 67 to 288 seeds per fruit. These results demonstrated that inducing loss-of-function mutations in *S-RNase* genes resulted in self-compatibility in potato. The growth vigour and plant morphology of the mutant lines did not differ from those of the wild type, indicating that they could be used directly for breeding. The proportion of T_1 plants in the selfed families without the Cas9 cassette varied from 3.6% to 24.5%, thus demonstrating that self-compatibility can be achieved without introducing any exogenous DNA. This is now an alternative route to the use of the *Sli* gene.

2.6 Self-Fertility and Potato Genetics

Hanneman (1985) found self-fertile plants in the wild diploid (EBN = 2) potato species *Solanum chacoense*, which is an extremely polymorphic species found in south Bolivia, north and central Argentina, Paraquay, Uruguay and south Brazil, often as a field weed in lowland pastures (Hawkes 1990; Ochoa 1990). Furthermore, he inbred several lines to the S_6 generation. Subsequently, Hosaka and Hanneman (1998a) studied the inheritance of self-compatibility. They found that the offspring of a cross between a self-compatible (SC) inbred clone (S_7) of *S. chacoense* and a self-incompatible (SI) clone of cultivated diploid Phureja segregated for SC (7 offspring) and SI (8). Diallel crosses were made among 15 offspring (F_1's), as well as selfing 10 of them and the SC parental clone, and making two further sib-matings. All plants were tested for self-compatibility. The results were consistent with a single dominant gene (*Sli*) acting sporophytically in the male parent to inhibit *S* gene expression in the pollen so that the pollen was compatible with its own parent and plants with similar *S* genes. The *Sli* gene had been maintained in a heterozygous condition through eight generations of selfing (S_8) implying that dominant homozygotes might be associated with lethality; for example, tight linkage of *Sli* to a recessive lethal gene.

Hosaka and Hanneman (1998b) went on to cross a highly inbred line of *S. chacoense* (chc 525-3, SC) as the female parent with a Group Phureja clone (1.22, SI). One clone of the F_1 progeny (F_1-1, SC) was then crossed as the male parent with a selected clone (93H100-1, SI) from a Stenotomum/Phureja bulk population. The resulting 116 offspring were assessed for SC by selfing each plant and determining berry set. The female-fertile plants (berries set after bulk-pollen pollinations) with no berry set after 10 self-pollinations were regarded as SI plants. The results were 66 SC and 35 SI plants with 15 not determined, a significant departure from the expected Mendelian ratio of 1:1. A genetic linkage map was constructed using unique alleles from the *S. chacoense* parent that were transmitted via F_1-1 and hence could be scored as present or absent in the segregating population. From a total of 28 RAPD and 127 RFLP markers, 109 markers mapped into 12 linkage groups, as expected. Distorted segregation ratios were found for all markers on linkage groups 2 and 11, with lower frequencies of marker bands, while localized distortion was found for linkage groups 1, 5, 7, 10 and 12. The dominant *Sli* gene for SC mapped to the distal end of (chromosome) linkage group 12, 10.0 cM away from two cosegregating markers, CT156E5120 and 89–1320. As the *S* locus itself is on chromosome 1, the *Sli* gene is independent of the *S* locus. These results demonstrated that it is possible to have both SC and SI plants that are both male and female fertile.

Another interesting and different example of self-fertility comes from the cross between the two diploid genotypes used in the potato genome sequencing project, Group Phureja DM 1–3 516 R44 (DM, male-sterile) and Group Tuberosum RH89–039-16 (RH, self-fertile) (Potato Genome Sequencing Consortium 2011). Peterson et al. (2016) observed a wide range of fruit set on many plants after self-pollination, suggesting that the self-fertility was not a strictly qualitative trait. Examination of pollen tube growth in self-fertile and self-unfruitful F_1 plants after controlled self-pollinations revealed no difference in the ability of pollen tubes to reach the ovary. Hence it is unlikely that the *Sli* gene was present in the germplasm. The F_1 population (103 plants) was genotyped using a genome-wide single-nucleotide polymorphism (SNP) array (Infinium 8303). All 85 SNPs segregating with self-fertility occurred on chromosome 12. Seeds obtained after self-pollination of self-fertile individuals were used to advance the population for four generations. Genotyping 46 self-fruitful and 46 self-unfruitful S_3 plants on the Infinium 8303 Potato SNP array then revealed eight SNPs segregating with self-fertility, located on chromosomes 4, 9 and 11 in addition to 12. Three times more heterozygosity than expected was found in the S_3 generation.

Two research groups have used the *Sli* gene in genetic studies which involve self-pollination. Endelman and Jansky (2016) were the first to use an inbred-line-derived diploid F_2 population for genetic mapping in potato. The female parent was the doubled monoploid Phureja potato DM1–3 (male sterile) used to sequence the potato genome. It has two unusual properties, namely a strong susceptibility to jelly end, a physiological condition associated with a high concentration of reducing sugars and low starch content, and its tendency to produce tubers instead of normal sprouts at the eyes of tubers. The male parent was an S_7 inbred line (M6) derived from the wild relative *S. chacoense*. A single F_1 plant was self-pollinated to produce the F_2 population of 103 to 211 individuals for phenotyping in two glasshouse and two fields, and also for genotyping with an expanded version of the original Infinium array of 8303 SNPs. A set of 178 progeny and 2264 markers were used for linkage analysis (12 linkage groups, each 60 to 100 cM, and total map length 883 cM). Overall, there was excellent agreement between the results of the linkage analysis and version 4.03 of the potato reference genome (Potato Genome Sequencing Consortium 2011; Sharma et al. 2013), which is based on the maternal inbred parent of this population (DM1–3) and integrates the potato genome with genetic and physical maps. The phenotypic data for nine traits allowed 10 major genes/QTLs to be mapped using R package *qtl* (Broman et al. 2003). The allele present in DM1–3 and the chromosome for the locus are given in brackets (flanking markers and candidate genes can be found in the paper): (1) tuber pigmentation as yes/no (*D* on 10); (2) pigment colour as red/purple (*p* on 11 and *R* on 2); (3) pigmented skin intensity on 0–10 scale (*pf* on 10); (4) pigmented flesh intensity on 0–10 scale (*Pf2* on 9 and *pf* on 10); (5) yellow flesh as yes/no (*Y* on 3); (6) tuber length/width ratio (*ro* on 10); (7) eye tubers as yes/no (*et* on 5); (8) jelly end as yes/no (*je* on 5); and (9) short anthers as yes/no (*Sa* on 6). For tuber skin colour it should be pointed out that M6 is white, DM1–3 is red and their F_1 purple. Furthermore, although M6 has normal anthers and sheds copious pollen, it was the source of the deleterious recessive allele for short anthers which suggested an interaction between the *sa* allele contributed by *S. chacoense* M6 and the Phureja cytoplasm contributed by DM1–3.

Meijer et al. (2018) did QTL mapping in an F_2 population from the cross *S. tuberosum* × *S. chacoense*. The *S. tuberosum* parent (D2) was a highly heterozygous diploid potato which lacked any wild species in its pedigree and which had been selected for cooking quality, round/oval tubers, yellow tuber flesh, shallow eye depth and no presence of anthocyanins in the tuber skin. The *S. chacoense* parent (DS) was a virtually homozygous diploid clone containing the dominant *Sli*-gene for 'self-compatibility' whose tubers were long, with deep eyes, white tuber flesh and slight anthocyanin levels in their skin. An F_2 population was generated by self-pollination of an F_1 plant and 272 out of 415 F_2 plants produced the minimum of six tubers required for a field trial with two replicates and three-plant plots. A total of 91 SNPs which were heterozygous in the F_1 was used to generate a linkage map. Subsequently, nine F_3 populations derived from another F_1 plant were used to validate the QTLs and monitor the overall increase in the homozygosity level. The distribution of loci in the F_2 population was 7537 homozygous DS, 15,870 heterozygous and 4578 homozygous D2 with 45% D2 alleles and 55% DS alleles. The segregation of markers on the short arm of chromosome 1 (near *S*-locus) and the long arm of chromosome 12 (near *Sli*-locus) was heavily skewed in the direction of DS. A major QTL was detected for tuber shape on chromosome 10, for curved tuber shape also on chromosome 10, for tuber skin colour on chromosome 2 and for tuber flesh colour on chromosome 3; with a minor QTL for tuber shape and minor QTLs for skin colour also detected.

Prinzenberg et al. (2018) studied the genetic variation in a diploid mapping population of size 225 for the steady-state quantum yield of photosystem II of photosynthesis [PSII (Φ_{PSII})] using chlorophyll fluorescence (CF) imaging, which is a rapid, precise and non-invasive method. The diploid germplasm was provided by Solynta (Wageningen, The Netherlands, www.solynta.com) from their diploid F_1

hybrid breeding programme. Using SNP markers, the researchers found two QTLs, one on chromosome 3 and the other on chromosome 5, that were associated with differences in Φ_{PSII}. The higher Φ_{PSII} values were associated with the allele from the DS parent (derived from *S. chacoense*). The researchers also tested the effects of salt stress, low temperature and deficiency in phosphate. For the last two stresses, significant decreases in photosynthetic activity could be shown, revealing the potential for stress detection with CF-based tools. These findings show the potential of high-throughput phenotyping for physiological research and breeding in potato.

These three mapping efforts are just the start. The longer-term aim is true breeding lines of diploid *S. tuberosum* which can readily be selfed and crossed and which display disomic inheritance for the kind of genetical studies that have proved possible with inbred lines of outbreeding maize. Furthermore, research on using such lines in potato breeding and for the production of diploid hybrid cultivars for propagation through true botanical seed is now well advanced (Lindhout et al. 2011, 2018; Jansky et al. 2016).

2.7 Unreduced (2*n*) Gametes from Diploid Potatoes

Breeding as well as genetical studies can be done at the diploid level with a return to the tetraploid level through unreduced (2*n*) gametes. It is therefore time to consider the production of 2*n* gametes in diploids as a way of returning to the tetraploid level for further breeding and cultivar production.

2.7.1 Unreduced (2n) Gametes and Sexual Polyploidization

A number of mechanisms during the pre-meiotic, meiotic and post-meiotic stages of gamete formation may produce 2*n* gametes (Veilleux 1985), referred to as diplandroid and diplogynoid for pollen and egg, respectively. They can be viewed as abnormal events resulting from genetic mutations that affect the outcome of meiosis. In potatoes they can be classified genetically into first division restitution (FDR) or second division restitution (SDR) (Peloquin et al. 1999), as shown in Fig. 2.1.

Starting with microsporogenesis, FDR can arise from an abnormal parallel orientation of the spindles during metaphase II and anaphase II that prevents cell division, and consequently two 2*n* microspores are formed following cytokinesis. The trait is simply inherited as a Mendelian recessive (*ps*), as reported by Mok and Peloquin (1975). FDR can also arise from fused spindles during the second meiotic division, but the genetic consequences are the same (Ramanna 1979). In contrast, SDR occurs when the first division of meiosis is followed by a premature cytokinesis so that the second division does not occur, but the chromatids do separate. Consequently, a dyad of two 2*n* microspores is formed. Premature cytokinesis is also simply inherited as a Mendelian recessive ($pc-1$ and $pc-2$) (Mok and

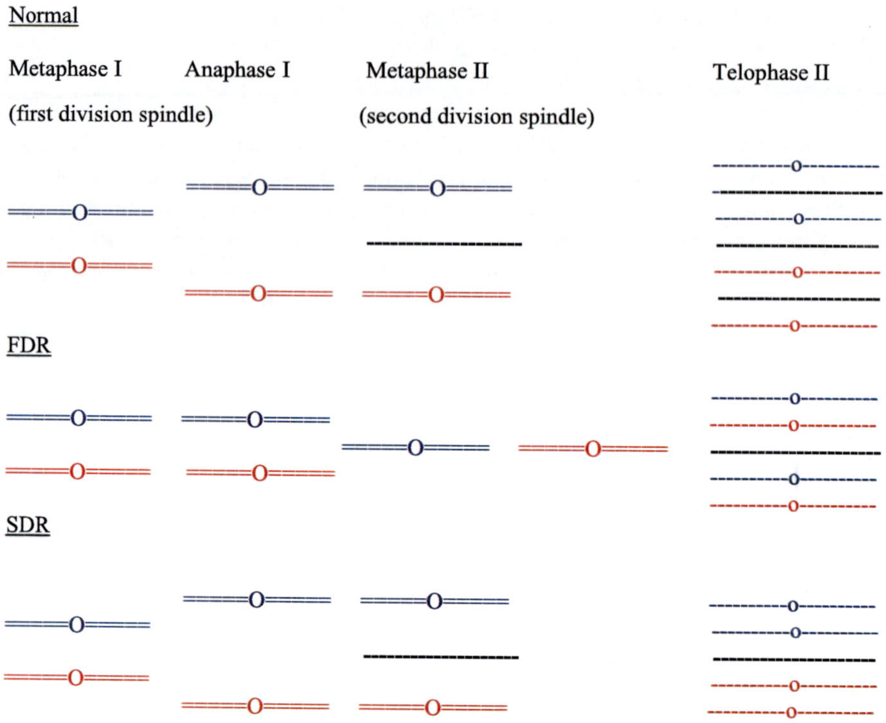

Fig. 2.1 Reduced gametes from normal meiosis in diploids and unreduced gametes from first division restitution (FDR) and second division restitution (SDR): diagrams show two homologous chromosomes, centromeres (O and O) and adjoining sections of chromosomes (chromatids ----- and -----) but no chiasmata, and cytokinesis (----) (from Bradshaw 2016 with permission)

Peloquin 1975). Synaptic mutants have been found which cause poor pairing and/or reduced chiasmata frequencies in microsporogenesis (Jongedijk and Ramanna 1988). When the synaptic mutant *sy*-3 is combined with 'parallel spindles', it can produce FDR 2*n* pollen with no crossing over (NCO), and hence the mechanism is called FDR-NCO. In megasporogenesis, the predominant mechanism of 2*n* egg formation is omission of the second meiotic division, again simply inherited as a Mendelian recessive (*os*), and resulting in two 2*n* megaspores of which one is functional (Werner and Peloquin 1990). Another simply inherited mechanism is failure of cytokinesis after the second meiotic division (*fc*). Both mechanisms are genetically second division restitutions. FDR 2*n* eggs can be produced through simply inherited desynapsis (*ds* − 1) resulting in a direct equational division of univalent chromosomes at anaphase I (pseudohomotypic division) (Jongedijk et al. 1991).

Diploid species and dihaploid-species hybrids in general produce both 2*n* and *n* gametes (Mendiburu and Peloquin 1977). Matings between 4*x* and 2*x* parents (4*x*-2*x* and 2*x*-4*x* crosses) give rise to almost entirely 4*x* progeny due to a 'triploid block' mechanism. Matings between 2*x* parents (2*x*-2*x* crosses) produce 2*x* and 4*x* offspring, the frequencies being cross dependent (Mendiburu and Peloquin 1977). Hence the 4*x* offspring have to be distinguished by chromosome counts or a ploidy correlated trait such as mean number of chloroplasts in stomata (Tai 1994).

2.8 The Genetic Consequences of FDR and SDR in Diploids

We will now briefly consider the genetic consequences of FDR and SDR. More details can be found in the reviews by Tai (1994) and Peloquin et al. (1999). We need to return to the diagram above, add a chiasma and look at the consequences for heterozygous loci between the centromere and chiasma (*A*/*a*) and beyond the chiasma (*B*/*b*) (Fig. 2.2).

With FDR, no-exchange tetrads (NET) are formed (strictly dyads are formed) when there is no crossing over between chromatids carrying *A* and *a*. All 2*n* gametes are of genotype *Aa*. Single-exchange tetrads (SET) are formed when there is crossing over between chromatids carrying *B* and *b*. Two types of chromosomal orientation are possible at metaphase II with SET, as shown in Fig. 2.2. The first orientation results in 2*n* gametes with genotypes *BB* and *bb* in a 1: 1 ratio, whereas the second orientation results in all *Bb* genotypes. If the two orientations are equally likely, then we expect ratios of $\frac{1}{4}$ *BB*: $\frac{1}{2}$ *Bb*: $\frac{1}{4}$ *bb*. If the frequency of SET is β, then the ratio of *BB*, *Bb* and *bb* gametes is $\frac{1}{4}\beta$: $(1-\frac{1}{2}\beta)$: $\frac{1}{4}\beta$. If one chiasma is equally likely to occur anywhere along a chromosome arm, the chromosome is 50 cM long and β increases from zero to one in a linear fashion as you move from centromere to the end of the chromosome arm. Furthermore, the average value of β for the chromosome arm is one half, and the average frequency of transmission of heterozygosity from diploid parent to 2*n* gamete is 0.75. If we extend considerations to two-, three- and four-strand double crossovers (DET), then for a heterozygous locus (*C*/*c*) beyond the second chiasma we find ratios of $\frac{1}{8}$ *CC*: $\frac{3}{4}$ *Cc*: $\frac{1}{8}$ *cc*; in other words a heterozygous frequency of 0.75, as shown by Cuenca et al. (2011). With two crossovers per chromosome arm, we have an arm length of 100 cM (one Morgan) and a problem. What is the distribution of these two crossovers? Do they occur at random or do they interfere with one another? Cuenca et al. (2011) considered a number of possibilities including interference. Here I am just going to look at one simple situation (Table 2.3). The level of heterozygosity falls from 100% at the centromere to 75% one-third the way along, to 66.7% two-thirds the way along, and back to 75% at the end of the chromosome arm; with an average value of near 75% for the whole arm.

With SDR and NET, the 2*n* gametes have genotypes *AA* and *aa* with equal frequency, whereas with SET, the chromosomal orientation at metaphase II does not alter the outcome of only *Bb* gametes. Again, assuming the frequency of SET is β, the ratio of *BB*, *Bb* and *bb* gametes is $\frac{1}{2}(1-\beta)$: β: $\frac{1}{2}(1-\beta)$. If one chiasma is equally

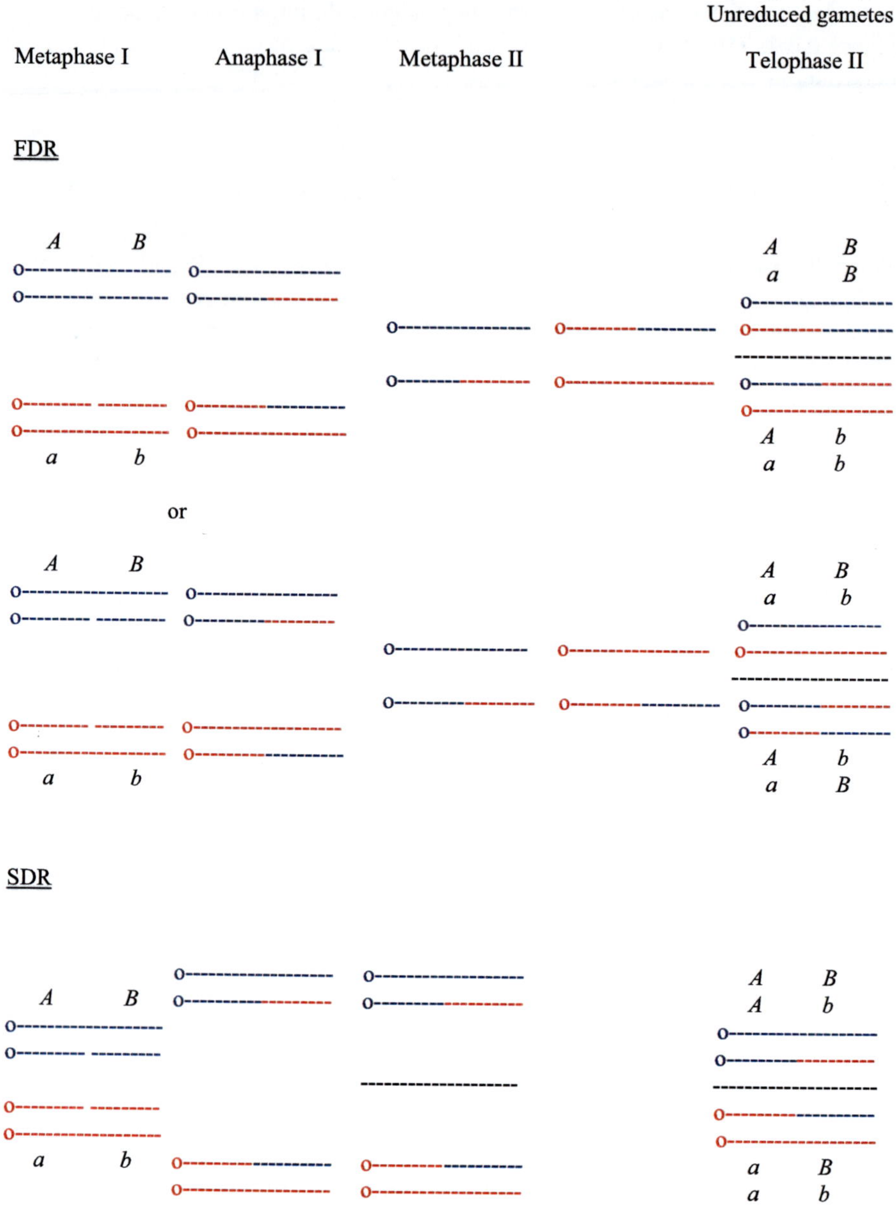

Fig. 2.2 The genetical consequences of producing unreduced (2*n*) gametes by FDR and SDR. (From Bradshaw 2016 with permission)

Table 2.3 Expected heterozygosity (H%) of locus *l*, which is *d* cM from centromere on chromosome arm of length 100 cM, in unreduced (2*n*) gametes produced by FDR and SDR, assuming that there are always two chiasmata (χ) distributed at random along chromosome arm where *P* is the probability of a chiasma between the centromere and locus *l*

Probability (*P*) of a χ	0.0	0.1	0.2	0.3	0.4	0.5	0.6	0.67	0.7	0.8	0.9	1.0
Average number of χ	0.0	0.2	0.4	0.6	0.8	1.0	1.2	1.33	1.4	1.6	1.8	2.0
Locus *l* position *d* cM	0	10	20	30	40	50	60	66.7	70	80	90	100
FDR H%	100	90.8	83.0	76.8	72.0	68.8	67.0	66.7	66.8	68.0	70.8	75.0
SDR H%	0	18.5	34.0	46.5	56.0	62.5	66.0	66.7	66.5	64.0	58.5	50.0

			H%	
			FDR	SDR
Probability of no chiasma	$(1-P)^2$	NET	100%	0%
Probability of one chiasma	$2P(1-P)$	SET	50%	100%
Probability of two chiasmata	P^2	DET	75%	50%

From Bradshaw (2016) with permission

likely to occur anywhere along a chromosome arm, again the average value of β for the chromosome arm is one half, but the average frequency of transmission of heterozygosity from diploid parent to 2*n* gamete is 0.50. When $\beta = \frac{2}{3}$, the ratio of *BB*, *Bb* and *bb* gametes is 1: 4: 1 for both FDR and SDR. If we extend considerations to two-, three- and four-strand double crossovers (DET) for SDR, then for a heterozygous locus (*C*/*c*) beyond the second chiasma we find ratios of $\frac{1}{4}$ *CC*: $\frac{1}{2}$ *Cc*: $\frac{1}{4}$ *cc*; in other words a heterozygous frequency of 0.50, as shown by Cuenca et al. (2011). Again, with two crossovers per chromosome arm, I am just going to consider the same situation as for FDR (Table 2.3). The level of heterozygosity increases from 0% at the centromere to 50% one-third the way along, to 66.7% two-thirds the way along, and back to 50% at the end of the chromosome arm; with an average value of near 50% for the whole arm. Different answers are obtained with different models (Cuenca et al. 2011), but in all cases the average frequency of transmission of heterozygosity from diploid parent to 2*n* gamete is much higher for FDR (about 75%) than for SDR (about 50%). Furthermore, from 0 cM at the centromere to 33 cM, it is less than 50% with SDR and more than 75% with FDR. FDR gametes also maintain and transmit a large amount of inter-locus epistatic interactions, which are important in maximizing heterosis, but are not considered here.

$$\text{FDRH\%} = 100(1-P)^2 + 100P(1-P) + 75P^2$$

$$\text{SDRH\%} = 200P(1-P) + 50P^2$$

Tai (1994) reviewed the genetic consequences of 2*n* gametes for sexual polyploidization in 2*x* × 2*x* crosses (FDR × FDR, FDR × SDR and SDR × SDR) and 4*x* × 2*x* crosses (4*x* × FDR and 4*x* × SDR), where the tetraploid genotype can have up to four different alleles ($A_1A_2A_3A_4$) at a locus. He concluded that the overall advantage of FDR in transmitting heterozygosity from 2*x* parents to 4*x* progeny depends on the allelic diversity present in the parents. In diploid by diploid crosses, the SDR × SDR cross is expected to be inferior for maintaining hybrid vigour in the progenies unless the diallelic genotypes have the same degree of heterosis as the tri- and tetra-allelic ones. Furthermore, the FDR × FDR cross is favoured for the possibility of creating a genetically more uniform hybrid population than the other two types of crosses, since it has a higher frequency of tetra-allelic genotypes ($A_1A_2A_3A_4$) for loci up to 30 cM (out of 50 cM) from the centromere, and a low frequency of diallelic genotypes for loci located anywhere in a chromosome. The use of $2n$ gametes in potato breeding has been reviewed by Tai (1994) and Ortiz (1998), and more recently by Carputo and Frusciante (2011).

2.9 Meiosis in Autotetraploids, Double Reduction and Tetrasomic Inheritance

In a true autotetraploid, the four sets of chromosomes are entirely homologous; meiotic pairing within each group of four homologous chromosomes is completely random and tetrasomic inheritance always occurs. DNA synthesis takes place before the start of the first meiotic division so that each homologous chromosome comprises two identical sister chromatids which are joined at the centromere (Fig. 2.3). The first meiotic division begins with competition between homologous chromosomes as they try to form pairs, followed by chiasma formation and crossing over. With four chromosomes there are a number of possible outcomes depending upon whether or not changes of pairing partner occur along the lengths of the chromosomes. In the simplest situation (chromosomal segregation), the four homologous chromosomes pair at random to give two bivalents (2II's) and the two divisions of meiosis proceed normally to give equally viable diploid gametes. Male and female gametes then unite at random to form tetraploid offspring. More complex pairing patterns can result in quadrivalents or various combinations of trivalents, bivalents and univalents. Hence quadrivalent formation is evidence for autotetraploidy, but its absence does not prove disomic inheritance. Quadrivalents can be a chain (CIV) or a ring (RIV), with three chiasmata sufficient for a chain and four for a ring. Five or more chiasmata would give a complex pattern but may not occur in practice, particularly with short chromosomes. Quadrivalent orientations on the spindle that result in 2 + 2 disjunction (i.e. the migration of two chromosomes to each pole) produce balanced diploid chromosome sets. Where a trivalent and univalent (CIII + I) occur, the trivalent will usually disjoin two and one, and the univalent to one or the other pole, or be left behind and lost. The simplest trivalent has two chiasmata and forms a chain, but more complex arrangements are possible, particularly if there are three or more chiasmata. One bivalent and two univalents (II + 2I's) and four univalents (4I's) are the remaining possibilities. The second division of meiosis is straightforward in that the centromeres lie on the spindle equator and then split into daughter centromeres which disjoin to opposite spindle poles. Overall, some but not all outcomes of meiosis in autotetraploids lead to balanced diploid chromosome sets and hence viable and fertile diploid gametes. Thus, some reduction in fertility must be anticipated, but may or may not occur. Back in 1953, Swaminathan and Howard

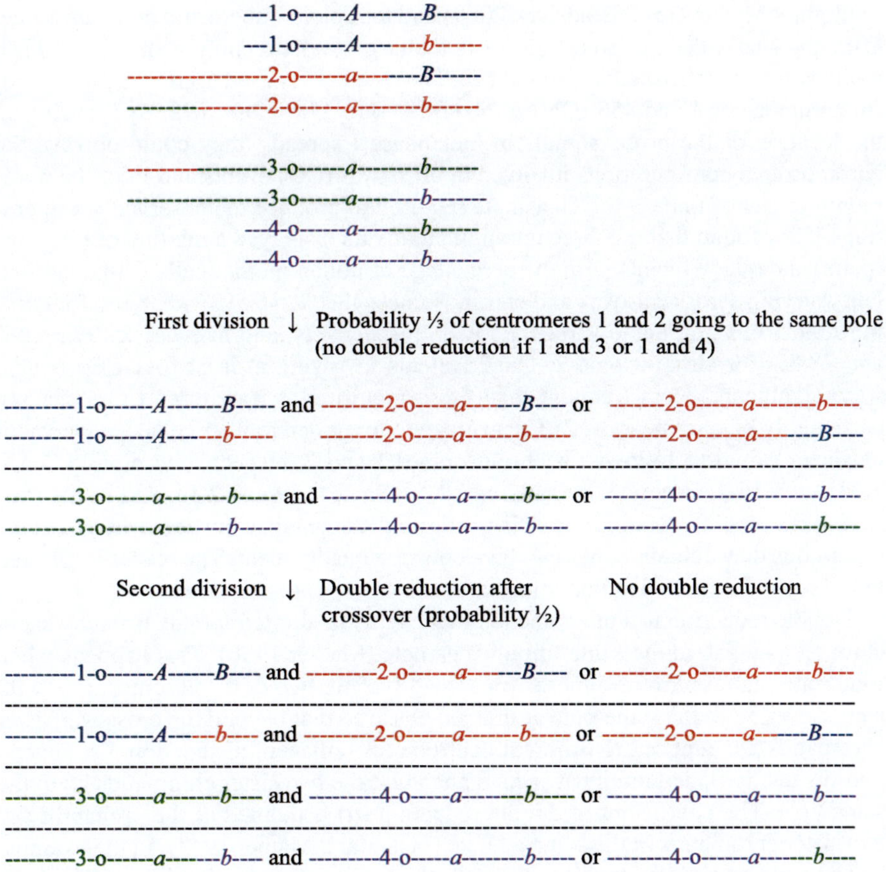

Fig. 2.3 Double reduction (*BB*, *bb*, *bb* and *bb* in same diploid gamete) in an autotetraploid (frequency 0 before and 1/6 after crossover between loci *A* and *B*). Four homologous chromosomes pair: 1 and 4 plus 2 and 3 before the centromeres; 1 and 2 plus 3 and 4 after the centromeres, with crossing over between pairs of chromatids (sister chromatids are shown in same colour); chromosome 1 carries alleles *A* and *B* and chromosomes 2, 3 and 4 carry alleles *a* and *b*. (Modified from Bradshaw 2016 with permission)

(1953) reported that low frequencies of quadrivalents, trivalents and univalents occur in potatoes, but bivalents predominate.

More recently, Bourke et al. (2015) concluded from the segregation of 3273 molecular markers (*Mmmm* × *mmmm*) that quadrivalents accounted for between 20 and 30% of all meiotic pairing configurations in their biparental tetraploid cross (235 offspring), which translates into between 2.4 and 3.6 quadrivalents per cell. Their conclusion was based on an estimate of 6% double reduction at the distal ends of chromosomes. Furthermore, they found no evidence of preferential pairing within the groups of four chromosomes. Then even more recently, Choudhary et al. (2020) have used Fluorescence In Situ Hybridization (FISH) to differentiate individual meiotic chromosomes 1 and 2 in diploid Phureja cultivar Scapa (Mayan

Twilight × Mayan Gold; Bradshaw 2009) and tetraploid Tuberosum cultivars Sante, Cara and Maris Peer. Chromosome 1 is the largest potato chromosome at ~90 Mb while potato 2 is the smallest at ~50 Mb. The researchers used 5S rDNA probe (red) for chromosome 1 and 45S rDNA probe (green) for chromosome 2. By focusing on the location of the probe signals in metaphase I spreads, they could observe the chromosomal configurations involved in crossover (CO) events and infer the minimum number of underlying chiasmata required to produce the observed configurations. They found that the three tetraploid cultivars displayed a mixture of bivalents (pairs) and quadrivalents with the percentage of pollen mother cells with quadrivalents varying across cultivars and chromosomes: 48.1%, 15.5% and 38.0% for chromosome 1 in Cara, Sante and Maris Peer, respectively, and likewise 25.9%, 7.0% and 28.5% for chromosome 2. Quadrivalents involving at least four COs (rings) were significantly less frequent than those requiring at least three COs (chains), particularly on chromosome 2. The average (over pollen mother cells) frequency of chismata varied as follows for chromosomes 1 and 2, respectively: diploid Scapa 1.47 and 1.15, tetraploid Cara 3.07 and 2.48, Sante 2.59 and 2.13, and Maris Peer 2.87 and 2.32. Pollen mother cells in the diploid all contain a bivalent whereas in the tetraploids they contain either two bivalents or a quadrivalent. The reader is referred to the paper for more detailed information.

Double reduction is a phenomenon that occurs in autotetraploids whereby sister chromatids finish in the same (diploid) gamete (Mather 1936). This happens when four homologous chromosomes form a quadrivalent, two of the chromosomes with a crossover go to the same pole at first anaphase so that beyond the crossover sister chromatids are attached to different centromeres (adjacent disjunction, i.e. genetic non-disjunction), and finally at second anaphase the two sister chromatids go to the same pole. The coefficient of double reduction (α) is defined as the probability of two sister chromatids in the same gamete (Fisher and Mather 1943). In the example just considered and illustrated in Fig. 2.3, the maximum value for the coefficient of double reduction is 1/6. However, if a double crossover is possible between the centromere and the section of chromatid under consideration, the maximum coefficient is 1/4, as pointed out by Luo et al. (2006). The value of α therefore depends upon the frequency and type of quadrivalent (ring or chain), the distance of the locus from the centromere, and the position and frequency of double crossovers (probably rare in potatoes on same side of centromere).

The gametic output of a single locus with four different alleles ($A_1A_2A_3A_4$) follows from the definition of α. Each of the four homologous chromosomes can provide a pair of sister chromatids so that the probability for any one chromosome is $\frac{1}{4}\alpha$. Then for the $(1 - \alpha)$ situations where this does not occur, two chromosomes (non-sister chromatids) are chosen at random so that the overall gametic outcome from ($A_1A_2A_3A_4$) is:

$$\frac{1}{4}\left(A_1A_1 + A_2A_2 + A_3A_3 + A_4A_4\right)\alpha + \frac{1}{6}\left(A_1A_2 + A_1A_3 + A_1A_4 + A_2A_3 + A_2A_4 + A_3A_4\right)(1-\alpha)$$

When the coefficient is zero, one has chromosomal segregation whereas when it is 1/7 one appears to have random chromatid segregation; but meiosis is always a two-division process with 'half-centromeres' and chromatids separating in the second division. Chromosomal segregation with two alleles *A* and *a* at a locus is shown in Table 2.4 for some of the commonly encountered crosses in potato breeding and genetics.

Hence for the cross *Aaaa* × *aaaa*, with chromosomal segregation one has a 1:1 ratio of *Aaaa*: *aaaa* whereas with chromatid segregation one has a 1:12:15 ratio of *AAaa*: *Aaaa*: *aaaa*. With $\alpha = \frac{1}{6}$, the ratio is 1:10:13 for gametes *AA*:*Aa*:*aa*, as can be seen in Fig. 2.3: $\frac{1}{4}$ *AA* multiplied by $\frac{1}{2}$ (second division) and then multiplied by $\frac{1}{3}$ (first division) gives $\frac{1}{24}$. With real data, deviations from these expected ratios occur by chance and we need a test of significance for the deviations. A commonly used one for categorical data is the chi-square (X^2) goodness-of-fit statistic devised by Karl Pearson in 1900 (Bulmer 1967). If a category has an observed count of *o* observations and an expected count *e* under some hypothesis, then the goodness-of-fit test statistic is formed by summing the quantity $(o - e)^2/e$ over categories:

$X^2 = \Sigma\ (o - e)^2/e$, where summation is over categories.

If the hypothesis is true, it is likely that *o* will differ from *e* by chance in each category, but not by too much. How far apart they may be by chance can be found from the properties of the chi-square distribution. Chi-square values can be looked up in tables such as those compiled by Fisher and Yates (1963). Genetical examples can be found in textbooks of genetics and I included one in Chap. 4 of my book *Plant breeding*: *past, present and future* (Bradshaw 2016).

An appreciation of tetrasomic inheritance allowed Mendelian analysis of major genes and such knowledge proved useful to breeders in designing introgression programmes and progeny testing for determining copy number of major genes in potential parents (Bradshaw and Mackay 1994; Mackay 2005), as can be seen in the first four crosses in Table 2.4. In anticipation of mapping work at the tetraploid level, Luo et al. (2000) presented a theoretical model and a statistical analysis for predicting the genotypes of a pair of tetraploid parents at a codominant or dominant molecular marker locus based on their own and their progeny's gel-band patterns at

Table 2.4 Chromosomal segregation with two alleles *A* and *a* at a locus (divide values by 36 to get proportions)

		Offspring					Ratio
Parent 1	Parent 2	*AAAA*	*AAAa*	*AAaa*	*Aaaa*	*aaaa*	*A---*: *aaaa*
Aaaa	*aaaa*	0	0	0	18	18	1:1
AAaa	*aaaa*	0	0	6	24	6	5:1
AAAa	*aaaa*	0	0	18	18	0	1:0
AAAA	*aaaa*	0	0	36	0	0	1:0
Aaaa	*Aaaa*	0	0	9	18	9	3:1
AAaa	*Aaaa*	0	3	15	15	3	11:1
AAaa	*AAaa*	1	8	18	8	1	35:1

that locus. The theory allows for null alleles and for any degree of double-reduction to be modelled. Computer simulations revealed that a satisfactory prediction of tetrasomic segregation can be achieved in most circumstances by using a full-sib progeny size of about 100.

2.10 Differences Between Tetrasomic and Disomic Inheritance

There are three important ways in which tetrasomic inheritance differs from disomic inheritance. Diploid gametes mean that inbreeding effects (A and A in AA can be identical by descent) and digenic (dominance effects) interactions (A and a interaction in Aa) can be transmitted from parents to offspring. Furthermore, in a random mating population, the single-locus equilibrium is not attained in one generation and the genotype frequencies at equilibrium are not the products of the gene frequencies unless $\alpha = 0$. We shall encounter the consequences for digenic interactions when considering quantitative traits, so let us take a brief look at the other two differences in reverse order.

The Hardy–Weinberg principle holds true under tetrasomic inheritance in the sense that the zygotic genotype frequencies are given by the products of the appropriate gametic frequencies. With multiple alleles at a locus, when $\alpha = 0$, the gametic and genotypic frequencies at equilibrium are $(\sum p_i A_i)^2$ and $(\sum p_i A_i)^4$, respectively, where p_i is the frequency of allele A_i at locus A. This is the assumption that we will make in the simple examples presented in this book. The general case and the rate of approach to equilibrium can be found in the book *An Introduction to Population Genetics* by Crow and Kimura (1970). Two specific results are as follows. First, at equilibrium with $\alpha = 1/7$, we have the equivalent of 20% inbreeding in the population. Second, the reduction in disequilibrium for gametes is $(1 - \alpha)/3$ each generation so that when $\alpha = 0$, three generations of random mating reduce the disequilibrium to 3.7% of its initial value. So, the comparison between tetrasomic and disomic inheritance is as follows.

With disomic inheritance, one generation of random mating establishes the equilibrium between gene and genotype frequencies at a single locus (the well-known Hardy–Weinberg equilibrium). For two alleles at a locus we have:

$(p_1A_1 + p_2A_2)^2 = p_1^2A_1A_1 + 2p_1p_2A_1A_2 + p_2^2A_2A_2$ where p_1 and p_2 are the frequencies of alleles A_1 and A_2 and $p_1 + p_2 = 1$.

With tetrasomic inheritance the single-locus equilibrium is not attained in one generation and the genotype frequencies at equilibrium are not the products of the gene frequencies unless there is no double reduction. When there is no double reduction, for four alleles at equilibrium we have:

$(p_1A_1 + p_2A_2 + p_3A_3 + p_4A_4)^4$ which results in 55 genotypes.

If we self-pollinate our populations to homozygosity, we obtain two diploid genotypes (A_1A_1 and A_2A_2, frequencies p_1 and p_2) and four tetraploid genotypes ($A_1A_1A_1A_1$, $A_2A_2A_2A_2$, $A_3A_3A_3A_3$ and $A_4A_4A_4A_4$, frequencies p_1, p_2, p_3 and p_4). The array of genotypes in an arbitrary generation arising from an arbitrary initial population under a particular system of inbreeding can be determined by the generation matrix approach of Fisher (1965). The generation matrix for selfing an autotetraploid can be found in the book chapter by Bradshaw (1994). The recurrence equation for the inbreeding coefficient (F) for selfing from generation (t − 1) to t is as follows (Kempthorne 1957).

$$\text{Diploids:} \quad F_t = \frac{1}{2} + \frac{1}{2}F_{t-1}$$

$$\text{Tetraploids (in absence of double reduction):} \quad F_t = \frac{1}{6} + \frac{5}{6}F_{t-1}$$

The inbreeding coefficient (F) for diploids is the probability that the two alleles at a locus are identical by descent and for tetraploids is the probability that a randomly chosen pair of alleles at a locus are identical by descent. Starting with a non-inbred population ($F_0 = 0$), seven generations of selfing increases F to 0.9922 in a diploid population and to 0.7209 in a tetraploid population, where $F = 1$ is a completely inbred population. Looked at another way, seven generations of selfing are required to make an initially heterozygous diploid 99% homozygous, whereas in tetraploids 27 generations are required in the absence of double reduction (i.e. chromosomal segregation) (Bradshaw 1994), and this would not be done in practice.

If we random mate (cross in all combinations) the two diploid homozygotes, we re-establish the equilibrium random mating: $p_1^2A_1A_1 + 2p_1p_2A_1A_2 + p_2^2A_2A_2$. Furthermore, the effects of inbreeding are removed. Hence, we have two ways of multiplying the 'best' genotype (say A_1A_2) in the population: either by vegetative reproduction or by sexual hybridization of the appropriate two inbred lines ($A_1A_1 \times A_2A_2$). Hence there is no genetical advantage or disadvantage in diploid TPS propagated cultivars compared with diploid tuber propagated cultivars, and cycles of hybridization and selection can improve both types of cultivar for the selected traits, as we shall see in Chap. 8.

In contrast, if we random mate (cross in all combinations) the four tetraploid homozygotes, we secure just 10 out of the 55 genotypes in an equilibrium population, namely four homozygotes ($A_1A_1A_1A_1$, $A_2A_2A_2A_2$, $A_3A_3A_3A_3$ and $A_4A_4A_4A_4$) and six heterozygotes each with two pairs of different alleles ($A_1A_1A_2A_2$, $A_1A_1A_3A_3$, $A_1A_1A_4A_4$, $A_2A_2A_3A_3$, $A_2A_2A_4A_4$ and $A_3A_3A_4A_4$), and the effects of inbreeding have not been completely removed. It takes another round of random mating of these 10 genotypes in the correct frequencies to secure the complete set of 55 genotypes also in the correct frequencies. The theory of inbreeding and crossbreeding in autotetraploids is therefore more complicated than in diploids, and there are differences which have consequences for breeding strategies aimed at exploiting heterosis (hybrid vigour) in hybrid cultivars. Hence in tetraploids we can only multiply genotypes with three or four different alleles by vegetative means as two appropriate inbred lines for sexual hybridization do not exist. Genotypes with three or four different alleles only exist in the genetically variable double-cross hybrids which can be superior in performance to the single-cross hybrids (Gallais 2003); for example, $A_1A_1A_2A_2 \times A_3A_3A_4A_4$ (composition 1 $A_1A_1A_3A_3$ 1 $A_1A_1A_4A_4$ 1 $A_2A_2A_3A_3$ 1 $A_2A_2A_4A_4$, 4 $A_1A_1A_3A_4$ 4 $A_2A_2A_3A_4$ 4 $A_1A_2A_3A_3$ 4 $A_1A_2A_4A_4$ and 16 $A_1A_2A_3A_4$) would be higher yielding than $A_1A_1A_1A_1 \times A_2A_2A_2A_2$ (composition $A_1A_1A_2A_2$) if genotypes with three or four different alleles were superior to those with two different alleles; but this is an unproved assumption. If tetraploids are inherently superior to diploids, one might want to produce tetraploid genotype $A_1A_1A_2A_2$ rather than diploid genotype A_1A_2 by hybridization of inbred lines $A_1A_1A_1A_1$ and $A_2A_2A_2A_2$. If tetraploid genotype $A_1A_2A_3A_4$ is superior to genotypes with fewer different alleles, one might accept a genetically variable TPS (true potato seed) cultivar with the highest possible frequency of genotype $A_1A_2A_3A_4$ or one might argue that every effort should be made to grow a clonally propagated cultivar with maximum heterozygosity.

As already mentioned, the inbreeding coefficient (F) in a diploid genotype is simply the probability that the two alleles at a locus are identical by descent, i.e. they are copies of a gene occurring previously in the ancestry, or one is a copy of the other (Malécot 1948). It depends on the mating system, with self-pollination the strongest form of inbreeding. Furthermore, for a diploid, the inbreeding coefficient of an individual equals the coancestry of its parents. The coefficient of coancestry (Φ), or parentage, is the probability (P) that a pair of alleles taken at random from each of the two individuals, X and Y, are identical by descent.

$$F = \Phi_{XY} = \tfrac{1}{4}\left[P(x_1 \equiv y_1) + P(x_1 \equiv y_2) + P(x_2 \equiv y_1) + P(x_2 \equiv y_2)\right]$$

The coancestry of an individual with itself (Φ_{11}) is defined as the probability that two genes taken at random, with replacement, from the individual are identical by descent (Mendoza and Haynes 1973).

$$\Phi_{11} = \tfrac{1}{2} + \tfrac{1}{2}F$$

In contrast, the inbreeding coefficient (F) in a tetraploid is the probability that a randomly chosen pair of alleles at a locus are identical by descent. Hence for an individual $X_iX_jY_kY_l$ formed from gametes X_iX_j and Y_kY_l we have the following.

$$F = \frac{1}{6}\left[P(x_i \equiv x_j) + P(x_i \equiv y_k) + P(x_i \equiv y_l) + P(x_j \equiv y_k) + P(x_j \equiv y_l) + P(y_k \equiv y_l)\right]$$
$$= \frac{1}{6}\left(F_x + 4\Phi_{XY} + F_y\right)$$

$$\Phi_{11} = \tfrac{1}{4} + \tfrac{3}{4}F$$

The probability of a gamete containing two alleles that are identical by descent [$P(x_i \equiv x_j)$] is by definition the inbreeding coefficient of the parent from which it came.

Mendoza and Haynes (1973, 1974) used these concepts of coancestry and inbreeding to discuss the effects of inbreeding tetraploid potatoes and their implications for potato breeding, and to determine the genetic relationships among potato cultivars grown in the USA. They concluded that the highly significant figures obtained in the statistical analysis of per cent yield reduction, relative to the inbreeding coefficient of successive generations of selfing, indicated that the tetraploid potato exhibits intense inbreeding depression. Furthermore, they concluded that the gene pool of North American cultivars in 1974 was rather limited and inbred due to close relationships between their parents and ancestors. For a full description of inbreeding further parameters are required, as reviewed briefly by Bradshaw (1994), and seen in the books by Kempthorne (1957), Fisher (1965) and Gallais (2003). Here it will suffice to point out another important difference between disomic and tetrasomic inheritance. In a diploid species the effects of inbreeding can be removed by one generation of random mating, whereas in a tetraploid, the inbreeding coefficient of the offspring depends on the inbreeding coefficients of the parents, and the probability that two pairs of alleles are identical by descent is greater than zero if both parents are inbred.

Finally, on inbreeding, Haynes and Potts (1993) looked at ways of minimizing inbreeding in tetraploids derived through sexual polyploidization involving $2n$ gametes in tetraploid × diploid, tetraploid × dihaploid-species hybrids or diploid × diploid hybridizations. For the two breeding strategies involving tetraploid female parents, they showed less inbreeding in the derived tetraploid from a diploid parent producing $2n$ pollen by first division restitution rather than by second division restitution, when the frequency of single exchange tetrads in the diploid is less than 2/3. In bilateral sexual polyploidization, they showed that the inbreeding coefficient of a derived tetraploid for a given set of parents from a first division × first division restitution cross is less than the inbreeding coefficient of a second division × first division restitution cross which in turn is less than the inbreeding coefficient of a second division × second division restitution cross when the frequency of single exchange tetrads is less than 2/3. In summary, the inbreeding coefficient of derived tetraploids is a complex function of the coancestry of the parents, the inbreeding of the parents, the coefficient of double reduction in the tetraploid parent and the frequency of single exchange tetrads in the diploid parent(s), and finally on the mechanism of $2n$ gamete formation.

2.11 Quantitative Genetics Theory

We now need to take a brief look at what might be called classical quantitative genetics theory. For much of the second half of the twentieth century, breeders had to rely on combining ability analysis for most of their genetic knowledge and the concepts of heritability, additive and non-additive genetic variation, genotype × environment interaction and population improvement in determining breeding strategies, choosing parents and in predicting and improving responses to selection and rates of progress (Bradshaw and Mackay 1994). We shall therefore look at some quantitative genetics theory before examining more recent work on linkage mapping and quantitative trait locus (QTL) analysis at the tetraploid level.

Once cultivars are available that are reasonably well-adapted to local conditions, breeders try to make further improvements in yield from crosses among such cultivars and advanced breeding clones. Although it is highly unlikely that the cultivars and advanced breeding clones will be a population of clones (genotypes) in genetic equilibrium, readers should be aware of some results from such simple theory for quantitative traits like yield. We are going to consider a single locus *A* with just two alleles, *A* (frequency *p*) and *a* (frequency *q*), where there are main effects of alleles and diallelic interactions but no triallelic and tetra-allelic interactions, and where there is chromosomal segregation (Table 2.5). Multiple alleles with triallelic and tetra-allelic interactions at a locus and non-allelic interactions between two loci were considered by Kempthorne (1957), Wright (1979), Wricke and Weber (1986) and Gallais (2003) with the results summarized by Bradshaw (1994). There are 14 components of genetic variance for a random mating population in equilibrium but it is not possible to obtain (accurate) estimates of so many components.

It is assumed that the population is in the tetraploid equivalent of Hardy–Weinberg equilibrium and that the next generation is produced by random-mating. Furthermore, it is assumed that there is no correlation between the genotypes assessed and their environment (i.e. assessment is done in a properly randomized trial). Under these circumstances, the covariance of allele frequencies and phenotypic values equals the covariance of allele frequencies and genotypic values (+ *e*'s and − *e*'s cancel out), whereas the phenotypic variance has an environmental component (+ *e*'s and − *e*'s both square to + e^2).

For my purposes, our simple model is adequate, but needs some explanation: 4*a* is simply the genetic difference between the two homozygotes *AAAA* and *aaaa*; but the additional 'dominance' parameter in the three heterozygotes *AAAa*, *AAaa* and *Aaaa* takes the values 3*d*, 4*d* and 3*d*, respectively (I have opted for *d* rather than *h*), because there are three, four and three digenic interactions (*A* with *a*) in these genotypes (e.g. in *AAAa*, *a* can interact with three *A*'s). Needless to say, different researchers have used different models over the years! The general combining abilities (GCAs) of genotypes are the means of their offspring from mating with gametes in equilibrium frequencies, minus the population mean, and the breeding values of the genotypes are twice the GCAs, at least when defined as the value of an individual when judged by the mean value of its progeny (the GCA is doubled because for GCA the parent provides only half the genes in the progeny, the other half coming at random from the population). This practical definition of the breeding value of a genotype is different from the theoretical one which is defined and calculated in terms of the average effects of its genes (the difference arises because average effects are across the whole population for single alleles whereas progeny are produced by combining diploid gametes in which two alleles may interact). This difference can be seen in the derivation of breeding value given by Endelman et al. (2018), based on Kempthorne (1957).

Table 2.5 Quantitative genetics model for five genotypes at a single locus in genetic equilibrium with two alleles *A* and *a* at frequencies *p* and *q*, respectively, and chromosomal segregation (GCA = general combining ability of genotype; α_t = the average effect of the allele substitution)

Genotype	*AAAA*	*AAAa*	*AAaa*	*Aaaa*	*aaaa*
Frequency	p^4	$4p^3q$	$6p^2q^2$	$4pq^3$	q^4
Frequency *A*	1	$\frac{3}{4}$	$\frac{1}{2}$	$\frac{1}{4}$	0
Genotypic value	$2a$	$a + 3d$	$4d$	$-a + 3d$	$-2a$
Phenotypic value	$2a \pm e$	$a + 3d \pm e$	$4d \pm e$	$-a + 3d \pm e$	$-2a \pm e$
Genotypic value - mean	$4aq - 12pqd$	$a(-p + 3q) - 3(4pq - 1)d$	$2a(-p + q) - 4(3pq - 1)d$	$-a(3p-q) - 3(4pq - 1)d$	$-4ap - 12pqd$
GCA	$2aq - 6pqd + 4q^2d$	$\frac{1}{2}a(-p + 3q) - 1\frac{1}{2}(4pq - 1)d - 2q(p-q)d$	$a(-p + q) - 2(3pq - 1)d + (\frac{2}{3} - 4pq)d$	$-\frac{1}{2}a(3p-q) - 1\frac{1}{2}(4pq - 1)d - 2p(q-p)d$	$-2ap - 6pqd + 4p^2d$
Breeding value (practical)	$4aq - 12pqd + 8q^2d$	$a(-p + 3q) - 3(4pq - 1)d - 4q(p-q)d$	$2a(-p + q) - 4(3pq - 1)d + 2(\frac{2}{3} - 4pq)d$	$-a(3p-q) - 3(4pq - 1)d - 4p(q-p)d$	$-4ap - 12pqd + 8p^2d$
Breeding value (theoretical)	$4aq - 12pqd + 12q^2d = 4q\alpha_t$	$a(-p + 3q) - 3(4pq - 1)d - 6q(p-q)d = (3q-p)\alpha_t$	$2a(-p + q) - 4(3pq - 1)d + 3(\frac{2}{3} - 4pq)d = (2q-2p)\alpha_t$	$-a(3p-q) - 3(4pq - 1)d - 6p(q-p)d = (q-3p)\alpha_t$	$-4ap - 12pqd + 12p^2d = -4p\alpha_t$
Selfed family	$2a$	$a + 2\frac{1}{2}d$	$3\frac{1}{3}d$	$-a + 2\frac{1}{2}d$	$-2a$

The average effects of alleles *A* and *a* are calculated as follows.

Allele *A* can combine with three (*AAA*), two (*AAa*, *AaA*, *aAA*), one (*Aaa*, *aAa*, *aaA*) or zero (*aaa*) *A*s at a locus with combined frequencies p^3, $3p^2q$, $3pq^2$ and q^3. The mean value of these genotypes minus the population mean is the average effect of *A* (α_A) in the population.

$$\begin{aligned}\alpha_A &= (2a)p^3 + (a+3d)3p^2q + (4d)3pq^2 + (-a+3d)q^3 - 2(p-q)a - 12pqd \\ &= q\left[a + 3(q-p)d\right] = q\alpha_t\end{aligned}$$

Likewise, allele *a* can combine with three (*AAA*), two (*AAa*, *AaA*, *aAA*), one (*Aaa*, *aAa*, *aaA*) or zero (*aaa*) *A*s at a locus with combined frequencies p^3, $3p^2q$, $3pq^2$ and q^3. The mean of these genotypes minus the population mean is the average effect of *a* (α_a) in the population.

$$\begin{aligned}\alpha_a &= (a+3d)p^3 + (4d)3p^2q + (-a+3d)3pq^2 + (-2a)q^3 - 2(p-q)a - 12pqd \\ &= -p\left[a + 3(q-p)d\right] = -p\alpha_t\end{aligned}$$

Hence the average effect of the gene (allele) substitution '*a*' to '*A*' is $\alpha_A - \alpha_a = q\alpha_t + p\alpha_t = \alpha_t$.

The theoretical breeding values of individual genotypes are the sum of the average effects of the alleles that they contain because these are the alleles that they transmit to the next generation (Table 2.5):

$$AAAA = 4q\alpha_t$$

$$AAAa = (3q - p)\alpha_t$$

$$AAaa = (2q - 2p)\alpha_t$$

$$Aaaa = (q - 3p)\alpha_t$$

$$aaaa = -4p\alpha_t$$

2.11.1 Simple Statistics

So, let us look at the genetic variation in our population of clones and determine the response in the next sexual generation to selection of the best clones for yield and also the effect of inbreeding on our population. We need the following statistics:

$$\begin{aligned}\text{Population mean}: \mu_x &= (\Sigma x)/n \\ &= \Sigma fx \text{ when } f \text{ is proportion (frequency) of } x \text{ and hence } \Sigma f = 1;\end{aligned}$$

$$\text{Variance } V_x = \sigma_x^{\,2} = \left(\Sigma(x-\mu_x)^2\right)/n = \Sigma f(x-\mu_x)^2 = \Sigma fx^2 - \mu_x^{\,2}; V\mu_x = V_x/n;$$

$$\text{Covariance}\,\text{cov}_{xy} = \left(\Sigma(x-\mu_x)(y-\mu_y)\right)/n = \Sigma f(x-\mu_x)(y-\mu_y) = \Sigma fxy - \mu_x\mu_y$$

Regression of y (random) on x (fixed): slope of regression line $\beta = Cov_{xy}/V_x$

2.11.2 Genetic Variation and Response to Selection

Applying the above statistics to our population we get the following (the algebra is tedious).

Frequency of '*A*' in population = *p*.

Mean of the population of clones (genotypes) $M = 2(p\text{-}q)a + 12pqd$.

Genetic variance of the population of clones $V_G = 4pq[a + 3(q\text{-}p)d]^2 + 24(pqd)^2 = V_A + V_D$, where V_A is the additive genetic variance ($4pq\alpha_t^2$) (variance of theoretical breeding values in Table 2.5), V_D is the component of variance for diallelic (dominance) interactions (variance of genotypic values minus the mean and minus the theoretical breeding values in Table 2.5) and $a + 3(q\text{-}p)d = \alpha_t$, the average effect of the allele substitution '*a*' to '*A*'.

If the population is produced from a mating design in which half-sib and full-sib families can be recognized (hierarchical, factorial or diallel set of crosses), V_A and V_D can be estimated as shown in Table 2.6, but V_D with less precision than V_A. The algebra is tedious and not shown, but can be found in Wricke and Weber (1986). The variation in the population is divided into between and within full-sib families, and the variation between full-sib families is further divided into between half-sib families (GCA) and a remainder which is the variation due to specific combining ability (SCA). Bradshaw and Mackay (1994) provided a table of references to publications on general and specific combining abilities for a range of traits which could be updated with more recent references, but there seems little point as they are germplasm specific and hence would not provide a particularly useful reference table for breeders. Nevertheless, we shall encounter examples throughout this book.

Table 2.6 Components of variance for population in genetic equilibrium under tetrasomic inheritance with chromosomal segregation in absence of triallelic, tetra-allelic and non-allelic interactions: V_A is the additive genetic variance, V_D is the variance for diallelic interactions, GCA is general combining ability and SCA is specific combining ability

	V_A	V_D
Population	1	1
Between full-sib families (FS)	$\frac{1}{2}$	2/9
Within full-sib families (FS)	$\frac{1}{2}$	7/9
Between half-sib families (HS) = GCA	$\frac{1}{4}$	1/36
FS – 2HS = SCA	–	1/6
$4\text{GCA} - \frac{2}{3}\text{SCA}$	1	–
6SCA	–	1

The response to selection in our population of clones is determined from Table 2.5 in two stages as follows: (1) the change in frequency of allele '*A*' as a result of phenotypic selection and (2) the change in population mean as a result of the change in allele frequency.

The covariance of allele '*A*' frequency and phenotypic value equals the covariance of allele '*A*' frequency and genotypic value (assuming clones assessed in properly randomized trial):

$$= pq\left[a+3(q-p)d\right] = pq\alpha_t.$$

The regression (β) of allele '*A*' frequency on phenotypic value is $\beta = pq\alpha_t/V_P$, where V_P is the variance of phenotypic values of the clones: $V_P = V_G + V_E/r = V_A + V_D + V_E/r$, where V_E is the environmental plot to plot variation in the assessment trial of clones and r is the number of replicates of the clones in the trial.

The selection differential $S = M_S - M$, where M_S is the mean of the selected group of clones, M is the mean of the unselected population of clones and the standardized selection differential S/σ_P is the intensity of selection i, where $\sigma_P = (V_P)^{\frac{1}{2}}$.

The expected change in allele '*A*' frequency (Δp) for selection differential S is simply determined from the slope (β) of the linear regression line when we plot the graph of allele '*A*' frequency against the phenotypic values of the clones: $\Delta p = \beta S$.

The effect of changes in allele '*A*' frequency (Δp) on the population mean (ΔM is the change in the population mean) can be derived using the differential calculus.

$$M = 2(p-q)a + 12pqd = 2(2p-1)a + 12p(1-p)d$$

$$dM / dp = 4\left[a+3(q-p)d\right] = 4\alpha_t$$

$$\Delta M = \Delta p.dM / dp = \Delta p.4\alpha_t$$

So finally putting all of this together, the response (R) to selection is the change in population mean (ΔM) for change in allele frequency Δp:

$$R = \Delta M = \Delta p.4\alpha_t = \beta S.4\alpha_t = (pq\alpha_t / V_P)S.4\alpha_t = (4pq\alpha_t^2 / V_P)S = (V_A / V_P)S$$

and we have the classic formula (breeders' equation) for the response to selection:

$$R = h_n^2 S, \text{where } h_n^2 \text{ is the narrow-sense heritability} (V_A / V_P), \text{or } R = h_n^2 i\sigma_P.$$

The response to selection can also be determined through the theoretical concept of breeding value. In tetraploids, the practical definition is only the same as the theoretical one in the absence of allelic as well as non-allelic interactions because allelic interactions can be transmitted through diploid gametes.

The covariance of breeding values and phenotypic values equals the covariance of breeding values and genotypic values = $4pq\alpha_t^2 = V_A$ (the additive genetic variance).

The regression (β_{AP}) of breeding values (A) on phenotypic values (P) is therefore:

$$\beta_{AP} = V_A / V_P = h_n^2$$

The change in breeding value ΔA as a result of change in phenotypic value ΔP is therefore:

$$\Delta A = \beta_{AP}\Delta P = (V_A / V_P)\Delta P = (V_A / V_P)S = R(\text{the response to selection})$$

As yield is a quantitative trait, we assume that many loci contribute to the variation in yield, and in our simple model we assume that we can sum over loci so, for example, the additive genetic variance (V_A) which we estimate is $\Sigma 4pq\alpha_t^2$. Finally, it should be pointed out that we have determined the equilibrium response to selection, not the immediate response to selection (one round of random mating does establish single-locus equilibrium under tetrasomic inheritance). The latter can be determined from the regression of offspring means on mid-parental values and for our simple model is $R = [(V_A + \frac{1}{3} V_D)/V_P]S$, where again for our simple model $(V_A + \frac{1}{3} V_D = 4\sigma^2_{GCA} + 1\frac{1}{3}\sigma^2_{SCA})$.

2.11.3 *Effect of Inbreeding*

Under tetrasomic inheritance, the effects of inbreeding are potentially complex, as can be seen in the book by Gallais (2003). Here, we will stick with our simple model. The selfed families of our population of clones are shown in Table 2.5.

$$\text{The mean of the selfed families is:} \qquad M_1 = 2(p-q)a + 10pqd$$

$$\text{The mean of the population of clones is:} \qquad M_0 = 2(p-q)a + 12pqd$$

One generation of selfing has reduced the mean by $2pqd$. If the diallelic interactions are all positive, summation over loci ($\Sigma 2pqd$) can result in a large reduction in yield on selfing, whereas if the interactions are equally likely to be positive or negative, there may be no net effect on the trait on selfing.

The formula for the change in mean after n generations of selfing is: $M_0 - M_n = 12pq[1 - (\frac{5}{6})^n]d$.

The formula for the mean after n generations of selfing is: $M_n = 2(p-q)a + 12pq(\frac{5}{6})^n d$.

The change over generations is shown in Fig. 2.4.

The inbreeding depression only increases slowly in our simple model, but Bennett (1976) pointed out that when p is much lower than q, triallelic and tetra-allelic interactions could contribute to a very rapid inbreeding depression on selfing.

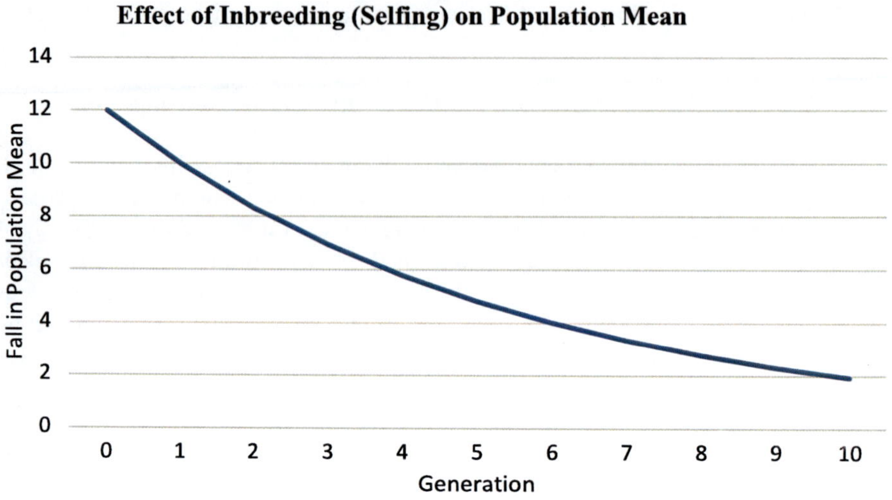

Fig. 2.4 Change in population mean on selfing ($p = q = \frac{1}{2}$, $d = 4$, for illustration)

Table 2.7 Means of parents and offspring

Parents	*AAAa*	*AAaa*	*Aaaa*
Parental means	$a + 3d$	$4d$	$-a + 3d$
Crosses	*AAAa* × *AAaa*	*AAAa* × *Aaaa*	*AAaa* × *Aaaa*
Cross means	$\frac{1}{2}a + 3\frac{1}{6}d$	$3\frac{1}{2}d$	$-\frac{1}{2}a + 3\frac{1}{6}d$

2.11.4 *Hybridization of Selected Clones*

As mentioned at the start of this section, it is highly unlikely that the cultivars and advanced breeding clones chosen as parents of the next generation will be a population of genotypes in genetic equilibrium. So, what will be the mean of the next generation compared with that of the parents? If $p = q = \frac{1}{2}$ in Table 2.5, the genotypes *AAAA*, *AAAa*, *AAaa*, *Aaaa* and *aaaa* occur in the ratio 1:4:6:4:1. Let us assume that selection for higher yield has been effective in eliminating genotype *aaaa* and that genotype *AAAA* is rare so that the parents comprise the three heterozygotes in the ratio 2:3:2 (Table 2.7).

If the parents *AAAa*, *AAaa* and *Aaaa* are in the ratio 2:3:2, their mean is $(24/7)d$.

Then their offspring mean (three crosses in ratio 3:2:3) is $3\frac{1}{4}d$, which is $(5/28)d$ lower.

Hence, the mean of the next generation is likely to be less than the mean of their parents in the presence of unidirectional (across loci) diallelic interactions.

2.12 Linkage Analysis with Chromosomal Segregation

Linkage map construction and QTL mapping in autotetraploids requires appropriate computer software which has, for example, been provided by Biomathematics and Statistics Scotland. The first software to be readily available was TetraploidMap for Windows (Hackett et al. 2007). It analyses presence/absence data for dominant and/or codominant markers scored on two parents and the full-sib offspring of the cross between them (Hackett et al. 1998). It was developed to analyse the data becoming available at the Scottish Crop Research Institute (SCRI), now the James Hutton Institute (JHI), using AFLP and SSR markers (Luo et al. 2001). The linkage analysis is based on chromosomal segregation in which the four homologous chromosomes pair at random to give two bivalents, the two divisions of meiosis proceed normally and give equally viable diploid gametes, and male and female gametes then unite at random to form tetraploid offspring. The linkage analysis includes a test for double reduction to assess the adequacy of this part of the model, but a significant result could occur from segregation distortion or genotyping problems. Nevertheless, the software proved suitable for linkage and QTL analysis in potato, the theory for interval mapping of QTLs having been extended from diploid to autotetraploid species by Hackett et al. (2001). A review of the work done up to 2008 can be found in the book chapter by Milbourne et al. (2008). With the advent and use of extensive SNP (dosage) data, the second software to be made readily available was TetraploidSNPMap (Hackett et al. 2017). It has two modules, one for linkage map estimation and the other for QTL mapping. It was developed for analysing data from the Infinium 8300 potato SNP array, but is equally applicable to other species and SNP dosage data from other technologies. It also analyses data scored on two parents and the full-sib offspring of a cross between them. The theory underlying the program is described in detail in the publications by Hackett et al. (2013, 2014). TetraploidSNPMap is distributed with a detailed user manual and test datasets from these papers (www.bioss.ac.uk/knowledge/tetraploidmap/).

2.12.1 Linkage Analysis in Autotetraploids

The reader unfamiliar with the complexities of linkage analysis in autotetraploids may find useful the five examples shown in Fig. 2.5. They consider two loci with two alleles at each locus (*A*, *a* and *B*, *b*) in one parent, where there are one (simplex) or two (duplex) copies of each dominant (or expressed) allele, namely *A* and *B*, which can be in coupling or repulsion linkage. The other parent has genotype *aaaa*/*bbbb* so that we are mapping loci in the first parent. We have to consider all of the possibilities that can occur during meiosis (not all are shown in Fig. 2.5). The three possible chromosome pairings are 12 + 34, 13 + 24 and 14 + 23. With pairing 12 + 34, centromeres 1 and 3 or 1 and 4 go to the same pole at anaphase I and then the centromeres divide and go to opposite poles at anaphase II, there having been two possible spindle orientations at metaphase II. With simplex/simplex coupling all three pairings give the same result (Fig. 2.5a), whereas with the other four

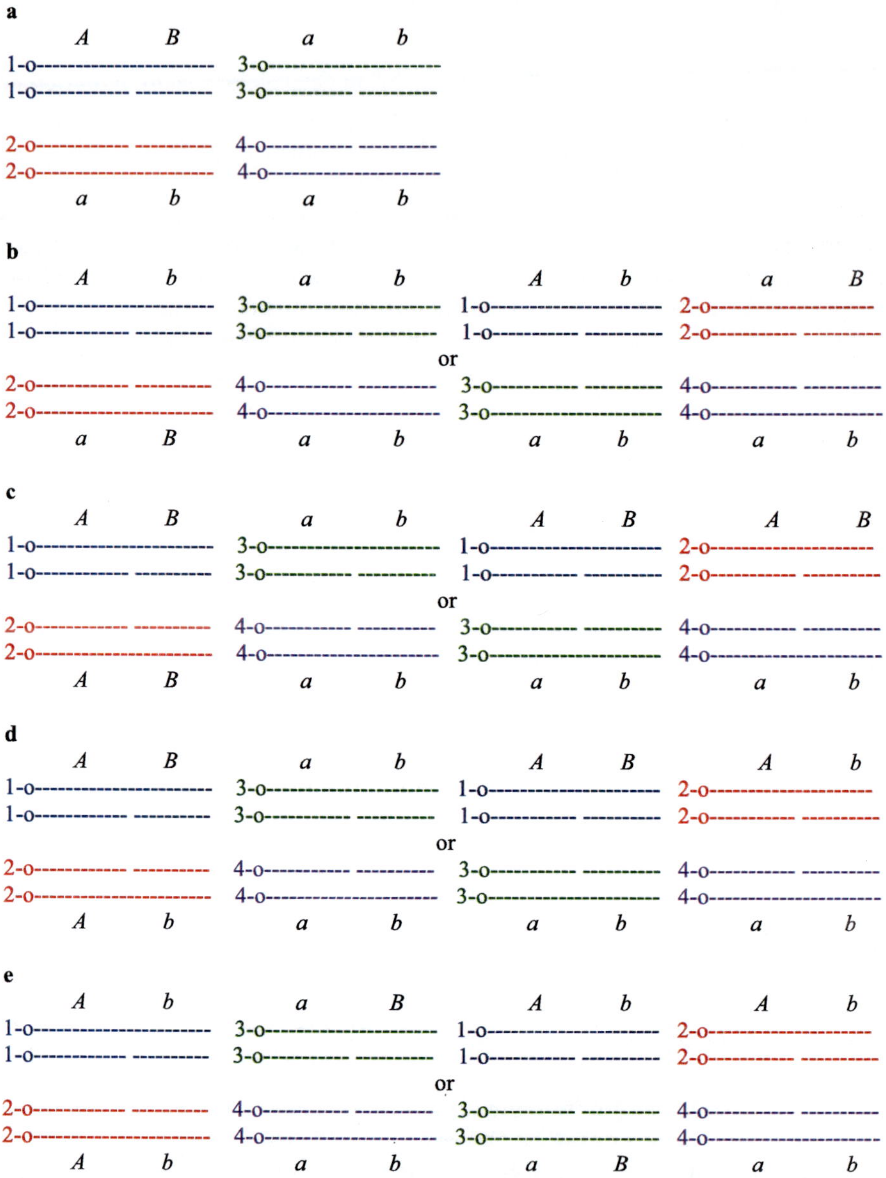

Fig. 2.5 Linkage analysis in autotetraploids with chromosomal segregation (two bivalents). **(a)** Simplex/simplex coupling (12 + 34) = (13 + 24) = (14 + 23). **(b)** Simplex/simplex repulsion (12 + 34), (13 + 24) = (14 + 23). **(c)** Duplex/duplex coupling (12 + 34), (13 + 24) = (14 + 23) **(d)** Duplex/simplex coupling (12 + 34), (13 + 24) = (14 + 23). **(e)** Duplex/simplex repulsion (12 + 34), (13 + 24) = (14 + 23). (From Bradshaw 2016 with permission)

examples, pairings 13 + 24 and 14 + 23 give the same result which is different from pairing 12 + 34 (Fig. 2.5b–e). The interested reader may like to go back to Fig. 2.3 and work out what happens for simplex/simplex coupling when a quadrivalent forms and double reduction occurs at locus *B*. I think that you will find that $(b + c)/n$ estimates $\frac{3}{4}r$ rather than *r*, so that a mixture of two bivalents and a quadrivalent will result in a biased estimate of *r* but should not affect the mapped order of loci.

The marker classes that can be recognized are AB, AO, OB and OO where A and B are expression of A and B, respectively, and O is absence of expression. The expected phenotypic frequencies for the four marker classes in the progeny can be determined for recombination fraction (frequency) *r* and the observed frequencies used to calculate the maximum likelihood estimate of *r*, the asymptotic variance of *r* and its standard error (Hackett et al. 1998). For simplex/simplex coupling the result is the same as for a diploid backcross (*Ab/ab* × *ab/ab*), which I used in Chap. 4 of my book on plant breeding (Bradshaw 2016) to illustrate estimation of recombination frequency by maximum likelihood. I also dealt with mapping functions (Haldane and Kosambi), test for linkage and membership of linkage group, and ordering loci in a linkage group. I have decided not to repeat these here, but to simply illustrate some of the complications of tetrasomic inheritance. The biggest one is the difference between coupling and repulsion linkages where the four phenotypes occur with the following frequencies:

Phenotypes	AB	AO	OB	OO	total
Coupling (12+34) = (13+24) = (14+23)	$\frac{1}{2}(1 - r)$	$\frac{1}{2}(r)$	$\frac{1}{2}(r)$	$\frac{1}{2}(1 - r)$	1
Repulsion (12+34)	$\frac{1}{2}(r)$	$\frac{1}{2}(1 - r)$	$\frac{1}{2}(1 - r)$	$\frac{1}{2}(r)$	1
Repulsion (13+24) = (14+23)	$\frac{1}{4}$	$\frac{1}{4}$	$\frac{1}{4}$	$\frac{1}{4}$	1
Repulsion Overall	$\frac{1}{6}(1 + r)$	$\frac{1}{6}(2 - r)$	$\frac{1}{6}(2 - r)$	$\frac{1}{6}(1 + r)$	1
Observed numbers	*a*	*b*	*c*	*d*	*n*

The maximum likelihood estimate of *r* and its standard error is (Hackett et al. 1998):

$$r = (b + c)/n \qquad SE = \left[r(1-r)/n\right]^{\frac{1}{2}} \qquad \text{for simplex / simplex coupling and}$$

$$r = \left[2(a+d)-(b+c)\right]/n \qquad SE = \left[(1+r)(2-r)/n\right]^{\frac{1}{2}} \qquad \text{for simplex / simplex repulsion.}$$

If $r = 0.10$, and $n = 100$, its SE with coupling linkage is 0.030, but 0.145 with repulsion linkage.

The method of maximum likelihood is a very general method of estimation which was developed by Ronald A. Fisher and leads to estimators with a number of desirable properties (Bulmer 1967). The method consists of taking as estimators, those values of the parameters which maximize the likelihood of the observations. In practice, for ease of computation, the logarithm (l) of the likelihood (L) is maximized as its maximum occurs at the same value of a parameter as that of the likelihood. Again, I have decided not to go into details here but refer readers to Chap. 4 in my book on plant breeding (Bradshaw 2016). The ratio of the likelihood (probability) that r equals the estimate of r ($\hat{r}$) [($L(r = \hat{r})$] over the likelihood (probability) that r equals $\frac{1}{2}$ (its maximum value) [($L(r = \frac{1}{2})$] is called the odds. The significance of the odds is commonly expressed as the LOD score, namely the common logarithm to base 10 of the likelihood ratio (LR). If the odds are greater than 1000 (LOD = 3) one has very good evidence for linkage, but when considering a large number of pairs of loci, a more stringent criterion may be required.

These results for simplex/simplex coupling and repulsion linkages demonstrate that it is difficult to detect repulsion linkages with tetrasomic inheritance. With simplex markers one is therefore likely to produce a linkage map comprising linkage groups equal in number to that of the tetraploid chromosomes ($4x = 48$), having failed to identify homologous chromosomes. However, the situation improves when we have dominant duplex markers and consider duplex/duplex coupling (Fig. 2.5c) and duplex/simplex coupling (Fig. 2.5d) and repulsion (Fig. 2.5e) linkages. With these linkages the 12 + 34 pairing is not informative whereas the other two possible pairings are informative about recombination. Estimation of r and its standard error can be found in the paper by Hackett et al. (1998). In the absence of explicit formulae for the recombination fractions, the log-likelihood is maximized numerically.

The standard errors for $r = 0.10$, and $n = 100$ are as follows.

$$\mathrm{SE} = \left[3r(2-r)(5-2r+r^2)\right]^{\frac{1}{2}} \Big/ \left[4n(5-6r+3r^2)\right]^{\frac{1}{2}} \text{ for duplex / duplex coupling} = 0.039$$

$$\mathrm{SE} = \left[r(1-r)(2+r)(3-r)\right]^{\frac{1}{2}} \Big/ \left[n(1+r-r^2)\right]^{\frac{1}{2}} \text{ for duplex / simplex coupling} = 0.071$$

$$\mathrm{SE} = \left[r(1-r)(2+r)(3-r)\right]^{\frac{1}{2}} \Big/ \left[n(1+r-r^2)\right]^{\frac{1}{2}} \text{ for duplex / simplex coupling} = 0.071$$

The graphs of SE against r (for $r = 0$ to 0.5) are shown in Fig. 2.6. For r between 0 and 0.1 the SE for duplex/duplex coupling is only slightly larger than that for simplex/simplex coupling, and the SE for duplex/simplex linkage (either coupling or repulsion) is at worst half that for simplex/simplex repulsion. Hackett et al. (1998) went on to show by computer simulation that these combinations of markers can be used to identify and map four homologous chromosomes in an autotetraploid. They concluded that a population size of at least 150 individuals should be used, and that a larger number, say 250, would provide a better chance of identifying homologous chromosomes. The steps of linkage map estimation can be summarized as segregation analysis, marker clustering, calculation of pairwise recombination fractions and LOD scores, marker ordering and identification of parental phase (Luo et al. 2001).

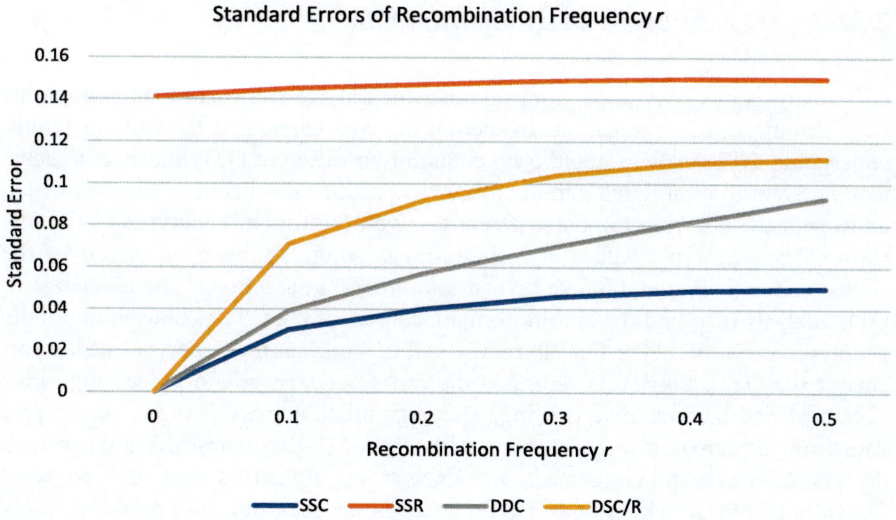

Fig. 2.6 Standard errors of the recombination frequency *r* for simplex/simplex coupling (SSC), simplex/simplex repulsion (SSR), duplex/duplex coupling (DDC) and duplex/simplex coupling and repulsion (DSC/R) when the full-sib family size is 100

Since 1998 further progress has been made in the theory and practice of linkage analysis in autotetraploids. The best tetraploid map of potatoes published to date is the one by Hackett et al. (2013) in which they modified the steps of linkage map estimation to use dosage data on SNPs in the parents and offspring. The two parents (processing clone 12601ab1 and table cultivar Stirling) and 190 F_1 offspring were genotyped using an Infinium 8300 Potato SNP Array. Overall, 3839 of the 5378 polymorphic SNPs were assigned putative genetic locations and a combined map of the two parents was produced. Between 74 and 152 SNPs were mapped to each of the 12 linkage groups (chromosomes) and between 100 and 300 further SNPs were allocated to approximate bins. The length of chromosomes, calculated using the mapped SNPs, varied from 71.9 cM to 121.6 cM, with a total length of 1087.5 cM. A low level of double reduction products was detected: on average 2.7 out of 190 individuals for Stirling as parent and 3.4 for 12601ab1. The SNP panel was selected to target single-copy regions of the genome. Hence for the majority of SNPs it was possible to assign a unique genomic location in the published potato genome sequence. This direct 'map to genome' link should facilitate the identification of candidate genes at trait loci. Another example of the use of TetraploidSNPMap can be found in the paper by Park et al. (2019). They produced a map for the cross between parents NY121 and NY115 using 2761 SNPs and a population size of 195 offspring, with a view to mapping QTL for resistance to cyst nematodes (see Chap. 5). Out of the 2761 SNPs, 1909 and 2262 segregated in NY121 and NY115, respectively, while 1410 SNPs segregated in both parents. The total genetic map lengths of NY121 and NY115 were 905.9 and 919.5 cM, respectively. The genetic map distances varied from 96.8 cM for chromosome 1 to 66 cM for chromosome 10.

2.12.2 QTL Analysis in Autotetraploids

QTL analysis has developed alongside that for linkage analysis in autotetraploids with chromosomal segregation. In theory, the two parents of the full-sib family undergoing QTL analysis could each contain four different QTL alleles at a locus, making eight in total. Each parent could thus produce six different gametes which unite with the six from the other parent to give 36 different genotypes in the offspring. The genotypic values at a single locus comprise the main effects of the alleles and their diallelic, triallelic and tetra-allelic interactions. The challenge in QTL analysis is to find the chromosomal locations of the QTLs contributing to the genotypic variation in the full-sib family and to work out the genotypes of the parents at the QTL. So, let us return to the cross between potato processing clone 12601ab1 and table cultivar Stirling. The most reliable analysis of the phenotypic data from the cross is the one by Hackett et al. (2014) using the linkage map described in the previous section (Hackett et al. 2013) and the software TetraploidSNPMap which does a QTL analysis for each linkage group separately (Hackett et al. 2017). These papers extend the theory and application of the interval mapping method of Hackett et al. (2001) to the use of dosage data on SNPs. For each offspring, Hackett et al. (2001) used the marker information on each chromosome to identify possible configurations of chromosomes inherited from the two parents and the locations of crossovers on these chromosomes. A branch and bound algorithm identified configurations with the minimum number of crossovers. From these configurations, the conditional probability of each possible QTL genotype was estimated for a series of positions along the chromosome. An iterative weighted regression then related the trait values to the QTL genotype probabilities.

The mapping steps subsequently used by Hackett et al. (2013, 2014) can be summarized as follows. The QTL genotype probabilities are estimated at each SNP from the parental genotypes and phases and the offspring dosages, using a Hidden Markov Model. QTL genotype probabilities between the SNPs are interpolated at 1 cM spacing along each chromosome, using a cubic smoothing spline. Trait values are modelled as a function of the QTL genotype, using a normal mixture model with a constant variance. As a result of computer simulations, Hackett et al. (2014) recommended fitting an additive QTL model without iterative re-weighting, and then exploring non-additive models for the genotype means estimated at the most likely position of the QTL. For a simplex QTL ($Qqqq \times qqqq$), there are 2 possible genotypes $qqqq$ or $Qqqq$, with separate means. However, for duplex or double-simplex QTLs ($QQqq \times qqqq$ and $Qqqq \times Qqqq$) there are 3 genotype means $qqqq$, $Qqqq$, $QQqq$ and these can be modelled as an additive effect with means (m, $m + a$, $m + 2a$), or as separate groups with means (m_1, m_2, m_3) or as a dominant effect with means (m, $m + d$, $m + d$). All possible models are fitted and compared using the Schwarz Information Criterion (SIC, Schwarz 1978), and the best 6 models are displayed. The SNP dosage information provides enough information to align the two parental maps so that the effects of alleles from both parents can be studied simultaneously.

a

```
Parent 1                                   Parent 2

---1-o------A------Q------B------          ---5-o------a------q------b------
---2-o------A------Q------B------          ---6-o------a------q------b------
---3-o------a-------q------b------         ---7-o------a------q------b------
---4-o------a-------q------b------         ---8-o------a------q------b------

Offspring
             1                                        4                                      1
---1-o------A------Q------B----   ---1-o------A-------Q------B---   ---3-o------a------q------b----
---2-o------A------Q------B----   ---3-o------a-------q-------b---  ---4-o------a------q------b------
---5-o------a-------q------b----  ---5-o------a-------q-------b---  ---5-o------a------q------b-----
---6-o------a-------q------b----  ---7-o------a-------q-------b---  ---8-o------a------q------b-----
```

b

```
Parent 1                                   Parent 2

---1-o------A------Q------B------          ---5-o------a------q------b------
---2-o------a------ q------b------         ---6-o------a------q------b------
---3-o------a-------q------b------         ---7-o------a------q------b------
---4-o------a-------q------b------         ---8-o------a------q------b------

Offspring
             1                                        1
---1-o------A------Q------B------          ---3-o------a------q------b------
---2-o------a------ q------b------         ---4-o------a------q------b------
---5-o------a-------q------b------         ---7-o------a------q------b------
---6-o------a-------q------b------         ---8-o------a------q------b------
```

Fig. 2.7 Marker and QTL locations in autotetraploid. (**a**) Duplex markers and QTL. (**b**) Simplex markers and QTL

The analysis of the 12601ab1 × Stirling phenotypic data (lowest heritability 0.87) revealed the following QTLs (per cent of trait variance): one for maturity (51.9), one for height (35.6), two for late blight resistance of foliage (33.4, 17.8), two for late blight resistance of tubers (20.7, 16.3) and two for resistance to the white potato cyst nematode (29.8, 8.6); as well as two loci for flower colour (51.9, 7.9) and one for *R*-gene resistance to late blight (57.7). Furthermore, the inferred genotypes of the two parents had just two different alleles and were duplex or simplex as shown in Fig. 2.7. Of particular interest was a region on chromosome 4 (20–34 cM) and another on chromosome 5 (13–27 cM). The section on chromosome 4 contained the two-LOD support interval (confidence interval) for a duplex QTL in Stirling for foliage and tuber resistance to late blight and a duplex QTL in 12601ab1 for resistance to the white potato cyst nematode. Two copies of the alleles gave greater resistance than one copy for both blight and cyst nematodes: *QQqq*,

Qqqq and *qqqq* was 6.3, 5.1 and 1.7 for foliage blight resistance, 38.2%, 57.1% and 84.1% for tuber blight susceptibility, and 2.9, 4.0 and 5.7 for the square roots of cyst counts, i.e. susceptibility. The expected frequency of all four alleles in an offspring is 1 in 36, making resistance to blight and to cyst nematodes a difficult combination to achieve in practice along with all of the other desirable traits in a new cultivar. When the effect of the large QTL for cyst nematode resistance was removed by regression on a candidate SNP for the QTL and the residuals analysed, a further duplex QTL in 12601ab1 was detected on chromosome 11 which explained 8.6% of the variation. Possession of each *Q* allele was associated with a mean decrease of 0.7 in the square root score of cysts. The section on chromosome 5 contained the two-LOD support interval for a simplex QTL in Stirling for early maturity, shorter height and foliage and tuber susceptibility to late blight. This QTL can be interpreted as the physiological effects of maturity on plant height and resistance to late blight.

All of the desired alleles had a large enough effect for them to be desirable in a new cultivar, a goal that can be achieved with the use of flanking molecular markers. The two-LOD support intervals on chromosomes 4 and 5 were both 14 cM, so flanking markers *A* and *B* would need to enclose this length of chromosome. A simulation study had revealed that the two-LOD support intervals were a reasonable approximation to a 95% confidence interval for the true position of the QTL. Where a candidate SNP can be confirmed as a causal gene for the trait, then the SNP can be used for marker-assisted selection. A major advantage of the SNP map was the ability to link directly to the potato genome sequence for the vast majority of SNP markers. Hackett et al. (2014) put forward tentative suggestions for possible causal genes.

The computer simulations (Hackett et al. 2014) revealed that the population size (190 offspring), and not the marker density, limited the size of QTLs that could be detected. A single QTL explaining 5% of the trait variance in a population of 200 was detected in only 29 out of 50 simulations, and the size of the effect was on average over-estimated. In contrast, with a population size of 400, such a QTL was detected in 41 out of 50 simulations, and the magnitude of the effect was estimated more accurately.

Bourke et al. (2019) quantified the power and precision of QTL analysis in autopolyploids under bivalent and multivalent genetic models that include double reduction (double reduction results in a lot of complications; for example, a parent with genotype *Qqqq* can produce *QQqq* offspring). Computer simulations revealed that only marginal gains in QTL detection power are achieved by using the double reduction model when multivalent pairing occurs. However, the researchers found that dense marker coverage on all four homologues of a chromosome (high genotypic information content) is required for both QTL detection power and precision (ability to locate true position). This effect far outweighed considerations regarding bivalent or multivalent pairing (and double reduction) during meiosis and ranks along adequate population size in importance for detection power. Heritability should not be an issue as broad-sense heritability can be increased by increased replication. The methods were checked with data from the cross Altus x Colomba.

The parents and 235 offspring clones were assessed for field maturity [from 1 (very late) to 9 (very early) in increments of 1] and tuber flesh colour on a five-point scale from white to dark yellow in the growing seasons 2012, 2013 and 2014 and genotyped for 6912 SNP markers. A single QTL peak for plant maturity was found on chromosome 5 around 18–20 cM with segregation of a single early allele from cultivar Colomba explaining 47% of the variance. A single peak for flesh colour was found on chromosome 3 at 60 cM with segregation of two copies of a yellow flesh allele from cultivar Colomba explaining 53% of the variance. These traits were chosen because these QTLs of large effect were known to be associated with the *StCDF1* locus on chromosome 5 (with gene annotation PGSC0003DMG400018408) and the *StChy2* locus on chromosome 3 (gene annotation PGSC0003DMT400026363), respectively, so that they could be used to compare the physical and genetic positions of the QTL peaks. The gene position fell within the LOD-2 intervals of the QTL peaks.

Chen et al. (2020) have developed a likelihood-based method (QvMethod) for mapping quantitative trait loci under tetrasomic inheritance that takes account of quadrivalents and double reduction. It uses orthogonal contrast-based genetic models. The authors used computer simulations and real data to demonstrate that their method outperforms bivalent-based methods both in terms of statistical power of QTL detection and in accuracy of locating QTLs. The real data were from Hackett et al. (2014) for 12 traits and chromosome 5 in the 12601ab1 × Stirling population described above. Both methods detected the same QTLs of large effect for plant height, maturity, resistance to foliage blight and resistance to tuber blight which were all associated with the same candidate gene *StCDF1* mentioned above.

2.13 Genome-Wide Association Studies (GWAS)

Genome-wide association studies (GWAS), or simply association mapping, are applied to a diverse set of genotypes in a collection of interest to the breeder. GWAS offers a number of advantages over the use of biparental populations, such as better mapping resolution, analysis of a greater number of traits in a single study, and an increased chance that marker-trait associations will hold across a wider range of germplasm. The idea behind GWAS can be seen by considering a simple example.

2.13.1 *A Simple Example*

Let us consider the set of 16 tetraploid clones in Table 2.8. They are from a much larger population that has been scored for a quantitative trait, such as yield, and also for the dosage of many SNP markers of which three (*A*, *B* and *C*) are shown in the table. The three are tightly linked on the same chromosome but are in equilibrium both within each locus and between loci. The example is artificial in the sense that

Table 2.8 Association genetics: yields of 16 clones (departures from mean) and SNP marker genotypes at three tightly linked loci *A*, *B* and *C* in equilibrium (number of *A*, *B* and *C* alleles coded 2, 1, 0, −1 and − 2 for 4, 3, 2, 1 and 0)

Clone	Yield *Y*	Locus *A*	Locus *B*	Locus *C*
1	2	*AAAa* 1	*BBBB* 2	*Cccc* −1
2	1	*AAAa* 1	*BBBb* 1	*CCcc* 0
3	1	*Aaaa* −1	*BBBb* 1	*CCCc* 1
4	1	*aaaa* −2	*BBBb* 1	*CCCc* 1
5	1	*AAaa* 0	*BBBb* 1	*CCcc* 0
6	0	*AAaa* 0	*BBbb* 0	*Cccc* −1
7	0	*AAAA* 2	*BBbb* 0	*CCcc* 0
8	0	*AAAa* 1	*BBbb* 0	CCCC 2
9	0	*AAaa* 0	*BBbb* 0	cccc −2
10	0	*Aaaa* −1	*BBbb* 0	*CCCc* 1
11	0	*Aaaa* −1	*BBbb* 0	*Cccc* −1
12	−1	*AAAa* 1	*Bbbb* −1	*CCCc* 1
13	−1	*AAaa* 0	*Bbbb* −1	*CCcc* 0
14	−1	*Aaaa* −1	*Bbbb* −1	*Cccc* −1
15	−1	*AAaa* 0	*Bbbb* −1	*CCcc* 0
16	−2	*AAaa* 0	*bbbb* −2	*CCcc* 0

I have ensured that the frequencies of the five genotypes at each locus are in the exact ratio of 1:4:6:4:1 and that there are no correlations between the numbers of alleles at the three loci (i.e. I have eliminated sampling variation). For simplicity, we assume equal frequencies (0.5) of the two SNP alleles at each locus. Furthermore, we assume that at locus *B*, the SNP occurs in a causal gene that affects yield. We now determine the yields (*Y*) of the 16 (*N*) clones and their marker allele composition and then perform the regression of phenotypic values on all available markers, as first proposed by Meuwissen et al. (2001).

In other words, we do a standard multiple regression analysis.

$$\hat{Y} = b_0 + b_A X_A + b_B X_B + b_C X_C$$

where b_0 is the intercept ($b_0 = \mu_{(Y)} - \sum b\mu_{(X)}$) and b_A, b_B and b_C are the regression coefficients for predictors X_A, X_B and X_C, respectively, which take values of 4, 3, 2, 1 and 0 for the number of copies of *A*, *B* and C (i.e. an additive genetic model). They are estimated to minimize $\sum(Y - \hat{Y})^2$ (least squares estimate), *Y* being the observed yield and $\hat{Y}$ the estimated yield. We will in fact work with departures from the means for the *Y* values ($y = Y - \mu_{(Y)}$) and the *X*'s (means of *X*'s are all 2 so values of *x*'s are 2, 1, 0, −1 and − 2):

$\hat{y} = b_A x_A + b_B x_B + b_C x_C$, which in matrix algebra is:

$$\hat{\mathbf{y}} = \mathbf{Xb} \text{ where } \hat{\mathbf{y}}(16 \text{ rows}) \text{ and } \mathbf{b}(3 \text{ rows}) \text{ are column vectors and } \mathbf{X} \text{ is a matrix } (16 \text{ rows and } 3 \text{ columns}).$$

The intercorrelation matrix for the pairs of columns (**y**, $\mathbf{x}_A$, $\mathbf{x}_B$ and $\mathbf{x}_C$) is:

$$\mathbf{R} = \begin{array}{c|cccc|} & y & x_A & x_B & x_C \\ y & 1.0 & 0.0 & 1.0 & 0.0 \\ x_A & 0.0 & 1.0 & 0.0 & 0.0 \\ x_B & 1.0 & 0.0 & 1.0 & 0.0 \\ x_C & 0.0 & 0.0 & 0.0 & 1.0 \end{array}$$

We therefore have a very simple intercorrelation matrix in which yield is correlated with marker dosage at locus B ($r = 1.0$) but not at A and C ($r = 0$), and there are no correlations between markers (all 0).

The least squares estimate of **b** (which equals the maximum likelihood estimate for regression models with normal errors) is:

$$\mathbf{b} = \left(\mathbf{X}'\mathbf{X}\right)^{-1}\mathbf{X}'\mathbf{y} \text{ where } \left(\mathbf{X}'\mathbf{X}\right)^{-1} \text{ is the inverse of } \left(\mathbf{X}'\mathbf{X}\right) \text{ and } \mathbf{X}' \text{ is the transpose of } \mathbf{X}.$$

The derivation can be found in any standard textbook on statistics. In our simple example:

$$\left(\mathbf{X}'\mathbf{X}\right)^{-1} = \begin{vmatrix} 1/16 & 0 & 0 \\ 0 & 1/16 & 0 \\ 0 & 0 & 1/16 \end{vmatrix} \quad \mathbf{X}'\mathbf{X} = \begin{vmatrix} 0 \\ 16 \\ 0 \end{vmatrix} \quad \mathbf{b} = \begin{vmatrix} 0 \\ 1 \\ 0 \end{vmatrix}$$

In other words, $b_A = b_C = 0.0$ and $b_B = 1.0$, and the predicted yields ($\hat{y}$) equal the actual yields (y). The b's are the (genetic) effects of changing an allele a to A, b to B and c to C, respectively, in any genotype (i.e. an additive genetic model).

The total sum of squares for yield is **y'y** = 16.

The sum of squares due to regression is **b'X'Xb** = 16, and hence accounts for all of the variation.

As **b'X'X** = [0 16 0] and **b'** = [0 1 0], marker B accounts for all of this variation (correlation between yield and marker was $r = 1.0$; hence $r^2 = 1.0$ for marker B).

Furthermore R^2 = **b'X'Xb/y'y** = 1.0 in our simple example, where R is the multiple correlation coefficient (the correlation between y and $\hat{y}$).

In practice, there will be environmental effects (e's) as well as genetic effects so the model for n genotypes ($i = 1 \ldots n$) and p markers ($j = 1 \ldots p$) will be:

$$y_i = g_i + e_i = \sum b_j x_{ij} + e_i, \text{ where summation is over the } p \text{ markers.}$$

The predicted yields ($\hat{y} = \sum b_j x_{ij}$) will no longer equal the actual yields (y).

In practice, a number of QTLs will make various genetic contributions to the trait of interest and there could be intra- and inter-locus interactions between alleles. Hence the question arises as to how many markers are required to cover the whole genome to allow us to detect all the QTLs, and the interactions of QTL alleles, at least in theory. Usually, thousands of markers are used, thus creating a computational problem in which the number of markers is considerably larger than the number of genotypes under consideration. A second question arises from the fact that we assumed that our collection of clones came from an equilibrium population. In practice, this is unlikely to be true and we need to ask if marker-trait associations could arise from causes other than tight linkage of markers and QTLs, such as population structure (a population including genotypes bred for different end uses) and kinship (a population including genotypes from different crossing schedules where the schedules included related parents). I discussed these issues with examples in my book on plant breeding (Bradshaw 2016) and hence have decided not to repeat them here.

2.13.2 Linkage Disequilibrium

The number of markers required for a GWAS analysis (i.e. resolution of mapping) depends on the extent of linkage disequilibrium (LD) between pairs of markers at different distances apart. Linkage disequilibrium (LD) is the non-random association of alleles at different loci in a breeding population of interest. It is useful to compare the situations in diploids and tetraploids. For simplicity, let us consider two loci each with two alleles at equal frequency (*A*, *a* each at frequency of $\frac{1}{2}$ and *B*, *b* each at frequency of $\frac{1}{2}$).

In diploids, with complete disequilibrium there would be just two gametes, either *AB* and *ab* or *Ab* and *aB*, whereas at equilibrium there would be equal frequencies of the four gametes *AB*, *ab*, *Ab* and *aB*. One measure of disequilibrium (*D*) is:

$$D = P(AB)P(ab) - P(Ab)P(aB)$$

where *P*(*AB*), *P*(*ab*), *P*(*Ab*) and *P*(*aB*) are the frequencies of gametes *AB*, *ab*, *Ab* and *aB*.

Hence with complete disequilibrium we have

$$D = \left(\frac{1}{2}\times\frac{1}{2}\right) - (0\times 0) = \frac{1}{4} \text{ or } D = (0\times 0) - \left(\frac{1}{2}\times\frac{1}{2}\right) = -\frac{1}{4}$$

and with equilibrium ½ we have

$$D = \left(\frac{1}{4}\times\frac{1}{4}\right) - \left(\frac{1}{4}\times\frac{1}{4}\right) = 0$$

Although two unlinked loci can be in disequilibrium, with GWAS we are interested in close linkage. By considering the change in gamete frequencies from one generation to the next it can be shown that $D_1 = (1 - c)D_0$, where *c* (below *r* is used for correlation) is the recombination frequency between the two loci. Hence the disequilibrium (D_n) after *n* generations of random mating is:

$$D_n = (1-c)^n D_0$$

The disequilibrium depends on the recombination frequency and the number of generations of random mating since it first arose. The interested reader will find the theory of linkage disequilibrium in any standard textbook of Population Genetics (e.g. *Genetics of Populations, Fourth Edition* by Hedrick (2011)).

There are, however, other measures of linkage disequilibrium such as the correlation (r) between alleles at two loci and its square (r^2). As can be seen in Tables 2.9 and 2.10, r^2 has values between 1 (complete disequilibrium) and 0 (linkage equilibrium). The relationship between r^2 and D is $r^2 = D^2/pquv$, where p, q, u and v are the frequencies of alleles A, a, B and b. Hence in our simple example D declines 1/4, 1/8, 1/16, 1/32, 1/64, 1/128, etc. and r^2 declines 1, 1/4, 1/16, 1/64, 1/256, 1/1024, etc. over generations.

Table 2.9 Complete disequilibrium: frequencies (copies) of alleles *A* and *B* (coded 1, 0 and − 1 for 2, 1 and 0 copies) in genotypes from random union of gametes *AB* and *ab* or *Ab* and *aB*

Frequency	*A*	*B*		*A*	*B*
1 *AB*/*AB*	1	1	1 *Ab*/*Ab*	1	−1
2 *AB*/*ab*	0	0	2 *Ab*/*aB*	0	0
1 *ab*/*ab*	−1	−1	1 *aB*/*aB*	−1	1
r		1			−1
r^2		1			1

Table 2.10 Equilibrium and partial disequilibrium: frequencies (copies) of alleles *A* and *B* (coded 1, 0 and − 1 for 2, 1 and 0 copies) in genotypes from random union of gametes *AB*, *ab*, *Ab* and *aB* and from random mating genotypes *AB*/*AB*, 2 *AB*/*ab* and *ab*/*ab* in Table 2.9 when $c = \frac{1}{2}$ ($n = 1$)

Equilibrium	*A*	*B*	$c = \frac{1}{2}$ ($n = 1$)	*A*	*B*
1 *AB*/*AB*	1	1	9 *AB*/*AB*	1	1
2 *AB*/*Ab*	1	0	6 *AB*/*Ab*	1	0
1 *Ab*/*Ab*	1	−1	1 *Ab*/*Ab*	1	−1
2 *AB*/*aB*	0	1	6 *AB*/*aB*	0	1
4 *AB*/*ab*	0	0	20 *AB*/*ab*	0	0
2 *Ab*/*ab*	0	−1	6 *Ab*/*ab*	0	−1
1 *aB*/*aB*	−1	1	1 *aB*/*aB*	−1	1
2 *aB*/*ab*	−1	0	6 *aB*/*ab*	−1	0
1 *ab*/*ab*	−1	−1	9 *ab*/*ab*	−1	−1
r		0			$\frac{1}{2}$
r^2		0			$\frac{1}{4}$

In tetraploids, we are interested in the frequencies of chromosomes (haplotypes) *AB*, *ab*, *Ab* and *aB* but actually observe the frequencies (copies) of alleles *A* and *B* in the tetraploid genotypes, coded 2, 1, 0, −1 and − 2 for 4, 3, 2, 1 and 0 copies. Hence for complete disequilibrium we can produce tetraploid offspring from the diploid gametes *AB*/*AB*, *AB*/*ab* and *ab*/*ab* in the ratio 1:2:1 or from the diploid gametes *Ab*/*Ab*, *Ab*/*aB* and *aB*/*aB* in the ratio 1:2:1, as shown in Table 2.11. For equilibrium, we can produce 25 different tetraploid offspring from the 10 diploid gametes in the ratios $(AB + ab + Ab + aB)^2$ as shown in Table 2.12.

We can also see that the decline in r^2 from the first generation of random mating for unlinked loci ($c = \frac{1}{2}$) is the same for tetraploids as for diploids, namely 1 to $\frac{1}{4}$. It should, however, be pointed out that in our simple example, I ensured that each locus was in equilibrium with the five genotypes in the ratio 1:4:6:4:1. Gallais (2003) considered the more general case of panmictic disequilibrium at each locus and linkage disequilibrium between two loci which he grouped together as gametic disequilibrium. He went on to show that for a given recombination frequency, the change in the chromosome association (correlation) of two genes is nearly the same in autotetraploids as in diploids.

Having given a brief explanation of the use of r^2 (the square of the Pearson correlation coefficient) as a measure of linkage disequilibrium (LD) it is time to return to the matter in hand, the number of markers required for a GWAS analysis. If the range of LD is very narrow then a large number of markers is required to pick up a QTL linked to a marker, but the QTL is precisely located (i.e. high resolution). On the other hand, if the range of LD is extensive, not so many markers are required to find linkage between a QTL and marker, but the QTL is not located with as much precision (i.e. low resolution). Vos et al. (2017) evaluated LD decay and various LD-decay estimators in simulated and SNP-array data for tetraploid potatoes. The real data came from a panel of 537 tetraploid cultivars (ancient and modern) and progenitor clones that was genotyped with a 20 K SNP array. For LD decay in the real data they looked at $LD_{\frac{1}{2},90}$ the distance at which the short-range LD (based on markers pairs within 100 kb) is halved when using the 90% percentile of r^2 at

Table 2.11 Complete disequilibrium: frequencies (copies) of alleles *A* and *B* (coded 2, 1, 0, −1 and − 2 for 4, 3, 2, 1 and 0 copies) in tetraploid genotypes from random union of diploid gametes *AB*/*AB*, *AB*/*ab* and *ab*/*ab* in the ratio 1:2:1 or from the diploid gametes *Ab*/*Ab*, *Ab*/*aB* and *aB*/*aB* in the ratio 1:2:1

Frequency	*A*	*B*		*A*	*B*
1 *AB*/*AB*/*AB*/*AB*	2	2	1 *Ab*/*Ab*/*Ab*/*Ab*	2	−2
4 *AB*/*AB*/*AB*/*ab*	1	1	4 *Ab*/*Ab*/*Ab*/*aB*	1	−1
6 *AB*/*AB*/*ab*/*ab*	0	0	6 *Ab*/*Ab*/*aB*/*aB*	0	0
4 *AB*/*ab*/*ab*/*ab*	−1	−1	4 *Ab*/*aB*/*aB*/*aB*	−1	1
1 *ab*/*ab*/*ab*/*ab*	−2	−2	1 *aB*/*aB*/*aB*/*aB*	−2	2
r		1			−1
r^2		1			1

Table 2.12 Equilibrium and partial disequilibrium: frequencies (copies) of alleles *A* and *B* (coded 2, 1, 0, −1 and − 2 for 4, 3, 2, 1 and 0 copies) in tetraploid genotypes in equilibrium and from random mating genotypes *AB/AB/AB/AB*, 4 *AB/AB/AB/ab*, 6 *AB/AB/ab/ab*, 4 *AB/ab/ab/ab* and *ab/ab/ab/ab* in Table 2.11 when $c = \frac{1}{2}$ $(n = 1)$

Equilibrium	*A*	*B*	$c = \frac{1}{2}$ $(n = 1)$	*A*	*B*
1 *AB/AB/AB/AB*	2	2	169 *AB/AB/AB/AB*	2	2
4 *AB/AB/AB/Ab*	2	1	260 *AB/AB/AB/Ab*	2	1
6 *AB/AB/Ab/Ab*	2	0	126 *AB/AB/Ab/Ab*	2	0
4 *AB/Ab/Ab/Ab*	2	−1	20 *AB/Ab/Ab/Ab*	2	−1
1 *Ab/Ab/Ab/Ab*	2	−2	1 *Ab/Ab/Ab/Ab*	2	−2
4 *AB/AB/AB/aB*	1	2	260 *AB/AB/AB/aB*	1	2
16 *AB/AB/AB/ab*	1	1	928 *AB/AB/AB/ab*	1	1
24 *AB/AB/Ab/ab*	1	0	840 *AB/AB/Ab/ab*	1	0
16 *AB/Ab/Ab/ab*	1	−1	256 *AB/Ab/Ab/ab*	1	−1
4 *Ab/Ab/Ab/ab*	1	−2	20 *Ab/Ab/Ab/ab*	1	−2
6 *AB/AB/aB/aB*	0	2	126 *AB/AB/aB/aB*	0	2
24 *AB/AB/aB/ab*	0	1	840 *AB/AB/aB/ab*	0	1
36 *AB/AB/ab/ab*	0	0	1524 *AB/AB/ab/ab*	0	0
24 *AB/Ab/ab/ab*	0	−1	840 *AB/Ab/ab/ab*	0	−1
6 *Ab/Ab/ab/ab*	0	−2	126 *Ab/Ab/ab/ab*	0	−2
4 *AB/aB/aB/aB*	−1	2	20 *AB/aB/aB/aB*	−1	2
16 *AB/aB/aB/ab*	−1	1	256 *AB/aB/aB/ab*	−1	1
24 *AB/aB/ab/ab*	−1	0	840 *AB/aB/ab/ab*	−1	0
16 *AB/ab/ab/ab*	−1	−1	928 *AB/ab/ab/ab*	−1	−1
4 *Ab/ab/ab/ab*	−1	−2	260 *Ab/ab/ab/ab*	−1	−2
1 *aB/aB/aB/aB*	−2	2	1 *aB/aB/aB/aB*	−2	2
4 *aB/aB/aB/ab*	−2	1	20 *aB/aB/aB/ab*	−2	1
6 *aB/aB/ab/ab*	−2	0	126 *aB/aB/ab/ab*	−2	0
4 *aB/ab/ab/ab*	−2	−1	260 *aB/ab/ab/ab*	−2	−1
1 *ab/ab/ab/ab*	−2	−2	169 *ab/ab/ab/ab*	−2	−2
256			9216		
r		0			$\frac{1}{2}$
r^2		0			$\frac{1}{4}$

short-range as the estimator for LD. They used their data to estimate lower and upper limits for the number of markers required for GWAS. They assumed a genome size of 400 Mb (including the gene-rich arms and excluding the pericentromeric heterochromatin). They divided this by 2 Mb (equivalent to a genetic distance of about 5 cM), the distance at which they found that LD had decayed to the very low threshold value for the approach to equilibrium (decay curves flatten). The result was 200 markers as the lower limit for the number required per haploid genome to detect QTLs. However, as they observed no LD decay within 100 kb, they concluded that the upper limit was 4000 markers. They went on to reason that the lower and upper limits (200 and 4000) need to be multiplied by the number of haplotypes per locus (genetic diversity) in the population of genotypes of interest. They assumed 10 founder haplotypes in the potato gene pool and hence came up with 40,000 SNPs as an upper bound for QTL discovery. They concluded that although their 20 K SolSTW array, with 14,530 SNPs, did not reach the upper bound, it should still be able to detect sufficient QTLs. Other interesting results were $LD_{\frac{1}{2},90}$ values of 1.5 Mb for cultivars released before 1945 and 0.6 Mb for cultivars released after 2005, a value of 2.5 Mb for introgressed regions, indicating large haplotype blocks, and negligible LD decay in pericentromeric heterochromatin. The short-range LD calculated based on markers pairs within 100 kb for the cultivar panel ranged from an r^2 of 0.19 to 0.22.

Sharma et al. (2018) examined LD in an association mapping panel of 341 tetraploid potato cultivars and breeding lines, largely comprising European founder and cultivated germplasm, but including 29 non-European cultivars. They genotyped the panel using an 8 k SNP array. They used two LD estimators, $LD_{1/2max,90}$ and $LD_{1/10,90}$, denoting the distances at which LD equals one-half of its maximum fitted r^2 value ($r^2_{max,90}$) and where r^2 equals 0.1 on the 90th percentile, respectively. Overall $LD_{1/10,90}$ for the short and long arms of chromosomes and the combined euchromatin were comparable with values of 2.52 Mb, 2.76 Mb and 2.73 Mb, respectively. LD decay in the pericentromeric heterochromatin for all chromosomes was negligible due to the suppression of recombination in this fraction of the genome. Despite this insignificant LD decay in heterochromatin, the overall $LD_{1/10,90}$ calculated at the whole chromosome level (3.27 Mb) was within twofold limits of that observed for the euchromatin (2.73 Mb). All chromosomes exhibited modest $LD_{1/10,90}$, ranging from 2.54 Mb (Chromosome 9) to 4.68 Mb (Chromosome 3), but with chromosome 8 displaying an extremely conserved $LD_{1/10,90}$ of 20.04 Mb. Sharma et al. (2018) also estimated the overall euchromatin $LD_{1/2max,90}$ at 0.91 Mb for cultivars in their panel released after 2005 compared with the value of 0.6 Mb obtained by Vos et al. (2017). They also estimated 1 Mb to correspond to about 1 cM to 2 cM, compared with Vos et al. (2017) quoting about 2.5 cM.

2.13.3 *Computer Software for GWAS Analysis*

GWAS analysis requires appropriate computer software, so I will start with the paper by Rosyara et al. (2016). However, the method of analysis also requires acquaintance with the general mixed model and matrix algebra. The model is familiar to animal breeders and geneticists, but until recently, less familiar to plant breeders and geneticists. It embraces the vast majority of estimation problems encountered in Quantitative Genetics, including the estimation of breeding values of potential parents (Lynch and Walsh 1998). I gave an introductory account with an example in Chap. 6 of my book on plant breeding (Bradshaw 2016). Here, I am simply going to state the model in order to show how it can be extended for GWAS analysis.

2.13.4 General Mixed Model

The general mixed model is mixed because it includes both random and fixed effects. Random effects are where treatments (e.g. genotypes from a cross) are a random sample from a large population about which we want to make inferences, whereas with fixed effects we are only interested in the actual treatments used (e.g. no fungicide compared with both the full and half the full recommended application). The general mixed model can be written as follows using matrices and analysed using matrix algebra (Lynch and Walsh 1998).

$$\mathbf{y} = \mathbf{X}\beta + \mathbf{Zu} + \mathbf{e} \begin{pmatrix} \text{b is sometimes used for the estimate of } \beta, \\ \text{and replaces } \beta \text{ in the model} \end{pmatrix}$$

y is a column vector ($n \times 1$) containing the phenotypic values of a trait (e.g. yield) measured on n individuals (or plots in yield trial).

The linear model comprises the following:

X is $n \times p$ design matrix with elements 1 (effect contributes) or 0 (effect does not contribute).
$\boldsymbol{\beta}$ is $p \times 1$ column vector of fixed effects; e.g. trial mean followed by experimental treatments.
Z is $n \times q$ incidence matrix with elements 1 (effect contributes) or 0 (effect does not).
u is $q \times 1$ column vector of random effects; e.g. genetic effects, breeding values.
e is $n \times 1$ column vector of residual deviations independent of random genetic effects.

A special case of the model is the general linear model, $\mathbf{y} = \mathbf{X}\boldsymbol{\beta} + \mathbf{e}$. When **X** is a design matrix of indicator variables (1 or 0), we have ANOVA (analysis of variance) models, whereas in multiple regression models **X** contains predictor variables which can take on a continuous range of values. Estimates of $\boldsymbol{\beta}$ are obtained by the method of least squares in which the residual sum of squares is minimized. The method of ordinary least squares assumes that the residual errors are homoscedastic ($\sigma^2(e_i) = \sigma_e^2$ for all i) and uncorrelated ($\sigma(e_i, e_j) = 0$ for $i \neq j$), whereas in generalized (weighted) least squares the residual errors can be heteroscedastic (some inherently more variable than others) and/or correlated. If the residuals follow a multivariate normal distribution, the ordinary least squares estimate is also the maximum-likelihood estimate.

The means and variances of the component vectors of the mixed model are as follows.

The means (expected values) of **u** and **e** are **0** by definition.

The mean (expected value) of **y** is therefore $\mathbf{X}\boldsymbol{\beta}$.

If **G** is the ($q \times q$) covariance matrix of **u** and **R** is the ($n \times n$) covariance matrix of **e**, and if **u** and **e** are uncorrelated, then **V**, the ($n \times n$) covariance matrix of **y,** is **ZGZ′** + **R** (fixed effects do not contribute). (**Z′** is the transpose of **Z**).

In summary, the random variable **y** has mean $\mathbf{X}\boldsymbol{\beta}$ and variance **V**.

The mixed-model analysis involves prediction of **u** (random effects) and estimation of $\boldsymbol{\beta}$ (fixed effects), **G** and **R** from **y**, **X** and **Z.** Fixed effects are estimated as best linear unbiased estimators (BLUEs) and random effects are predicted by best linear unbiased predictors (BLUPs). They are best because they minimize the sampling variance, linear because they are linear functions of **y** (the observed phenotypes) and unbiased because their expected (mean) values are $\boldsymbol{\beta}$ and **u**, respectively.

$$\mathrm{BLUE}(\boldsymbol{\beta}) = \left(\mathbf{X}'\mathbf{V}^{-1}\mathbf{X}\right)^{-1}\mathbf{X}'\mathbf{V}^{-1}\mathbf{y}\left(\mathbf{V}^{-1}\,\text{is the inverse of}\,\mathbf{V}\right)$$

$$\mathrm{BLUP}(\mathbf{u}) = \mathbf{G}\mathbf{Z}'\mathbf{V}^{-1}\left(\mathbf{y}-\mathbf{X}\boldsymbol{\beta}\right)$$

G and **R** are generally assumed to be functions of a few unknown variance components and need to be estimated prior to the BLUP analysis by ANOVA or by restricted maximum likelihood (REML). BLUP estimates remain unbiased when estimates of genetic variances are used in place of actual values (as is usual) but are no longer guaranteed to be the best of all unbiased linear estimators. Unlike ANOVA (unbiased) estimation of variance components, maximum likelihood estimation can cope with unbalanced experimental designs. In practice, restricted maximum likelihood (REML) is used to eliminate bias in the estimates of variance components that arises when fixed effects are subject to error. REML estimators maximize only the portion of the likelihood that does not depend on the fixed effects. The estimation is done in an iterative fashion from BLUP estimates of random effects assuming that the appropriate variance components are known. Details can be found in the book by Lynch and Walsh (1998). BLUP is primarily used for the identification of individuals (clones of potato genotypes) with maximum genetic merit in selection programmes and for monitoring the response to selection. In practice, the computation of $\mathbf{V}^{-1}$ is difficult for very large **V** so $\boldsymbol{\beta}$ and **u** are jointly obtained from mixed model equations (Henderson 1950):

$$\begin{pmatrix}\mathbf{X}'\mathbf{R}^{-1}\mathbf{X} & \mathbf{X}'\mathbf{R}^{-1}\mathbf{Z} \\ \mathbf{Z}'\mathbf{R}^{-1}\mathbf{X} & \mathbf{Z}'\mathbf{R}^{-1}\mathbf{Z}+\mathbf{G}^{-1}\end{pmatrix}\begin{pmatrix}\boldsymbol{\beta} \\ \mathbf{u}\end{pmatrix} = \begin{pmatrix}\mathbf{X}'\mathbf{R}^{-1}\mathbf{y} \\ \mathbf{Z}'\mathbf{R}^{-1}\mathbf{y}\end{pmatrix}$$

While the general mixed model forms the fundamental framework for BLUP analysis, there are numerous ways in which it can be formulated, including ones to estimate breeding values (Lynch and Walsh 1998). Later in this book, when considering genomic selection, we shall need to refer to animal models that estimate the breeding values of individuals (which will be potato clones). So, let us take a brief look at such a model before returning to GWAS analysis.

$$y_i = \mu + a_i + e_i$$

where a_i is the additive genetic (breeding) value of individual i. With k individuals, the model can be expressed as

$$\mathbf{y} = \mathbf{X}\beta + \mathbf{Zu} + \mathbf{e}$$

where $\mathbf{X} = \begin{vmatrix} 1 \\ 1 \\ \\ 1 \end{vmatrix}, \boldsymbol{\beta} = \mu, \mathbf{u} = \begin{vmatrix} a_1 \\ a_2 \\ \\ a_k \end{vmatrix}$

The matrix **G** ($k \times k$) describing the covariances among random effects **u** (breeding values) follows from standard results for the covariances between relatives which can be found in Lynch and Walsh (1998). The additive genetic covariance between two relatives i and j is $2\Phi_{ij}\sigma_A^2$, where Φ_{ij} is the coefficient of coancestry (given earlier in this chapter) and σ_A^2 is the additive genetic variance in the base population under consideration. Hence $\mathbf{G} = \sigma_A^2\mathbf{A}$, where the additive genetic (or numerator) relationship matrix **A** has elements $A_{ij} = 2\Phi_{ij}$ (e.g. in diploids Φij is $\frac{1}{8}$ for half-sibs and $\frac{1}{4}\sigma_A^2$ is the covariance for half-sibs). The standard assumption for the covariance matrix **R** for the vector of residual errors is $\mathbf{R} = \sigma_e^2\mathbf{I}$ (**I** is the identity matrix) where the residual error for each observation has the same variance σ_e^2 and is uncorrelated with all other residual errors. However, this assumption will not be true if the trait displays allelic and non-allelic interactions. If the total genetic merit (additive plus non-additive genetic values) of an individual (potato clone) is of interest, the model can be extended to include non-additive genetic values as well as additive genetic values as shown by Lynch and Walsh (1998). If non-additive effects are present but not taken into account, the estimates of additive effects will be biased.

2.13.5 GWAS Analysis

Rosyara et al. (2016) reported on the development and validation of the R package GWASpoly, which was designed for GWAS with biallelic single-nucleotide polymorphisms (SNPs) in autopolyploids using the Q (or P) + K method to take account of population structure and kinship, respectively. The Q + K mixed model for GWAS can be written as follows.

$$\mathbf{y} = \mathbf{X}\beta + \mathbf{ZSt} + \mathbf{ZQv} + \mathbf{Zu} + \mathbf{e}$$

y is a column vector ($w \times 1$) containing the observed phenotypic values for the trait for w potato clones (in our context). [Rosyara et al. (2016) use w rather than n.]

X is $w \times p$ incidence matrix to model environmental effects or other covariate effects.

$\boldsymbol{\beta}$ is $p \times 1$ column vector of fixed (environmental) effects.

Z is $w \times n$ incidence matrix to map genotypes to observations.

S is $n \times d$ incidence matrix which depends on genetic model (**ZS** is incidence matrix $w \times d$).
t is $d \times 1$ column vector of SNP effects which depend on genetic model.
Q is $n \times q$ incidence matrix for population size n (**ZQ** is incidence matrix $w \times q$).
v is $q \times 1$ column vector of effects for the subpopulations.
u is $n \times 1$ column vector of (random background) polygenic effects.
e is $w \times 1$ column vector of residuals.

X, **ZS**, **ZQ** and **Z** are incidence matrices of 1 s and 0 s relating fixed environmental effects, SNPs, populations and genotypes to the phenotypes **y**.

Var(**u**) = σ_g^2**K,** where **K** is kinship matrix ($n \times n$), and Var(**e**) = **I**σ_e^2.

The reader is referred to the paper by Rosyara et al. (2016) for details of the estimation of the variance components σ_g^2 and σ_e^2, for the three kinship matrices (**K**) considered, and for the three different methods available in GWASpoly for establishing a p-value detection threshold for statistical significance.

For biallelic SNPs in autotetraploids, there are five genotype classes which can be parameterized by the dosage of the minor allele (A): (0, 1, 2, 3, 4). In addition to the general model, the authors present results for four different single-parameter genetic models (**S** matrix): additive (genetic value increases in equal steps from *aaaa* to *AAAA*), diploidized additive (*Aaaa*, *AAaa* and *AAAa* have same genetic value halfway between *aaaa* and *AAAA*) and simplex dominant and duplex dominant (*Aaaa*, *AAaa* and *AAAa* have same value as either *aaaa* or *AAAA* according to dominance).

Using a simulated tetraploid population, the authors confirmed their hypothesis that statistical power is higher when the assumed gene action in the GWAS model matches the gene action at unobserved quantitative trait loci (QTL). Furthermore, with 300 markers per 100 cM chromosome and 200 individuals, for the same proportional increase, population size had a bigger effect than marker density on statistical power.

The authors then analysed 13 traits in the Solanaceae Coordinated Agricultural Project (SolCAP) potato diversity panel. Their analysis of population structure was conducted using all 221 tetraploid clones in the panel while their GWAS results were based on the 187 tetraploid clones with both marker and phenotypic data. The tetraploid clones comprised potatoes from six different market categories [chip (69), fry (34), round white table (38), table russet (13), pigmented (32) and yellow (27)] as well as three wild species and five genetic stocks. The markers were 3441 SNPs with minor allele frequency greater than 0.05. The phenotypic data came from a randomized complete block trial with two replicates in each of four environments (New York-2010, Wisconsin-2010, New York-2011 and Wisconsin-2011), although not all traits were measured in every environment. The 13 traits covered agronomic (total yield, vine maturity 95 and 120 days after planting), morphological (tuber size, shape, length, width and eye depth), biochemical (tuber fructose, glucose, sucrose and malic acid content) and quality (chip colour) ones. The broad-sense heritability on an entry-mean basis for each trait ranged from 0.60 for tuber malic acid content to 0.94 for tuber shape. The phenotypic values used in the analysis were the best linear unbiased estimators (BLUEs) for each tetraploid clone for each trait (i.e. treating them as fixed effects). The GWAS analyses were conducted using the Q_{DAPC} + K_{RR} model (i.e. discriminant analysis of principal components and realized relationship, respectively).

Surprisingly, just nine significant QTLs for 6 out of the 13 traits were detected: tuber yield (1), eye depth (2), tuber shape (1), tuber length (1), vine maturity at 95 days (3) and vine maturity at 120 days (1). The two most significant QTLs were for tuber shape and tuber eye depth, both at 48.9 Mb on chromosome 10, as found in earlier biparental linkage mapping studies (see Chap. 4). The scarcity of major QTLs detected was attributed to the lack of statistical power. A power simulation using simulated QTL and phenotypes, but with the actual marker data, revealed that for a monogenic trait with a heritability of 0.3, the genome-wide median for the probability of QTL detection was only 0.01. This was even lower than expected with a small population size of 187. The simulation also revealed that power was lower in regions of lower marker density. The authors concluded that in addition to increasing the population size, higher marker density could also improve future GWAS studies in potato.

Sharma et al. (2018) developed an association mapping panel of 341 tetraploid potato cultivars and breeding clones largely comprising European founder and cultivated germplasm but including 29 non-European cultivars. They genotyped the panel using an 8 k SNP array and examined various GWAS models (combinations of presence and absence of **Q** and **K**) using a set of 20 quantitative traits from field trials in 2012 and 2013 at two sites (Cambridge and York, UK) with two nitrogen levels (100 and 200 kg/ha), making eight environments in total. Trials had two replicates and an alpha design. Genotypes were treated as a fixed effect and the best linear unbiased estimates (BLUEs) for all traits were calculated using REML. As these estimates contain an environmental component in addition to the 'true' genotypic value, they are regarded as phenotypic values. There was sufficient phenotypic data to include 290 of the tetraploid cultivars in the GWAS analyses. The SNP array was produced from only six potato cultivars (Hamilton et al. 2011), but proved highly polymorphic. SNP genotypes were called using the R package fitTetra (Voorrips et al. 2011) which employs a mixture model based on the allele signal ratio to classify the individuals into one of the five possible genotypic classes in the tetraploid genotype model, giving dosage information (0, 1, 2, 3 or 4) for each SNP. For performing GWAS, SNPs with 20% or more missing data and 5% or less minor allele frequency (MAF) were excluded, together with SNPs with inconsistent calls and mapping to more than one location, leaving 5718 usable SNPs, of which 5157 were used in the GWAS analyses. Physical positions for all SNPs were derived from the chromosome-scale pseudomolecules for potato integrating the potato genome with genetic and physical maps (Sharma et al. 2013). Like Rosyara et al. (2016), the authors used the GWASpoly R package (additive genetic model) for their GWAS analyses.

Kinship analysis of the 341 cultivars identified ten clusters (Kinship groups) in which groups 1 and 2 together were the most divergent from the rest of the tetraploid germplasm and quite different from each other. Kinship group 1 largely comprised cultivars from the USA and Canada, whereas group 2 was mainly a collection of 'salad' and early potato cultivars from various European countries. In contrast, there was a lack of any strong population structure in the panel of 341 cultivars, although five putative subpopulations were weakly identified but showed low levels of population differentiation. Nevertheless, there was a high correspondence of kinship groups and subpopulations with six kinship groups nesting into their respective subpopulations as single large blocks, with the remaining kinship groups assigned across different subpopulations but in cohesive blocks of at least three genotypes (shared lineage). Non-metric multidimensional scaling (nMDS) proved the most useful method for investigating population structure. However, the method used for population structure estimation was not critical for performing GWAS when correction for kinship effects was included. Indeed, the overall conclusion from all of the analyses was that kinship (**K**) alone is sufficient to control type I error (false positives) for conducting GWAS in cultivated potato.

Surprisingly, just 6 significant QTLs (marker-trait associations) for 6 out of the 20 traits were detected: eye depth, flesh colour, foliage, plant height, stolon attachment and tuber shape. For tuber shape and eye depth, the significant association (solcap_snp_c1_8019) was located at 48.9 Mb on chromosome 10; for stolon attachment, foliage (height to breadth ratio at 80% ground cover) and plant height the significant association (solcap_snp_c2_50302) was on chromosome 5, around 0.5 Mb away from the *StCDF1* gene for maturity; and for flesh colour the significant association (solcap_snp_c2_20285) was on chromosome 3 close to the *Y* locus for yellow/white flesh colour. No significant marker-trait associations were found for yield and its components, dry matter content, after-cooking blackening, enzymic browning, glucose content after storage and dormancy. Nevertheless, Sharma et al. (2018) thought from their genome-wide survey of linkage disequilibrium (LD) in their GWAS panel that the 8 k SNP array was suitable for performing GWAS in potato.

Aliche et al. (2019) also used the Q + K mixed model for GWAS in their association mapping of tuber yield, tuber size distribution and marketable yield under irrigated and non-irrigated (drought stress) field conditions, in France (Connantre) and the Netherlands (Zeeland) in 3 and 2 years, respectively. The mapping procedure was implemented in R Studio version 3.2.3, using the GWASpoly package for autotetraploids (Rosyara et al. 2016). They analysed 95 out of 103 European cultivars (10 late, 46 intermediate and 47 early maturing) using a 14 K SNP array. In 2015, the unirrigated Connantre trial experienced a drought during the later stages of plant development, as a result of a total rainfall of only 42 mm, which resulted in a 51% reduction in the marketable fractions of tuber fresh weight. In contrast, at the same location the previous year there had been a total rainfall of 258.2 mm, but drought during the early stages of plant growth which resulted in a 13% reduction in the marketable fractions of tuber number. Under irrigated conditions in Connantre in 2015, the GWAS analysis revealed significant associations of tuber fresh weight with three SNP markers in close proximity to the well-known StCDF/maturity locus on Chromosome 5. In contrast, in the trial without irrigation (drought stress), there was a significant association of an SNP marker with size distribution spread of tuber number, size class where maximum tuber number occurred, marketable fractions of tuber number and marketable fractions of tuber fresh weight. The marker, PotVar0030768, is found at position 55,657,256 bp on the scaffold PGSC0003DMB000000062 of Chromosome 3 of the potato genome. The allele dosage classes for the SNP were quadruplex (5), triplex (20), duplex (38), simplex (20) and nulliplex (13). The allele frequency was therefore 0.4583 so that the expected numbers in the dosage classes for an equilibrium population with chromosomal segregation were 4, 20, 36, 28 and 8, numbers that were not significantly different from those observed ($P > 0.05$ on chi-square goodness-of-fit test). Increases in the allelic dosage improved the marketable fraction, indicating additive effects. Among the candidate genes present in the region of the SNP was one coding for a pentatricopeptide repeat-containing protein, but more research is required to establish a causal connection with the phenotypes. The reader is referred to the original paper for more detail.

Yuan et al. (2020) likewise used the Q + K mixed model implemented in the GWASpoly package for autotetraploids (Rosyara et al. 2016) for their association mapping of resistance to common scab (*Streptomyces scabies* Thaxter) in Canadian germplasm. Their association panel comprised 123 advanced breeding clones and 20 commercial cultivars. The panel displayed both significant population structure and familial relationships and clustered into five distinct subpopulations. It was genotyped with 10,392 polymorphic SNPs, out of 12 K SolCAP SNPs, and phenotyped for resistance in each of 3 years in a scab nursery maintained in the same field. The scab scores were approximately normally distributed and the broad-sense heritability for the means of the genotypes was 0.81. Three resistance QTLs were identified on chromosomes 2, 4 and 12 and explained 21%, 19% and 26%, respectively, of the variation in scores. The QTL on chromosome 2 was simplex-dominant whereas those on chromosomes 4 and 12 were duplex-dominant QTLs.

In summary, the method used by Rosyara et al. (2016) has become established for GWAS using SNPs in tetraploid potato but its success so far has been confined to detecting QTLs of large effect. This is in contrast to earlier studies with AFLP and SSR markers such as the one by D'hoop et al. (2014). They phenotyped 19 tuber quality traits in 205 historical and contemporary cultivars, and also in 299 clones from Dutch breeders, and did association analysis using 3364 AFLP loci and 653 SSR alleles. Even after a marked reduction in marker-trait associations following correction for population structure and genetic relatedness (using a kinship matrix estimated from marker information), the use of 720 mapped AFLP markers identified 43 and 24 QTLs in the 205 cultivars and 299 clones, respectively. Nevertheless, as we shall see in a later chapter, genomic selection is probably the way forward for polygenic traits with many alleles of small effect where the breeder wishes to use molecular markers to combine as many alleles with a positive effect as possible, in contrast to combining particular alleles of large effect.

2.14 Major Genes, QTL Alleles of Large Effect and Many Alleles of Small Effect

The aim of this chapter has been to provide the knowledge required for understanding the theoretical basis of breeding methods. It has also led to recognition of three categories of genes of interest to plant breeders in general and potato breeders in particular. It is worth finishing this chapter with a further consideration of these categories which are major genes, QTL alleles of large effect and many alleles of small effect that affect economically important traits. Since 1988, it has been possible to map these genes to the 12 chromosomes of potatoes through their linkages to mapped molecular markers. In this chapter, we have seen the progress that has been made over the last 30 years in developing linkage maps. The first map had 135 markers whereas today we can work with thousands, or even millions, of SNPs (single-nucleotide polymorphism markers). Mapping can be done with diploid inbred lines and tetraploid heterozygous clones as well as with diploid heterozygous clones. However, the most important developments followed the publication on 14 July 2011, in *Nature*, of 86% of the sequence of the 844 Mb-genome of the potato, with a prediction of 39,031 protein-coding genes (Potato Genome Sequencing Consortium 2011). Today genetic (linkage) and physical maps can be integrated with the potato reference genome, causal genes sought through candidate SNPs, and the SNPs used for marker-assisted selection. However, despite all of this progress, there is one striking feature of all of the published work considered in this chapter, the small sizes of the mapping populations: from 49 to 272 genotypes in biparental populations and from 95 to 537 genotypes in panels for genome-wide association studies (GWAS). In other words, there is a lack of statistical power for detecting alleles of small effect.

2.14.1 *Major Genes for Qualitative Traits*

Throughout this book we will encounter examples of major genes affecting qualitative traits such as tuber skin and flesh colour and disease resistance. Here one example will suffice. Three key loci are known to affect skin colour as determined by anthocyanins: locus *R* (chromosome 2) affects production of red anthocyanins, locus *P* (chromosome 11) purple anthocyanins and locus *I* (chromosome 10) determines whether or not *R* and *P* are expressed in the tubers. If we focus on locus *P* (in the presence of *I* allele(s)), we find that *Pppp* × *pppp* segregates 1:1 for purple or non-purple skin colour and the locus can be mapped to chromosome 11. It is tempting to be lazy and say *P* is the gene for purple colour. More accurately we should say that the change from *pppp* to *Pppp* results in a change of skin colour from white or red to purple. Today we know enough about the biosynthesis (biochemistry and genetics) of anthocyanins to say that allele *P* encodes a functional version of the enzyme (protein) flavonoid 3′, 5′–hydroxylase which converts dihydrokaempferol to dihydromyricetin, a precursor for the production of purple-coloured delphinidin-3-glucoside. In contrast, allele *p* encodes a non-functional version of the enzyme. When Jung et al. (2005) genetically transformed the red-skinned cultivar Desiree with the *P* allele (i.e. introduced *P* into Desiree), the result was purple-skinned tubers. The researchers also showed that the functional and non-functional alleles differed by 10 amino acids, of which one was within a highly conserved motif and hence probably critical for function. The mapping study of Endelman and Jansky (2016) was also consistent with *P* encoding flavonoid 3′, 5′–hydroxylase. We shall consider in a later chapter the logistics of combining major genes, and QTL alleles of large effect, in a single cultivar. Finally, we saw earlier in this chapter that population sizes of 100 are adequate for determining the genotypes of parents when a major gene is segregating (Luo et al. 2000).

2.14.2 Quantitative Trait Locus (QTL) Alleles of Large Effect

Throughout this book, we will also encounter examples of QTL alleles of large effect affecting quantitative traits. My example here is maturity in the tetraploid cross between processing clone 12601ab1 and table cultivar Stirling which we encountered earlier in this chapter, where maturity was scored on a 1 (early) to 9 (late) scale. Hackett et al. (2014) confirmed earlier analyses showing that segregation of SNP c2_47609 (*Mmmm*) on chromosome 5 in Stirling explained 55.5% of the continuous variation in maturity scores among the 222 offspring. The heritability of the scores was 0.916 so there was little environmental variation. Allele *M* was associated with early maturity and the average maturity scores for two groups of offspring (*Mmmm* and *mmmm*) were 3.53 and 5.87. The difference of 2.34 had a standard error of 0.15 and hence was a statistically significant difference. More importantly, as the lowest and highest scores in the offspring were 1.654 to 7.506 (difference 5.852), the QTL accounted for 40% of the phenotypic range (2.34/5.85) or 44% of the genotypic range (2.34/5.36); clearly a QTL allele of large effect. The SNP maps to a genome super scaffold which is about 1.5 Mb from the *StCDF1* gene coding for plant maturity (Kloosterman et al. 2013); clearly the candidate gene and a satisfying result.

I now want to turn this practical example into a theoretical one. I am going to consider the segregation of two alleles at a number of unlinked marker loci in three populations that give the same result: at the tetraploid level *Mmmm* × *mmmm* and at the diploid level *Mm* × *mm* and the inbred lines from selfing *Mm* (*MM* and *mm*), i.e. all the offspring are *M*: m in a 1: 1 ratio (*Mmmm*: *mmmm*; *Mm*: *mm* and *MM*: *mm*). Let us start with five such loci segregating where one locus has a large effect (difference between M and m phenotype equals 2) and the other four loci each have half the effect (difference of 1). Figure 2.8 shows a population of 32 clones where all of the variation comes from segregation at the five loci (i.e. there is no environmental variation). The population has a mean of 5, a variance of 2, and lowest and highest values of 2 and 8. One locus (*M*) can be detected by a molecular marker that is either

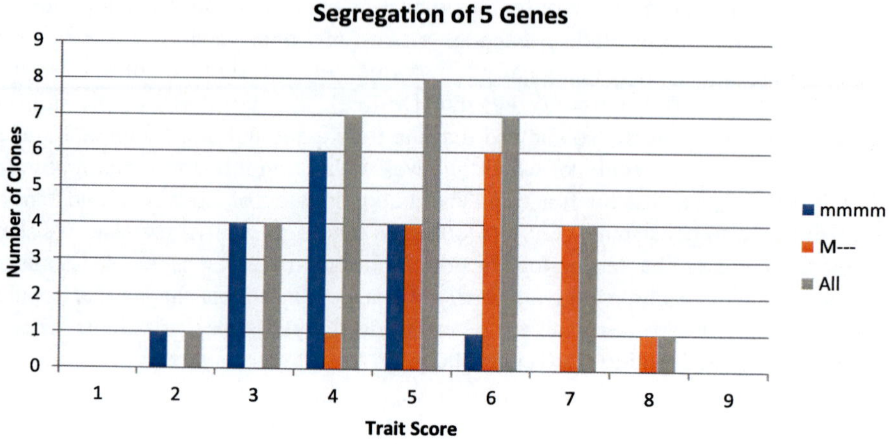

Fig. 2.8 Population of 32 clones where the variation in the trait scores is generated by segregation at 5 loci, one of which is detected through a molecular marker M and explains 50% of the variation

in the causal gene or very tightly linked to it. Segregation at this marker locus splits the population into two groups (*Mmmm* and *mmmm*) with means are 4 and 6 and explains 50% of the variance: both the between- and within-group variances are 1. The difference between the groups (6–4 = 2) is one-third of the difference between the clones with the highest and lowest values (8–2 = 6). The variance of the difference between the group means is the sum of the variances of the two group means each of which is the variance of the 16 clones divided by the number of clones ($V\mu_x = V_x/n$) (1/16). Hence the variance of the difference is $\frac{1}{8}$ and the standard error of the difference is 0.354. The difference of 2 ± 0.354 is statistically significant as determined by a *t*-test ($t = 5.65$, $P < 0.001$, 30 df; Bulmer 1967).

The key points from this simple example are as follows. There is no way of knowing from the continuous distribution how many loci are segregating. Nevertheless, segregation of a molecular marker has explained 50% of the variation and selection of the clones with this marker (M) would increase the population mean from 5 to 6. Breeders could use this marker for marker-assisted selection in their breeding programmes and assuming the marker was an SNP of known location in the potato genome, geneticists could seek the causal gene and a better understanding of the genetic basis of the trait.

Chen et al. (2018) have developed orthogonal contrast-based models for quantitative genetic analysis in autotetraploid species and applied them to a segregating full-sib family of 304 individuals, derived from the cross between American cultivar Atlantic and Chinese cultivar Longshu-3, which was assessed in three field trials in 2015. They found a major QTL for flowering time that explained 39.7% of the phenotypic and 50.1% of the genotypic variation and a major QTL for plant height that explained 29.7% of the phenotypic and 40.8% of the genotypic variation. They also showed that the statistical power for detecting a major QTL is low when the heritability is below 0.35 and the population size is below 500. Bourke et al. (2019) also examined maturity, as considered earlier in this chapter, and found a single QTL peak for plant maturity on chromosome 5 around 18–20 cM with segregation of a single early allele from cultivar Colomba explaining 47% of the variance.

Molla et al. (2018) were primarily interested in the cadmium (Cd) and zinc (Zn) contents of potato tubers in their tetraploid mapping population, cadmium being potentially toxic to humans, whereas zinc is a desirable micronutrient. Potatoes grown in soil with high levels of Cd can concentrate Cd in the tubers to levels higher than the maximum permissible level in many countries of 0.1 mg kg^{-1} fresh weight. Hence breeding for lower Cd but higher Zn contents is desirable. The mapping population comprised 188 F_1 progeny from the cross of Lady Rosetta by Cara, the former being early maturing with a determinate growth habit and a short canopy height, and the latter late maturing with an indeterminate growth habit and a tall canopy height. Lady Rosetta is also the parent with the higher Cd and Zn contents. The parents and offspring were grown in pots in soil with naturally high levels of Cd, in experiments with three replicates in 2014 and 2015, and their tubers assessed for Cd and Zn content (mg kg^{-1} dry-weight) by inductively coupled plasma mass spectrometry (ICP-MS). The mapping population was also scored for maturity (1 to 9 scale with 9 very early) in 2014 and 2015 and canopy height in 2015 in field maintenance plots of 15 tubers without replication. Leaf material was genotyped using the SolCap 8303 SNP array and linkage and QTL mapping performed using the TetraploidSNPMap software mentioned earlier. The final genetic map comprised 2893 SNP markers and had a total length of 985.7 cM. The QTL of largest effect was for maturity on chromosome 5 and explained 53% of the variation with a simplex allele from Lady Rosetta associated with early maturity (marker present gave score of 6.68 and marker absent a score of 3.4, a large difference of 3.28). The closest SNP marker with simplex configuration was solcap_snp_c2_11829 (PGSC0003DMG400030549) at 25 cM on chromosome 5, located about 0.50 Mb from the *StCDF1* gene. This gene also explained 37% of the variation in canopy height, 38% of the variation in Zn content and 33% of the variation in Cd content, with early maturity associated with short canopy height and high Zn and Cd contents. When Cd and Zn contents were corrected for maturity, non-maturity related QTLs of small effect were detected for Cd and Zn concentrations. For Cd, they were on chromosomes 3 (8% of variation), 6 (12.5%) and 7 (10%) and for Zn on chromosomes 1 (7%), 3 (5.3%) and 6 (7.0%), with the locations being similar on chromosomes 3 and 6 (there was possibly a second QTL for Zn on chromosome 6). Overall, the results indicated partially overlapping genetic control of tuber-Cd and tuber-Zn concentration in the cross, involving both maturity-related and non-maturity-related mechanisms. The interested reader is referred to the original paper for more details and a discussion of the implications for potato breeding.

2.14.3 Many Alleles of Small Effect

Now let us look at the examples in Figs. 2.9 and 2.10 where the population size is 256 clones.

In Fig. 2.9 the variation in the trait scores is generated by segregation at 8 loci, each with the same effect, one of which is detected through a molecular marker M and explains 12.5% of the variation. The means of the two groups are 4.5 and 5.5 (difference 1 ± 0.165), the between-group variance is 0.25 and the within-group variance is 1.75 (total 2.0). The difference between the two groups is $\frac{1}{8}$ of the difference between the highest and lowest clones (difference of 8), which are the two extreme genotypes with all of the alleles of increasing and decreasing effects, respectively, and which each occur at a frequency of 1 in 256. The difference between the two groups is statistically significant ($t = 6.05$, $P < 0.001$, 254 df; Bulmer 1967).

In Fig. 2.10 the variation in the trait scores is generated by segregation at 16 loci, each with the same effect, one of which is detected through a molecular marker M and explains 6.25% of the variation. The means of the two groups are 6.0103 and 6.7174 (difference 0.7071 ± 0.171), the between-group variance is 0.125 and the within-group variance is 1.875 (total 2.0). The difference between the two groups is 1/16 of the difference between the two extreme genotypes with all of the alleles of increasing and decreasing effects, respectively. However, these genotypes do not occur in the sample of 256 clones because their frequencies are each only 1 in 65,536. The difference between the highest and lowest clones, which each occurs at

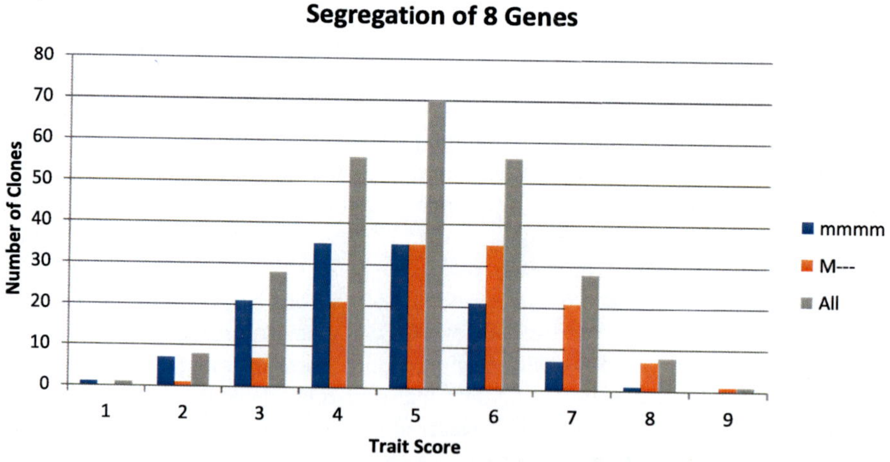

Fig. 2.9 Population of 256 clones where the variation in the trait scores is generated by segregation at 8 loci, one of which is detected through a molecular marker M and explains 12.5% of the variation

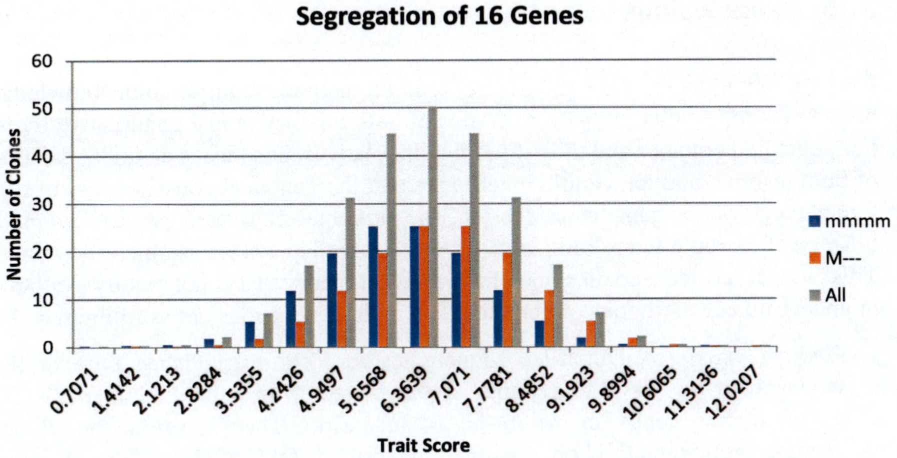

Fig. 2.10 Population of 256 clones where the variation in the trait scores is generated by segregation at 16 loci, one of which is detected through a molecular marker M and explains 6.25% of the variation

a frequency of about 2 in 256, is 7.071. The difference between the two groups is statistically significant ($t = 4.13$, $P < 0.001$, 254 df; Bulmer 1967). It is also worth noting that the best clones that occur in the populations segregating at 8 and 16 loci are 2.83 and 2.5 standard deviations, respectively, from their population means; the standard deviation being 1.414 in both populations. However, whereas the former is the best achievable genotype, in the population segregating for 16 loci, the best achievable genotype (clone) is 4 standard deviations from the mean.

If we keep the population size and variance the same, at 256 and 2, respectively, and increase the number of loci to 128, each with the same effect, our molecular marker now explains just 0.78% of the variance. The between-group and within-group variances are 0.015625 and 1.984375, respectively, and the difference between the means of the two groups is 0.25 ± 0.176, which is not statistically significant ($t = 1.42$, $P > 0.10$, 254 df; Bulmer 1967). The genetic difference is too small to be statistically significant in a population of size 256. In other words, we lack the power to detect the difference. Nevertheless, if over sexual generations we select for higher frequencies of all the alleles of increasing effect, we should eventually produce the genotype with all 128 desirable alleles, which will have a value of 11.32 standard deviations from the original population mean. The eventual selection limit depends on the number of loci segregating, but there is no way of telling this from looking at a continuous distribution of scores. The way forward for breeders for many quantitative traits, such as yield, is a GWAS analysis using molecular markers, not to identify causal genes of small effect and to understand the genetics of the trait, but to produce robust prediction equations to estimate breeding values (genomic estimated breeding values), as we shall see in a later chapter.

2.15 Conclusions

Understanding the theoretical basis of breeding methods requires some knowledge of genetics for predicting the inheritance of both qualitative and quantitative traits. For cultivated potatoes and their diploid wild relatives, this means an understanding of both disomic and tetrasomic inheritance and the genetical consequences of different types of $2n$ gamete production, the principal cultivated potato (*Solanum tuberosum*) being a tetraploid ($2n = 4x = 48$) that displays tetrasomic inheritance. This chapter has focused on some of the relevant theory that is not readily available in undergraduate textbooks of genetics. The main conclusions are as follows.

1. *Potato Genome*: A Dutch-led sequencing consortium published 86% of the sequence of the 844 Mb-genome of the potato, with a prediction of 39,031 protein-coding genes, in *Nature* on 14 July 2011. They had used the whole-genome shotgun method on a doubled monoploid (DM) of Group Phureja potatoes. DM has also been used to define a core set (pan-genome) of 30,401 genes (77.4%) required for the growth and development of potatoes, the remainder being dispensable genes that can be found missing in individual genotypes of cultivated potatoes. The sequence should aid gene discovery; for example, by linking single-nucleotide polymorphism (SNP) markers to the sequence and hence to candidate genes.
2. *Dihaploids*: These diploid *S. tuberosum* can be produced from tetraploid *S. tuberosum* by pollination with elite haploid-inducing clones of Group Phureja. They can be crossed with many diploid wild species which also have an endosperm balance number of two (EBN = 2). However, primary dihaploids are usually male sterile, and most dihaploids, diploid landraces and wild species, and their hybrids, display gametophytic self-incompatibility. Nevertheless, crosses involving diploid *S. tuberosum* and crossable wild species have proved useful for mapping major genes and quantitative trait loci (QTLs).
3. *Self-incompatibility and Self-compatibility*: Outcrossing in diploids is enforced by a single multiallelic *S*-locus, gametophytic self-incompatibility, S-RNase-based system, with the *S*-locus located on chromosome 1. Self-compatibility can be achieved by using the CRISPR–Cas9 system to knock-out the self-incompatibility *S-RNase* gene. It can also be achieved through use of a single dominant self-incompatibility inhibitor gene (*Sli*) from *Solanum chacoense* located on chromosome 12. It is therefore now possible to produce self-fertile, inbred lines of diploid potatoes for genetical research and breeding purposes.
4. *2n Gametes*: Breeding as well as genetical studies can be done at the diploid level with a return to the tetraploid level through unreduced ($2n$) gametes (sexual polyploidization). The mechanisms that produce $2n$ gametes can be classified genetically as first division restitution (FDR) or second division restitution (SDR). The average frequency of transmission of heterozygosity from the diploid parent to $2n$ gamete is much higher for FDR (about 75%) than for SDR (about 50%).

5. *Double Reduction*: Mendelian ratios under tetrasomic inheritance require an understanding of meiosis in autotetraploids and the possibility of double reduction, with the coefficient of double reduction (α) defined as the probability of two sister chromatids in the same diploid gamete.
6. *Tetrasomic and Disomic Inheritance*: There are three important ways in which tetrasomic inheritance differs from disomic inheritance. Inbreeding effects and digenic (dominance) interactions can be transmitted from parents to offspring in diploid gametes. Furthermore, in a random mating population, the single-locus equilibrium is not attained in one generation and the genotype frequencies at equilibrium are not the products of the gene frequencies unless $\alpha = 0$.
7. *General and Specific Combining Abilities*: At the start of a breeding programme, for quantitative traits, analysis of the parental material for general (GCA) and specific (SCA) combining ability can still be of value in predicting likely progress from cycles of crossing and selection, but the results will be specific to the breeding programme and its germplasm.
8. *Linkage and QTL Analysis in Biparental Populations*: The theory for autotetraploids has been developed and put into practice through appropriate computer software. Compared with diploids, it is difficult to detect repulsion linkages with tetrasomic inheritance, but duplex markers can be used to identify homologous chromosomes. With a population size of 400, about 80% of QTLs each explaining 5% of the trait variance can be detected. Hence only QTLs of large effect are likely to be discovered, say those explaining at least 10% of the trait variance.
9. *GWAS*: Genome-wide association studies (GWAS), or simply association mapping, can be used to detect QTLs in a diverse collection of genotypes of interest to the breeder. Computer software is available for autotetraploids and takes appropriate account of population structure and kinship in the collection of genotypes. Kinship has proved more important than population structure in potatoes. Although GWAS has widened the scope and resolution of analysis over biparental populations, statistical power is an issue and to date only QTLs of large effect have been detected. It is perhaps sobering to reflect that it came as a surprise to human geneticists studying psychological traits with GWAS, that there were no associations that account for more than 1% of the heritable differences between individuals and that the average value is less than 0.01%. This means that thousands of DNA differences contribute to the heritability of psychological traits (often around 40%) and that huge GWAS sample sizes (100,000 to 300,000 individuals) are needed to detect these tiny associations (Plomin 2018). Whatever the eventual results for potatoes, breeders are turning to genomic selection of quantitative traits to complement selection of major genes and QTL alleles of large effect.

References

Aliche EB, Oortwijn M, Theeuwen TPJM, Bachem CWB, van Eck HJ, Visser RGF, van der Linden CG (2019) Genetic mapping of tuber size distribution and marketable tuber yield under drought stress in potatoes. Euphytica 215:186. https://doi.org/10.1007/s10681-019-2508-0

Bennett JH (1976) Expectations for inbreeding depression on self-fertilization of tetraploids. Biometrics 32:449–452

Bonierbale MW, Plaisted RL, Tanksley SD (1988) RFLP maps based on a common set of clones reveal modes of chromosomal evolution in potato and tomato. Genetics 120(4):1095–1103

Bourke PM, Voorrips RE, Visser RGF, Maliepaard C (2015) The double-reduction landscape in tetraploid potato as revealed by a high-density linkage map. Genetics 201:853–863. https://doi.org/10.1534/genetics.115.181008

Bourke PM, Hackett CA, Voorrips RE, Visser RGF, Maliepaard C (2019) Quantifying the power and precision of QTL analysis in autopolyploids under bivalent and multivalent genetic models. G3: Genes|Genomes|Genetics 9:2107–2122. https://doi.org/10.1534/g3.119.400269

Bradshaw JE (1994) Quantitative genetics theory for tetrasomic inheritance. In: Bradshaw JE, Mackay GR (eds) Potato genetics. CAB International, Wallingford, pp 71–99

Bradshaw JE (2009) Potato breeding at the Scottish plant Breeding Station and the Scottish crop research institute: 1920-2008. Potato Res 52:141–172

Bradshaw JE (2016) Plant breeding: past, present and future. Springer, Cham

Bradshaw JE, Mackay GR (1994) Breeding strategies for clonally propagated potatoes. In: Bradshaw JE, Mackay GR (eds) Potato genetics. CAB International, Wallingford, pp 467–497

Broman K, Wu H, Sen S, Churchill GA (2003) R/QTL: QTL mapping in experimental crosses. Bioinform 19:889–890

Bulmer MG (1967) Principles of statistics, 2nd edn. Oliver & Boyd, Edinburgh. 252p

Cadman CH (1942) Autotetraploid inheritance in the potato: some new evidence. J Genet 44:33–52

Carputo D, Frusciante L (2011) Classical genetics and traditional breeding. In: Bradeen JM, Kole C (eds) Genetics, genomics and breeding of potato. Science Publishers, Enfield, pp 20–40

Chen J, Zhang F, Wang L, Leach L, Luo Z (2018) Orthogonal contrast-based models for quantitative genetic analysis in autotetraploid species. New Phytol 220:332–346. https://doi.org/10.1111/nph.15284

Chen J, Leach L, Yang J, Zhang F, Tao Q, Dang Z et al (2020) A tetrasomic inheritance model and likelihood-based method for mapping quantitative trait loci in autotetraploid species. N Phytol. https://doi.org/10.1111/nph.16413

Choudhary A, Wright L, Ponce O, Chen J, Prashar A, Sanchez-Moran E, Luo Z, Compton L (2020) Varietal variation and chromosome behaviour during meiosis in *Solanum tuberosum*. Heredity. https://doi.org/10.1038/s41437-020-0328-6

Cipar MS, Peloquin SJ, Hougas RW (1964) Inheritance of incompatibility in hybrids between *Solanum tuberosum* haploids and diploid species. Euphytica 13:163–172

Crow JF, Kimura M (1970) An introduction to population genetics theory. Harper and Row, New York

Cuenca J, Froelicher Y, Aleza P, Juárez J, Navarro L, Ollitrault P (2011) Multilocus half-tetrad analysis and centromere mapping in citrus: evidence of SDR mechanism for 2n megagametophyte production and partial chiasma interference in mandarin cv 'Fortune'. Heredity 107:462–470

D'hoop BB, Keizer PL, Paulo MJ, Visser RG, van Eeuwijk FA, van Eck HJ (2014) Identification of agronomically important QTL in tetraploid potato cultivars using a marker-trait association analysis. Theor Appl Genet 127:731–748. https://doi.org/10.1007/s00122-013-2254-y

Dodds KS (1962) Classification of cultivated potatoes. In: Correll DS (ed) The potato and its wild relatives. Texas Research Foundation, Renner, pp 517–539

Dodds KS (1965) The history and relationships of cultivated potatoes. In: Hutchinson JB (ed) Essays in crop plant evolution. Cambridge University Press, Cambridge, pp 123–141

Endelman JB, Jansky SH (2016) Genetic mapping with an inbred line-derived F2 population in potato. Theor Appl Genet 129(5):935–943

Endelman JB, Schmitz Carley CA, Bethke PC, Coombs JJ, Clough ME, da Silva WL, De Jong WS, Douches DS, Frederick CM, Haynes KG, Holm DG, Miller JC, Muñoz PR, Navarro FM, Novy RG, Palta JP, Porter GA, Rak KT, Sathuvalli VR, Thompson AL, Craig Yencho G (2018) Genetic variance partitioning and genome-wide prediction with allele dosage information in autotetraploid potato. Genetics 209(1):77–87. https://doi.org/10.1534/genetics.118.300685

Fisher RA (1965) The theory of inbreeding, 2nd edn. Oliver and Boyd, London

Fisher RA, Mather K (1943) The inheritance of style length in *Lythrum salicaria*. Ann Eugenics 12:1–23

Fisher RA, Yates F (1963) Statistical tables, 6th edn. Oliver and Boyd, Edinburgh. 146p

Gallais A (2003) Quantitative genetics and breeding methods in autopolyploid plants. INRA, Paris. 515p

Gebhardt C, Ritter E, Debener T, Schachtschabel U, Walkemeier B, Uhrig H, Salamini F (1989) RFLP analysis and linkage mapping in *Solanum tuberosum*. Theor Appl Genet 78(1):65–75

Gebhardt C, Ritter E, Barone A, Debener T, Walkemeier B, Schachtschabel U, Kaufmann H, Thompson RD, Bonierbale MW, Ganal MW, Tanksley SD, Salamini F (1991) RFLP maps of potato and their alignment with the homoeologous tomato genome. Theor Appl Genet 83(1):49–57

Hackett CA, Bradshaw JE, Meyer RC, McNicol JW, Milbourne D, Waugh R (1998) Linkage analysis in tetraploid species: a simulation study. Genet Res, Camb 71:143–154

Hackett CA, Bradshaw JE, McNicol JW (2001) Interval mapping of quantitative trait loci in autotetraploid species. Genetics 159:1819–1832

Hackett CA, Milne I, Bradshaw JE, Luo Z (2007) TetraploidMap for windows: linkage map construction and QTL mapping in autotetraploid species. J Hered 98:727–729

Hackett CA, McLean K, Bryan GJ (2013) Linkage analysis and QTL mapping using SNP dosage data in a tetraploid potato mapping population. PLoS One 8(5):e63939. https://doi.org/10.1371/journal.pone.0063939

Hackett CA, Bradshaw JE, Bryan GJ (2014) QTL mapping in autotetraploids using SNP dosage information. Theor Appl Genet 127:1885–1904. https://doi.org/10.1007/s00122-014-2347-2. Epub 2014 Jul 1

Hackett CA, Boskamp B, Vogogias A, Preedy K, Milne I (2017) TetraploidSNPMap: software for linkage analysis and QTL mapping in autotetraploid populations using SNP dosage data. J Hered 108:438–442

Hamilton JP, Hansey CN, Whitty BR, Stoffel K, Massa AN et al (2011) Single nucleotide polymorphism discovery in elite North American potato germplasm. BMC Genomics 12:302. https://doi.org/10.1186/1471-2164-12-302

Hanneman RE (1985) Self fertility in *Solanum chacoense*. Am Potato J 62:428–429

Hardigan MA, Crisovan E, Hamilton JP, Kim J, Laimbeer P, Leisner CP, Manrique-Carpintero NC, Newton L, Pham GM, Vaillancourt B, Yang X, Zeng Z, Douches DS, Jiang J, Veilleux RE, Buell CR (2016) Genome reduction uncovers a large dispensable genome and adaptive role for copy number variation in asexually propagated *Solanum tuberosum*. Plant Cell 28:388–405. https://doi.org/10.1105/tpc.15.00538

Hawkes JG (1990) The potato: evolution, biodiversity & genetic resources. Belhaven Press, London

Hawkes JG, Jackson MT (1992) Taxonomic and evolutionary implications of the endosperm balance number hypothesis in potatoes. Theor Appl Genet 84:180–185

Haynes KG, Potts WE (1993) Minimizing inbreeding in tetraploids derived through sexual polyploidization. Am Potato J 70(9):617–624

Hedrick PW (2011) Genetics of populations, 4th edn. Jones and Bartlett, Sudbury. 675p

Henderson CR (1950) Estimation of genetic parameters. Ann Math Stat 21:309–310

Hermsen JGT (1994) Introgression of genes from wild species, including molecular and cellular approaches. In: Bradshaw JE, Mackay GR (eds) Potato genetics. CAB International, Wallingford, pp 515–538

Hosaka K, Hanneman RE (1998a) Genetics of self-compatibility in a self-incompatible wild diploid potato species *Solanum chacoense*. 1. Detection of an S-locus inhibitor (Sli) gene. Euphytica 99:191–197. https://doi.org/10.1023/A:1018353613431

Hosaka K, Hanneman RE (1998b) Genetics of self-compatibility in a self-incompatible wild diploid potato species *Solanum chacoense*. 2. Localization of an S-locus inhibitor (*Sli*) gene on the potato genome using DNA markers. Euphytica 103:265–271. https://doi.org/10.1023/A:1018380725160

Hougas RW, Peloquin SJ (1958) The potential of potato haploids in breeding and genetic research. Amer Potato J 35:701–707

Hougas RW, Peloquin SJ, Ross RW (1958) Haploids of the common potato. J Hered 49:103–107

Hougas RW, Peloquin SJ, Gabert AC (1964) Effect of seed-parent and pollinator on the frequency of haploids in *Solanum tuberosum*. Crop Sci 4:593–595

Ivanovskajya EV (1939) A haploid plant of *Solanum tuberosum*. Comptes Rendus de L'Acadẻmie de Science. URSS 24:517–520

Jansky S (2006) Overcoming hybridization barriers in potato. Plant Breed 125:1–12

Jansky SH, Charkowski AO, Douches DS, Gusmini G, Richael C, Bethke PC, Spooner DM, Novy RG, De Jong H, De Jong WS, Bamberg JB, Thompson AL, Bizimungu B, Holm DG, Brown CR, Haynes KG, Sathuvalli VR, Veilleux RE, Miller JC, Bradeen JM, Jiang J (2016) Reinventing potato as a diploid inbred line–based crop. Crop Sci 56:1412–1422. https://doi.org/10.2135/cropsci2015.12.0740

Johnston SA, der Nijs TPM, Peloquin SJ, Hanneman RE Jr (1980) The significance of genic balance to endosperm development in interspecific crosses. Theor Appl Genet 57:5–9

Jongedijk E, Ramanna MS (1988) Synaptic mutants in potato, *Solanum tuberosum* L. I. Expression and identity of genes for desynapsis. Genome 30:664–670

Jongedijk E, Ramanna MS, Sawor Z, Hermsen JGT (1991) Formation of first division restitution (FDR) 2*n*-megaspores through pseudohomotypic division in *ds-1* (desynapsis) mutants of diploid potato: routine production of tetraploid progeny from 2*x* FDR × 2*x* FDR crosses. Theor Appl Genet 82:645–656

Jung CS, Griffiths HM, De Jong DM, Cheng S, Bodis M, De Jong WS (2005) The potato P locus codes for flavonoid 3',5'–hydroxylase. Theor Appl Genet 110:269–275

Kaufmann H, Salamini F, Thompson RD (1991) Sequence variability and gene structure at the self-incompatibility locus of *Solanum tuberosum*. Mol Gen Genet 226:457–466

Kempthorne O (1957) An introduction to genetic statistics. Wiley, New York

Kloosterman B, Abelenda JA, Carretero Gomez MDM, Oortwijn M, de Boer JM, Kowitwanich K, Horvath BM, van Eck HJ, Smaczniak C, Prat S, Visser RGF, Bachem CWB (2013) Naturally occurring allele diversity allows potato cultivation in northern latitudes. Nature 495(7440):246–250

Lewis D (1943) The physiology of incompatibility in plants. III Autopolyploids J Genet 44:171–185

Lewis D (1954) Comparative incompatibility in angiosperms and fungi. Adv Genet 6:235–285

Lindhout P, Meijer D, Schotte T, Hutten RCB, Visser RGF, van Eck HJ (2011) Towards F1 hybrid seed potato breeding. Potato Res 54:301–312

Lindhout P, de Vries M, ter Maat M, Ying S, Viquez-Zamora M, van Heusden S (2018) Hybrid potato breeding for improved varieties. In: Wang-Pruski G (ed) Achieving sustainable cultivation of potatoes. Burleigh Dodds, Cambridge, pp 99–122

Lunden AP (1937) Arvelighetsundersokelser i potet. (Inheritance studies in the potato). Meldinger fra Norges Landbrukshøishkole 17:1–156

Luo ZW, Hackett CA, Bradshaw JE, McNicol JW, Milbourne D (2000) Predicting parental genotypes and gene segregation for tetrasomic inheritance. Theor Appl Genet 100:1067–1073

Luo ZW, Hackett CA, Bradshaw JE, McNicol JW, Milbourne D (2001) Construction of a genetic linkage map in tetraploid species using molecular markers. Genetics 157:1369–1385

Luo ZW, Zhang Z, Leach L, Zhang RM, Bradshaw JE, Kearsey MJ (2006) Constructing genetic linkage maps under a tetrasomic model. Genetics 172:2635–2645

Lynch M, Walsh B (1998) Genetics and analysis of quantitative traits. Sinauer Associates, Sunderland. 980p

Mackay GR (2005) Propagation by traditional breeding methods. In: Razdan MK, Mattoo AK (eds) Genetic improvement of Solanaceous crops volume I: potato. Science Publishers, Enfield, pp 65–81

Malécot G (1948) Les mathématiques de l'hérédité. Masson, Paris

Mann H, Iorizzo M, Gao L, D'Agostino N, Carputo D, Chiusano ML, Bradeen JM (2011) Molecular linkage maps: strategies, resources and achievements. In: Bradeen JM, Kole C (eds) Genetics, genomics and breeding of potato. Science Publishers, Enfield, pp 68–89

Mather K (1936) Segregation and linkage in autotetraploids. J Genet 32:287–314

McClure B, Cruz-García F, Romero C (2011) Compatibility and incompatibility in S-RNase-based systems. Ann Bot (Lond) 108:647–658. https://doi.org/10.1093/aob/mcr179

Meijer D, Viquez-Zamora M, van Eck HJ, Hutten RCB, Su Y, Rothengatter R, Visser RGF, Lindhout WH, van Heusden AW (2018) QTL mapping in diploid potato by using selfed progenies of the cross *S. tuberosum* × *S. chacoense*. Euphytica 214(7):121

Mendiburu AO, Peloquin SJ (1977) The significance of 2*n* gametes in potato breeding. Theor Appl Genet 49:53–61

Mendoza HA, Haynes FL (1973) Some aspects of breeding and inbreeding in potatoes. Am Potato J 50:216–222

Mendoza HA, Haynes FL (1974) Genetic relationship among potato cultivars grown in the United States. HortSci 9:328–330

Meuwissen TH, Hayes BJ, Goddard ME (2001) Prediction of total genetic value using genome-wide dense marker maps. Genetics 157:1819–1829

Milbourne D, Bradshaw JE, Hackett CA (2008) Molecular mapping and breeding in polyploid crop plants. In: Kole C, Abbott AG (eds) Principles and practices of plant genomics, volume 2 molecular breeding. Science Publishers, Enfield, pp 355–394

Mok DWS, Peloquin SJ (1975) Three mechanisms of 2*n* pollen formation in diploid potatoes. Can J Genet Cytol 17:217–225

Molla F, Mengist MF, Alves S, Griffin D, Creedon J, McLaughlin MJ, Jones PW, Milbourne D (2018) Genetic mapping of quantitative trait loci for tuber-cadmium and zinc concentration in potato reveals associations with maturity and both overlapping and independent components of genetic control. Theor Appl Genet 131:929–945

Ochoa CM (1990) The potatoes of South America: Bolivia. Translated by Ugent D, Cambridge University Press, Cambridge

Ortiz R (1998) Potato breeding via ploidy manipulations. In: Janick J (ed) Plant breed revs, vol 16. Wiley, New York, pp 15–86

Ortiz R (2001) The state of the use of potato genetic diversity. In: Cooper HD, Spillane C, Hodgkin T (eds) Broadening the genetic base of crop production. CAB International, Wallingford, pp 181–200

Ortiz R, Mihovilovich E (2020) Genetics and cytogenetics of the potato. In: Campos H, Ortiz O (eds) The potato crop: its agricultural, nutritional and social contribution to humankind. Springer, ebook, pp 219–247

Park J, Hackett CA, Dandurand L-M, Wang X, Jong WS (2019) QTL for resistance to *Globodera rostochiensis* pathotype Ro2 and *G. pallida* pathotype Pa2/3 in autotetraploid potato. Am J Potato Res. https://doi.org/10.1007/s12230-019-09745-4

Peloquin SJ, Boiteux LS, Carputo D (1999) Meiotic mutants in potato: valuable variants. Genetics 153:1493–1499

Peterson BA, Holt SH, Parker F, Laimbeer E, Doulis AG, Coombs J, Douches DS, Hardigan MA, Buell CR, Veilleux RE (2016) Self-fertility in a cultivated diploid potato population examined with the Infinium 8303 potato single-nucleotide polymorphism array. Plant Genome 9. https://doi.org/10.3835/plantgenome2016.01.0003

Plomin R (2018) Blueprint: how DNA makes us who we are. Allen Lane, Milton Keynes. 266p

Potato Genome Sequencing Consortium (2011) Genome sequence and analysis of the tuber crop potato. Nature 475:189–195

Prinzenberg AE, Víquez-Zamora M, Harbinson J, Lindhout P, van Heusden S (2018) Chlorophyll fluorescence imaging reveals genetic variation and loci for a photosynthetic trait in diploid potato. Physiol Plant 164(2):163–175

Ramanna MS (1979) A re-examination of the mechanisms of 2*n* gamete formation in potato and its implications for breeding. Euphytica 28:537–561

Ross H (1986) Potato breeding – problems and perspectives. Advs in Plant Breed 13, Paul Parey, Berlin and Hamburg

Rosyara U, De Jong W, Douches D, Endelman J (2016) Software for genome-wide association studies in autopolyploids and its application to potato. Plant Genome 9:1–10

Schwarz GE (1978) Estimating the dimension of a model. Ann Stat 6:461–464

Semagn K, Babu R, Hearne S, Olsen M (2014) Single nucleotide polymorphism genotyping using Kompetitive Allele Specific PCR (KASP): overview of the technology and its application in crop improvement. Mol Breed 33:1–14. https://doi.org/10.1007/s11032-013-9917-x

Sharma SK, Bolser D, de Boer J, Sønderkaer M, Amoros W, Carboni MF, D'Ambrosio JM, de la Cruz G, Di Genova A, Douches DS, Eguiluz M, Guo X, Guzman F, Hackett CA, Hamilton JP, Li G, Li Y, Lozano R, Maass A, Marshall D, Martinez D, McLean K, Mejía N, Milne L, Munive S, Nagy I, Ponce O, Ramirez M, Simon R, Thomson SJ, Torres Y, Waugh R, Zhang Z, Huang S, Visser RGF, Bachem CWB, Sagredo B, Feingold SE, Orjeda G, Veilleux RE, Bonierbale M, Jacobs JME, Milbourne D, Martin DMA, Bryan GJ (2013) Construction of reference chromosome-scale pseudomolecules for potato: integrating the potato genome with genetic and physical maps. G3 (Bethesda) 3:2031–2047

Sharma SK, MacKenzie K, McLean K, Dale F, Daniels S, Bryan GJ (2018) Linkage disequilibrium and evaluation of genome-wide association mapping models in tetraploid potato. G3 8:3185–3202. https://doi.org/10.1534/g3.118.20037

Spooner DM (2016) Species delimitations in plants: lessons learned from potato taxonomy by a practicing taxonomist. J Syst Evol 54:191–203

Spooner DM, Nunez J, Trujillo G, Herrera MR, Guzman F, Ghislain M (2007) Extensive simple sequence repeat genotyping of potato landraces supports a major reevaluation of their gene pool structure and classification. Proc Nat Acad Sci 104:19398–19404

Spooner DM, Ghislain M, Simon R, Jansky SH, Gavrilenko T (2014) Systematics, diversity, genetics, and evolution of wild and cultivated potatoes. Bot Rev 80:283–383

Swaminathan MS, Howard HW (1953) The cytology and genetics of the potato (*Solanum tuberosum*) and related species. Bibliogr Genet 16:1–192

Tai GCC (1994) Use of 2*n* gametes. In: Bradshaw JE, Mackay GR (eds) Potato genetics. CAB International, Wallingford, pp 109–132

Tanksley SD, Ganal MW, Prince JP, de-Vicente MC, Bonierbale MW, Broun P, Fulton TM, Giovannoni JJ, Grandillo S, Martin GB, Messeguer R, Miller JC, Miller L, Paterson AH, Pineda O, Roder MS, Wing RA, Wu W, Young ND (1992) High density molecular linkage maps of the tomato and potato genomes. Genetics 132(4):1141–1160

Van Ooijen JW, Jansen J (2013) Genetic mapping in experimental populations. Cambridge University Press, Cambridge, 157p

van Os H, Stam P, Visser RGF, van Eck HJ (2005a) RECORD: a novel method for ordering loci on a genetic linkage map. Theor Appl Genet 112(1):30–40

van Os H, Stam P, Visser RGF, van Eck HJ (2005b) SMOOTH: a statistical method for successful removal of genotyping errors from high-density genetic linkage data. Theor Appl Genet 112(1):187–194

van Os H, Andrzejewski S, Bakker E, Barrena I, Bryan GJ, Caromel B, Ghareeb B, Isidore E, de Jong W, van Koert P, Lefebvre V, Milbourne D, Ritter E, Rouppe van der Voort JNAM, Rousselle-Bourgeois F, van Vliet J, Waugh R, Visser RGF, Bakker J, van Eck HJ (2006) Construction of a 10,000-marker ultradense genetic recombination map of potato: providing a framework for accelerated gene isolation and a genomewide physical map. Genetics 173(2):1075–1087. https://doi.org/10.1534/genetics.106.055871

Veilleux RE (1985) Diploid and polyploid gametes in crop plants: mechanisms of formation and utilization in plant breeding. Plant Breed Rev 3:253–288

Voorrips RE, Gort G, Vosman B (2011) Genotype calling in tetraploid species from bi-allelic marker data using mixture models. BMC Bioinformatics 12:172. https://doi.org/10.1186/1471-2105-12-172

Vos PG, Paulo MJ, Voorrips RE, Visser RGF, van Eck HJ, van Eeuwijk FA (2017) Evaluation of LD decay and various LD-decay estimators in simulated and SNP-array data of tetraploid potato. Theor Appl Genet 130:123–135. https://doi.org/10.1007/s00122-016-2798-8

Werner JE, Peloquin SJ (1990) Inheritance and two mechanisms of $2n$ egg formation in $2x$ potatoes. J Hered 81:371–374

Wricke G, Weber WE (1986) Quantitative genetics and selection in plant breeding. Walter de Gruyter, Berlin, New York. 406p

Wright AJ (1979) The use of differential coefficients in the development and interpretation of quantitative genetic models. Heredity 43:1–8

Ye M, Peng Z, Tang D, Yang Z, Li D, Xu Y, Zhang C, Huang S (2018) Generation of self-compatible diploid potato by knockout of S-RNase. Nat Plants 4:651–654

Yuan J, Bizimungu B, De Koeyer D, Rosyara U, Wen Z, Lagüe M (2020) Genome-wide association study of resistance to potato common scab. Potato Res 63(2):253–266

Part II
Breeding Objectives

Chapter 3
Increasing Potato Yields: A Conundrum

3.1 Introduction

Increasing yield per hectare is an obvious breeding objective for all uses of potatoes, whether as a staple food for local consumption, as a vegetable for cash or as a crop for processing into food or starch. However, the potato is a global crop which is grown in a number of agroecological zones defined by length of growing season(s), day lengths, temperatures and water availability. Hence, we need to be more specific about increasing yield as a breeding objective, both in terms of end use of crop and in terms of the targeted agroecological zone, because we are likely to need a separate breeding programme for each combination of end use and zone. Having decided the targeted agroecological zone, we can ask how well or badly existing cultivars are adapted to the zone, or a more specific region within the zone, and hence how much effort needs to go into breeding new cultivars with higher yields. As yield is a quantitative trait, variation in yield can result from environmental differences as well as genetical ones, and also from genotype × environment interactions. Hence, we shall need to assess potential parents and potential cultivars in a range of environments, as well as seeking to generate and utilize genetical variation for higher yields in our breeding programmes. In this chapter, we discover that increasing potato yields has proved a conundrum for potato breeders. I start with some theory and an example of the conundrum before going on to consider the potential yields of potatoes, the gaps between these and actual yields and the extent to which the gaps can be reduced by improved agronomy and new cultivars. Finally, we consider how to assess and select potential cultivars for higher yields, including the importance of choosing the right environments in which to do the assessments. Then, in the next chapter, we will consider the target end user of new cultivars and add quality criteria to that of yield in the target environment.

J. E. Bradshaw, *Potato Breeding: Theory and Practice*,
https://doi.org/10.1007/978-3-030-64414-7_3

3.2 Crosses Among Cultivars and Breeders' Clones

My first example in this chapter is from the potato programme at the Scottish Crop Research Institute (SCRI) which combined research with breeding finished cultivars. As part of the programme, a 6 × 6 half diallel set of crosses with selfs was assessed at Mylnefield and Gourdie Farms, Dundee (56.46°N, 2.97°W) in 1993 and 1994, respectively (Bradshaw 2009a). The two trials each had randomized complete block (RCB) designs with two replicates and five-plant plots. They were grown and evaluated in the way described by Bradshaw et al. (1998) for ware trials. The trial in 1993 was planted on 27 April and harvested on 5 October and the one in 1994 on 27 April and 14 September, respectively. Two of the six parents, cultivars Maris Piper (M. Piper) and Desiree, were still in the top twenty cultivars in Great Britain (GB) in 2017, at positions 1 and 17, respectively, and cultivars Pentland Squire (P. Squire) and Cara had been in the top ten in 1986. Cultivar Katahdin, released in 1932, was still in the top ten cultivars in the USA in 1984 (Plaisted and Hoopes 1989), as well as being a major contributing ancestor to 44 prominent North American cultivars (Love 1999). Lastly, clone 3683a2 was a parent of SCRI cultivar Kirsty, released in 1982 (Bradshaw 2009b). It was intended to include the six parents four times in each replicate, but M. Piper was included six times and Katahdin twice. Each of the 294 offspring clones was included just one in each replicate. There were 76 clones from the self-pollinations of the six parents and 218 clones from the 15 hybridizations of the six parents.

The results of the analysis of variation among the 218 hybrid clones were encouraging for improving yield. Let us start with the analysis of variance and components of variance (the 'true' amounts of variation) shown in Table 3.1, where years, replicates and clones were treated as random effects. Yields were higher in 1993 (10.75 kg/plot) than in 1994 (6.39 kg/plot), but our interest is in the clones. The differences between clones were statistically significant when tested against the years × clones interaction which was also significant, but smaller in size. The

Table 3.1 Components of variance for yields of 218 clones in two trials with two replicates (reps), with the trials grown in separate years (1993 and 1994) (number of missing values shown in brackets under df): df = degrees of freedom and MS = mean squares

Items	df	MS	Expected MS
Years	1	4137.684**	
Residual (reps)	2	12.643	
Clones	217	20.611***	$\sigma^2_e + 2\sigma^2_{YC} + 4\sigma^2_C$
Years × Clones	213(4)	4.424**	$\sigma^2_e + 2\sigma^2_{YC}$
Residual (plots)	419(15)	3.168	σ^2_e
Components		Value	
Clones	σ^2_C	4.047	
Years × Clones	σ^2_{YC}	0.628	
Residual (plots)	σ^2_e	3.168	

*** $P < 0.001$, ** $P < 0.01$, * $P < 0.05$

components of variance for clones and years × clones interactions were 4.047 and 0.628, respectively. The proportion of the variation between the means of clones (averaged over years and replicates) that was genetical ($\sigma^2_C/(\sigma^2_C + \sigma^2_{YC}/2 + \sigma^2_e/4)$) was 0.785, and is called the broad-sense heritability. In other words, 78.5% of the difference between the yield of a clone selected as a potential new cultivar and the mean yield of all 218 clones should be a true difference that will be seen in subsequent clonal generations. There were in fact 17 clones (mean yield 12.914 kg/plot) out of the 218 (7.8%) that out-yielded the highest yielding parent which was Cara (11.872 ± 0.526 kg/plot; standard error of mean based on years × clones interaction and four plots of Cara per replicate). Hence, selection of these clones should result in a set of clones which is 3.409 kg/plot (4.343 × 0.785) higher yielding than the mean of all 218 clones (8.571 kg/plot); that is 11.98 kg/plot.

In Table 3.2, the variation between clones has been partitioned into between and within the 15 progenies. The components were estimated using the harmonic mean (14.31) of the numbers of clones in each of the progenies. The progeny mean square was significant when tested against the clones within progenies mean square, but it can be seen from the expected mean squares that this is a test of significance for progenies and years × progenies combined. Nevertheless, the component of variance for progenies was larger than the component for the interaction with years, which was statistically significant. There was more variation within (3.168) than between progenies (0.954) and also smaller amounts of variation for years × progenies interactions (0.232) and years × clones within progenies interactions (0.410). We could use these components to compare the responses to selection between individual clones and also between whole progenies with index selection which combines between and within progeny selection. However, we will assume that the high broad-sense heritability for individual clones (0.785) means that individual clone

Table 3.2 Components of variance for yields of 218 clones from 15 progenies in two trials with two replicates (reps), with the trials grown in separate years (1993 and 1994) (number of missing values shown in brackets under df): df = degrees of freedom and MS = mean squares

Items	df	MS	Expected MS
Progenies	14	77.912***	$\sigma^2_e + 2\sigma^2_{YC/P} + 4\sigma^2_{C/P} + 28.62\sigma^2_{YP} + 57.24\sigma^2_P$
Clones/progenies	203	16.660***	$\sigma^2_e + 2\sigma^2_{YC/P} + 4\sigma^2_{C/P}$
Years × progenies	14	10.635**	$\sigma^2_e + 2\sigma^2_{YC/P} + 28.62\sigma^2_{YP}$
Years × clones/progenies	199(4)	3.987*	$\sigma^2_e + 2\sigma^2_{YC/P}$
Residual (plots)	419(15)	3.168	σ^2_e
Components		Value	
Progenies	σ^2_P	0.954	
Clones/progenies	$\sigma^2_{C/P}$	3.168	
Years × progenies	σ^2_{YP}	0.232	
Years × clones/progenies	$\sigma^2_{YC/P}$	0.410	
Residual (plots)	σ^2_e	3.168	

*** $P < 0.001$, ** $P < 0.01$, * $P < 0.05$

selection will be superior to progeny selection and almost as good as the more complicated index selection.

Finally, in Table 3.3, the variation between progenies has been partitioned into general (GCA) and specific combining ability (SCA) using Experimental Method 4 and Model II of Griffing (1956). The selfed progenies were excluded from the analysis (Method 4) as they cannot be considered part of the implied (hypothetical) random mating population from which we assume the parents and progeny were drawn at random (Model II). The six selfed progenies were in fact much lower yielding than their parents, 6.21 compared with 10.27 kg/plot, a fall of 36%. The 218 offspring clones were also lower yielding than their parents, 8.57 compared with 10.27 kg/plot, a fall of 16.5%. Both of these falls were expected from our simple theory of the previous chapter in the presence of unidirectional diallelic interactions. Griffing (1956) showed that the expected mean square (MS) for GCA includes $[\sigma^2_{\mathrm{SCA}} + (p\text{-}2)\sigma^2_{\mathrm{GCA}}]$, where p is the number of parents, a result that is not obvious. The SCA item was statistically significant when tested against the clones within progenies item, but not the GCA item when tested against the SCA item due to the small numbers of degrees of freedom involved. Nevertheless, it can be seen that the components of variance for GCA and SCA were larger than those for their interactions with years, and that SCA accounted for slightly more variation than GCA. Additive genetic variance (GCA) is required for progress over a number of generations, whereas within any one generation, breeders can make use of all the genetic variation, both between and within crosses. In other words, not all of the superiority of clones selected for their high yield will be transmitted to their offspring if some of this is due to non-additive genetic variation in the form of SCA.

There were 17 clones (mean yield 12.914 kg/plot) out of the 218 (mean yield 8.571 kg/plot) that out-yielded the highest yielding parent, Cara (yield 11.872 ± 0.526 kg/plot), and we concluded from using the broad-sense heritability value of 0.785 that selection of these clones should result in a set of clones which is 3.409 kg/plot (4.343 × 0.785) higher yielding than the mean of all 218 clones. Now let us consider the yield increase (over the mean of the 218 clones) we can expect in the next generation from inter-crossing the 17 superior clones. The formula for the response to selection (R), derived in the previous chapter, is $R = h_n^2 S$, where h_n^2 is the narrow-sense heritability and S is the selection differential, the difference (4.343 kg/plot) between the mean of the 17 clones and the mean of all 218 clones. Whereas the broad-sense heritability was $\sigma^2_C/(\sigma^2_C + \sigma^2_{YC}/2 + \sigma^2_e/4)$, the narrow-sense heritability is $\sigma^2_A/(\sigma^2_C + \sigma^2_{YC}/2 + \sigma^2_e/4)$, where σ^2_A ($=V_A$) is the additive genetic variance. Based on the components of variance for a random mating population (Wricke and Weber 1986; Bradshaw 1994a), and in the absence of triallelic and tetra-allelic interactions, $4\sigma^2_{GCA} - \frac{2}{3}\sigma^2_{SCA}$ can be used as the estimate of the additive genetic variance (σ^2_A).

I am going to use $4\sigma^2_{GCA} - \frac{2}{3}\sigma^2_{SCA}$ (1.111) as my estimate of the additive genetic variance and $6\sigma^2_{SCA}$ (2.742) as my estimate of the non-additive genetic variance and take comfort from the fact that their sum (3.853) is close to σ^2_C (4.047), the total genetical variation between clones. Hence, my estimate of the narrow-sense heritability is 0.216 (1.111/5.153), giving a response to selection of 0.938 kg/plot (4.343 × 0.216). This estimate serves my purpose, but it is very unlikely that the

Table 3.3 Components of variance for yields of 218 clones in 15 progenies from a 6 × 6 half diallel set of crosses grown in a trial with two replicates in each of two years (number of missing values shown in brackets under df): GCA = general combining ability, SCA = specific combining ability, df = degrees of freedom and MS = mean squares

Items	df	MS	Expected MS
Progenies			
GCA	5	135.81	$\sigma^2_e + 2\sigma^2_{YC/P} + 4\sigma^2_{C/P} + 28.62\sigma^2_{YSCA} + 114.48\sigma^2_{YGCA} + 57.24\sigma^2_{SCA} + 228.96\sigma^2_{GCA}$
SCA	9	45.75**	$\sigma^2_e + 2\sigma^2_{YC/P} + 4\sigma^2_{C/P} + 28.62\sigma^2_{YSCA} + 57.24\sigma^2_{SCA}$
Clones/progenies	203	16.660***	$\sigma^2_e + 2\sigma^2_{YC/P} + 4\sigma^2_{C/P}$
Years × progenies			
Years × GCA	5	15.942	$\sigma^2_e + 2\sigma^2_{YC/P} + 28.62\sigma^2_{YSCA} + 114.48\sigma^2_{YGCA}$
Years × SCA	9	6.901	$\sigma^2_e + 2\sigma^2_{YC/P} + 28.62\sigma^2_{YSCA}$
Years × clones/progenies	199(4)	3.987*	$\sigma^2_e + 2\sigma^2_{YC/P}$
Residual (plots)	419(15)	3.168	σ^2_e
Components		Value	
Progenies			
GCA	σ^2_{GCA}	0.354	
SCA	σ^2_{SCA}	0.457	
Clones/progenies	$\sigma^2_{C/P}$	3.168	
Years × progenies			
Years × GCA	σ^2_{YGCA}	0.079	
Years × SCA	σ^2_{YSCA}	0.102	
Years × clones/progenies	$\sigma^2_{YC/P}$	0.410	
Residual (plots)	σ^2_e	3.168	

*** $P < 0.001$, ** $P < 0.01$, * $P < 0.05$

parents were a random sample of genotypes from a random mating population in equilibrium, and hence my estimates of additive and non-additive genetic variance are probably inaccurate.

We can summarize the conclusions of this lengthy analysis as follows. If we select the 17 highest yielding clones out of the total of 218, we can expect them to be 3.409 kg/plot higher yielding than the population mean of 8.571 kg/plot. If we inter-cross the 17 clones to produce a new generation of 218 clones, we can expect an increase of 0.938 kg/plot from 8.571 to 9.509 kg/plot, an 11% increase. Furthermore, although we cannot predict the long-term response to selection, we might reasonably expect to find that similar increases occur in the next four generations without reducing the variation among the clones in each generation. If this happens, the population mean will increase as follows (Fig. 3.1): 8.571, 9.509, 10.447, 11.385, 12.323 and 13.261 (11.7% greater than Cara). After two generations of selection, the population mean will be greater than the average of the six original parents (10.266), and after four generations of selection, greater than Cara, which means that so will over half of the clones. In other words, it is the improvement in yield over generations that is the driver of higher yielding clones in each generation. The improvement in yield over generations exploits additive genetic variation, whereas the additional improvement seen in the best clones in each generation exploits both additive and non-additive genetic variation.

But what happened in practice at SCRI? The 17 highest yielding clones were assessed as potential cultivars in the SCRI potato breeding programme, but none of the 17 was good enough for other traits to be selected as a new cultivar. Furthermore, they were not inter-crossed because in 1991 a new breeding programme had been started to deal with the priority of combining improvements in yield with better processing quality and disease and pest resistance (Bradshaw et al. 2009). In a later chapter, we shall see that clones combining better fry colour with better resistance

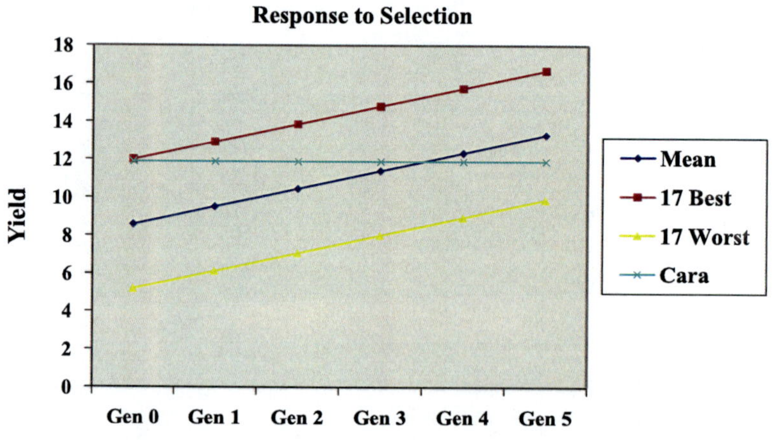

Fig. 3.1 Response to selecting 17 highest yielding clones out of 218 in each generation on population mean (218 clones) and means of 17 highest and 17 lowest yielding clones in next generation (yield in kg/plot); Cara is highest yielding original parent

to late blight and to the white potato cyst nematode than found in cultivar Maris Piper were produced, but were lower yielding than this cultivar, which was still the most widely grown one in Great Britain in 2017, having been released in 1966 and having held the number one spot for over 40 years. Hence, despite good prospects for improving yield, British farmers did not benefit from higher yielding cultivars from the 6 × 6 half diallel set of crosses, but nor did they benefit from higher yielding cultivars from much larger breeding efforts at SCRI and elsewhere. We can conclude that combining yield increases with other commercially desirable traits is difficult.

In later chapters, we will return to the question of combining high yield with other desirable traits in new cultivars. We will look at the theory of the response to selection in more detail and ask how selection for high yield and other traits can be made more effective. We will need to consider the intensity of selection, both the broad-sense and narrow-sense heritability, and the amount of additive and non-additive genetic variation. However, I cannot emphasize enough at this stage that the prerequisite for improving yield, or any other trait, over generations of selection, is adequate additive genetic variation. The amount of additive genetic variation is unique to any breeding programme (i.e. the parents used to initiate the programme) and therefore has to be determined at the start of any new programme. Breeders have certainly found the relative proportions of GCA and SCA to be specific to their particular populations of genotypes under investigation; for example, Tai (1976) and Killick (1977) found that all of their variation in tuber yield was due to SCA, whereas Brown and Caligari (1989) found only GCA which was significant when tested against the GCA × sites interaction which was present. Neele et al. (1991) found both significant GCA and SCA which were of similar magnitudes. Now, however, I want to consider another aspect of the conundrum of increasing potato yields, namely the extent to which current potato cultivars are already potentially high yielding and hence the real challenge is to reduce the gap between actual and potential yields through improved agronomy and new cultivars.

3.3 Potential Yields

Crop physiologists have developed models of crop growth which allow them to predict potential yields based primarily on the effects of length of growing season, day lengths, temperatures and water availability on resource capture for photosynthesis (other factors such as nutrients can be added to the models). Sometimes, potential yields are approached in practice, as seen in the following two examples, one from Scotland and one from North America. In Scotland, MacKerron (1985) reported achieved yields for four crops that were 102, 101, 94 and 85% of their calculated potential yields. The achieved yields of 80 to 64 t/ha compared with UK average yields of around 40 t/ha. In the Columbia Basin, Washington, USA, Kunkel and Campbell (1987) reported an even higher maximum measured yield of 124 t/ha for cultivar Russet Burbank, compared with predicted yields of 140 and 160 t/ha based on intercepted radiation and length of growing season, respectively (the former is more accurate). Cultivar Russet Burbank had been released back in 1914 and is still widely grown in North America (Ortiz 2001). Hence, there may not be much scope to increase the potential yields of potatoes if cultivars are available which match their growth cycle from planting until crop maturity with the length of the available growing season (Haverkort and Struik 2015). There are three reasons for high potential yields: a high harvest index, adaptation to tuberization in different day-lengths and tuber initiation and rate of bulking matched to the available growing season.

3.4 Harvest Index

The harvest index of a potato is the proportion of the whole plant's dry-weight that is harvestable tuber. Haverkort and Struik (2015) quote 0.75 as the typically high harvest index (HI) of potato compared with 0.4 to 0.6 for cereals. It is easy to find examples of high values in the literature, but two will suffice. Mazurczyk et al. (2009), in Poland (52°29′N 21°03′E), found that diverse water supplies and organic or mineral nitrogen did not change the harvest index values of mid-early cultivar Triada, which varied between 0.7 and 0.8 at the end of the growing period. In contrast, Jamaati-e-Somarin et al. (2010), in Iran, found that plant density and nitrogen fertilizer did affect the harvest index of cultivar Agria. The harvest index was highest (0.76) at an intermediate nitrogen level of 160 kg ha^{-1} and a low plant density of 5.5 plants m^{-2} and lowest (0.66) with no applied nitrogen at a high plant density of 11 plants m^{-2}. Increasing plant density from 5.5 to 7.5 to 11 plants m^{-2} reduced the harvest index at all nitrogen levels for which the order for harvest index was $160 > 80 > 0 > 200$ kg ha^{-1}.

Variation in harvest index does exist as Victorio et al. (1986) found in Peru. At a site typical of Andean potato farming, with a season of 150 days, 9 cultivated genotypes grew well and at maturity had a high harvest index that ranged from 0.73 to 0.85. However, at a site in the warm coastal desert region of Peru, despite good growth, the harvest index dropped and ranged from 0.33 (for a triploid Group Chaucha genotype) to 0.75 (for a tetraploid Tuberosum × Andigena hybrid). The night temperature was cooler and the day-night temperature differences were greater in the Andes (Sierra) than on the Coast. However, the only consistent differences in potato growth between the two sites were shorter plants in the Andes with an increased tuber dry matter percentage (20–28% compared with 14–21%). These results raise the question of whether or not worldwide there is still scope to select potatoes for a higher harvest index through better adaptation to the environments in which they are now being grown.

Neele (1990) thought that the complex inheritance of tuber yield might be clarified by describing tuber yield as a function of total light interception, light utilization efficiency and harvest index, of which total light interception and harvest index were the most promising as they had been shown to have a high heritability (Spitters et al. 1989). He also thought that harvest index could be used to identify genotypes with an early tuber initiation and a relatively low tuber bulking rate so that they have a long tuber bulking period which makes full use of the growing season. He pointed out that the development of the harvest index during the growing season is characterized by three parameters: tuber initiation, the rate of increase during tuber bulking and the maximum reached in the mature crop. The maximum is difficult to determine as the total biomass includes that lost through the season as rotten leaves, stems and stolons. Fortunately, Spitters et al. (1989) had found that differences between genotypes in harvest index at the end of the growing period were closely related to differences found in mid- July. The inheritance study done by Neele (1990) in The Netherlands over two years comprised 20 progenies from a partial 6 × 6 diallel set of crosses in which the six parents were four commercial cultivars and two breeding lines which differed in maturity. Although inheritance studies will be dealt with in greater detail later in this book, Neele's results are worth mentioning here in the context of the importance of harvest index as a contributor to high yield. There were large differences between progenies in harvest index estimated in mid-July (0.31–0.58 and 0.21–0.56 in the two years) with a very high correlation ($r = 0.97$) between years. Furthermore, most of the variation between progenies was due to differences in general combining ability (GCA) which were 10 times larger than the differences in specific combining ability (SCA). In contrast, for tuber yield at harvest in mid-September, the GCA and SCA variances were similar in magnitude (the former slightly larger than the latter). Mid-parent values were poor predictors of progeny means for tuber yield ($r = 0.16$ and 0.24 in the two years), an issue we will need to deal with later in this book. However, differences in harvest index between progenies could be predicted with a high degree of precision by using the harvest index parameters of the parents. This estimated harvest index of progenies showed a reasonable correlation to the tuber yield of the progenies in mid-September ($r = 0.67$ and 0.82 in the two years; and 0.68 and 0.69 from each of the two years to a third year) and hence can be used in the prediction of the yield of the progeny, which is an interesting result for potato breeders. It should also be pointed out that there was a moderate correlation ($r = 0.66$) between parental GCAs for yield from the two years and progeny means in a third year. We shall return to the prediction of progeny performance later in this book. For now, our conclusion is that genetic variation exists for harvest index and has a high heritability, but the variation is often for low rather than high harvest index, possibly because the latter is near the physiological limit.

3.5 Adaptation to Tuberization in Different Day-Lengths

By 2001, potatoes were being grown in 149 countries from latitudes 65°N to 50°S (Hijmans 2001), having been domesticated in the highlands of southern Peru (Spooner et al. 2005) some 7000 years ago. The success of the potato as a global crop was dependent on the selection of genotypes (cultivars) that were adapted to producing tubers in the day-lengths of the regions and seasons where they were being grown. Two contrasting growing seasons, for example, occur in the lowland temperate and lowland subtropical regions of the world (Govindakrishnan and Haverkort 2006). In the temperate regions, crop duration is fairly long (120–150 days), planting time coincides with increasing temperatures after cold winters and is followed by long days (more than 14 h) with temperatures suitable for haulm growth as well as for tuber formation and bulking. Hence, yields are potentially high as we have seen for achieved yields in Scotland of 64–80 t/ha (MacKerron 1985). In contrast, in the lowland tropics and sub-tropics, the growing season is quite short (100–120 days). Where winters are too severe for potatoes, planting is done in the autumn during relatively warm conditions and the temperature decreases as the season progresses and day length also decreases. With early tuber initiation and rapid bulking, high yields (40–50 t/ha) can nevertheless be achieved in 100–120 days; for example, in the Indo-Gangetic plains. Where winters are milder and frosts are infrequent, planting can be done in winter so that crop emergence, growth and development take place with increasing temperature and day length (Govindakrishnan and Haverkort 2006). The situation in the Indo-Gangetic plains is in fact more complicated than this as short-duration autumn (September to mid-December) and spring crops (mid-January to April) are grown in the North-West Plains, where the period mid-December to mid-January is too cold, a long-duration crop (October to April) is grown in the Central Plains and a short-duration, early maturing crop (November to February) is grown in the Eastern Plains (Dua et al. 2018). One of the success stories of Indian potato breeding is the production of cultivars suitable for the short-day, winter conditions of the subtropical plains. Kawar et al. (2018) described the successful breeding of an early-maturing, moderately late-blight resistant cultivar, Kufri Khyati, for the Indian plains which was released in 2008 from a cross made in 1992. In on-farm trials for two seasons at 14 sites across the Northern, Central and Eastern Plains, marketable yields of Kufri Khyati at 75 days after planting averaged 23.9 t ha^{-1} (range over 28 trials: 9.3–39.9 t ha^{-1}), compared with 20.3 t ha^{-1} for the best control (Kufri Pukhraj at eight locations and Kufri Ashoka at six locations), a 17.7% increase. Respectable total yields of 20 t ha^{-1} had been obtained in earlier trials just 60 days after planting.

We now know from the work of Kloosterman et al. (2013) that maturity in potato is controlled by the *StCDF1* locus on chromosome 5, and that selection of genetic variant *StCDF1.2* at this locus results in early maturity and the ability to tuber in long days. They used fine mapping to locate *StCDF1* as a possible candidate gene and then used gene insertion by genetic transformation for proof of function. The early-maturing/tuberizing *StCDF1.2* allele, under control of the cauliflower mosaic virus 35S promoter (CaMV 35S), was introduced into a short-day-dependent *S. tuberosum* Group Andigena genotype and into a very late maturing diploid potato genotype. Under long-day conditions, transformed Andigena plants had formed tubers by four weeks after planting, whereas the untransformed controls were unable to tuberize in long days. Overexpression of the *StCDF1.2* allele also resulted in early tuber formation in the late maturing diploid potato. Overexpression of *StCDF1.2* had no effect on flowering in potato, indicating a separation of signal transduction pathways for tuberization and flowering. Kloosterman et al. (2013) found that *StCDF1* belongs to the family of DOF (DNA-binding with one finger) transcription factors and regulates tuberization and plant life cycle length by acting as a mediator between the circadian clock and a mobile tuberization signal (StSP6A). In *StCDF1*.2 an insertion results in a frame-shift which introduces a premature stop codon and results in a truncated protein with impaired ability to interact with other proteins, thus affecting post-translational light regulation. The outcome is expression of the mobile StSP6A signal and induction of tuber development at the stolon termini in long days as well as short ones. Hardigan et al. (2017) explored haplotype diversity at the *StCDF1* locus and found introgression of truncated alleles encoding shortened StCDF1 proteins from wild species, particularly *S. microdontum*, in long-day–adapted cultivars. Zhang et al. (2020) have shown that reduced expression of *StCEN* (a member of the TERMINAL FLOWER 1/CENTRORADIALIS gene family) accelerates tuber formation, whereas transgenic lines over-expressing this gene display delayed tuberization and reduced tuber yield. They concluded that expression of *StCEN* suppresses tuberization by directly antagonizing StSP6A function in stolons, identifying *StCEN* as a breeding marker to improve tuber initiation and yield, through the selection of genotypes with reduced *StCEN* expression.

Adaptation to long days can occur quickly, as seen in the breeding of long-day adapted Andigena potatoes (Neotuberosum), where recurrent selection for tuber production in long days resulted in large increases in yield. After five cycles of selection, Glendinning (1975) achieved mean yields per plant of 1.15 kg for Neotuberosum compared with 0.33 kg for unadapted Andigena and 0.76 kg for Tuberosum, albeit from a late harvest. Likewise, Munoz and Plaisted (1981) achieved an increase in yield over six cycles of selection from 0.32 to 1.37 kg per plant (hill). Whilst such dramatic increases in yield seem unlikely today in breeding for local adaptation, some improvements may be possible and some may be needed to accommodate climate change. In other words, today, in most potato growing areas, reasonably well-adapted cultivars with a high harvest index are usually available.

3.6 Adaptation to Growing Season: Ideotypes

Crop physiologists have also used their models of crop growth to suggest ideotypes to potato breeders. Haverkort and Grashoff (2004) defined an ideotype as the ideal genotype for a particular environment. The ideotype has a growth cycle with a length that matches that of the available growing season, which is characterized by a temperature window (neither too hot nor too cold) and resources such as solar radiation, water and nutrients. They argued that the optimal genetic constitution of a new cultivar can be determined from an appropriate crop-growth model before starting a breeding programme for any target environment. Their approach aims to quantify, for example, the benefits of the following.

1. Frost tolerance for earlier planting
2. Greater sprout growth rate for earlier emergence
3. A lower base temperature for crop growth
4. Higher initial leaf areas at emergence and/or higher leaf area development rates
5. Thinner leaves in temperate climates with less intense solar radiation than in the tropics
6. Changes in maximum leaf area index
7. Changes in patterns of senescence
8. Deeper rooting
9. Increased water-use efficiency

Optimization of nine parameter values in simulation runs resulted in predicted increases of tuber dry matter production of 78% for unirrigated crops grown in dry and sunny years and over 40% for crops with irrigation. Hence, the results of their simulations of ideotyping exercises were encouraging but need testing in practice. The outcome may be the use of physiological selection criteria for adaptation to target environments in ways that have been elusive in the past. A more likely outcome is help in the choice of parents with complementary traits for crossing followed by selection for yield in the target environment and growing season.

More recently, Kleinwechter et al. (2016) have used the SUBSTOR potato model to help to identify, assess and prioritize potential traits for yield improvement and environmental adaptation (including adaptation to climate change) by creating 'virtual' genotypes to explore the relationship between physiological traits and yield and genotype-environment interactions. In other words, cultivar parameters in the model are varied to mimic new genotypes and to explore the impact of cultivar variation on potato yields in contrasting climatic regions. The model assumes that pests and diseases are controlled and nitrogen is the only limiting soil nutrient. Five phenological stages of plant development are considered: preplanting, planting to sprout germination, sprout germination to emergence, emergence to tuber induction and tuber induction to maturity. Vegetative growth and biomass accumulation take place in the tuber from emergence to maturity. The model has five cultivar parameters that can be explored to evaluate the possible effects of virtual traits on crop yields:

- G2: Potential leaf area growth rate which determines leaf expansion (higher values should lead to higher yields).
- TC: Critical temperature for tuber induction (higher values can be interpreted as improved heat tolerance).
- P2: Sensitivity to long photoperiods (it takes high values for early cultivars with high sensitivity to long photoperiods and low values for less sensitive late cultivars so that lower values should lead to higher yields in long day environments).
- G3: Potential tuber growth rate (higher values should lead to higher yields).
- PD: Index for determinacy (a determinate habit genotype has a value of 1, causing the partitioning to tubers to begin more rapidly and vine growth to decrease more rapidly than those with fractions smaller than 1, so that higher values will

lead to higher yields provided enhanced early tuber growth still allows the development of a sufficiently large canopy).

The model was calibrated with real data from yield trials from five locations: Kalengyere (cultivar Asante = Victoria), in the tropical highlands of Uganda; Gisozi (cultivar Asante = Victoria) in the tropical highlands of Burundi; La Molina (cultivar Amarilis), in the subtropical coastal lowlands of Peru; Modipuram (cultivar Kufri Bahar), in the subtropical lowlands of India; and temperate Ithaca (cultivar Kathadin) in New York State, USA.

The authors then explored the responses of simulated fresh-weight yields to changes in each of the cultivar parameters of the model. Their results showed that yield responses to changes in cultivar parameters are specific to the environment. Some changes are less effective in subtropical and tropical environments and more effective in increasing yields in temperate environments. Solar radiation, day length and temperature are the environmental factors that constrain the effectiveness of cultivar parameters in changing yields. The simulated variation in yields among environments was larger than the variation from changes in cultivar parameters. The impact of cultivar parameter changes on yield also varied with the cultivar parameter. Potential tuber growth rate was the cultivar parameter with the strongest effect on yields. Changes in the potential tuber growth rate parameter can lead to large yield changes in tropical highlands and temperate environments that have high solar radiation to ensure sufficient assimilate production for a larger sink. The results also suggested that improving crop management, particularly optimizing water and nitrogen input, is more important for increasing yields than the potential for cultivar improvement.

3.7 Ecophysiological Models

In a further development of the modelling approach, Khan et al. (2014) used an ecophysiological crop growth model 'GECROS' to analyse yield differences among 100 diploid genotypes from an F_1 segregating population, their parents and a set of standard potato cultivars. The reader is referred to the paper for details of the model which includes genotype-specific parameters: (1) thermal days from plant emergence to onset of tuber bulking (m_v), (2) thermal days from onset of tuber bulking to maturity (m_R), (3) maximum plant height (H_{max}), (4) maximum tuber nitrogen concentration (n_{so}) and (5) total crop nitrogen uptake at crop maturity (N_{max}). Data sets came from six field experiments (two in each of years 2002, 2004 and 2005), at Wageningen (52°N 5.7°E), the Netherlands. The F_1 population displayed a wide variation for all the parameters, with distributions that were approximately normal, apart from m_R and H_{max} which had bimodal distributions. When averaged across four of the field experiments for which the most data were available, there was a good correlation (R^2 = 0.86) between observed and predicted tuber yields for the F_1

genotypes. By dropping parameters in turn, it was concluded that their relative importance for determining tuber yield was $N_{max} > n_{SO} > m_R > m_V$ and H_{max}. In other words, genotypes with high N_{max} and low n_{SO} tend to be higher yielding. Although these results are interesting, more research is needed before model-based approaches can be used to design strategies for potato ideotype breeding for specific environments. The researchers also concluded that modelling differences in crop yield among individuals in a genetic population under field conditions remains a major challenge.

The main conclusion so far from the work of crop physiologists is that if available cultivars have a high harvest index, are adapted to tuberization in the daylengths of the target growing season, and if their tuber initiation and rate of bulking match the available growing season, then potential yields will be high. If not, there may be scope to improve potential yields by choosing appropriate parents and making appropriate crosses. Crop-growth models should help to identify desirable parents but it will probably still be easier to select offspring for yield per se rather than for physiological traits that affect yield. Before going on to compare potential yields with actual yields, I am going to insert a short section on path coefficient analysis. Although I have never used such an analysis as a potato breeder, I still get papers to review that claim that the technique is useful.

3.8 Path Coefficient Analysis

Path coefficient analysis was developed by Wright (1921, 1923) and subsequently applied in a plant breeding context by Dewey and Lu (1959) on crested wheat. It can be viewed as an alternative way of doing multiple linear regression analysis, as shown by Li (1975) in his book on the subject. In multiple linear regression, we explore how much of the variation in yield (y, the dependent variable) can be explained by p fixed or random independent variables (the latter are assumed to be from a multivariate-normal distribution) (x_1 to x_p), such as tuber number and average tuber weight (the components of fresh-weight yield), number of leaves and stems, plant height and other morphological traits. Our model assumes that each observed value of yield y (mean of all y's is $\bar{y}$) is a linear function of x_1 to x_p plus an uncorrelated residual term e:

$$y = b_0 + b_1x_1 + b_2x_2 + \ldots + b_px_p + e$$

y will be correlated with some or all of the x's and the x's may be correlated with each other.

The corresponding regression equation is:

$$\hat{y} = b_0 + b_1x_1 + b_2x_2 + \ldots + b_px_p$$

$\hat{y}$ is the predicted yield, b_0 is the intercept and $b_1, b_2 \ldots b_p$ are the partial regression coefficients for predictors $x_1, x_2 \ldots x_p$, respectively, and are estimated to minimize $\sum(y - \hat{y})^2$ (least squares estimate).

The multiple correlation coefficient R is the correlation between y and $\hat{y}$ (R has values between 0 and 1). R^2 is the percentage of variation accounted for by the x's, but the sample estimate is not an unbiased estimate of the population parameter which needs correction for p and n, the number of measurements of yield (observations). In other words, the total sum of squares $\sum(y - \bar{y})^2$ is partitioned into two components, the sum of squares due to regression $\sum(\hat{y} - \bar{y})^2$ and the residual sum of squares $\sum(y - \hat{y})^2$, so that $R^2 = \sum(\hat{y} - \bar{y})^2 / \sum(y - \bar{y})^2$.

If all of the variables are standardized (Y and X's) so that they are measured from their mean values in units of their own standard deviations (they will have been measured in different units), they will each have a mean ($\bar{Y}$ for Y) of zero and unit variance ($V_Y = \sigma_Y^2$ for Y).

$$Y = (y - \bar{y}) / \sigma_y \text{ and } X = (x - \bar{x}) / \sigma_x$$

$\bar{Y} = 0$ and $V_Y = \sigma_Y^2 = \sum(Y - \bar{Y})^2/n = 1$ where n is the number of observations.

The transformation to standardized variables does not affect the correlations (r) between variables:

$$r(X,Y) = r(x,y)$$

Path analysis deals with the relationships among the standardized variables. In the standardized model for regression the partial regression coefficients are replaced by the standardized partial regression coefficients and are represented by the symbol ρ so that b_1 above becomes ρ_{01} and is called the path coefficient from X_1 to Y.

The standardized model for regression on three variables, for example, is (Fig. 3.2):

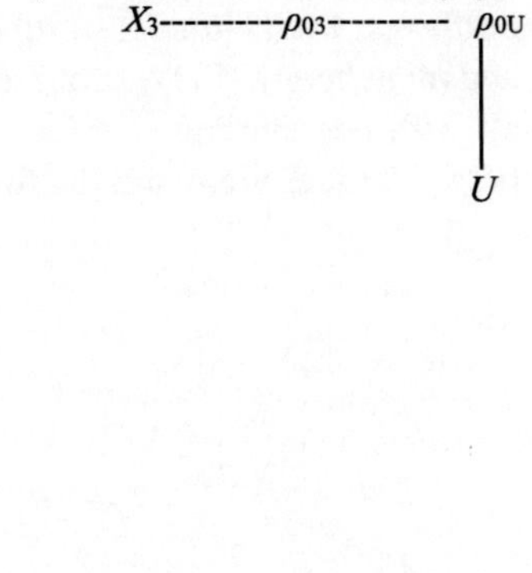

Fig. 3.2 Path coefficient analysis for three variables

$$Y = \rho_{01}X_1 + \rho_{02}X_2 + \rho_{03}X_3 + U$$

The regression equation is:

$$\hat{Y} = \rho_{01}X_1 + \rho_{02}X_2 + \rho_{03}X_3$$

The residual U is uncorrelated with any of the X's, hence:

$$\sigma_Y^2 = \sigma_{\hat{Y}}^2 + \sigma_U^2 = 1$$

We now have a path diagram that summarizes the essential features of the relationship between the dependent variable (yield), the independent variables and the residual which includes all other variables that may affect yield but are not included in the data. Not shown in the path diagram are the correlations between Y and the X's [$r(y, x_1) = r_{01}$ etc.].

Least squares estimation of the standardized partial regression coefficients leads to the following normal equations:

$$\rho_{01} + \rho_{02}r_{12} + \rho_{03}r_{13} = r_{01}$$

$$\rho_{01}r_{12} + \rho_{02} + \rho_{03}r_{23} = r_{02}$$

$$\rho_{01}r_{13} + \rho_{02}r_{23} + \rho_{03} = r_{03}$$

We can calculate the r's from our data and estimate the ρ's by standard least squares procedures. The outcome is a partitioning of the correlations between Y and the X's into direct (e.g. ρ_{01}) and indirect effects (e.g. $\rho_{02}r_{12} + \rho_{03}r_{13}$) of the X's. The correlation between Y and X_1 for example is the direct effect of X_1 plus the indirect effects of X_2 and X_3. The indirect effect of X_2, for example, is the direct effect of X_2 multiplied by the correlation between X_1 and X_2. It can be shown (Li 1975) that there is a simple relationship between path coefficients and partial regression coefficients:

$$\rho_{01} = b_1\left(\sigma_{x1} / \sigma_y\right) \text{ etc.}$$

Can path analysis be useful? Let us take a look at an example. Asghari-Zakaria et al. (2006) recorded yield and its components for 120 potato clones from their breeding programme, grown in an augmented randomized complete block design with three replications and five check cultivars in 2004 in the Ardabil region of northwest Iran. Number of tubers per plant (NTP), average weight per tuber (AWT) and plant height (PH) accounted for nearly 89% of the variation in tuber yield (TY), all with positive direct effects on tuber yield. The partitioning of the correlations between tuber yield and the three traits is shown in Table 3.4.

Table 3.4 Partitioning of the correlations between tuber yield (TY) and number of tubers per plant (NTP), plant height (PH) and average weight per tuber (AWT) into direct and indirect effects (Asghari-Zakaria et al. 2006, with permission)

Correlation	r	Direct effect	Indirect effect	Indirect effect
TY and NTP	0.714	0.729	0.229 × 0.531 = 0.122	0.502 × −0.273 = −0.137
TY and PH	0.779	0.229	0.729 × 0.531 = 0.387	0.502 × 0.325 = 0.163
TY and AWT	0.377	0.502	0.729 × −0.273 = −0.199	0.229 × 0.325 = 0.074

Correlations: NTP and PH $r = 0.531$, NTP and AWT $r = -0.273$, PH and AWT $r = 0.325$

It can be seen that for TY and NTP the indirect effects cancelled each other leaving a moderate direct effect of similar value to the correlation. In contrast, for TY and PH small values for the direct and indirect effects combined to give a moderately large correlation. In contrast again, for TY and AWT, a negative indirect effect from NTP reduced the direct effect of AWT to give a small correlation. The authors went on to do a sequential path analysis in which they showed that 40% of the total variation for plant height was explained by three characters, namely, number of main stems per plant (NMS), number of secondary stems per plant (NSS) and diameter of main stems (DMS). Likewise, 17% of the total variation for number of tubers per plant was explained by number of main stems per plant and number of secondary stems per plant. Finally, diameter of main stems and tuber dry matter concentration, respectively, had significant positive and negative direct effects on average weight per tuber.

Although these results are interesting, it is not obvious that measuring the components of yield and other traits affecting yield will involve less work than measuring yield itself, nor is it obvious that selection for these traits will give a greater increase in yield than simply selecting for yield. Furthermore, the causal connection between height and yield is not obvious. The same comments apply to the paper by Sattar et al. (2007) on a trial of 28 genotypes (24 from CIP and 4 cultivars from the Netherlands) at the Bangladesh Agricultural Research Institute (BARI), Joydebpur, Gazipur during the winter season of 2004–2005. They also apply to the paper by Shubha and Singh (2018) on a trial of 28 diverse potato genotypes, representing the agro-climatic zones of India, and four Indian cultivars, at the Vegetable Research Centre of G. B. Pant University of Agriculture and Technology, Pantnagar, Uttarakhand, India. Hence, my reservations about path analysis; but readers will need to draw their own conclusions. In a different approach, Yuan et al. (2016) in Canada studied the effectiveness of breeders' field selection parameters (nine) on potato yield through linear mixed models and principal component analysis (PCA).

3.9 Actual Yields

It is now time to consider actual yields. Globally during the 50 years from 1966 to 2016, potato yields increased by 50%, or in simple terms by 1% per annum (http://faostat.fao.org). Although this increase was not as great as that for the major cereals, in 2016, the average yield of potatoes on a dry-weight basis (3.92 t/ha) compared favourably with those of maize (4.79), wheat (2.90) and rice (3.94). However, potato growers are less interested in dry-weight yield than in high fresh-weight yields with a suitable dry-matter content (estimated by specific gravity) for the targeted end use. High dry-matter content, for example, is needed for making crisps because it is associated with an increased yield of product and lower oil absorption. On a fresh-weight basis, in 2016 there was a large variation in potato yields between countries, from 0.8 (Central African Republic) to 49.0 t/ha (New Zealand), with a global average of 19.6 t/ha. Furthermore, in 2016, actual yields typically ranged over climatic regions from 24% to 72% (globally about 50%) of potential yields, which varied according to the length of growing season, day-lengths and temperatures (Beukema and van der Waag 1990; Haverkort and Struik 2015). Hence, we need to consider the extent to which new cultivars contributed to the yield increases from 1966 to 2016 and the extent to which new cultivars can reduce the gap between actual and potential yields. The answers will have major implications for our breeding objectives.

3.10 Past Increases in Yields

Let us consider three examples, one each from Great Britain, North America and India. Yields of over 40 t/ha and a high-quality product are now achieved in Great Britain, but 100 years ago, yields were only around 16 t/ha. Then after modest increases, yields almost doubled from 22 t/ha in 1960 to 40 t/ha in 2000, with an average of 46.5 t/ha for years 2013–2016 (British Potato Council and AHDB Statistics). A number of factors were involved in these yield increases. Modern chemical fertilizers were already available by 1900; disease-free seed was guaranteed from the 1920s through cultivar registration and seed certification; modern chemical herbicides, insecticides and fungicides became available from the 1950s; mechanization and larger fields have been trends since the 1960s; and extensive irrigation has been used since the 1980s (Bradshaw 2009a). Furthermore, post-harvest losses have been reduced by the use of controlled environment stores since the 1960s. But did new cultivars contribute to the yield increases? Simmonds (1981) estimated that of the 8.9 t/ha increase in yield in Great Britain from 1964 to 1976, 4.0 t/ha was environmental (better husbandry and management), 5.5 t/ha was due to the near replacement of cultivars King Edward and Majestic with three modern cultivars, Pentland Crown, Maris Piper and Desiree, and that the difference of 0.6 t/ha was a small but negative genotype × environment interaction. Simmonds (1980) also estimated that in agriculture for the six years 1971–1976, Pentland Crown (the new most widely grown cultivar in Great Britain) out yielded Majestic (the former most widely grown cultivar) by 10.1 t/ha. However, cultivars Maris Piper and Desiree were still in the top twenty cultivars in Great Britain in 2017, at positions 1 and 17, respectively, so new cultivars probably did not contribute to the further increases in yield.

In the USA, national potato yields increased from 5.6 t/ha in 1920 to over 33.6 t/ha in 1989 (Lucier et al. 1990). These increases have also been attributed to factors such as improved cultural practices, use of fertilizer, irrigation, pest management and a shift in production to the western USA. New cultivars were introduced through public sector breeding efforts, but did they contribute to the yield increases? Douches et al. (1996) found a lack of improvement in yield and specific gravity when cultivars released over the period 1861 to 1991 were compared for three years in Michigan in a high-yielding environment (50 t/ha), but they did find trends to earlier maturity (high yield and late maturity were correlated: $r = 0.55$) and improved tuber appearance. They also suggested that a factor contributing to yield stasis has been the increasing number of quality traits for which selection is practised in breeding programmes in North America. Indeed, Love et al. (1998) have provided evidence of significant progress since 1960 in North America in developing cultivars with good processing quality, including an increase in specific gravity between 1960 and 1970. Douches et al. (1996) also suggested that lack of yield improvement might be due to the narrow genetic base of the cultivated potato in North America (i.e. lack of additive genetic variance).

Before leaving the USA, it is worth considering the situation in Northern Maine which has a climate with average temperatures ideally suited for potato production (16.8–18.6 °C average daily temperature from June to August) and ample rainfall (930 mm per annum), but a short growing season of between 100 and 110 days. Surprisingly, potato yields remained fairly constant from 1940 to 2010, fluctuating between 23 and 35 t/ha. Frantz et al. (2012) used a simple light-driven mechanistic model to identify potential yield limitations. They found that meteorologically limited productivity based on typical cloud cover was about 55 t/ha, some 10% greater than maximum measured yields (in small areas on commercial farms in some years) and 80% higher than historical averages. Yield differences between actual and potential could be attributed to differences between actual radiation capture and optimal radiation capture efficiency that should be possible in the absence of stresses. Most yield increases in the specific case studies examined (five different crop management systems) were due to enhancement of radiation capture, which was achievable either by improved water management or disease suppression. Hence, the authors concluded that strategies for sustained yield improvements should focus on improved radiation capture, either by increasing the peak radiation capture potential, prolonging the radiation capture duration, or by shifting radiation capture to coincide with available light. They pointed out that selecting cultivars that can emerge in cooler soil conditions or pre-treating the seed tubers with a longer dormancy breaking period after cold storage are, theoretically, approaches to improve early potato stand establishment. Their model predicts a 10% and 17%

increase in yield from planting two and four weeks earlier in the season, respectively, without changing the length of the growing season; an example of adaptation to the growing season.

Finally, in India, potato production during 2017–2018 was 51.31 million tonnes from 2.14 million hectares with a productivity of 23.97 t/ha. Luthra et al. (2020) point out that this compares with 1.54 million tonnes from 0.23 million hectares with a productivity of 6.6 t/ha in 1949. They attributed 50% of the 3.63-fold increase in yield per hectare to the release of improved, high-yielding cultivars possessing resistance/tolerance to various biotic and abiotic stresses. Although they did not provide experimental evidence for this claim, it does not seem unreasonable given that the expansion in area was to a new set of environmental conditions in the Plains of India, namely those of the lowland subtropics in winter.

3.10.1 Irrigation

Globally, the potato is grown in a number of agroecological zones defined by length of growing season(s), day lengths, temperatures and water availability. Summer crops in the north-west of the USA and north-east of China, autumn crops in the Mediterranean and winter crops where rice is grown in the rainy summer season, all need irrigation (Haverkort and Struik 2015). In other areas and production systems, potato crops may benefit from some irrigation. In 2009, farming as a whole was using 70% of the available fresh water, and with human population growth, there is going to be increased competition for water resources between agricultural, industrial and domestic users (Bell 2009). Furthermore, climate change is expected to exacerbate drought events throughout the world and yield losses for potatoes of between 18% and 32% over the 30 years from 2003 have been predicted by Hijmans (2003).

Not surprisingly, potato growers and researchers are exploring irrigation management with a view to making better use of the available water. In China, for example, half of the potato crop is grown in a semi-arid region frequently affected by water restriction (Fig. 3.3). Thus, Qin et al. (2018), at the experimental station of Zhangjiakou Academy of Agricultural Sciences in Zhangbei County (41°04′N, 114°42′E, 1505 masl), compared partial root-zone drying furrow irrigation (PRD) and the commonly used drip irrigation (DI) with furrow full irrigation (FI) as control. They found that PRD at 50% of FI was more promising than DI for maintaining yields and increasing revenues while saving water and reducing CO_2 emissions. In the Indian Punjab (30°56′N and 75°52′E at 247 masl), Brar et al. (2019) compared drip irrigation with furrow irrigation for both bed and ridge planting, with and without straw mulching, for cultivar Pukhraj in dry winter growing season from 21 October to 28 January, which followed summer monsoon season (July to September). The straw mulching resulted in 19.0% higher yields as a result of 36.2 mm higher transpiration and 44.2 mm lower soil evaporation. Drip irrigation of beds and ridges resulted in 34.9% and 26.4% higher yields, respectively, than furrow irrigation

Fig. 3.3 Seed potato crop being grown under drip irrigation near Jining City, Inner Mongolia, China (41.03°N 113.11°E, 1409 masl) (6 August 2010)

along with savings of 23.3 mm and 53.3 mm irrigation water, respectively. Averaged over two winter seasons, yields for bed planted potatoes with mulch and drip irrigation were 32.31 t/ha compared with 18.28 t/ha for ridge planted potatoes with furrow irrigation but without mulch.

It is therefore important for potato breeders to be aware of the irrigation system that is going to be used and to assess and select potential new cultivars for yielding ability (adaptation) under such growing conditions. Furthermore, in many semi-arid and arid regions where irrigation is obligatory, transient wilting under high temperature and low air humidity is a common phenomenon so that selection for heat tolerance is also desirable (Levy et al. 2013). In contrast, where potatoes are primarily rainfed and hence grown under a variable water supply, periods of drought can be an issue. In temperate climates, for example, where potatoes are grown during the rainy summer season, dry spells during various growth stages may lead to transient water shortage and drought stress and result in loss of yield and tuber quality.

3.10.2 Weed Control

Finally, the importance of weed control should be mentioned. For example, in Egypt, Shehata et al. (2019) found that the total fresh-weight yield of cultivar Spunta at harvest (110 days after planting) was 37.6 t/ha in plots where weeds had been removed by hoeing twice, compared with 17.9 t/ha in unweeded plots, when averaged over two trials in different autumn growing seasons (2015/16 and 2016/17). Application of metribuzin herbicide resulted in a slightly lower yield of 33.8 t/ha, whereas slightly higher yields were achieved with the use of biodegradable (38.3 t/ha), infra-red transmitting (38.1), black (40.9) and blue (40.5) plastic mulches. Weed control efficacy at 90 days after planting had been 93.8% (biodegradable), 88.1% (infra-red transmitting), 98.1% (black) and 83.7% (blue), compared with 83.1% from hoeing and almost 98% from the herbicide. The researchers concluded from an economic analysis that black and blue plastic mulches were viable alternative weed management methods to hoeing and synthetic herbicides, especially for newly reclaimed areas of land and for organic farms.

3.11 Current Gap Between Actual and Potential Yields

Despite the increases in potato yields over the last 50 years, average (actual) potato yields in many regions are still somewhat below potential yields. Beukema and van der Waag (1990) estimated that potential yields in their chosen tropical, sub-tropical and temperate countries/regions could vary from 40 to 140 t/ha depending upon the length of growing season and temperatures. However, average fresh-weight yields in these countries in 2016 were still much lower than the potential yields (about 50% of potential on average; range 24% to 65%). Similar figures have been quoted by Haverkort and Struik (2015) for average over potential yields: 72% (65 compared with 90 t/ha) in the most productive areas of the Netherlands; 35% in the less productive areas of Chile; and 60% (31 compared with 55 t/ha) in Northern Maine, USA. A further interesting example has been provided by Deguchi et al. (2016), who compared actual and potential yield levels in different production systems in Japan, which change from a winter crop free from frost in Okinawa in the south (approximately 25°N; actual yields 15.5 t/ha) to a summer crop free from heat in Hokkaido in the north (approximately 45°N; actual yields over 30 t/ha) that accounts for 78% of national production. Potential yields were calculated using the crop growth model LINTUL-POTATO-DSS (Haverkort et al. 2015). The actual/potential yield ratio varied between 0.21 and 0.63 with an unweighted (by area) average of 0.40. Actual yields of the Hokkaido summer crops decreased from 35 to 32 t/ha over the period 1993–2013 because of an increase in higher dry matter cultivars for processing and starch, whereas potential yield remained stable at 56 t/ha. The most widely grown table cultivars in 2011 were still Irish Cobbler and May Queen which had been introduced in 1908 and 1912, respectively! Climate change predictions were as follows: a 2 °C increase in temperature would reduce potential yield from 56 to 53 t/ha but a 2 week increase in growing season would increase it from 53 to 60 t/ha. However, with an estimated yield increase of 28% as CO_2 concentrations increase from 380 ppm in 2000 to 550 ppm, yields of 77 t/ha could be achieved by 2050.

Likewise, using the LINTUL crop growth model, Yagiz et al. (2020) were able to make predictions about potato yields in 2050 in the 15 potato-producing regions of Turkey, based on meteorological data and the available periods of daily average temperatures between 13 °C and 22 °C for potato growth. Current average yields for Turkey are 34.93 t/ha, which is close to the predicted water limited yield of 36.31 t/ha, but way below (37%) the potential yield of 94.42 t/ha. However, by 2050, with CO_2 concentrations at 550 ppm, temperatures +2 °C higher, a higher radiation-use efficiency (RUE) and planting and harvest dates adjusted to the new periods between 13 °C and 22 °C, water limited yields could increase to 54.41 t/ha and potential yields to 120.63 t/ha.

In order to consider ways of reducing the current yield gap, I am going to consider the reasons for low yields in parts of Sub-Saharan Africa. In 2016, yields in Burundi (9.8 t/ha) were half the global average (19.6 t/ha), having risen above 5 t/ha for the first time in 2013 (6.2 t/ha) (http://faostat.fao.org). Bararyenya et al. (2018) published an assessment of opportunities for Burundian small-scale potato farmers to increase their productivity and income. They did on-farm experiments at five farms in each of three provinces surrounding Bujumbura town in two seasons during the three years 2014, 2015 and 2016, making 30 'experiments' and 580 plots in total. They hypothesized that in rain-fed conditions, improved cultivars, healthier seed tubers, pre-sprouting in light, proper fertilization and protection from late blight attack would improve yield and quality and lower the incidence of brown rot (*Ralstonia solanacearum* cause of 'brown rot' of tubers and 'bacterial wilt' of foliage). They compared the following treatments with the current practice of growers: introduction of two new cultivars, use of early generation seed from a rapid multiplication scheme, earlier harvesting of a seed crop, seed storage in a diffused light store, and research-based timing and dose rate of fertilizers and fungicides. They found that a new cultivar contributed up to 20% yield increase and healthier seed 80% (yield 13.89 t/ha compared with 7.69 t/ha from farm saved seed). Early harvesting reduced yield by 30% and reduced the incidence of brown rot in the current season but increased it (from 21% to 39%) in the following season when tubers were replanted. Diffused light storage of seed tubers resulted in yields of 11.76 t/ha compared with 9.15 t/ha from dark storage. Alternating contact and systemic fungicide application to control late blight resulted in yields of 9.94 t/ha compared with 6.68 t/ha from farmers lack of appropriate methods of control. Application of chemical fertilizers (N, P, K at 60, 90, 60 kg/ha) resulted in yields of 14.4 t/ha compared with 9.05 t/ha from farmer application of manure and compost. The best average yields achieved in a set of treatments was 15.5 t/ha. The authors concluded that it is possible to double yields and economic returns under the growing conditions in Burundi when growers plant healthy pre-sprouted seed of a new cultivar and apply chemical fertilizers and fungicides.

Wang'ombe and van Dijk (2013) assessed the relative impacts of the adoption of clean seed tubers, fertilizers, fungicides and irrigation on potato yields in Kenya. The potato growing areas in Kenya are all high altitude (1400 to 2700 masl), high-rainfall zones (mean annual rainfalls above 1000 mm but areas varying from 500 to 2600 mm per year) with temperatures ranging from a minimum of 2 °C to a maximum of 26 °C (i.e. temperate). Soils can be volcanic, red clay or loamy. Nearly 800,000 households were producing potatoes. Wang'ombe and van Dijk (2013) analysed 402 (a random sample) completed questionnaires from farmers in 2009/2010. They found that clean seeds (seed tubers), proper use of fertilizers, use of fungicides and irrigation (in lower rainfall areas) all increased yield, but that the combination of clean seeds and proper use of fertilizer achieved the maximum average yield of 18.35 t/ha compared with the overall average of 12.17 t/ha. However, they also found that clean seeds had the lowest adoption rate of only 4.5%, despite a high proportion of farmers (79%) being aware of the need to use clean seeds (more clean seed was not in fact available). The adoption rate of irrigation was also low at

23%, whereas there was widespread use of fungicides (92%), to control late blight, and of fertilizers (96%). Farmers are in fact discouraged from using animal manure because it may be contaminated with bacterial wilt. The authors went on to analyse possible factors leading to the adoption or otherwise of the four factors.

In both of the previous examples, yield increases came primarily from better agronomy rather than new cultivars, and more generally this is the way to achieve yield increases in the immediate future. In the longer term, new cultivars could help prevent yield losses through better tolerance of abiotic stresses and better resistance to pests and diseases, which would also allow a reduced use of chemical control. In a later chapter, we will also explore propagation of new cultivars through true potato seed as an alternative way of dealing with the problem of farm saved seed tubers being infected with diseases. The big unanswered question is whether or not a better understanding of both carbon and nitrogen assimilation and partitioning and the control of storage organ development can be translated into new breeding strategies and methods. The interested reader is referred to the review by Fernie et al. (2020) on the synchronization of developmental, molecular and metabolic aspects of source–sink interactions, which includes potato and cassava as models to illustrate the signalling control of storage sink initiation. For the rest of this chapter, we are going to deal with the issues that arise in conducting yield trials and the most appropriate environment(s) in which to practise selection for increased yield. We will need to consider genotype × environment interactions and the optimum biological and economic levels of resources such as the use of water and fertilizers for improving yield.

3.12 Yield Trials

Potential cultivars (genotypes) are assessed in yield trials, the aim being to accurately predict their relative yields and other attributes when grown under agricultural or horticultural conditions. Simmonds (1980) compared the yields of four potato cultivars (two old and two new) in official trials and in an annual survey of farm crops in Great Britain over the years 1969–1977. He found that yields in trials (39.67 t/ha) were higher than in agriculture (36.52 t/ha) and that the trials underestimated the mean yield differences between cultivars in agriculture; for example, the Pentland Crown minus Majestic yield difference was 4.7 compared with 10.1 t/ha. However, the differences between paired cultivars were reasonably well correlated between trials and farms. In contrast, it was not possible to predict yield stability from trials to farms. These results are just one set of data, but they remind us that it is important for their relevance that yield assessments mirror farmers' fields and management procedures, for example, fertilizer application and irrigation regimes.

Let us start by considering a yield trial in a field at one location in one year. The trial area will be divided into a rectangular grid of plots in which to grow the entries (Fig. 3.4). More than one plot of each entry is required to assess the environmental

Fig. 3.4 Replicated yield trials of potato at ware site in Dundee, Scotland (from Bradshaw and Bonierbale 2010 with permission). (Source: SCRI)

variation because the differences between entries in different plots will be due to environment (σ^2) as well as genotype (σ^2_G). Good experimental techniques are required to minimize the increase in 'environmental' variation that can occur from lack of uniformity in plant spacing and the application of fertilizers, herbicides, fungicides and pesticides. In the early generations of a breeding programme, and in trials for genetic analysis, there will be a large number of entries, often several hundred. Furthermore, plots will be small in size because only limited amounts of planting material will be available and consequently only a limited number of tubers will be available after harvest for assessments of quality traits. At most, plots will comprise several rows, but sometimes a single row or even a single plant may be used. For a given total trial area, statistical considerations suggest that the use of many smaller plots is preferable to fewer larger ones (Mead 1997), but shape and size will often be dictated by agricultural practice and machinery. Furthermore, resources usually permit relatively few replicates of each entry, often two, three or four. The simplest experimental layout would be to allocate all of the entries and their replicates at random to the trial plots; that is complete individual randomization. Randomization is necessary for a valid estimate of the precision of a trial which is the standard error of the difference between the means (over replicates) of two entries; the smaller the standard error the better.

A slightly more complicated, but standard layout, is to have each replicate in a separate block so that each block contains a single plot of each entry (randomized complete block design). A different randomization of entries is used in each block. This leads to a simple additive model of plot yield (y):

$$y = \mu + g + b + e$$

where μ is the trial mean, g the genotype effect, b the block effect and e the plot error.

The assumption of additivity implies that both the differences between blocks and the differences between genotypes remain consistent over a trial. A second assumption is usually made in the analysis of trials, namely that plot errors are homogeneous in size and unaffected by genotype; that is the yield variation between plots is similar for all of the genotypes in the trial. The standard method of analysis is based on least squares estimates for general linear models (Mead 1997). One can estimate the components of variance from the mean squares in an analysis of variance (ANOVA) of plot yields. Of particular interest are the plot-to-plot variation (soil heterogeneity) for replicates of the same genotype (σ^2) within a block (although in practice genotypes are in different blocks) and the variation between the mean (averaged over replicates) yields of genotypes ($\sigma^2_G + \sigma^2/r$), where r is the number of replicates and σ^2_G is the true genotypic variation. The ratio $\sigma^2_G/(\sigma^2_G + \sigma^2/r)$ is called the broad-sense heritability (h^2_b), and measures the proportion of the observed (phenotypic) variation between genotype means that is due to genetic (genotypic) differences. The standard error of the difference (SED) between two genotype means is $(2\sigma^2/r)^{½}$. Hence, the broad-sense heritability can be increased, and the standard error of the difference can be decreased, by increasing the number of replicates.

Two approaches can be taken to reduce the problem of within-trial heterogeneity and hence to reduce σ^2. One is to group plots into small blocks for which the assumption of uniform conditions within blocks is more realistic (Mead 1997). The other is the use of spatial analysis (nearest neighbour) which models the observed similarity of yields from adjacent plots (Gleeson 1997), although one must be aware of possible negative inter-plot interference (see below). Computer-generated trial designs and software for their analysis are available for both approaches, for example, AGROBASE Generation II™ from Agronomix Software (http://www.agronomix.com), which has been linked to the popular R-Statistical Software (http://www.r--project.org/). Blocking is highly desirable and can be supplemented with spatial analysis, although the latter will not be considered further in this book.

Failure to use blocks will usually result in a mixture of two disadvantages. Firstly, the distortion of relative means, because by chance all replicates of some genotypes are on high fertility plots, whereas for others all replicates are on low ones; hence the differences between genotype means will be biased. Secondly, the inflation of standard errors, because by chance the replicates of some genotypes are on an equal number of high and low fertility plots; hence the plot-to-plot variation is large for these genotypes. The effective control of variation through blocking usually requires fairly small block sizes, whilst most selection trials use large numbers of genotypes. This leads almost inevitably to the use of incomplete block designs. Thus, a genotype trial will be designed in complete replicate blocks, each replicate being split

Replicate 1

1	2	3	4	5	6	7	8	9	10
11	12	13	14	15	16	17	18	19	20
21	22	23	24	25	26	27	28	29	30
31	32	33	34	35	36	37	38	39	40
41	42	43	44	45	46	47	48	49	50
51	52	53	54	55	56	57	58	59	60
61	62	63	64	65	66	67	68	69	70
71	72	73	74	75	76	77	78	79	80
81	82	83	84	85	86	87	88	89	90
91	92	93	94	95	96	97	98	99	100

Replicate 2

1	20	28	36	49	57	64	72	83	95
17	65	43	89	6	100	78	54	21	32
26	4	52	80	15	88	99	63	37	41
40	48	85	51	74	2	16	97	69	23
44	22	7	98	33	79	90	11	55	66
59	73	96	47	30	61	5	38	12	84
68	87	19	3	92	24	31	45	76	60
75	91	34	62	58	13	27	86	50	9
82	39	71	25	67	46	53	10	94	18
93	56	70	14	81	35	42	29	8	77

Fig. 3.5 Two replicates, 10 rows by 10 columns design for field trial comparing 100 genotypes (numbered 1–100) in small plots (Patterson and Robinson 1989; from Bradshaw 2016 with permission)

into a number of incomplete blocks. Comparisons between genotype means will then require an adjustment to allow for differences between blocks, as explained in books on the subject (Mead et al. 2002). The subset of genotypes in each incomplete block in one replicate should be evenly spread among blocks in each of the other replicates.

Alpha designs, also called generalized lattices (Patterson et al. 1978; Patterson et al. 1988), extended the lattice designs of Yates (1936) and have provided a well-understood system for producing appropriate and efficient incomplete block designs. Analysis of data from past genotype trials suggests that allowing for block differences in two directions can further improve precision. The lattice squares of Yates (1937) required the number of genotypes to be a perfect square p^2 and the number of replicates to be $p + 1$, and hence were rather restrictive. Today, other row-and-column designs are available which place fewer restrictions on the numbers of genotypes and replicates (Patterson and Robinson 1989). A 10×10 design with two replicates is shown in Fig. 3.5, where it is easily seen that cultivars 1–10 appear in different rows in replicate 2 and share their columns with nine different cultivars.

3.13 Interference Between Plots

One potential complication in yield trials is interference between plots (Kempton 1997). This occurs when plot yield is affected not only by the genotype grown in the plot, but also by the particular genotypes in neighbouring plots. Interference is most likely to occur when the plots are small and not bordered, as in the early generations of a breeding programme. Interference error can lead to a systematic bias in genotype effects that persists across trials and is not reduced by randomization and replication. Interference sometimes affects only the scale of responses but can affect the ranking of genotypes. Interference arises from competition for resources, such as light, nutrients and moisture. In tuber crops such as potato (Connolly et al. 1993; Bradshaw 1994b), an important factor is the tuber yields of neighbouring plots. Interference may also be caused by the differential spread of plant pests or pathogens between plots. Thus, in a disease screening trial, a highly susceptible genotype may act as a secondary source of infection for other genotypes in the trial, so that the effectiveness of more resistant genotypes is underestimated. Kempton (1997) gave as an example late blight scores for four potato genotypes when grown in pure and mixed stands. Although the ordering of genotypes for resistance was the same (Shelagh > Cara > Record > Pentland Crown), the difference between Shelagh and Pentland Crown was much reduced in the mixture.

An example of interference is provided in Fig. 3.6 and Table 3.5 from Bradshaw (1994b). Five cultivars of potato (Cara, Estima, Pentland Dell, Pentland Javelin and Torridon) were assessed for two years (1992 and 1993) in all possible pairings (a competition diallel) in three consecutive drills (rows), such that one member of the pair (the cultivar) was the centre-drill, and the other member (the neighbour) was in the two adjacent drills. Analyses of variance were done on data from the centre-drills with cultivars and neighbours as fixed effects and years and replicates as random effects. For ware yield (45–85 mm), inter-cultivar competition was present and the rankings of cultivars in pure-stands were different from those under competition. As there was year × cultivar and year × neighbour interactions, but no cultivar × neighbour interactions, the results for the two years are shown separately and include the two-sided neighbour effect. It can be seen that Cara and Torridon are

						Neighbour
	CCC	CDC	CEC	CJC	CTC	C
	DCD	DDD	DED	DJD	DTD	D
	ECE	EDE	EEE	EJE	ETE	E
	JCJ	JDJ	JEJ	JJJ	JTJ	J
	TCT	TDT	TET	TJT	TTT	T
Cultivar	C	D	E	J	T	Mean

Fig. 3.6 Example of a competition diallel with three-row plots in potato (Bradshaw 1994b, and 2016 with permission): cultivars are Cara (C), Pentland Dell (D), Estima (E), Pentland Javelin (J) and Torridon (T). The 25 combinations of cultivar and neighbour had different randomizations in each of four replicates in each of two years

Table 3.5 Potato cultivar fresh-weight ware yields (kg/plot) under competition (comp) and in pure-stands (pure), together with cultivar means (direct effects) and estimated two-sided neighbour effects (neighbour); direct + neighbour = predicted pure (Bradshaw 1994b, and 2016 with permission)

	1992				1993			
	Comp	Pure	Direct	Neighbour	Comp	Pure	Direct	Neighbour
Cara	9.91	9.67	9.86	−0.51	14.36	12.21	13.93	−1.86
Dell	9.49	9.53	9.49	0.04	10.38	12.44	10.79	0.79
Estima	9.94	11.23	10.20	0.44	11.72	14.06	12.19	1.11
Javelin	7.84	8.35	7.94	0.67	9.82	11.22	10.10	1.05
Torridon	10.41	9.77	10.28	−0.60	11.86	11.36	11.76	−1.08
SED	0.54	1.08	0.48	0.48	0.54	1.08	0.48	0.48

particularly strong competitors, reducing the yields of neighbouring cultivars. Their own yields were higher under competition than in pure stands. The correlation between cultivar yields under competition and in pure-stands was moderate ($r = 0.73$) in 1992 and low ($r = 0.18$) in 1993.

In a competition diallel, the yield of the centre plot (drill) of genotype A with genotype B as neighbour is modelled as:

Yield A with B = replicate mean + direct effect A + interference effect B + interaction A with B + plot error.

Thus, g in our model for a randomized complete block trial now becomes:

$$g = v_r + 2a_s + 2i_{rs}$$

where v_r is the direct effect of cultivar r, a_s the effect of cultivar s as a one-sided neighbour, and i_{rs} the non-additive interaction of cultivar r and neighbour s.

Hence, pure-stand yields are $v_r + 2a_r$.

There are various ways of dealing with interference (Kempton 1997). In the advanced generations of a breeding programme, when the number of potential cultivars has been reduced and more planting material is available, multiple-row plots can be used and the outer rows discarded. There may also be scope to group similar genotypes (e.g. early-maturing potatoes) and thus avoid interference, provided prior knowledge is available about the trait values for the genotypes. Finally, it may be possible to adjust for interference using an appropriate plant covariate (e.g. plot yield) derived from neighbouring plots (Kempton 1985; Connolly et al. 1993). If the covariate is the mean yield of neighbouring left and right plots, the statistical model for a cultivar (genotype) trial with a randomized complete block design becomes:

$$y_{ij} = \mu + g_i + b_j + \frac{1}{2}\beta\left(y_{left} + y_{right}\right) + e_{ij}$$

where y_{ij} is the yield of genotype i in block j, μ is the trial mean, g_i the genotype effect, b_j the block effect and e_{ij} the plot error. Genotype yields in pure-stands are then estimated by imagining that the neighbouring plots both contain genotype i so that $½\beta(y_{left} + y_{right}) = ½\beta(y_{ij} + y_{ij}) = \beta y_{ij}$. Hence,

$$y_{ij} - \beta y_{ij} = \mu + g_i + b_j + e_{ij}$$

$$y_{ij} = \left(\mu + g_i + b_j + e_{ij}\right) / \left(1 - \beta\right)$$

and the genotype yields in pure stands are $g_i/(1\text{-}\beta)$. In other words, high yields (positive g_i) are corrected for over estimation by the presence of low-yielding neighbours and low yields (negative g_i) for under estimation by the presence of high-yielding neighbours as β is negative in the presence of competition. The estimation of the parameters in the model is more complicated than implied by this simple presentation and the procedure can be found in the paper by Connolly et al. (1993) which refers to the original statistical literature. Such interference models are useful as they can be used to test for the presence and size of interference effects and hence alert the breeder to potential problems. For example, Connolly et al. (1993) found that values for the common competition coefficient (β) for yield in 31 breeders' trials at SCRI (East of Scotland) from 1988 to 1991 ranged from −0.28 (competition) to 0.37 (a positive association between plots and their neighbours for which spatial analysis would be appropriate: see above). They concluded that it is unlikely that competition will seriously affect a breeder's ability to select clones that are high yielding under monoculture. Nevertheless, it is desirable to use multiple-row plots for the final stages of yield assessment of new cultivars because modelling and grouping may not be completely effective. In the potato diallel, for example, individual neighbour effects, but not covariate adjustment, markedly improved the correlations between observed and predicted pure-stand yields (r = 0.94, 0.77 and 0.73 in 1992 and r = 0.84, 0.28 and 0.18 in 1993 for individual neighbour effects, covariate adjustment and no adjustment, respectively). The regression coefficients for covariate adjustment were – 0.19 and – 0.33 in 1992 and 1993, respectively. Although competitive ability in potatoes is associated with 'tuber' yield, other factors must also be involved; but determining individual neighbour effects is not a practical option when assessing a large number of potential cultivars.

3.14 Genotype × Environment Interaction (G × E)

Within any agroecological zone targeted by a breeding programme, yields will vary between locations and seasons. It may not be obvious if genotypes can be found with general adaptability and yield stability across these environments. For example, can cultivars be bred with tolerance (little reduction in yield) of abiotic stresses but which also yield well in the absence of such stresses, or are two different types of cultivar required? Likewise, can cultivars be bred with high resource-use efficiency under both high and low inputs of the resource? For example, can cultivars be bred for higher levels of production than other cultivars on organic farms, low-input farms and high-input farms, or are different cultivars needed for each farming system? Ultimately, an understanding of the physiological and biochemical bases of stress tolerance and resource-use efficiency, and their implications for yield potential, are required if the breeder is to be provided with selection criteria for manipulating G × E. In the meantime, an assessment is required of genotype × environment interactions at the start of breeding programmes in order to determine objectives and to choose parents for hybridization, and at the end of programmes in order to make recommendations to farmers and growers about which cultivars to grow. Equally important for the success of a breeding programme is the choice of environment at every intermediate stage where clones are assessed and selected. Hence, we are now going to look at the concept of genotype × environment interactions, how they are assessed and their consequences for potato breeding.

3.15 Concept of Genotype × Environment Interactions

Genotype × environment interactions can occur between genotypes and specific macro-environments such as locations, seasons and fertilizer regimes. The interaction may or may not involve changes in the ranking of genotypes over environments, something that Baker (1988) has shown can be assessed by a technique developed by Azzalini and Cox (1984). The literature on genotype × environment interactions is vast and the methods of analysis are numerous, but there are relatively few underlying concepts. Even so, the topic has attracted controversy and debate, as seen by two reviews in *Crop Science* (Yan et al. 2007; Gauch et al. 2008).

A simple (artificial) example is shown in Fig. 3.7 for the yields of three genotypes in three environments. Genotype 1 is best in environment 1 (E1), genotype 2 in environment 2 (E2), and genotype 3 in environment 3 (E3). Hence, the results of any genetic analysis will depend upon the environment in which the genotypes

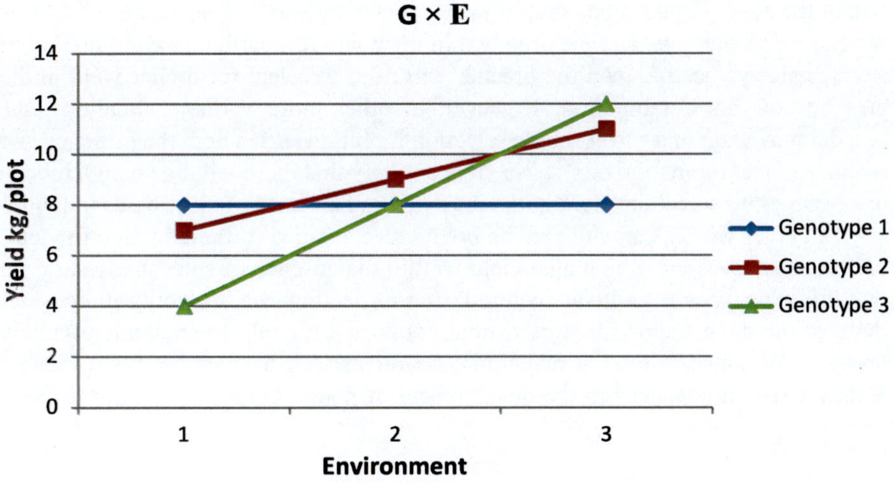

Fig. 3.7 Yields of three genotypes grown in three environments

are assessed. Likewise, if a breeding programme was conducted in low-yielding environment 1, genotype 1 would be selected but would be the worst if grown in environment 3. In contrast, genotype 3 would be selected in high-yielding environment 3 but would be the worst if grown in environment 1. Interestingly, genotype 2 which would be selected in environment 2 would rank second in both environments 1 and 3, and first overall. It would also be selected on the basis of the average of yields in environments 1 and 3. Genotype 2 is high yielding (average yield of 9) with general adaptability, whereas genotypes 1 and 3 have specific adaptability to low- and high-yielding environments, but the same average yield of 8. Ironically, we would also predict from Fig. 3.7 that selection for low yield in below-average environments (E1) should result in genotypes (genotype 3) with a high sensitivity to environmental change, whereas selection for low yield in above-average environments (E3) should result in genotypes (genotype 1) with a low sensitivity to environmental change. These predictions have been confirmed in selection experiments with the fungus *Schizophyllum commune* (Jinks and Connolly 1973, 1975) and with the tobacco *Nicotiana rustica* (Jinks and Pooni 1982). An example of the issue of general and specific adaptability and selection environment arises when we consider breeding for tolerance of abiotic stresses.

3.16 Tolerance of Abiotic Stress

The limited numbers of seed tubers available in the early and intermediate generations of a programme means that the breeding material will be assessed at very few sites, usually just one or two. We now need to consider if high yield potential (and quality) in a particular agroecological zone can be combined with tolerance of any likely abiotic stresses in order to achieve yield stability across locations and seasons within the zone. Potato production may encounter various combinations of stresses depending on the zone and the area within the zone. If a particular combination of stresses always occurs, then the breeder will need to select for higher yield in the presence of this combination. However, in other more variable situations, the breeder may want to try to combine selection for high yield in both the presence and absence of a combination of stresses (it is unlikely that there will be enough tubers, or resources, to select for yield in the presence and absence of all stresses individually). In other words, can cultivars be bred with tolerance (little reduction in yield) of abiotic stresses but which also yield well in the absence of such stresses, or are two different types of cultivar required? In this section, we are going to consider three of the most important stresses that potatoes are likely to encounter, namely drought, heat and salinity. The reader may be interested in a recent review by Prashar et al. (2018) on advances in the development of potato cultivars resistant to these stresses.

3.16.1 Drought

Potatoes, like all crops, can deal with drought by three mechanisms, namely escape, avoidance and tolerance. So, should breeders select new cultivars for earlier maturity to escape later occurring droughts in short season areas, or should breeders select new cultivars to have a different root system that can reach deeper moisture zones and thus avoid the drought? This would probably mean using a suitable glasshouse test to select genotypes with a desired root system under field conditions along the lines developed by Iwama (2008) and Wishart et al. (2013) (Fig. 3.8). Or finally, should breeders select for true tolerance by selecting new cultivars that tolerate drought by reducing transpiration losses through leaf rolling, early closure of stomatal openings and cuticular resistance? This would almost certainly result in sub-optimal yield by reducing CO_2 fixation.

Hence, one can ask if selection for yield under water stress results in dehydration-avoidance (e.g. by osmotic adjustment) to maintain high water status under stress rather than tolerance of dehydration, and would this also lower yield potential? For the practical breeder, the issue becomes one of selection for general or specific adaptability. In other words, can new cultivars be selected that are tolerant of drought conditions but also yield well in its absence (general adaptability), or are different cultivars required for drought and non-drought environments (specific adaptability)? These questions need to be answered at the beginning of a breeding programme in order to make sensible choices of parents to cross and environments in which to grow and assess potential cultivars. Selection for wide adaptability will usually

Fig. 3.8 Root system on field grown plant and glasshouse test at SCRI, Dundee. (Source: SCRI)

involve selection in more than one environment and for more than one physiological component of crop growth and tuber production. Crop physiologists can provide some guidance, as seen in reviews such as the ones by Levy et al. (2013) and Monneveux et al. (2013), including suggestions for genes to use in the genetic transformation (modification) of potatoes, but more research is also required. For example, in their association mapping study of tuber size distribution and marketable tuber yield under drought stress, Aliche et al. (2019) found that the most severe effect of drought on the marketable fractions of tuber fresh weight was a 51% reduction at Connantre in France in 2015, when drought occurred during the later stages of plant development as a result of a total rainfall of only 42 mm, whereas the previous year there had been a 13% reduction in the marketable fractions of tuber number as a result of drought during the early stages of plant growth. While more research is being done, assessment of diverse germplasm for drought tolerance is of value to breeders globally.

Cabello et al. (2012) assessed a set of 918 potato accessions from the CIP genebank for drought tolerance over the five-year period 2004–2008 at CIP's La Molina experimental station (12.1°S 76.9°W, 240 masl) on the coast of Peru, characterized by absence of rainfall, high relative humidity and average annual minimum and maximum temperatures of 14.9 °C and 18.0 °C, respectively. The accessions comprised 226 improved cultivars [(including Late Blight Resistant (LBR) and Lowland Tropic Virus Resistant (LTVR) clones from populations developed at CIP and also foreign ones)], 74 genetic stocks (including two diploid mapping populations) and 618 landraces [*S. ajanhuiri* (15), *S. curtilobum* (15), *S. juzepczukii* (22) and *S. tuberosum* Group Andigena (261), Chaucha (113), Goniocalyx (45), Phureja (81), Stenotomum (43) and (Chilean) Tuberosum (23)]. Three control cultivars were included: Revolucion (drought-susceptible), Achirana (drought-tolerant) and Compis (Andigena landrace). Different accessions were evaluated each year but with the controls included each year. The improved cultivars, genetic stocks and landraces were grown in separate trials, all with three replicates and all planted the first week of August and harvested either 90 (improved cultivars) or 120 (genetic stocks and landraces) days later. The total quantity of water received by each trial over the five years varied between 375 and 414 mm for the irrigated treatment (water applied every 10 days) and between 128 and 175 mm for the drought treatment (furrow irrigation stopped either 35 days (improved cultivars) or 42 days after planting (genetic stocks and landraces). The interaction between irrigation treatments and accessions was highly significant for tuber yield, tuber number (except for accessions in subset of genetic stocks), average tuber weight and tuber dry matter, thus revealing differences in drought tolerance. Average tuber yield per plant under irrigated conditions was 813.7, 392.1 and 475.9 g for the improved cultivars, genetic stocks and landraces, respectively, whereas under drought conditions, tuber yield decreased to 420.2, 223.0, and 299.7 g, respectively. Despite the interactions, there were significant correlations for yield between irrigated and drought conditions ($r = 0.78$, 0.43, and 0.74 for improved cultivars, genetic stocks and landraces, respectively). Within the landraces, on average, the *S. curtilobum* accessions had the highest yields under irrigated (606.7) and drought (379.0) conditions. For each accession, a Drought Susceptibility Index (DSI) was calculated as DSI = (1 – Yd/Yw)/D where Yd = mean yield of the accession under drought, Yw = mean yield of the same accession under well-watered conditions and D = a measure of environmental stress intensity. Although really low DSI (drought tolerance) was associated with low yield under irrigation, the authors were able to identify accessions combining above median high yield under irrigated conditions with below median DSI. Furthermore, there were 13 accessions (5 LBR clones, 3 LTVR clones and 5 Andigena landraces) with yields greater than 800 g plant^{-1} under both irrigated (average 1154) and drought (883 = 77% of irrigated) conditions, and these were superior to the controls. Interestingly, the 13 accessions with the highest yields under irrigation (average 1278) had relatively lower yields under drought (808 = 63% of irrigated) conditions, whereas yields were less changed for the 13 highest yielding accessions under drought conditions (886 = 80%) and their yields under irrigation (average 1115). This suggests that the highest yielding accessions under drought conditions had yield potential rather than drought tolerance under these conditions. Readers are referred to the paper for more details, and can draw their own conclusions, but my conclusion is that clones (potential cultivars) can be found with general (wide) adaptability by assessment for yield in two contrasting environments and that selecting for drought tolerance (DSI) rather than yield potential under drought conditions was not helpful in this respect.

3.16.2 Heat

Average daily temperatures of above 21 °C have an adverse effect on plant growth and tuber formation (Govindakrishan and Haverkort 2006). Hence, heat is a major limiting factor for potato production in the lowland tropics and subtropics. As potato cultivars being grown in Brazil were native to Europe and not fully adapted to the tropical conditions, a breeding programme for heat tolerance was undertaken at Ufla, Lavras (21°14′S 45°00′W, 919 masl), as described by Benites and Pinto (2011). The base population comprised five Brazilian and five heat-tolerant clones, four of which came from CIP. After three cycles of recurrent selection with seedling populations of up to 6000 in size, 103 clones were evaluated in the winter of 2006 and the rainy growing season of 2007: eight clones from the base population and 29, 32 and 30 from the three cycles of selection, respectively, together with four controls (Agata, Asterix, Atlantic and Monalisa). The best eight clones (on index primarily of yield, yield components and visual appearance) in each generation were used to calculate the responses to selection. There were statistically significant differences between clones and clone × season interactions for yield. The rainy season is the hot one, where for example, in 2007 mean daily temperatures were above 20 °C for about 55% of the growing season and reached levels higher than 25 °C in 25% of this period. Compared with the winter season, mean tuber yield was reduced by 56.9% due to reductions in the percentage of large tubers (26.4%) and tuber weight (20.4%). The mean grade for tuber appearance also decreased by 7.1%, due to the high temperatures which, in turn, increased the incidence of physiological disorders. Averaged over the two seasons, the three cycles of selection resulted in a 37.85% increase in yield, most of which was achieved in the first cycle of selection. The best eight clones were all higher yielding than the best control. Readers are referred to the paper for more details, and can draw their own conclusions, but my conclusion is that selection in two contrasting environments, one favourable for potato growing and the other resulting in heat stress, has resulted in general adaptability.

Khan et al. (2015) developed and tested an early generation in vitro assay to identify potato populations and clones tolerant to heat. Identification of populations (full-sib families) was more successful ($r = 0.57$ for correlation with field results) than identification of clones ($r < 0.30$). Twenty families were produced from parents (clones) from CIP's lowland tropics virus resistant (LTVR) population, their B3 (late blight resistant) population, and *Solanum tuberosum* cultivars (TBR) adapted to long-day conditions. The LTRV population is characterized by virus resistance, heat tolerance, sub-tropical lowland adaptation and earliness, whereas the B3 population is characterized by late blight resistance and cool, highland adaptation. Families were tested for tuber production as seedling progenies in vitro in the dark at 18, 25 and 32 °C and as tuber progenies in a field trial at Tacna (439 masl), an arid sub-tropical environment in the South of Peru, in the summer of 2008 when night temperatures ranged from 10 to 21 °C and day temperatures ranged from 26 to 32 °C. The 25 °C dark treatment provided the best discrimination between families. The percentage of genotypes per family with tubers ranged from 4.7% (a B3 × B3 family) to 82% (a TBR × LTVR family). There was a moderate correlation ($r = 0.57$) between these results and those from the field trial. Hence, the in vitro test (see paper for details) can be used as an initial screen in a breeding programme to identify promising families for more detailed field evaluation of yield and other desirable traits.

Breeding cultivars for heat tolerance is also important in India where for example, high temperatures, particularly at night (above 20 °C), were restricting early planting in the North Western Plains. Mite/leafhopper burn was also a problem. ICAR-CPRI started breeding work in 1984 with the screening of their germplasm collection (Luthra et al. 2020). Genotypes with desirable characteristics for heat tolerance were crossed with well adapted genotypes of the region. Initial screening of seedlings was done in pots in a glasshouse at Shimla (in the hills) where they were exposed to 22 °C night temperature, created using heat convectors. Observations on tuberization were recorded four and eight weeks after planting and seedlings that tuberized under these conditions were selected as early and heat tolerant, usually around 300 out of 20,000 planted. Multiplication over two years of selected genotypes took place in the plains during the (aphid) vector-free period with rejection of clones producing undesirable tuber types. Just 40 clones would enter more extensive trialling (Fig. 3.9) over an eight-year period with a view to selecting one new cultivar capable of yielding in a 75-day growing season (assessments were also done at 60 and 90 days).

Fig. 3.9 Heat-tolerant cultivars (Kufri Lauvkar, released 1972, is one of the controls) on display at Central Potato Research Institute, Modipuram, Meerut, Uttar Pradesh, India (29°3′29″N 77°42′10″E, 225 masl) (13 December 2008)

Kufri Surya was the first heat-tolerant cultivar to be produced from the programme and was released in 2006. The most recent cultivar, Kufri Lima (released in 2018) was developed from a collaboration with the International Potato Centre (CIP) in Lima, Peru (https://cipotato.org/blog/new-potato-variety-kufri-lima-take-heat-virus-india/). It is an early heat-tolerant cultivar with extreme virus resistance (PVX and PVY), suitable for early planting in the North Indian Plains (Fig. 3.10). It was selected in Peru from CIP's LTRV population as CIP397065.28 and then evaluated in India from 2006 (Bonierbale et al. 2020).

With a quantitative trait such as heat tolerance, the underlying loci (QTLs) can be sought through associations between trait scores and molecular markers. Sometimes, desirable alleles of large effect can be found and incorporated into new cultivars by marker-assisted selection. Thus, Trapero-Mozos et al. (2018) identified a locus (on chromosome 4) associated with tuber yield (fresh-weight) at both normal (22 °C) and elevated temperature (28 °C) in a diploid, long-day adapted, Phureja–Tuberosum mapping population (2157 SNP markers). They used a nodal cutting assay to assess the tuber yield of 170 genotypes (12 cuttings of each) under controlled environment, growth-room conditions (the reader is referred to the paper for details of growing conditions and heat treatments). The population exhibited fairly late and uniform foliage maturity and did not segregate for variation at the maturity locus on chromosome 5 (Kloosterman et al. 2013). Inspection of the genomic region corresponding to about 5 cM either side of the QTL peak on

Fig. 3.10 Heat-tolerant cultivars Kufri Lima and Kufri Surya on display at Field Day, Global Potato Conclave 2020, Mahatma Mandir, Gandhinagar, Gujarat, India (31 January 2020)

chromosome 4 revealed a strong candidate gene encoding a heat-shock cognate protein HSc70 which is a constitutively expressed molecular chaperone. Sequence analysis of the PCR products from a region of the *HSc70* gene in the two parents of the mapping population identified four distinct alleles (unique differences were present in their promoters). One of these alleles (*A2* with presence of ten rather than four TA repeats in promoter) was linked to elevated yield in the progeny. Expression of this allelic variant was much higher than other alleles, particularly on exposure to moderately elevated temperature. Furthermore, expression of the *HSc70* allelic variant under its native promoter in the potato cultivar Desiree resulted in enhanced *HSc70* expression at elevated temperature (28 °C). The researchers established that elevated transient expression of *A2 HSc70* provided a temperature-dependent protective effect as assessed by an electrolyte leakage assay. Furthermore, in nodal cutting tuber yield assays, there was no difference in fresh-weight yield between three transgenic lines and wild-type Desiree at 20 °C, but yield in wild-type Desiree decreased by 75% at 28 °C compared to 22 °C, whereas yield reductions were less in the transgenic lines (about 50%). In plants grown from stem cuttings, with a maximum day temperature of 20 °C, just one of the three transgenic lines exhibited a significant increase in both tuber fresh- and dry-weight over wild-type Desiree, whereas at 28 °C, tuber yields were at least twice as high in all of the transgenic lines relative to the controls. The transgenic lines were also about 50% higher yielding at 28 °C than at 20 °C, whereas there were no significant differences in the wild-type Desiree (values were actually lower). In summary, the researchers had demonstrated that the presence of the *A2 HSc70* allele confers tolerance to elevated temperature, both in a mapping population and in transgenic lines where *HSc70* expression is driven by the *A2* promoter. In the nodal cutting system and in plants generated from stem cuttings, the presence of the *A2* allele results in enhanced yield at elevated temperature. In transgenic lines, the yield was up to two-fold greater than in wild-type plants at 28 °C, but with no significant effect on yield at 20 °C. In the mapping population genotypes, however, yield was enhanced at both 22 °C and 28 °C. In both situations, the researchers proposed that the rapid response of the *A2* allele to temperature perturbation underlies the yield effect. They also thought that use of the *A2* allele in potato breeding might provide a strategy for enhancing yield either under temperate conditions or under periods of abiotic stress.

Finally, Lehretz et al. (2019) have discovered a small RNA that comprises around 19 nucleotides and that blocks tuber formation at day and night temperatures of 29 °C and 27 °C, respectively, by effectively repressing *SP6A* mRNA accumulation and the formation of SP6A (SELF-PRUNING 6A) protein (*SP6A* is a homologue of the floral inductor *FLOWERING LOCUS T* and controls tuber formation). At more normal day and night temperatures for potato growth of 21 °C and 18 °C, respectively, the RNA is inactive and SP6A is formed and induces the formation of tubers. The researchers successfully switched off this small RNA in plants being grown in a glasshouse at the higher temperatures and found that good-quality tubers were produced. The researchers now need to test the potato plants under field conditions to see if they can withstand high temperatures in these circumstances.

3.16.3 Salinity

Soil salinity is a major environmental constraint to the production of crops including potato. Qadir et al. (2014) estimated that by 2014 about 62 million hectares (20%) of the world's irrigated lands had been degraded by salt, up from 45 million in the early 1990s. This area is expected to increase due to global climate change and as a consequence of many irrigation practices (Roy et al. 2014). Salinity is thus a major threat to the sustainable irrigation required to maintain the yield and quality of many crops including potato (Bell 2009). Rain and surface water dissolve the earth's natural minerals as they make their way to the sea, which is salty as a result. Salt-degradation occurs in arid and semi-arid regions where rainfall is too low to maintain regular percolation of rainwater through the soil and where irrigation is practised without a natural or artificial drainage system. Water settles on the irrigated land and evaporates, leaving the salt which can accumulate to dangerous levels in the root zone of crops and reduce their productivity. In other words, the irrigation has the effect of stopping the minerals running off to the ocean. If sufficient cheap energy were available (e.g. solar energy), desalination of irrigation water could be the ultimate solution to the problem. In the meantime, breeding for salt tolerance may be necessary.

Potato breeders have sought in vitro screens for salt tolerance which translate into yield responses to salt stress in whole plants under field conditions. One example is provided by the research of Shaterian et al. (2008) on methodologies and traits for evaluating the salt tolerance of diploid potato clones. Their aim was to develop a simple, yet accurate, method for segregating salt-tolerant and sensitive genotypes and to identify phenotypic and physiological parameters indicative of yield tolerance to salt stress. The germplasm assessed was diverse diploid *Solanum* genotypes (22 clones) known to vary in salt tolerance from preliminary screening. The clones were grown in pots in a glasshouse under conditions in which they all tuberized. At tuber initiation, using a hydroponic system, each genotype was challenged with 100 mM salt (10 dS cm^{-1}) for two weeks followed by an additional two weeks at 150 mM (total four weeks). Non-stressed control plants were watered with a standard nutrient solution. A number of traits were measured after the salt treatment including a growth index which was calculated as: (shoot height after four weeks salt stress – height at the beginning of the stress treatment) / (shoot height after four weeks under non-stressed control conditions – height at the beginning of the control treatment) × 100. In a second experiment, seven diploid clones differing in salt tolerance were chosen to study the effect of short-term salt stress on tuber yields at maturity. Beginning at tuber initiation, half the plants of each clone were challenged by adding 150 mM NaCl (16.3 dS m^{-1}) salt to the nutrient solution for seven days, after which standard watering and fertilization were resumed. The control plants were watered with a standard nutrient solution throughout the experiment. Traits assessed included shoot weights just prior to initiation of the salt stress and shoot and tuber fresh weight at the final harvest, which was either at senescence or after 120 days. A growth index was also calculated: (shoot height after salt stress – height at the beginning of the stress treatment) / (shoot height after treatment period under control conditions – height at the beginning of the treatment period) × 100. The key findings were that the growth index in both experiments was a good predictor of absolute tuber yield but not of tuber yield tolerance to salt stress (tuber yield under stress/tuber yield under non-stress conditions). Tuber yield tolerance to salt stress was however positively correlated with time to maturity. When the genotypes were ranked based on cluster analysis of a range of readily measured phenotypic and physiological responses to salt stress, the ranking was effective at predicting the relative impact of a salt stress event on tuber yield at maturity. The highest yielding clone at maturity after salt stress was also the highest yielding clone in the absence of stress, although the stress reduced its yield to 63% of that in the absence of stress. In other words, the best salt tolerance did not mean the best yield under salt stress. Again, my interpretation of the results is that one can find clones that yield well in the presence and in the absence of the stress and hence can be considered to have general adaptability; but readers should consult the original paper and draw their own conclusions.

Other researchers have explored the mechanism of tolerance (osmotic tolerance, ion exclusion and tissue tolerance) with a view to informing breeders about selection criteria for salt tolerance. Jaarsma et al. (2013), for example, looked at the effect of salt stress on growth, sodium ion (Na^+) accumulation and proline metabolism in six potato cultivars differing in tolerance. The more tolerant cultivars had the capacity to accumulate Na^+ ions from the transpiration stream into their stem tissue, whereas the salt-sensitive cultivars transported relatively more Na^+ to leaf tissue. Roy et al. (2014) reviewed the genetic basis of salt tolerance in a wide range of crop species. They presented a compilation of transgenes and their effects on salt tolerance in crops (but only one reference to potato) and explained that temporal and spatial (specific cell types) control of gene-expression are often important for increasing tolerance. Importantly, the effects of introduced genes need to be evaluated in the field to confirm their effect on salt tolerance and yield improvement.

3.17 Detection and Analysis of G × E

It is now time to turn to the detection and analysis of G × E. Provided genotypes (*G*) are replicated in each environment (*e* is plot-to-plot variation) as well as over environments (*E*), genotype × environment interactions (*GE*) can be detected through a two-way analysis of variance, and their relative importance determined from the components of phenotypic (yield) variance (σ^2_P):

$$\sigma^2_P = \sigma^2_G + \sigma^2_E + \sigma^2_{GE} + \sigma^2_e$$

The observed variance of genotype means over *n* environments with *r* replicates in each environment is:

$$\sigma^2_G + \sigma^2_{GE} / n + \sigma^2_e / (nr)$$

The standard error of the difference (SED) between two genotype means is:

$$\text{SED} = (2\text{VAR})^{½} \text{ where VAR} = \sigma^2_{GE} / n + \sigma^2_e / (nr).$$

The critical difference (d_α) between two genotypes is the difference that can occur by chance in a proportion α of trials when the true difference is zero.

$d_\alpha = t_\alpha \times$ SED (or as a percentage of the mean yield of the trials μ; $d_\alpha = t_\alpha \times \text{SED} \times 100/\mu$), where t_α is the value from the normal distribution that is exceeded with probability α.

Potential new cultivars are usually evaluated in a series of trials conducted in an appropriate range of locations (*L*) for more than one year (*Y*), before making recommendations to farmers and end users. Locations will differ in factors such as soil type and agronomic management practices, as well as day length, altitude and climate, and years will differ primarily in weather. Differences in disease levels may occur both between locations and years.

The standard error of the difference between the means of a new genotype and c checks evaluated in trials with r replicates at l locations for y years is:

$\text{SED} = [(1 + c)\text{VAR}/c]^{1/2}$, where $\text{VAR} = \sigma^2_{GL}/l + \sigma^2_{GY}/y + \sigma^2_{GLY}/(ly) + \sigma^2_e/(lyr)$.

Estimates of the components of variance from a series of trials can be used to evaluate the efficiency and effectiveness of the trials system in terms of the number of locations, years and replicates used; and to make improvements where necessary (Talbot 1984, 1997). The residual (restricted) maximum likelihood (REML) approach to estimation can be used on incomplete 'genotypes × locations × years' data matrices (Talbot 1997). Talbot's (1984) analysis of data from the Official UK Variety Testing System for 1969–1978 included 178 potato trials in which 36 varieties (cultivars) had been assessed at 20 test centres over 10 sowing years. The mean fresh-weight yield over all trials was high at 45.2 t/ha. The components of variance ((t/ha)^2) were as follows:

Varieties	9.78
Centres	11.07
Years	44.25
Centres × Years	75.78
Varieties × Centres	2.98
Varieties × Years	2.63
Varieties × Centres × Years	14.96
Plots	18.79

Potatoes were the least accurately assessed of all the agricultural crops (cereals, forage crops, potatoes, sugar beet and spring beans), except forage maize, with a CV% of 9.6 (coefficient of variation). The environmental differences were large compared with those between varieties, with the largest component the Centres × Years interaction (75.78). The component of variance for Varieties × Centres × Years interaction (14.96) was larger than that for Varieties (9.78). The critical percentage difference (mean over trials) for a new variety compared with a control was 9.8% for 2 years of trials with 2 replicates at 12 centres (i.e. difference between two varieties which will be exceeded in 0.025 proportion of cases if there is no real difference). In other words, a new variety needs to be almost 10% higher yielding than best control to be declared higher yielding; quite a big difference.

In addition to being used to make recommendations to farmers, data from such a set of trials will help the breeder to choose cultivars for use as parents in further crosses. However, at the start of new programmes, breeders will want to assess additional germplasm of interest in their own trials. They will also need more detailed information about genotype × environment interactions.

3.18 Principal Component Analysis

3.18.1 Theory

In general, the complexity or otherwise of genotype × environment interactions can be seen from a principal component analysis (PCA). If the overall (additive) effects of genotypes (*G*) and environments (*E*) are first removed from the data, the analysis results in multiplicative terms for the interaction, the so-called additive main effects and multiplicative interaction (AMMI) model (Gauch 1988; Gauch et al. 2008):

$$Y_{ij} = \mu + G_i + E_j + \Sigma\theta_k u_{ki} v_{kj} + \bar{e}_{ij}$$

where Y_{ij} is the yield of *i*th genotype in *j*th environment, summation is from $k = 1$ to K, the number of principal component axes (PCA can also stand for principal component axis) retained in the model, and θ, u and v are the singular values (i.e. square roots of the eigenvalues), normalized eigenvectors for genotypes and normalized eigenvectors for environments, respectively. Finally, $\bar{e}_{ij}$ is the residual variation for the mean of the *i*th genotype in *j*th environment. Axis *k* is assigned $(G + E - 1 - 2\,k)$ degrees of freedom, where *G* and *E* are the number of genotypes and environments, respectively. The normalized eigenvectors are usually scaled by multiplying by $\theta^{½}$ to give $\theta^{½}u$ (PCA scores for genotypes) and $\theta^{½}v$ (PCA scores for environments) so that $\Sigma\theta_k u_{ki} v_{kj}$ becomes $\Sigma\theta^{½}{}_k u_{ki} \theta^{½}{}_k v_{kj}$. A simple artificial example is shown in Table 3.6 for three genotypes grown in three environments where the first two principal components account for all of the G × E variation (i.e. all of the sum of squares for G × E).

For readers familiar with matrix algebra, the PCA analysis was done as follows, where **Z** is the G × E matrix, with elements z_{ij}, shown in Table 3.6b.

Z	(1 0 −1)
	(−1 −1 2)
	(0 1 −1)

We obtain the eigenvalues (λ) and eigenvectors (PCA scores) for genotypes from the matrix **ZZ'**, where **Z** has been post-multiplied by its transpose **Z'**.

ZZ'	(2 −3 1)
	(−3 6 −3)
	(1 −3 2)

The trace of the matrix (sum of elements in the leading diagonal) is 10 which is the sum of squares for G × E $(z_{11}^2 + z_{12}^2 \ldots z_{33}^2)$.

We now consider the following set of homogeneous equations: $(\mathbf{ZZ'} - \lambda\mathbf{I})\mathbf{U} = \mathbf{0}$, where **I** is the unit matrix of the same order as **ZZ'** and the λs are the eigenvalues of **ZZ'** for which the corresponding **U**s, called the eigenvectors, are the non-trivial

Table 3.6 Principal component analysis of yields of three genotypes grown in three environments (from Bradshaw 2016 with permission)

A Yields (Y_{ij}) of three genotypes grown in three environments where Y_{ij} is yield of ith genotype in jth environment

	Environment 1	Environment 2	Environment 3	Mean
Genotype 1	8	9	10	9
Genotype 2	5	7	12	8
Genotype 3	5	8	8	7
Mean	6	8	10	8

B G × E (Z_{ij}) for three genotypes in three environments ($Z_{ij} = Y_{ij} - Y_{i.} - Y_{.j} + Y_{..}$, where $Y_{i.}$ is the mean yield of genotype i, $Y_{.j}$ is the mean yield of environment j and $Y_{..}$ is the overall mean yield)

	Environment 1	Environment 2	Environment 3	PCA 1	PCA 2
Genotype 1	1	0	−1	$-1/\sqrt{2}$	$-1/\sqrt{2}$
Genotype 2	−1	−1	2	$2/\sqrt{2}$	0
Genotype 3	0	1	−1	$-1/\sqrt{2}$	$1/\sqrt{2}$
PCA 1	$-1/\sqrt{2}$	$-1/\sqrt{2}$	$2/\sqrt{2}$		
PCA 2	$-1/\sqrt{2}$	$1/\sqrt{2}$	0		

C Analysis of variance

Source	df	SS	MS
Environment	2	24	12
Genotype	2	6	3
G × E	4	10	2½
PCA 1	3	9	3
PCA 2	1	1	1
Total	8	40	

solutions (i.e. $\mathbf{U} \neq \mathbf{0}$) of the equations which occur when the determinant $|\mathbf{ZZ'} - \lambda\mathbf{I}| = 0$.

There are two non-trivial solutions of the eq. $(\mathbf{ZZ'} - \lambda\mathbf{I})\mathbf{U} = \mathbf{0}$, with $\lambda = 9$ and 1 (which sum to the trace = 10), and correspondingly normalized (squares of elements sum to 1) eigenvectors for genotypes:

Vector1	Vector2
$(-1/\sqrt{6})$	$(-1/\sqrt{2})$
$(\ 2/\sqrt{6})$	$(\ 0\ )$
$(-1/\sqrt{6})$	$(\ 1/\sqrt{2})$

These are usually scaled by multiplying by $\theta^{½}$ $(= \lambda^{¼}) = \sqrt{3}$ and 1, respectively, to give the scaled and normalized PCA scores for genotypes:

PCA1(G)	PCA2(G)
$(-1/\sqrt{2})$	$(-1/\sqrt{2})$
$(\ 2/\sqrt{2})$	$(\ 0\ )$
$(-1/\sqrt{2})$	$(\ 1/\sqrt{2})$

We obtain the eigenvalues and eigenvectors (scores) for environments from the matrix **Z'Z**, where **Z** is now pre-multiplied by its transpose **Z'**.

Z'Z	(2 1 −3)
	(1 2 −3)
	(−3 −3 6)

The trace of this matrix is again 10, the sum of squares for G × E.

There are two non-trivial solutions of the eq. $(\mathbf{Z'Z} - \lambda\mathbf{I})\mathbf{V} = \mathbf{0}$, with $\lambda = 9$ and 1 (which sum to the trace = 10), and corresponding normalized eigenvectors for environments:

Vector1	Vector2
$(-1/\sqrt{6})$	$(-1/\sqrt{2})$
$(-1/\sqrt{6})$	$(\ 1/\sqrt{2})$
$(\ 2/\sqrt{6})$	$(\ 0\ )$

These are usually scaled by multiplying by $\theta^{1/2}$ $(= \lambda^{1/4}) = \sqrt{3}$ and 1, respectively, to give the scaled and normalized PCA scores for environments:

PCA1(E)	PCA2(E)
$(-1/\sqrt{2})$	$(-1/\sqrt{2})$
$(-1/\sqrt{2})$	$(\ 1/\sqrt{2})$
$(\ 2/\sqrt{2})$	$(\ 0\ )$

The rescaled PCA scores for genotypes multiplied by the transpose of the rescaled PCA scores for environments reproduce the genotype × environment interactions.

The contribution of PCA1 to the genotype × environment interactions is PCA1(G) PCA1(E)'.

$(-1/\sqrt{2})\ (-1/\sqrt{2}\ -1/\sqrt{2}\ 2/\sqrt{2})$ =	$(\frac{1}{2}\ \frac{1}{2}\ -1)$
$(\ 2/\sqrt{2})$	$(-1\ -1\ 2)$
$(-1/\sqrt{2})$	$(\frac{1}{2}\ \frac{1}{2}\ -1)$

The contribution of PCA2 to the genotype × environment interactions is PCA2(G) PCA2(E)'

$(-1/\sqrt{2})\ (-1/\sqrt{2}\ 1/\sqrt{2}\ 0)$ =	$(\frac{1}{2}\ -\frac{1}{2}\ 0)$
$(\ 0\ )$	$(0\ \ 0\ 0)$
$(\ 1/\sqrt{2})$	$(-\frac{1}{2}\ \frac{1}{2}\ 0)$

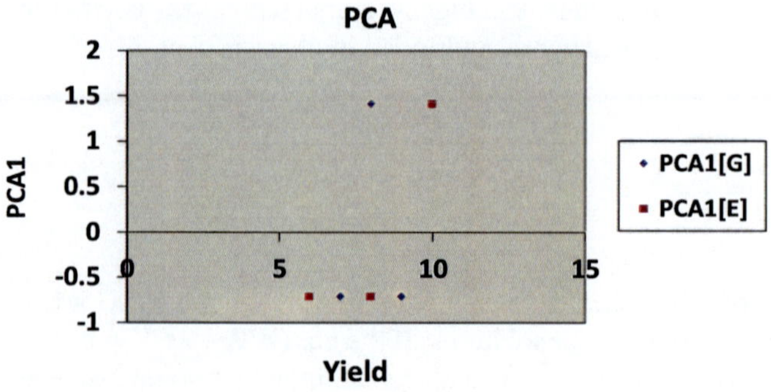

Fig. 3.11 PCA1 scores for genotypes [G, genotypes 1 to 3 are shown right to left] and environments [E, environments 1 to 3 are shown left to right] plotted against the additive means for genotypes ($Y_{i.}$) and environments ($Y_{.j}$): the largest interaction is between genotype 2 and environment 3 (1.414 × 1.414 = 2; PCA2 does not contribute to this interaction)

The sum of the contributions from PCA1 and PCA2 equals the G × E values shown in Table 3.6:

$$\left(\frac{1}{2}\ \frac{1}{2} -1\right) \quad \left(\frac{1}{2} -\frac{1}{2}\ 0\right) \quad (\ 1\ \ 0 -1)$$

$$(-1 -1\ \ 2) + (\ 0\ \ 0\ \ 0\ ) = (-1 -1\ \ 2)$$

$$\left(\frac{1}{2}\ \frac{1}{2} -1\right) \quad \left(-\frac{1}{2}\ \frac{1}{2}\ 0\right) \quad (\ 0\ \ 1 -1)$$

The first principal component accounts for 90% (first eigenvalue) of the G × E sum of squares (trace of matrix) and the second component for the remaining 10% (second eigenvalue). Furthermore, the first component provides a good prediction of the interactions, being out by only ½ in just four out of nine interactions. When the first principal component explains all or most of the variation, the PCA1 scores for genotypes and environments are commonly plotted (y-axis) against the additive means for genotypes ($Y_{i.}$) and environments ($Y_{.j}$) and the overall mean ($Y_{..}$) (x-axis). Thus, in Fig. 3.11, one can see that genotype 1 has the highest yield but a low PCA1 score. This means that it has relatively small interactions with the three environments and is therefore high yielding with general adaptability and a stable yield across environments (Table 3.6a). In contrast, genotype 2 has an intermediate yield but a high PCA1 score. This means that it has relatively large interactions with the three environments, including a large positive interaction with environment 3 which is the highest yielding environment. Genotype 2 is therefore responsive to environmental change and has specific adaptability to high-yielding environments (Table 3.6).

When the first two principal components explain all or most of the variation, the PCA2 scores are commonly plotted against the PCA1 scores. Furthermore, they may have a biological meaning, in which case the nature of the interaction can be visualized in biplots (Kempton 1984). The breeder can thus identify interesting interactions and hence genotypes and possible aspects of the environment (e.g. drier conditions or poorer soil) for further research. As more components are added, the predicted G × E becomes closer to the observed G × E, but equally it is no longer possible to give simple biological interpretations of the interactions. As the principal component analysis accounts for all of the G × E, it can reveal the complexity or simplicity of the interactions. If the interactions are spread evenly across all of the components then one has a complex situation, whereas if the first or first and second components account for most of the interactions, one has a simple situation.

An alternative analysis of a set of yield trials combines and analyses the total variation between genotype main effects and genotype × environment interactions (GGE model) but will not be considered here. The interested reader can consult the reviews in *Crop Science* mentioned earlier (Yan et al. 2007; Gauch et al. 2008).

3.18.2 Potato Example

Mulema et al. (2008) used an AMMI model to analyse yield data for nine clones from the International Potato Centre (CIP) and three control cultivars which had been evaluated for three cropping seasons (long rains of 2001 and 2002 and short rains of 2002) at four locations (1524 to 2450 masl), one in eastern and three in southwestern Uganda (very close to the equator). Seven of the clones were from CIP's population B (quantitative resistance to late blight) and had been selected in Uganda for good yields and acceptable agronomic characteristics, as well as for their late blight resistance. The three control cultivars were susceptible (Kabale), moderately resistant (Victoria) and resistant (Rutuku) to late blight. Each of the 12 trials had a randomized complete block design with 3 replicates and 60 planted tubers per plot. The trials were subjected to normal agronomic and cultural practices, except that no attempt was made to control late blight with fungicides. The AMMI model was the one considered above:

$$Y_{ij} = \mu + G_i + E_j + \Sigma\theta_k u_{ki} v_{kj} + \overline{e}_{ij}$$

The 144 means (12 genotypes × 12 environments) had a very wide range from 0.8 to 87.0 t/ha (converted from kg per plot). There were statistically significant differences between genotypes (range 22.6 to 43.1 t/ha) and between environments (3.7 to 52.3 t/ha), as well as statistically significant genotype × environment interactions. The components of variance for genotypes, environments and G × E were 37, 229 and 98, respectively. The first principal component for G × E (21 df) was statistically significant, whereas the second principal component was not significant when tested against the residual G × E. The first component accounted for 40% of the G × E sum of squares (121 df). The authors plotted the PCA1 scores for genotypes and for environments (both on y-axis) against the mean yields for genotypes and environments (both on x-axis) as described above. They went on to use the figure to discuss yield stability across environments. However, I am now going to use a simpler form of analysis to discuss yield stability.

3.19 Regression Analysis

Sometimes, all or most of the genotype × environment interactions can be accounted for by the heterogeneity of regressions (β_i) in a joint regression analysis of the interactions on to an environmental index. In the absence of a physical assessment of the environments, the mean value of all genotypes in a given environment (E_j) is used as a biological assessment of that environment (Finlay and Wilkinson 1963) and is expressed as a deviation from the overall mean (μ) where $\bar{e}_{ij}$ is the average plot error for r replicates.

$$Y_{ij} = \mu + G_i + E_j + (GE)_{ij} + \bar{e}_{ij}$$

$$= \mu + G_i + E_j + \beta_i E_j + d_{ij} + \bar{e}_{ij}$$

Where the heterogeneity of regressions accounts for all of the interactions (i.e. the *d*s are all zero), the analysis is equivalent to a principal component analysis in which the first component accounts for all of the interactions. Furthermore, the regression coefficients are properties of the genotypes per se and measure their responsiveness to environmental change, as we saw and discussed in the simple example shown in Fig. 3.7.

Where the heterogeneity of regressions accounts for most but not all of the interactions, the regression coefficients still have a predictive value, but they are no longer properties of the genotypes per se, nor are the deviations from regressions (the only property of the genotype per se is its variation over environments which will vary over genotypes). The joint regression analysis for our PCA example is shown in Table 3.7. The heterogeneity of regressions account for 70% of the sum of squares (less than PCA1) and the predictions are not quite so good as those for PCA1, being out by 1 in two interactions in environment 2 and also getting the rankings of genotypes 2 and 3 wrong in this environment (Fig. 3.12). We shall also see in the next section, when we consider response curves and surfaces of genotypes to changes in their physical environment (Knight 1970, 1973), that a regression analysis could be inappropriate and misleading even if it accounts for much of the variation due to genotype × environment interactions.

3.20 Response Curves

3.20.1 Introduction

Response curves have been worked out for the application of fertilizers to crops so that recommendations can be made to farmers about the amounts to apply for maximum yield and, perhaps more importantly, the economically optimum fertilization rates, particularly for nitrogen. Consideration will also be given to reducing the

Table 3.7 Joint regression analysis of yields of three genotypes grown in three environments (from Bradshaw 2016 with permission)

A Yields (Y_{ij}) of three genotypes grown in three environments where Y_{ij} is yield of *i*th genotype in *j*th environment

	Environment 1	Environment 2	Environment 3	Mean
Genotype 1	8	9	10	9
Genotype 2	5	7	12	8
Genotype 3	5	8	8	7
Mean	6	8	10	8

B G × E (Z_{ij}) for three genotypes in three environments ($Z_{ij} = Y_{ij} - Y_{i.} - Y_{.j} + Y_{..}$, where $Y_{i.}$ is the mean yield of genotype i, $Y_{.j}$ is the mean yield of environment j and $Y_{..}$ is the overall mean yield) together with E_j, the environmental index, β_i, where $(1 + \beta_i)$ is the slope of the regression line for genotype i, and Σd^2_i, the sum of squared deviations from regression for genotype i

	Environment 1	Environment 2	Environment 3	β_i	Σd^2_i
Genotype 1	1	0	−1	−0.50	0.0
Genotype 2	−1	−1	2	0.75	1.5
Genotype 3	0	1	−1	−0.25	1.5
E_j	−2	0	2		

C Analysis of variance

Source	df	SS	MS
Environment	2	24	12
Genotype	2	6	3
G × E	4	10	2½
Regression	2	7	3½
Residual	2	3	1½
Total	8	40	

environmental risk from leaching of excess nitrogen. Such agricultural experimentation features in Fisher's famous book *The Design of Experiment* (Fisher 1935), where, for example, he considers an experiment done at Rothamsted Experimental Station in England in 1931 to test the effect on potato yield of three levels of nitrogenous manure in combination with three similar levels of potassic manures. Details of earlier experiments done in 1925 and 1926 on cultivar Kerr's Pink and in 1927 on cultivar Arran Comrade can be found in the paper by Eden and Fisher (1929). They provide a fascinating insight into how Fisher thought about improving the design of agricultural experiments. Over 80 years on, experiments are still being done to determine appropriate levels of fertilizers for potato production; for example, to achieve high yield and acceptable tuber quality for the potato processing (French fry) industry in the Argentinian pampas, where fully irrigated crops are grown on sandy-loam soils and N, P and S levels are considered critical factors (Caldiz et al. 2018).

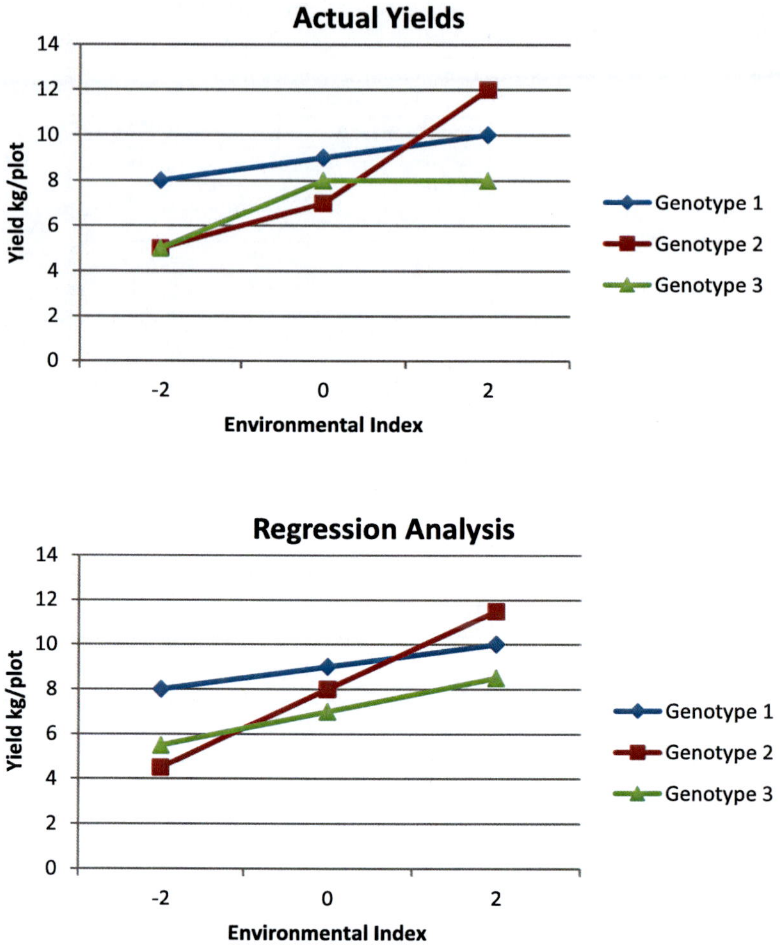

Fig. 3.12 Joint regression analysis of yields of three genotypes grown in three environments (from Bradshaw 2016 with permission)

3.20.2 *Essential Mineral Elements*

A useful recent review on improving nutrient management in potato cultivation is by White (2018). He reminds readers that plants can synthesize all of the organic compounds they require provided they have access to sufficient carbon dioxide, water and 14 essential mineral elements. These comprise six macroelements at tissue concentrations greater than 1 mg g^{-1} dry matter (DM) [nitrogen (N), phosphorus (P), potassium (K), calcium (Ca), magnesium (Mg) and sulphur (S)] and eight micronutrients in smaller amounts [chlorine (Cl), boron (B), iron (Fe), zinc (Zn), manganese (Mn), copper (Cu), nickel (Ni) and molybdenum (Mo)]. He also reminds readers that a deficiency of any essential mineral element will result in slower growth, impaired development and loss of yield, even if the supply of all other elements is sufficient, a clear message for ensuring that actual yields approach potential yields of new cultivars. White (2018) describes nutrient management practices for the potato crop, concluding with how modern agronomic strategies can be complemented by breeding cultivars with greater acquisition of mineral nutrients and better physiological utilization of nutrients for tuber yields. The former can be achieved by focusing on characteristics of the root system and the latter by focusing on canopy architecture and nutrient or biomass partitioning within the plant. Here, I am going to concentrate on nitrogen response curves and phosphorus-use efficiency.

3.20.3 Nitrogen Response Curves

A simple quadratic response curve is commonly used to explore nitrogen fertilizer rates:

$$y = a + bx - cx^2,$$

where y is the yield response, x the level of one environmental factor and the coefficients a, b and c are constants for any one genotype. Hence,

$$y_{ij} = a_i + b_i x_j - c_i x^2{}_j$$

where y_{ij} is the yield of genotype i in environment j.

Better fitting models can sometimes be found, as discussed by Fontes et al. (2012).

Let us first show that a Finlay and Wilkinson (1963) regression analysis is not particularly helpful. If M is the response curve for the mean of many genotypes (Fig. 3.13), then in a Finlay and Wilkinson (1963) regression analysis, it has a line of slope 1, with no deviations, and passes through the origin. We now consider the regression analysis for genotypes A and B, where M is the biological environmental index (i.e. the x-values). We ignore the curves above environmental level 10 on the grounds that all genotypes display a reduction in yield and hence this is the maximum level of interest. Genotype A has a higher maximum yield than M but at the same level of the physical index. Its regression line has a slope of 1.25 ($\beta_A = 0.25$) with no deviations. In contrast, genotype B has the same maximum yield as M but at a lower level of the physical index. Its regression line has a slope of 0.98 ($\beta_B = -0.02$), which is very close to 1, but there are also deviations from regression. The regression line accounts for 97% of the total sum of squares for genotype B over environments, and hence has good predictive ability. However, it would not make sense to say that B is less stable than A; it is simply different (Lin et al. 1986). In fact, the total sum of squares for A (185.3) is higher than that for B (117.6), reflecting its greater variability. The regression analysis would reveal that genotype

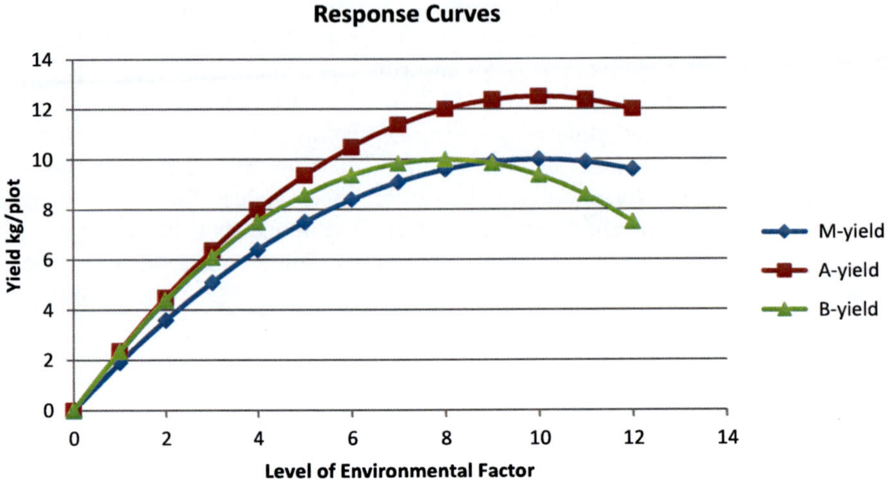

Fig. 3.13 Yield responses of genotypes A, B and M (mean of all genotypes) to an environmental factor such as nitrogen fertilizer (from Bradshaw 2016 with permission)

A has the highest yield at all fertilizer levels but fail to reveal that genotype B requires less fertilizer for maximum yield than the average over all genotypes. So, let us return to response curves.

A good example for potatoes is provided by the experiments of Fontes et al. (2010). They were interested in nitrogen fertilizer rates for potatoes being grown in South-eastern Brazil and conducted four experiments from 3 April to 21 July 2003 in fields close to the University of Federal Viçosa (UFV), in Viçosa, Minas Gerais State (20°45′S 42°51′W 693 masl). Each experiment explored the response to nitrogen fertilizer of a different cultivar (Ágata, Asterix, Atlantic and Monalisa). The researchers used a quadratic model to describe the yield response to increasing fertilizer rates (0, 50, 100, 200 and 300 kg N ha^{-1}). The estimated maximum total tuber fresh-weight yield of Ágata, Asterix, Atlantic and Monalisa occurred at 180, 201, 175 and 176 kg N ha^{-1}, respectively, and the maximum marketable yields at slightly different values of 168, 212, 175 and 193 kg N ha^{-1}, respectively. The predicted maximum marketable yields were 33.1, 32.3, 33.3 and 25.9 t ha^{-1}, respectively, where non-marketable yield was tubers of size smaller than or equal to 3.3 cm and those that were rotten, green, cracked or had secondary growth. When the calculations were done on a dry-weight basis, the maximum marketable yields were 6.03, 6.05, 8.68 and 4.97 t ha^{-1}. The economic optimum N fertilization rates (the point where the last increment of N returns a yield large enough to pay for the additional N) were slightly lower and ranged from 147 to 201 kg N ha^{-1} depending upon cultivar and relative prices of nitrogen fertilizer and potato tubers. For the median relative prices, the rates for the four cultivars were 151, 198, 156 and 172 kg N ha^{-1}, respectively. Hence, depending on the cultivar, under favourable price conditions (low N price and high tuber price), the economic optimum N fertilization rates for growers were 92–95% of the estimated N fertilization rate for obtaining the maximum potato yield, whereas under unfavourable conditions (high N price and low potato tuber price), they fell to 86–92% of the rates for maximum yield. Finally, it is interesting to look at (agronomic) nitrogen-use efficiency (total and marketable tuber yield divided by the N fertilization rate in kg tuber kg^{-1} N). Nitrogen-use efficiency (NUE) decreased (457–614 to 81–102 kg tuber kg^{-1} N depending on cultivar) as the N fertilization rate increased (50 to 300 kg N ha^{-1}). The NUE calculated for the maximum marketable tuber yield was 197, 152, 190 and 134 kg tuber kg^{-1} N for Ágata, Asterix, Atlantic and Monalisa, respectively. So, in conclusion, based on just marketable yield and no other factors, which is the best cultivar? Cultivars Agata and Atlantic have the highest marketable yields (33.1 and 33.3 t ha^{-1}) at the lowest fertilizer (168 and 175 kg N ha^{-1}) and economic fertilizer (151 and 156 kg N ha^{-1}) rates, and also the highest nitrogen-use efficiency (197 and 190 kg tuber kg^{-1} N). Experiments in other countries on nitrogen fertilizer rates for potatoes can easily be found in the scientific literature (for example, in Potato Research), but are not considered here as they do not allow me to make additional points about potato breeding.

If we now return to Fig. 3.13, we can argue that genotype A is superior to M because its maximum yield is higher than the maximum yield of M at the same level of the environmental factor. We can also argue that genotype B is superior to M because it has the same maximum yield but at a lower level of the environmental factor. However, A is always superior to B because it is higher yielding at all levels of the environmental factor. Hence, in practice, a breeder will need to seek the advice of agronomists on the appropriate level of nitrogen fertilizer and then select higher yielding genotypes (potential cultivars) in field trials with that amount of applied nitrogen fertilizer (I will not go into the issue of a split application). If agronomists are trying to encourage farmers to apply more fertilizer to achieve higher yields, the breeder may want to assess potential cultivars at two nitrogen levels, current farmer practice (low-input) and desirable farmer practice (high-input). New cultivars that yield better than other potential (rejected) cultivars at both levels of input would best serve both current and future practice and hence the farming community as a whole.

3.20.4 Nitrogen-Use Efficiency

In summarizing research results on breeding for better acquisition and utilization of nitrogen, White (2018) made the following points. Physiological N utilization efficiency (NUtE, DM production per unit N accumulated by a crop) often contributes more than N uptake efficiency (NUpE, amount of N accumulated by a crop per unit N supplied) to agronomic N-use efficiency (NUE, DM production per unit N supplied) when plants are grown with low N supply, although NUpE can contribute significantly to NUE when more N fertilizer is applied. Although genetic variation among potato cultivars exists for both components, variation in NUE is strongly associated with maturity class, with early maturing, determinate cultivars generally having lower NUE than later maturing, indeterminate cultivars. Within a maturity class, variation in NUpE is likely to be associated with root traits, whilst variation in NUtE might be attributed in part to optimizing leaf canopy architecture, maintaining N in the shoot, and maximizing DM harvest index. Care is therefore needed in setting breeding objectives which are best done in consultation with crop physiologists.

3.20.5 Phosphorus-Use Efficiency

The potato is regarded as a crop with an inability to acquire phosphorus (P) effectively from the soil. As a result, White et al. (2018) point out that the crop recovers less than 10% of the P applied as fertilizer in the year of application with a consequent high loss of P from fields and environmental pollution. For example, in Great Britain, in 2016, the potato occupied 3% of arable land but was responsible for greater than 12% of all inorganic P-fertilizer applied to tillage crops. Hence, there is a need for new cultivars that use P-fertilizer inputs more efficiently. Agronomic phosphorus-use efficiency (PUE) is commonly defined as crop dry matter (DW) yield per unit of P available in the soil (g DW g^{-1} P soil), which is numerically equal to the product of P acquisition (uptake) efficiency (PUpE) and crop physiological utilization efficiency (PUtE), where PUpE is the P acquired by the crop per unit of available P (g P crop g^{-1} P soil) and PUtE is the yield per unit P acquired by a crop (g DW g^{-1} P crop). White et al. (2018) concluded from a review of the literature that differences in yield responses to P-fertilizer applications between crop genotypes, including potato, are often correlated with PUpE, but rarely with PUeT. They tested the hypothesis that juvenile root vigour improves PUE of potato in a series of trials in Dundee, Scotland (56°28'N 03°07'W) over the period 2006–2011, using a wide range of *S. tuberosum* germplasm (in total 26 tetraploid Tuberosum, 36 diploid Phureja and 3 diploid hybrids). They demonstrated that increased juvenile root vigour accelerated P-acquisition and initial canopy cover and hence increased tuber yields. As juvenile root vigour is a heritable trait, it can be selected to improve P-fertilizer-use efficiency of potato, particularly for early cultivars and short season crops in soils with low P-availability. Another aspect of their results is of interest in the context of potato breeding. A subset of 32 genotypes (23 tetraploid Tuberosum, 7 diploid Phureja and 2 diploid hybrids) was assessed for two years in replicated field trials with (147 kg P ha^{-1}) and without P-fertilizer where soil P-levels before fertilization in the two years were 43 and 40 mg kg^{-1} soil, respectively. The P-fertilization increased average tuber fresh-weight yields from 7.28 to 8.04 kg $plot^{-1}$ for the seven Phurejas (10.4% increase) and from 12.36 to 13.96 kg $plot^{-1}$ for the 23 Tuberosums (12.9% increase). Yields of genotypes with and without fertilizer were highly correlated in both years ($r = 0.94$ and 0.84, respectively) and the four highest yielding cultivars without fertilizer (14.92 kg $plot^{-1}$) were also the highest yielding with fertilizer (17.41, an increase of 16.7%). Hence, by assessing potato genotypes in two contrasting environments, ones can be found that are high yielding in both environments (general adaptability).

As P deficiency is one of the most widespread nutrient constraints limiting yields in the tropics, my second example is from Brazil. Fernandes and Soratto (2016) evaluated the effect of P fertilization rates on P uptake, fertilizer P recovery (FPR), tuber yield and applied P-use efficiency (APUE) for five cultivars grown in tropical soils with low, medium and high P availability. The five cultivars were Agata, Asterix, Atlantic, Markies and Mondial, which together accounted for more than 80% of the potatoes grown in Brazil. The replicated field experiments were done during the hot, rainy summer season of 2011 (114 to 129 day growing season from April to September) in Sao Paulo State, south-eastern Brazil (23°S 49°W 649–744 masl) in soils with 14, 36 and 70 mg P dm^{-3} soil (strictly, resin-extractable P). Five P fertilization rates were used: 0, 55.5, 111, 222 and 444 kg P ha^{-1} (0, 125, 250, 500 and 1000 kg P_2O_5 ha^{-1}). In other respects, regional cultural practices were followed for crop management including N and K fertilization, sprinkler irrigation and chemical control of pests and diseases. P uptake was determined from the P concentrations of each plant organ (tubers, shoots and roots) and the accumulated dry-matter (yield). Fertilizer P recovery (%) was calculated as [(P uptake from soil treated with P fertilizer – P uptake from the untreated control)/(amount of P fertilizer applied)] × 100. Finally, applied phosphorus-use efficiency (APUE) was calculated as APUE = (tuber yield for P fertilization rate – tuber yield without P fertilization)/ (P_2O_5 fertilization rate). The reader is referred to the paper for details, but some of the key results were as follows. The highest dry matter accumulation (yield) in the tubers occurred in cultivar Mondial. Phosphate fertilizer increased the DM accumulation in the tubers up to applications of 500 kg P_2O_5 ha^{-1} in the soil with low initial P availability and up to 250 kg P_2O_5 ha^{-1} in the soils with medium and high initial P availability. Importantly, in the soil with low P availability, less DM accumulated regardless of P application rate, i.e. the plateau was much lower than for medium and high availability (<4 compared with >5 t ha^{-1}). The percentage fertilizer P recovery was highest for cultivar Mondial, with values falling from 9.3% at 125 kg P_2O_5 ha^{-1} to 2.3% at 1000 kg P_2O_5 ha^{-1}, thus confirming the reported low values for potatoes. Likewise, the applied phosphorus-use efficiency was highest for cultivar Mondial and fell from 100 kg tubers kg^{-1} P_2O_5 at 125 kg P_2O_5 ha^{-1} to 15 kg tubers kg^{-1} P_2O_5 at 1000 kg P_2O_5 ha^{-1}. Application of phosphate fertilizer increased the total fresh-weight yield of tubers up to application rates of 500, 250 and 125 kg P_2O_5 ha^{-1} in the soils with low, medium and high P availability, respectively. The soils with medium and high P availability did not differ in tuber yield, whereas yields were lower in soils with low P availability (20 compared with 35 t ha^{-1}). Mondial was the highest yielding cultivar. There were interactions between cultivars and soil P availability but not between cultivars and application rates. Fernandes and Soratto (2016) therefore concluded that it is not necessary to tailor phosphate fertilizer recommendations for individual cultivars because they respond similarly to the same rates of P application under field conditions. However, potato cultivars for soils with high P-fixing capacity (i.e. low plant availability) should have considerable tolerance to low P availability to improve applied phosphorus-use efficiency. In summary, with an application rate of 500 kg P_2O_5 ha^{-1}, the two contrasting environments in which to practise selection for increased yield were soil P availability of 14 and ≥ 36 mg dm^{-3} soil, respectively. In other words, the choice of site for the assessments was more important than the application rate of P fertilizer.

My third example is from Egypt, where the potato crop is grown in arid and semi-arid regions in soils with a sandy texture and low fertility and under irrigation because there is no rainfall during the growing season. Although the research involved only a single cultivar Diamant, it resulted in recommendations to growers about changes to their production system. I have therefore included it as an example of a different kind of production system for potatoes and hence a different kind of environment in which a breeder may need to practise selection for yield. Eissa (2019) studied the effect of P-fertigation frequency through a drip-irrigation system and the form of P-fertilizer on the uptake of P and yield of potatoes grown on calcareous sandy soils in two seasons, mid-October to beginning of February 2017 and 2018, respectively. There were three frequencies of P-fertigation (daily, every two days and every three days) and two forms of P-fertilizer (acidic urea phosphate (UP) and monopotassium phosphate (MPP)), with all treatments delivering a total of 150 kg P_2O_5 ha^{-1} and no P-fertigation as the control. Adequate N (345 kg N ha^{-1}) and K (230 kg K_2O ha^{-1}) were also supplied through the irrigation system and water was added every day according to the calculated evapotranspiration. The best results came from daily P-fertigation with urea phosphate from one week to two months after planting. This treatment reduced the soil pH from 8.10 to 7.43, increased P-availability from a very low 3.5 to around 22 mg kg^{-1} soil and increased the fresh-weight yield of tubers at harvest from 17 to around 30 t ha^{-1}, as well as increasing the P concentration in the petioles and both phosphorus and water-use efficiency.

3.21 G × E as a Correlated Response to Selection

A breeder may opt for indirect selection for yield by selection for a different root system to avoid drought, increased juvenile root vigour to improve phosphorus acquisition, in vitro tuber production at 25 °C to improve heat tolerance and use in vitro screens for salt tolerance. Using the concept of correlated response to selection a breeder can assess whether or not indirect selection is better than direct selection for yield. Furthermore, the concept can be extended to selection problems involving genotype × environment interactions by regarding a trait measured in two environments as two different but correlated traits. For example, to what extent will selection for higher yields in one production system, say non-organic, result in higher yields in a different production system, say organic; or selection in a temperate climate result in higher yields in a Mediterranean climate? The improvement of performance in one environment, as a result of selection in another, is seen as a correlated response which depends on the square roots of the heritabilities of performance in each environment and the genetic correlation between the two performances (Falconer and Mackay 1996).

The correlated response in environment y (CR_y) is the response in environment x ($i_x h_x \sigma_{Gx}$) multiplied by the regression of genotypic values in y on those in x, the latter being the genetic correlation (r_G) multiplied by σ_{Gy}/σ_{Gx}. In the current context, we do not need to worry about whether the heritabilities are narrow or broad sense, and whether or not the genotypic effects are in fact breeding values.

$$CR_y = i_x h_x \sigma_{Gx} \times r_G \sigma_{Gy} / \sigma_{Gx} = i_x h_x r_G \sigma_{Gy} = i_x h_x h_y r_G \sigma_{Py}$$

i is intensity of selection.
h is square root of heritability.
r_G is genetic correlation.
σ_{Gy} is square root of genotypic variance in y.
σ_{Py} is square root of phenotypic variance in y.
The response to direct selection in environment y (R_y) is:

$$R_y = i_y h_y \sigma_{Gy}$$

Hence, the correlated response can be compared with the response (R_y) expected from direct selection in environment y:

$$CR_y / R_y = r_G i_x h_x / i_y h_y$$

If the intensity of selection in a glasshouse screening test for some root trait, say, is much higher than for direct selection for yield in a field trial, and the heritabilities are similar, the indirect selection will be superior providing $r_G i_x > i_y$. On the other hand, if the intensities of selection are similar, indirect selection in environment x will be superior to direct selection in environment y providing $r_G h_x > h_y$.

Let us look at three examples, conventional compared with organic farming, temperate compared with Mediterranean climate and breeding for the Indian plains.

3.21.1 Conventional Versus Organic Farming

In 2018, Lady Balfour was still one of the most successful organically grown cultivars for consumption in the UK, having been released in 2001; but it was not bred for organic production. It came from a cross aimed at combining field resistance to late blight (*Phytophthora infestans*) with quantitative resistance to potato cyst nematodes (*Globodera rostochiensis* and *G. pallida*) (Bradshaw 2009b). Subsequently it was found to yield well under organic growing conditions. This raises the question of whether or not breeding cultivars for organic farming can be justified in terms of specific adaptability as well as from an economic perspective. In other words, is organic farming so different from conventional farming that cultivars need to be bred specifically for organic farming; for example, in order to cope with competition from weeds and limited nutrient availability from less readily available nutrient sources? Experimental data are needed to determine the correlation between selection under conventional and organic farming. If the heritabilities and intensities of selection are comparable under organic and conventional selection environments, then direct selection under organic farming is more efficient by an amount depending on how much r_G is less than one. If the breeder decides that a separate selection programme is justified for organic farming, there are two ways to proceed: direct selection under organic farming or indirect selection under conventional farming for adaptation to organic farming. The indirect selection could involve traits thought to be important for organic farming such as capacity for shading and hence suppressing weeds. If the same selection intensity is applied, the efficiency of indirect selection for a given trait depends on the heritabilities under the different farming systems and the genotypic correlation between the two farming systems. Early work that compared organic and conventional farming, such as that by Varis et al. (1996) and Maggio et al. (2008), involved too few cultivars (three and two, respectively) to provide answers to these particular questions. Hence, answers will be found in future experiments, not in this book, but I can point the reader in the direction of some interesting literature.

Tiemens-Hulscher et al. (2012) proposed an organic crop ideotype for potatoes in their review of the topic. They quote organic potato yields in Europe of 12 to 35 t/ha compared with 36 to 43 t/ha for conventional potato production and attribute the low and variable yields to the timing of late blight attacks and low nitrogen input (availability). They also point out that quality and appearance traits are the same for organic and conventional production but the requirements for disease and pest resistance differ. Cyst nematodes can be managed by a long crop rotation in organic farming, whereas pests and diseases with a wider host range are difficult to control, and diseases such as early blight (*Alternaria solani*) cause more yield loss under low-input and stress conditions. Other diseases and pests of particular concern to organic growers are *Rhizoctonia* stem canker and black scurf, silver scurf (*Helminthosporium solani*), *Verticilium* wilt and the dreaded Colorado potato beetle (*Leptinotarsa decemlineata*). Breeding methods for organic production are currently similar to those for conventional production, but more restricted as somatic hybridization and genetic modification are non-permissible techniques. A good account with lots of coloured photographs can be found in the book *Potato Breeding* edited by Tiemens-Hulscher et al. (2013). As potential cultivars need to be assessed under organic conditions, participation of organic potato growers as farmer breeders in collaboration with breeding companies has its appeal. If farmers select cultivars adapted to their own areas, regions, or even farms, rather than ones with more general adaptability, a consequence will be the diversity of cultivars desired by advocates of organic farming. Interestingly, when Khakbazan et al. (2015) looked at the economics of organically managed and conventional potato production systems in Atlantic Canada, they found that without an organic price premium, significant net revenue losses are expected from organic production because of lower yields and high costs. In other words, higher yielding cultivars are desirable for organic production systems. Finally, it is important to emphasize a key point made by Döring and Lynch (2018) in their review: organic potato cultivation must be viewed and judged as part of a whole organic farming system with a particular focus on rotation design.

3.21.2 Temperate Compared with Mediterranean Climate

Potato breeding at SPBS and SCRI (Scottish Plant Breeding Station and Scottish Crop Research Institute) was always done primarily in the context of providing new cultivars better adapted to seed production in Scotland, ware growing in Great Britain, and the requirements of end users. However, from 1980 overseas trialling identified cultivars suitable for seed export to the Mediterranean and North Africa (Bradshaw 2009b). Brown et al. (1996) analysed some of the trial results and found a greater correlation for total marketable yield between the Scottish ware site where clones were selected and sites in England (r = 0.43 to 0.70) than with sites in the Mediterranean (r = 0.00 to 0.67). Although the very best clones from the Scottish ware site performed reasonably well in the Mediterranean (general adaptability), the results supported the view that selection would be optimized by selecting in environments more similar to those in which the cultivars were to be grown. However, the problem faced by potato breeders is the logistics of seed-tuber multiplication. This means that potato breeders are likely to select their early generations at local seed and ware sites and then test relatively few potential cultivars in a wider range of environments. Given the global importance of potatoes, it would seem sensible for potato-growing countries to have their own breeding programmes targeted at local agroecological environments, notwithstanding an understandable desire for breeders and commercial companies to see their new cultivars grown as widely as possible.

3.21.3 Breeding for the Indian Plains

In India a major breeding objective is early-maturing, short-duration cultivars suited for cultivation in the plains. Kawar et al. (2018) described the successful breeding of such a cultivar, Kufri Khyati, which was released in 2008 from a cross made in 1992 at the Central Potato Research Station at Kufri (Shimla, 31.10°N 77.25°E 2630 masl). It provides an example of where a good choice of selection environment produced a cultivar adapted to a wider range of similar environments. Kufri Khyati was selected from seedling and clonal generations evaluated at the Central Potato Research Station at Jalandhar (31.33°N 75.58°E 228 masl) in the north western plains over the period 1993 to 1996. In four years of replicated yield trials at Jalandhar from 1997 to 2001, the total and marketable yields of Kufri Khyati were higher than those of the control cultivar Kufri Ashoka at 60 days after planting, with averages of 25.2 compared with 22.1 (14.0% higher) and 24.5 compared with 21.2 t ha^{-1} (15.6% higher), respectively. As a result of these promising yields, Kufri Khyati progressed to three years (2002 to 2005) of multilocational trials at 14 sites across the northern, central and eastern plains. Total yields of Kufri Khyati at 60 and 75 days after planting were on average higher than those of the best control (Kufri Ashoka or Kufri Jawahar), namely 19.99 compared with 18.57 t ha^{-1} (7.6% higher) and 27.74 compared with 24.59 t ha^{-1} (12.8% higher), respectively. Finally, two years (2005 to 2007) of on-farm trials were done at the same 14 sites across the northern, central and eastern plains. The marketable yield of Kufri Khyati at 75 days after planting was on average higher than the best control (Kufri Ashoka at six sites and Kufri Pukhraj at eight sites) at 23.9 compared with 20.3 t ha^{-1} (17.7% higher). Kufri Khyati was released in 2008 for general cultivation in the plains of India, having also established at Kufri that it has moderate resistance to late blight (unusual in an early cultivar) and at Jalandhar that it has better keeping quality at ambient temperature than Kufri Ashoka and Kufri Pukhraj.

3.22 Conclusions

Increasing yield per hectare is an obvious breeding objective for all uses of potatoes, whether as a staple food for local consumption, as a vegetable for cash or as a crop for processing into food or starch. The end use will determine the dry-matter concentration which is desirable in a new cultivar, but selection will be for high fresh-weight yield at this level of dry-matter rather than for high dry-weight yield. In this chapter, we have reviewed the need to define the target environment in which new cultivars will be grown and hence the importance of choosing the right environments in which to select for higher yield. As yield is a quantitative trait, we have reviewed the assessment of yield in appropriate trials and the analysis and implications of genotype × environment interactions. We have reached some important conclusions as follows.

1. *Target Environments*: Breeders needs to define the target environments of their programmes in terms of time of year and length of growing season(s) available for potato growth, day lengths and temperatures during growing season(s) and water availability (crop grown under irrigation, crop rain-fed with irrigation, crop rain-fed without irrigation).
2. *Potential Yields*: Breeders need to consult crop physiologists about potential yields in target environments and how close the best and average farmers get to these yields with current cultivars. Potential yields have been approached in practice with cultivars having a high harvest index, adaptation to tuber formation in the day lengths of the growing season, and tuber initiation and rate of bulking matched to the available growing season. Failure to achieve potential yields because currently available cultivars lack these traits leads to selection criteria for improving potential yields and the search for parental genotypes that possess these traits. Crop physiologists can help by defining ideotypes derived from eco-physiological models.

3. *Path Coefficient Analysis*: Path coefficient analysis is presented as a way of investigating the effects of morphological and physiological components of yield on yield itself. However, it is not obvious that measuring the components of yield and other traits affecting yield will involve less work than measuring yield itself, nor is it obvious that selection for these traits will give a greater increase in yield than simply selecting for yield.
4. *Yield Gap*: The gap between potential and achieved yields of potato crops has narrowed over the 50 years from 1966 to 2016 as globally yields increased by 50% to an average of 19.6 t/ha, but with large variations between countries from 0.8 (Central African Republic) to 49.0 t/ha (New Zealand). Nevertheless, in 2016, achieved yields still ranged over climatic regions from 24% to 72% of potential yields with a global average of around 50%.
5. *Increasing Yields in the Short Term*: Where yields are currently low, compared with global average, they can probably be doubled in the immediate future by planting seed tubers that are free of disease and in an appropriate physiological state (e.g. pre-sprouted) and by applying adequate levels of fertilizers, fungicides (and herbicides and pesticides) and irrigation (in low rainfall areas). Furthermore, post-harvest losses can be reduced by the use of appropriate storage facilities. In the longer term, new cultivars should help to prevent yield losses, particularly through inbuilt resistance to pests and diseases and tolerance of abiotic stresses.
6. *Drought, Heat and Salt Stress*: In different agroecological zones, and areas within zones, potato production may encounter various combinations of drought, heat and salt stress. If a particular combination of stresses always occurs, then the breeder will need to select for higher yield in the presence of this combination. However, in other more variable situations, the breeder may be best advised to select for high yield in both the presence and absence of the combination of stresses (the breeder is unlikely to have enough tubers, or resources, to select for yield in the presence and absence of each individual stress).
7. *Increased Crop Inputs*: Response curves for crop management inputs are used by agronomists to make recommendations to farmers about the amounts of fertilizer (and irrigation) to apply to their crops for maximum yield, or perhaps more importantly, recommendations about economically optimum fertilization rates, possibly also including reducing the environmental impact from leaching. If all farmers in a target region for potato breeding follow these recommendations, the breeder will also need to follow them in their yield trials. However, these recommendations may be aspirations for the future with many farmers currently unable or unwilling to follow them. In these circumstances, the breeder may be best advised to select for high yields in both high and low input trials. There is, however, one complication to bear in mind. With a nutrient like nitrogen, selection

under optimum and below optimum rates of nitrogen fertilizer is appropriate. In contrast, with phosphorus, trials sites are required with high and low soil availability of the nutrient rather than simply high and low rates of fertilizer application.

8. *Field Trials for Yield Assessment*: In a potato breeding programme, field trials for yield assessment will have a large number of entries and small sizes of plots. The problem of within-trial heterogeneity of environment can be reduced by the use of small blocks and spatial analysis (nearest neighbour). Interference between small plots with different genotypes may be a problem which can be reduced by modelling and grouping of phenotypically similar trial entries. As interference is unlikely to be completely eliminated, multiple-row plots are desirable for the final stages of yield assessment of new cultivars.

9. *Data Analysis*: In this age of modern computers, analysis of data from yield trials is likely to use the general mixed model and restricted maximum likelihood estimation (REML). The more familiar (to potato breeders) general linear model, analysis of variance and multiple linear regression are special cases of the general mixed model.

10. *Genotype × Environment Interactions*: An assessment of genotype × environment interactions is required at the start of a breeding programme in order to determine objectives and to choose parents for hybridization, and at the end of the programme in order to make recommendations to farmers and growers about which cultivars to grow. Equally important for the success of a breeding programme is the choice of environments at every intermediate stage where clones are assessed and selected. With limited resources, particularly number of tubers, the breeder will be restricted to selection in relatively few contrasting environments which will need to be carefully chosen in terms of agronomy inputs and abiotic stresses.

11. *Analysis of Genotype × Environment Interactions*: Principal component analysis is one of the most informative analyses of genotype × environment interactions as it accounts for all of the interactions. When the interactions are spread evenly across all of the components, one has a complex situation, whereas if the first or first and second components account for most of the interactions, one has a simple situation with the possibility of a biological interpretation of the interactions. A principal component analysis in which the first component accounts for all of the interactions is equivalent to a joint regression analysis in which the heterogeneity of regressions accounts for all of the interactions.

12. *Correlated Response to Selection*: It is sometimes useful to consider genotype × environment interactions in the context of a correlated response to selection. A breeder might opt for indirect selection for yield by selection for a different root system to avoid drought, increased juvenile root vigour to improve phosphorus acquisition, in vitro tuber production at 25 °C to improve heat tolerance and use in vitro screens for salt tolerance. Using the concept of correlated

response to selection, a breeder can assess whether or not indirect selection is better than direct selection for yield. Furthermore, a breeder might consider it useful to ask to what extent selection for higher yields in one production system, say non-organic, results in higher yields in a different production system, say organic; or selection in a temperate climate result in higher yields in a Mediterranean climate? The answer should help the breeder decide if more than one breeding programme can be justified.

13. *Response to Selection*: Once a breeder has defined all aspects of the target environment for their programme, chosen appropriate parents and made appropriate crosses, selection for higher yields and other desirable traits can begin. A conventional, now traditional, combining ability analysis allows the breeder to predict the response to selection over generations (uses additive genetic variation) and within generations (uses all genetic variation, both additive and non-additive). Improvement in yield over generations is in fact the driver for producing higher yielding clones in each generation. Relatively few major genes and quantitative trait loci (QTLs) alleles of large effect have been identified and used by breeders in selecting for higher yields.
14. *Combining Higher Yields with Improvement in Quality Traits*: Despite good prospects for improving yield, breeders have often found it difficult to combine higher yields with the quality traits required in a successful new cultivar.

In the next chapter we will consider the target end user of new cultivars and add quality criteria to that of yield in the target environment.

References

Aliche EB, Oortwijn M, Theeuwen TPJM, Bachem CWB, van Eck HJ, Visser RGF, van der Linden CG (2019) Genetic mapping of tuber size distribution and marketable tuber yield under drought stress in potatoes. Euphytica 215:186. https://doi.org/10.1007/s10681-019-2508-0

Asghari-Zakaria R, Fathi M, Hasan-Panah D (2006) Sequential path analysis of yield components in potato. Potato Res 49(4):273–279. https://doi.org/10.1007/s11540-007-9022-9

Azzalini A, Cox DR (1984) Two new tests associated with analysis of variance. J R Statist Soc B 46:335–343

Baker RJ (1988) Tests for crossover genotype-environmental interactions. Can J Plant Sci 68:405–410

Bararyenya A, Nahayo P-C, Nduwimana A, Niyonzima P, Nyawakira D, Sindihebura J-P, Vyizigiro E, Ndayishimiye V, Haverkort AJ (2018) Assessment of opportunities for Burundian small-scale potato farmers to increase productivity and income. Potato Res 61(1):73–88

Bell A (2009) Peak water. Luath Press Limited, Edinburgh, 207 p

Benites FRG, Pinto CABP (2011) Genetic gains for heat tolerance in potato in three cycles of recurrent selection. Crop Breed & Appl Biotechnol 11:133–140

Beukema HP, van der Waag DE (1990) Introduction to potato production. Pudoc, Wageningen, pp 23–24

Bonierbale MW, Amoros WR, Salas E, de Jong W (2020) Potato breeding. In: Campos H, Ortiz O (eds) The potato crop: its agricultural, nutritional and social contribution to humankind. Springer, ebook, Cham, pp 163–218

Bradshaw JE (1994a) Quantitative genetics theory for tetrasomic inheritance. In: Bradshaw JE, Mackay GR (eds) Potato genetics. CAB International, Wallingford, pp 71–99

Bradshaw JE (1994b) Assessment of five cultivars of potato (*Solanum tuberosum* L.) in a competition diallel. Ann Appl Biol 125:533–540

Bradshaw JE (2009a) A genetic perspective on yield plateau in potato. Potato J 36:79–94

Bradshaw JE (2009b) Potato breeding at the Scottish Plant Breeding Station and the Scottish Crop Research Institute: 1920–2008. Potato Res 52:141–172

Bradshaw JE (2016) Plant breeding: past, present and future. Springer, Cham

Bradshaw JE, Bonierbale M (2010) Potatoes. In: Bradshaw JE (ed) Root and tuber crops, handbook of plant breeding 7. Springer, New York, pp 1–52

Bradshaw JE, Dale MFB, Swan GEL, Todd D, Wilson RN (1998) Early-generation selection between and within pair crosses in a potato (*Solanum tuberosum* subsp. *tuberosum*) breeding programme. Theor Appl Genet 97:1331–1339

Bradshaw JE, Dale MFB, Mackay GR (2009) Improving the yield, processing quality and disease and pest resistance of potatoes by genotypic recurrent selection. Euphytica 170:215–227

Brar AS, Buttar GS, Thind HS, Singh KB (2019) Improvement of water productivity, economics and energetics of potato through straw mulching and irrigation scheduling in Indian Punjab. Potato Res 62(4):465–484

Brown J, Caligari PDS (1989) Cross prediction in a potato breeding programme by evaluation of parental material. Theor Appl Genet 77:246–252

Brown J, Dale MFB, Mackay GR (1996) General adaptability of potato genotypes selected in the UK for the Mediterranean region. J Agric Sci Camb 126:441–448

Cabello R, De Mendiburu F, Bonierbale M, Monneveux P, Roca W, Chujoy E (2012) Large-scale evaluation of potato improved varieties, genetic stocks and landraces for drought tolerance. Amer J Potato Res 89:400–410

Caldiz DO, Viani PG, Giletto CM, Zamuner EC, Echeverría HE (2018) Improving yield and quality of processing potato crops grown in the Argentinian Pampas: the role of N, P and S and their impact on CO_2 emissions. Potato Res 61(2):147–168

Connolly T, Currie ID, Bradshaw JE, McNicol JW (1993) Interplot competition in yield trials of potatoes (*Solanum tuberosum* L.) with single-drill plots. Ann Appl Biol 123:367–377

Deguchi T, Iwama K, Haverkort AJ (2016) Actual and potential yield levels of potato in different production systems of Japan. Potato Res 59:207–225

Dewey DR, Lu KH (1959) A correlation and path-coefficient analysis of components of crested wheatgrass seed production. Agron J 51:515–518

Döring TF, Lynch DH (2018) Organic potato cultivation. In: Wale S (ed) Achieving sustainable cultivation of potatoes volume 2: production, storage and crop protection. Burleigh Dodds Science Publishing, Cambridge, pp 89–118

Douches DS, Maas D, Jastrzebski K, Chase RW (1996) Assessment of potato breeding progress in the USA over the last century. Crop Sci 36:1544–1552

Dua VK, Rawal S, Singh S, Sharma J (2018) Improving potato cultivation practices: an overview. In: Wale S (ed) Achieving sustainable cultivation of potatoes volume 2: production, storage and crop protection. Burleigh Dodds Science Publishing, Cambridge, pp 23–43

Eden T, Fisher RA (1929) Studies in crop variation: VI. Experiments on the response of the potato to potash and nitrogen. J Agric Sci 19:201–213

Eissa MA (2019) Efficiency of P-fertigation for drip irrigated potato grown on calcareous sandy soils. Potato Res 62(1):97–108

Falconer DS, Mackay TFC (1996) Introduction to quantitative genetics, 4th edn. Longman, Harlow, 464p

Fernandes AM, Soratto RP (2016) Response of potato cultivars to phosphate fertilization in tropical soils with different phosphorus availabilities. Potato Res 59:259–227

Fernie AR, Bachem CWB, Helariutta Y, Neuhaus HE, Prat S, Ruan Y-L, Stitt M, Sweetlove LJ, Tegeder M, Wahl V, Sonnewald S, Sonnewald U (2020) Synchronization of developmental, molecular and metabolic aspects of source–sink interactions. Nat Plants 6:55–66. https://doi.org/10.1038/s41477-020-0590-x

Finlay KW, Wilkinson GN (1963) The analysis of adaptation in a plant-breeding programme. Aust J Agric Res 14:742–754

Fisher RA (1935) The design of experiment. Oliver and Boyd, Edinburgh, 252 p

Fontes PCR, Braun H, Busato C (2010) Economic optimum nitrogen fertilization rates and nitrogen fertilization rate effects on tuber characteristics of potato cultivars. Potato Res 53(2):1–13

Fontes PCR, de Castro Silva MC, Miranda GV (2012) Statistical models in plant diagnosis and calculating recommended nitrogen rates. In: He Z, Larkin RP, Honeycutt W (eds) Sustainable potato production: global case studies. Springer, New York, pp 391–402

Frantz JM, Larkin RP, Trusty G, Honeycutt CW, He Z, Olanya OM, Halloran JM (2012) Comparing modelled productivity to historical data in New England potato production systems. In: He Z, Larkin RP, Honeycutt W (eds) Sustainable potato production: global case studies. Springer, New York, pp 81–95

Gauch HG Jr (1988) Model selection and validation for yield trials with interaction. Biometrics 44:705–715

Gauch HG Jr, Piepho HP, Annicchiarico P (2008) Statistical analysis of yield trials by AMMI and GGE: further considerations. Crop Sci 48:866–889

Gleeson AC (1997) Spatial analysis. In: Kempton RA, Fox PN (eds) Statistical methods for plant variety evaluation. Chapman & Hall, London, pp 68–85

Glendinning DR (1975) Neo-Tuberosum: new potato breeding material. 2. A comparison of Neo-Tuberosum with unselected Andigena and with Tuberosum. Potato Res 18:343–350

Govindakrishnan PM, Haverkort AJ (2006) In: Gopal J, Khurana SMP (eds) Handbook of potato production, improvement and postharvest management. The Haworth Press, Binghampton, pp 179–229

Griffing B (1956) Concept of general and specific combining ability in relation to diallel crossing systems. Aust J Biol Sci 9:463–493

Hardigan MA, Laimbeer FPE, Newton L, Crisovan E, Hamilton JP, Vaillancourt B, Wiegert-Rininger K, Wood JC, Douches DS, Farré EM, Veilleux RE, Buell CR (2017) Genome diversity of tuber-bearing *Solanum* uncovers complex evolutionary history and targets of domestication in the cultivated potato. PNAS 114(46):E9999–E10008. https://doi.org/10.1073/pnas.1714380114

Haverkort AJ, Grashoff C (2004) Ideotyping-potato a modelling approach to genotype performance. In: MacKerron DKL, Haverkort AJ (eds) Decision support systems in potato production. Wagengingen Academic Publishers, Wageningen, pp 199–211

Haverkort AJ, Struik PC (2015) Yield levels of potato crops: recent achievements and future prospects. Field Crop Res 182:76–85

Haverkort AJ, Franke AC, Steyn JM, Pronk AA, Kooman PL (2015) A robust potato model: LINTUL-POTATO-DSS. Potato Res 58:313–327

Hijmans RJ (2001) Global distribution of the potato crop. Am J Potato Res 78:403–412

Hijmans RJ (2003) The effect of climate change on global potato production. Amer J Potato Res 80:271–280

Iwama K (2008) Physiology of the potato: new insights into root system and repercussions for crop management. Potato Res 51:333–353

Jaarsma R, de Vries RSM, de Boer AH (2013) Effect of salt stress on growth, Na^+ accumulation and proline metabolism in potato (*Solanum tuberosum*) cultivars. PLoS One 8(3):e60183. https://doi.org/10.1371/journal.pone.0060183

Jamaati-e-Somarin S, Zabihi-e-Mahmoodabad R, Yari A (2010) Yield and yield components of potato (*Solanum tuberosum* L.) tuber as affected by nitrogen fertilizer and plant density. AJBAS 4(8):3128–3131

Jinks JL, Connolly V (1973) Selection for specific and general response to environmental differences. Heredity 30:33–40

Jinks JL, Connolly V (1975) Determination of the environmental sensitivity of selection lines by the selection environment. Heredity 34:401–406

Jinks JL, Pooni HS (1982) Determination of the environmental sensitivity of selection lines of *Nicotiana rustica* by the selection environment. Heredity 49:291–294

Kawar PG, Kardile HB, Raja S, Dutt S, Kumar R, Manivel P, Bhardwaj V, Singh BP, Govindakrishnan PM, Chakrabarti SK (2018) Developing early-maturing and stress-resistant potato varieties. In: Wang-Pruski G (ed) Achieving sustainable cultivation of potatoes. Burleigh Dodds, Cambridge, pp 143–167

Kempton RA (1984) The use of biplots in interpreting variety by environment interactions. J Agric Sci (Camb) 103:123–135

Kempton RA (1985) Statistical models for inter-plot competition. Asp Appl Biol 10:111–120

Kempton RA (1997) Interference between plots. In: Kempton RA, Fox PN (eds) Statistical methods for plant variety evaluation. Chapman & Hall, London, pp 101–116

Khakbazan M, Henry R, Haung J, Mohr R, Peters R, Fillmore S, Rodd V, Mills A (2015) Economics of organically managed and conventional potato production systems in Atlantic Canada. Can J Plant Sci 95(1):161–174. https://doi.org/10.4141/cjps-2014-050

Khan MS, Yin X, van der Putten PEL, Struik PD (2014) An ecophysiological model analysis of yield differences within a set of contrasting cultivars and an F1 segregating population of potato (*Solanum tuberosum* L.) grown under diverse environments. Ecol Model 290:146–154

Khan MA, Munive S, Bonierbale M (2015) Early generation in vitro assay to identify potato populations and clones tolerant to heat. Plant Cell Tissue Organ Cult 121:45–52. https://doi.org/10.1007/s11240-014-0677-z

Killick RJ (1977) Genetic analysis of several traits in potatoes by means of a diallel cross. Ann Appl Biol 86:279–289

Kleinwechter U, Gastelo M, Ritchie J, Nelson G, Asseng S (2016) Simulating cultivar variations in potato yields for contrasting environments. Agr Syst 145:51–63

Kloosterman B, Abelenda JA, Carretero Gomez MM, Oortwijn M, de Boer JM, Kowitwanich K, Horvath BM, van Eck HJ, Smaczniak C, Prat S, Visser RGF, Bachem CWB (2013) Naturally occurring allele diversity allows potato cultivation in northern latitudes. Nature 495(7440):246–250

Knight R (1970) The measurement and interpretation of genotype-environment interactions. Euphytica 19:225–235

Knight R (1973) The relation between hybrid vigour and genotype-environment interactions. Theor Appl Genet 43:311–318

Kunkel R, Campbell GS (1987) Maximum potential potato yield in the Columbia Basin, USA: model and measured values. Am Potato J 64:355–366. https://doi.org/10.1007/BF02853597

Lehretz GG, Sonnewald S, Hornyik C, Corral JM, Sonnewald U (2019) Post-transcriptional regulation of *FLOWERING LOCUS T* modulates heat-dependent source-sink development of potato. Curr Biol 29(10):1614–1624. https://doi.org/10.1016/j.cub.2019.04.027

Levy D, Coleman WK, Veilleux RE (2013) Adaptation of potato to water shortage: irrigation management and enhancement of tolerance to drought and salinity. Am J Potato Res 90:186–206. https://doi.org/10.1007/s12230-012-9291-y

Li CC (1975) Path analysis – a primer. The Boxwood Press, Pacific Grove

Lin CS, Binns MR, Lefkovitch LP (1986) Stability analysis: where do we stand? Crop Sci 26:894–900

Love SL (1999) Founding clones, major contributing ancestors, and exotic progenitors of prominent North American cultivars. Am J Potato Res 76:263–272

Love SL, Pavek JJ, Thompson-Johns A, Bohl W (1998) Breeding progress for potato chip quality in North American cultivars. Am J Potato Res 75:27–36

Lucier G, Budge A, Plummer C, Spurgeon C (1990) U.S. potato statistics, 1949-89. USDA, Economic Research Service 829:1-8. U.S. Gov Print Office, Washington, DC

Luthra SK, Gupta VK, Tiwari JK, Kumar V, Bhardwaj V, Sood S, Dalamu, Kaur RP, Kumar R, Vanishree G, Kumar D, Mhatre P, Chakrabarti SK (2020) Potato breeding in India. CPRI Technical Bulletin No 74 (revised), ICAR-Central Potato Research Institute, Shimla, Himachal Pradesh, India

MacKerron DKL (1985) A simple model of potato growth and yield: part II. Validation and external sensitivity. Agric Forest Meteorol 34:285–300

Maggio A, Carillo P, Bulmetti GS, Fuggi A, Barbieri G, De Pascale S (2008) Potato yield and metabolic profiling under conventional and organic farming. Eur J Agron 28(3):343–350. https://doi.org/10.1016/j.eja.2007.10.003

Mazurczyk W, Wierzbicka A, Trawczyński C (2009) Harvest index of potato crop grown under different nitrogen and water supply. Acta Sci Pol Agricultura 8(4):15–21

Mead R (1997) Design of plant breeding trials. In: Kempton RA, Fox PN (eds) Statistical methods for plant variety evaluation. Chapman & Hall, London, pp 40–67

Mead R, Curnow RN, Hasted AM (2002) Statistical methods in agriculture and experimental biology, 3rd edn. Chapman and Hall/CRC, Boca Raton, 488p

Monneveux P, Ramirez DA, Pino M-T (2013) Drought tolerance in potato (*S. tuberosum* L.): can we learn from drought tolerance research in cereals? Plant Sci 205-206:76–86

Mulema JMK, Adipala E, Olanya OM, Wagoires W (2008) Yield stability analysis of late blight resistant potato selections. Exp Agric 44:145–155

Munoz FJ, Plaisted RL (1981) Yield and combining abilities in Andigena potatoes after six cycles of recurrent phenotypic selection for adaptation to long day conditions. Am Potato J 58:469–479

Neele AEF (1990) Study on the inheritance of potato tuber yield by means of harvest index components and its consequences for choice of parental material. Euphytica 48:159–166

Neele AEF, Nab HJ, Louwes KM (1991) Identification of superior parents in a potato breeding programme. Theor Appl Genet 82:264–272

Ortiz R (2001) The state of the use of potato genetic diversity. In: Cooper HD, Spillane C, Hodgkin T (eds) Broadening the genetic base of crop production. CAB International, Wallingford, pp 181–200

Patterson HD, Robinson DL (1989) Row-and-column designs with two replicates. J Agric Sci (Camb) 112:73–77

Patterson HD, Williams ER, Hunter EA (1978) Block designs for variety trials. J Agric Sci (Camb) 90:395–400

Patterson LJ, Wild P, Williams ER (1988) An algorithm to generate designs for variety trials. J Agric Sci (Camb) 111:133–136

Plaisted RL, Hoopes RW (1989) The past record and future prospects for the use of exotic potato germplasm. Am Potato J 66:603–627

Prashar A, Colwell Fde J, Hornyik C, Bryan GJ (2018) Advances in development of potato varieties resistant to abiotic stress. In: Wang-Pruski G (ed) Achieving sustainable cultivation of potatoes. Burleigh Dodds, Cambridge, pp 125–142

Qadir M, Quillérou E, Nangia V, Murtaza G, Singh M, Thomas RJ, Drechsel P, Noble AD (2014) Economics of salt-induced land degradation and restoration. Nat Resour Forum. https://doi.org/10.1111/1477-8947.12054

Qin J, Antonio DCR, Xie K, Li W, Yactayo W, Jin L, Quiroz R (2018) Is the partial root-zone drying more appropriate than drip irrigation to save water in China?: a preliminary comparative analyses for potato cultivation. Potato Res 61(4):391–406

Roy SJ, Negrão S, Tester M (2014) Salt resistant crop plants. Curr Opin Biotechnol 26:115–124

Sattar MA, Sultana N, Hossain MM, Rashid MH, Islam AA (2007) Genetic variability, correlation and path analysis in potato (*Solanum tuberosum* L.). Ban J Pl Brd and Gen 20(1):33–38

Shaterian J, Waterer DR, de Jong H, Tanino KK (2008) Methodologies and traits for evaluating the salt tolerance in diploid potato clones. Am J Potato Res 85:93–100. https://doi.org/10.1007/s12230-008-9003-9

Shehata SA, Abouziena HF, Abdelgawad KF, Elkhawaga FA (2019) Weed control efficacy, growth and yield of potato (*Solanum tuberosum* L.) as affected by alternative weed control methods. Potato Res 62(2):139–155

Shubha K, Singh D (2018) Selection of yield-associated morphological and biochemical traits using correlation and path coefficient analysis in potato (*Solanum tuberosum* L.) in the foothills of North-Western Himalayas. Pot Res 61(3):273–281. https://doi.org/10.1007/s11540-018-9376-1

Simmonds NW (1980) Comparisons of the yields of four potato varieties in trials and in agriculture. Exp Agric 16:393–398

Simmonds NW (1981) Genotype (*G*), Environmemt (*E*) and *GE* components of crop yields. Exp Agric 17:355–362

Spitters CJT, Neele AEF, Schapendonk AHCM (1989) Crop physiological components of tuber yield and their use in potato breeding. In: Louwes KM, Toussaint HAJM, Dellaert LMW (eds) Parental line breeding and selection in potato breeding. Proceedings of the joint conference of the EAPR breeding section and the EUCARPIA potato section. Pudoc, Wageningen, The Netherlands, pp 51–60

Spooner DM, McLean K, Ramsay G, Waugh R, Bryan GJ (2005) A single domestication for potato based on multilocus amplified fragment length polymorphism genotyping. Proc Natl Acad Sci 102:14694–14699

Tai GCC (1976) Estimation of general and specific combining abilities in potato. Can J Genet Cytol 18:463–470

Talbot M (1984) Yield variability of crop varieties in the U.K. J Agric Sci (Camb) 102:315–321

Talbot M (1997) Resource allocation for selection systems. In: Kempton RA, Fox PN (eds) Statistical methods for plant variety evaluation. Chapman and Hall, London, pp 162–174

Tiemens-Hulscher M, van Bueren ETL, Hutten CB (2012) Potato: perspectives to breed for an organic crop ideotype. In: van Bueren ETL, Myers JR (eds) Organic crop breeding. Wiley-Blackwell, Chichester, pp 227–237

Tiemens-Hulscher M, Delleman J, Eising J, van Bueren ETL (2013) Potato breeding. Aardappelwereld BV, The Hague, 171 p

Trapero-Mozos A, Morris WL, Ducreux LJM, McLean K, Stephens J, Torrance L, Bryan GJ, Hancock RD, Taylor MA (2018) Engineering heat tolerance in potato by temperature-dependent expression of a specific allele of *HEAT-SHOCK COGNATE 70*. Plant Biotechnol J 16:197–207

Varis E, Pietilä L, Koikkalainen K (1996) Comparison of conventional, integrated and organic potato production in field experiments in Finland. Acta Agric Scandinav Sec B-Soil & Plant Sci 46(1):41–48

Victorio RG, Moreno U, Black CC Jr (1986) Growth, partitioning and harvest index of tuber bearing *Solanum* genotypes grown in two contrasting Peruvian environments. Plant Physiol 82:103–108

Wang'ombe JG, van Dijk MP (2013) Low potato yields in Kenya: do conventional input innovations account for the yields disparity? Agric Food Secur 2:14. https://doi.org/10.1186/2048-7010-2-14

White PJ (2018) Improving nutrient management in potato cultivation. In: Wale S (ed) Achieving sustainable cultivation of potatoes volume 2: production, storage and crop protection. Burleigh Dodds Science Publishing, Cambridge, pp 45–67

White PJ, Bradshaw JE, Brown LK, Dale FMB, Dupuy LX, George TS, Hammond JP, Subramanian NK, Thompson JA, Wishart J, Wright G (2018) Juvenile root vigour improves phosphorus use efficiency of potato. Plant and Soil 432:45–63. https://doi.org/10.1007/s11104-018-3776-5

Wishart J, George TS, Brown LK, Ramsay G, Bradshaw JE, White PJ, Gregory PJ (2013) Measuring variation in potato roots in both field and glasshouse: the search for useful yield predictors and a simple screen for root traits. Plant and Soil 368:231–249

Wricke G, Weber WE (1986) Quantitative genetics and selection in plant breeding. Walter de Gruyter, Berlin/New York, 406p

Wright S (1921) Correlation and causation. J. Agric Res 20:557–587

Wright S (1923) Theory of path coefficients. Genetics 8:239–255

Yagiz AK, Cakici M, Aydogan N, Omezli S, Yerlikaya BA, Ayten S, Maqbool A, Haverkort AJ (2020) Exploration of climate change effects on shifting potato seasons, yields and water use employing NASA and National long-term weather data. Potato Res 63:565–577

Yan W, Kang MS, Ma B, Woods S, Cornelius PL (2007) GGE biplot vs. AMMI analysis of genotype-by-environment data. Crop Sci 47:643–653

Yates F (1936) A new method of arranging variety trials involving a large number of varieties. J Agric Sci (Camb) 26:424–455

Yates F (1937) A further note on the arrangement of variety trials: quasi-Latin squares. Ann Eugenics 7:121–140

Yuan JZ, Murphy AM, De Koeyer D, Lague M, Bizimungu B (2016) Effectiveness of the field selection parameters on potato yield in Atlantic Canada. Can J Plant Sci 96:701–710

Zhang X, Campbell R, Ducreux LJM, Morris J, Hedley PE, Mellado-Ortaga E, Roberts A, Stephens J, Bryan G, Torrance L, Chapman S, Prat S, Taylor MA (2020) TERMINAL FLOWER-1/ CENTRORADIALIS inhibits tuberization via protein interaction with the tuberigen activation complex. Plant J. https://doi.org/10.1111/tpj.14898

Chapter 4
Improving Potato Quality: A Problem of Definition and Measurement

4.1 Introduction

The potato is grown as a staple food for local consumption, as a vegetable for cash, or as a crop for processing into food or starch. The intended end uses of new cultivars determine the quality criteria that need to be added as breeding objectives to that of yield in the production environments. Nutritional value will be important for the potato as a staple food; cooking quality, texture, taste and flavour will be important for the potato as a vegetable sold for cash and processing quality will be defined by the manufacturers of French fries, chips (crisps) and other processed products. Finally, the composition of the potato starch will determine its value for different uses by the starch industry. A separate breeding programme will probably be needed for each combination of end use and targeted production environment. In the last chapter, we considered the importance of choosing the right environments in which to select for higher yield. We now need to take up the challenges of defining breeding objectives for improving potato quality that can be translated into selection criteria for use on the tubers from assessment plots and yield trials, both immediately after harvest and after storage. We also need to give some consideration to the germplasm available for improving quality and what is known about the genetics of quality traits.

Let us consider quality traits in the broadest sense in the order and the way in which they can be assessed after harvest. The interested reader can find more information in the reviews by Storey and Davies (1992) and Dale and Mackay (1994) and in the book *Potato biology and biotechnology advances and perspectives* edited by Vreugdenhil et al. (2007). Details of chemical composition can be found in the book *Advances in potato chemistry and technology, 2nd Edition*, edited by Singh and Kaur (2016), along with analytical techniques. Readers may also like to refer to the protocols in the book *Potato breeding a practical manual for the potato chain* edited by Tiemens-Hulscher et al. (2013) as they are illustrated with excellent colour photographs together with tables of scores for standard cultivars. Breeders will need

J. E. Bradshaw, *Potato Breeding: Theory and Practice*,
https://doi.org/10.1007/978-3-030-64414-7_4

to draw up their own lists of selection criteria for the intended end use(s) of their new cultivars. Broad-end uses such as vegetable for table use and potato for processing will need to be divided into subcategories; for example, table use into cooking method such as baking, boiling, deep-fat frying, microwaving, roasting and steaming and processing use into French fries, chips (crisps) and other processed products.

4.2 Tuber Appearance and Dry Matter

After harvest, the tubers from each field plot are taken into store in bags or boxes and weighed as soon as possible. With small plots (e.g. 5 plants) the total yield will be recorded whereas with larger plots (e.g. 10 plants) the produce will first be graded and each grade weighed separately before calculating marketable yield. At the same time, the tubers can be scored for aspects of their appearance. Consumers in some countries may have a preference for a particular skin colour (e.g. white or red) and appearance (e.g. russet or non-russet) and for a particular flesh colour (e.g. white or yellow), thus making these selection criteria. Processors may also have a preference for a particular flesh colour (e.g. white or yellow for French fries, yellow for chips (crisps)) as well as for a particular tuber shape such as round or round-oval for chips (crisps) and long-oval or long-oblong for French fries, again making these selection criteria for end use. Round potatoes are preferred for canned new potatoes, round-oval for pre-packing and oval to long-oval for baking. Tuber size and tuber size distribution may also be important for end use (e.g. small size for baby new potatoes or large tubers for baking and French fries). For most commercial purposes, an even-sized sample and a uniform tuber shape are preferred. Eye depth can be visually assessed with deep eyes considered undesirable for most uses because of peeling losses. Growth cracking and greening are also considered serious defects capable of leading to substantial marketable yield losses through rejections by processors and pre-packers. Growth cracking has been attributed in part to fluctuating water stress, but a breeder will discard any clone which is particularly prone to this defect. Greening is caused by the formation of chlorophylls a and b as a result of exposure to light before, at and after harvest during storage and retailing. Although chlorophylls are harmless and tasteless to humans, their production may be accompanied by increased levels of glycoalkaloids, which are toxic. Again, a breeder will discard any clone which is particularly prone to this defect. In the next chapter we will consider skin blemish diseases as a blemish-free skin is desirable in a vegetable destined for washed pre-packs and bakers.

The genetics of skin and flesh colour will be considered latter in this chapter. Here it is worth mentioning two mapped genes, one for tuber shape and the other for eye depth. The tuber length/width (L/W) ratio was determined by Endelman and Jansky (2016) in their diploid F_2 mapping population. The trait was scored in replicated field trials for 2 years, from which they estimated a broad-sense heritability of 0.79 for genotype means. One parent (DM) was long (L/W = 3.5) while the other parent (M6) and the F_1 hybrid were oblong (1.5 and 1.8, respectively). A single Quantitative Trait Loci (QTL) (designated *Ro*) was detected on chromosome 10 that accounted for 41% of the variation. Gene action was partially dominant for oblong ($d/a = 0.29/0.59$). Furthermore, the 1.5-LOD (logarithm of the odds) interval for *Ro* also contained the *D(I)* locus for pigmentation, thus confirming earlier reports that these loci are linked (De Jong and Burns 1993). Several previous studies had mapped a QTL for tuber shape, designated *Ro*, to this region (van Eck et al. 1994; Li et al. 2005; Prashar et al. 2014). Interestingly, in the first paper by van Eck et al. (1994), tuber shape was treated as both a qualitative and a quantitative trait. The experimental material was a cross between two diploid potato clones: the female parent (clone USW5337.3) was a hybrid between Group Phureja P1225696.1 and the dihaploid USW42 extracted from cv. Chippewa and the male parent (clone 77.2102.37) was a hybrid between a *Solanum vernei-Solanum tuberosum* backcross) and the USW5337.3 clone. Both parental clones had round tubers but the cross segregated for tuber shape. Inheritance of qualitative tuber shape (round or long) was explained by a single locus *Ro*, round being dominant to long, and mapped to chromosome 10 using restriction fragment length polymorphism (RFLP) markers. Tuber shape was also studied as a quantitative trait using the length/width ratio as the trait value. The estimated broad-sense heritability was 0.80 and the morphologically mapped *Ro* locus explained 75% of the genetic variation, indicating the presence of a major QTL at the *Ro* locus and minor genetic factors. More detailed analysis provided evidence for multi-allelism at the *Ro* locus. The researchers pointed out that previously published studies at the tetraploid level had failed to find a clear genetic model for tuber shape (De Jong and Burns 1993). However, in a genome-wide association study on a diversity panel of 187 tetraploid clones by Rosyara et al. (2016), the two most significant QTLs were for tuber shape and tuber eye depth, both at 48.9 Mb on chromosome 10.

Li et al. (2005) examined tuber eye depth (deep or shallow) in a cultivated diploid potato family which was also segregating for tuber shape (round or long). They came to the following conclusions: 1) there is a major locus controlling eye depth, 2) deep eye (*Eyd*) is dominant to shallow eye (*eyd*), 3) the *Eyd/eyd* locus is located on chromosome 10 and 4) the *Eyd/eyd* locus is closely linked with the major locus for tuber shape (*Ro/ro*), at a distance of about 4 cM. Again, classification of eye depth is not so clear in tetraploids, whereas in the diploid family there were very few potatoes with intermediate eye depth. However, this did not apply in the diploid population of Prashar et al. (2014).

Prashar et al. (2014) also examined tuber shape and eye depth in a highly heterozygous full-sib diploid population (06H1). Each parent was a hybrid between a Group Tuberosum dihaploid and a Group Phureja clone. The genetic map was constructed from a set of 8303 single nucleotide polymorphism (SNP) markers. High levels of collinearity were found between the 12 chromosomal maps and an improved version of the potato genome assembly, with the expected genetic clustering in centromeric regions (Sharma et al. 2013). QTL analysis was done on tuber shape and eye depth data from two growing seasons. Tuber shape was assessed using the length/width ratio and eye depth was visually assessed on a 1–9 scale where 1 was very deep eyes and 9 was very shallow eyes. Tuber shape displayed a continuum of scores and eye depth was approximately normally distributed. The major loci segregating for tuber shape were on chromosomes 2 and 10 (*Ro*), with the latter having the larger effect (0.52 difference between two marker classes compared with 0.36). Three smaller effects at the limit of detection mapped to chromosomes 1, 3 and 9. A major locus for tuber eye depth co-located with the tuber shape locus on chromosome 10 (*Eyd*), with two minor QTLs on chromosomes 2 and 3. In contrast to the results of Li et al. (2005), it was not possible to separate the two loci on chromosome 10 and Prashar et al. (2014) concluded that round tubers and deep eyes were the pleiotropic effects of a single locus. The locus on chromosome 10 for tuber eye depth and on chromosome 2 for tuber shape had previously been reported by Sliwka et al. (2008), but they only accounted for 14.7% and 8.0%, respectively, of what were traits showing continuous variation in their population. More recently, about 250 clones from the 06H1 population were stored at 10 °C and days recorded to >2 mm sprout length for all eyes per tuber. Sprout length of the apical bud was also assessed at set intervals to determine sprout growth rate (Taylor 2020). The largest effect QTL for rapid tuber sprout growth was found on chromosome 3 and a candidate gene underlying the QTL was identified as TERMINAL FLOWER 1/ CENTRORADIALIS orthologue (PGSC0003DMG400014322).

Finally, it is worth mentioning that one of the parents (DM) used by Endelman and Jansky (2016) had two unusual properties, namely, a strong susceptibility to jelly end, a physiological condition associated with a high concentration of reducing sugars and low starch content, and a tendency to produce tubers instead of normal sprouts at the eyes of tubers. Both the eye tuber (*Et*) and jelly end (*Je*) traits mapped to the end of chromosome 5 with the 1.5-LOD interval (1.8 Mb) for *Et* contained within the interval for *Je* (3.2 Mb). Because of the differences in physiology for the traits, *Et* and *Je* are probably distinct but tightly linked loci; but pleiotropy cannot be ruled out from the mapping data.

Dry-matter (DM) content can also be determined at this stage by measuring the specific gravity of a sample of tubers [e.g. we used 3.63 kg at Scottish Crop Research Institute (SCRI)] with a hydrometer or by weighing tubers (can be as few as five) in air (x) and then underwater (y) [$SG = x/(x-y)$]. Dry-matter content is then estimated from the relationship with specific gravity found by von Scheele et al. (1937). Some breeders use sample sizes as large as 5 kg to determine under water weight (Tiemens-Hulscher et al. 2013) (Fig. 4.1). The relationships between specific gravity, dry-matter content and starch content of potatoes were reviewed by Simmonds (1977), starch being the major component of the dry matter (65–75%). The DM content of potato tubers is an extremely important trait which has been associated with many other quality traits such as texture of the cooked potato, suitability for processing and susceptibility to mechanical damage and bruising (Dale and Mackay 1994). For example, DM content is used to assess suitability for processing purposes because high content increases process efficiency (less energy to remove water) and product yield per unit of fresh-weight (FW) and lowers oil absorption (Storey and Davies 1992). Dry-matter content, like fresh-weight yield, is a typical quantitative trait in being affected by genotype and environment and their interaction (Dale and Mackay

Fig. 4.1 Specific gravity being determined through weight in air and under water at HZPC (21 January 2020)

1994), although Killick and Simmonds (1974) reported little interaction for specific gravity of tubers from their review of data available at the time.

4.3 Damage Resistance

Mechanical damage and bruising are major causes of yield losses in potatoes, particularly as a result of mechanized harvesting and handling (e.g. grading). Packers and processors can reject or downgrade deliveries of potatoes from growers. External damage ranges from slight scuffing to splits, cuts or cracks, while internal damage includes shattering and blackspot with localized discolouration due to melanin formation from enzymic browning (Dale and Mackay 1994). The damage resistance of clones can be assessed in tests ranging from shaker tests which simulate riddling conditions to various types of drop tests in which tubers are impacted by a bolt or pendulum type of weight (Storey and Davies 1992). De Maine et al. (1992) showed that external mechanical damage can be assessed on the produce of yield trials by sorting the tubers from each plot into four categories: (1) undamaged, (2) scuffed, (3) peeler damaged and (4) severely damaged. Each category is weighed and a damage index calculated: total of the percentages of scuffed, peeler and severely damaged tubers, weighted in the ratios 1:3:7. The potatoes had been kept for 3 months in an ambient temperature store with frost protection. Reasonably consistent results were obtained for 9 cultivars over 5 years (rank correlations of 0.667 to 0.993). Furthermore, in an 8 × 8 half-diallel set of crosses with selfs, assessed in 2 years, all of the variation could be attributed to general combining ability (GCA) with the GCA × year interactions not statistically significant. Parental scores were highly correlated with parental GCAs (rank correlation of 0.905). These results indicate that the best parents for breeding to improve damage resistance can be predicted using their own damage index. The damage index can be determined on the tubers from small plots (10 plants) but is laborious and slow to use. It may therefore be confined to the assessment of potential parents at the start of a breeding programme and potential cultivars at the end of a programme. Tiemens-Hulscher et al. (2013) describe a shaker test for assessing resistance to damage and black spot bruising that starts with 40 undamaged tubers in the 40–60 mm size range (smaller batches of 20 tubers can be assessed in a modified test).

In a later chapter we will encounter a different solution to the blackspot bruising problem. The formation of the pigment melanin results from initial oxidation of the substrate tyrosine to 3,4-dihydroxyphenylalanine (DOPA) by the enzyme polyphenol oxidase (PPO). Hence, enzymic browning, which also occurs in cut or prepared potatoes, can be prevented by inhibition of PPO activity or prevention of tyrosine oxidation. This can be achieved chemically by adding bisulphite or hypochlorite compounds (Storey 2007) or by silencing of the *Ppo5* gene (polyphenol oxidase-5). Interestingly, Gubb et al. (1989) found that 18 out of 50 genotypes of the wild Mexican potato species *Solanum hjertingii* were non-browning when sliced compared with a range of browning in 5 commercial cultivars, including cultivar Maris Piper which had a low level of browning. The authors adopted potential browning of a rehydrated freeze-dried powder as the most efficient technique for screening enzymic browning. Tuber slices 4 mm thick were cut transversely across the long axis of the tuber. Slices for fresh and frozen reflectance measurements and visual assessment were taken from the centre of the tuber, whilst the slices adjacent to these were used to produce a freeze-dried powder. Slices for freeze-drying (20 slices per genotype) were cut and immediately immersed into liquid nitrogen, freeze-dried and brought back to equilibrium at 20 °C, finely ground and stored at −70 °C until required. A spectrophotometric analysis of the freeze-dried potato macerate of each genotype provided the measure of potential browning. This is a measure of the total enzymic browning resulting from the breakdown of cellular compartmentalization and the release of endogenous substrates and water-soluble enzymes in tuber tissue.

4.4 Internal Defects

Tubers can suffer from internal defects such as hollow heart, brown centre and internal rust spot, which are of particular concern for processors as well as being unwelcome in a vegetable for table use. Clearly, the breeder will need to cut open a sample of harvested tubers to see if any clones are particularly susceptible to any of these defects. At SCRI we cut open as few as three large tubers per plot. Hollow heart is a physiological defect resulting in an internal cavity of varying dimensions. Brown centre is often a precursor to hollow heart and is sometimes referred to as incipient hollow heart. Its characteristic browning is due to cell death in the pith area. Development of hollow heart is often associated with periods of rapid tuber growth, commonly preceded by water or nutrient stress. Breeders tend to select against particularly susceptible clones as and when environmental factors allow expression of the defect (Dale and Mackay 1994). Internal rust (or brown) spot is also a physiological disorder, characterized by rust-coloured spots distributed irregularly throughout the tubers. Its development is associated with restricted water supply, high temperatures, uneven growth rates and a low calcium concentration in affected tubers (see section on mineral biofortification for more on calcium). Again, breeders tend to select against particularly susceptible clones as and when environmental factors allow expression of the defect (Dale and Mackay 1994), as can be seen in a recent set of field studies in Italy (Raimo et al. 2018).

4.5 Storage Decisions

The traits considered so far can be assessed immediately after harvest or as soon as practical thereafter. The same is true of cooking quality, processing quality and nutritional value. However, where only one potato crop is grown each year, the produce needs to be stored for as much as 6 months or more, and this raises the issue of the effect of the storage regime on quality traits and nutritional value (Gottschalk and Ezekiel 2006). Physiological weight loss occurs during storage as a result of respiration and transpiration; the former is minimized through temperature control (2–4 °C maintains a low respiration rate) and the latter is reduced by maintaining high relative humidity (95%). Many other physiological and biochemical changes occur during storage, including cold-induced sweetening (the production of reducing sugars, glucose and fructose), which is particularly acute at 4 °C or less, but which can be reversed to varying degrees by increasing the temperature to 10–20 °C, a process known as reconditioning. Furthermore, clones being assessed in a breeding programme will vary in their dormancy period, typically from 5 to 9 weeks. As table and processing potatoes are in demand throughout the year, both short- and long-term storage are needed. Refrigerated storage is the best option for long-term storage whereas non-refrigerated, on-farm methods are cheap alternatives for short-term storage, particularly in tropical and subtropical countries. In Kenya, for example, there is interest in modified atmosphere packaging (MAP) of ware potato tubers for short-term storage at ambient temperatures. Nyankanga et al. (2018) found that perforated low-density black polyethylene bags were the best of seven packaging materials due to low sprouting, reduced weight loss, low rate of tuber greening and reduced rate of tuber decay over the 5 weeks of their experiment. Interestingly, Scott et al. (2019) have identified the need for 16–24 million tonnes more of cold storage capacity in India to meet a projected increase in potato demand of 20–30 million tonnes by 2030.

Fig. 4.2 Box storage and use of ethylene in the Netherlands (22 January 2020)

Optimal storage conditions are based on the purpose for which potatoes are stored and the cultivar, particularly ones that have been bred for processing. Cultivars for table use are typically stored at 4–6 °C, for French fries at 6–8 °C and for crisping at 7–12 °C (Gottschalk and Ezekiel 2006). When potatoes are stored above 4 °C, sprout growth occurs and warm storage (above 7 °C) of processing potatoes requires their treatment with sprout inhibitors (their dormancy is usually too short to avoid this). For over 50 years, the major global sprout suppressant has been chlorpropham (CIPC), but the European Commission has decided not to renew its approval for use in the European Union from 1 January 2020 because of concerns over residues in tubers. Hence, there is currently much interest in alternatives such as the use of ethylene (Fig. 4.2), maleic hydrazide, 1,4-dimethylnaphthalene and spearmint oil (Briddon et al. 2018). More details on storage can be found in the reviews by Gottschalk and Ezekiel (2006) and Briddon et al. (2018), but it should be clear that breeders need to make important decisions about the storage and assessment of their clones after harvest, preferably in consultation with the targeted end users. Not surprisingly, resistance to cold-induced sweetening is commonly a breeding objective for processing cultivars. At the Scottish Crop Research Institute (SCRI), for example, when the plots were weighed after harvest, samples of three and five tubers were taken from each plot and stored at 4 and 10 °C, respectively, for a minimum of 3 months. Sprouting (dormancy), keeping quality and cooking quality were determined on the samples stored at 10 °C and fry colour was assessed after storage at both at 4 °C and 10 °C.

4.6 Cooking Quality

4.6.1 After-Cooking Blackening and Sloughing

Two cooking quality traits that are relatively easy to determine on just a few tubers are after-cooking blackening (ACB) and sloughing. For example, at SCRI, after storage at 10 °C, two tubers from each plot were steamed for 45 min and scored for sloughing on a 1 (severe disintegration) to 9 (none) scale before being cut open. ACB was scored 4 to 5 hours later on a 1 (completely black) to 9 (none) scale. Likewise, in their breeding programme in Canada, Yuan et al. (2019) were able to bake or boil three tubers of 237 clones for culinary evaluation in each of 3 years. It is also common practice when assessing potential cultivars for cooking quality to divide them into four broad categories A, B, C and D, with A (quite firm salad type) and D (very mealy, mushy type) the extremes (Tiemens-Hulscher et al. 2013).

After-cooking blackening is a non-enzymic discolouration of the tuber which is usually more obvious at the stolon end (Storey and Davies 1992). Its undesirable presence is most obvious in boiled or steamed potatoes, but discolouration may also be seen following cooking of baked potatoes, French fries, canned and dehydrated potato products. During cooking chlorogenic acid combines with iron to form a complex which oxidizes on cooling to the coloured ferri-dichlorogenic acid responsible for the blue-grey discolouration at normal pH (different colours occur above and below normal pH). Chelating agents in potato tubers such as citric acid compete with the chlorogenic acid for the iron to form a colourless complex. The incidences of ACB depend on cultivar susceptibility as well as growing conditions (high N to K ratio), seasonal factors and storage length and conditions. Cultivar differences have been correlated with their contents of chlorogenic acid, and to a lesser extent, citric acid (Hughes and Evans 1967). Selection for lack of ACB is considered a realistic breeding objective for this quantitative trait (Dale and Mackay 1994).

4.6.2 Flavour and Texture

Flavour is important to the consumer but is likely to be assessed during the final stages of a breeding programme when the number of potential cultivars has been reduced to 40 or less and it is possible to use a professionally trained taste panel for organoleptic tests, an inherently slow process. In research at SCRI, for example, sensory and volatile analyses were done on 22 potato clones (16 Group Phureja and 6 Group Tuberosum) from a yield trial with two replicates, in which a panel of 14 sensory assessors (provided by Charis Innovative Food Services Limited) were able to score 22 sensory attributes (Winfield et al. 2005). Rapid screening of large numbers of clones for flavour is still not possible, to the best of my knowledge. Hence, there is a need to progress understanding of the relationship between metabolite production and human perception of flavour in order to define and select desirable

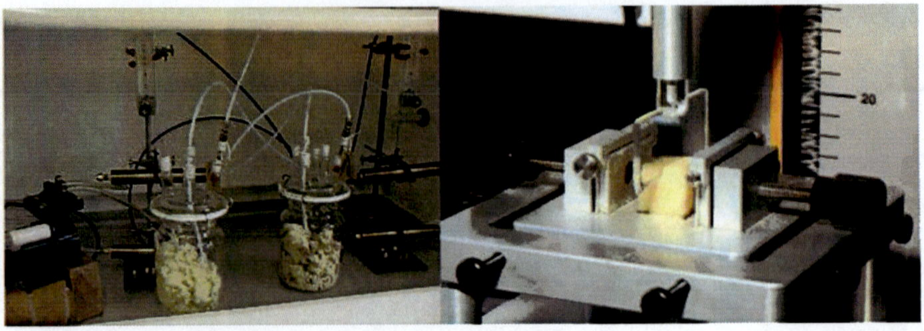

Fig. 4.3 Entrainment of headspace volatile organic compounds (left) and texture profile analysis using probe and load cell (right). (Source: SCRI)

flavour profiles. One of the more recent reviews of potato flavour is by McKenzie and Corrigan (2016). They point out that flavour is defined as the combination of the sense of taste (sweet, sour, salt, bitter and umami), which is perceived on the tongue, and the sense of smell when volatiles come into contact with receptors located in the nasal passages, either by sniffing or chewing. Raw potatoes have a limited but distinctive aroma whereas cooking potato tubers releases a wide range of volatile compounds, with the cooking method (baking, steaming, boiling, roasting or frying) determining the type and extent of the flavour-producing reactions. Common flavour precursors are sugars, lipids and amino acids. McKenzie and Corrigan (2016) list over 100 key compounds involved in potato flavour production. Potato agronomy and post-harvest storage (e.g. cold sweetening) as well as cultivar characteristics can influence the final perception of flavour. Sensitive volatile analysis of boiled potatoes can be done by entrainment of headspace volatile organic compounds (Fig. 4.3) followed by various types of gas chromatography (GC). McKenzie and Corrigan (2016) briefly discuss studies comparing metabolomic content with sensory analysis, including interesting differences found between 'flavoursome' Group Phureja potatoes and generally 'blander' Group Tuberosum. Turning to taste, the five basic human tastes (sweet, sour, salty, bitter and umami) are all naturally found in potato, although saltiness appears very limited (McKenzie and Corrigan 2016). Morris et al. (2007) assessed the levels of the major umami compounds in steamed potatoes and found that both glutamate and 5′-nucleotides were more than twofold higher in mature tubers of Phureja cultivars Mayan Gold and Inca Sun compared with Tuberosum cultivars Pentland Dell and Montrose, both at harvest and after 6 weeks storage at 4 °C. The 5′-nucleotides were guanosine 5′-monophosphate (5′-GMP) and adenosine 5′-monophosphate (5′-AMP). The equivalent umami concentration (based on levels of glutamate, aspartate, 5′-GMP and 5′-AMP) was calculated for 5 cultivars (3 Phureja and 2 Tuberosum), and strong positive correlations ($R^2 \geq 0.79$) found with flavour attributes and acceptability scores of boiled tubers from a trained evaluation panel, suggesting that umami is an important component of potato flavour. Finally, Sato et al. (2019) have suggested that the egumi-taste which can occur in potatoes and is disliked by Japanese consumers because it is unpleasantly harsh, bitter and astringent, may be due to bitter-tasting free amino acids or to a metabolite of tyrosine, homogentisic acid.

Texture is also an extremely important sensory attribute in potato as it has a direct influence on consumer preference as well as influencing cooking time (McKenzie and Corrigan 2016; Singh et al. 2016a). Texture can be related to the potato tuber's resistance to an applied force and depends on many structural and chemical properties of the tuber, including starch and dry-matter content (Singh et al. 2016a). However, it is often difficult to relate these to consumers' perceptions of a particular texture and to their preferences (Storey and Davies 1992). Hence, selection for a particular texture often results indirectly from selection for other traits such as the quality category of the cooked tuber or lack of disintegration on cooking. Disintegration (none to complete), consistency (firm to very soft), mealiness (waxy/not mealy to very mealy/floury), dryness (humid to dry) and structure (fine to coarse) can all be considered aspects of cooked potato texture (Storey and Davies 1992). The mealiness/waxiness continuum is one of the components of texture most readily described by sensory panellists, although the full list of attributes for panellists would include hardness, firmness, springiness, adhesiveness, graininess, mealiness, moistness and chewiness (McKenzie and Corrigan 2016). Correlations of varying degree are often found between these traits and dry-matter content; for example, high content and increased disintegration (and hence shorter cooking time) and high content and mealiness. Singh et al. (2016a) have reviewed the factors that influence the cooking, textual and rheological properties of potatoes. Textural qualities can be visualized using scanning electron microscopy and confocal scanning laser microscopy and analysed using probes and suitable load cells that measure compression, puncture and shear forces (Fig. 4.3). The parameters derived from texture profile analysis are fracturability, hardness, adhesiveness, cohesiveness, gumminess, springiness and chewiness. Although these approaches are useful for research purposes, breeders still regard texture as a quantitative trait with a major genetic component, but often treat it as a descriptive trait rather than a selection criterion (Dale and Mackay 1994).

4.7 Processing Quality

4.7.1 Fry Colour and Reducing Sugar Content After Storage

We have already dealt with high dry-matter (DM) content as an important trait in a cultivar for processing, as well as mentioning processors' preferences for tuber flesh colour, size and shape. Processors may also have preferences for dry-matter distribution and uniformity of texture, but the most important quality criterion not yet covered is fry colour of the fresh and stored potato. Fry colour after storage is a relatively easy trait to assess in a breeding programme. At SCRI, for example, fry colour assessment after storage of tubers at 4 °C (Fry4) and 10 °C (Fry10) was done on three tubers. Each tuber was cut in half and a 1 to 1.5 mm-thick slice was taken and fried in vegetable oil at 175 °C in a thermostatically controlled chip fryer until all of the water had boiled off (bubbling ceased). Colour was assessed by comparison with a standardized chart on a 1 (extremely dark) to 9 (extremely pale) scale, where five and above is an acceptably light colour (Mackay and Dale 1990). Again, in their book, Tiemens-Hulscher et al. (2013) provide protocols and colour photographs for assessing French fry and crisp quality on a 1 to 9 scale of increasing lightness of colour.

Whilst starch is the major carbohydrate in potatoes, the small but varying amounts of sugars (sucrose, glucose and fructose) are critical for processing quality, as reviewed by Storey and Davies (1992). The production of unacceptably dark and bitter tasting fry products by non-enzymic browning is associated with a high percentage of reducing sugars (glucose and fructose) and has been the subject of much research. Frying at high temperatures (175 °C) results in a typical Maillard reaction between the reducing sugars and the α-amino acid groups of nitrogenous compounds (including free amino acids) also present in the tuber (Schallenberg et al. 1959). A reducing sugar content of 0.1% fresh-weight is ideal for the manufacture of crisps whereas higher than 0.33% is unacceptable (Storey and Davies 1992). A major environmental factor in determining reducing sugar content is the storage temperature, with cold storage (4 °C and less) resulting in cold-induced sweetening (i.e. higher levels). Other types of sweetening result from tuber immaturity at harvest, rapid sprout growth at higher storage temperatures in the absence of sprout-suppressors and senescent sweetening in unsprouted tubers following storage at high temperatures. The biochemistry is complex and the regulation of starch mobilization is not fully understood, but targets for genetic reduction (modification) of cold-induced sweetening by genetic transformation have been identified, explored and some shown to work; for example, lowering the expression of tuber-expressed starch-associated *R1* and phosphorylase-L (*PhL*) genes in cultivar Ranger Russet by gene silencing (Rommens et al. 2006). Reduction of cold-induced sweetening has also been achieved by conventional breeding (Brown et al. 1990; Mackay et al. 1990; Mackay 2005) and shown to be a highly heritable trait (broad-sense heritability for FRY4 of 0.79) in the cross between processing clone 12601ab1 and table cultivar Stirling (Bradshaw et al. 2008) (Fig. 4.4), despite no QTLs (Quantitative Trait Loci) of large effect being found (4 QTLs explained about 30% of the variation). Hence, breeding cultivars resistant to cold-induced sweetening is a realistic objective and will negate the need to store potatoes at high temperatures and use sprout suppressants or to try to breed ones with a longer dormancy. Nevertheless, Raigond et al. (2018) still recommended that the Indian processing cultivar Kufri Chipsona-1 should be stored at 12 °C with sprout suppressants, rather than at 4 °C, because its reducing sugar content increased from 180 mg/100 g fresh weight (0.18%) to 712 mg/100 g fresh weight (0.71%) after 180 days of storage at 4 °C, but remained below 0.33% at 12 °C. In contrast, cultivars Kufri Jyoti and Kufri Pukhraj, which are not suitable for processing, exceeded this level at 12 °C as well as at 4 °C.

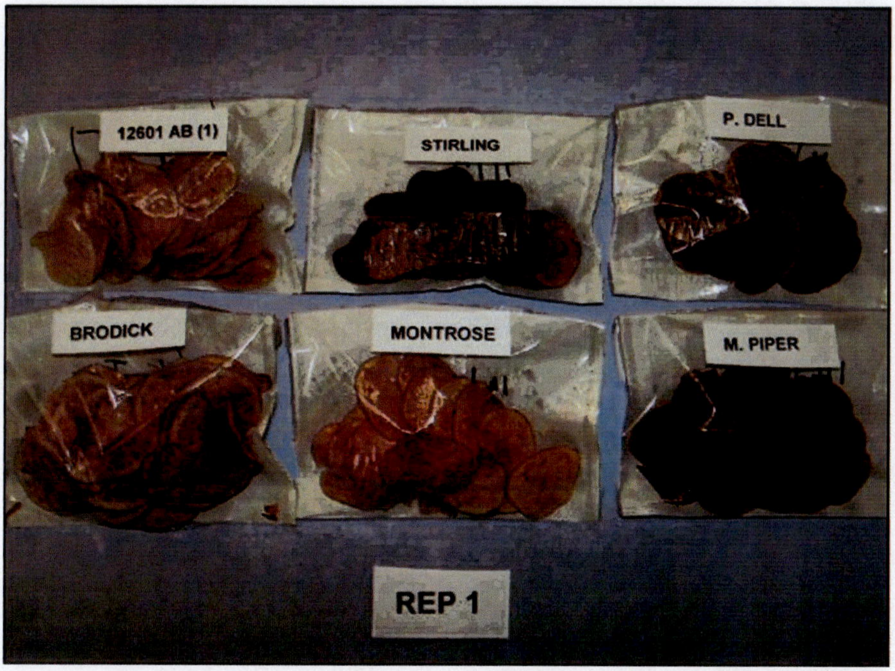

Fig. 4.4 Fry colour of crisps (chips) after potatoes had been stored at 4 °C for 3 months: Stirling is a table cultivar, Maris Piper is a widely grown table cultivar that can be used for French fries, Pentland Dell has been extensively used for processing into French fries and Brodick, Montrose and 12601ab1 are cultivars/clone with resistance to cold-sweetening. (Source: SCRI)

4.7.2 Acrylamide

Today, processors are under pressure to reduce acrylamide formation in crisps (chips) and French fries because of concerns about its effects on human health (Amrein et al. 2003; Pedreschi 2009; Pinhero et al. 2009; De Meulenaer et al. 2018). Acrylamide is classified as a Group 2A (probably carcinogenic to humans) carcinogen. Mandatory mitigation measures for reducing acrylamide in food came into force in the European Union on 11 April 2018 with benchmark levels for potato crisps and French fries of 750 and 500 μg kg^{-1}, respectively (European Commission).

Acrylamide is formed during high-temperature cooking (frying, baking and roasting) and food processing by Maillard browning reactions between reducing sugars and free asparagine. Lower levels of these compounds are obvious targets for conventional breeding and genetic engineering. Muttucumaru et al. (2017) analysed the acrylamide-forming potential of 20 potato cultivars grown at two sites (Doncaster and Woburn) in the United Kingdom (UK). They modelled the relationship between the ratio of free asparagine to reducing sugars and the levels of acrylamide. They identified a value of 2.26 as the tipping point in the ratio below which free asparagine concentration could affect acrylamide formation. This is an important result as reducing sugars have been the main determinant of acrylamide-forming potential in most datasets (e.g. De Wilde et al. 2006), but with free asparagine sometimes contributing to the variation (e.g. Shepherd et al. 2010). In contrast, acrylamide formation in heated wheat and rye flour is determined by the free asparagine concentration unless the flour is derived from sulphur-deprived wheat (Muttucumaru et al. 2017).

The J R Simplot Company in the USA is taking the genetic transformation route to improving processing cultivars being grown in North America, but it is not yet clear if they will be acceptable to companies such as McDonald's. Chawla et al. (2012) of Simplot Plant Sciences, have shown that tuber-specific silencing of *asparagine synthetase 1* (by RNA interference) reduces the acrylamide-forming potential of potatoes grown in the field without affecting tuber shape and yield. Zhu et al. (2016) found that fried potato products from cultivar Russet Burbank silenced for the vacuolar invertase gene *VInv* and the asparagine synthetase genes *StAS1* and *StAS2* contained only one-fifteenth of the acrylamide content of the controls. Interestingly, the extent of acrylamide reduction of the best triple-silenced line was similar to that of the best *VInv* single-silenced line, suggesting that the production of cultivars with low-reducing sugars may be an effective and sufficient approach for minimizing the acrylamide-forming potential of French fry processing potatoes. We shall take a further look at these two research papers in a later chapter on genetically modified potatoes.

Acrylamide-producing potential can also be lowered by conventional breeding. Shepherd et al. (2010) examined the genetic variation released in the cross mentioned above between processing clone 12601ab1 and table cultivar Stirling. The processing clone was known to be resistant to cold-temperature sweetening (Bradshaw et al. 2008). A subset of 43 offspring clones from a replicated yield trial in 2007 were stored at 4 °C for 4 months. Fry colour and acrylamide content of their crisps were determined after frying at 180 °C, as well as the concentrations of reducing sugars and asparagine in their tubers. Most of the variation (>88%) between the 43 clones and 5 controls was genetical. Multiple regression analysis revealed that fry colour was determined by reducing sugar content (71.4% of variation), whereas acrylamide content was determined by both reducing sugars and asparagine (in total 53.4% of variation). There was no correlation between reducing sugars and asparagine. The clone with the lowest acrylamide content was third lowest for reducing sugars, fourth lowest for asparagine and second best for fry colour (light colour). The researchers concluded that both reducing sugars and asparagine should be targeted in conventional breeding.

Novy et al. (2017) reported the release in the USA in 2015 of cultivar Payette Russet, which was notable for its cold-sweetening resistance and associated low-acrylamide formation, making it ideally suited for processing into French fries and other potato products. Low asparagine and reducing sugar concentrations contributed to an 81% reduction in acrylamide content in French fries relative to cultivars Ranger Russet and Russet Burbank following 8 months storage at 9 °C.

In conclusion, lowering acrylamide formation is a realistic breeding objective for both conventional breeding and genetically modified potatoes. Furthermore, in production, nitrogen and sulphur (if applied) fertilizer levels can affect acrylamide-forming potential, but in a complicated way (Muttucumaru et al. 2013). It is also known that many additives can reduce acrylamide formation in potato products on a laboratory scale but they are still to be proved on an industrial scale (De Meulenaer et al. 2018).

4.8 Glycoalkaloids

Potato tubers contain small quantities of steroidal glycoalkaloids which are usually regarded as undesirable toxic compounds, despite having some anticarcinogenic properties (Friedman and McDonald 1997; Friedman and Levin 2016). About 95% of total content is accounted for by α-solanine and α-chaconine, which are differently glycosylated forms (galactose and glucose, respectively) of the aglycone, solanidine. Other plant parts have higher concentrations and it is generally thought that glycoalkaloids confer some protection against various pathogens. Tuber concentrations above 15 mg 100 g^{-1} FW (fresh-weight) confer bitterness and above 20 mg are considered unsafe for human consumption, resulting in symptoms typically associated with food poisoning. Factors which can affect tuber concentrations are cultivar, climate, storage environment, maturity, damage, temperature and exposure to light (Sinden et al. 1984; Dale and Mackay 1994). Hence, most causes of excessive glycoalkaloid levels are under the control of the breeder, the grower or those involved in processing and marketing. Potato cultivars vary in content, but commercial potatoes rarely exceed 20 mg 100 g^{-1} FW. However, some wild species used in breeding have higher contents.

Van Gelder et al. (1988) determined the compositions of glycosidic-bound steroidal alkaloids of *Solanum* species used in potato breeding by capillary gas chromatography using simultaneous nitrogen-specific (NPD) and flame ionization detection (FID). Total glycoalkaloid contents varied from 12.3 to 734.8 mg 100 g^{-1} FW. Similar small-sized tubers of cultivars grown under the same conditions showed lower contents of 12.6–72.1 mg 100 g^{-1} FW, with values of 4–36 mg for normal-sized field grown tubers. The highest contents were found in *S. vernei*, a source of resistance to potato cyst nematodes.

Today, there is concern that introgression breeding from wild species may inadvertently result in unacceptably high concentrations of tuber glycoalkaloids, and hence new cultivars are commonly tested to ensure that their contents are below the accepted safe limit of 20 mg 100 g^{-1} FW. American cultivar Lenape was removed from commerce in 1970 due to its high glycoalkaloid levels which averaged 29 mg 100 g^{-1} FW over 39 locations and rose to 65 mg in stressed conditions. It is thought that the high levels were due to *Solanum chacoense* in its immediate ancestry (Zitnak and Johnston 1970; Sinden et al. 1984). Nevertheless, glycoalkaloid contents are highly heritable and the breeder can control the levels in new cultivars by appropriate selection of wild species and parental cultivars and by testing selected hybrids and potential cultivars (Sinden et al. 1984). It may also be necessary to check wild species for unusual forms of glycoalkaloids when used for the first time.

Although not all of the steps in the biosynthesis of the major glycoalkaloids (α-solanine and α-chaconine) are known, virtual elimination of tuber glycoalkaloids by gene silencing is feasible, as we shall see in the chapter on genetically modified potatoes (Itkin et al. 2013; Shepherd et al. 2015; Kaminski et al. 2016; Umemoto et al. 2016).

4.9 Nutritional Value

As well as being a major supplier of carbohydrate (starch) in the human diet, the potato also provides significant amounts of protein, with a good amino acid balance, minerals (particularly potassium), vitamins (particularly C), micronutrients and phytonutrients, which include antioxidants. It is high in dietary fibre, especially when eaten unpeeled with its skin. Fresh potatoes are virtually free of fat and cholesterol and have a water content of about 80%.

As a major food crop, the potato has an important role to play in the United Nations '2030 Agenda for Sustainable Development' which started on 1 January 2016 (http://faostat.fao.org). The agenda includes 17 goals, the second of which is to end hunger, achieve food security and improved nutrition, and promote sustainable agriculture. The background to the agenda is as follows: In 2015, out of a world population of 7.3 billion, around 800 million people suffered from hunger (821 million in 2017, www.fao.org/state-of-food-security-nutrition/en/) and more than 2 billion from one or more micronutrient deficiencies, known as 'hidden hunger', especially in women, infants and children. Ironically about 1.9 billion people were overweight, of whom 600 million were obese (672 million adults in 2017). Furthermore, the world population is expected to reach 8.5 billion by 2030 and 9.7 billion in 2050. By 2030 the aim of the agenda is to 'ensure access by all people, in particular the poor and people in vulnerable situations, including infants, to safe, nutritious and sufficient food all year round'. For potatoes, the need is to improve nutritional value as well as to increase production.

The prerequisites to improving nutritional value are usable genetic variation, of which there appears to be plenty, and rapid and reliable screens for the chosen targets such as near infrared spectroscopy (NIRS). This was used, for example, by Escuredo et al. (2018) to predict with good accuracy the phenol and flavonoid content, antioxidant capacity, dry-matter, texture and soluble solid content of 73 potato tuber samples involving 35 cultivars. In contrast, screening techniques involving a gas chromatograph and mass spectrometer (GC-MS) or high-performance liquid chromatography (HPLC) are more time consuming and hence a constraint in breeding, and so can be spectroscopic and spectrophotometric methods. The complications in deciding breeding targets for improved nutritional value stem from the fact that before being eaten, potatoes are stored at various temperatures for varying lengths of time and then cooked or processed in various ways, all of which can affect chemical composition (Fig. 4.5).

Fig. 4.5 Mayan Gold roast potatoes with parsnips: the potatoes have been stored and cooked and are about to be eaten. (Kindly supplied by Shiona Mackie 25 December 2018)

Potatoes need to be cooked because of the indigestibility of their ungelatinized starch (Burton 1989). Cooking results in physical, chemical and enzymic changes in the potato tuber and these affect its nutritional value to different degrees (Tian et al. 2016). Furthermore, following food digestion, the nutrients must be taken up by the human body (i.e. bioavailable) in order to be utilized. For example, Macdonald-Clarke et al. (2016) found that the bioavailability of potassium (K) is as high from potatoes as from potassium gluconate supplements. However, iron (Fe) concentration provides a good example of the complexity of bioaccessibility and bioavailability and of issues in deciding breeding objectives (Andre et al. 2014). The concentration of Fe in potato is low compared with cereals and legumes, but its bioavailability is greater due to the presence of high levels of ascorbic acid (promoter of Fe absorption) and low levels of phytic acid (inhibitor of Fe absorption) (Fairweather-Tait 1983). Andre et al. (2015) estimated the bioaccessibility (potential bioavailability) and bioavailability of Fe from 12 Andean potato clones. Between 64% and 79% of the Fe was released from the tuber matrix during in vitro gastrointestinal digestion and was therefore available at the intestinal level. Furthermore, on average, 32% and 24.5% of the ascorbic acid and chlorogenic acid (the principal phenolic acid), respectively, were also bioaccessible from boiled tubers. There were negative relationships between bioavailable Fe values and phenolic concentrations, whereas ascorbic acid concentrations were positively associated with the ferritin values. However, intestinal absorption of intrinsic Fe from potato tubers could not be detected by the method used, so further work is required to conclusively determine the bioavailability of intrinsic Fe from potato tubers. Nevertheless, these results reveal that the consequences for human nutrition of increasing phenolic concentrations as well as iron content require further research to determine the optimum balance for potato breeding objectives.

Only after all of these kinds of considerations can the chemical composition of potato tubers be related to recommended daily allowances (RDA) and intakes (RDI) of nutrients and sensible targets set for improving nutritional value. In other words, the breeder needs evidence that selecting for altered chemical composition will actually improve nutritional value and human health. The latter is particularly difficult to prove as seen in the review by Beals (2019) which has the following conclusions. 'Research suggests that potato nutrients and components may have favourable impacts on blood pressure, satiety and gut health; however, data from observational studies examining the effects of consuming whole potatoes on body weight and disease risk remain controversial. There is currently a lack of experimental data regarding the impact of potato consumption on obesity, weight management and/or diabetes; thus, randomized controlled intervention trials investigating the effect of potato consumption on various health outcomes and disease states are needed to better isolate potato consumption from other known risk factors'.

In April 2019, Volume 96(2) of the American Journal of Potato Research was a special issue on the contribution of potatoes to human nutrition (Jansky et al. 2019) to which the reader is referred for the many papers of interest. In addition, I recently reviewed improving the nutritional value of potatoes by conventional breeding and genetic modification (Bradshaw 2019). What follows are my conclusions, with justification, for realistic breeding objectives, together with the genetics of tuber skin and flesh colour. Other breeders may come to different conclusions about priorities.

4.9.1 Resistant Starch and Glycaemic Index

The form of energy storage in potato tubers is almost entirely starch, which is made up of two polymers: amylose (20–30%), which is a linear molecule, and amylopectin (70–80%), which is branched. In considering the nutritional value of potatoes we need to consider their glycaemic index (GI) (Monro and Mishra 2009). This is a ranking system that allows carbohydrate foods to be classified on a scale of 0–100 as either low (<55), medium (56–69) or high (>70), where glucose is used as the reference food and assigned a score of 100. GI measures the food's effects on postprandial blood glucose levels over 2 h. High GI foods result in a rapid increase and corresponding insulin responses. They can therefore contribute to obesity, glucose intolerance, type-2 diabetes and cardiovascular disease, but could be considered beneficial to sports persons who need high-quality (nutrient-dense) carbohydrate sources to enhance muscle glycogen storage and to deliver carbohydrate to muscle and other organs during strenuous exercise (Kanter and Elkin 2019). However, nutritional advice based on the GI of a fixed amount (50 g) of available carbohydrate, may not be representative of postprandial satiety and glycaemia when carbohydrates are consumed freely in a mixed meal (Akilen et al. 2016; Borch et al. 2016). With these reservations in mind, we will now look at the GI of potatoes.

GI values reported for potatoes by Foster-Powell et al. (2002) range from 23 to 111, where the glucose control is 100. In Great Britain (GB), Henry et al. (2005) reported GI values from 56 to 94 for 8 potato cultivars. Potatoes with waxy textures produced medium GI values whilst floury potatoes had high values. There is also some evidence that coloured potatoes have a lower GI than white ones (78.7 v 93.0), and that this is possibly due to an inhibitory effect of anthocyanins (polyphenols) on intestinal α-glucosidase (Ramdath et al. 2014). Despite the wide range of values, potatoes are usually considered a high GI food and this has raised concerns about their nutritional value and led to the search for low-GI potatoes (Pinhero et al. 2016). However, it must be remembered that GI is affected by cooking method and potatoes are usually eaten in mixed meals. Compared to boiling or mashing, for example, some cooking methods (e.g. frying, microwaving and baking) reduce the postprandial glycaemic response significantly, and a cooling step or co-digestion with protein, lipid and vinegar are effective ways to decrease the GI of potato (Tian et al. 2016). More detailed information on the effect of cooking methods on the glycaemic index of potatoes can be found in the reviews by Nayak et al. (2014) and Tian et al. (2016). Here, however, we are concerned with genetic variation in GI of potato.

Clinical evaluation of GI is expensive. Hence, effective, quick and low-cost alternatives have been sought and validated, such as in vitro starch digestibility as determined by the rate and extent of glucose release from enzymatic digestion under controlled conditions. Working with 7 potato cultivars (GI 53–103), Ek et al. (2014a) found a strong positive correlation ($r = 0.91$) between in vivo and in vitro GI methods, particularly in vitro starch hydrolysis of cooked potatoes at 120 min. In contrast, amylose, dietary fibre and total starch content were not correlated with either in vitro starch digestibility or GI. The researchers concluded that low-GI potato cultivars can be identified by screening using a high-throughput in vitro digestion procedure. They classified cultivar Carisma™ (a waxy type with GI = 53) as low-GI, cultivar Nicola (GI = 69) as medium-GI and the other 5 cultivars as high-GI. They explained the low GI of cultivar Carisma™ in terms of the physical and chemical properties of its starch (Ek et al. 2014b). Cultivar Carisma™ was launched as a low-GI potato in Australia in 2010 and is now marketed and grown in North America. Another new conventionally bred cultivar (currently unnamed) developed by Agriculture and Agri-Food Canada (www.agr.gc.ca/potato-cultivars) is being trialled as a low-GI potato. It is reported to have a moderate index of 65 when served hot and a very low one of 34 when served cold.

Total starch can be classified by its digestibility into nutritionally important fractions: rapidly digestible starch (RDS), slowly digestible starch (SDS) and resistant starch (RS) (Pinhero et al. 2016). Rapidly digestible starch is the amount of starch hydrolysed in the first 20 min of in vitro incubation with a-amylase, SDS is the amount hydrolysed between 20 and 120 min and RS is the starch left after 120 min, calculated by subtracting RDS and SDS from the total. Resistant starch reaches the large intestine essentially intact and has similar physiological effects and health benefits as fibre (Birt et al. 2013; Lockyer and Nugent 2017). Pinhero et al. (2016) evaluated 14 early potato cultivars grown in Canada for dietary fibre, total starch,

RDS, SDS and RS as well as for estimated GI (eGI). They found a strong positive correlation ($r > 0.96$) between eGI and RDS and a strong negative one ($r < -0.96$) between eGI and RS, irrespective of whether the potatoes were uncooked, cooked or retrograded (boiled and cooled for 48 h at 4 °C). Hence, RDS and RS were major factors contributing to eGI. Monro et al. (2009) also found genetic variation for starch digestibility among 37 New Zealand potato breeding lines. Slowly digestible starch ranged from 7% to 37% of total starch and RS from 12% to 27% after a post-cooking cooling treatment that increased SDS and RS. The researchers concluded that there was sufficient genetic variation for selection in conventional breeding, and that the glycaemic impact of some potatoes may be substantially reduced by cool storing after cooking.

The overall conclusion from the studies reported in this section is that low-GI potato clones can be identified in a conventional breeding programme by screening using a high-throughput in vitro digestion procedure; but it remains to be seen if such selection will be for speciality potatoes rather than for more general table use. Finally, it is worth mentioning that the introduction of inulin to potato is another way to improve its nutritional value by reducing its energy density through increased dietary fibre. Hellwege et al. (2000) have developed transgenic potato tubers that synthesize the full range of inulin molecules naturally occurring in globe artichoke (*Cynara scolymus*), as we shall see in the chapter on genetically modified potatoes.

4.9.2 Protein Content and Amino Acid Balance

On a dry weight (DW) basis, potatoes contain about 10% protein (2% on FW basis), equal to that of most cereals such as rice or wheat. The protein is considered good quality in the sense that it has a biological value of between 90 and 100 (proportion of absorbed protein from a food which becomes incorporated into the proteins of the human body relative to a standard), which is greater than most other non-animal sources of protein and close to whole egg protein which is given a value of 100 (Kanter and Elkin 2019). The proteins are often classified into three groups, namely, patatin, protease inhibitors and high-molecular weight proteins and are in fact associated with several health benefits (Waglay and Karboune 2016). They are a good source of lysine, but low contents of sulphur amino acids (methionine and cysteine) limit their nutritive value (Friedman 1996). This was confirmed by Bártová et al. (2015) in glasshouse produced tubers of four Andigenum accessions and one Tuberosum cultivar (Desiree): protein content was 5.2–8.0% DW, patatin (the major storage protein) content was 17.1–41.7% of protein and methionine and cysteine were the amino acids with the lowest concentrations in all five genotypes. Where potatoes are an important dietary constituent, breeding for increased protein quantity and quality can be justified, but it is difficult to find good examples in the scientific literature. In contrast, differences between cultivars have often been reported (e.g. Galdón et al. 2010; Pęksa et al. 2013). Protein quantity and quality can also be improved by genetic modification. Chakraborty et al. (2010) developed transgenic potatoes with enhanced nutritive value by tuber-specific expression of a seed protein, Amaranth Albumin 1 (AmA1), from *Amaranthus hypochondriacus*, in 7 Indian cultivars suitable for cultivation in different agro-climatic regions. As AmA1 is a non-allergenic protein that originated from an edible crop, transgenic potato crops expressing AmA1 should receive public acceptability. We shall consider this example in more detail in the chapter on genetically modified potatoes.

4.9.3 Mineral Biofortification

White et al. (2009) reviewed the mineral elements required by humans in their diet, the mineral composition of potatoes, and the effects of fertilizer application and genetic variation on tuber mineral concentrations. They concluded that it should be possible to increase the mineral content of tubers (biofortification) by combining genotypes that have naturally higher concentrations (genetic biofortification) with appropriate fertilization strategies (agronomic biofortification) to deliver greater quantities of essential minerals to the human diet without compromising yield.

White et al. (2009) identified considerable genetic variation between modern cultivars for N, P, K, S, Ca, Mg, Cu, Fe, Mn and Zn, and also between Andean landraces for Ca, Fe and Zn and between wild species for Ca. Further information on the mineral content of North American cultivars and potential cultivars can be found in the papers by Brown et al. (2010, 2011, 2012b, 2013, 2014) and the paper by Nassar et al. (2012) which also includes Se. The authors also consider potential breeding objectives. Subramanian et al. (2017) evaluated the variation in mineral content of wild and cultivated potato accessions in the Commonwealth Potato Collection (CPC) and thought that breeding for higher tuber concentrations of Ca, Fe and Zn was worthwhile. Haynes et al. (2012) looked at the Cu, Fe, Mn and Zn contents of cultivar Atlantic and 17 hybrids between tetraploid Tuberosum and diploid hybrids of Phureja-Stenotomum, grown for 2 years at three locations in the East of the USA. Genetic variation was large for all four micronutrients, suggesting that contents could be improved through breeding. But what are the urgent needs for genetic biofortification of minerals?

4.9.4 Genetic Biofortification of Iron and Zinc and Significance of Calcium

In 2004, the International Potato Centre (CIP) chose Fe and Zn as potato targets in the HarvestPlus Biofortification Challenge Program (Paget et al. 2014). More recently, Kromann et al. (2017) have summarized the continuing need for Fe and Zn biofortification in a world where more than 60% of the global population is thought to be dietary deficient in Fe and more than 30% in Zn. In contrast to Fe, Zn is not significantly stored in the human body and must be ingested daily.

The research of Burgos et al. (2007) and Paget et al. (2014) has shown that the cultivated diploid potatoes of the Andes are a good source of high concentrations of Fe and Zn which can be increased further by conventional breeding. Mineral content was determined by inductively coupled plasma–optical emission spectrophotometry (ICPOES). Concentrations in raw, peeled tubers of native Andean potato accessions ranged from 9 to 37 mg Fe kg^{-1} DW and from 8 to 20 mg Zn kg^{-1} DW, with diploid Phureja accession 703274 and diploid Stenotomum accession 701165 showing the highest levels of Fe and Zn, respectively (Burgos et al. 2007). In a derived diploid breeding population, Paget et al. (2014) found additive genetic variation and moderate narrow-sense heritabilities for individual plants on a DW basis (0.42 for Fe and 0.37 for Zn, averaged over two generations). Furthermore, the maximum mineral contents found in the second generation of 1329 clones were 42.5 mg Fe kg^{-1} DW and 38.9 mg Zn kg^{-1} DW. The researchers concluded that the improved material should provide a useful source of dietary Fe and Zn. They estimated that in Peru, the best material would provide 56% of the estimated average requirement (EAR) of Fe for children aged 4–8 years and 28% for female adults aged 19–30 years. Similarly, it would provide 37% of the EAR of Zn for children aged 4–8 years and 22% for female adults aged 19–30 years. We will return to this breeding work in a later chapter.

Genetic biofortification of Zn, but not Fe, can be complemented with agronomic biofortification using foliar or soil fertilizers (White et al. 2012, 2017; Kromann et al. 2017). The highest tuber concentrations of both Fe and Zn found by Kromann et al. (2017) were in the diploid Phureja cultivars Chaucha amarilla and Chaucha roja. Kromann et al. (2017) also reported that CIP has more than 50 genetically biofortified potato clones with the capacity to absorb and contain more than 35 mg Fe and 35 mg Zn kg^{-1} tuber flesh DW.

Paget et al. (2014) also found genetic variation for calcium (Ca) content in their diploid population derived from Stenotomum, Goniocalyx and Phureja accessions studied by Burgos et al. (2007). However, they did not consider Ca a high priority compared with Fe and Zn. Although the potato is not a rich source of Ca, high tuber Ca content from Ca fertilization is associated with reduced incidences of disease (common scab) and physiological disorders (hollow heart) and hence is of interest to breeders (Zorrilla et al. 2014; Chung et al. 2016). For example, Zorrilla et al. (2014) wanted to combine the high yield and specific gravity of chipping (crisping) cultivar Atlantic with the internal tuber quality and moderate resistance to pitted (common) scab of cultivar Superior, which also has a high tuber Ca content compared with Atlantic. After hybridization of these cultivars, they were able to identify eight promising offspring clones that had similar chipping quality to Atlantic, but higher tuber Ca content and lower incidences of hollow heart and pitted scab. Palta (2020) has provided a summary, with references, of the importance of calcium for healthy cell membranes and strong cell walls, as well as for the role of calcium as a signalling molecule mediating plant response to environmental stresses and hormones. Other key points are as follows: Tubers accumulate much less calcium than leaves and stems because calcium moves with the water in the xylem, and the tuber is a low-transpiring organ. Furthermore, it is the roots on the tuber and at the tuber-stolon junction that supply water to the tuber, not the main root system. Tuber calcium content can therefore be increased by placing calcium in the tuber and stolon areas during tuber bulking. Total calcium applied at 100–200 kg/ha has been shown to reduce storage rot and internal defects (brown centre, internal brown spot and hollow heart), and also black spot bruise injury. Transgenic potato plants of cultivar Atlantic which overexpressed a short, deregulated version of the cation exchanger 1 gene (*sCAX1*, a vacuolar calcium transporter gene) resulted in calcium deficiency symptoms in potato leaves and tubers, and a high incidence of hollow heart (Zorrilla et al. 2019). These symptoms could be mitigated through higher levels of Ca^{2+} in the media or soil nutrient solution. The researchers concluded that cellular calcium was being transported to the vacuole and sequestered in the form of calcium oxalate. Finally, Palta (2020) reported that genetic variability exists for tuber calcium accumulation ability among wild potato species (*Solanum microdontum* has high ability), as well as among cultivars, and that tuber calcium is a heritable trait.

4.9.5 Biofortification of B Vitamins

The five most abundant B vitamins in potato are B_1 (thiamin), B_3 (niacin), B_5 (pantothenic acid), B_6 (pyridoxine) and B_9 (folate). A summary of their chemical structures, recommended daily uptakes for humans, biological functions and contents in potato can be found in the book chapter by Navarre et al. (2018). Here we are going to look at whether or not increasing contents can be justified as breeding objectives and whether or not genetic variation for contents has been found.

Thiamin (vitamin B_1: thiamine diphosphate) deficiency can cause cardiovascular and neurological problems commonly referred to as beriberi (Goyer and Haynes 2011; Navarre et al. 2018). Severe thiamin deficiency is still common in developing countries, where the main food source is low in thiamin and rich in carbohydrate (e.g. white rice), and marginal deficiency remains a health concern in industrialized countries. Potato is potentially a good source of thiamin despite some loss (up to 12%) during cooking and processing because thiamin is heat sensitive (Goyer and Haynes 2011). Navarre et al. (2018) reported that 100 g of boiled potatoes contains around 10% of the Recommended Dietary Allowance (RDA) of thiamin and French fries and baked potatoes slightly less. Thus, developing potatoes with increased thiamin content could have a positive impact on human health and seems a worthwhile breeding objective.

Goyer and Haynes (2011) provided evidence of genetical variation for thiamin content among 39 potato cultivars and clones, mainly from the Pacific Northwest (USA) Potato Development Program. Thiamin was extracted by combining acid digestion and enzymatic hydrolysis and concentrations were measured by a microbiological assay employing *Lactobacillus viridescens* grown in thiamine-deficient medium. The researchers considered this a simple, fast, low cost and very sensitive method that can be performed in any standard laboratory without the need for expensive equipment such as HPLC. Thiamin concentrations ranged from 292 to

1317 ng g^{-1} fresh-weight (FW). Broad-sense heritability of the genotype means over the four environments was 0.49 (95% confidence interval of 0.21–0.72), but there were significant clone x environment interactions. Tubers harvested at a mature stage late in the growing season had higher amounts of thiamin than tubers harvested at a young stage early in the season. Thiamin concentrations increased slightly in some genotypes during storage at 7 °C for 1 to 3 months.

Goyer and Sweek (2011) examined a wider range of germplasm for thiamin concentrations, namely, 54 field-grown clones from 33 accessions of primitive Andigenum cultivars and 3 modern potato cultivars, in 2009 and 2010, together with 64 glasshouse-grown accessions of 25 wild potato species in 2010. Thiamin concentrations ranged from 490 to 2325 ng g^{-1} FW (2908 to 16,401 ng g^{-1} DW in 2010, where DW is dry weight). The majority of clones in both years had thiamin concentrations of 1200 to 1600 ng g^{-1} FW. The researchers identified Andigenum accession PI 320377 (a Phureja from Colombia) as promising material to integrate in breeding programmes for thiamin enhancement. Among the wild species, *S. vernei* subsp. *vernei* and *Solanum acaule* subsp. *aemulans* had the highest thiamin concentrations ($\geq$8633 ng g^{-1} DW).

Niacin (vitamin B_3) deficiency can lead to pellagra, a disease associated with dementia, diarrhoea and dermatitis, and is common in countries where the population relies on a corn-based diet (Navarre et al. 2018). Although 100 g of baked or boiled potatoes contains around 10% of RDA of niacin, little research has been done on genetic variation in potato and improving content is not currently a breeding objective.

Panthothenic acid (vitamin B_5) deficiency is uncommon in humans so increasing its content in potato is not currently a breeding objective (Navarre et al. 2018).

Pyridoxine (vitamin B_6: pyridoxal-phosphate) deficiency has wide effects on human health with symptoms ranging from neurological disorders to anaemia, certain dermatological problems and weakening of the immune system (Navarre et al. 2018). Potatoes are an excellent source of pyridoxine with baked potatoes and potato chips containing 23 and 60% of the RDA per 100 g, respectively (Navarre et al. 2018). Hence, developing potatoes with increased pyridoxine content could have a positive impact on human health and seems a worthwhile breeding objective. Juhaszi et al. (2014) have developed a novel method for simultaneous detection of vitamins B_5, B_6 and C (see next section) by GC–MS (gas chromatograph–mass spectrometer) that can handle 30–40 samples per day. It is faster than high-performance liquid chromatography (HPLC) and insensitive to tuber colour, unlike some spectrophotometric methods. They found genetical variation for all three vitamins in two diploid potato populations, each grown at two sites in the Netherlands. At Washington State University, USA, Mooney et al. (2013) also provided evidence of genetical variation among cultivars for pyridoxine content that could be used in a breeding programme. Contents ranged from 16.9 to 27.0 μg B_6/g DW. Furthermore, concentrations increased significantly over a 227-day storage period at either 6.7 or 8.9 °C, with the greatest increases (23–42%) at the higher temperature.

Folates (vitamin B_9: tetrahydrofolate and its derivatives) are essential micronutrients in the human diet whose deficiency is associated with increased risk of birth defects, neural tube defects, cardiovascular diseases, anaemia and some cancers (Goyer and Navarre 2007; Robinson et al. 2015). Current folate intake is suboptimal in most of the world's populations, even in developed countries. Although folic acid supplements and food fortification have proved efficient, they are expensive and hard to implement in developing countries. Increasing the folate concentration in potatoes by breeding should have a positive benefit and currently more breeding work is being done on folate than other vitamins (Goyer and Navarre 2007, 2009; Goyer and Sweek 2011; Robinson et al. 2015). Robinson et al. (2015) thought that doubling the folate content of modern cultivars was a realistic target. Based on the current per capita consumption of 137 g d^{-1} in the United States, such a potato would provide around 11% of the recommended daily need of 400 μg, assuming 20% dry matter (DM) and 80% retention during cooking. They also pointed out that further increases could be achieved by optimizing the time of harvest and storage conditions, as young tubers contain up to twofold the amount found in mature tubers and folates accumulate up to twofold in tubers stored at cold temperature.

Goyer and Navarre (2007) analysed 67 genotypes including white-, purple-, red-, blue- and yellow-fleshed potatoes and several wild species. Folate concentrations varied from 521 to 1373 ng g^{-1} DW, with white-fleshed cultivar Ranger Russet at 1037 ng g^{-1} DW. The five genotypes with the highest folate levels were colour-fleshed, of which four were yellow. Skin contained about 30% higher folate concentrations than flesh.

Goyer and Sweek (2011) examined a wider range of germplasm for folate concentrations, namely, 54 field-grown clones from 33 accessions of primitive Andigenum cultivars and 3 modern potato cultivars, in 2009 and 2010, together with 64 glasshouse-grown accessions of 25 wild potato species in 2010. Folate concentrations ranged from 15 to 416 ng g^{-1} FW (388 to 2098 ng g^{-1} DW in 2010). The majority of clones in both years had folate concentrations of 100 to 200 ng g^{-1} FW. The researchers identified clone RN 104.02 (from Andigenum accession PI 473260) as promising material to integrate in breeding programmes for folate enhancement. Among the wild species, *Solanum boliviense* accession 597736 had the highest folate concentration (3031 ng g^{-1} DW), but the other *S. boliviense* accession had a much lower amount, suggesting variability within species. *S. vernei* subsp. *vernei* also had relatively high folate concentrations.

Robinson et al. (2015) continued the search for high folate potato germplasm for breeding purposes. They screened 250 individual plants from 77 accessions and 10 *Solanum* species (*Solanum stipuloideum*, *S. chacoense* subsp. *chacoense*, *Solanum candolleanum*, *S. acaule*, *Solanum demissum*, *Solanum microdontum*, *Solanum okadae*, *S. tuberosum* Andigenum Group, *S. boliviense* and *S. vernei*) and found a 10.5-fold range of folate concentrations (221 to 2336 ng g^{-1} DW). Some individuals from *S. boliviense* (PI 597736), *S. tuberosum* Andigenum Group (PI 225710 and PI 320377, both Phureja from Colombia) and *S. vernei* (PI 558149 and PI 230468) had double the folate concentrations of commercial cultivars such as Russet Burbank (1040 ng g^{-1} DW).

Currently, screening germplasm for folate is a time consuming and tedious analytical process that involves a tri-enzyme extraction and microbial assay. Furthermore, wild species may produce small or even no tubers in the field so that they need to be grown in winter in greenhouses or crossed with adapted cultivated potatoes. Hence, Bali et al. (2018) sought molecular markers (SNPs) linked to genes (alleles) for high folate content with a view to practising marker-assisted selection for high folate on a large number of individuals in subsequent breeding work. They crossed a high-folate clone (Fol 1.6) from diploid *S. boliviense* (PI 597736) and a low-folate diploid *S. tuberosum* clone and then intermated the progeny to generate a segregating 'F_2' population of 94 individuals which was assessed in a greenhouse. Folate content ranged from 304 to 2952 ng g^{-1} DW. Associations were sought with 4174 polymorphic SNPs. The authors concluded that there were two major QTLs located in chromosomes 4 and 6, and additional minor QTLs in other chromosomes, that contribute to folate biosynthesis but acknowledged that further work was required to confirm their results and develop markers for use in breeding.

4.9.6 Biofortification of Vitamin C (Ascorbic Acid)

Vitamins C (L-ascorbic acid) is an essential micronutrient in the human diet with scurvy a well-known symptom of vitamin C deficiency. As a staple food, the potato can be a more important source of vitamin C than fruits and vegetables that have higher contents but are only complementary components of the diet. Love and Pavek (2008) argued that a three-pronged approach should be taken to increase the ascorbic acid content of potatoes, involving breeding, improved crop management and modification of cooking processes. They reported that tuber ascorbic acid concentrations ranged from 11.5 to 29.8 mg 100 g^{-1} FW among parental clones in North American breeding programmes and concluded that adequate variability exists within adapted, high-quality potato germplasm for rapid progress in breeding for higher ascorbic acid concentration of tubers.

Burgos et al. (2009b) identified considerable genetic variation in ascorbic acid content among 25 Andean potato cultivars grown in three environments, despite some genotype × environment interaction. Concentrations were evaluated by a spectrophotometric method based on the ability of ascorbic acid to reduce the dye 2,6-dichloroindophenol. They ranged in freshly harvested, raw, peeled tubers from 22.2 to 121.4 mg 100 g^{-1} DW (6.5 to 36.9 mg 100 g^{-1} FW), with diploid cultivar (CIP) 704393 (Goniocalyx Group) showing the highest levels in all three locations (105.1 to 121.4 mg 100 g^{-1} DW or 33.6 to 34.0 mg 100 g^{-1} FW). Burgos et al. (2009b) point out that this value is higher than those reported for Indian, Canadian and Norwegian potato cultivars but lower than the highest value reported for Korean cultivars, albeit in the central part of the tuber. Cultivar 704393 also had the highest level after cooking (54% retention on boiling) and after storage for 26 weeks under farmers' conditions (34% retention). In conclusion, 100 g of fresh-harvested and boiled potatoes of cultivar 704393 could provide adults with 17% to 20% of their RDA (100–120 mg d^{-1}) of ascorbic acid.

Cho et al. (2013) identified 10 out of 268 advanced clones in their Korean programme as having high ascorbic acid contents of 32.3 to 38.9 mg 100 g^{-1} FW. Among them, four double-cropping clones with short dormancy were selected for further analysis, with cultivars Dejima and Chubaek as controls. Ascorbic acid content was higher in younger tubers than in mature ones, and its content was highest in the selected clones 90 days after planting. During storage, ascorbic acid content significantly declined in all clones. The decrease was highest in H06035-4 and H06085-2 compared to cultivars Chubaek and Dejima, so the differences between selected clones and control cultivars narrowed after 4 months of storage. Nevertheless, H06035-4 was selected as a promising high ascorbic acid clone because of its agronomic characteristics and consistency of high ascorbic content.

4.10 Beneficial Phytochemicals

Potatoes contain a number of beneficial phytochemicals, including carotenoids and phenolic compounds such as phenolic acids and anthocyanins, many of which are antioxidants (Brown 2005; Ezekiel et al. 2013). The health-promoting effects of carotenoids and phenolic compounds have been reviewed by Lachman et al. (2016) and Akyol et al. (2016), respectively, along with the effects on them of growing conditions, tuber development, post-harvest storage and cooking processes. Short-term feeding of foods rich in the retinal carotenoids lutein and zeaxanthin can substantially increase macula pigment density in human subjects (Hammond et al. 1997).

All potatoes contain chlorogenic acid as the principal phenolic acid. Yellow and orange flesh are due to carotenoids, whereas the pink, red, blue and purple colours in potato tubers are due to anthocyanins whose antioxidant properties have been demonstrated in red- and purple-fleshed potatoes (Brown et al. 2003; Reyes et al. 2005). Kaspar et al. (2011) confirmed that yellow-fleshed potatoes had higher concentrations of phenolic acids and carotenoids than white-fleshed ones and that purple-fleshed potatoes had higher concentrations of phenolic acids and anthocyanins. Tierno et al. (2015) found that losses of total phenolics, anthocyanins and carotenoids occurred on boiling peeled tubers of highly pigmented potatoes, with big differences between genotypes in retention percentages: 45% to 74% for phenolics, 27% to 82% for anthocyanins and 32% to 100% for carotenoids.

The enhancement of anthocyanins and carotenoids, by conventional breeding and genetic modification, provides an opportunity to improve the nutritional value of potatoes and their processed products; but it will be important to assess new cultivars for their total and individual phytochemicals after cooking and processing.

4.10.1 Anthocyanins

Lewis et al. (1998a, b) used HPLC to quantify the anthocyanins, flavonoids and phenolic acids in tetraploid cultivars of *S. tuberosum* with coloured tubers, and in wild, tuberous *Solanum* species, the tubers of which were mostly white or light purple. When the flesh was also pigmented, the same anthocyanins were found as in the skin, but at much lower concentrations. The flesh tissues in which anthocyanin pigmentation can be observed are the vascular bundle, the cortex, the peri-medullary zone and the central pith (van Eck 2007). Combinations of different pigments and pigment patterns occur. Whereas red- and purple-skinned potatoes are relatively common, potatoes with varying distributions of anthocyanins in the flesh (rings, arcs and radiating stars) are rare outside the native cultivars of the Andes. Solidly coloured red or purple tuber flesh is unusual throughout the world but occurs as rare segregants in certain red- and purple-skinned breeding populations, as found by Brown et al. (2003). Total anthocyanin content ranged from 6.9 to 35 mg 100 g^{-1} FW in the red-fleshed clones and from 5.5 to 17.1 in the purple-fleshed ones. A few pigmented flesh types are available commercially as specialty potatoes, for example, cultivars Cranberry Red and All Blue (Brown et al. 2003). Interest now centres on increasing the concentrations of anthocyanins as a way of improving the levels of antioxidants in potato tubers, but first it is useful to consider some genetics, particularly as already mentioned, consumers in some countries may have a preference for a particular skin colour (e.g. white or red).

The inheritance of anthocyanin pigmentation in cultivated potatoes (diploid and tetraploid) was reviewed by van Eck (2007). Key genes/alleles are at locus *R* (chromosome 2) for the production of red anthocyanins, locus *P* (chromosome 11) for the production of purple (and blue) anthocyanins and locus *I* (chromosome 10) which determines whether or not *R* and *P* are expressed in the tubers (both skin and flesh). Two alleles are recognized at the *I* locus: *Iep* giving pigmentation in the epidermis of the tuber and *Ico* in the outer cortex. Three alleles are recognized at the *R* locus: *R*, R^{pw} and *r*, with R^{pw} giving pink tubers. De Jong et al. (2003) were able to show that although the dosage of *R* in tetraploids contributes to the intensity of skin colour, it is not a major factor. White potatoes lack either dominant allele *I* or both *R* and *P*; pink and red potatoes lack *P* but have *I* with R^{pw} and *R*, respectively; and purple potatoes have *P* and *I*. The *I* locus is in fact part of the *B-I-F* linkage group whose alleles determine the tissue-specific regulation of anthocyanin biosynthesis (van Eck 2007). In tetraploid potatoes, the *I* locus was originally referred to as *D* for developer by Salaman (1910). Candidate genes for the loci were proposed by De Jong et al. (2004) and subsequently confirmed (Fig. 4.6). The *P* locus on chromosome 11 codes for flavonoid 3′, 5′–hydroxylase (Jung et al. 2005); the *R* locus (*drf*) on chromosome 2 codes for dihydroflavonol 4-reductase (Zhang et al. 2009a) and the *D* (*I*) locus on chromosome 10 encodes an R2R3 MYB transcription factor that regulates expression of multiple anthocyanin structural genes in tuber skin (Jung et al. 2009). As purple-skinned potatoes can have genotype *R*--- or *rrrr*, both the 'red' and 'not-red' alleles of *drf* must code for catalytically active enzymes. The naringenin

↓ F3H

← dihydrokaempferol →

↓ ↓ F3'H ↓ F3'5'H

↓ dihydroquercetin → dihydromyricetin

↓ F3'5'H

↓ DFR ↓ DFR ↓ DFR

leucopelargonidin leucocyanidin leucodelphinidin

↓ ↓ ↓

pelargonidin-3-glucoside cyanidin-3-glucoside delphinidin-3-glucoside

(red) (pink) (purple)

↓ ↓ ↓

F3H	flavanone 3–hydroxylase
F3'H	flavonoid 3'–hydroxylase
F3'5'H	flavonoid 3', 5'–hydroxylase
DFR	dihydroflavonol 4-reductase

Fig. 4.6 Part of anthocyanin biosynthetic pathway, simplified. (Holton and Cornish 1995; from Bradshaw 2016 with permission)

enzyme encoded by the red allele appears capable of reducing dihydromyricetin as well as dihydrokaempferol (and dihydroquercetin), whereas the non-red allele(s) appears capable of reducing the former, but not the latter. It is assumed that similar arguments apply to the pink allele and dihydroquercetin. In contrast, the *pppp* genotype results in a loss of enzyme activity. The dominant R2R3 MYB domain allele is transcribed in red and purple tubers, but the recessive allele is not transcribed in white ones with genotype *iiii* (Jung et al. 2009).

In the inbred-line-derived diploid F_2 mapping population of Endelman and Jansky (2016), major genes/QTLs were segregating for tuber pigmentation, colour and intensity. For four of them, the published candidate gene was within the 1.5-LOD interval of the corresponding locus in the mapping population. Hence, they were able to provide confirmatory evidence for *D*(*I*) encoding the R2R3-MYB transcription factor Stan2 (*S. tuberosum* anthocyanin 2), *P* encoding flavonoid 3′, 5′–hydroxylase, *R* encoding dihydroflavonol 4-reductase and *Pf2* on chromosome 9 being a regulatory gene for flesh pigment intensity, namely, the basic helix-loop-helix (bHLH) regulatory gene *S. tuberosum* anthocyanin 1 (Stan1). The *Pf2* locus corresponds to one of the three QTLs detected by Zhang et al. (2009b) which influenced the extent (none to complete) of flesh pigmentation (purple) in the progeny of a diploid cross segregating for the trait. The other two QTLs were on chromosomes 5 and 8. The other QTL affecting both pigmented skin and flesh intensity detected by Endelman and Jansky (2016) was *Pf* on chromosome 10, which was linked to but separate from *D(I)*, and which they postulated to be the *Pf* (pigmented flesh) locus described by De Jong (1987).

Let us now turn to quantitative variation in the concentrations of anthocyanins. Proof of the existence of such variation has come from the breeding and release in North America of speciality potatoes such as AmaRosa (Brown et al. 2012a) and Purple Pelisse (Vales et al. 2012), which are remarkable for their unusual colours. Cultivar AmaRosa resulted from a seedling from a cross made in 2000 between a red-fleshed clone and a bulk of pollen from other red-fleshed clones. The red-skinned, red-fleshed seedling was multiplied and evaluated over a number of years, including extensive trials in 2006 and 2007. AmaRosa tubers had higher total anthocyanin and hydrophilic oxygen radical absorption capacity (H-ORAC) than the control cultivar All Blue, while total carotenoids and lipophilic oxygen radical absorption capacity (L-ORAC) were equivalent. Hence, overall antioxidant capacity was greater. Cultivar Purple Pelisse resulted from a cross made in 2000 between a red-skinned, red-fleshed clone and a bulk of pollen from red-fleshed potatoes. It was multiplied from a seedling identified in 2001 as having purple-skinned, purple-fleshed tubers. It was released as a cultivar in 2009 after extensive trials in 2006 and 2007. Purple Pelisse tubers had higher total anthocyanin and hydrophilic oxygen radical absorption capacity (H-ORAC) than the control cultivar All Blue, and higher total antioxidant values than cultivars Russet Burbank and Snowden. More recently, in 2018, ICAR-CPRI in India have released a new speciality potato cultivar Kufri Neelkanth which is high in antioxidants. Its tubers have a dark purple skin and cream flesh (Luthra et al. 2020).

The overall conclusion from this section is that red- and purple-fleshed potatoes (Fig. 4.7) can be identified from appropriate crosses in a conventional breeding programme but it remains to be seen if such selection will continue to be for speciality potatoes rather than for more general table use.

4.10.2 Carotenoids

Structurally, carotenoids are terpenoids, derived by condensation of prenyl pyrophosphates (Sandmann 2001), whose conjugated double-bond system determines their colour and is responsible for their biological actions as antioxidants.

The conversion of geranylgeranyl pyrophosphate to phytoene by phytoene synthase is the first dedicated step in the formation of carotenoids (Fig. 4.8). We shall see that the main carotenoids present in cultivated potatoes are lutein, lutein-5,6-epoxide, zeaxanthin, antheraxanthin, violaxanthin and neoxanthin, and that

Fig. 4.7 Red- and purple-fleshed potatoes on display in AgriExpo at Global Potato Conclave 2020, Mahatma Mandir, Gandhinagar, Gujarat, India, 28–31 January 2020 (30 January 2020)

beta-carotene, the precursor of vitamin A, is either absent or present in low concentrations. Orozco et al. (2013) confirmed that carotenoids can be present in potatoes in esterified as well as free form, and in their profiles of 60 cultivars found a direct correlation between the carotenoid content and the esterified fraction. They thought that esterification facilitated accumulation of carotenoids and hence the presence of esters should be included as a breeding trait. Sulli et al. (2017) confirmed the presence of esters in their core set of 10 cultivars and commented on their positive contribution to carotenoid stability, but also their negative effect on bioavailability.

Burgos et al. (2012) found that changes in concentrations of individual carotenoids due to boiling varied significantly among accessions of native Andean potatoes with diverse intensities of yellow flesh. Accessions (CIP) 705821 (Phureja) and 705172 (Phureja) were light yellow, 704393 (Goniocalix) and 701862 (Goniocalix) were intermediate yellow, and 702472 (Goniocalix), 705799 (Phureja) and 704218 (Phureja) were deep yellow. Boiling significantly reduced the violaxanthin and antheraxanthin concentrations of all accessions but lutein and zeaxanthin concentrations were not affected or were higher than in raw tubers. Burgos et al. (2013) then estimated the lutein and zeaxanthin concentrations in boiled, freeze dried and milled samples of seven yellow-fleshed accessions (including four from previous study) before and after different steps (gastric, duodenal and micellar phase) of in vitro digestion. For all accessions, the amounts of lutein and zeaxanthin in the micelles after micellarization (i.e. the bioaccessible amounts) were significantly lower than the original amounts found in the boiled samples. The accession 701862 showed the highest bioaccessible lutein concentration (281 out of 396 μg 100 g^{-1} FW) and the accessions 703566 and 704218 showed the highest bioaccessible zeaxanthin concentrations (665 out of 1196 and 608 out of 1106 μg 100 g^{-1} FW, respectively).

Geranylgeranyl pyrophosphate

↓ Phytoene synthase (*CrtB* gene)

Phytoene (colourless)

↓

↓ Phytoene desaturase (*CrtI* gene)

Lycopene

Lycopene epsilon-cyclase ↓ ↓ Lycopene beta-cyclase (*CrtY* gene)

(*Lcye* gene) ↓ ↓

δ-Carotene (γ-Carotene)

↓ ↓

α-Carotene β-Carotene (red) → → → → Astaxanthin (red)

↓ ↓ (β-Carotene) ketolase (*bkt1* gene)

↓ ↓ β-Carotene hydroxylase (*Chy2* gene)

α-Cryptoxanthin β-Cryptoxanthin

↓ ↓ β-Carotene hydroxylase (*Chy2* gene)

Lutein (yellow) Zeaxanthin (orange) → → → → Astaxanthin (red)

↓ ↓ (Zeaxanthin) ketolase

↓ ↓ Zeaxanthin epoxidase (*Zep* gene)

Lutein-5,6-epoxide Antheraxanthin (yellow)

↓ Zeaxanthin epoxidase (*Zep* gene)

Violaxanthin (yellow)

↓

Neoxanthin (yellow)

Fig. 4.8 Carotenoid biosynthetic pathway (simplified)

Considering the mean potato intake in the Andes (500 g d^{-1}), the accession 701862 would provide 14% of the lutein intake suggested for health benefits and the accessions 703566 and 704218 would provide 50% more than the suggested zeaxanthin intake. Hence, higher levels of lutein and zeaxanthin are sensible targets when breeding new cultivars with higher levels of beneficial carotenoids, given their role in protection against macular eye degeneration (Snodderly 1995). Furthermore, there are relatively few dietary sources of zeaxanthin (Morris et al. 2004).

Bonierbale et al. (2009) confirmed that diploid Phureja is a good source of yellow-fleshed potatoes with high levels of lutein or zeaxanthin. They developed, validated and applied near-infrared reflectance spectroscopy (NIRS) to the characterization of

152 Phureja accessions (colour range: cream, light yellow and deep yellow) for total and individual carotenoid contents of raw tubers. NIRS estimated total carotenoids (103 to 2135 μg 100 g^{-1} FW) and zeaxanthin concentrations (trace to 1290 μg 100 g^{-1} FW) with high accuracy and could differentiate accessions with low, medium and high concentrations of violaxanthin (trace to 278 μg 100 g^{-1} FW), antheraxanthin (3 to 354 μg 100 g^{-1} FW), lutein (55 to 189 μg 100 g^{-1} FW) and β-carotene (trace to 18 μg 100 g^{-1} FW). NIRS can therefore be used to assess large numbers of genotypes (clones) in a breeding programme in preference to costlier and more time consuming HPLC. Accessions 701025 and 703566 showed the highest estimated zeaxanthin concentrations (above 1000 μg 100 g^{-1} FW). Forty-three accessions showed relatively high concentrations of β-carotene (above 10 μg 100 g^{-1} FW) with 703552, 706782, 701570, 706787, 706759, 706767, 706763, 706829, 704213, 704227 and 703581 showing the highest values. Burgos et al. (2009a) took a more detailed look at total and individual carotenoid concentrations in 23 of the 152 accessions, using spectrophotometry and HPLC. They confirmed that total carotenoid concentration was positively correlated with antheraxanthin and zeaxanthin concentrations, but negatively correlated with β-carotene concentration. They also pointed out that the concentrations of carotenoids in the diploids they studied were much higher than those reported for tetraploid cultivars. Breithaupt and Bamedi (2002), for example, reported total carotenoid contents of 58 to 175 μg 100 g^{-1} FW in four yellow-fleshed cultivars and 38 to 62 μg 100 g^{-1} FW in four white-fleshed ones.

It is worth mentioning here a more general method of chemical analysis. Drapal et al. (2020) used metabolite profiling to identify biochemical compounds with phytochemical, nutritional and flavour properties in raw and cooked (boiled) potato tubers from a panel of 32 advanced tetraploid breeding clones, 15 diploid landraces from Groups Phureja and Stenotomum, 15 diploid biofortified bred clones and 10 wild genotypes from species indicated as ancestors of cultivated potato. The potatoes were grown in 2016 in a field trial using local agronomic practices in the highlands of Peru at 4100 masl. The researchers identified 77 metabolites including amino acids, sugars, intermediates of the tricarboxylic acid cycle (Krebs cycle), phenolics, glycoalkaloids and carotenoids. The study highlighted which metabolites with nutritional properties are available in potatoes and how they change after a cooking treatment. It confirmed that total carotenoid content of the diploid bred clones had changed as intended through the biofortification crossing programme of CIP. The reader is referred to the paper for details of the profiling and the results.

High heritabilities for flesh colour intensity have been reported for tetraploid breeding clones (0.93 on clonal mean basis) (Haynes et al. 1996) and for long-day-adapted, diploid breeding families (0.99 for half-sib families) (Haynes 2000), as well as a negative correlation with tuber size, as measured by weight (Haynes et al. 1994). Hence, in a practical breeding programme, selection for intense yellow flesh in similar sized tubers should result in higher contents of total carotenoids without a reduction in tuber weight. If necessary, the different degrees of yellowness can be quantified by spectrophotometry (Burgos et al. 2012). However, the breeder may also be interested in the concentrations of individual carotenoids, so let us look at some more results.

Individual carotenoids can be separated and quantified by HPLC with photodiode array (PDA) detection, the eluting compounds being identified by their absorbance spectra and co-elution with commercially available authentic standards (Lu et al. 2001; Wolters et al. 2010). Eleven clones from the Phureja/Stenotomum (diploid) breeding population of Haynes (2000) were analysed by Lu et al. (2001) along with two tetraploid controls. They detected six major carotenoids in all of the material: neoxanthin, violaxanthin, lutein-5,6-epoxide, lutein, zeaxanthin and an unknown carotenoid. Total carotenoid content in the diploid clones ranged from 136 (white flesh) to 1435 μg 100 g^{-1} FW (intense yellow flesh), and the main carotenoids were lutein-5,6-epoxide (34%), violaxanthin (33%) and lutein (26%). In contrast, tetraploid cultivars Superior (white flesh) and Yukon Gold (yellow flesh) contained 64 and 111 μg 100 g^{-1} FW of total carotenoids, and the main components were violaxanthin (40%) and lutein (22%). Both total and individual carotenoid contents were positively correlated with intensity of yellow. Again, this suggests that selecting for more intense yellow flesh will result in higher levels of carotenoids.

The clones analysed by Lu et al. (2001) were chosen because they produced at least 5% of 2*n* pollen, and hence could be used to improve the nutritional value of tetraploid potatoes through 4*x* by 2*x* hybridizations, as done with cultivar Yukon Gold. It was this Canadian cultivar that opened the way to acceptance of yellow-fleshed potatoes by American consumers. Haynes et al. (2011) investigated the inheritance of carotenoid content in 4*x* by 2*x* hybridizations. They crossed a light yellow-fleshed tetraploid breeding clone with three diploid clones that produced 2*n* pollen and had high (dark yellow flesh), medium (moderate yellow flesh) and low (white/cream flesh) carotenoid levels. In both of the yellow-fleshed diploids, the proportion of zeaxanthin was low at 3.1% or less of the total carotenoid content. After harvest, in each of 2 years, carotenoids were extracted and quantified by HPLC in 13 or 14 randomly selected clones from each family, as well as cultivar Yukon Gold as control. A continuous distribution of carotenoid concentration with high- and low-segregants was observed in all three families. There were significant differences among clones and high broad-sense heritabilities for zeaxanthin, total lycopene beta cyclase (LCY-b) pathway carotenoids (zeaxanthin + antheraxanthin + violaxanthin + neoxanthin), lutein and total carotenoids (LCY-b + lutein) despite clone × year interactions. Average contents in the clones ranged from 57 to 280 μg zeaxanthin 100 g^{-1} FW (Yukon Gold 103), 177 to 548 μg LCY-b 100 g^{-1} FW (Yukon Gold 262) and 62 to 169 μg lutein 100 g^{-1} FW (Yukon Gold 81). Based on flesh colour segregation, the tetraploid parent was duplex (*Chy2Chy2chy2chy2*) for the *Chy2* allele 3 (*β-carotene hydroxylase 2*) governing yellow-flesh, the white-fleshed diploid was homozygous recessive (*chy2chy2*) and the two yellow-fleshed diploid parents were heterozygous (*Chy2chy2*) and produced 2*n* gametes by a second division restitution mechanism. The researchers concluded that tetraploid clones can be bred with higher levels of individual and total carotenoids than Yukon Gold.

Brown et al. (1993) studied orange-fleshed potatoes from the diploid breeding population of Haynes (2000) and showed that this colour was associated with large amounts of zeaxanthin. When averaged over a number of backcrosses, zeaxanthin levels were much higher in orange than white segregants (1426 compared with 304 μg 100 g^{-1} FW), whereas lutein levels were identical (146 μg 100 g^{-1} FW). Zeaxanthin also accounted for slightly more than half (51%) of the carotenoid content of a deep yellow, long-day-adapted Phureja cultivar, Inca Dawn (DB375/1), bred at the Scottish Crop Research Institute (Morris et al. 2004). Antheraxanthin (25%), violaxanthin (11%), an unidentified carotenoid (8%) and lutein (3%) were the other major carotenoids present. About 15% of the carotenoids were esterified. During tuber storage at 4 °C for 9 months in the dark, esterified carotenoids increased by 73%, zeaxanthin and antheraxanthin levels decreased by 11% and 7%, respectively, and the levels of lutein and an unidentified carotenoid both increased by 6%. Morris et al. (2004) found that there was an inverse trend between the level of zeaxanthin epoxidase transcript level and tuber carotenoid content in a range of potato germplasm, with implications for the regulation of carotenogenesis in potato tubers.

Romer et al. (2002) inhibited zeaxanthin conversion to violaxanthin in cultivars Baltica and Freya by tuber-specific down-regulation of the *zeaxanthin epoxidase* gene, using transformation with both sense and antisense constructs encoding the gene. As a consequence, zeaxanthin content was elevated 4–130-fold, reaching values up to 40 μg g^{-1} DW, while the amount of violaxanthin was dramatically reduced. Surprisingly, total carotenoid levels increased by up to 5.7-fold compared with controls. The transformed potatoes had dark yellow to orange flesh.

Wolters et al. (2010) subsequently demonstrated that only genotypes combining the presence of dominant *β-carotene hydroxylase 2* (*Chy2*) allele 3 with homozygosity for recessive *zeaxanthin epoxidase* (*zep*) allele 1 (locus on chromosome 2) produced orange-fleshed tubers that accumulated large amounts of zeaxanthin (>250 μg 100 g^{-1} FW). They performed association analysis between SNP haplotypes and flesh colour phenotypes in a wide range of diploid (including orange) and tetraploid (none were orange) potato genotypes. Just 1 out of 11 *Chy2* alleles (3) changed white into yellow flesh (in diploids and tetraploids), whereas none of the *lycopene epsilon cyclase* (*Lcye*) alleles (locus on chromosome 12) had a large effect on flesh colour. The *Lcye* gene product is required for the synthesis of α-carotene, the precursor of lutein (Fig. 4.8). The *Chy2* allele is the most likely candidate for the yellow flesh allele (*Y*) at the locus on chromosome 3 (Bonierbale et al. 1988; Brown et al. 2006), and not an allele of the phytoene synthase gene (Thorup et al. 2000). The F_2 population of Endelman and Jansky (2016) also segregated for yellow (DM and F_1 hybrid) and white (M6) flesh colour, and the 1.5-LOD interval for the single dominant locus on chromosome 3 contained the beta-carotene hydroxylase gene *Chy2*. Genes other than *Chy2* and *Lcye* determine the intensity of the yellow flesh colour. Wolters et al. (2010) also found that the recessive *zep* allele 1 contained a non-long-terminal-repeat (LTR) retrotransposon sequence in intron 1, which was absent in all other *Zep* alleles investigated, and had a lower expression level (lower steady state mRNA level). Only 5 out of 230 tetraploid cultivars contained the *zep* allele 1, all in simplex (single copy), thus confirming that this allele is rare in the tetraploid potato gene pool. In contrast, five diploid orange-fleshed potatoes were homozygous for *zep* allele 1: cultivars Papa Pura, Andean Sunrise and Yema de Huevo and clones IvP92-030-11 and IvP01-84-19. The researchers concluded that allele 1 is probably a Phureja *Zep* allele, as it is almost absent in the tetraploid Tuberosum gene pool apart from a few genotypes with Phureja in their ancestry. This means that breeding of a tetraploid potato cultivar with orange tuber flesh (with high zeaxanthin content) is a challenging task.

More recently, Sulli et al. (2017) were able to divide a collection of five tetraploid (*S. tuberosum*) and five diploid (four Phureja and one *S. chacoense*) genotypes into three groups based on *Chy2* allele 3 and *zep* allele 1. The first group comprised four tetraploids and Phureja cultivar Mayan Gold, heterozygous *Chy2*--- and homozygous *ZepZepZepZep* and *Chy2chy2* and *Zep***zep*, respectively (* novel allele), with yellow flesh and high levels of xanthophylls (antheraxanthin, violaxanthin, neoxanthin and lutein) and xanthophyll esters (23% to 41% of xanthophylls). The second group comprised one tetraploid and *S. chacoense*, homozygous recessive *chy2chy2chy2chy2* and *chy2chy2*, respectively, with white-flesh and low levels of carotenoids. Finally, the third group comprised three Phureja, heterozygous *Chy2chy2* and homozygous recessive *zepzep*, with orange flesh and high levels of non-esterified zeaxanthin.

It might be concluded from the previous paragraph that conventional breeding for high zeaxanthin levels and orange flesh would prove difficult. Nevertheless, in Japan, Sakamoto et al. (2017) have reported breeding a new tetraploid potato cultivar Nagasaki Kogane with good eating quality, high carotenoid content (yellowish/orange flesh) and resistance to pests (cyst nematodes) and diseases (bacterial wilt and *Potato virus Y* (PVY)). It was selected from a cross between cultivars Saikai 35 (Phureja cytoplasm) and Saikai 33, where the former was a cross between chromosome-doubled (tuber-disc method) cultivar Inca-no-mezame and cultivar Sakurafubuki. Diploid cultivar Inca-no-mezame was derived from crosses involving a Phureja clone and has a high carotenoid content. Cultivar Nagasaki Kogane contained lutein and zeaxanthin, but at lower concentrations than cultivar Inca-no-mezame (43.8 compared with 79.7 μg lutein 100 g^{-1} FW and 801.5 compared with 1131.7 μg zeaxanthin 100 g^{-1} FW). By way of comparison, the popular double-cropping tetraploid cultivar Dejima contained 48.1 μg lutein 100 g^{-1} FW and no zeaxanthin. Finally, diploid cultivar Inca-no-hitomi was selected from the open-pollinated progeny of cultivar Inca-no-mezame for its yellowish-orange flesh colour and other desirable agronomic characteristics (Kobayashi et al. 2008). They had similar total carotenoid contents which were four times higher than yellow-fleshed tetraploid cultivar Kita-akari, which in turn was ten times higher than white-fleshed tetraploid cultivar Irish Cobbler.

4.10.3 β-Carotene and Astaxanthin

Vitamin A deficiency remains a major health problem in some parts of the world where it is the major cause of blindness in children. Beta-carotene (provitamin A), the precursor of vitamin A, has been a target for genetic engineering of carotenoids in potatoes as it is either absent or present only in low concentrations in *S. tuberosum* and closely related *Solanum* species. Diretto et al. (2007), for example, transformed potato cultivar Desirée with a mini-pathway of key genes (enzymes) for the synthesis of β-carotene from geranylgeranyl diphosphate (Fig. 4.8). Expression of three bacterial genes from *Erwinia herbicola* encoding phytoene synthase (*CrtB*), phytoene desaturase (*CrtI*) and lycopene beta-cyclase (*CrtY*), under tuber-specific (patatin) promoter control, resulted in tubers with deep yellow flesh without any adverse leaf phenotypes. Carotenoid levels in one transformant had increased from 5.8 to 114.4 μg g^{-1} DW and β-carotene from 0.013 to 47.4 μg g^{-1} DW (and α-carotene 0 to 6.2, lutein 1.0 to 23.1, violaxanthin 0.7 to 5.6), compared with 'Golden Rice 2' at 31 μg g^{-1} DW. The transformants are known as 'Golden Potatoes' (GP). Chitchumroonchokchai et al. (2017) have provided data on bioavailability from transgenic lines GP17 and GP30. They concluded that for children and women of reproductive age, a 150-g serving of boiled Golden Potatoes has the potential to contribute 42% and 23%, respectively, of their daily requirement for provitamin A (measured as retinol activity equivalents). We shall look at this approach in more detail in the chapter on genetically modified potatoes.

There is also interest in engineering potatoes to accumulate ketocarotenoids. Astaxanthin, for example, has been associated with a range of health benefits and is used extensively as a feed additive in aquaculture (Morris et al. 2006). In nature, the main sources of astaxanthin are marine bacteria and microalgae. The conversion of β-carotene to astaxanthin requires the action of a hydroxylase as well as a ketolase. Morris et al. (2006) described the levels and types of carotenoid that accumulate in both tetraploid Tuberosum cultivar Desiree and diploid Phureja cultivar Mayan Gold on transformation with an algal *β-carotene oxygenase* gene (*bkt1* encoding a β-ketolase from *Haematococcus pluvialis*). A patatin promoter was used for tuber-specific expression. Two major ketocarotenoids, ketolutein and astaxanthin, absent in untransformed controls, were detected and gave tubers a reddish colour. Campbell et al. (2015) investigated approaches to optimizing tuber astaxanthin content. They achieved a nutritionally significant concentration of 77 μg g^{-1} DW, which has the potential to provide the recommended 5 mg daily intake of astaxanthin from 55 g DW of potato (ca. 220 g FW). Again, we shall look at this approach in more detail in the chapter on genetically modified potatoes.

4.11 Starch

In countries with a starch industry there are considerable opportunities to breed or genetically engineer potatoes with novel starches for use in food and non-food products. Potato starch has numerous useful functional properties such as thickening, coating, gelling, adhesion and encapsulation. Some of these functionalities are unique to the polymer as a result of the structure and organization of its linear amylose (17–25% of starch) and highly branched amylopectin (Li et al. 2006). Genetic engineering has already generated the two extreme types of potato starch. High amylopectin starch was produced by the down-regulation of the granule-bound starch synthase gene that controls amylose synthesis (Visser et al. 1991) and high amylose starch by concurrently down-regulating two starch branching enzymes, A and B (Schwall et al. 2000). Recently, Avebe (https://www.avebe.com/producten/eliane/) have produced a waxy potato starch (Eliane™) containing more than 99% amylopectin through traditional breeding. More information on potato starch and its modifications can be found in the review by Singh et al. (2016b). Finally, it is important to point out that starch extraction leaves the problem of what to do with the potato wastewater. The removal of potato proteins helps to deal with the polluting capacity of the wastewater and to offset the economic cost, but still leaves a use to be found for the deproteinated potato wastewater (Waglay and Karboune 2016).

4.12 Conclusions

The end uses of new cultivars determine the quality criteria that need to be added as breeding objectives to yield in the production environment. Nutritional value is important for the potato as a staple food; cooking quality, texture, taste and flavour are important for the potato as a vegetable sold for cash; and processing quality is defined by the manufacturers of French fries, chips (crisps) and other processed products. Finally, the composition of the potato starch determines its value for different uses in the starch industry. Broad-end uses can be divided into sub-categories such as market class, cooking method and type of processed product. Separate breeding programmes are required for each combination of end use and production environment, with some breeding objectives in common but others distinct. Substantial progress can only be made for a small number of traits so that identifying the most important is crucial for success. This will become clearer later in the book when we consider breeding methods. Furthermore, breeding objectives for quality traits need to be translated into selection criteria for use on the limited number of tubers from assessment plots and yield trials, immediately after harvest, and for some traits and after storage and cooking/processing. Assessing a few potential cultivars at the end of a breeding programme through taste panels or sophisticated chemical analyses does not exert much selection pressure but is important for marketing the chosen cultivar. The following conclusions apply to the quality traits considered in this chapter.

1. *Tuber Appearance*: This is a selection criterion when consumers and processors have a preference for a particular tuber shape, size, skin colour, skin finish or flesh colour. Deep eyes cause peeling losses in most uses and growth cracking and greening are also considered serious defects. Purple and red pigmentation are due to anthocyanins and yellow and orange flesh are due to carotenoids. The eye depth locus *Eyd/eyd* is closely linked to the tuber shape locus *Ro/ro* on chromosome 10 with deep dominant to shallow eye and round dominant to long shape.
2. *Dry Matter*: Dry-matter (DM) content is a quantitative trait that can be estimated from the specific gravity of tubers measured with a hydrometer or by weighing in air and then underwater. The major component is starch (65–75%). DM content affects the texture of the cooked potato, suitability for processing (high value desirable) and susceptibility to mechanical damage and bruising.
3. *Damage Resistance*: Mechanical damage and bruising result from mechanized harvesting and handling and can result in rejection by packers and processors. External damage ranges from slight scuffing to splits, cuts or cracks, while internal damage includes shattering and blackspot with localized discolouration due to melanin formation from enzymic browning. Damage resistance can be assessed in shaker tests which simulate riddling conditions and in various types of drop tests where tubers are impacted by a bolt or pendulum type of weight. Enzymic browning, which also occurs in cut or prepared potatoes, can be prevented by inhibition of polyphenol oxidase (PPO) activity by silencing the *Ppo5* gene or by prevention of tyrosine oxidation with addition of bisulphite or hypochlorite compounds during processing.
4. *Internal Defects*: Hollow heart, brown centre and internal rust spot are of particular concern for processors as well as being unwelcome in a vegetable for table use. Breeders select against particularly susceptible clones as and when environmental factors allow expression of the defect.
5. *After-Cooking Blackening (ACB) and Sloughing*: Disintegration from sloughing can be assessed on tubers steamed for 45 min and undesirable ACB once tubers have been cut and left for 4 to 5 h. During cooking, it is chlorogenic acid that combines with iron to form a complex which oxidizes on cooling to the coloured ferri-dichlorogenic acid responsible for the blue-grey discolouration at normal pH. It is common practice to place cooked tubers into four broad categories: A, B, C and D, with A (quite firm salad type) and D (very mealy, mushy type) the extremes.
6. *Flavour and Texture*: Flavour is important to the consumer but likely to be assessed on 40 or less potential cultivars by a professionally trained taste panel. Flavour is defined as the combination of the sense of taste (sweet, sour, salt, bitter and umami), which is perceived on the tongue, and the sense of smell when volatiles come into contact with receptors located in the nasal passages, either by sniffing or chewing. Raw potatoes have a limited but distinctive aroma whereas cooking potato tubers releases a wide range of volatile compounds. Sensitive volatile analysis of boiled potatoes can be done by entrainment of headspace volatile organic compounds followed by various types of

gas chromatography, but has not resulted in commonly used selection criteria. Texture is also an important sensory attribute as it has a direct influence on consumer preference as well as cooking time. Selection for a particular texture often results indirectly from selection for other traits such as cooking quality category or lack of disintegration on cooking. Textural qualities can be analysed for research purposes using probes and suitable load cells that measure compression, puncture and shear forces, but breeders still treat texture as a descriptive trait rather than a selection criterion.

7. *Fry Colour*: The production of unacceptably dark and bitter tasting fry products by non-enzymic browning is associated with a high percentage of reducing sugars (glucose and fructose). Frying at high temperatures (175 °C) results in a typical Maillard reaction between the reducing sugars and the α-amino acid groups of nitrogenous compounds (including free amino acids). Cold storage (4 °C and less) results in cold-induced sweetening (i.e. higher levels of reducing sugars). Lower levels of reducing sugars and hence sweetening (and acrylamide formation) can be achieved by conventional breeding and gene silencing through genetic transformation of tuber-expressed starch-associated *R1* and phosphorylase-L (*PhL*) genes. It is common practice to assess tubers for fry colour after storage at both at 4 °C and 10 °C, as well as sprouting (dormancy), keeping quality and cooking quality on the samples stored at 10 °C.
8. *Acrylamide*: Acrylamide is classified as a Group 2A carcinogen (probably carcinogenic to humans). It is formed during high-temperature cooking (frying, baking and roasting) and food processing by Maillard browning reactions between reducing sugars and free asparagine. Lower levels of these compounds can be achieved by conventional breeding and genetic engineering with European Union benchmark levels for potato chips (crisps) and French fries of 750 and 500 μg kg^{-1}, respectively, being realistic targets. Tuber-specific silencing of the *asparagine synthetase 1* gene by RNA interference reduces the acrylamide-forming potential of potatoes grown in the field without affecting tuber shape and yield.
9. *Glycoalkaloids*: Potato tubers contain small quantities of steroidal glycoalkaloids which are usually regarded as undesirable toxic compounds, despite having some anticarcinogenic properties. About 95% of total content is accounted for by α-solanine and α-chaconine. Concern that introgression breeding from wild species may inadvertently result in unacceptably high concentrations of tuber glycoalkaloids means that new cultivars are commonly tested to ensure that their contents are below the accepted safe limit of 20 mg 100 g^{-1} FW. Sufficient is known about the biosynthesis of α-solanine and α-chaconine for their virtual elimination by gene silencing to be feasible.
10. *Nutritional Value*: Deciding breeding targets for improved nutritional value is complicated by the fact that before being eaten, potatoes are stored at various temperatures for varying lengths of time and then cooked or processed in various ways, all of which can affect chemical composition. Furthermore, to be of value following food digestion, the nutrients must be bioavailable and taken up and utilized by the human body. There are, however, still issues over the

bioavailability of some nutrients and hence some uncertainty in relating chemical composition to recommended daily allowances (RDA) and intakes (RDI), and in setting sensible targets for improving nutritional value.

11. *Glycaemic Index* (*GI*): Despite a wide range of values, potatoes are usually considered a high GI food that can contribute to obesity, glucose intolerance, type-2 diabetes and cardiovascular disease. This has led to the search for low-GI potatoes (value below 70 or even 55 on scale with glucose 100). Such potatoes can be bred by selection for an increased amount of resistant starch by screening using a high-throughput in vitro starch digestion procedure. It remains to be seen if low-GI potatoes will be for speciality rather than more general table use. The introduction of inulin to potato is another way to improve its nutritional value by reducing its energy density through increased dietary fibre. Potatoes can be transformed so that their tubers synthesize the full range of inulin molecules naturally occurring in globe artichoke (*Cynara scolymus*).
12. *Protein Quantity and Quality*: Potatoes contain about 10% protein on a dry-weight basis (2% on fresh-weight basis), equal to that of most cereals such as rice or wheat. The protein is considered good quality, despite being low in methionine and cysteine, with a biological value of between 90 and 100. Where potatoes are an important dietary constituent, breeding for increased protein quantity and quality can be justified, but is not currently a high priority. Potatoes with enhanced nutritive value can be achieved by tuber-specific expression of a seed protein, Amaranth Albumin 1 (AmA1), from *Amaranthus hypochondriacus.*
13. *Mineral Content*: Increasing iron (Fe) and zinc (Zn) content to over 35 mg Fe and 35 mg Zn kg^{-1} tuber flesh dry weight, as being done by CIP as part of the HarvestPlus Biofortification Challenge Program, is feasible and can be justified as 60% of the global population is thought to be dietary deficient in Fe and more than 30% in Zn. The cultivated diploid potatoes of the Andes are a good source of high concentrations of Fe and Zn which can be increased further by conventional breeding. Genetic biofortification of Zn, but not Fe, can be complemented with agronomic biofortification using foliar or soil fertilizers. Although the potato is not a rich source of Ca, high tuber Ca content from Ca fertilization is associated with reduced incidences of disease (common scab) and physiological disorders (hollow heart) and hence is also of interest to breeders.
14. *Vitamin Content*: The five most abundant B vitamins in potato are B_1 (thiamin), B_3 (niacin), B_5 (pantothenic acid), B_6 (pyridoxine) and B_9 (folate). Current folate intake is suboptimal in most of the world's populations, even in developed countries. Although folic acid supplements and food fortification have proved efficient, they are expensive and hard to implement in developing countries. Currently, more breeding work is being done on folate than other vitamins and doubling the folate content of modern cultivars from around 1000 to 2000 ng g^{-1} dry weight (DW) is a realistic target, using accessions of Phureja potatoes, *S. boliviense* and *S. vernei*. Furthermore, as a staple food, the potato can be a more important source of vitamin C (ascorbic acid) than fruits and vegetables that have higher contents but are only complementary components

of the diet. Diploid Goniocalyx potatoes with contents of 105.1 to 121.4 mg 100 g^{-1} DW (or 33.6 to 34.0 mg 100 g^{-1} fresh-weight (FW)) have been found. Furthermore, tetraploid clones have been identified in a Korean breeding programme with contents of 32.3 to 38.9 mg 100 g^{-1} FW.

15. *Anthocyanins*: These are phenolic compounds which are antioxidants as well as conferring purple and red pigmentation to tubers, usually the skin but also in rare instances to the flesh. Key genes/alleles are locus *R* (chromosome 2) for the production of red anthocyanins, locus *P* (chromosome 11) for purple (and blue) anthocyanins, locus *I* (chromosome 10) for determining whether or not *R* and *P* are expressed in the tubers and locus *Pf2* (chromosome 9) for pigment intensity in the flesh. Both *I* and *Pf2* are regulatory genes. Red- and purple-fleshed potatoes can be identified from appropriate crosses in a conventional breeding programme but it remains to be seen if such potatoes will be for speciality rather than more general table use.
16. *Carotenoids*: The main carotenoids present in cultivated potatoes are lutein, lutein-5,6-epoxide, zeaxanthin, antheraxanthin, violaxanthin and neoxanthin, with beta-carotene, the precursor of vitamin A, either absent or present in low concentrations. In a practical breeding programme, selection for intense yellow flesh in similar-sized tubers results in higher contents of total carotenoids without a reduction in tuber weight. Lutein (yellow-flesh) and zeaxanthin (deep yellow/orange-flesh) have a role in protection against macular eye degeneration and are sensible targets for breeding new cultivars with higher levels of beneficial carotenoids. Only genotypes combining the presence of dominant *β-carotene hydroxylase 2* (*Chy2*) allele 3 with homozygosity for recessive *zeaxanthin epoxidase* (*zep*) allele 1 (chromosome 2) produce orange-fleshed tubers that accumulate large amounts of zeaxanthin (>250 μg 100 g^{-1} FW). Diploid Phureja is a good source of yellow-fleshed potatoes with high levels of lutein or zeaxanthin: Mayan Gold, heterozygous *Chy2chy2* and *Zep*zep*, respectively (* novel allele), has yellow flesh and high levels of xanthophylls (antheraxanthin, violaxanthin, neoxanthin and lutein) and xanthophyll esters (23% to 41% of xanthophylls). The *Chy2* allele is the most likely candidate for the well-known yellow flesh allele (*Y*) (chromosome 3). Phurejas, heterozygous *Chy2chy2* and homozygous recessive *zepzep*, have orange flesh and high levels of non-esterified zeaxanthin (e.g. cultivars Papa Pura, Andean Sunrise and Yema de Huevo). Tetraploid cultivar Nagasaki Kogane contains lutein and zeaxanthin, but at lower concentrations than in one of its progenitors, diploid cultivar Inca-no-mezame (43.8 compared with 79.7 μg lutein 100 g^{-1} FW and 801.5 compared with 1131.7 μg zeaxanthin 100 g^{-1} FW).
17. *Beta-carotene and Astaxanthin*: Vitamin A deficiency remains a major health problem in some parts of the world where it is the major cause of blindness in children. Potatoes can be genetically transformed to produce beta-carotene (provitamin A), the precursor of vitamin A, by insertion of three bacterial genes from *Erwinia herbicola* encoding phytoene synthase (*CrtB*), phytoene desaturase (*CrtI*) and lycopene beta-cyclase (*CrtY*), under tuber-specific (patatin) promoter control. The tubers have deep-yellow flesh without any adverse leaf phenotypes. One ‘Golden Potato’ transformant has 47.4 μg beta-carotene g^{-1}

DW compared with 'Golden Rice 2' at 31 μg g^{-1} DW. Carotenoid biosynthesis can also be engineered to produce astaxanthin, a feed additive in aquaculture.

18. *Starch*: In countries with a starch industry, there are considerable opportunities to breed or genetically engineer potatoes with novel starches for use in food and non-food products. Potato starch properties depend on the structure and organization of its linear amylose (17–25% of starch) and highly branched amylopectin. High amylopectin starch can be produced by down-regulation of the granule-bound starch synthase gene that controls amylose synthesis and high amylose starch by concurrent down-regulation of two genes encoding starch branching enzymes, A and B.

In the next chapter we need to consider the diseases and pests of potatoes and add resistance to these to yield and quality traits as overall breeding objectives.

References

Akilen R, Deljoomanesh N, Hunschede S, Smith CE, Arshad MU, Kubant R, Anderson GH (2016) The effects of potatoes and other carbohydrate side dishes consumed with meat on food intake, glycemia and satiety response in children. Nutr Diabetes 6:e195. https://doi.org/10.1038/nutd.2016.1

Akyol H, Riciputi Y, Capanoglu E, Caboni MF, Verardo V (2016) Phenolic compounds in the potato and its byproducts: an overview. Int J Mol Sci 17:835. https://doi.org/10.3390/ijms17060835

Amrein TM, Bachmann S, Noti A, Biedermann M, Barbosa MF, Biedermann-Brem S, Grob K, Keiser A, Realini P, Escher F, Amadò R (2003) Potential of acrylamide formation, sugars, and free asparagine in potatoes: a comparison of cultivars and farming systems. J Agric Food Chem 51:5556–5560

Andre CM, Legay S, Iammarino C, Ziebel J, Guignard C, Larondelle Y, Hausman J-F, Evers D, Miranda LM (2014) The potato in the human diet: a complex matrix with potential health benefits. Potato Res 57:201–214

Andre CM, Evers D, Ziebel J, Guignard C, Hausman J-F, Bonierbale M, zum Felde T, Burgos G (2015) In vitro bioaccessibility and bioavailability of iron from potatoes with varying vitamin C, carotenoid, and phenolic concentrations. J Agric Food Chem 63:9012–9021. https://doi.org/10.1021/acs.jafc.5b02904

Bali S, Robinson BR, Sathuvalli V, Bamberg J, Goyer A (2018) Single Nucleotide Polymorphism (SNP) markers associated with high folate content in wild potato species. PLoS One 13(2):e0193415. https://doi.org/10.1371/journal.pone.0193415

Bártová V, Bárta J, Brabcová A, Zdráhal Z, Horáčková V (2015) Amino acid composition and nutritional value of four cultivated South American potato species. J Food Compost Anal 40:78–85. https://doi.org/10.1016/j.jfca.2014.12.006

Beals KA (2019) Potatoes, nutrition and health. Am J Potato Res 96(2):102–110

Birt DF, Boylston T, Hendrich S, Jane JL, Hollis J, Li L, McClelland J, Moore S, Phillips GJ, Rowling M, Schalinske K, Whitley EM (2013) Resistant starch: promise for improving human health. Adv Nutr 4:587–601

Bonierbale MW, Plaisted RL, Tanksley SD (1988) RFLP maps based on a common set of clones reveal modes of chromosomal evolution in potato and tomato. Genetics 120(4):1095–1103

Bonierbale M, Grüneberg W, Amoros W, Burgos G, Salas E, Porras E, zum Felde T (2009) Total and individual carotenoid profiles in *Solanum phureja* cultivated potatoes: II. Development and application of near-infrared reflectance spectroscopy (NIRS) calibrations for germplasm characterization. J Food Compost Anal 22:509–516

Borch D, Juul-Hindsgaul N, Veller M, Astrup A, Jaskolowski J, Raben A (2016) Potatoes and risk of obesity, type 2 diabetes, and cardiovascular disease in apparently healthy adults: a systematic review of clinical intervention and observational studies. Am J Clin Nutr 104:489–498. https://doi.org/10.3945/ajcn.116.132332

Bradshaw JE (2019) Improving the nutritional value of potatoes by conventional breeding and genetic modification. In: Qureshi AMI, Dar ZA, Wani SH (eds) Quality breeding in field crops. Springer, Cham, pp 41–84

Bradshaw JE, Hackett CA, Pande B, Waugh R, Bryan GJ (2008) QTL mapping of yield, agronomic and quality traits in tetraploid potato (*Solanum tuberosum* subsp. *tuberosum*). Theor Appl Genet 116:193–211

Breithaupt DE, Bamedi A (2002) Carotenoids and carotenoid esters in potatoes (*Solanum tuberosum* L.): new insights into an ancient vegetable. J Agric Food Chem 50:7175–7181

Briddon A, Cunnington A, Harper G (2018) Post-harvest storage of potatoes. In: Wale S (ed) Achieving sustainable cultivation of potatoes volume 2: production, storage and crop protection. Burleigh Dodds Science Publishing, Cambridge, pp 119–135

Brown CR (2005) Antioxidants in potato. Am J Potato Res 82:163–172

Brown J, Mackay GR, Bain H, Griffith DW, Allison MJ (1990) The processing potential of tubers of the cultivated potato, *Solanum tuberosum* L., after storage at low temperature. 2. Sugar concentration. Potato Res 33:219–227

Brown CR, Edwards CG, Yang CP, Dean BB (1993) Orange flesh trait in potato: inheritance and carotenoid content. J Am Soc Hortic Sci 118:145–150

Brown CR, Wrolstad R, Durst R, Yang CP, Clevidence B (2003) Breeding studies in potatoes containing high concentrations of anthocyanins. Am J Potato Res 80:241–250

Brown CR, Kim TS, Ganga Z, Haynes K, De Jong D, Jahn M, Paran I, De Jong W (2006) Segregation of total carotenoid in high level potato germplasm and its relationship to beta-carotene hydroxylase polymorphism. Am J Potato Res 83:365–372

Brown CR, Haynes KG, Moore M, Pavek MJ, Hane DC, Love SL, Novy RG, Miller JC (2010) Stability and broad-sense heritability of mineral content in potato. Iron. Am J Potato Res 87:390–396. https://doi.org/10.1007/s12230-010-9145-4

Brown CR, Haynes KG, Moore M, Pavek MJ, Hane DC, Love SL, Novy RG, Miller JC (2011) Stability and broad-sense heritability of mineral content in potato. Zinc. Am J Potato Res 88:238–244. https://doi.org/10.1007/s12230-011-9188-1

Brown CR, Vales I, Yilma S, James S, Charlton B, Culp D, Hane D, Shock C, Feibert E, Pavek M, Knowles R, Novy R, Whitworth J, Stark J, Miller JC, Holm D, Quick R, Navarre R (2012a) "AmaRosa," a red skinned, red fleshed fingerling with high phytonutrient value. Am J Potato Res 89:249–254. https://doi.org/10.1007/s12230-012-9248-1

Brown CR, Haynes KG, Moore M, Pavek MJ, Hane DC, Love SL, Novy RG, Miller JC (2012b) Stability and broad-sense heritability of mineral content in potato: calcium and magnesium. Am J Potato Res 89:255–261. https://doi.org/10.1007/s12230-012-9240-9

Brown CR, Haynes KG, Moore M, Pavek MJ, Hane DC, Love SL, Novy RG (2013) Stability and broad-sense heritability of mineral content in potato: potassium and phosphorus. Am J Potato Res 90:516–523

Brown CR, Haynes KG, Moore M, Pavek MJ, Hane DC, Love SL, Novy RG, Miller JC (2014) Stability and broad-sense heritability of mineral content in potato: copper and sulfur. Am J Potato Res 91:618–624

Burgos G, Amoros W, Morote M, Stangoulis J, Bonierbale M (2007) Iron and zinc concentration of native Andean potato cultivars from a human nutrition perspective. J Sci Food Agric 87:668–675. https://doi.org/10.1002/jsfa.2765

Burgos G, Salas E, Amoros W, Auqui M, Muñoa L, Kimura M, Bonierbale M (2009a) Total and individual carotenoid profiles in *Solanum phureja* of cultivated potatoes: I. Concentrations and relationships as determined by spectrophotometry and HPLC. J Food Compost Anal 22:503–508

Burgos G, Auqui S, Amoros W, Salas E, Bonierbale M (2009b) Ascorbic acid concentration of native Andean potato varieties as affected by environment, cooking and storage. J Food Compost Anal 22:533–538

Burgos G, Amoros W, Salas E, Muñoa L, Sosa P, Díaz C, Bonierbale M (2012) Carotenoid concentrations of native Andean potatoes as affected by cooking. Food Chem 133:1131–1137

Burgos G, Muñoa L, Sosa P, Bonierbale M, zum Felde T, Díaz C (2013) In vitro bioaccessibility of lutein and zeaxanthin of yellow fleshed boiled potatoes. Plant Foods Hum Nutr 68:385–390

Burton WG (1989) The potato, 3rd edn. Longman Scientific and Technical, Harlow

Campbell R, Morris WL, Mortimer CL, Misawa N, Ducreux LJ, Morris JA, Hedley PE, Fraser PD, Taylor MA (2015) Optimising ketocarotenoid production in potato tubers: Effect of genetic background, transgene combinations and environment. Plant Sci 234:27–37

Chakraborty S, Chakraborty N, Agrawal L, Ghosh S, Narula K, Shekhar S, Naik PS, Pande PC, Chakrborti SK, Datta A (2010) Next-generation protein-rich potato expressing the seed protein gene *AmA1* is a result of proteome rebalancing in transgenic tuber. Proc Natl Acad Sci U S A 107:17533–17538. https://doi.org/10.1073/pnas.1006265107

Chawla R, Shakya R, Rommens CM (2012) Tuber-specific silencing of *asparagine synthetase-1* reduces the acrylamide-forming potential of potatoes grown in the field without affecting tuber shape and yield. Plant Biotechnol J 10:913–924

Chitchumroonchokchai C, Diretto G, Parisi B, Giuliano G, Failla ML (2017) Potential of golden potatoes to improve vitamin A and vitamin E status in developing countries. PLoS One 12(11):e0187102. https://doi.org/10.1371/journal.pone.0187102

Cho K-S, Jeong H-J, Cho J-H, Park Y-E, Hong S-Y, Won H-S, Kim H-J (2013) Vitamin C content of potato clones from Korean breeding lines and compositional changes during growth and after storage. Hortic Environ Biotechnol 54:70–75

Chung YS, Palta P, Bamberg J, Jansky S (2016) Potential molecular markers associated with tuber calcium content in wild potato germplasm. Crop Sci 56:576–584

Dale MFB, Mackay GR (1994) Inheritance of table and processing quality. In: Bradshaw JE, Mackay GR (eds) Potato genetics. CAB International, Wallingford, pp 285–315

De Jong H (1987) Inheritance of pigmented tuber flesh in cultivated diploid potatoes. Am Potato J 64:337–343

De Jong H, Burns VJ (1993) Inheritance of tuber shape in cultivated diploid potatoes. Am Potato J 70:267–283

De Jong WS, De Jong DM, Bodis M (2003) A fluorogenic 5′ nuclease (TaqMan) assay to assess dosage of a marker tightly linked to red skin color in autotetraploid potato. Theor Appl Genet 107:1384–1390. https://doi.org/10.1007/s00122-003-1420-z

De Jong WS, Eannetta NT, De Jong DM, Bodis M (2004) Candidate gene analysis of anthocyanin pigmentation loci in the *Solanaceae*. Theor Appl Genet 108:423–432. https://doi.org/10.1007/s00122-003-1455-1

De Maine MJ, Bradshaw JE, Caligari PDS (1992) Inheritance of the external mechanical damage resistance of potato cultivars. Ann Appl Biol 121:379–384

De Meulenaer B, Vinci RM, Mestdagh F (2018) Acrylamide formation in fried potato products and its mitigation. In: Wale S (ed) Achieving sustainable cultivation of potatoes volume 2: production, storage and crop protection. Burleigh Dodds Science Publishing, Cambridge, pp 137–176

De Wilde T, De Meulenaer B, Mestdagh F, Govaert Y, Ooghe W, Fraselle S, Demeulemeester K, Van Peteghem C, Calus A, Degroodt J-M, Verhé R (2006) Selection criteria for potato tubers to minimize acrylamide formation during frying. J Agric Food Chem 54(6):2199–2205

Diretto G, Al-Babili S, Tavazza R, Papacchioli V, Beyer P, Giuliano G (2007) Metabolic engineering of potato carotenoid content through tuber-specific overexpression of a bacterial minipathway. PLoS One 2(4):e350

Drapal M, Lindqvist-Kreuze H, Mihovilovich E, Aponte M, Bonierbale M, Fraser PD (2020) Cooking dependent loss of metabolites in potato breeding lines and their wild and landrace relatives. J Food Compost Anal 88:103432. https://doi.org/10.1016/j.jfca.2020.103432

Ek KL, Wang S, Copeland L, Brand-Miller JC (2014a) Discovery of a low-glycaemic index potato and relationship with starch digestion *in vitro*. Br J Nutr 111:699–705

Ek KL, Wang S, Brand-Miller JC, Copeland L (2014b) Properties of starch from potatoes differing in glycemic index. Food Funct 10:2509–2515

Endelman JB, Jansky SH (2016) Genetic mapping with an inbred line-derived F2 population in potato. Theor Appl Genet 129(5):935–943

Escuredo O, Seijo-Rodríguez A, González-Martín MI, Rodríguez-Flores MS, Seijo MC (2018) Potential of near infrared spectroscopy for predicting the physicochemical properties on potato flesh. Microchem J 141:451–457

Ezekiel R, Singh N, Sharma S, Kaur A (2013) Beneficial phytochemicals in potato – a review. Food Res Int 50:487–496

Fairweather-Tait S (1983) Studies on the availability of iron in potatoes. Br J Nutr 50:15–23

Foster-Powell K, Holt SHA, Brand-Miller JC (2002) International table of glycemic index and glycemic load values: 2002. Am J Clin Nutr 76:5–56

Friedman M (1996) Nutritional value of proteins from different food sources. A review. J Agric Food Chem 44:6–29

Friedman M, Levin CE (2016) Glycoalkaloids and calystegine alkaloids in potatoes. In: Singh J, Kaur L (eds) Advances in potato chemistry and technology, 2nd edn. Elsevier, Amsterdam, pp 167–194

Friedman M, McDonald GM (1997) Potato glycoalkaloids: chemistry, analysis, safety, and plant physiology. Crit Rev Plant Sci 16:55–132

Galdón BR, Mesa DR, Rodríguez EMR, Romero CD (2010) Amino acid content in traditional potato cultivars from the Canary Islands. J Food Compost Anal 23:148–153. https://doi.org/10.1016/j.jfca.2009.08.009

Gottschalk K, Ezekiel R (2006) Storage. In: Gopal J, Khurana SMP (eds) Handbook of potato production, improvement and postharvest management. The Haworth Press, Binghampton, pp 489–522

Goyer A, Haynes KG (2011) Vitamin B1 content in potato: effect of genotype, tuber enlargement, and storage, and estimation of stability and broad-sense heritability. Am J Potato Res 88:374–385. https://doi.org/10.1007/s12230-011-9203-6

Goyer A, Navarre DA (2007) Determination of folate concentrations in diverse potato germplasm using a trienzyme extraction and a microbiological assay. J Agric Food Chem 55:3523–3528

Goyer A, Navarre DA (2009) Folate is higher in developmentally younger potato tubers. J Sci Food Agric 89:579–583

Goyer A, Sweek K (2011) Genetic diversity of thiamin and folate in primitive cultivated and wild potato (*Solanum*) species. J Agric Food Chem 59:13072–13080. https://doi.org/10.1021/jf203736e

Gubb IR, Hughes JC, Jackson MT, Callow JA (1989) The lack of enzymic browning in the wild potato species *Solanum hjertingii* Hawkes compared with commercial *Solanum tuberosum* varieties. Ann Appl Biol 114:579–586

Hammond BR, Johnson EJ, Russell RM, Krinsky NI, Yeum KJ, Edwards RB, Snodderly DM (1997) Dietary modification of human macular pigment density. Invest Ophthalmol Vis Sci 38:1795–1801

Haynes KG (2000) Inheritance of yellow-flesh intensity in diploid potatoes. J Am Soc Hortic Sci 125:63–65

Haynes KG, Potts WE, Chittams JL, Fleck DL (1994) Determining yellow-flesh intensity in potatoes. J Am Soc Hortic Sci 119:1057–1059

Haynes KG, Sieczka JB, Henninger MR, Fleck DL (1996) Clone × environment interactions for yellow-flesh intensity in tetraploid potatoes. J Am Soc Hortic Sci 121:175–177

Haynes KG, Clevidence BA, Rao D, Vinyard BT (2011) Inheritance of carotenoid content in tetraploid × diploid potato crosses. J Am Soc Hortic Sci 136:265–272

Haynes K, Yencho G, Clough M, Henninger M, Sterrett S (2012) Genetic variation for potato tuber micronutrient content and implications for biofortification of potatoes to reduce micronutrient malnutrition. Am J Potato Res 89:192–198

Hellwege EM, Czapla S, Jahnke A, Willmitzer L, Heyer AG (2000) Transgenic potato (*Solanum tuberosum*) tubers synthesise the full spectrum of inulin molecules naturally occurring in globe artichoke (*Cynara scolymus*). Proc Natl Acad Sci U S A 97:8699–8704

Henry CJK, Lightowler H, Strik C, Storey M (2005) Glycaemic index values for commercially available potatoes in Great Britain. Br J Nutr 94:917–921

Holton TA, Cornish EC (1995) Genetics and biochemistry of anthocyanin biosynthesis. Plant Cell 7:1071–1083

Hughes JC, Evans JL (1967) Studies on after-cooking blackening in potatoes. IV Field experiments. Eur Potato J 10:16–36

Itkin M, Heinig U, Tzfadia O, Bhide AJ, Shinde B, Cardenas PD, Bocobza SE, Unger T, Malitsky S, Finkers R, Tikunov Y (2013) Biosynthesis of antinutritional alkaloids in solanaceous crops is mediated by clustered genes. Science 341:175–179

Jansky S, Navarre R, Bamberg J (2019) Introduction to the special issue on the nutritional value of potato. Am J Potato Res 96(2):95–97

Juhaszi Z, Dancs G, Marincs F, Vossen M, Allefs S, Banfalvi Z (2014) Vitamin C, B5, and B6 contents of segregating potato populations detected by GC-MS: a method facilitating breeding potatoes with improved vitamin content. Plant Breed 133:515–520. https://doi.org/10.1111/pbr.12169

Jung CS, Griffiths HM, De Jong DM, Cheng S, Bodis M, De Jong WS (2005) The potato *P* locus codes for flavonoid 3′,5′–hydroxylase. Theor Appl Genet 110:269–275

Jung CS, Griffiths HM, De Jong DM, Cheng S, Bodis M, Kim TS, De Jong WS (2009) The potato *developer* (*D*) locus encodes an R2R3 MYB transcription factor that regulates expression of multiple anthocyanin structural genes in tuber skin. Theor Appl Genet 120:45–57

Kaminski KP, Kørup K, Andersen MN, Sønderkær M, Andersen MS, Kirk HG, Nielsen KL (2016) Next generation sequencing bulk segregant analysis of potato support that differential flux into the cholesterol and stigmasterol metabolite pools is important for steroidal glycoalkaloid content. Potato Res 59:81–97. https://doi.org/10.1007/s11540-015-9314-4

Kanter M, Elkin C (2019) Potato as a source of nutrition for physical performance. Am J Potato Res 96(2):201–205

Kaspar KL, Park JS, Brown CR, Mathison BD, Navarre DA, Chew BP (2011) Pigmented potato consumption alters oxidative stress and inflammatory damage in men. J Nutr 141:108–111

Killick RJ, Simmonds NW (1974) Specific gravity of potato tubers as a character showing small genotype-environment interactions. Heredity 32:109–112

Kobayashi A, Ohara-Takada A, Tsuda S, Matsuura-Endo C, Takada N, Umemura Y, Nakao T, Yoshida T, Hayashi K, Mori M (2008) Breeding of potato variety "Inca-no-hitomi" with a very high carotenoid content. Breed Sci 58:77–82

Kromann P, Valverde F, Alvarado S, Vélez R, Pisuña J, Potosí B, Taipe A, Caballero D, Cabezas A, Devaux A (2017) Can Andean potatoes be agronomically biofortified with iron and zinc fertilizers? Plant Soil 411:121–138. https://doi.org/10.1007/s11104-016-3065-0

Lachman J, Hamouz K, Orsák M, Kotíková Z (2016) Carotenoids in potatoes–a short overview. Plant Soil Environ 62:474–481

Lewis CE, Walker JRL, Lancaster JE, Sutton KH (1998a) Determination of anthocyanins, flavonoids and phenolic acids in potatoes. I: Coloured cultivars of *Solanum tuberosum* L. J Sci Food Agric 77:45–57

Lewis CE, Walker JRL, Lancaster JE, Sutton KH (1998b) Determination of anthocyanins, flavonoids and phenolic acids in potatoes. II: Wild, tuberous *Solanum* species. J Sci Food Agric 77:58–63

Li XQ, De Jong H, De Jong DM, De Jong WS (2005) Inheritance and genetic mapping of tuber eye depth in cultivated diploid potatoes. Theor Appl Genet 110:1068–1073

Li X-Q, Scanlon MG, Liu Q, Coleman WK (2006) Processing and value addition. In: Gopal J, Khurana SMP (eds) Handbook of potato production, improvement, and postharvest management. Food Products Press, New York, pp 523–555

Lockyer S, Nugent AP (2017) Health effects of resistant starch. Nutr Bull 42:10–41

Love SL, Pavek JJ (2008) Positioning the potato as a primary food source of vitamin C. Am J Potato Res 85:277–285

Lu W, Haynes K, Wiley E, Clevidence B (2001) Carotenoid content and color in diploid potatoes. J Am Soc Hortic Sci 126:722–726

Luthra SK, Gupta VK, Tiwari JK, Kumar V, Bhardwaj V, Sood S, Dalamu KRP, Kumar R, Vanishree G, Kumar D, Mhatre P, Chakrabarti SK (2020) Potato breeding in India. CPRI Technical Bulletin No 74 (Revised). ICAR-Central Potato Research Institute, Shimla

Macdonald-Clarke CJ, Martin BR, McCabe LD, McCabe GP, Lachcik PJ, Wastney M, Weaver CM (2016) Bioavailability of potassium from potatoes and potassium gluconate: a randomized dose response trial. Am J Clin Nutr 104:346–353. https://doi.org/10.3945/ajcn.115.127225

Mackay GR, Dale MFB (1990) Quality in potatoes. SCRI Annual Report 1990, pp 9–12

Mackay GR (2005) Propagation by traditional breeding methods. In: Razdan MK, Mattoo AK (eds) Genetic improvement of Solanaceous crops volume I: potato. Science Publishers, Enfield, pp 65–81

Mackay GR, Brown J, Torrance CJW (1990) The processing potential of tubers of the cultivated potato, *Solanum tuberosum* L., after storage at low temperature. 1. Fry colour. Potato Res 33:211–218

McKenzie M, Corrigan V (2016) Potato flavour. In: Singh J, Kaur L (eds) Advances in potato chemistry and technology, 2nd edn. Elsevier, Amsterdam, pp 339–368

Monro J, Mishra S (2009) Nutritional value of potatoes: digestibility, glycemic index, and glycemic impact. In: Singh J, Kaur L (eds) Advances in potato chemistry and technology. Academic, Burlington, pp 371–394

Monro J, Mishra S, Blandford E, Anderson J, Genet R (2009) Potato genotype differences in nutritionally distinct starch fractions after cooking, and cooking plus storing cool. J Food Compost Anal 22:539–545

Mooney S, Chen L, Kühn C, Navarre R, Knowles NR, Hellmann H (2013) Genotype-specific changes in vitamin B_6 content and the PDX family in potato. Biomed Res Int. https://doi.org/10.1155/2013/389723

Morris WL, Ducreux L, Griffiths DW, Stewart D, Davies HV, Taylor MA (2004) Carotenogenesis during tuber development and storage in potato. J Exp Bot 55:975–982. https://doi.org/10.1093/jxb/erh121

Morris WL, Ducreux LJM, Fraser PD, Millam S, Taylor MA (2006) Engineering ketocarotenoid biosynthesis in potato tubers. Metab Eng 8:253–263

Morris WL, Ross HA, Ducreux LJM, Bradshaw JE, Bryan GJ, Taylor MA (2007) Umami compounds are a determinant of the flavour of potato (*Solanum tuberosum* L.). J Agric Food Chem 55:9627–9633

Muttucumaru N, Powers SJ, Elmore JS, Mottram DS, Halford NG (2013) Effects of nitrogen and sulfur fertilization on free amino acids, sugars, and acrylamide-forming potential in potato. J Agric Food Chem 61:6734–6742. https://doi.org/10.1021/jf401570x

Muttucumaru N, Powers SJ, Elmore JS, Dodson A, Briddon A, Mottram DS, Halford NG (2017) Acrylamide-forming potential of potatoes grown at different locations, and the ratio of free asparagine to reducing sugars at which free asparagine becomes a limiting factor for acrylamide formation. Food Chem 220:76–86

Nassar AMK, Sabally K, Kubow S, Leclerc YN, Donnelly DJ (2012) Some Canadian-grown potato cultivars contribute to a substantial content of essential dietary minerals. J Agric Food Chem 60:4688–4696

Navarre DA, Moehninsi M, Lin S, Hellman H (2018) Nutritional properties and enhancement/biofortification of potatoes. In: Wang-Pruski G (ed) Achieving sustainable cultivation of potatoes. Volume 1: Breeding improved varieties. Burleigh Dodds Science Publishing, Cambridge, pp 191–221

Nayak B, Berrios JDJ, Tang J (2014) Impact of food processing on the glycemic index (GI) of potato products. Food Res Int 56:35–46

Novy RG, Whitworth JL, Stark JC, Schneider BL, Knowles NR, Pavek MJ, Knowles LO, Charlton BA, Sathuvalli V, Yilma S, Brown CR, Thornton M, Brandt TL, Olsen N (2017) Payette Russet: a dual-purpose potato cultivar with cold-sweetening resistance, low acrylamide formation, and resistance to late blight and potato virus Y. Am J Potato Res 94:38–53

Nyankanga RO, Murigi WW, Shibairo SI (2018) Effect of packaging material on shelf life and quality of ware potato tubers stored at ambient tropical temperatures. Potato Res 61:283–296

Orozco RF, Gallardo-Guerrero L, Hornero-Mendez D (2013) Carotenoid profiling in tubers of different potato (*Solanum* sp) cultivars: accumulation of carotenoids mediated by xanthophyll esterification. Food Chem 141:2864–2872. https://doi.org/10.1016/j.foodchem.2013.05.016

Paget M, Amoros W, Salas E, Eyzaguirre R, Alspach P, Apiolaza L, Noble A, Bonierbale M (2014) Genetic evaluation of micronutrient traits in diploid potato from a base population of Andean Landrace Cultivars. Crop Sci 54:1949–1959

Palta JP (2020) Improving potato tuber quality, seed piece performance and reducing storage rot by targeted calcium nutrition. In: Souvenir, global potato conclave 2020. Mahatma Mandir, Gandhinagar, pp 211–216

Pedreschi F (2009) Fried and dehydrated potato products. In: Singh J, Kaur L (eds) Adavances in potato chemistry and technology. Academic, Burlington, pp 319–337

Pęksa A, Kita A, Kułakowska K, Aniołowska M, Hamouz K, Nemś A (2013) The quality of protein of coloured fleshed potatoes. Food Chem 141:2960–2966. https://doi.org/10.1016/j.foodchem.2013.05.125

Pinhero RG, Coffin R, Yada RY (2009) Post-harvest storage of potatoes. In: Singh J, Kaur L (eds) Adavances in potato chemistry and technology. Academic Press, Burlington, pp 339–370

Pinhero RG, Waduge RN, Liu Q, Sullivan JA, Tsao R, Bizimungu B, Yada RY (2016) Evaluation of nutritional profiles of starch and dry matter from early potato varieties and its estimated glycemic impact. Food Chem 203:356–366

Prashar A, Hornyik C, Young V, McLean K, Sharma SK, Dale MFB, Bryan GJ (2014) Construction of a dense SNP map of a highly heterozygous diploid potato population and QTL analysis of tuber shape and eye depth. Theor Appl Genet 127:2159–2171

Raigond P, Mehta A, Singh B (2018) Sweetening during low-temperature and long-term storage of Indian potatoes. Potato Res 61:207–217

Raimo F, Pentangelo A, Pane C, Parisi B, Mandolino G (2018) Relationships between internal brown spot and skin roughness in potato tubers under field conditions. Potato Res 61:327–339

Ramdath DD, Padhi E, Hawke A, Sivaramalingam T, Tsao R (2014) The glycemic index of pigmented potatoes is related to their polyphenol content. Food Funct 5:909–915

Reyes LF, Miller JC, Cisneros-Zevallos L (2005) Antioxidant capacity, anthocyanins and total phenolics in purple- and red-fleshed potato (*Solanum tuberosum* L.) genotypes. Am J Potato Res 82:271–277

Robinson BR, Sathuvalli V, Bamberg J, Goyer A (2015) Exploring folate diversity in wild and primitive potatoes for modern crop improvement. Genes 6:1300–1314. https://doi.org/10.3390/genes6041300

Römer S, Lübeck J, Kauder F, Steiger S, Adomat C, Sandmann G (2002) Genetic engineering of a zeaxanthin-rich potato by antisense inactivation and co-suppression of carotenoid epoxidation. Metab Eng 4:263–272. https://doi.org/10.1006/mben.2002.0234

Rommens CM, Ye J, Richael C, Swords K (2006) Improving potato storage and processing characteristics through all-native DNA transformation. J Agric Food Chem 54:9882–9887

Rosyara UR, De Jong WS, Douches DS, Endelman JB (2016) Software for genome-wide association studies in autopolyploids and its application to potato. Plant Genome 9. https://doi.org/10.3835/plantgenome2015.08.0073

Sakamoto Y, Mori K, Matsuo Y, Mukojima N, Watanabe W, Sobaru N, Tamiya S, Nakao T, Hayashi K, Watanuki H, Nara K, Yamazaki K, Chaya M (2017) Breeding of a new potato variety ‘Nagasaki Kogane’ with high eating quality, high carotenoid content, and resistance to diseases and pests. Breed Sci. https://doi.org/10.1270/jsbbs.16168

Salaman RN (1910) The inheritance of colour and other characters in the potato. J Genet 1:7–46

Sandmann G (2001) Genetic manipulation of carotenoid biosynthesis: strategies, problems and achievements. Trends Plant Sci 6:14–17

Sato H, Koizumi R, Itoyama R, Ichisawa M, Negishi J, Sakuma R, Furusho T, Sagane Y, Takano K (2019) Free amino acids in potato (*Solanum tuberosum*) may cause *egumi*-taste in food products. Potato Res 62(3):305–314

Schallenberg RS, Smith O, Treadaway RH (1959) Role of sugars in the browning reaction in potato chips. J Agric Food Chem 7:274–277

Schwall GP, Safford R, Westcott RJ, Jeffcoat R, Tayal A, Shi Y-C, Gidley MJ, Jobling SA (2000) Production of very high amylose potato starch by inhibition of SBE A and B. Nat Biotech 18:551–554

Scott GJ, Petsakos A, Suarez V (2019) Not by bread alone: estimating potato demand in India in 2030. Potato Res 62(3):281–304

Sharma SK, Bolser D, de Boer J, Sønderkaer M, Amoros W, Carboni MF, D’Ambrosio JM, de la Cruz G, Di Genova A, Douches DS, Eguiluz M, Guo X, Guzman F, Hackett CA, Hamilton JP, Li G, Li Y, Lozano R, Maass A, Marshall D, Martinez D, McLean K, Mejía N, Milne L, Munive S, Nagy I, Ponce O, Ramirez M, Simon R, Thomson SJ, Torres Y, Waugh R, Zhang Z, Huang S, Visser RGF, Bachem CWB, Sagredo B, Feingold SE, Orjeda G, Veilleux RE, Bonierbale M, Jacobs JME, Milbourne D, Martin DMA, Bryan GJ (2013) Construction of reference chromosome-scale pseudomolecules for potato: integrating the potato genome with genetic and physical maps. G3 (Bethesda) 3:2031–2047

Shepherd LVT, Bradshaw JE, Dale MFB, McNicol JW, Pont SDA, Mottram DS, Davies HV (2010) Variation in acrylamide producing potential in potato: segregation of the trait in a breeding population. Food Chem 123:568–573

Shepherd LVT, Hackett CA, Alexander CJ, McNicol JW, Sungurtas JA, Stewart D, McCue KF, Belknap WR, Davies HV (2015) Modifying glycoalkaloid content in transgenic potato–metabolome impacts. Food Chem 187:437–443

Simmonds NW (1977) Relations between specific gravity, dry matter content and starch content of potatoes. Potato Res 20:137–140

Sinden SL, Sanford LL, Webb RE (1984) Genetic and environmental control of potato glycoalkaloids. Am Potato J 61:141–156

Singh J, Kaur L (2016) Advances in potato chemistry and technology, 2nd edn. Elsevier, Amsterdam

Singh J, Kaur L, Rao MA (2016a) Textural characteristics of raw and cooked potatoes. In: Singh J, Kaur L (eds) Advances in potato chemistry and technology, 2nd edn. Elsevier, Amsterdam, pp 475–501

Singh J, Colussi R, McCarthy OJ, Kaur L (2016b) Potato starch and its modification. In: Singh J, Kaur L (eds) Advances in potato chemistry and technology, 2nd edn. Elsevier, Amsterdam, pp 195–247

Sliwka J, Wasilewicz-Flis I, Jakuczun H, Gebhardt C (2008) Tagging quantitative trait loci for dormancy, tuber shape, regularity of tuber shape, eye depth and flesh colour in diploid potato originated from six *Solanum* species. Plant Breed 127:49–55

Snodderly DM (1995) Evidence for protection against age-related macular degeneration by carotenoids and antioxidant vitamins. Am J Clin Nutr 62(Suppl):1448–1461

Storey M (2007) The harvested crop. In: Vreugdenhil D (ed) Potato biology and biotechnology advances and perspectives. Elsevier, Oxford, pp 441–470

Storey RMJ, Davies HV (1992) Tuber quality. In: Harris P (ed) The potato crop. Chapman & Hall, London, pp 507–569

Subramanian NK, White PJ, Broadly MR, Ramsay G (2017) Variation in tuber mineral concentrations among accessions of *Solanum* species held in the Commonwealth Potato Collection. Genet Resour Crop Evol 64:1927–1935

Sulli M, Mandolino G, Sturaro M, Onofri C, Diretto G, Parisi B, Giuliano G (2017) Molecular and biochemical characterization of a potato collection with contrasting tuber carotenoid content. PLoS One 12(9):e0184143. https://doi.org/10.1371/journal.pone.0184143

Taylor MA (2020) The genetic control of sprout growth in potato tubers during storage. In: Souvenir, global potato conclave 2020. Mahatma Mandir, Gandhinagar, pp 208–210

Thorup TA, Tanyolac B, Livingstone KD, Popovsky S, Paran I, Jahn M (2000) Candidate gene analysis of organ pigmentation loci in the Solanaceae. PNAS 97:11192–11197

Tian J, Chen J, Ye X, Chen S (2016) Health benefits of the potato affected by domestic cooking: a review. Food Chem 202:165–175

Tiemens-Hulscher M, Delleman J, Eising J, van Bueren ETL (2013) Potato breeding. Aardappelwereld BV, The Hague, 171p

Tierno R, Hornero-Méndez D, Gallardo-Guerrero L, López-Pardo R, Ruiz de Gallareta JI (2015) Effect of boiling on the total phenolic, anthocyanin and carotenoid concentrations of potato tubers from selected cultivars and introgressed breeding lines from native potato species. J Food Compost Anal 41:58–65

Umemoto N, Nakayasu M, Ohyama K, Yotsu-Yamashita M, Mizutani M, Seki H, Saito K, Muranaka T (2016) Two cytochrome P450 monooxygenases catalyze early hydroxylation steps in the potato steroid glycoalkaloid biosynthetic pathway. Plant Physiol 171:2458–2467

Vales MI, Brown CR, Yilma S, Hane DC, James SR, Shock CC, Charlton BA, Karaagac E, Mosley AR, Culp D, Feibert E, Stark JC, Pavek MJ, Knowles NR, Novy RG, Whitworth JL (2012) Purple Pelisse: a specialty 'fingerling' potato with purple skin and flesh and medium specific gravity. Am J Potato Res 89:306–314

van Eck HJ (2007) Genetics of morphological and tuber traits. In: Vreugdenhil D (ed) Potato biology and biotechnology advances and perspectives. Elsevier, Oxford, pp 91–115

van Eck HJ, Jacobs JME, Stam P, Ton J, Stiekema WJ, Jacobsen E (1994) Multiple alleles for tuber shape in diploid potato detected by qualitative and quantitative genetic analysis using RFLPs. Genetics 137:303–309

Van Gelder WMJ, Vinket JH, Scheffer JJC (1988) Steroidal glycoalkaloids in tubers and leaves of *Solanum* species used in potato breeding. Euphytica S:147–158

Visser RGF, Somhorst I, Kuipers GJ, Ruys NJ, Feenstra WJ, Jacobsen E (1991) Inhibition of the expression of the gene for granule-bound starch synthase in potato by antisense constructs. Mol Gen Genet 225:289–296

Von Scheele C, Svensson G, Rasmussen J (1937) Die bestimmung des starkegehalts und der trockensubstanz der kartoffel mit hilfe des spezifischen gewichts. Landw Vers Sta 127:67–96

Vreugdenhil D, Bradshaw J, Gebhardt C, Govers F, MacKerron DKL, Taylor MA, Ross HA (eds) (2007) Potato biology and biotechnology advances and perspectives. Elsevier, Oxford, 823p

Waglay A, Karboune S (2016) Potato proteins: functional food ingredients. In: Singh J, Kaur L (eds) Advances in potato chemistry and technology, 2nd edn. Elsevier, Amsterdam, pp 75–104

White PJ, Bradshaw JE, Dale MFB, Ramsay G, Hammond JP, Broadley MR (2009) Relationships between yield and mineral concentrations in potato tubers. HortScience 44:6–11

White PJ, Broadley MR, Hammond JP, Ramsay G, Subramanian NK, Thompson J, Wright G (2012) Bio-fortifcation of potato tubers using foliar zinc-fertiliser. J Hortic Sci Biotechnol 87:123–129

White PJ, Thompson JA, Wright G, Rasmussen SK (2017) Biofortifying Scottish potatoes with zinc. Plant Soil 411:151–165. https://doi.org/10.1007/s11104-016-2903-4

Winfield M, Lloyd D, Griffiths W, Bradshaw J, Muir D, Nevison I, Bryan G (2005) Assessing organoleptic attributes of *Solanum tuberosum* and *S. phureja* potatoes. Asp Appl Biol 76:127–135

Wolters AMA, Uitdewilligen JGAML, Kloosterman BA, Hutten RCB, Visser RGF, van Eck HJ (2010) Identification of alleles of carotenoid pathway genes important for zeaxanthin accumulation in potato tubers. Plant Mol Biol 73:659–671. https://doi.org/10.1007/s11103-010-9647-y

Yuan J, Bizimungu B, Leblanc D, Lague M (2019) Effects of field selection parameters and specific gravity on culinary evaluation traits in a potato breeding program. Potato Res 62(4):361–377

Zhang Y, Cheng S, De Jong D, Griffiths H, Halitschke R, De Jong W (2009a) The potato *R* locus codes for dihydroflavonol 4-reductase. Theor Appl Genet 119:931–937

Zhang Y, Jung CS, De Jong WS (2009b) Genetic analysis of pigmented tuber flesh in potato. Theor Appl Genet 119:143–150

Zhu X, Gong H, He Q, Zeng Z, Busse JS, Jin W, Bethke PC, Jiang J (2016) Silencing of vacuolar invertase and asparagine synthetase genes and its impact on acrylamide formation of fried potato products. Plant Biotechnol J 14(2):709–718

Zitnak A, Johnston GR (1970) Glycoalkaloid content of B5141-6 potatoes. Am Potato J 47:256–260

Zorrilla C, Navarro F, Vega S, Bamberg J, Palta J (2014) Identification and selection for tuber calcium, internal quality and pitted scab in segregating 'Atlantic' × 'Superior' reciprocal tetraploid populations. Am J Potato Res 91:673–687

Zorrilla C, Schabow J, Chernov V, Palta J (2019) CAX1 vacuolar antiporter overexpression in potato results in calcium deficiency in leaves and tubers by sequestering calcium as calcium oxalate. Crop Sci 59:176–189

Chapter 5
Improving Resistance to Diseases and Pests: A Dynamic Situation

5.1 Introduction

Finally, in deciding breeding objectives, we need to add the goal of resistance to diseases and pests to those already considered of high and stable yield and quality. It is sometimes useful to distinguish between pathogen and parasite. A parasite lives in or on another organism from which it obtains nutrient, whereas a pathogen is a parasite that produces a disease in its host (Day 1974). Obligate parasites cannot multiply in nature without their hosts, whereas facultative ones can grow and live on substrates other than a living host. For major disease and pest problems in a targeted agroecological zone, new and durable forms of resistance need to be actively sought, together with ways of combining them with other desirable traits as quickly as possible. Breeders commonly opt to chemically control major diseases and pests in their assessments of yield and quality traits and to assess resistance in separate special tests. Hence, the number of healthy tubers available for resistance testing is a practical issue, as some tests require more tubers than others. For less important diseases and pests, potential cultivars will simply be screened to avoid extreme susceptibility. This may be done in special tests or when natural epidemics of minor diseases occur. The prerequisites for successful breeding are a source of resistance, a reliable screen for resistance using the most appropriate isolate(s) of the pathogen or population(s) of the pest, and finally, an understanding of the inheritance of resistance. It is easiest to use resistances already present in cultivars, but new sources may be required from landraces or wild relatives to provide adequate levels of resistance. Furthermore, strategies are needed for achieving durability of resistance because this is not a property of new cultivars per se. It also depends on the population genetics of the pathogen or pest and the epidemiology of the disease. Appropriate management of resistance in the environment requires knowledge of the relative fitnesses of pathogen and pest genotypes that can and cannot overcome the resistance genes being used. Durability of resistance is still difficult to predict with certainty and there have been mixed fortunes in the past. Hence, durability is often

J. E. Bradshaw, *Potato Breeding: Theory and Practice*,
https://doi.org/10.1007/978-3-030-64414-7_5

either an historical fact or a hope for the future. Furthermore, priorities in resistance breeding may alter as a result of climate change because pathogens and pests have specific temperature and moisture requirements for their population growth. So, bearing these points in mind, let us take a closer look at breeding for resistance to some of the major diseases and pests of potatoes, first having considered terminology and a few general points about screening for resistance. Still relevant are the chapters on diseases and pests of potato in *Potato Biology and Biotechnology: Advances and Perspectives* (Vreugdenhil et al. 2007), including the one on genetics of resistance (Simko et al. 2007). Also relevant is the book *Insect Pests of Potato: Global Perspectives on Biology and Management* (Alyokhin et al. 2012). Finally, there are useful chapters in *The Potato Crop: Its Agricultural, Nutritional and Social Contribution to Humankind* (Campos and Ortiz 2020).

5.2 Types of Resistance

I find that the most useful classifications of resistance are qualitative versus quantitative and specific versus broad spectrum or general resistance; but I acknowledge that my generation of plant breeders were influenced by Van der Plank's (1968) book *Disease Resistance in Plants*, and his terms vertical and horizontal resistance. He developed what proved to be a controversial but testable theory of vertical and horizontal resistance, but is acknowledged as a founding father of modern quantitative epidemiology (Thresh 1998). Simmonds (1991), who was director of the Scottish Plant Breeding Station when I started work there in 1975, was enthusiastic about horizontal resistance, particularly for late blight of potato, but not everyone shared his views!

Qualitative resistance is where plants from a cross can be classified unambiguously as resistant or susceptible and the number in each category can be counted. The difference between resistant and susceptible plants is usually large with resistant ones free, or showing just traces of disease, and susceptible ones showing extensive symptoms (infection). Mendelian ratios and the corresponding genes can be sought. Qualitative resistance has proved to be oligogenic, controlled by one or a few major genes. In contrast, quantitative resistance is where there are no unambiguous categories, rather degrees of resistance or susceptibility which are measured on some scale. Genes are identified indirectly through their associations with molecular markers and have proved to be both few and many in number. Hence, it seems reasonable to talk about qualitative oligogenic resistance versus quantitative resistance. Both types of resistance can be assessed on plant parts (e.g. detached leaves), seedlings or whole plants in controlled environment facilities such as growth chambers or specialized glasshouses. Under these circumstances, the breeder or pathologist, for example, may apply inoculum of the pathogen containing a known concentration of spores and wait to see how many lesions develop, how quickly they spread, how soon and for how long they sporulate and how many spores they produce. Some disease tests will give clear-cut qualitative differences:

plants or detached leaves are either susceptible, producing sporulating lesions, or resistant, showing complete immunity to infection or a hypersensitive reaction in which infected cells die. Other tests will give quantitative results, showing variation in the number of lesions, their spread and spore production. Both types of resistance can also be assessed on plants in the field, often in special nurseries, where the progress of the disease can be followed under more natural but less controlled conditions. Hence care is needed in using the term field resistance, as it need not necessarily imply quantitative resistance, although often it will. Disease progress is usually the result of a number of cycles of infection and sporulation, with the outcome referred to as an epidemic. Van der Plank (1968) argued that vertical resistance delays the start of an epidemic because it reduces the number of spores that can infect a cultivar, whereas horizontal resistance slows an epidemic down after it has started because it slows the rate of infection. Together they determine the amount of disease. Similar considerations apply to pest resistance.

The pathogen (or pest) may be differentiated into species, subspecies, biovars or physiological races. A cultivar (or potential cultivar) may be resistant to some isolates of a pathogen but susceptible to other isolates (isolate specific) or it may be resistant to most or all known isolates of a pathogen (broad spectrum). A cultivar may be more or less susceptible to different isolates of the pathogen, the isolates thus varying in their aggressiveness. Once many potential cultivars are considered, cultivar × isolate interactions may occur which are relevant to the programme's objectives, and which determine the most appropriate isolate or isolates of the pathogen for use in screening. The results are usually clearer when both the potential cultivars and pathogen isolates are genetically homogeneous. Broad-spectrum quantitative resistance equates with horizontal resistance and race-specific qualitative oligogenic resistance with vertical resistance, as shown in Table 5.1.

In the example of vertical resistance, one cultivar (3) is susceptible to isolate 1 but two cultivars (2 and 3) are susceptible to isolate 2, so we say that isolate 2 is more virulent than isolate 1. We also say that cultivar 2 displays isolate-specific resistance, which is termed race-specific if the isolates are physiological races of the pathogen, whereas cultivar 1 has broad-spectrum resistance, albeit to two isolates in our example. In contrast, there are no cultivar × pathogen isolate interactions in the example of horizontal resistance. We say that cultivar 4 is more resistant than cultivar 5 which in turn is more resistant (less susceptible) than cultivar 6, and that

Table 5.1 Vertical (race-specific, qualitative) and horizontal (broad-spectrum, quantitative) resistance. (Modified from Bradshaw 2016 with permission)

Vertical			Horizontal		
	Pathogen			Pathogen	
Host	Isolate 1	Isolate 2	Host	Isolate 3	Isolate 4
Cultivar 1	Resistant	Resistant	Cultivar 4	10% disease	30% disease
Cultivar 2	Resistant	Susceptible	Cultivar 5	30% disease	50% disease
Cultivar 3	Susceptible	Susceptible	Cultivar 6	50% disease	70% disease

isolate 4 is more aggressive than isolate 3. The term pathogenicity (ability to cause disease) includes both aggressiveness and virulence.

Whisson et al. (2011) reviewed the molecular basis of pathogenicity and how pathogens and plants interact and coevolve. In brief, pathogens produce surface or secreted molecules that are needed for their growth. These are referred to as pathogen-associated molecular patterns (PAMPs). Both protein and non-protein PAMPs have been identified. Perception of PAMPs by pattern recognition receptors on plant cells initiates pattern-triggered immunity (PTI) which includes responses such as callose deposition. Successful pathogens target and suppress PTI through the action of effector molecules, leading to effector-triggered susceptibility (ETS). Resistance (R) proteins in plants detect the presence or action of specific pathogen effectors. Recognized effectors are referred to as avirulence (Avr) effectors. R proteins initiate effector-triggered immunity (ETI) which may include a hypersensitive (HR) response to contain pathogen ingress, as well as a PTI response. The host (*R*) and pathogen (*Avr*) genes are the genetic basis of race-specific resistance and hence also the basis of Flor's original gene-for-gene (concept) hypothesis (Flor 1942) for the genetics of resistance to the rust *Melampsora lini* in flax (*Linum usitatissimum*). The gene-for-gene concept has been demonstrated for a wide range of host–parasite systems, including rusts, smuts, bunts, mildews, other parasitic fungi, nematodes, insects, bacteria, viruses and higher plant parasites of economically important crops (Day 1974). Thus, if *R* and *r* are the resistant and susceptible alleles at a locus in the host, and *Avr* and *avr* the avirulent and virulent alleles at a locus in the pathogen, we have:

		Pathogen	
		Avr	
Host	*R*	R	S
	r	S	

where R is an incompatible interaction resulting in resistance and S is a compatible one resulting in susceptibility. In today's terminology, we say that the *R* gene encodes a receptor for the pathogen ligand (effector protein) specified by the *Avr* gene.

5.3 Screening for Resistance

Reliable testing is required both for screening germplasm collections for new sources of resistance and for assessing material under selection in a breeding programme. A reliable test for qualitative resistance is one where there are no escapes; susceptible control plants should all be susceptible. A reliable test for quantitative resistance is one in which there is little environmental variation or errors of measurement. A statistically significant offspring-midparent regression, for example, is evidence of heritable resistance, and the slope of the regression line is a measure of (narrow-sense) heritability. When low heritabilities result from environmental and genotype × environmental interaction variation, heritability and response to selection can be increased through increased replication in individual tests and an increased number of tests, respectively, if this is thought to be a good use of additional resources. Initial tests for qualitative and quantitative resistance will probably be done in a glasshouse or screenhouse, but confirmation of resistance under field conditions should be sought. For any specific disease test, the best stage of plant growth for inoculation and scoring, inoculation concentration, method of inoculation and environmental conditions (temperature, day-length and humidity) are all factors to be determined by experiment, if not already known in the literature. Testing will need to use the most appropriate isolate or isolates of the pathogen (or population of pest), taking into account its differentiation into species, subspecies, biovars or physiological races; and the distribution of these in the target environments of the breeding programme. For some pathogens, a mixture of isolates/inoculum from different geographical locations may be appropriate; for others, the most virulent or most aggressive isolate may be required. The literature on screening for resistance is vast and a breeder will need to consult it for those diseases and pests of particular interest in their is gender neutral programme.

5.4 Late Blight of Foliage, Stems and Tubers

Late blight of potato foliage, stems and tubers is caused by the oomycete (algal-like) pathogen *Phytophthora infestans* (Mont.) de Bary which is a diploid heterothallic species, requiring the association of A1 and A2 mating types to form long-lived sexual oospores. *P. infestans* spreads through sporangia which are blown or splashed to new infection sites where growth and development within host tissue are followed by sporulation and further dispersal. Its life history can be found in reviews such as the one by Fry (2008). It is still potentially the most serious disease of potatoes worldwide, over 170 years after its first impact outside of Mexico when severe epidemics swept through North America and Europe in the mid-1840s. In Ireland, in 1845, 1846, 1848 and 1849, the result was complete failure of a staple crop and a famine with serious political and societal consequences which have been the subject of much debate (Nally 2011). The primary cause of the crop failure was a susceptible cultivar attacked by a virulent and aggressive pathogen in environmental conditions favourable to disease development at a time when chemical control was unavailable. Up until the 1980s, the epidemics of late blight outside Mexico were caused by the asexual reproduction of strains of the A1 mating type of *P. infestans* which therefore had clonal lineages (Yoshida et al. 2013). However, since the late 1970s, new populations of *P. infestans*, comprising both the A1 and A2 mating types, have spread from Mexico to the rest of the world through trade in seed tubers, and possibly also in young plants (Hohl and Iselin 1984; Goodwin and Drenth 1997). There is therefore the possibility of sexual reproduction resulting in oospores

Fig. 5.1 Late blight of potato in the foliage and stems (left) and tubers (right). (Source: SCRI)

which can over-winter in the soil and start epidemics earlier each season. In contrast, with asexual reproduction, *P. infestans* survives between seasons as living mycelia in late-blight-infected tubers. Furthermore, sexual reproduction may also allow the faster evolution of more virulent and aggressive strains of the pathogen. Chemical control is available and necessary for the many popular but susceptible cultivars which are widely grown. However, spraying is expensive, not always effective, and the new populations of *P. infestans* are often resistant to the widely used systemic fungicide metalaxyl (phenylamide is active ingredient) (GILB 1999). This is particularly true in many areas of the developing world where the disease is a serious threat to small-scale potato producers, for example, in the tropical highlands where it coincides with year-round potato production (Forbes 2012). Hence, the need for resistant cultivars is as great today as it was after the epidemics of the mid-1840s (Fig. 5.1).

It is important to appreciate that resistance in the foliage does not guarantee resistance in the tubers, and vice versa, although correlations can be found in some germplasm (Stewart et al. 1994b). The numbers of tubers required for assessments mean that the most effective way of selecting for resistance to both aspects of the disease in a breeding programme is to select those clones resistant in the foliage and then screen them for tuber resistance. Furthermore, resistance is often associated with late maturity and hence it is important to score both if one wants to break the association, and probably best to test early- and late-maturing germplasm separately (Bradshaw et al. 2004). Methods of assessing the foliage and tubers for resistance can be found in the Materials and Methods sections of papers on the breeding and genetics of resistance. At SPBS/SCRI, we developed and used glasshouse progeny tests for assessments of seedlings, 5–7 weeks after sowing (Stewart et al. 1983c), and tubers from flowering plants (Wastie et al. 1987); glasshouse tests for assessing the foliage of whole plants just prior to flowering (Stewart et al. 1983a) and for assessing tubers from glasshouse-grown flowering plants (Wastie et al. 1987); and a field trial for assessing foliage resistance

(Malcolmson 1976) and a test for field-grown immature tubers (Stewart et al. 1983b). Only in the field trial for assessing foliage resistance is an epidemic followed. In the other tests, foliage and tuber resistance are assessed 1 and 2 weeks, respectively, after inoculation. Details of testing can be found in the cited references, but three additional points are worth making. Firstly, at SPBS/SCRI, we found glasshouse whole plant tests more reliable than the traditional detached leaflet tests for identifying *R*-genes and the virulence characteristics of *P. infestans* isolates. Secondly, Stewart and Bradshaw (2001) found that the use of mixed inoculum (two races) led to an over-estimation of quantitative foliage resistance of genotypes with which only one of the races was compatible in field trials, and likewise to a lesser extent in glasshouse tests, but that assessment of tuber resistance was not significantly affected. In other words, assessment of genotypes which may possess *R*-genes using mixed inoculum should be interpreted with caution, and apparent high levels of quantitative field resistance should be confirmed through genetic studies of the inheritance of resistance. Thirdly, quantitative foliage resistance can be assessed both in a whole-plant glasshouse test and in a field assessment, as done by Bradshaw et al. (2006) who detected a QTL of large effect on chromosome 4 in both tests. However, it is common practice to assess such quantitative resistance in the field.

5.4.1 Populations of P. infestans

When considering breeding strategies, it is important for the breeder to know the genetic make-up of the *P. infestans* populations present in the target environments of the breeding programme, and how they are changing with time. So, let us take a brief look at the history of populations of *P. infestans*.

Goss et al. (2014) have provided evidence that *P. infestans* originated in central Mexico rather than the Andes and hence the populations found in the Andes are descendants of the Mexican populations. Hence, the first hosts of *P. infestans* were wild relatives of our cultivated potatoes that were found in a secondary centre of host diversity. The authors used phylogeographic and approximate Bayesian computation analyses to unravel complex demographic histories. The analyses used microsatellite markers and sequences of four nuclear genes sampled from populations in the Andes, Mexico (Toluca valley) and elsewhere. The Toluca populations were sexual, with both mating types, whereas the South American populations were clonal.

Yoshida et al. (2013) compared the genomes of 11 herbarium and 15 modern strains of *P. infestans* and concluded that the original nineteenth-century pandemic was caused by a single clonal lineage, but that this lineage was not the direct ancestor of the one (US-1) that came to dominate the global *P. infestans* population during the twentieth century until the 1980s. Rather, a small *P. infestans* metapopulation was established at the periphery of its Mexican centre of origin, or possibly in North America, sometime before the first global *P. infestans* pandemic. The first lineage to spread from there was HERB-1, which persisted globally for at least half a century. Subsequently, the US-1 lineage expanded and spread from the metapopulation, replacing HERB-1. An interesting major genomic difference between the HERB-1 and US-1 lineages is the shift in ploidy, from diploid to triploid and even tetraploid.

The new populations of *P. infestans* that have been spreading from Mexico to the rest of the world since the end of the 1970s comprise both the A1 and A2 mating types (Hohl and Iselin 1984; Goodwin and Drenth 1997). Mating type can be determined using a molecular marker (CAPs) assay developed by Judelson et al. (1995). Martin et al. (2019) have provided insights into the evolution of these genetically diverse global populations of *P. infestans*, within which clonal lineages have emerged and become locally dominant, such as US-8 and more recently, US-22 and US-23 in North America and 13_A2 in Europe. The authors used new complementary mitochondrial (mtDNA) haplotype markers (maternally inherited) and nuclear SSRs to analyse over 100 *P. infestans* isolates representative of key clonal lineages found globally. A novel set of PCR primers amplified five target regions (loci) for mitochondrial DNA sequence analysis and increased the number of mtDNA haplotypes resolved from four with the PCR-RFLP method to 37 (17, 6, 8 and 4 for Ia, Ib, IIa and IIb haplotypes, respectively, plus 2 HERB-1 haplotypes). A 12-plex nuclear SSR assay (12 loci) revealed 90 multilocus genotypes and provided accurate discrimination of dominant clonal lineages and other genetically diverse isolates (earlier, RG-57 was a commonly used nuclear RFPL probe). These markers form a key element of monitoring *P. infestans* populations by EuroBlight (www.euroblight.net) and USAblight (www.usablight.org).

Breeders need to be aware of changes in *P. infestans* populations, but sometimes, the changes occur too quickly for the timescale on which breeders operate. One example will suffice. In Britain, there was a dramatic change over the period 2003–2007: SSR genotype 13_A2 (A2 mating type) was absent in 2003 but comprised 71.4% of samples (1023) in 2007, whereas genotypes 2_A1, 7_A1 and 8_A1 (A1 mating type) comprised 67.5% in 2003, but only 7.3% in 2007 (Cooke et al. 2008). Cultivar Lady Balfour was National Listed in Britain in 2001 and quickly became established as the number one organic cultivar, partly because of its field resistance to late blight. It proved relatively more susceptible to 13_A2 than might have been expected from its field resistance to other genotypes (White and Shaw 2009; Lees and Stewart 2012), but in 2019 was still being widely grown for the organic market! In contrast, cultivar Sarpo Mira retained its high resistance rating. Another example of the spread of genotype 13_A2 is in Spain between surveys done in 2003–2005 and 2011–2014 (Alor et al. 2019). As a consequence of the rise of genotype 13_A2 in Britain, the Commonwealth Potato Collection was screened in 2008 for new sources of resistance to 13_A2 (Fig. 5.2), but will they prove durable?

Fig. 5.2 Screening the Commonwealth Potato Collection in 2008 for sources of resistance to *Phytophthora infestans* genotype 13_A2: resistance found in *S. bulbocastanum*, *S. chacoense*, *S. commersonii*, *S. demissum*, *S. okadae*, *S. polyadenium*, *S. stoloniferum* and *S. verrucosum* (10 June 2008)

5.4.2 Breeding Strategies for Resistance: Introgressing R*-Genes from Wild Species (*Solanum demissum *and* S. bulbocastanum*)*

The devastation of potato crops in North America, Ireland and NW Europe after the arrival of late blight in the 1840s led to greater breeding efforts throughout Europe and North America. Although no highly resistant or immune cultivars had been produced by the early years of the twentieth century, sufficient resistance was supposedly built up to prevent further disasters, no doubt helped from 1885 by the use of Bordeaux mixture as a copper sulphate-based fungicide (Forbes 2012). The apparent breakthrough was the discovery of the *R*-genes for high levels of resistance in *Solanum demissum*. Introgression (backcrossing) breeding started in 1908 in Germany and in 1909 in Great Britain (Muller and Black 1951). Cultivar Craigs Snow White was released from the Scottish Plant Breeding Station (SPBS) in 1948 from a cross made in 1939 between Craigs Defiance and clone W800(2), which was bred by Dr. J. H. Wilson of St Andrews, Scotland, from germplasm including *S. demissum*, and given to the newly founded SPBS after his death in January 1920. Craigs Snow White contained what became known as the *R1* (strictly *Rpi-R1*) gene. As reviewed by Van der Plank (1968), it failed to provide durable resistance. By 1955, cultivars with the *R1* gene were being grown widely in West Germany and races virulent on *R1* were frequent enough for the gene to delay a blight epidemic by only 5 days. In contrast, in the Netherlands, in 1956, there were no commercial cultivars of any importance with the *R1* gene, virulent races were rare and the gene delayed blight epidemics by about a month. However, the potential rate of increase of virulent races could be seen in the USA. Cultivar Kennebec was the first one with the *R1* gene to be grown widely in the USA. When released in 1948, it either escaped blight altogether or had a few lesions too late in the season to do any harm. By 1954, it had increased in popularity, but virulent races were now abundant, and it was no longer very resistant. Cultivar Pentland Ace, with gene *R3a* and *R3b* variant $R3b^{G1696/G3111}$ (Armstrong et al. 2019), was released from SPBS in 1951, but soon succumbed to blight. The commercially more successful cultivar Pentland Dell, used for making French fries, entered production in Britain in 1963 when race 4 was the prevalent race of *P. infestans.* It had three genes, thought to be *R1*, *R2* and *R3*, each of which conferred resistance to race 4. Recent research by Armstrong et al. (2019) has shown that it actually has *R1*, *R3a*, $R3b^{G1696/G3111}$ and *R2* family member $abpt^{T86}$. It succumbed to blight in 1967, but nevertheless went on to become one of SPBS's most successful cultivars and was still the 12th most widely grown cultivar in Britain in 2019, albeit under the protection of fungicides. Interestingly, in the 1960s, SPBS took part in an International Collaborative Programme with CPRI in India and the Rockefeller Foundation Programme in Mexico. Under this programme, segregating material was exchanged. The initial screening for resistance was done in India and selected genotypes were sent to Edinburgh or Mexico for confirmation of resistance. Genotypes showing a high degree of resistance in Mexico and Edinburgh were evaluated for adaptation under Indian conditions. One cultivar developed from the same germplasm as Lady Balfour was Kufri Jyoti (*R3*, *R4* and *R7*) which was released in 1968. It is still in 2020 the most popular cultivar in the North-western and North-eastern hills of India and parts of the North Indian plains. It remained almost immune to late blight until the year 1974, but subsequently started to get infected and, by 1986, had completely succumbed to the disease (Luthra et al. 2020). I find it remarkable that germplasm in a late blight resistance breeding programme could deliver one of the most successful organic cultivars in the UK and one of the most successful table cultivars in India, two different uses in two very different environments. Perhaps we need to think of *Solanum demissum* as a source of adaptability to different environments and end uses rather than for failed *R*-genes for late blight resistance.

By 1968, the number of *R*-genes identified had risen to the 11 currently recognized ones, but races of *P. infestans* which could overcome the more recently discovered ones (*R5* to *R11*) were already widely distributed in Britain, despite the *R*-genes not being present in common commercial cultivars (Malcolmson 1969). Breeders began to doubt that these *R*-genes would provide durable resistance, either singly or in combination, due to the evolution of new races of *P. infestans* which arose by mutation.

Following Black et al. (1953), the physiological races of *P. infestans* were numbered to indicate their virulence towards the *R* genes. Thus, race 0 was avirulent on all 11 *R*-genes, whereas race 1,2,3 was virulent on any combination of the genes *R1*, *R2* and *R3*, and so on. Two sets of 11 late blight differentials were established by Black (Bl) and Mastenbroek (Ma), respectively, and are still maintained for the benefit of breeders and pathologists, for example, at SASA, Edinburgh, and Wageningen, The Netherlands, respectively. Ideally, the 11 differentials should have been monogenic for the same 11 *R*-genes, but unsatisfactorily, this is certainly not true. Malcolmson and Black (1966) concluded that the BlR9 differential contained genes *R1*, *R2*, *R3* and *R9*; Huang et al. (2005) concluded that the BlR3 differential contained *R3a,* whereas the MaR3 differential contained two resistance genes, *R3a* and *R3b*; and Kim et al. (2012) concluded that the MaR8 and MaR9 differentials contained at least four (*R3a*, *R3b*, *R4* and *R8*) and seven (*R1*, *Rpi-abpt1*, *R3a*, *R3b*, *R4*, *R8* and *R9*) *R*-genes, respectively. More recently, Armstrong et al. (2019) have shown that the BlR3 differential Pentland Ace contains *R3a* and *R3b* variant $R3b^{G1696/G3111}$, and the BlR9 differential 2573(2) contains *R1* variant $R1^{\Delta T4109}$, $abpt^{T86}$, *R3a*, $R3b^{G1696/G3111}$, *R8* and *R9a*. Hence, there was a strong case for a new set of more useful differentials such as the one developed by Zhu et al. (2015). They extended the Mastenbroek differential set with plants harbouring reduced *R*-gene complexity and with plants containing newly identified *R*-genes from related but different *Solanum* species. In addition, they produced a differential set of ten genetically modified (GM) plants with single late blight *R*-genes in the same genetic background (Desiree), but legislative restrictions (on GM potatoes) are currently limiting the use of the GM Desiree differential set.

Interestingly, an extensive review of available data by Świeżynski et al. (2000) revealed that the resistance of differentials BlR5, BlR8 and BlR9 was overcome least frequently, and hence, their *R*-genes might be of value after all. Subsequently, when investigating the resistance of cultivar Sarpo Mira, in field studies performed in Denmark with locally collected isolates between 2004 and 2011, Orłowska et al. (2012) found that Sarpo Mira was fully resistant and Bintje was fully susceptible in all 7 years (no data for 2009). The resistance of six (R1, R3, R4, R7, R10 and R11) of the 11 differentials of Black was broken (susceptible) in all years studied; R2 and R6 were broken in all but one of the tested years; and R5 was broken in three out of the 7 years. In contrast, R8 and R9 were fully resistant in all years tested (7 and 6, respectively). Likewise, Kim et al. (2012) found that differentials MaR8 and MaR9 had a high level of resistance in the field in three areas of the Netherlands over the 5-year period 2006–2010. They were also able to demonstrate that a plant (*R8-18*) that contained only the *R8* gene delayed the start of a blight epidemic by a similar number of days (50) to MaR8 (50) and to blight-resistant cultivar Sarpo Mira (51). Furthermore, whereas MaR8 and MaR9 were resistant to *P. infestans* strain IPO-C in both the field and detached leaf tests, *R8–18* was only resistant in the field.

Rietman et al. (2012) found that Sarpo Mira contains five *R*-genes, namely *R3a*, *R3b*, *R4*, *Rpi-smira1* and *Rpi-smira2*. The four genes *R3a*, *R3b*, *R4* and *Rpi-Smira1* conferred qualitative resistance that could be detected in detached-leaf tests. *P. infestans* strain IPO-C could fully infect Sarpo Mira under detached-leaf test conditions but not in the field where Sarpo Mira had a certain level of resistance which was conferred by gene *Rpi-Smira2*. In a cross with a susceptible parent, field resistance to strain IPO-C segregated even though the plants were fully susceptible to this strain in detached-leaf assays. Rietman et al. (2012) therefore suggested that field resistance may represent a weaker form of *R*-gene-based resistance that does not cause a full-blown hypersensitive response (HR) and resistance phenotype upon recognition of the pathogen. Orłowska et al. (2012) had also found that in Sarpo Mira, the progress of the late blight infection was different in whole plants compared with detached leaves, and that in segregating populations in the field, one gene was providing the resistance (1:1 segregation).

Tomczynska et al. (2014) examined 137 offspring from the cross between Sárpo Mira and susceptible cultivar Maris Piper in detached leaflet assays using complex races of *P. infestans* from Poland. They mapped a marker (45/XI) at the end of Sárpo Mira's chromosome 11, in the vicinity of the *R3* locus, which explained between 55.8% and 67.9% of the variation in the mean scores for late blight resistance. They concluded that they had probably found a new *R*-gene homologue, located in the *R3* gene cluster, that may correspond to gene *Rpi-Smira1* identified by Rietman et al. (2012). In contrast, the *Rpi-Smira2* gene has been localized in the *R8* locus on potato chromosome 9 by Jo (2013) who provided additional support for the likely identity of the *Rpi-Smira2* and *R8* genes. Furthermore, when MaR8 was crossed to a susceptible cultivar, a 1:1 segregation of resistance to susceptibility was observed in the F1 population in field trials, but not in detached leaf assays. In other words, a single gene was providing resistance in the field.

There is still much to learn about the components of resistance of Sárpo Mira and their durability, and hence, for how long and in what circumstances it will remain a useful source of resistance genes. The same is true of the MaR8 and MaR9 differentials. Nevertheless, we have seen that high levels of broad-spectrum resistance may result from individual *R*-genes, although stacking of such *R*-genes and an appropriate genetic background are expected to prolong durability.

One other backcrossing programme is worthy of mention. A start was made in 1959 to introgress resistance from *Solanum bulbocastanum*, but it involved bridging crosses with *S. acaule* and *S. phureja* (strictly Group Phureja) before three successive backcrosses to *S. tuberosum* finally yielded cultivars Toluca and Bionica in 2006 and 2008, respectively, with the *Rpi-blb2* resistance gene. Armstrong et al. (2019) have shown that Bionica also has *Rpi-abpt*^{T86}, *Rpi-R3a* and *Rpi-R3b*$^{G1696/G3111}$. Identification of functionally equivalent loci in more closely related species may allow breeders to transfer the desired *R*-genes into *S. tuberosum* without the use of transgenic technologies or laborious bridge crosses and somatic hybridization.

5.4.3 Breeding Strategies for Resistance: Cloning and Deploying R-Genes

The arrival of new populations of *P. infestans* throughout the potato-growing world prompted new screenings of germplasm collections for novel sources of resistance, as we saw earlier (Fig. 5.2). As a result, new *R*-genes have been found, mapped, cloned and deployed through a transgenic (cisgenic) route which will be dealt with in more detail in the chapter on genetic transformation. There has been much interest and debate over whether or not these new *R*-genes will be more durable per se or can be deployed in a more durable way, for example, in multi-lines or by pyramiding (stacking) in a single genotype (Goverse and Struik 2009; Forbes 2012). Incidentally, Van der Plank (1968) thought that the best use of *R*-genes was in a multiline-pathogen system because the genes could not be mismanaged.

Table 5.2 Cloned *R*-genes, species of origin, chromosome location (LG1 to LG12, equivalent to chromosomes 1 to 12) and spectrum of resistance (to isolates/races of *P. infestans*) in the Netherlands (modified from Hein et al. 2009 with permission; Bradeen 2011; Haverkort et al. 2016)

Species of origin	Gene	LG	Spectrum of resistance	References
S. berthaultii	*Rpi-ber*	10	Broad	Rauscher et al. (2006), Haverkort et al. (2016)
S. bulbocastanum	*RB/Rpi-blb1*	8	Broad	Naess et al. (2000), Song et al. (2003), van der Vossen et al. (2003)
	Rpi-blb2	6	Broad	van der Vossen et al. (2005)
	Rpi-blb3	4	Intermediate	Lokossou et al. (2009), Haverkort et al. (2016)
	Rpi-abpt	4	Intermediate	Lokossou et al. (2009), Haverkort et al. (2016)
S. chacoense	*Rpi-chc1*	10	Broad	Vossen et al. (2011)
	Rpi-chc2	10	Narrow	Haverkort et al. (2016)
S. demissum	*R1*	5	Narrow	Leonards-Schippers et al. (1992), Ballvora et al. (2002)
	R2	4	Intermediate	Li et al. (1998), Park et al. (2005a, b), Lokossou et al. (2009)
	R3a	11	Narrow	Huang et al. (2004, 2005)
	R3b	11	Narrow	Li et al. (2011)
	R8 (Rpi-smira2)	9	Broad	Haverkort et al. (2016)
	R9a	9	Broad	Haverkort et al. (2016)
	Rpi-edn1	4	Intermediate	Haverkort et al. (2016)
	Rpi-edn2	9	Broad	Haverkort et al. (2016)
S. mochiquense	*Rpi-mcq1*	9	Intermediate	Smilde et al. (2005), Haverkort et al. (2016)
S. papita	*Rpi-pta1*	8	Broad	Vleeshouwers et al. (2008), Wang et al. (2008), Haverkort et al. (2016)
S. stoloniferum	*Rpi-sto1*	8	Broad	Vleeshouwers et al. (2008), Haverkort et al. (2016)
S. tarijense	*Rpi-tar1*	10	Broad	Haverkort et al. (2016)
S. venturii	*Rpi-vnt1.1*	9	Broad	Foster et al. (2009), Pel et al. (2009)

Hein et al. (2009), Bradeen (2011) and Haverkort et al. (2016) reviewed progress in mapping and cloning the *R*-genes shown in Table 5.2, including *R*-genes from *S. demissum.* Haverkort et al. (2016) also attempted to identify ones with broad-spectrum resistance. Later, in the chapter on genetic transformation, we shall return to how Haverkort et al. (2016) stacked combinations of two and three *R*-genes using marker-free transformation. At one field-monitoring plot in The Netherlands in 2011, they found that narrow-spectrum *R*-genes alone in cultivar Desiree gave only a short delay to first infection, compared with untransformed Desiree, whereas broader-spectrum *R*-genes resulted in delays of up to 70 days and stacks of two or three *R*-genes remained uninfected at 90 days.

Other species known to contain resistance to late blight are: *S. avilesii* (*Rpi-avl1*, LG11, Verzaux et al. 2011), *S. microdontum* (*Rpi-mcd*, LG4, LG10, Sandbrink et al. 2000; *Rpi-mcd1*, LG4, Tan et al. 2008; QTLs on LG5 and LG6, Meade et al. 2020), *S. pampasense* (QTLs on LG11 and LG12, Meade et al. 2020), *S. paucissectum* (Villamon et al. 2005), *S. pinnatisectum* (*Rpi-pin1*, LG7, Kuhl et al. 2001), *S. polytrichon* (*Rpi-plt1*, LG8, Wang et al. 2008), *S. tuberosum* Group Phureja, LG9, Sliwka et al. 2006) and *S. verrucosum* (*RB*$_{ver}$, Liu and Halterman 2006). It should also be pointed out that cultivar Alouette contains *Rpi-vnt1.3* (closely related to *Rpi-vnt1.1*) from *S. venturii* and cultivar Innovator contains *Rpi-R2-like* (Armstrong et al. 2019).

5.4.4 Breeding Strategies for Resistance: Importance of Pathogen Effectors

The discovery of pathogen effectors in *P. infestans* led Vleeshouwers et al. (2008, 2011) to propose the use of candidate *P. infestans* effector libraries to catalogue *R*-genes based on which effectors are detected and hence allow the stacking of *R*-genes that detect different ones. This should increase the likelihood of achieving more durable resistance to late blight. *R*-genes and putative effectors can be transiently co-expressed in leaves of *Nicotiana benthamiana* to see if a hypersensitive reaction occurs, having been delivered into the leaf tissue via *Agrobacterium tumefaciens* (agroinfiltration). Effector-based studies by Vleeshouwers et al. (2008) have shown that *Rpi-sto1* and *Rpi-pta1* are functionally equivalent to *Rpi-blb1* (i.e. they are orthologous genes: homologous genes between species), the *Rpi-blb1* locus being present prior to the divergence of tuber-bearing and non-tuber-bearing *Solanum* species. Foster et al. (2009) and Pel et al. (2009) found that *Rpi-phu1* was orthologous with *Rpi-vnt1.1* and Liu and Halterman (2006) found that the *RB* genes from *S. bulbocastanum* and *S. verrucosum* were orthologues. Hence, the number of distinct functional *Rpi* genes is less than the apparent number present in diverse wild species. Further results can be found in Hein et al. (2009). In the DuRPh Project described by Haverkort et al. (2016), 10 out of 20 *R*-genes remained after elimination of functional redundancy, but these were reduced to five *R*-genes after further elimination of ones with a narrow spectrum of resistance in The Netherlands. The five remaining were: *Rpi-edn2, Rpi-blb3, Rpi-vnt1.1, Rpi-chc1* and *Rpi-sto1* (Table 5.2).

In a more recent review of the interaction between effector *Avr*-genes and *R*-genes, Jones et al. (2014) used late blight of potatoes (*Phytophthora infestans*) as an example (*R*-genes are designated *Rpi*-genes). Hosts and pathogens have co-evolved large *R*-gene and effector repertoires, respectively, potato and *P. infestans* being no exception. The sequenced doubled-monoploid potato clone DM and *P. infestans* have more than 438 *R*-genes and 563 effector candidates, respectively. Jones et al. (2014) explain that some *R*-genes are more durable than others because some effectors are more indispensable to the pathogen than others. Effector molecules are used by pathogens to suppress host defence mechanisms. They are usually delivered into host cells but can act in the intercellular spaces of the leaf. Effectors interfere with defence processes directly, or with their activation, by disrupting host-signalling mechanisms. However, plants have evolved the capacity to recognize specific effectors, either directly or indirectly by detecting their effects on host components. This recognition is usually mediated by intracellular receptors encoded by disease resistance (*R*) genes. Advances in DNA sequencing methods mean that the effector repertoire of pathogens can be analysed and effectors predicted from their protein sequence motifs. For example, *P. infestans* effectors have a signal peptide and a so-called RXLR (arginine-any amino acid-leucine-arginine) amino acid sequence motif close to the signal peptide cleavage site in the secreted effector. If a particular effector is shared between multiple races of a pathogen, then it is more likely to be indispensable for the pathogen than an effector that is present in some but not all races. This knowledge provides a crucial tool for discovering the most indispensable pathogen effectors, and prioritizing *Rpi*-genes that recognize these effectors for deployment in crops. The importance of the *P. infestans* Avr3a effector was directly examined by testing the virulence of races in which the *Avr3a* gene had been silenced. Remarkably, *P. infestans* lines with reduced levels of *Avr3a* expression are either weakly virulent or completely non-virulent. In other words, an effector needs to be produced for virulence. *Avr3a* is found in two allelic forms, $Avr3a^{KI}$ and $Avr3a^{EM}$ (varying in two amino acid positions). The *Rpi* gene *R3a* confers recognition of the $Avr3a^{KI}$ form but not the $Avr3a^{EM}$ form. The $Avr3a^{EM}$ form is present in the widespread virulent race 13_A2, which is why it can overcome *R3a*. Efforts are underway in at least two laboratories to identify novel forms of *R3a* that can recognize both $Avr3a^{KI}$ and $Avr3a^{EM}$, for example, by screening wild *Solanum* populations followed by transient *Agrobacterium*-mediated assays of *R3a* and $Avr3a^{EM}$ alleles.

Recently, Du et al. (2015) demonstrated that the receptor-like protein elicitin response (ELR) from the wild potato *Solanum microdontum* mediates extracellular recognition of the elicitin domain, a molecular pattern that is conserved in *Phytophthora* species. ELR associates with the immune co-receptor BAK1/SERK3 and mediates broad-spectrum recognition of elicitin proteins from several *Phytophthora* species, including four diverse elicitins from *P. infestans*. Transfer of ELR into cultivated potato resulted in enhanced resistance to *P. infestans*. Hence, pyramiding cell surface pattern recognition receptors with intracellular immune receptors could maximize the potential of generating a broader and potentially more durable resistance to *P. infestans*.

5.4.5 Breeding Strategies for Resistance: Quantitative Field Resistance

As breeders began to doubt the durability of *R*-gene resistance, many started to select for quantitative field resistance, either by using races of *P. infestans* compatible with the *R*-genes present in their material, or by creating *R*-gene-free germplasm so that screening could be done with any race (Toxopeus 1964; Black 1970;

Wastie 1991; Ortiz 2001). The latter approach was found useful by CIP in developing their potato Populations B1, B2 and B3. Race-specific resistance was eliminated from these populations where necessary by progeny testing with crosses to a susceptible tester that did not possess *S. demissum*-derived *R*-genes for resistance (Trognitz et al. 2001). In contrast, the former approach was used at the Scottish Plant Breeding Station (SPBS). The breeding was started in 1954 by Black (1970) and continued after his retirement in 1968. Further progress from 1968 to 1990 was reviewed by Wastie (1991), including selection for quantitative field resistance in tubers as well as foliage because tubers can be infected by spores from the slowly spreading, sporulating lesions of a foliage-resistant cultivar. Finally, Bradshaw et al. (2009) reported progress from 1991 to 2008 in combining quantitative resistance to late blight (in foliage and tubers) with quantitative resistance to the white potato cyst nematode (*Globodera pallida*), a modest increase in yield, and acceptable fry colour for processing. The pedigrees of the late blight resistant material are shown in the paper by Bradshaw (2009a) and trace back to crosses made with *S. demissum* in 1932 and 1937. In fact, *S. demissum* features in the pedigrees of 58 out of the 72 cultivars bred at the Scottish Plant Breeding Station and the Scottish Crop Research Institute from 1920 to 2008 (Bradshaw 2009a). In terms of breeding strategy, however, it is most instructive to look at the results of the detailed genetic analyses done on the blight resistance of cultivar Stirling, which was released in 1991.

The late blight resistance of cultivar Stirling was analysed before the advent of *P. infestans* genotype 13_A2 (Blue-13) by assessing the resistance of offspring from the cross with the susceptible clone 12601ab1 (Stewart et al. 2003; Bradshaw et al. 2004). The provisional map positions and effects of resistance genes were confirmed by further analyses of the results (Hackett et al. 2014). The results can be summarized as follows.

Whole-plant glasshouse tests revealed that Stirling has a single copy of a major *R*-gene on linkage group XI (chromosome 11) between 81 and 87 cM (distal end) which provided foliage resistance to simple race 1,4 (restricted lesions or spreading lesions on the oldest leaves alone) but was overcome by complex race 1,2,3,4,6,7 (at least a few spreading lesions on leaves of any age). The gene is close to or part of a large cluster of NB-LRR genes including *R3a*. Stirling does have the R7 differential in its pedigree but the *R*-gene has not been confirmed as *R7*. Further results from the cross of Stirling with susceptible cultivar Maris Piper confirmed that the *R*-gene provides resistance to the simple race but not the complex race (Soloman-Blackburn et al. 2007).

The defeated *R*-gene, or a closely linked gene(s), provided a small but statistically significant contribution to the foliage field resistance of Stirling, as assessed in Ayrshire in 1998 (average of five scoring dates). The 137 clones with the *R*-gene had an average score of 6.27 compared with 5.78 (difference 0.49 ± 0.170) for the 78 clones without the *R*-gene; this difference being 9.4% of the range found in the progeny clones (2.70 to 7.90, on a 1 to 9 scale of increasing resistance). Results from the cross of Stirling with susceptible cultivar Maris Piper and the backcross of clone 94B13A57 to Maris Piper provided further evidence of the defeated *R*-gene contributing to resistance, this time in a glasshouse test (Soloman-Blackburn et al.

2007). The average scores of the backcross clones with and without the *R*-gene were 4.23 and 2.52.

Stirling has a single copy of a gene for early maturity on linkage group V (chromosome 5) between 14 and 22 cM that is associated with susceptibility to blight. Clones with the gene had an average score of 3.77 compared with 5.63 for those without the gene on the fourth scoring date in Ayrshire in 1998. They were also more susceptible to blight in their tubers (74.63% versus 45.17% susceptible) in a glasshouse test in 1999.

Stirling possesses two copies (duplex) of a resistance allele (*R*) on linkage group IV (chromosome 4) between 22 and 30 cM. Offspring with genotype *RRrr* had an average score of 6.3, those with *Rrrr* 5.1, and those with *rrrr* 1.7, on the fourth scoring date in Ayrshire in 1998. The corresponding (glasshouse) means for tuber blight susceptibility were 38.2%, 57.1% and 84.1%. Maturity did not affect these results for chromosome 4 which confirmed that the gene was a true resistance gene. Segregation at this locus explained 43.6% of the variation in foliage blight scores and 19.2% of the variation in tuber blight scores.

A clone (94B13A61) from the cross of Stirling with Maris Piper inherited field resistance but not the major *R*-gene from Stirling. It is reasonable to assume that the field resistance was due to a resistant allele from chromosome 4 (Soloman-Blackburn et al. 2007). When 11 offspring selected for resistance from the cross between 94B13A61 and Maris Piper were assessed in Ayrshire in 2007, they were on average as susceptible as Maris Piper and none was more resistant than Stirling. That year *P. infestans* genotype 13_A2 (Blue-13) was naturally present, dominated the blight trial, and proved virulent (compatible) on the allele from chromosome 4. We thus appear to have a QTL × isolate interaction (Bradshaw 2009b).

The results of the genetic analyses of cultivar Stirling and cultivar Sárpo Mira, considered earlier, suggest that the relationships between qualitative resistance and quantitative field resistance and their durability are not as clear-cut as once thought, and that breeders need to think in terms of assembling the best combinations of *R*-genes and major QTLs for resistance, in the light of predictions about the likely responses of populations of *P. infestans* to deployment of these combinations. Both cultivars were selected for field resistance, Stirling in Scotland and Sárpo Mira in Hungary. Stirling's QTL on chromosome 4 is not effective against *P. infestans* genotype 13_A2, whereas the *Rpi-Smira2* gene on chromosome 9 in Sárpo Mira does provide resistance. Neither the analysis of resistance in Stirling nor in Sárpo Mira provides evidence of field resistance that does not display resistance × isolate interactions, that is horizontal resistance as proposed by Van der Plank (1968). Such horizontal resistance is meant to result from the action of many minor genes on morphological and developmental processes and on basal host defence, and hence not select for increased pathogen adaptation and aggressiveness (Mundt 2014). It remains to be seen if analysis of other sources of field resistance, such as CIP's Populations B1 to B3, can provide the missing evidence. High-yielding clones with good field resistance have been selected from population B3C2 in Uganda, after two cultivars from population B3C1 succumbed to late blight (Byarugaba et al. 2013).

Interestingly, Li et al. (2010) reported that by 2009, a new blight-resistant cultivar Cooperation-88 (C88) was the most widely grown cultivar in Yunnan Province of China (186,667 ha; 29.44% of total area) and that the total area of production in this and neighbouring provinces was 390,000 ha, making it one of the most widely grown cultivars in the world. It was bred over the period 1986–1996 from a cross between I-1085 and a bulk of pollen from a highly late blight resistant population with its origin in Andigena potatoes from CIP in Peru. I-1085 was a clone from India which had been selected for late blight resistance in Mexico and distributed by the CIP. The authors report that Cooperation-88 was resistant to late blight under local conditions from 2002 to 2009 and also in inoculation tests to race 1,2,3,4,5,6,7,8,9,10,11, with small and restricted lesions with few mycelia. Cooperation-88 was also highly resistance to viruses PVY^{O} and PVX, and moderately resistant to PLRV, when inoculated in a greenhouse. It is late maturing in the long-day summer growing season at over 2000 m above sea level in Yunnan Province, but earlier maturing in the warmer winter season at lower elevations. It is used both for the fresh table market and by the chipping industry. More recently, Tung et al. (2018) included Cooperation-88 as one of three cultivars with different levels of resistance in their study of combining ability for foliar resistance to a complex race (3, 4, 6, 7, 8, 10, 11) of late blight in 8-week-old plants (seedlings) in a greenhouse at Dalat, Vietnam (1500 m above sea level) in July 2015. Although Cooperation-88 had the highest combining ability, it also contains the major genes *R1*, *R2*, *R3a* and *Rpi-blb1*, thus leaving questions unanswered about the inheritance of its historical durable resistance, particularly as *R2* and *Rpi-blb1* would appear to be a good combination of genes in Vietnam.

5.4.6 *Assessment of Late Blight in the Foliage of Potatoes in Field Nurseries*

The previous section raises the issue of what breeders are actually assessing when it comes to late blight in the foliage of potatoes in field nurseries. The Scottish Plant Breeding Station and Scottish Crop Research Institute used a site in Ayrshire on the western seaboard of Scotland where the moist weather favours the development of late blight (Fig. 5.3).

The trial was planted in early May. Every third drill was the susceptible cultivar King Edward so that every test drill was adjacent to one spreader drill of King Edward. A randomized complete block design was used for each set of test material, with two or three replicates of two- or three-plant plots. Infector plants of susceptible cultivar King Edward were raised in pots in a glasshouse and inoculated in mid-July with an appropriate complex race of *P. infestans*, assuming assessment was for quantitative resistance. Three to four days later, these infector plants were placed at metre intervals along the spreader drills. The amount of blight present on each test plot was scored on a 1–9 scale of increasing resistance (Cruickshank et al. 1982) at

Fig. 5.3 Assessment of resistance to late blight in the foliage of potatoes in a field trial in Ayrshire, Scotland (every third drill is the susceptible cultivar King Edward). (**a**) Genetic experiment in August 1998. (**b**) Breeding material in 2008 (from Bradshaw 2016 with permission; Source: SCRI and 5 August 2008)

intervals of a few days during August (Fig. 5.3). The best scoring date was chosen retrospectively as either the one when the standards were showing their expected reaction to the disease on past experience of the trial, or the one that best discriminated between the material under test. The test was quite severe as the spreader rows can produce a lot of inoculum and inter-plot interference is a potential problem, as a clone with good quantitative resistance can by chance have very susceptible ones as neighbours. Nevertheless, the test proved its worth over many years. For research purposes, the infection rate and the area under the disease progress curve can be estimated, but how do you interpret the results?

5.4.7 *Theory and Practice of Disease Progress Curves*

Inoculum of *P. infestans* can survive winter in cull piles of diseased tubers and then start an epidemic in the following crop. In our simple theory, we are going to relate the infection rate of an epidemic to the proportion of infected tissue. Although this will prove adequate for our purpose, for late blight, we should really be estimating infection rate based on infectious tissue. We will assume that an epidemic is started by a low concentration of spores so that they act independently and the number of lesions produced is proportional to the number of spores. As a lesion of potato blight grows through a leaf following successful infection by a spore, there is an outer zone of mycelium working its way into healthy tissue. Behind it is a zone in which sporangia are formed if the humidity is high enough. In turn, behind this sporulating zone is a sterile, inner zone, containing perhaps some undispersed sporangia but producing no new sporangia even under optimal conditions of temperature and humidity. Thus, there is a latent period before newly infected tissue becomes infectious. Then follows an infectious period before the tissue is removed from the epidemic because it neither produces inoculum nor is it susceptible to re-infection by *P. infestans*. Considerable necrosis in the centre of a lesion and a narrower zone of spores is a known characteristic of many resistant potato cultivars. Hence Van der Plank (1968) thought that resistance in the form of a reduced infectious period, as well as increased latent period, was probably important for late blight of potatoes.

The infection rate (r) of an epidemic is defined by the following differential equation:

$$\mathrm{d}y / \mathrm{d}t = ry(1-y)$$

where y is the proportion of disease (e.g. leaf area of plants covered by lesions such as those of late blight of potatoes) at time t and $(1-y)$ is the proportion of tissue still healthy and available for infection. (Van der Plank used x instead of y, but the reader will probably be used to y as the vertical axis in a graph). The differential equation is a mathematical model of an epidemic in which y increases with t, and the rate of increase depends on both the amount of disease present and the amount of healthy tissue remaining to be infected. We solve the differential equation as follows.

$$\int \mathrm{d}y / \left[y(1-y)\right] = \int \mathrm{d}y / y + \int \mathrm{d}y / (1-y) = r \int \mathrm{d}t$$

Therefore

$\ln(y) - \ln(1-y) = rt + k$ where k is a constant, and

$$\ln\left(y / (1-y)\right) = rt + k$$

$$y / (1-y) = e^{rt+k}$$

$$y = e^{rt+k} / \left(1 + e^{rt+k}\right)$$

Hence, the graph of $\ln(y/(1-y))$ against t is a straight line of slope r and intercept k, whereas the graph of y against t is a curve, the disease progress curve.

Early in an epidemic, y will be small and $(1-y)$ will be very close to 1, so that

$$\ln(y) \approx rt + k$$

Hence, $\ln(y_1) - \ln(y_0) = rt_1 - rt_0$, where y_0 and y_1 are the proportions of disease at times t_0 and t_1, and

$\ln(y_1/y_0) = r(t_1 - t_0)$ or $(t_1 - t_0) = (1/r)\ln(y_1/y_0)$, so that $(t_1 - t_0)$ is the time required for the amount of disease to increase from y_0 to y_1.

Also

$$y_1 / y_0 = e^{r(t1-t0)} \quad y_1 = y_0 e^{r(t1-t0)}$$

In other words, early in the epidemic, we have exponential growth in which infections are independent of one another (i.e. no interference), as observed with late blight of potatoes. Furthermore, these equations show that the longer the duration of an epidemic (the bigger t-t_0), the greater the effect of a given value of r on the final value of y for a given initial value y_0.

As an example, we can use the data of Bradshaw et al. (2004) for field assessments of foliage resistance to late blight in Ayrshire in 1998. On July 20, glasshouse-grown plants of cultivar King Edward were inoculated in the laboratory with a zoospore suspension of the complex race 1, 2, 3, 4, 6, 7 of *P. infestans*. Three days later, on July 23, they were placed in the spreader rows of the trial. The amount of blight present on the two (or three) plants in each plot was scored on a 1–9 scale of increasing resistance on July 30 and on August 3, 7, 11 and 14. The 1–9 scale corresponded to the percentage of necrotic tissue (100, 90, 80, 70, 60, 40, 25, 10 and 0). Two replicates of the Black *R*-gene differentials, with the exception of R_9, which was not available, were planted adjacent to the trial in three-plant plots and scored on the same dates. The scores confirmed that the trial had been inoculated with the intended race. Despite some differences, I am going to assume that the R1, R2, R3, R4, R6 and R7 differentials were equally susceptible to race 1, 2, 3, 4, 6, 7 of *P. infestans* and work with their mean scores. Figure 5.4 shows the plot of $\ln(y/(1-y))$ against t in days from July 23. It is a straight line of slope (r) 0.2725 with intercept (k) −3.299.

$$\ln\left(y/\left(1-y\right)\right) = 0.2725\,t - 3.299$$

These values for r and k can be used to plot the disease progress curve shown in Fig. 5.5.

$$y = e^{rt+k} / \left(1 + e^{rt+k}\right)$$

where y is multiplied by 100 to get the percentage of disease at time t in days.

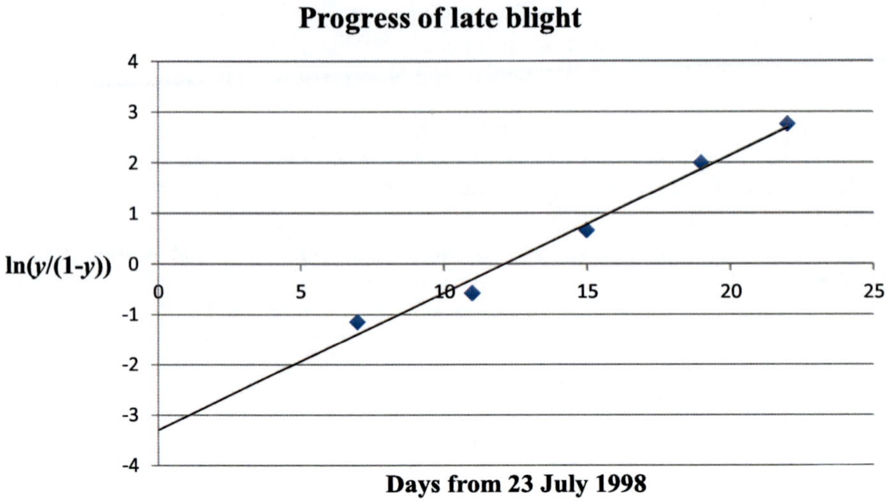

Fig. 5.4 Progress of complex race 1, 2, 3, 4, 6, 7 of *Phytophthora infestans* on six *R*-gene differentials (mean of R1, R2, R3, R4, R6 and R7) of potato in Ayrshire, Scotland in 1998, where *y* is proportion of disease (from Bradshaw 2016 with permission)

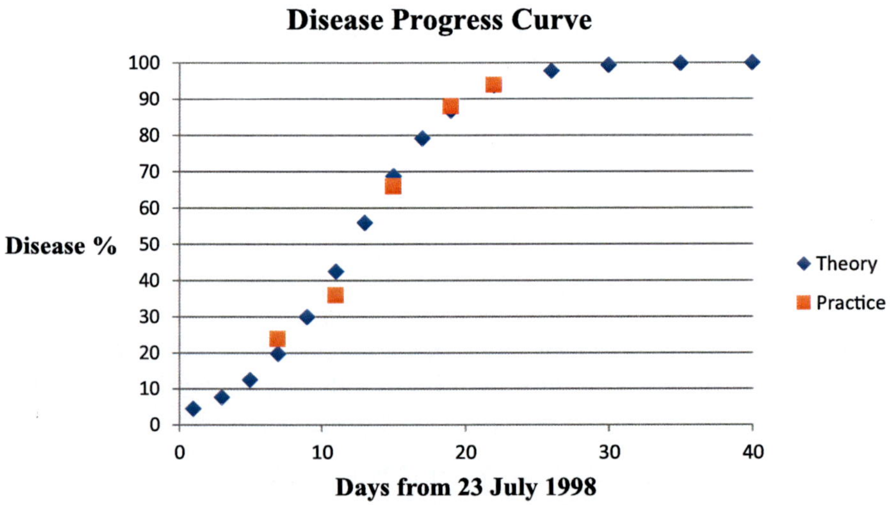

Fig. 5.5 Disease Progress Curve with $r = 0.2725$ and $k = -3.299$ and five actual values for complex race 1, 2, 3, 4, 6, 7 of *Phytophthora infestans* on six *R*-gene differentials (mean of R1, R2, R3, R4, R6 and R7) of potato in Ayrshire, Scotland in 1998 (from Bradshaw 2016 with permission)

Despite only having five actual values, the disease progress curve is typical of late blight and shows why it is potentially one of the most destructive diseases in the world. Under favourable conditions, it took about 3 weeks to destroy the susceptible plants.

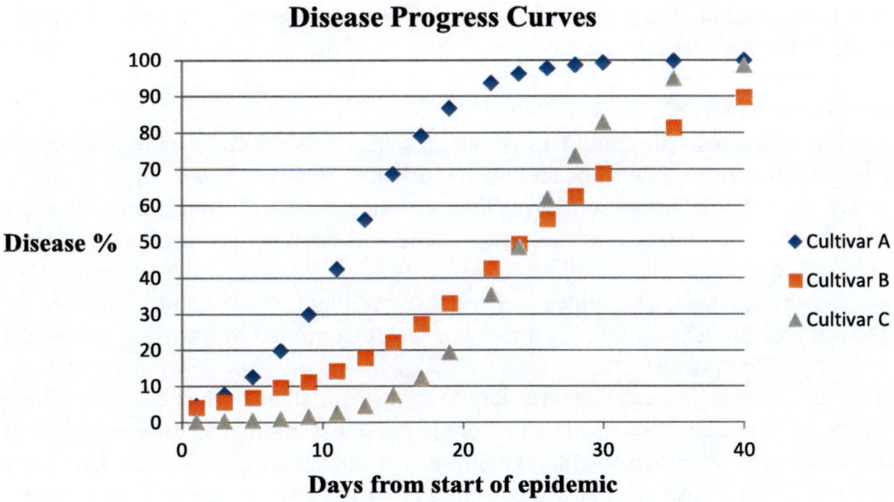

Fig. 5.6 Theoretical Disease Progress Curves differing in r and k (cultivar A: $r = 0.2725$ and $k = -3.299$; cultivar B: $r = 0.13625$ and $k = -3.299$; cultivar C: $r = 0.2725$ and $k = -6.598$) (from Bradshaw 2016 with permission)

We can now explore two questions about disease progress curves, the effect of changing r and the effect of changing k. In Fig. 5.6, the rate of infection on cultivar B is half that on cultivar A ($r = 0.13625$ compared with $r = 0.2725$; $k = -3.299$ for both) and k for cultivar C is double that of cultivar A ($k_C = -6.598$ compared with $k_A = -3.299$; $r = 0.2725$ for both). Halving the rate of infection halves the rate of progress of the disease, whereas the change in k (k_A-k_C) has delayed the epidemic by 12 days [$(k_A-k_C)/r = 12.1$]. So, what does the change in k represent? The answer is the amount of disease at the start of the epidemic when $t = 0$ and y is very small. Then, we have $\ln(y_A) = k_A$ and $\ln(y_C) = k_C$ so that $(k_A-k_C) = \ln(y_A) - \ln(y_C) = \ln(y_A/y_C)$. But we have shown above that with the same rate of infection r, $\ln(y_A/y_C)/r$ is the time required for the level of disease to increase from y_C to y_A. Furthermore, if we assume that the number of lesions produced (amount of disease) is proportional to the number of spores starting the epidemic, we conclude that the epidemic is delayed by fewer spores infecting cultivar C than cultivar A. The ratio is in fact 27.1 to 1. This situation could arise if a natural epidemic was started by a mixture of races (1, 2, 3, 4, 6, 7) and (1, 2, 3, 4, 6, 7, 10) in the ratio of 27 to 1 and cultivar A possessed *R*-genes *R1R2R3* and cultivar C possessed *R10*. All spores could infect cultivar A but only 1 in 28 could infect cultivar C. The number of spores infecting a cultivar is determined by the race composition of the pathogen and whether or not the cultivar possesses resistance to any of the races.

In summary, the amount of disease is determined by the number of spores infecting a cultivar and the subsequent rate of infection. Van der Plank (1968) gave examples for late blight of potatoes from the 1950s. He presented data from West Germany in 1955 showing that the *R1* gene delayed epidemics by 5 days and the

R3R4 combination delayed them by 15 days. He also presented data from 117 potato fields in the Netherlands that compared the rate of disease progress on susceptible cultivar Bintje ($r = 0.42$ per day), intermediate cultivar Eigenheimer ($r = 0.21$ per day) and resistant cultivar Voran ($r = 0.16$ per day). Van der Plank (1968) interpreted r as horizontal resistance and k as vertical resistance. What one sees in practice for a number of cultivars depends on how the disease nursery is managed. It is important to distinguish disease nurseries that rely on natural epidemics, those that are inoculated with a mixture of races and those inoculated with a single race. This is because mixtures of races will give biased results unless each can overcome all of the *R*-genes present, as pointed out by Van der Plank (1968) and confirmed by Stewart and Bradshaw (2001). Furthermore, care is needed in reporting the results from disease nurseries.

It is common practice for breeders to assess resistance when the susceptible control(s) is clearly susceptible and there is good discrimination between resistant and susceptible germplasm, although Simmonds and Wastie (1987) argued in favour of the mean resistance over all scoring dates as a simple and accurate way of making use of all of the data. In Fig. 5.6, on day 19 (fourth score on August 11), the disease percentages of cultivars A, B and C are 87%, 33% and 19%, respectively. Cultivar A is clearly susceptible and cultivars B and C resistant, with C apparently more resistant than B. But if the assessment is delayed to day 24, cultivars B and C now both have 49% disease compared with 96% for cultivar A and appear equally more resistant (or less susceptible) than the susceptible control.

A common way to try to make use of the information from a number of scoring dates is through the area under the disease progress curve (AUDPC). Again, the areas from day 0 to day 19 or day 24 would show the order A > B > C, but extending the areas beyond day 24 would reduce the difference between B and C. The exact areas can be calculated from the definite integral of the disease progress curve from time zero to t, but a good approximation with only a few scores such as the five in Fig. 5.6 is as follows (Shaner and Finney 1977).

$$\text{AUDPC} = \Sigma[(y_{i+1} + y_i/2)][t_{i+1} - t_i] \text{ from } i = 1 \text{ to } n-1$$

where y is the disease percentage (or other unit of severity) at the ith observation, t is the time (in days) at the ith observation and n is the total number of observations. The first observation in Fig. 5.6 would be $y = 0$ at $t = 0$. Sometimes, areas are expressed relative to the maximum possible area.

Neither the best date for discriminating between resistant and susceptible genotypes nor the area under the disease progress curve can discriminate between horizontal (r) and vertical resistance (k) in the way achieved by the graph of $\ln(y/(1-y))$ against t. Even here, there are potential complications when comparing genotypes. Genotypes may combine both types of resistance, changes in environmental conditions and the physiological state of the plants may occur during an epidemic so that the rate of infection changes with time, and a new race may arrive during the course of an epidemic. Nevertheless, Andrivon et al. (2006) have provided evidence from natural late blight infections in potato breeding nurseries in France, that the graph of $\ln(y/(1-y))$ against t can be used to evaluate potential cultivars for their ability to delay the onset of an epidemic and also their ability to reduce infection rates. Likewise, the disease progress curves for potential cultivars at a site conducive to late blight on the Isle of Anglesey, North Wales, from 2005 to 2008 could be interpreted in these terms, but not the areas under the curves (White and Shaw 2009). In 2005 and 2006, the trials relied on natural infection and the relative areas appeared to largely reflect vertical resistance (k), whereas in 2007 and 2008, the trials were inoculated with an isolate of genotype 13_A2 and the relative areas appeared to largely reflect horizontal resistance (r).

5.5 Fungal Diseases of Foliage, Stems and Tubers in Warm Climates

5.5.1 Early Blight

Early blight, caused by the fungus *Alternaria solani* Sor., is a serious early dying disease of potatoes grown in warm climates, especially where crops are irrigated. As the plants mature, necrotic lesions develop on the leaves and coalesce, causing defoliation and loss of yield. The disease can be controlled with fungicides, but this can be costly and ineffective, and hence, resistant cultivars are desirable. Varietal differences in field resistance were reported in the 1980s and shown to be heritable, and hence, a high level of resistance was a realistic breeding objective provided true resistance was distinguished from late maturity (Pavek and Corsini 1994).

One set of examples of assessing material in a breeding programme for resistance to early blight is provided by the papers of Caligari and Nachmias (1988), Nachmias et al. (1988), Stewart and Bradshaw (1993) and Stewart et al. (1994a) which also give references to the earlier work of other researchers. The assessments were as follows: (1) a field assessment of clones (eight-tuber plots) at the Gilat Regional Experiment Station, Negev, Israel, which had been sprayed with a spore suspension of *A. solani* (10^5 spores/ml) four times, at weekly intervals, from 45 days after planting, and scored for percentage of necrotic foliage at weekly intervals from 75 days at planting; (2) a glasshouse assessment of individual plants which had been grown from tubers in their own pots and inoculated at flowering time with a spore suspension (2.5×10^3 spores/ml) of *A. solani*, the inoculation being done in an enclosed chamber at 25 °C with the humidity high enough to keep the leaves damp, and the plants returned to the glasshouse after 24 h; the plants were scored for leaf area covered at 4- or 5-day intervals from 7 days after inoculation; (3) a glasshouse progeny test in which samples of 25 seedlings of each progeny were raised in 10 cm pots and inoculated with a spore suspension of *A. solani* containing 15×10^3 spores/ml, either 35 (young), 45 (middle) or 55 (old) days post sowing (plus uninoculated control), scored for level of resistance, and surviving, resistant plants transplanted into individual pots, grown to maturity, and their tubers harvested; and (4) a field progeny test in which tubers from the glasshouse progeny test were grown in the Negev (one tuber from each of ten clones of a progeny in a ten-plant plot), the trial being inoculated and scored the same way as for clones. The results of the experiments can be summarized as follows. Glasshouse and field assessments of clones were in good agreement ($r = 0.765$), with the best agreement with field results coming from glasshouse plants inoculated at flowering time with a spore suspension containing 2.5×10^3 spores/ml and the foliage kept damp for 24 h post inoculation. There was no correlation between 'young' and 'middle' aged seedlings and the field scores, but evidence of a small correlation ($r = 0.46$) between the 'old' seedlings and the field scores. In contrast, there was much better agreement between glasshouse and field assessment of progenies ($r = 0.75$ for progeny means) when adult plants from tubers were used in the glasshouse test. Furthermore, selection of 68 resistant seedlings in the glasshouse was effective; the overall level of susceptibility in the field decreased from 66.9% necrosis (120 unselected seedlings) to 35.4% necrosis for the 68 selected seedlings. The mid parent scores and the progeny means of 12 progenies were correlated both in the field ($r = 0.81$) and in the glasshouse ($r = 0.875$), thus confirming that the resistance was heritable. The overall conclusion from this work was that selecting and saving resistant individuals at the seedling stage in the glasshouse were the most effective way of selecting for resistance (66.9% to 35.4% necrosis) and preferable to choosing the most resistant progenies from an adult plant glasshouse test a year later. The seedling test could therefore be used to identify resistant individuals for multiplication and further assessment in the field. However, in 1994, early blight was not important enough for SCRI's commercial breeding partners to include it as a breeding objective and hence resistant cultivars were not deliberately bred. Now, 25 years later, there is renewed interest in breeding for resistance to early blight by other groups around the world.

Wolters et al. (2019) have developed a rapid method to screen wild *Solanum* germplasm for resistance to early blight. Plants from micro-propagation were transferred to pots in a climate-regulated greenhouse under long-day conditions (16 h light and 22 °C, 8 h dark and 18 °C). When 5 weeks old, plants were transferred to a climate cell (25 °C, relative humidity >70%, 16 h light and 8 h dark) the day before inoculation. The plants were placed inside a tent that was constructed by covering a frame with transparent plastic sheeting. An atomizer was placed inside the tent together with the plants to create a fine mist and increase humidity after inoculation of the plants. The upper three fully expanded compound leaves were drop-inoculated with 10 μl droplets of conidial suspension (1×10^5 conidia/ml). Directly after inoculating the plants, the lights were switched off and the atomizer was turned on. The normal light regime was resumed the next day and the atomizer was only switched on during subsequent nights. This mimics the alternating leaf wetness during the night and gradual drying of the leaves during the course of the day that can occur outdoors. The diameters of lesions produced on all the inoculated spots were recorded 5 days after inoculation using a digital calliper. Isolates of *A. grandis* and *A. protenta* were used as well as *A. solani* as all are large-spored *Alternaria* species that can cause early blight.

Xue et al. (2019) characterized resistance to early blight in 217 old and modern tetraploid potato cultivars in field experiments in Pennsylvania in 2016 and 2017. Relative area under the disease progress curve (RAUDPC) was calculated based on visual assessment of foliar disease during the growing season each year. RAUDPC ranged from 0.0090 to 0.7372 in 2016 and from 0.0215 to 0.7889 in 2017, respectively, with significant differences between cultivars and a significant cultivar × environment interaction. Broad-sense heritability of resistance was estimated as 0.89 with a 95% confidence interval of 0.86–0.92. All cultivars were also evaluated for foliage maturity in separate field trials in 2016 and 2017, and a strong correlation was found between resistance and late maturity. Maturity-adjusted RAUDPC was calculated by regressing maturity on RAUDPC. Although most resistant and moderately resistant cultivars showed late maturity and most susceptible cultivars showed early maturity, a few exceptions were found.

Jones and Perez (2019) have made a start to exploring candidate genes which could be involved in resistance through toxin resistance, pathogen recognition and response, and reduced cell death; but more research is required.

5.5.2 *Verticillium Wilt*

Verticillium wilt, caused by *Verticillium dahliae* Kelb., is a common soil-borne disease of potatoes under irrigation in warm-growing season areas, and often occurs concurrently with early blight. *V.albo-atrum* Reinke and Berthold also causes Verticillium wilt but prefers cooler conditions (e.g. northern USA and southern Canada). The pathogen infects the vascular tissue of potatoes through the roots, interfering with the transport of water and nutrition, and thus reducing both the yield and quality of tubers. The microsclerotia produced by the fungus are capable of long-term survival in soil in the absence of a susceptible host plant (Simko and Haynes 2017). The following summary is based on the review by Pavek and Corsini (1994). The disease results in early dying, smaller tubers and lower dry-matter content. Juvenile plants in the rapid vegetative stage of growth are resistant to colonization by the fungus, but once tuber bulking occurs, considerable differences in resistance to colonization and symptom development can be seen. Apparent resistance is correlated with late maturity, but not with true resistance to colonization. Some tolerant cultivars yield well despite infection and colonization of the vascular system. Chemical control (e.g. soil fumigation) of the disease is effective but expensive with a negative environmental impact, thus making breeding for resistance an attractive proposition; for example, the breeding programme started in 1949 at Aberdeen, Idaho (USA) by John McLean. Clone USDA 41956 was one important source of resistance which was also used as the source of Rx_{adg} for immunity to PVX. The third generation of crosses from McLean's starting material produced two highly resistant (not immune) clones, A66107-51 and A68113-4, with good tuber maturity characteristics, high yield and processing qualities for French fries (Corsini et al. 1990). Mohan et al. (1990) reported on how to maintain and prepare single conidial isolates of *V. dahlia*; how to do glasshouse evaluations of clones of cultivated potatoes and seedlings of non-cultivated tuber-bearing *Solanum* species; how to do field evaluations of resistance; how to prepare crude toxin (acetone precipitates) from *V. dahlia* cultures and how to test the reaction of excised leaves to injections of the toxin preparations (chlorosis and necrosis 72 h after inoculation). The details can be found in the paper, but perhaps it is worth noting the following. The virulence of isolates can be verified on four-week-old glasshouse-grown plants by injecting a conidial/microsclerotial suspension at the petiole-stem axils of ten leaves, incubating the plants in a glasshouse at 24 °C day and 16 °C night temperature, and observing wilt symptoms for up to 6 weeks after inoculation. Clones can be evaluated in pots in a glasshouse by placing sprouted seed pieces between two layers of inoculum, filling the pot with a mixture of sand-vermiculite-peat, and observing wilt severity at intervals until 56 days after planting. Germinated seedlings can be transplanted into pots containing a known concentration of inoculum and the development of disease symptoms followed over time. In field evaluations, with sprinkler irrigation, tubers are planted in plots in soil with a known concentration of conidia forming units and the incidence of wilt and severe wilt recorded periodically, followed by tuber yield at harvest, as well as determining the degree of stem colonization. The main conclusions from the work of Mohan et al. (1990) on 40 potato clones and six accessions of *Solanum* species were as follows. There was a highly significant correlation between the wilt reaction of the clones in the glasshouse, the incidence and progress of wilt and severe wilt in the field ($r = 0.90$), and the degree of colonization of stem apices by *V. dahlia* ($r = 0.82$). As already mentioned, clones A66107-51 and A68113-4 had high levels of resistance, as did the species *S. chacoense*, *S. sparsipilum* and *S. tarijense*. The reaction of genotypes to the acetone precipitates of *V. dahliae* toxin in an excised leaf bioassay was not correlated with their reaction to *Verticillium wilt* in the field or glasshouse. Yield differences could not be attributed entirely to differences in wilt resistance. Nevertheless, the yields of clones A66107-51 and A68113-4 were as high as those of the best cultivars. However, it was moderately resistant clone A7411-2 which was released as cultivar Ranger Russet in 1991.

Nearly 20 years later, Jansky (2009) assessed the Verticillium wilt resistance in US potato breeding programmes by evaluating 14 advanced clones and 11 cultivars over 2 years in a field at the Hancock, Wisconsin, Agricultural Experiment Station that had been fumigated and then inoculated with *V. dahliae*. She concluded that measurable progress had been made towards resistance as the six clones identified as resistant in both years were five breeding clones and a recent release (Megachip) which were more consistently resistant than the moderately resistant cultivar standard Ranger Russet. She also advocated a selection strategy that involved three consecutive measures of resistance on a reducing number of clones to separate true resistance from tolerance as follows. First, retain clones if they are early maturing or if they are at least as resistant as Ranger Russet based on symptom scores on a single date (showing best discrimination). Then, retain remaining clones if their mean sap colony count score (levels of conidia) is less than 200 cfu/100 μl sap. Finally, retain clones remaining after the first two steps if their mean dried stem score (levels of microsclerotia) is less than 200 cfu/50 mg dried stem material. This selection method reduces the number of samples processed for costly sap and dried stem assays, while consistently identifying resistant clones. However, because of the large variability among and within plants for stem colonization scores, it is important to sample a sufficient number of stems (e.g. 12 per clone).

Bae et al. (2009) provided some evidence for effective selection of clones in single-plant and four-plant plots for maturity, symptom expression, yield and stem colonization; but once again, resistance based on symptom expression was correlated ($r = 0.6$) with late maturity. Simko and Haynes (2017) confirmed a significant correlation ($r = 0.4$) between plant maturity and partial resistance to the pathogen when they evaluated 274 cultivars and 9 breeding selections for resistance under greenhouse conditions, with late-maturing clones being generally more resistant. However, when they estimated maturity-adjusted resistance from the residuals of the linear regression between the two traits, they found that a substantial part of the resistance could not be explained by the effect of maturity. Interestingly, clone USDA 41956 was in the group of germplasm with the best maturity-adjusted resistance, the highest resistance being found in cultivar Navajo. The reader is referred to the paper for details, but two points are worth mentioning. First, two-week-old plants were inoculated by adding spore suspension to each of three holes made 3 cm from the base of the stem and passing from the soil surface through the root zone; and 4 weeks later, the plant reaction was scored on a 1–5 scale of increasing susceptibility. Second, control plants were used to score foliage maturity based on the development of floral buds and flowering. The researchers also mentioned that their previous work showed that the homologue of tomato gene *Ve1* (*StVe1*) is associated with partial resistance to Verticillium wilt in potato, but more work is required to develop robust molecular markers that can be used by breeders.

Over the past few years, Verticillium wilt has been expanding into most potato production areas of China and is now one of the most serious soil-borne diseases affecting potatoes in China. Li et al. (2019) have therefore evaluated 30 commercially popular potato cultivars for resistance to *V. dahliae* strain Vdp83 and *V. nonalfalfae* strain Vnp24, two well-characterized strains causing Verticillium wilt of potato in China, the latter a relatively new species of increasing importance. Ten plants of each cultivar were inoculated with the *V. dahliae* strain and incubated on greenhouse benches. Symptoms were rated at weekly intervals and the relative area under the disease progress curve was calculated. Five resistant and four susceptible cultivars identified from the analyses were then challenged with the *V. nonalfalfae* strain. Cultivar responses to *V. nonalfalfae* were like those to *V. dahliae*, except for one cultivar. The cultivars resistant to both species were Qingshu 9, Zhongshu 18, Longshu 8 and Zhongshu 19. Hence, resistance among potato cultivars exists in China and can be deployed to manage the disease until breeding programmes in China develop new cultivars with resistance from the sources identified.

5.5.3 *Fusarium Wilt*

Fusarium wilt, caused by *Fusarium solani* (Mart.) App. and Wr. var. *eumartii* (Carp.) Wr. and *F. oxysporum* Schlecht, is also a soil-borne disease, which has been reported in many potato-growing areas, but causes little loss in yield and appears to have been mainly a localized problem in the past. Hence, little effort appears to have gone into breeding for resistance, although differences in susceptibility have been reported among potato germplasm (Pavek and Corsini 1994).

5.6 Fungal Diseases of Tubers

5.6.1 *Wart*

Wart is a persistent, soil-borne, disfiguring, blemish-forming disease of potato tubers caused by the obligate biotrophic fungus *Synchytrium endobioticum* (Schilb.) Perc. whose persistent resting spores can remain infectious for more than 20 years. Chemical control of the disease is not possible. It is assumed that *S. endobioticum* spread at the end of the nineteenth century from its centre of origin in the Andes, first to Europe and North America, and subsequently across the whole potato-growing regions of Asia, Africa and Oceania (Obidiegwu et al. 2014). It provides an early success story in breeding for resistance. Back in 1920, when the Scottish Plant Breeding Station (SPBS) was established in Edinburgh, wart was considered a serious soil-borne disease. Some of the cultivars available for use as parents in 1920 were resistant to wart, which was inherited in a simple dominant manner. Wart was largely eliminated from Britain during the first half of the twentieth century by coupling the breeding of cultivars with immunity/resistance to pathotype 1 (the only pathotype found in Britain and synonymous with pathotype 1(D1)) with scheduling of land, which prohibited the cultivation of susceptible cultivars on land known to be infested (Simmonds 1969). The only complication in breeding for resistance was the fact that wart was a quarantine disease. This meant that SPBS could not do the testing, so potential cultivars in the advanced generations of the breeding programme were tested by the Quarantine Authority in Scotland. Nevertheless, until as recently as 1990, all cultivars bred at SPBS (and its successor SCRI) were resistant to wart. However, in other European countries, wart is still a problem, and since 1941, other pathotypes of the fungus have been found and are evolving. Hence, there is still a need for research to ensure a continuation of effective and sustainable management of the disease, including quarantine measures and resistant cultivars, as reviewed by Obidiegwu et al. (2014). A few of the key points from their review are as follows. Since 1941, more than 30 new pathotypes have been identified, of which pathotypes 1, 2, 6, 8 and 18 are the most important ones in Europe. They can be differentiated using ten potato cultivars: Tomensa and Deodara are susceptible to all five pathotypes, whereas Karolin, Ulme and Belita are resistant to all five. Wart resistance testing is done under field or greenhouse conditions to complement laboratory bioassays, details of which can be found in the review by Obidiegwu et al. (2014). The reliable assessment of resistance is laborious and time consuming and requires the inoculation of at least 20 tubers per pathotype. Hence, it can only be done in the later stages of a breeding programme. Breeding for resistance has also been hampered by the lack of single genes for resistance and the complexity of resistance screening with several pathotypes. A single dominant gene (*Sen1*) for resistance to pathotype 1 has been mapped to chromosome 11 (Hehl et al. 1999) and another single dominant gene (*Sen1-4*), also for resistance to pathotype 1, has been mapped to chromosome 4 (Brugmans et al. 2006). Groth et al. (2013) found a major quantitative resistance locus (QRL) near *Sen1* that explained between 46% and 56% of the phenotypic variation for resistance against pathotype 1. Other QRL of smaller effect for resistance to combinations of pathotypes 1, 2, 6 and 18 have also been mapped (Ballvora et al. 2011; Groth et al. 2013) and their positions shown on the 12 chromosomes of potato by Obidiegwu et al. (2014). However, of more significance is the discovery of a major gene (*Sen2*) on chromosome 11, distinct from *Sen1*, which provides resistance to eight pathotypes of wart: pathotype 1(D1), four pathotypes also common in Western Europe, 2(G1), 6(O1), 8(F1), 18(T1), and three pathotypes found in Poland and named 2(Ch1), 3(M1) and 39(P1) (Plich et al. 2018). Gene *Sen2* was present in a complex diploid hybrid of various *Solanum* species, bred at IHAR-PIB, Młochów Research Centre, Poland. The hybrid was the resistant parent of the mapping population of 176 offspring clones and also had *Sen1*, as did the other diploid parent which lacked *Sen2*. DArTseq markers were used for the mapping. The genetic and physical distances between the *Sen1* and *Sen2* loci were indirectly estimated at 63 cM and 32 Mbp, respectively. The work involved testing more than 58,000 tubers by the Glynne–Lemmerzahl method, as modified by Przetakiewicz (2008). The authors developed PCR markers co-segregating with the *Sen2* locus that can be applied in marker-assisted selection of potatoes resistant to eight important pathotypes of *S. endobioticum*. They concluded that the wide spectrum of the *Sen2* resistance may be an indication of durability which can be enhanced by the pyramiding of the *Sen2* and *Sen1* genes, as occurred in 61 clones in their mapping population.

5.6.2 Powdery Scab

Powdery scab is a persistent, soil-borne, disfiguring, blemish-forming disease of tubers caused by the fungus *Spongospora subterranean* (Wallr.) Lagerh. Roots can also be infected. The pathogen thrives, and powdery scab becomes severe, in cool, damp, soil conditions. Hence, the disease is still a problem for growers in many countries in the cool, temperate regions of the world, resulting in loss of tuber quality and yield, as can be seen from the Proceedings of the Third International Powdery Scab Workshop, held in Einsiedeln, Switzerland, from July 18–21, 2016 (Merz and Falloon 2017). The review by Falloon (2008) of integrated disease management to control powdery scab is still relevant, and so is breeding cultivars with better resistance as an important component of such a strategy. No potato clones had been identified that were immune to infection, so crops of resistant cultivars were likely to support pathogen multiplication and survival and maintain soil infestation. This also means that resistance was quantitative rather than qualitative in nature. Seed potatoes with powdery scab are commercially unacceptable for establishment of new crops because the disease can be transmitted on infected and infested seed tubers, and seed certification systems usually include low tolerance levels for powdery scab incidence. Falloon (2008) mentioned that resistance to powdery scab was an important breeding objective in New Zealand, with Gladiator an example of a cultivar with a high degree of resistance. Russell Genet and colleagues in New Zealand assessed 138 potato cultivars and 24 breeding clones over 25 consecutive growing seasons (1991/2 to 2015/16) in powdery scab trials (Merz and Falloon 2017). The trials were in soil inoculated with macerated potato tubers which were severely infected with powdery scab. The trials had randomized complete block designs with six replicates and five-plant plots. They were planted in mid-September and harvested by hand in early March. Plants were regularly irrigated during growth, especially at and just after tuber initiation, to encourage disease development. At harvest, tubers >30 g were washed free of soil and assessed for powdery scab severity using a 0–3 scale (0 = no infection, 1 = <5% of tuber surface area infected, 2 = 5–25% and 3 = >25%). Mean tuber scores for each cultivar/clone were linearly scaled from 1 (all tubers heavily infected) to 9 (complete resistance) through a generalized linear model. Merz et al. (2012) looked at ten reference cultivars, ranging from resistant to susceptible, in more detail over 4 years in four to six locations across Europe. The ten cultivars performed similarly across the different environments. The authors suggested that the lack of a genotype × location interaction was due to low genetic variation in the pathogen population. However, there was no relationship between tuber and root susceptibility. The authors thought that cultivars with low tuber susceptibility but high root susceptibility were likely to be a particular risk factor in potato production because sporosori produced in root galls contribute unnoticed to a large part of the long-term soil inoculum.

5.6.3 *Fusarium Dry Rot*

Fusarium dry rot is an important cause of spoilage of tubers in storage worldwide. Lees (2018) deals with it as a case study of fungal diseases affecting potato storage. She points out that as many as 13 species of *Fusarium* can cause dry rot of potato and the species composition appears to vary according to climate and geographical location, cultivar preference, and possibly agronomic practice. *Fusarium coeruleum* (Libert) Sacc. is the most commonly isolated species in Great Britain and the Nordic countries, whereas *F. sambucinum* (Fuckel) (synonymous with *F. sulphureum* and *Gibberella cyanogena*) is the most significant species in other parts of Europe, China and North America. *F. avenaceum* Sacc. (Corda ex. Fr.) occurs in Scotland

and Norway. *Fusarium* can be both soil- and tuber-borne, with inoculum transmitted by infected or contaminated seed tubers. Although disease control can be achieved by good agronomic practice to minimize damage, the use of fungicides (but resistance to thiabendazole does occur), in-store ventilation and host resistance is also important. However, cultivars vary in their response to different *Fusarium* spp. Wastie et al. (1989), for example, examined the susceptibility of 15 potato cultivars to dry rot caused by *F. sulphureum* and *F. coeruleum* over 8 years. Field-grown tubers from a frost-free store (6–10 °C) were wounded (with a rounded-end glass rod), inoculated (with a 4-week-old maize-meal + sand culture of the *Fusarium* species), incubated at 10 °C and high (90–95%) relative humidity for 7–8 weeks, and the size of the rot was assessed. The number of tubers in each of the six infection classes (0, 1/16, 1/8, ¼, ½ and > ½ of exposed tissue colonized) was multiplied by 0, 1, 3, 6, 12 and 24, respectively, the products added and the sum divided by the total number of tubers. A mean score on the 0–24 scale was calculated for each cultivar in each year. The researchers found little correlation ($r = 0.371$) between the rank order of susceptibility of the cultivars to the two pathogens but a highly significant correlation between years (there was no cultivar × year interaction). They concluded that 2 years of tests are sufficient to assess susceptibility to both species. However, as the tests required three samples of 10 even-sized undamaged tubers of each cultivar, they are more suited to the final stages of a breeding programme than to the early generations when tubers are in short supply. Subsequent work on 22 tuber progenies from glasshouse- and field-grown (2 years) plants by Wastie and Bradshaw (1993, 1995) found that for *F. coeruleum,* variation due to general combining ability (GCA) was significant when tested against the variation due to specific combining ability (SCA), which was also significant for the field-grown progenies. Progeny means for the glasshouse- and field-grown plants (two years) were highly correlated ($r \geq 0.79$). Furthermore, parental and GCA values were highly correlated ($r \geq 0.88$), as were progeny means and mid-parental values ($r \geq 0.71$). In contrast, no such relationships were found for *F. sulphureum.* In later work with 71 different progenies, Lees and Bradshaw (2001) did find heritable resistance to *F. sulphureum* which was better assessed on tubers from field-grown (1 year) than from glasshouse-grown (2 years) plants, although the two were largely in agreement. Variation due to GCA was significant in the three tests when compared with the variation due to SCA, which was also significant in the two glasshouse tests. The GCA values from the three tests were in reasonable agreement ($r = 0.61$ and 0.69 between glasshouse and field, and 0.85 between two 'glasshouse' tests). The GCA values from all three tests were highly correlated ($r \geq 0.79$) with the 13 (one parent was missing) parental values from field-grown plants but not with the 11 parental values from the second year of glasshouse tests ($r \leq 0.49$), hence the superiority of field-grown tubers for progeny testing in a breeding programme. Valluru et al. (2006) also reported genetic differences for resistance to *F. sambucinum* (*F. sulphureum*) between clones from four tetraploid Tuberosum crosses. Broad-sense heritability in 2 years was high (≥0.77) for both lesion depth and diameter. However, in the 2 years, 24% and 38%, respectively, of the clones were unstable (high variance) over six assessments (experiments) of resistance. The researchers concluded that resistance to *F. sambucinum* has a large genetic component and thought that lesion diameter was more informative than lesion depth for evaluating resistance.

In conclusion, resistance to *F. coeruleum* is different and separate from resistance to *F. sulphureum* and hence, a breeder needs to decide if resistance is required to one or both species. Furthermore, the indications are that breeding for resistance to the former is easier than to the latter. This view is reinforced by the work of Burkhart et al. (2007) on resistance to a mixture of two isolates of *F. sambucinum* (*F. sulphureum*) and one isolate of *F. solani* in a North American diploid Phureja-Stenotomum population. Although more than half of the variation for resistance was genetic, the additive genetic component was small. Thus, higher levels of resistance to *F. sambucinum* were unlikely to be achieved by conventional breeding, so the authors advocated using a few highly resistant clones in 4*x*-2*x* crosses to transfer much of the dominance and epistatic variance governing resistance to the tetraploid level.

5.6.4 Gangrene

Gangrene, caused by *Phoma foveata* Foister, is a tuber-borne disease which causes spoilage of tubers during storage. As it was considered a problem in northern Europe in the 1980s, research on assessing and breeding for resistance was done at the Scottish Crop Research Institute. Wastie et al. (1988, 1990) developed methods for assessing clones and progenies for their resistance to gangrene. Appropriate-sized plots of the test material were raised free of disease in replicated trials in a field on an experimental farm and tubers harvested at maturity in early September. For clones, plots would comprise enough plants (e.g. four) to give a sample of 10 tubers, whereas for progenies, plots would comprise a row of 18 (or 20) plants to give a bulk of tubers, one from each plant. Tubers were stored in boxes, one for each plot, at 6–10 °C until mid-November each year, when they were washed and inoculated with a mixture of isolates (three to five, known to be aggressive) of *Phoma foveata* by a cornmeal-sand method. The tubers were dampened by being lightly watered and then rolled in a box of the diluted inoculum and hence, they were completely covered in sand, some of which adhered to them when they were removed. They were then replaced in the boxes and immediately incubated at 4 °C and 100% relative humidity for 8–10 weeks after which the level of infection was scored. In order to register a wide difference in score between resistant and susceptible clones or progenies, the number of tubers infected to a predetermined amount was recorded. An amount of either one-sixteenth, one-eighth, or one-quarter of the surface area of the tuber covered with gangrene lesions was chosen according to the rapidity with which the infection had developed in the control cultivars at the time of scoring. The percentage of infected tubers in each sample was then calculated (Stewart and Wastie 1989). Wastie et al. (1988) assessed the resistance of 36 tuber progenies from 10 parents differing in resistance/susceptibility and found a narrow-sense heritability of 0.49. They concluded that the mean resistance of the parents would give an indication of the likely resistance of their progeny. In a further experiment with 11 of the 36 progenies, Wastie et al. (1990) found a good correlation ($r = 0.80$) for the mean percentage of infected tubers between tubers from glasshouse-grown seedlings and those from plants grown in the field from tubers. They advocated the use of the seedling progeny test to identify progenies with useful levels of resistance, as well as providing information on the inheritance of resistance. Subsequently, Bradshaw et al. (1996) used the seedling progeny test to evaluate progenies from a 15 × 15 half diallel set of crosses originally made to investigate the inheritance of resistance to late blight and the white potato cyst nematode. An incubation period of 14 weeks at 4 °C was used. All of the variation between progenies could be attributed to differences in the general combining abilities (GCAs) of their parents. Interestingly, the four parents with the best GCAs for resistance had sizeable contributions from Group Andigena potatoes in their pedigrees. The overall conclusion from this work is that improved resistance to gangrene is relatively easy to achieve, should this be important enough as a breeding objective.

5.6.5 Black Scurf

Black scurf is a soil- and tuber-borne skin blemish disease caused by the fungus, *Rhizoctonia solani* (Kühn.), which also causes stem canker. The black scurf symptom is made up of the sclerotia (resting bodies) of the fungus. A comprehensive review of black scurf has been provided by Kumar et al. (2017), but does not include resistant cultivars. In countries like Great Britain, potential new cultivars are tested for resistance/susceptibility (NIAB TAG 2019), but resistance is not a high priority breeding objective. In contrast, there is now interest in breeding for resistance in Inner Mongolia, where *Rhizoctonia solani* AG2-1 is the main Anastomosis Group. Zhang et al. (2014) have developed a two-step method for evaluating and screening germplasm for resistance. Preliminary evaluation is based on tuber piece inoculation in the laboratory where relatively resistant cultivars are selected. These are then assessed in a field test using artificial inoculation of cut seed tubers with 2–4 g of wheat bran containing hyphae and sclerotinia. The researchers found that lesion areas on tuber pieces positively correlated ($r = 0.46$) with the disease indices of potato tubers during 2 years of field tests. Furthermore, the disease indices showed a positive correlation ($r = 0.66$) between the 2 years of field tests. The results also revealed differences in resistance between 20 potato cultivars/clones, but none of them was completely resistant, and most were susceptible.

Another common symptom of the disease is ‘damping-off’ of infected seedlings. Bidondo et al. (2019) have shown that the overexpression of antifungal genes enhances resistance to *Rhizoctonia solani* AG-3 in transgenic potato plants of cultivar Kennebec without affecting arbuscular mycorrhizal (AM) symbiosis. Rooted plants from micropropagation were transferred to trays filled with a soil-vermiculite-sand (70–15–15%) substrate infected with strain AG-3 and placed in a growing chamber. Out of six transgenic lines, at 7 days post infection (DPI), the three lines expressing high levels of a ribosome inactivating protein had the best pathogen resistance (infection reduction index of 65% to 87% for plant height relative to control). Then at 18 DPI, the line which also expressed high levels of chitinase protein exhibited 100% inhibition of infection. In addition, resistant GM plants showed no visual phenotypic differences from untransformed control plants under greenhouse conditions. Finally, the researchers demonstrated that arbuscular mycorrhizal symbiosis (isolate GA5 of *Rhizophagus intraradices* and a mix of indigenous AM fungi) was not adversely affected in the transgenic plants (assays under in vitro and microcosm conditions). Furthermore, the transgenic line with high levels of chitinase protein had improved mycorrhization levels.

5.6.6 Skin Spot, Silver Scurf and Black Dot

Skin spot (*Polyscytalum pustulans* (Owen and Wakef.) M.B. Ellis), silver scurf (*Helminthosporium solani* Dur. and Mont.) and black dot (*Colletotrichum coccodes* (Wallr.) Hughes) are three tuber-blemishing diseases. All three pathogens cause a deterioration of tubers in store; all are mainly or partly tuber-borne as well as soil-borne; and all are relatively common (Wastie 1994). More recent research in the UK (Lees et al. 2010) showed that soil-borne inoculum of *C. coccodes* was more important than tuber-borne inoculum in causing black dot. Potential new cultivars are tested for resistance/susceptibility in Britain (NIAB TAG 2019), but resistance to these diseases is not a high-priority breeding objective. Wastie et al. (1997) did develop a glasshouse progeny test that could be used for assessing large numbers of progenies for resistance to skin spot in genetical studies and for eliminating highly susceptible progenies in the early stages of a breeding programme. Furthermore, they found high correlations ($r > 0.864$) between mid-parent values and progeny means (i.e. a high narrow-sense heritability) for nine progenies from six parents for glasshouse-grown and field-grown tubers over 2 years. However, the test was not used in practice because other traits were given a higher priority. Nevertheless, it is worth mentioning that in the test, tubers are inoculated by dipping in a suspension of mycelium and spores of three isolates of *P. pustulans* and then placed in storage at 4 °C and 95% relative humidity for 4 months.

Similar work could be done on silver scurf and black dot if considered important. In fact, silver scurf did emerge as an economically important disease in the 1990s and the review by Errampalli et al. (2001) is still relevant. They pointed out that control of the disease by chemical and cultural practices is difficult and that *H. solani* isolates resistant to the postharvest fungicide thiabendazole (TBZ) have arisen. They reported methods of screening cultivars for susceptibility to silver scurf, but at their time of writing, no cultivars with high levels of resistance had been identified, at least in North America. Likewise, black dot has increased in importance over the last 20 years, and in the UK, Brierley et al. (2015) investigated factors affecting the development and control of the disease in field trials over the four-year period,

2004–2007. Cultivar resistance was shown to be an effective method of reducing black dot disease severity on tubers, with cultivar Sante more resistant than Maris Piper (17.0% compared with 43.8% of tubers unmarketable). Johnson et al. (2018) have reviewed the pathogen, disease development and management. They pointed out that potato cultivars with high levels of resistance to *C. coccodes* are not commercially available, but cultivars vary in susceptibility and extremely susceptible ones should be avoided for production unless providing market advantages because they require extra disease management.

5.7 Actinomycete Disease of Tubers

5.7.1 Common Scab

Common scab (CS) is caused by the saprophytic, filamentous, gram-positive, soil-borne bacterium *Streptomyces scabies* (Thaxt.) Waksman and Henrici and other pathogenic *Streptomyces* species. They have a wide host range and global distribution. Symptoms of the disease are scab-like (surface/raised/pitted) lesions on tubers which can cause a substantial reduction in marketable yield. Braun et al. (2017) reviewed the causal pathogens, management practices to control the disease, host resistance and screening methods which can be used by breeders, all made difficult by the diversity of pathogenic isolates. Key points can be summarized as follows. The bacterium is dispersed by spores and survives on seed, in soil and in soil water. Infection is aided by the production of a phytotoxin, thaxtomin A, which inhibits cellulose biosynthesis. The *txtAB* operon, responsible for thaxtomin A production, can be used as a marker to identify pathogenic strains of the bacterium. Tubers are most susceptible to infection up until 3–4 weeks after initiation, and with a soil moisture below field capacity. Hence, one control measure for growers is to maintain soil moisture in the 80–85% range. However, one of the best options for managing common scab is host plant resistance, and breeding for resistance has been a high priority among North American potato breeders. Braun et al. (2017) list the following as new cultivars with resistance: Alta Crown, Freedom Russet, GemStar Russet, Kalkaska, Liberator, Marcy, McBride, Megachip, Millennium Russet, Owyhee Russet, Premier Russet, Summit Russet, Teton Russet and Western Russet. Braun et al. (2017) also give references to potato wild relatives and landraces with reported resistance. To date, major genes for scab resistance have not been identified or mapped, and most of the literature treats scab resistance as a quantitative trait. By conducting a genome-wide association analysis (GWAS) of 20 commercial cultivars and 123 advanced Canadian breeding clones, Yuan et al. (2020) identified three QTLs for resistance on chromosomes 2, 4 and 12 which explained 21%, 19% and 26%, respectively, of the variation (assessments were done in each of 3 years in an on-farm scab nursery). Finally, Tegg and Wilson (2010) found no association between resistance to common scab disease and tolerance to thaxtomin A toxicity.

They tested a range of genetically diverse cultivars with varying resistances to common scab disease in both pot- and field-based studies and measured their mean necrosis response to thaxtomin A.

A number of assays have been developed to screen for resistance to common scab. In field assessments, the potatoes are grown to maturity in a naturally infested field (sandy soil) and the harvested tubers scored for scab symptoms. In greenhouse and growth chamber assessments, the potatoes are grown in inoculated potting medium (or in naturally infested soil) with controlled inoculum dose and composition and environmental conditions. Initial screening of seedlings with thaxtomin A has met with limited success. However, screening methods in breeding programmes are still time consuming and can produce variable results. At SCRI, for example, Caligari and Wastie (1985) used a glasshouse pot test in which a naturally infested sandy soil was diluted by 50% with river sand. Disease assessments by directly scoring the produce of the pots on a 1–9 scale of increasing resistance were as satisfactory as calculating the percentage of infected tubers after estimating the amount of common scab on each tuber. The data for 19 cultivars had a good correlation ($r = 0.73$) with official tests. The authors calculated that the use of 12 replicate plants would permit the detection of a difference in mean resistance of 1.3 points on a 9-point scale where resistant cultivar Pentland Crown scored 8 and susceptible cultivar Maris Piper scored 2. Ironically, Maris Piper replaced Pentland Crown as the most widely grown cultivar in the UK. More recently, Navarro et al. (2015) have provided an account of selecting for resistance to common scab in the Wisconsin (USA) potato breeding programme. From 2006 to 2013, they tested advanced clones and check varieties in 18 field environments with high levels of inoculum of pathogenic *Streptomyces* spp. During the same period, a similar set of potato clones were evaluated in 18 standard breeding trials which followed a potato-corn-corn-soybean-potato rotation to reduce soil-borne diseases and other sources of field variation. Typically, between 60 and 160 distinct genotypes were evaluated each year in these trials. In each trial, each clone was evaluated using three replicates of four-plant (hill) plots. The dedicated common scab trials proved superior for selecting for resistance, with higher heritabilities for clone means in a trial (average 0.83 compared with 0.53 for standard breeding trials) and the ability to separate resistant from susceptible cultivars, as well as an assessment of stability of resistance across environments. The researchers recommended testing advanced clones for 3 years in two dedicated common scab trials each year, making a total of six trials. Given the work involved in such trials, genomic selection for common scab resistance is an attractive prospect as shown by Enciso-Rodriguez et al. (2018). They assessed 370 tetraploid genotypes, primarily from Michigan State University's breeding programme, over 9 years (2009–2017) in a field disease nursery. The field had been inoculated with common scab from aggressive Michigan isolates and had been cultivated for high disease pressure. The researchers reported little genotype × year interaction and a moderately high genomic heritability estimate using an additive model (0.45 ± 0.017). Interestingly, they located a major-effect SNP-marker on chromosome 9 which was associated with a WRKY transcription factor.

5.8 Bacterial Diseases of Foliage, Stems and Tubers

Numerous bacterial pathogens can affect potatoes and be spread by the movement of infected seed, cultivation in infested soil or transmitted by insect vectors (Sjolund et al. 2018). The most serious diseases are bacterial wilt and brown rot, ring rot, blackleg, common scab (dealt with in the previous section) and a novel threat called zebra chip disease. The review by Zimnoch-Guzowska et al. (2005) of resistance to bacterial pathogens is still relevant along with more recent publications.

5.8.1 Bacterial Wilt of Plant and Brown Rot of Tuber

Ralstonia solanacearum Smith is the cause of bacterial wilt of the plant and brown rot of the tuber, which shows as progressive browning of the vascular ring tissue, its pathogenicity depending on a type III secretion system that delivers effectors that manipulate plant immunity (Sjolund et al. 2018). It is a soil-borne bacterium that enters plant's roots, invades xylem vessels and spreads rapidly to aerial parts of the plant through the vascular system where its high level of multiplication leads to wilting (Huet 2014). Fuglie (2007), in a review of research priorities for potatoes in developing countries, highlighted the control of *R. solanacearum* through cultivar resistance and crop management alongside control of late blight, improved yield potential, seed supply and virus resistance. More recently, Guchi (2015) identified bacterial wilt as a major production constraint in Ethiopia. The following information is taken from the review by Sjolund et al. (2018). *R. solanacearum* can be divided into five races on the basis of differences in host range, with race 3 infecting potato and tomato, and into six biovars on the basis of biochemical properties. Four major phylotypes, based on variation in the internal transcribed spacer region of ribosomal DNA and the endoglucanase gene, can be distinguished in relation to the phylogenetic and geographic origin of strains (I from Asia, II from the Americas, III from Africa and IV from Indonesia). Phylotype II has two sub clusters, A and B, with strains belonging to IIB responsible for bacterial wilt of potatoes in cold and temperate regions (Janse et al. 2004). *R. solanacearum* can spread through contaminated soil and debris, weed hosts and irrigation water, and over longer distances by latently infected seed potatoes. Interestingly, infection of a relatively common aquatic Solanaceous weed, *Solanum dulcamara*, growing on riverbanks, has been detected in several European countries and can result in large numbers of the bacteria being liberated in water (Zimnoch-Guzowska et al. 2005). *R. solanacearum* is endemic in many developing countries in inter-tropical (equatorial) regions where disease management remains limited to the use of pathogen-free seed tubers and adequate crop rotations with non-susceptible crops as chemical controls are not available.

Assessment of resistance can be done on plants grown in naturally contaminated fields with susceptible and more resistant cultivars interplanted as controls, as done in Brazil by Lopes et al. (1993). Tuber latent infection can be assessed by post-enrichment enzyme-linked immunosorbent assay on nitrocellulose membrane (NCM-ELISA) of 30 tubers per clone (Priou et al. 1999). It is a reliable and sensitive low-cost method that is rapid and easy to use, making it suitable for assessing susceptibility of breeding lines to bacterial wilt. Selection for resistance in the field and to tuber latent infection need to be considered in breeding programmes as the one does not guarantee the other. Lima et al. (1996) used a seedling test for early screening in a breeding programme in which seedlings grown in trays from true botanical seed were inoculated by dipping in a bacterial suspension for 10 min in a glasshouse.

Breeding for resistance has been hampered by the broad diversity of the pathogenic strains, the difficulties being highlighted in the review by Huet (2014). Patil et al. (2012) did a comprehensive review of improvement for bacterial wilt resistance in potato by conventional and biotechnological approaches, including all of the reported sources of resistance. They concluded that despite reports of a number of sources of resistance (e.g. in Group Phureja), breeding for resistance had only been reasonably successful to a very small extent in potato in South America and Asia, that the resistant cultivars developed were not adapted to different agroclimatic zones and were not effective against all strains of the pathogen. Hence, development of transgenics, containing genes imparting stable resistance, seemed to be the only effective alternative. In summary, despite limited success, breeding for resistance to bacterial wilt is likely to remain a priority where the disease is endemic.

5.8.2 Blackleg (Stem) and Soft Rot (Tuber in Storage)

As explained by Zimnoch-Guzowska et al. (2005) and Sjolund et al. (2018), blackleg of potato plants is a tuber-borne disease caused by a number of pectolytic bacteria species, most commonly *Pectobacterium atrosepticum* (synonym *Erwinia carotovora* subsp. *atroseptica*). The disease results in brown-black discolouration of the stem and stunting and wilting, as well as some initial non-emergence of seed tubers. Other bacterial species such as *Pectobacterium carotovorum* (synonym *E. carotovora* subsp. *carotovora*) and *Dickeya solani* can also cause field symptoms, but are more commonly associated with soft rots of tubers in storage. These can be cream to light brown in colour on the tuber surface with cut tubers showing a darker, clear margin between rotted and sound tissue. The *Dickeya* genus was previously classified as *E. chrysanthemi*. The species *D. solani* was first discovered in 2006 in crops showing blackleg in Northern Europe and Israel and has now been reported over a wide geographic range in Europe and Israel. Blackleg in north-western Europe and North America is increasing in incidence and severity. Which *Pectobacterium* and *Dickeya* pathogens dominate depends on the location and prevailing environmental conditions. Blackleg and soft rot can be controlled by seed management (seed free of pathogen), irrigation management (avoiding prolonged periods of high soil moisture), store management (proper ventilation) and cultivars with high levels of resistance, but not by chemical means. More information on the control of blackleg and tuber soft rot can be found in the review by Czajkowski et al. (2011), which includes a section on breeding for resistance.

As resistance to tuber soft rot and blackleg are not that closely related, assessment of both is required in breeding for resistance. The review by Zimnoch-Guzowska et al. (2005) of progress in breeding for resistance covered: (1) methods of screening for resistance (laboratory-based inoculation by various methods of whole tubers or tuber slices for soft rot; and for blackleg, in the greenhouse using stem-inoculated potted plants or under field conditions using inoculated seed tubers); (2) sources of resistance (relatively few cultivars, some wild and primitive cultivated species); (3) inheritance of resistance (quantitative and polygenic with both additive and non-additive genetic variation for soft rot; QTL analysis confirmed polygenic inheritance); and (4) breeding methods (conventional breeding, use of $2n$ gametes in $4x \times 2x$ crosses, protoplast fusion and transgenesis). Subsequently, Lebecka (2017) has provided a detailed account, together with colour photographs, of two methods of assessing tuber and stem tissue resistance to soft rot and blackleg, respectively. The former requires 3–10 tubers per genotype for each of two replications in each of two seasons and the latter requires from 20 to 30 pot-plants in each of the two replicates of the test.

The breeding methods (Zimnoch-Guzowska et al. 2005) have been demonstrated to work, but it is not obvious from a literature search that we are seeing more new cultivars with higher levels of resistance to blackleg and soft rot. Nevertheless, in a quantitative proteomic analysis comparing soft rot resistance and susceptibility to *D. solani*, Lebecka et al. (2019) were able to work with two resistant cultivars, Humalda and Bea, and a resistant diploid clone DG 00-270 derived by inter-crossing diploid *Solanum tuberosum* with *Solanum phureja* and other wild species: *S. chacoense*, *S. yungasense*, *S. verrucosum* and *S. microdontum*. Hence, it is briefly worth giving some breeding examples from around the year 2000. McGrath et al. (2002) partially transferred resistance to soft rot from *Solanum brevidens* (a diploid non-tuberous wild species) to tetraploid *S. tuberosum* over four backcross generations starting from a somatic hybrid (A206, hexaploid). Soft rot reactions were determined via a tuber plug inoculation method: the somatic hybrid had 20% of tubers and plugs showing evidence of severe rotting compared with 60% in the second to fourth backcrosses and 80% in control crosses between cultivated potatoes. Carputo et al. (2000) found resistance to soft rot in S. *tarijense* which they were able to transfer to tetraploid Tuberosum through a $2x \times 4x$ cross, where the diploid parent was a fertile interspecific hybrid between a Tuberosum haploid (dihaploid) and *S. tarijense* that produced $2n$ eggs. In contrast, they were less successful in transferring resistance from *S. commersonii* via a tetraploid somatic hybrid of a Tuberosum haploid (dihaploid) and *S. commersonii*. Zimnoch-Guzowska et al. (1999) identified 5 out of 28 potato diploid hybrids with both resistance to blackleg and soft rot as well as other desirable traits for use as parents in future breeding work. *S. chacoense* occurred frequently in the pedigree of these hybrids. Finally, Lees et al. (2000) identified long-day-adapted *Solanum phureja* as a source of resistance to blackleg caused by *Erwinia carotovora* subsp. *atroseptica* (*Pectobacterium atrosepticum*).

Two years of field tests with inoculated tubers and two controlled environment tests using stem inoculation were done on 19 clones from SCRI's long-day-adapted *S. phureja* population (Carroll 1982) which had not been selected for resistance to blackleg. One Group Phureja cultivar Mayan Twilight, which is commercially available, scored 9 for blackleg in official trials on a 1–9 scale of increasing resistance (NIAB TAG 2019).

5.8.3 Bacterial Ring Rot

Clavibater michiganensis subsp. *sepedonicus* (Spieck and Kotth.) Davis et al. is the cause of potato ring rot, a recurring seed-tuber-transmitted disease problem in temperate regions, despite many countries treating it as a quarantine disease and having a zero-tolerance policy for the import of infected seed potatoes. The disease results in wilting of the lower leaves and discolouration of the vascular ring tissue of tubers (Sjolund et al. 2018). Severely infected seed tubers may fail to sprout. The disease is spread mainly via infected seed tubers and one needs to be aware of symptomless infections. Both Jansky (2000) and Zimnoch-Guzowska et al. (2005) mentioned tetraploid *S. acaule* (disomic inheritance) as a source of extreme resistance (immunity to infection) which was lacking in current cultivars, as well as reviewing methods of screening for resistance. Kriel et al. (1995) concluded from their hybridizations that immunity was simply inherited, being mainly due to two complementary dominant alleles at two loci. They made four immune × immune, four immune × non-immune, four non-immune × immune and four non-immune × non-immune crosses, and noted the proportion of each progeny that was immune to root inoculations. Symptoms were noted at 6 weeks after inoculation and modified indirect immunofluorescence antibody staining (IFAS) was used to determine stem populations of *Clavibater michiganensis* subsp. *sepedonicus.* However, little breeding work on resistance appears to have been done.

5.8.4 Zebra Chip Disease

A brief account of zebra chip disease can be found in the review by Sjolund et al. (2018). The name zebra chip comes from the characteristic dark stripes produced when infected tubers are fried. The disease was first observed in Mexico in 1994 and then spread to the southern parts of the USA, other countries in Central America, and New Zealand. The causal agent of the disease is believed to be the unculturable bacterium *Candidatus* Liberibacter solanacearum (Lso) which resides in the plant phloem and is primarily vectored by the potato psyllid (lice) *Bactericera cockerelli.* Pesticides are the most common method of control but insecticide resistance has been reported. The reader is referred to the review of Sjolund et al. (2018) for more details.

There is interest in breeding for resistance as can be seen in the paper by Rashidi et al. (2017), which provides detailed methods for screening for resistance as well as a good literature review. At their time of writing, there were no reports of resistance to zebra chip disease in cultivated potatoes. The authors point out that resistance or low susceptibility represents a plant's ability to limit or reduce infection by the pathogen (Lso), whereas tolerance means relatively low symptom expression in tuber tissue despite infection. The authors conducted a two-year greenhouse study to evaluate Lso transmission success and the susceptibility of 11 potato breeding clones, representing a diverse genetic background, to zebra chip, relative to the susceptible commercial cultivar, Russet Burbank. When individual plants were exposed to two Lso-positive potato psyllids for 48 hours, the percentage of successful Lso transmission varied between 7% and 57%. Freshly-cut and fried tubers showed significant variation in symptom severity with several clones expressing relative tolerance compared to Russet Burbank. None of the clones showed statistically lower Lso titres than Russet Burbank with the exception of one clone in the second year of the study. However, the presence of a non-significant relationship between average symptom severity and Lso titre indicated variations in tolerance among the breeding clones, three of which had relatively lower Lso titre (low susceptibility) and tuber symptom expression (high tolerance). The reader is referred to the original paper for details, but a few aspects of the experiment are worth mentioning here. Under controlled greenhouse conditions, the temperatures fluctuated between 23 °C (day) and 18 °C (night) with a 14/10-hour, light/dark photoperiod. The plants were grown in two-gallon pots with 12 replicates of each clone including two non-infected controls. Inoculations were done on a single fully developed leaflet approximately 3 weeks after planting, with the release of two Lso-positive potato psyllids into a one-inch clip cage. Plants, including the non-infested controls, were covered individually with large mesh bags for the 48-hour inoculation access period to prevent any potential escapes into the greenhouse. Tubers were harvested approximately 7–8 weeks after the onset of vine senescence, sliced, and scored for the visual symptoms of disease both before and after frying in oil. Core samples were also taken from tubers for later Lso quantification. Clearly, a lot of work is involved in all of these assessments which means that they will probably be confined to seeking parents for crossing and then to the later stages of breeding programmes.

5.9 Viral Diseases

In 1921, Salaman (1921) recognized that the 'degeneration' of potato seed stocks which worsened with successive generations was the result of virus infection. Almost 100 years later, Jeffries and Lacomme (2018) list 54 viruses that can affect potatoes. They point out that virus infection can cause a wide range of symptoms in the foliage (leaf rolling, mosaic, chlorosis, necrosis and ringspot) and in the tubers (external necrosis, internal necrosis and growth cracks), affecting yield and tuber quality, and hence marketability. Infection can be primary (current season) or secondary (tuber-borne). Viruses can be transmitted by insects (aphids, thrips, leaf-hoppers, whitefly and beetles), nematodes, fungi and contact; and a few in true botanical seed and pollen. Although the genomes of most viruses comprise single-stranded RNA (ssRNA), there are a few DNA ones. Some viruses are common in potato and have a worldwide distribution, whereas others are either rare in potatoes or confined to particular geographic areas. Viruses can be controlled through: (1) seed certification schemes that produce virus-free planting material; (2) management of inoculum sources by roguing out diseased plants, avoiding proximity of ware to seed crops and control of weeds and alternative hosts); (3) cultural practices such as limiting the number of field generations of seed crops, early planting date and early haulm destruction, border crops to potatoes and mulches; (4) chemical control with insecticides and mineral oils to control aphids; and (5) resistant cultivars. The latter are particularly important where there are problems with the production of virus-free seed potatoes for planting. The paper by Priegnitz et al. (2019) reports the problem in Uganda of seed tubers which are highly degenerated with viruses as a result of smallholder farmers commonly using their own recycled seed potatoes or ones from the informal sector of family, neighbours and local market. The authors provided evidence of the benefits of more positive selection of seed tubers from healthy ware plants to reduce levels in the next ware crop of PLRV (*Potato leafroll virus*) and PVX (*Potato virus X*) [little infection with PVY (*Potato virus Y*) and PVA (*Potato virus A*) was found]. CIP (the International Potato Centre) on its establishment in 1971 recognized the need to make broad-based germplasm from its collection of South American landraces available to National Programmes in developing countries for breeding and selection, including germplasm with virus-resistance. Recurrent selection with evaluations in a range of environments on the Peruvian coast led to the development of germplasm with extreme resistance to PVY and PVX, and the selection of virus resistant clones better adapted to the warm conditions of the lowland tropics. Later, germplasm was incorporated to improve resistance to PLRV, while enhancing earliness and market traits (Bonierbale et al. 2020).

True resistance prevents virus infection, spread of the virus in the plant, and/or reduces accumulation of the virus in infected cells and tissues (Valkonen 2015). When, for example, potato breeding started at the Scottish Plant Breeding Station in 1920, strain-specific resistances to the mosaic viruses were present in the cultivars available for use as parents, and the second cultivar to be released, Craigs Defiance in 1939, was resistant to *Potexvirus Potato virus X* (PVX) and its variant *Potato virus B*, and to *Potyvirus* members *Potato virus Y* (PVY) strain C and *Potato virus A* (Bradshaw 2009a). The resistance was conferred by four Group Tuberosum genes, Nx_{tbr}, Nb_{tbr}, Nc_{tbr} and Na_{tbr} (Davidson 1980). Subsequently, genes conferring comprehensive resistance to all strains of these viruses became preferred, and introgression of resistance genes from potato wild relatives and landraces began in 1941 (Bradshaw 2009a). The same was true of other breeding programmes in Europe and North America. Today, potato breeders globally will need to seek the advice of virologists in deciding if breeding for resistance makes sense in terms of the importance of the virus for a particular target environment and the likelihood of success.

Here, I am going to consider nine viruses (all ssRNA) and one viroid that have attracted the attention of potato breeders and were reviewed by Solomon-Blackburn and Barker (2001b) who considered both traditional and molecular approaches. Here, we are going to concentrate on traditional approaches. Pathogen-derived transgenic resistance will be dealt with in the final chapter on genetically modified potatoes. The review by Solomon-Blackburn and Barker (2001b) also deals with methods of screening for resistance. We shall see that for some forms of resistance, glasshouse or screenhouse testing can be used in the initial stages of breeding, but a field exposure trial will be required to demonstrate that the resistance is operating under crop conditions. Furthermore, we shall see that some forms of resistance can only be assessed in the field. Details of methods can be found in the quoted scientific papers.

*5.9.1 PVY (*Potato Virus Y*)*

PVY (a *Potyvirus* giving severe mosaic symptoms) is transmitted by about 65 aphid species and is considered the most economically damaging virus in cultivated potatoes (Jeffries and Lacomme 2018). It is a non-persistently transmitted virus in which the aphid quickly (seconds to minutes) acquires virus on probing infected plants and may immediately infect other plants. The aphid usually loses its infectiousness within an hour. These attributes make chemical control with insecticides ineffective because they do not kill the aphids quickly enough. Most of the aphid species do not colonize potato plants.

A number of PVY strains have been characterized, as explained by Jeffries and Lacomme (2018), based on differential ability to trigger a hypersensitive response to specific resistance genes, symptoms on tobacco (vein necrosis or mosaic), development of potato tuber necrotic ringspot disease (PTNRD) on tubers, serology (ELISA) and genome phylogeny (molecular methods: RT-PCR and sequencing). The strains are PVY^{C}, PVY^{O}, PVY^{N}, PVY^{Z}, PVY^{E}, PVY^{NTN} (associated with PTNRD) and $PVY^{N\text{-}Wi}$. Strains PVY^{NTN} and $PVY^{N\text{-}Wi}$ are recombinants between the genomes of PVY^{O} and PVY^{N}. The relative incidence of PVY strains has been changing with PVY^{O} replaced by PVY^{N}, PVY^{NTN} or $PVY^{N\text{-}Wi}$ in some areas, no doubt due to a number of biological and epidemiological factors contributing to a fitness advantage. For example, Dupuis et al. (2019) reported from a survey in 2012, a slow yet progressive increase in the prevalence of $PVY^{N\text{-}Wi}$ strains in Swiss seed-potato crops and simultaneously a gradual decrease in the prevalence of PVY^{NTN} and PVY^{O} strains. They explained the increase by way of the interactions of several factors such as the evolution of the populations of vectors and the fitness of the PVY strains in the plant, as well as the resistance profile of the potato cultivars grown in the country.

In breeding for resistance, breeders have utilized major dominant *R*-genes for a hypersensitive resistance response (HR) or for extreme resistance (ER). ER is epistatic to HR, indicating that the genes for ER act earlier and more efficiently than the genes for HR. As explained by Valkonen (2015), HR prevents virus loading to phloem in the initially infected leaf and translocation to other parts of the plant but virus replication and initial cell-to-cell movement are not affected. Later, however, most of the infected cells die and a necrotic local lesion is seen at the site of infection. In contrast, ER suppresses virus multiplication in the initially infected cells so that no visible signs of infection can be observed apart from possible tiny, necrotic pin-point lesions, and no virus is detected in the inoculated tissue. The *R*-genes produce R proteins that stop viral proteins from suppressing RNA silencing, the most important basal antiviral defence system in plants, and thus elicit a quick and powerful defence response (the molecular details are quite complicated). We thus have another example of Flor's gene-for-gene concept (Flor 1942). Solomon-Blackburn and Barker (2001a) provided a comprehensive list of resistance genes identified by 2001.

Although PVY is naturally transmitted by aphids, in breeding for resistance, large numbers of young seedlings in seed pans at the cotyledon stage can be inoculated with PVY using infective plant extracts mixed with an abrasive in a spray gun. Susceptible seedlings showing mosaic symptoms can be discarded from about 2 weeks after inoculation, but to minimize the possibility of 'escapes', an additional manual inoculation 2 weeks later is advisable (Solomon-Blackburn and Barker 2001b). For more detailed breeding and genetical studies, young plants can be tested in a glasshouse by spray-inoculation with infected tobacco sap containing celite powder as an abrasive. Susceptible seedlings develop mottle symptoms within a few weeks. Field exposure trials can be used to confirm the resistance of clones selected for major-gene resistance; most clones with ER or HR genes to PVY remain healthy.

Virus strain-specificity of most HR genes makes resistance vulnerable to new variants of the virus which are not recognized by the gene and can overcome resistance (Valkonen 2015). PVY strains recognized by products of genes *Ny*, *Nc* and *Nz* are in strain groups PVY^{O} (ordinary strains), PVY^{C} (C strains) and PVY^{Z} (Z strains), respectively. Gene Ny_{tbr} (Tuberosum) for resistance to PVY^{O} strains proved useful for many years, having been found in clone 11-79 from Australia [USDA 41956 × (Katahdin × Snowflake)] (Bradshaw 2009a). The gene was present in cultivar Pentland Crown, which was released in 1958 and became the number one cultivar in Britain for a few years. Most of the strains that can overcome all three HR genes are in group PVY^{N}, and since the 1980s, more severe variants of PVY^{N} have become prevalent in potato crops throughout the world. These variants can cause severe symptoms of mosaic and yellowing in leaves and necrosis in tubers. Genome sequences have revealed that they have evolved via recombination between the PVY^{O} and PVY^{N} strains. The ones that can overcome all three HR genes but do not induce veinal necrosis in tobacco represent a new strain group PVY^{E}. Another problem with HR genes is that some are temperature-sensitive and hence, higher temperatures render the resistance less efficient, which results in systemic spread of the virus and severe, even lethal necrotic symptoms in the plant. A recent example is plants with gene *Ny-DG* (chromosome 9). Szajko et al. (2019) found that at 20 °C, strain PVY^{NTN} was localized in the inoculated leaves without symptoms, but at

28 °C, necrotic lesions were observed in both the inoculated leaves and the non-inoculated upper leaves so that the virus had spread systemically, which indicates that it had overcome the resistance. Finally, van Eck et al. (2017) mapped a *Ny* gene from *S. stoloniferum* ($Ny_{(o,n)sto}$) to the distal end of chromosome 11. The cross between resistant cultivar Festien and susceptible cultivar Seresta resulted in a 1:1 segregation of resistant to susceptible individuals (98:81) when tested with strain PVY^{NTN}. Cultivar Festien is also resistant to PVY^{O} and PVA, its resistance derived from a *S. stoloniferum* accession in the Commonwealth Potato Collection (CPC 2093) via the Dutch progenitor clone Y 66-13-636. However, the resistance does not prevent the spread of necrosis following graft inoculation.

In contrast to the HR genes, the ER ones (*Ry* genes) are currently effective against all strains of the virus (comprehensive resistance) and hence are the desired type of resistance to control the currently widely spread and severe recombinant PVY strains (Valkonen 2015). The genes Ry_{adg} (chromosome 11, Hamalainen et al. 1997), Ry_{sto} and $Ry\text{-}f_{sto}$ (chromosome 12, Song et al. 2005; Flis et al. 2005) and Ry_{chc} (chromosome 9, Sato et al. 2006) were derived from *S. tuberosum* Group Andigena, *S. stoloniferum* and *S. chacoense*, respectively. In European breeding programmes, a commonly used source of the Ry_{sto} gene was segregating progeny R49/272 from R.W. Ross in Cologne in 1954 [(*S. antipoviczii* = *S. stoloniferum* × Erika) selfed × Tuberosum hybrid]. Another commonly used source was the hybrid MPI 61.303/4 which had Group Andigena, *S. acaule*, *S. demissum* and *S. spegazzini* as well as *S. stoloniferum* in its pedigree (Ross 1986). The *Ry* cultivars listed by Ross were derived from *S. stoloniferum* by three to seven generations of backcrossing. Such cultivars have been produced in Germany, Holland, Poland and Hungary (Solomon-Blackburn and Barker 2001b). In Japan, Mori et al. (2012) were able to select a breeding line, Saikai 35, that combined good taste and bacterial wilt resistance from female parent TD0101 (from chromosome doubling of diploid cultivar Incano-mezame) with the *H1* (resistance to *Globodera rostochiensis* pathotype Ro1) and Ry_{chc} genes from male parent Sakurafubuki. They did so from a relatively small progeny size of 159 tuber-setting seedlings from which 14 clones were assessed for *H1* and Ry_{chc} using molecular markers and five were found to have both genes. They also did this in a short period of 3 years (2002–2004) with two cropping seasons per year. Saikai 35 is simplex for both genes and has Group Phureja-derived cytoplasm. In North America, where PVY is now a major pathogen because seed certification has become less effective, Fulladolsa et al. (2019) have described a *Solanum chacoense* clone (M19 = 39-7 from accession 275138) that is homozygous for a PVY resistance gene that is linked to a previously published marker for Ry_{chc}. Furthermore, M19 had been crossed with a diploid *S. tuberosum* clone (US-W4) to produce an adapted hybrid clone (XD3 = M20) carrying the resistance gene. M20 tuberizes in the field, producing round tubers with white skin and flesh and moderate size, is resistant to PVY^{O}, $PVY^{N:O}$ and PVY^{NTN}, and like M19 is female and male fertile and also self-compatible (M20 is considerably more self-fertile than M19). Finally, M20 does not produce 2*n* gametes to any appreciable extent. Interestingly, glycoalkaloids were not detected in the small tubers of M19.

5.9.2 *PVA (Potato Virus A) and PVV (Potato Virus V)*

PVA and PVV are species in the genus *Potyvirus* that are related to PVY but less commonly reported (Solomon-Blackburn and Barker 2001a; Jeffries and Lacomme 2018). PVA has a worldwide distribution apart from the Andes, whereas PVV has been reported in the Bolivian highlands, Peru, northern Europe and the USA. PVA causes symptoms that range from very mild mosaic to fairly severe, whereas PVV tends to cause few symptoms. Like PVY, they are aphid transmitted in a non-persistent manner. Major dominant genes for HR and ER are known. Interestingly, Ry_{sto} provides ER to PVA and PVV as well as to PVY strains PVY^{O}, PVY^{N} and PVY^{NTN} (Barker 1997) whereas Barker (1996) showed that Ra_{sto} is an unlinked ER gene specific to PVA. Likewise, Hamalainen et al. (1998) showed that Ra_{adg} is an ER gene specific to PVA that maps about 6.8 cM distal to Ry_{adg} on chromosome 11, and that Ry_{adg} provided ER to the two strains of PVY that they used (PVY^{O} and PVY^{N}). Interestingly, there was some evidence that the ER phenotype of Ra_{adg} depended on the genetic background.

5.9.3 *PVX (Potato Virus X)*

PVX is a *Potexvirus* with a worldwide distribution (Jeffries and Lacomme 2018). Its symptoms tend to be mild mosaic but mixed infections of PVX with other viruses such as PVY or PVA are very damaging. PVX is one of the few viruses that can be transmitted by contact under natural conditions in the field. In other words, plant infection occurs after infective sap contaminates wounds in healthy plants. Transmission may occur through abrasions caused during leaf contact during wind or rain, or during tuber storage when sprouts are developing. PVX can also be spread on clothing when walking through a crop containing infected plants and on farm machinery working such a crop.

According to Solomon-Blackburn and Barker (2001a), the main strain groups of PVX are groups 1–4, defined by Cockerham (1955) in terms of resistance genes in potato, and the resistant breaking strain PVX^{HB} (Moreira et al. 1980). Group 2 is commonly called B and Group 3 is the 'common strain'. However, Ohbayashi (2019) refers to known strains as PVX^{X3}, PVX^{S}, PVX^{UK3} and PVX^{Roth1} in Europe, PVX^{CP}, PVX^{CP4} and PVX^{HB} in South America and PVX^{O} (common strain) and PVX^{B} (necrosis strain) in Japan (Ohbayashi 2019). In the A-6 line of *S. demissum*, PVX^{O} causes small spots on inoculated leaves, spreads systemically in the plant and produces mosaic symptoms on upper leaves. In contrast, PVX^{B} causes severe necrotic spots on inoculated leaves but does not usually spread systemically and hence, no symptoms are observed on other leaves (Ohbayashi 2019).

In breeding for resistance, major dominant HR (*Nx*) and ER (*Rx*) genes have been found in potato wild relatives and landraces and introgressed into modern cultivars. Not surprisingly for a contact-transmitted virus, young seedlings or young plants can be tested in a glasshouse by spray-inoculation with infected sap containing celite powder (or carborundum) as an abrasive. Although the Nx_{tbr} gene provided useful resistance in cultivars such as Pentland Dell and Maris Piper (both also contain the Nb_{tbr} gene), some evolution of resistance-breaking strains was found (Jones 1982, 1985), and hence, *Rx* genes were preferred. The introgression of *Rx* genes for immunity to all strains of PVX started at the Scottish Plant Breeding Station in the 1950s (Bradshaw 2009a). USDA 41956, a parent of 11-79, was the source of one *Rx* gene (see below) and an Andigena accession in the Commonwealth Potato Collection (CPC 1673) was the source of another (Rx_{adg}, chromosome 12). Additionally, the most widely used source of Rx_{acl} (chromosome 5), from *S. acaule*, came from R.W. Ross in 1955 (44/1016/10), although it was also present in an *S. acaule* accession in the Commonwealth Potato Collection (CPC 379). Today Rx_{adg} is immune to all strains except South American PVX^{HB} (Ohbayashi 2019). Both Rx_{adg} (= *Rx1*) and Rx_{acl} (= *Rx2*) have been cloned (Bendahmane et al. 1997, 1999, 2000). Cultivars with these *Rx* genes have been produced in several countries including the USA, Germany, Argentina, the UK and Ireland (Solomon-Blackburn and Barker 2001b). Solomon-Blackburn and Barker (2001a) provided a comprehensive list of resistance genes identified by 2001.

Recently, Ohbayashi (2019) has shown that the *Rx* gene derived from USDA 41956 and the *Rx1* (= Rx_{adg}) gene derived from CPC 1673 confer equal resistance to the migration of *Potato virus X* from potato leaves to tubers, as well as leaf to leaf resistance, and that they are probably the same gene.

5.9.4 PVM (Potato Virus M) *and PVS* (Potato Virus S)

PVM and PVS are species in the genus *Carlavirus* which are found worldwide (Solomon-Blackburn and Barker 2001b; Jeffries and Lacomme 2018). The majority of infections are symptomless, but chlorotic mottling of leaves and rugosity have been reported in some cultivars. The lack of symptoms is one explanation for why Priegnitz et al. (2019) in Uganda found that some seed lots were initially highly infected with PVM and PVS and showed no reduction in virus incidence through positive selection of apparently healthy seed tubers over three clonal generations. The Andean strain of PVS (PVS^{A}) may cause more severe symptoms than the ordinary strain (PVS^{O}). PVM can be devastating in the seed and ware potato production areas of Central and Eastern Europe. It is mainly transmitted by aphids in a non-persistent manner, but some isolates are transmitted by contact. PVS is mainly transmitted by contact although some isolates are transmitted by aphids in a non-persistent manner. Dominant major genes for HR and ER are known: Gm_{grl} (chromosome 9) and Rm_{mga} (chromosome 11) (Marczewski et al. 2006) for PVM and Ns_{adg} (chromosome 8) (Marczewski et al. 2002) for PVS. Ristić et al. (2019) found that the incidence of PVS in Serbian seed potato crops increased over the three-year period 2014–2016 from 1.52% to 8.84%. They confirmed that Serbian PVS belongs to PVS^{O}. Although the incidence of infected plants varied with cultivar, infected samples of all 14 cultivars were found. The authors comment that only a few potato cultivars have been produced containing the Ns_{adg} gene, such as Szignal in Hungary and Fantasia in Germany.

5.9.5 PLRV (Potato Leafroll Virus*)*

PLRV is a *Polerovirus* that causes economic damage in cultivated potatoes throughout the world (Jeffries and Lacomme 2018). As implied in its name, leaf rolling is a typical symptom. It is the only potato virus which is aphid-transmitted in a persistent manner; but it does not propagate within the vector. The initial acquisition phase in which the aphid feeds on an infected plant's phloem may last for minutes, but days are required for maximum transmission efficiency. Acquisition is followed by a latent period of about 24–72 h during which the virus passes through the haemocoel to the salivary gland. After this, the virus can be transmitted to healthy plants throughout the aphid's life, including moults. Because of its relatively long acquisition phase and latent period, colonizing aphids are responsible for the spread of PLRV, and controlling aphid population development within the crop with an aphicide can be an effective control strategy. However, some clones of *M. persicae* are resistant to a number of major insecticide classes; for example, MACE (Modified Acetyl Cholinesterase) clones that arrived and spread in the UK from the year 2000 (Jeffries and Lacomme 2018). The most important and efficient vector of PLRV is *Myzus persicae* although *Aphis gossypii* is also important in subtropical areas. *Macrosiphum euphorbiae* and *Aulacorthum solani* are also vectors, albeit less efficient ones. Khaled et al. (2018) found that of the most common potato colonizing aphid species to acquire and transmit PLRV in Tunisia in 2015, the transmission potential was greatest in *M. persicae* (90%) followed by *M. euphorbiae* (80%), *Aphis fabae* (50%) and *A. gossypii* (30%). All four of the aphid species had acquired the virus within 2 h. *Aulacorthum solani* occurred at low frequencies in Tunisia but was not investigated. Interestingly, the greatest frequencies of viral infections were PVY and PVS, not PLRV.

In contrast to the viruses already considered, breeding for resistance to PLRV has involved selection for resistance to PLRV infection, accumulation or movement, rather than for HR or ER. A brief review of methods and references to the original literature can be found in the paper by Solomon-Blackburn and Barker (2001b). Resistance to infection is where the likelihood of infection by natural means from vector inoculation is reduced. It is usually assessed in field exposure trials where the virus is spread from infector plants to exposed plants by aphids, predominantly *M. persicae.* The frequency of secondary infection in tuber progeny plants is assessed by symptoms, ELISA (enzyme-linked immunosorbent assay) or a combination of the two in comparison with standard cultivars. Resistance to infection has been called field resistance. The plants can be exposed to PVY at the same time so that they can be selected for resistance to both viruses. The field trials are grown in areas of high natural aphid infestation which can be supplemented by applying cultured aphids to the infector plants. If these trials are separate from those for yield and other traits, they will put additional demands for tubers on the overall breeding programme. Results from field exposure trials can be variable but the most resistant clones are expected to remain virus-free. Brown et al. (1997), for example, reported a high narrow-sense heritability of 0.72 for field resistance in line × tester crosses of advanced tetraploid breeding material that were assessed in field exposure trials in 1994 and 1995 at Washington State University, USA.

Resistance to virus accumulation is where plants are infectible but the virus reaches only a relatively low concentration in the plant, for example, 1–5% of the concentration found in susceptible clones. Virus is less likely to be acquired by aphids from plants with this type of resistance and transmitted to other plants, thus reducing virus spread within the crop. Resistance to PLRV accumulation can be assessed by measuring the virus titre using a quantitative ELISA of tissue from glasshouse- or field-grown plants. In one method, glasshouse-grown plants are first graft-inoculated and then the virus titre is determined in the leaves of glasshouse-grown daughter tubers the following year in comparison with resistant and susceptible controls. Resistance to PLRV accumulation has been found in *S. brevidens* (Jones 1979), *S. chacoense* (a single dominant major gene Rlr_{chc} for extreme resistance: Brown and Thomas 1994), Group Phureja cv. 'Yema de huevo (Egg Yolk) clone 5010' (strong resistance in leaves: Franco-Lara and Barker 1999) and tetraploid breeding lines (Barker and Harrison 1985). Resistance to accumulation may in part be due to restricted movement as well as restricted virus multiplication per se. Resistance to virus movement can be expressed in a number of ways; for example, a lower percentage of PLRV-infected daughter tubers from an infected plant than susceptible controls. Wilson and Jones (1992) found that resistance to phloem transport in cultivar Bismark was separate from resistance to infection and accumulation.

Solomon-Blackburn and Barker (1993) found that resistance to PLRV can be greatly improved by combining two independent types of heritable resistance, namely resistance to infection and resistance to accumulation. Such a combination should greatly reduce the spread of PLRV. Both types of resistance have been identified in diploid breeding material such as DW.84-1457 (Dziewonska and Was 1994) and tetraploid clones such as G7445(1) (Barker 1987) and G8107(1) (Solomon-Blackburn et al. 2008). Dziewonska and Was (1994) demonstrated that DW.84-1457 is resistant to infection, both natural and by aphid inoculation, and resistant to PLRV multiplication with only very low levels of virus in graft-inoculated plants and their tuber-progeny plants. Furthermore, the clone produces 2*n* pollen, and hence can be used in 4*x*-2*x* crosses, and has extreme resistance to PVX and high resistance to PVM (the likely sources can be seen in its complicated pedigree involving wild species). Barker (1987) found that clone G7445(1) combined resistance to infection in field- and glasshouse-grown plants with resistance to movement of the virus from foliage to tubers. Later, Barker and Solomon (1990) provided evidence for the simple monogenetic control of the resistance in G7445(1). The cross with susceptible cultivar Maris Piper segregated in a 1:1 (21:19) ratio for low to high concentration of PLRV, as measured by quantitative ELISA, in the foliage of plants with secondary infection. Although the pattern of segregation in some subsequent crosses and selfs was a better fit to two unlinked dominant complementary genes for resistance, the simplest hypothesis is that resistance is conferred by a single dominant gene (Barker et al. 1994; Solomon-Blackburn et al. 2003/4). Solomon-Blackburn and Barker (1993) found that clone G8107(1) remained virus-free in field exposure trials. Later, Solomon-Blackburn et al. (2008) showed that clone G8107(1) was highly resistant to PLRV accumulation following graft inoculation and to PLRV infection by aphids. They suggested that an important element of the very strong resistance of this clone to PLRV infection is resistance to virus movement within leaves or from leaves to petioles, thus preventing spread to the rest of the plant.

Mihovilovich et al. (2007) studied Andigena potatoes as a source of resistance to PLRV. A preliminary evaluation of a group of 2518 clonal accessions from the in-trust germplasm collection held at the International Potato Centre (CIP) identified 46 that did not become infected with PLRV under natural field exposure and that were difficult to infect by graft inoculation. Working with this resistant material, Mihovilovich et al. (2007) analysed the inheritance of resistance, characterized the components and levels of resistance, and the expression of resistance in the genetic background of commercial Tuberosum potatoes. Analysis of a diallel set of crosses among Andigena cultivars revealed the importance of additive genetic variation, all the genetic variation being attributed to GCA (General Combining Ability) and none to SCA (Specific Combining Ability). Over 80% of the progeny of cultivar LOP-868 were uninfected. Phenotypic assessment of components of resistance showed that Andigena cultivars with high levels of resistance to infection were also resistant to the accumulation of PLRV and able to restrict translocation of the virus to their tubers. Three out of 19 Andigena cultivars (LOP-868, HUA-332 and OCH-7643) had these features and were as resistant as control DW.84-1457, which we met in the previous paragraph. The three resistant Andigena cultivars together with two susceptible ones were crossed with six Tuberosum cultivars. There were differences in the percentage of uninfected offspring between the Andigena parents and between the Tuberosum parents but no Andigena × Tuberosum interactions. The three resistant Andigena parents had the highest GCAs with LOP-868 the best of the three. The lack of interaction in the expression of Andigena resistance in Tuberosum backgrounds means that the resistance should be of general use in breeding programmes. The authors suggested that the multiple components of resistance in a single genotype could be the result of a common mechanism controlled by several genes. The remarkably high frequency (>70%) of uninfected individuals in all progenies derived from the highly resistant Andigena cultivar LOP-868 and the occurrence of highly resistant clones selected from this parent suggest the presence of an allele with large effects.

In summary, it looks as though major genes or QTL alleles of large effect exist for resistance to PLRV infection and accumulation or movement, and that their incorporation into new potato cultivars would greatly reduce the spread of the virus in potato crops and hence be an effective control measure.

5.9.6 *TRV (*Tobacco Rattle Virus*)*

TRV is a *Tobravirus* which can cause serious losses because of necrotic symptoms in the tuber flesh, known as spraing or corky ringspot, and which is distributed in cooler climates such as northern Europe, with significant variation in incidence between countries (Jeffries and Lacomme 2018). As few as 2–5% spraing-affected tubers may lead to rejection of consignments by processors and supermarkets. TRV is transmitted by free-living nematodes in the genera *Paratrichodorus* and *Trichodorus.* There is no evidence that the virus multiplies in the vector, is retained through the moult or is passed to the progeny. TRV-infected weeds perpetuate the virus within sites and within nematode populations. Distribution of TRV in soil reflects the distribution of the nematode vectors which prefer light sandy or peaty soils. Two types of TRV isolates are found, the M or normal multiplying type which consists of both RNA-1 and RNA-2 genomes, and the NM or non-multiplying type which consists of the RNA-1 genome and hence lacks the gene for the capsid protein and non-structural genes probably involved in nematode transmission. Resistance to TRV can be assessed in field trials on infective land (with trichodorid nematodes) by scoring tubers for frequency and severity of spraing symptoms. However, an uneven distribution of the viruliferous vector in the soil and variable weather conditions can make such trials difficult. Hence, Dale and Solomon (1988) developed a glasshouse test in which tubers are grown in pots of tested field soil containing viruliferous nematodes, and daughter tubers are scored for the severity but not the frequency of spraing symptoms. Symptomless tubers need to be tested for infection to distinguish desirable extreme resistance from undesirable tolerance which provides a means for new virus-free sites to become infected if the vector nematodes are present. Spraing disease is in fact a hypersensitive response to TRV. Examples of processing cultivars with resistance to spraing are Markies and Shepody for French fries and Hermes, Lady Rosetta, Record and Saturna for crisping (chipping).

5.9.7 *PMTV (*Potato Mop-Top Virus*)*

PMTV is a *Pomovirus* which is fungal-transmitted in a persistent manner in soil by *Spongospora subterranea*, which causes powdery scab (Jeffries and Lacomme 2018). PMTV causes spraing symptoms in tubers and is now widely distributed in potato-growing areas in South and North America, north-western Europe, Japan, China and Pakistan. PMTV can be disseminated to new fields in infected seed potatoes or in PMTV-containing resting spores of *S. subterranea* adhering to the surface of seed tubers. Treatment of soil with chemicals does not kill the resting spores and hence is ineffective as well as environmentally damaging. Cultivars with resistance to *S. subterranea* appear not to be protected against infection with PMTV. Hence, resistance of potato cultivars to PMTV is highly desirable but has proved elusive

(Valkonen 2015). Therefore, a wider screening of germplasm would seem desirable as well as exploring transgenic approaches. Once found, resistance to PMTV could be assessed in field trials on infective land in much the same way as TRV.

5.9.8 *PSTVd* (Potato Spindle Tuber Viroid)

The viroid PSTVd is the most important pathogen transmitted by True Potato Seed (TPS) (Jeffries and Lacomme 2018). Seed transmission frequency varies with a reported average of over 30% (Fernow et al. 1970). PSTVd is spread through contact between plants or contaminated machinery and can be pollen-transmitted under glasshouse and natural conditions. It is treated as a quarantine disease in countries where it is not endemic such as the UK. Given the lack of sources of resistance, Dianqiu and Xiaobao (2018) have recently advocated trying CRISPR-Cas13 as a new tool to achieve resistance through RNA knockdown with better specificity than RNAi.

5.10 Insect Pests

In their review of insect pests of potato, Radcliffe and Lagnaoui (2007) considered those of worldwide importance to be aphids, Colorado potato beetle, potato tuber moths and leafminers; and those of regional importance to be leafhoppers, potato psyllid, thrips, white grubs, wireworms, ladybird beetles, flea beetles, Andean potato weevils and cutworms. They looked at yield and quality reduction in terms of the effects of defoliators, sap feeders, root and tuber feeding and pathogen transmission. They considered insect control in terms of insecticides, biological control and host plant resistance. In a more recent review, Kroschel et al. (2020) have provided detailed information for a total of 49 insect pests of potato and the status quo of their management around the world. They thought that climate change will potentially contribute to the expansion of their geographical range of distribution, and that increasing populations will lead to greater crop and post-harvest losses. They also thought that farmers in tropical and subtropical countries, compared with those in temperate regions, have to contend with a higher number of pest species, with a minimum of two to four pests often reaching a status requiring the application of control measures. Here, we are going to focus on four insect pests of worldwide importance and host plant resistance. Andean potato weevils (APW, *Premnotrypes* spp.), a complex of tuber-boring herbivores, will not be considered despite being the most important pests of potatoes in the high Andes because resistant cultivars had not been found by 2012 (Parsa et al. 2012). In deciding objectives, breeders need to be aware of the control measures being used by farmers and growers and how host plant resistance could complement or even replace some of them.

5.10.1 Potato Tuber Moth

Potato tuber moth (PTM, *Phthorimaea operculella* Zeller) is the most serious insect pest of potatoes worldwide, particularly in warm tropical and sub-tropical climates. It originated in South America along with the potato, its preferred host, and spread with the potato around the world, the tubers providing enough food for its multiplication and sustenance. Its larvae mine the foliage, stems and tubers in the field, and in storage, with most economic losses occurring in storage, particularly non-refrigerated storage. Kroschel and Schaub (2013) reviewed the biology, ecology and management of potato tuber moths from a global perspective, and more recently, Gao (2018) from a Chinese and Chandel et al. (2020) from an Indian perspective. Control measures are the use of healthy seed, deep planting, proper ridging, removal of host weeds and potato volunteers, irrigation, harvesting soon after maturity, and use of insecticides both in the field and in storage. More recently, there has been interest in biological control: for example, Khorrami et al. (2018) considered the potential of entomopathogenic fungi and Gallego et al. (2020) of the *Blattisocius mali* mite (Acari: Blattisociidae) as biological control agents. Host resistance has been sought and transgenic approaches developed.

One example on breeding for resistance is the work of Ortiz et al. (1990) in diploid potatoes at the International Potato Centre (CIP) which has given a high priority to developing integrated methods of controlling potato tuber moth. Sixty-two $2x$ families were generated by appropriate mating of 16, $2x$ clones (resistant and susceptible), and tubers from pot-grown plants evaluated for resistance under natural infestation in storage at San Ramon, Peru. The trial was conducted in single natural diffused light stores maintained at ambient temperature as the simplest way of screening large number of genotypes for major differences in tuber damage. Tubers of clones were placed in paper bags and kept open for tuber moth entry (insects have choice). After 150 days of storage, the material was evaluated by counting the number of holes in the tubers and the amount of damage using a 1–4 scale. Then, a sample of 254 resistant or moderately resistant genotypes (clones) obtained from the evaluation in storage were tested for antibiosis (resistance to feeding by pest when no choice). Tubers from these clones were planted in pots at Huancayo, Peru and five tubers per genotype were harvested and used for laboratory evaluation. Each tuber was placed in a closed plastic container with five larvae of PTM. After 5 weeks, the evaluation was done using the same 1–4 scale. The number of adults per container was also counted. The main findings were as follows. A high frequency of resistant progeny was obtained from crosses between resistant parents, both the original *Solanum sparsipilum* (spl) accession and derived hybrids with diploid Tuberosum. Although there was a strong indication of reciprocal effects, spl cytoplasm was not essential for the expression of nuclear resistance genes. Both antibiosis and antixenosis (non-preference, i.e. tuber unattractive to pest) were mechanisms of PTM resistance, but the high percentage of resistant clones in the closed container test suggested that antibiosis was the major mechanism. Nevertheless, the two-stage strategy of assessing resistance would seem desirable in a breeding programme. Finally, the researchers were confident that $4x$ x $2x$ crosses could be used to transfer the resistance into commercial cultivars as it should be possible to select potential $2x$ progenitors with both resistance and $2n$ pollen production. Since then, there have been reports in the literature of resistant cultivars, but more effort appears to have gone into the transgenic route.

Since the early 1990s, efforts to engineer insect-resistant potatoes have focused on genes derived from *Bacillus thuringiensis* (Bt), a soil bacterium that accumulates high levels of insecticidal crystal (Cry) proteins, also known as δ-endotoxins, during sporulation (see review by Douches and Grafius 2005). Cry proteins have been used as biological insecticides for many years. They are host specific which is an advantage over chemical insecticides in terms of safety to non-target organisms. We shall consider the use of Bt toxin genes for insect control in more detail in the chapter on genetically modified potatoes. In summary, transgenic resistance to the potato tuber moth has been provided by the *Bt* proteins encoded by the *cry5* (equivalent to *cry1Ia1*) (Mohammed et al. 2000; Douches et al. 2004) and *cry1Ac9* (Davidson et al. 2004) genes using the constitutive CaMV 35S promoter. Effective resistance has been demonstrated in Egypt in field trials under natural infestations and under non-refrigerated storage (Douches et al. 2004), under laboratory and field conditions in the Pacific Northwest of the USA (Douches et al. 2011), and in a laboratory larval growth analysis in New Zealand (Davidson et al. 2004). Despite all of this potential, to the best of my knowledge, transgenic resistance is yet to be deployed by farmers in their fields.

5.10.2 Colorado Potato Beetle

Colorado potato beetle (CPB, *Leptinotarsa decemlineata* Say) is the best known and most destructive of the leaf-feeding insects attacking potatoes, being capable of defoliating a crop in the field with both adults and larvae feeding on the leaves. Cingel et al. (2016) described it as a 'ten-striped spearman' in the era of biotechnological warfare. It is a major pest in northern latitudes, particularly in North America, Europe, Russia and Central Asia. A relatively recent review on managing Colorado potato beetle and the need for resistance breeding is the one by Maharijaya and Vosman (2015), which contains extensive references to the original literature, although much of the information of relevance to breeding was also present in the earlier review by Douches and Grafius (2005). The key points from these reviews are as follows. Past and present control strategies rely on pesticides, but CPB has been able to develop resistance to most of the registered insecticides. Despite some promising experimental results, the use of biological control agents to suppress CPB in the field is still limited. CPB populations can be reduced through appropriate crop rotation but the limitations and constraints of other control measures are still high, and CPB is still difficult to control. It is known from the biology of CPB that feeding and oviposition behaviour are strongly affected by stimulants and deterrents released by potato leaves, and that volatiles released by the leaves affect host plant choice by the adult CPB. The presence of toxic compounds affects life history parameters during the feeding stage (adult/larvae survival, number of eggs deposited and development rate of adult/larvae). It is perhaps surprising and disappointing to find a review as recent as 2015 stating that there are no commercial cultivars on the market with a strong level of resistance to CPB, despite a number of resistance factors being identified in the wild relatives of potato that may be exploited in breeding programmes. The reviewers list 16 wild relatives and resistance factors that include the two well-known groups, namely glandular trichomes (types A and B) and leptine glycoalkaloids (Douches and Grafius 2005).

The best-known species with high densities of glandular trichomes is *S. berthaultii*, with derived breeding clone NYL 235-4 from Cornell University, NY, USA having resistance to the CPB (Plaisted et al. 1992). An initial cross made in 1980 between cultivar Hudson (4*x*) as female parent and the diploid species *S. berthaultii* (PI 310925) was followed by six backcross-type generations. There was strong selection for A trichome traits and some selection for B trichome traits, but the type B trichomes of NYL 235-4 are non-secretory and epicuticular sucrose esters are not detectable. Field resistance of NYL 235-4 is expressed in reduced egg laying by both overwintered (in the soil) and summer adults, increased mortality of larvae during the first and second instars and by increased mortality during pupation. NYL 235-4 is also resistant to potato leafhopper and has a total glycoalkaloid content comparable to cultivar Atlantic (i.e. acceptable for human consumption).

Likewise, the best-known species with leptine glycoalkaloids is *S. chacoense*. Yencho et al. (2000) summarized what was known in relation to CPB at the time of their inheritance studies as follows: *S. chacoense* was the only wild species known to produce leptines, leptines I and II being the acetylated analogues of α-chaconine and α-solanine, respectively, the commonly occurring glycoalkaloids present in all Tuberosum cultivars. A unique characteristic of the leptines is that, unlike other *Solanum* glycoalkaloids, they are synthesized only in the foliage, not in the tubers. This is advantageous for a breeding programme because glycoalkaloids present in high concentrations in tubers are toxic to humans, and high tuber glycoalkaloid levels are selected against by breeders. Introgressing the genes for leptine accumulation into a more adapted Tuberosum background had been slow, partly because most crosses between Tuberosum and *S. chacoense* had failed because of insufficient unreduced gametes. Eventually, tetraploid *S. chacoense* had been produced from a clone (8380-1 from PI 458310) that produces high levels of leptines. Yencho et al. (2000) evaluated seven tetraploid F_2 families derived from *Solanum tuberosum* (4x) × *S. chacoense* (4x) and determined resistance to CPB in the laboratory and field, and also foliar glycoalkaloid concentrations. Resistance was correlated with the concentrations of glycoalkaloids on a family and an individual basis. All of the progeny produced solanine and chaconine but the family with the highest mean leptine concentration had the lowest mean leaf disk feeding and CPB defoliation levels. Regressions of leaf disc consumption and field defoliation on type of glycoalkaloid revealed that only increased foliar levels of leptines resulted in decreased CPB feeding, with the regression models accounting for 17% and 26% of the variation in consumption and defoliation, respectively.

Transgenic resistance to the Colorado potato beetle has been provided by gene *cry3A*, which encodes the Cry3A protein from the bacterium *Bacillus thuringiensis* var. *tenebrionis*, using the constitutive CaMV 35S promoter (see review by Douches and Grafius 2005). The first genetically modified (GM) cultivars with this *Bt* gene to be commercialized by Monsanto in North America from 1995 onwards were Russet Burbank, Atlantic, Snowden and Superior (Duncan et al. 2002). Monsanto advised growers to plant 20% of their field area with a non-Bt-containing potato as a refuge for susceptible beetles and to rotate crops and fields in which transgenic potatoes were grown to extend the durability of resistance (Duncan et al. 2002). More generally, programmes to manage adaptation to Bt transgenic crops rely on a high dose of Bt toxin and structured refuges which provide sources of large numbers of susceptible individuals that readily mix in time and space with resistant individuals appearing in the transgenic crop. The theory is as follows: Bt resistance is inherited recessively with respect to the Bt dose present in the crop. The number of Bt resistant individuals will initially be very low. The refuges will produce a large number of susceptible individuals. These susceptible individuals will mix with resistant individuals arising in the crop before mating has occurred. The result will be heterozygous offspring, susceptible to Bt at the levels expressed in the transgenic crop. As explored by Roush (1998), pyramiding of *cry* genes would be even more effective at extending the durability of resistance. However, despite extensive evaluation, leading processing and fast-food outlet companies in North America were reluctant to purchase GM potatoes, because of consumer concerns over GM potato products, and Monsanto stopped marketing GM potatoes in 2001. We shall return to the subject of genetically modified potatoes in the last chapter of this book.

Coombs et al. (2003) and Douches and Grafius (2005) presented results from field studies in Michigan and Long Island, NY in 1998, 1999 and 2000 that compared natural (leptine glycoalkaloids and glandular trichomes), engineered (*Bt-cry3A* and *Bt-cry5* transgenic potato clones) and combined [(*Bt-cry3A* or *Bt-cry5*) + glandular trichomes and *Bt-cry3A* + leptine glycoalkaloids] plant resistance mechanisms of potato for control of CPB. The field trials were a choice situation under natural infestation with CPB. The researchers considered defoliation over time to be the best estimate of the host plant's resistance against CPB and therefore used the relative area under the defoliation curve (RAUDC) as the measure of host plant resistance. The results showed that the *Bt-cry3A* transgenic clones, the high leptine clone (USDA8380-1) and the high foliar glycoalkaloid clone (ND5873-15) were most effective for controlling defoliation by CPB adults and larvae. The *Bt-cry5* clone (SPc5-G2) was not as effective as the *Bt-cry3A* transgenic clones ('Russet Burbank Newleaf', RBN15 and YGc3.1). The glandular trichome (NYL235-4) and combined (*Bt-cry5* + glandular trichome) clones proved to be ineffective, which was surprising given the results mentioned earlier. Douches and Grafius (2005) concluded that the ability to integrate conventional breeding and transgenic strategies (including constitutively expressed plant-based proteins) may provide the best opportunity to build durable host resistance to insects such as CPB. However, the challenge still facing breeders and biotechnologists can be seen in review by Cingel et al. (2016) which, as already mentioned, has the title Colorado potato beetle: 'ten-striped spearman' in the era of biotechnological warfare.

5.10.3 Aphids

We earlier encountered aphids as vectors of viruses. Now, we need to briefly consider breeding for resistance to aphids rather than resistance to viruses which has received a lot more attention over the years. Davis et al. (2007) made a comprehensive field and glasshouse study of 49 cultivars, primarily of North American origin, for resistance to the green peach aphid (GPA), *Myzus persicae* (Sulzer), and the potato aphid, *Macrosiphum euphorbiae* (Thomas). The cultivars differed consistently in resistance to both aphids under greenhouse conditions and the mechanism of resistance appeared to be antibiosis (reduced reproduction and survival rather than reduced colonization). However, the resistances to the two aphids were not significantly correlated. In contrast, the field studies failed to give consistent results and differed from year to year. Perhaps not surprisingly, the field and greenhouse data were not correlated. The researchers concluded that the resistance found in cultivars, in combination with biological control or insecticide use, could provide substantial control while reducing the reliance on pesticides. Furthermore, they also thought that breeders could further improve the level of resistance through crosses between the more resistant cultivars currently available. Today, however, the emphasis in breeding research still appears to be on the effect of trichome morphology on aphid resistance in wild potato species, for example, by using Cryo Scanning Electron Microscopy (Cho et al. 2017).

5.10.4 Leafminer Flies

Pea leafminer (LMF, *Liriomyza huidobrensis* (Blanchard)) is the most destructive leafminer on potato. Adult females puncture the leaves to feed or lay eggs, which hatch into larvae that tunnel into the leaflets. Pea leafminer is indigenous to South America but did not become an important pest until the 1970s. By the time Radcliffe and Lagnaoui (2007) wrote their review, it was present in China and Indonesia, and had a foothold in Africa. Control is by natural enemies and chemical insecticides, but populations have developed insecticide resistance. Radcliffe and Lagnaoui (2007) mention that breeding work at CIP (International Potato Centre, Lima, Peru) produced cultivar Maria Tambeña with good levels of tolerance to leafminer fly infestations. This cultivar was included in the study by Mujica and Kroschel (2013) on the relationships between pest intensity and crop loss in different cultivars grown in coastal Peru, as a critical component for decision-making in integrated pest management (IPM). The results are also of value in determining breeding objectives (i.e. the level of resistance/tolerance required). The researchers used insecticides with different control efficiencies or the use of varying numbers of insecticide applications to obtain various levels of leafminer fly infestation intensities in their field trials over 3 years. Early cultivars showed higher foliar injury and yield loss than late potato cultivars with the exception of partially resistant Maria Tambeña. The researchers concluded that potato cultivars can tolerate considerable levels of foliar injury by the leafminer fly before control measures are needed.

5.11 Nematodes

*5.11.1 Cyst Nematodes (***Globodera rostochiensis** *Woll. and* **G. pallida** *Stone)*

Nematodes are typically small, slender, non-segmented worms of 0.1–2.5 mm in length which occur in the soil in great numbers. The nematode pests of potatoes have been reviewed by Davie and Pickup (2018), including their life cycles. Although 70 species from 24 genera have been reported to cause damage on potatoes, they list 20 species and explain the legislation in place to control them through quarantine and phytosanitary approaches in different countries. These depend on the country's long-term phytosanitary objectives for particular species of nematodes (e.g. establishment and maintenance of pest-free status, eradication of existing outbreaks and management of known infestations). The most highly regulated species are the cyst nematodes (PCN), *Globodera pallida* (pale/white potato cyst nematode) and *Globodera rostochiensis* (golden potato cyst nematode), which are treated as quarantine species in most areas of the world because of their economic impact. They have become widespread outside of South America since the 1840s, their introduction to other parts of the world predominantly associated with the movement of infested seed tubers or soil through the trade in seed potatoes. Hence, the management of known infestations has become common practice and involves loss of land for seed production, increased rotation periods, chemical or biological control and the use of resistant cultivars. A prerequisite of such actions is diagnosis of the nematode problem, an estimate of the nematode population density, knowledge of the threshold levels at which the nematode species causes damage to the potato crop, and the rate of decline of nematode populations in the absence of host crops. Interestingly, China is the area of the world that still appears to be free of cyst nematodes (Davie and Pickup 2018).

Cyst nematodes are the most damaging nematodes worldwide. A comprehensive account of the biology and management of cyst nematodes can be found in the book *Cyst Nematodes* edited by Perry et al. (2018). Their second-stage juveniles hatch from viable eggs in response to chemicals exuded from the roots of growing potato plants. The juveniles penetrate and feed on the roots and thus reduce tuber yield, size and quality. The nematode life cycle is completed with the formation of males and females, the latter each producing 200–600 eggs which following fertilization are protected within a cyst that can survive in the soil for 20 years or longer. A small proportion of eggs hatch each season and in the absence of any host will die. Both species of cyst nematode are diploid outbreeders and thus show variation both within and between populations (Phillips 1994). This variability makes it difficult to divide the species into pathogenic variants known as pathotypes, but we will encounter the scheme proposed by Kort et al. (1977) which divides *G. rostochiensis* into five pathotypes, Ro1 to Ro5, and *G. pallida* into three pathotypes, Pa1, Pa2 and Pa3 (for more details, see Phillips 1994).

Cyst nematodes can be controlled with chemicals, but nematicides are highly toxic pesticides for which there are both environmental and health concerns. They can persist long enough for run-off from field applications to contaminate watercourses and to leach through field soils into groundwater; and their residues can get into the harvested potatoes, particularly with carbamates that have both a contact and systemic mode of action (Davie and Pickup 2018). Hence, there is political pressure in many countries to reduce the use of these nematicides; for example, in the European Union (EU), Directive 2009/128/EC aims to achieve a sustainable use of pesticides in the EU by reducing the risks and impacts of pesticide use on human health and the environment and promoting the use of Integrated Pest Management (IPM) and alternative approaches or techniques, such as non-chemical alternatives to pesticides. Seenivasan (2020), for example, showed that liquid bioformulations of *Pseudomonas fluorescens* and *Purpureocillium lilacinum* applied as a seed treatment and as a soil drenching, respectively, were as effective as carbofuran in controlling cyst nematodes (reducing population densities).

Not surprisingly, breeding resistant cultivars is a high priority in many programmes worldwide. It is important, however, to distinguish resistance and tolerance. Host genotypes that are resistant to potato cyst nematode attack are those that inhibit or reduce nematode multiplication rates. The roots of such genotypes are nevertheless invaded by juvenile nematodes and can suffer damage, with large yield losses being sustained when they are grown in infested soil. Genotypes differ in the degree of yield loss sustained in comparable situations, and these differences are referred to as differences in tolerance (Dale et al. 1988). Higher priority has, however, been given to resistance breeding and work on tolerance limited to identifying methods of screening potential new cultivars in their final stages of evaluation. Such screening commonly requires expensive yield trials on infested soils, with and without nematicide control. In contrast, clonal tests for PCN resistance have been developed that allow the rapid and reliable assessment of large numbers of individual potato genotypes, for example, in clear closed containers such as 60 ml Clearopac canisters (Phillips et al. 1980). Replicate tubers of each genotype are planted in separate canisters containing sterilized soil and inoculated with the nematodes (*G. rostochiensis* or *G. pallida*) at around 20 eggs/gram of air-dried soil, with a final moisture level of 30%. The canisters are stored in the dark at a constant temperature of 20 °C. The number of cysts visible through the transparent walls of the containers is counted after 7–8 weeks to determine the level of resistance. Clonal material can be assessed at any stage of a breeding programme. Other screening tests will be found in the literature. They all allow a relatively large number of genotypes to be assessed using a relatively small number of tubers/plants.

The cyst nematodes co-evolved with wild potatoes in the Andes of South America (Davie and Pickup 2018), so it was to the wild relatives and Andean landraces of potatoes that breeders looked for sources of resistance by screening their germplasm collections. In Europe, for example, both the Commonwealth Potato Collection (CPC) in the UK (Turner 1989; Castelli et al. 2003) and the 'German-Netherlands Potato Collection' (BCRG) (Van Soest et al. 1983) have been extensively screened for resistance (summarized in Bradshaw and Ramsay 2005). Some of the genes found in some of the wild and cultivated species are summarized in Table 5.3. The following is a brief consideration of their use or potential use in potato breeding.

The cyst nematodes started to become a serious problem in Europe in the early 1950s. The first source of resistance to be successfully used was found by Ellenby in 1952 (Ellenby 1952) in an Andigena accession of the CPC (CPC 1673). It proved to be a simply inherited dominant major gene (Toxopeus and Huijsman 1953) which was named *H1*, from the old name for the nematode, *Heterodera*, and which was effective against what became known as pathotypes Ro1 and Ro4 of *G. rostochiensis* (screening had been done with a population of Ro1). A gene-for-gene relationship was shown for its interaction with a dominant avirulence gene in *G. rostochiensis* (Parrott 1981; Janssen et al. 1991) and it was subsequently mapped to chromosome 5 (Gebhardt et al. 1993). The *H1* gene was quickly backcrossed into the European cultivated potato, with cultivar Maris Piper released from the Plant Breeding Institute, Cambridge in 1966 and Pentland Javelin from the Scottish Plant Breeding Station, Edinburgh in 1967. The gene has been extensively used in potato breeding worldwide. It provides an extremely high level of resistance which is effective at reducing the size of PCN populations. However, as it inhibits PCN juveniles in potato roots at various stages of their development prior to cyst formation, tolerance of root damage is also required, as found in the Irish bred cultivar Cara. The resistance has proved durable in the UK where Ro1 appears to be the only pathotype, but pathotypes such as Ro3, which can overcome *H1*, have been found in other countries.

The PCN populations which were found in the UK that overcame the *H1* gene proved to be what we now know as pathotypes Pa1 and Pa2/3 of *G. pallida.* In a survey in 2002 (Minnis et al. 2002), PCN was detected in 64% of 484 potato fields in England and Wales, with 66% of these populations being wholly *G. pallida.* A major gene (*H2*) for resistance to Pa1 was found in the diploid wild species *Solanum multidissectum* (Dunnett 1960), but it has not been widely used in breeding as Pa2/3 became the main problem in the UK and elsewhere in Europe. Parrott (1981) provided evidence for a gene-for-gene relationship between resistance gene *H2* and a gene in *G. pallida.* The *H2* gene was eventually mapped to chromosome 5 by Strachan et al. (2019) in a cross between *H2*-containing tetraploid breeding clone P55/7 (Kort et al. 1977) and susceptible cultivar Picasso. Another major dominant gene *Gpa2* which provides resistance to Pa2 is thought to have been introgressed from CPC 1673 along with *Rx1* for resistance to PVX, rather than along with *H1*, as

Table 5.3 Major genes and QTLs of large effect for resistance to pathotypes of *Globodera rostochiensis* (Ro) and *G. pallida* (Pa) where LG is linkage group (chromosome)

Species of origin	Gene/QTL	LG	Spectrum of resistance	References
Group Andigena CPC 1673	*H1*	5	Ro1 and Ro4	Toxopeus and Huijsman (1953), Gebhardt et al. (1993)
Group Andigena CPC 1673	*Gpa2*	12	Pa2	Rouppe van der Voort et al. (1997)
Group Andigena CPC 2802	*H3* = *Gpa4* + QTL +?	4 11	Pa2/3	Howard et al. (1970), Fuller and Howard (1974), Bryan et al. (2004)
S. kurtzianum	*K1*	?	Ro1 and Ro2	Huijsman (1960), Kort et al. (1977)
S. multidissectum	*H2*	5	Pa1 (partial to Pa2/3)	Dunnett (1960), Strachan et al. (2019)
S. sparsipilum	QTLs = $GpaV^{s}_{spl}$ $GpaXI^{s}_{spl}$	5 11	Pa2/3 for mapping	Caromel et al. (2005)
S. spegazzinii	*Gro1–4*	7	Ro1 (for mapping), Ro2 to Ro5	Barone et al. (1990), Ballvora et al. (1995), Paal et al. (2004)
S. spegazzinii	*Fa*		Ro1 and Ro2?	Ross (1962)
S. spegazzinii	*Fb* (= *Gro1–4*)	(=7)	Ro1 to Ro5	Ross (1962)
S. spegazzinii	QTLs = *Gpa* + QTL	5 7	Pa2 and Pa3	Kreike et al. (1994)
S. spegazzinii	QTLs = *GpaM1* (*Gpa*) *GpaM2* *GpaM3*	5 6 12	Pa2/3	Caromel et al. (2003)
S. tarijense	QTLs = $GpaXI_{tar}^{l}$ + QTL	11 9	Pa3	Tan et al. (2009)
S. vernei	*GroVI*	5	Ro1 and Ro4	Jacobs et al. (1996)
S. vernei	QTLs = *Gpa5* + *Gpa6* +?	5 9	Ro and Pa; Ro1 and Pa2/3 and Pa2/3 for mapping	Ellenby (1948), Rouppe van der Voort et al. (2000), Bryan et al. 2002
Complex hybrid, including *S. vernei*	*Grp1*	5	Ro5 and Pa2 (partial to Pa3)	Rouppe van der Voort et al. (1998)
Complex hybrid, including group Andigena and *S. vernei*	QTLs Ro2_A = *Grp1?* Ro2_B	5 5	Ro2 and Pa2/3 Ro2	Park et al. (2019)

Gpa2 and *Rx1* are closely linked on chromosome 12 (Rouppe van der Voort et al. 1997). Armstrong et al. (2019) were able to confirm the presence of *Gpa2* $^{\Delta C2922}$ (functionally *Gpa2*) and *Rx1* in cultivars Cara and Picasso using diagnostic resistance gene enrichment sequencing (dRenSeq).

Other major dominant genes for resistance to *G. rostochiensis* are *Gro1* from *S. spegazzinii* (Barone et al. 1990; Ballvora et al. 1995), *GroVI* from *S. vernei* (Jacobs et al. 1996) and *K1* from *S. kurtzianum* (Huijsman 1960; Kort et al. 1977). Paal et al. (2004) were able to show that *Gro1* is a closely related candidate gene family, and by stable potato transformation of a susceptible cultivar, that the gene they named *Gro1-4* conferred resistance to *G. rostochiensis* pathotype Ro1. Jacobs et al. (1996) found that the *GroVI* locus mapped to chromosome 5, close to the *H1* gene, and provided resistance to pathotype Ro1 (and Ro4). The source of the *GroVI* resistance was a tetraploid hybrid between *S. tuberosum* and *S. vernei* $(VT^n)^2$62-33-3. The source of the *K1* gene was a hybrid (KTT 60.21.19) between *S. tuberosum* and *S. kurtzianum* which proved resistant to Ro1 and Ro2 (Kort et al. 1977). Huijsman (1960) thought that it would be straightforward to use *K1* in potato breeding.

Quantitative resistance to both *G. pallida* and *G. rostochiensis* was found in accessions of the diploid wild species *S. vernei* (Ellenby 1948) and in the pathotype scheme of Kort et al. (1977), hybrid 69.1377/94 was resistant to Ro1 to Ro5 and Pa1 to Pa3. *S. vernei*-derived resistance was used in breeding programmes in the UK from 1957 (CPC 2487 and CPC 2488), but very few successful cultivars with high levels of *S. vernei*-derived resistance have been produced in the UK, or elsewhere in Europe. Cultivars Lady Balfour from the Scottish Crop Research Institute, Sante from the Netherlands, and Nadine from Caithness Breeders, for example, were classified as partially resistant to *G. pallida* by NIAB TAG (2019). Subsequent genetical analysis at the diploid level revealed a QTL for resistance to *G. pallida* Pa2/3 on chromosome 5 and another on 9 which explained 24% and 17.6% of the variation in PCN scores (square root of cyst counts = number of females), respectively (Bryan et al. 2002). There was possibly a second QTL on chromosome 5 which explained a further 16% of the variation. Clones with all three QTLs had an average PCN score of 3.08 and those lacking all three QTLs had a score of 9.65, the three loci acting in a simple additive fashion. Rouppe van der Voort et al. (2000) also detected a QTL on each of chromosomes 5 (*Gpa5*) and 9 (*Gpa6*) in their '*S. vernei*' (other wild species in pedigree) derived resistance, the loci acting in an additive manner. Loci *Gpa5* and *Gpa6* explained 61% and 24%, respectively, of the variation.

Quantitative resistance to *G. pallida*, but not to *G. rostochiensis*, was found in some Andigena accessions (e.g. CPC 2775 and CPC 2802) and named *H3,* although it is not a single gene (Howard et al. 1970; Fuller and Howard 1974). It has been used in breeding programmes in the UK, and elsewhere, from 1970 (Bradshaw 2009a). A high level of partial resistance was achieved in cultivar Vales Everest, but the cultivar has not been widely grown since its release in the UK in 2005. Subsequent genetical analysis revealed that for the *H3* source, in the tetraploid cross of clone 12601ab1 and cultivar Stirling, 34.4% of the variation in resistance to *G. pallida* Pa2/3 could be explained by a QTL of large effect on chromosome 4 (square root of cyst count: *QQqq* 2.9 ± 0.25, *Qqqq* 4.0 ± 0.11, *qqqq* 5.7 ± 0.21) and a further 17.1% by a QTL of smaller effect on chromosome 11 (Bryan et al. 2004; Hackett et al. 2014). Caromel et al. (2005) also found two QTLs in *S. sparsipilum* that combined to give a high level of resistance to *G. pallida.* The QTL on chromosome 5 ($GpaV^s_{spl}$) explained 76.6% of the variation and the one on chromosome 11

(*GpaXI*$^{s}_{spl}$) explained 12.7%. The average numbers of cysts were 475.0 (no QTL), 126.6 (*GpaXI*$^{s}_{spl}$), 14.5 (*GpaV*$^{s}_{spl}$) and 1.7 (*GpaV*$^{s}_{spl}$ + *GpaXI*$^{s}_{spl}$). Both QTLs independently modified the sex ratio towards males, with the largest effect attributed to *GpaV*$^{s}_{spl}$, which increased the number of males two- to three-fold more than did *GpaXI*$^{s}_{spl}$. Furthermore, the two QTLs together but not alone, provoked a strong necrotic reaction in infected roots, near the head of the nematode. It should be pointed out that the naming of these QTLs is informative rather than historic: pathogen species (*Gpa*), potato linkage group (chromosome) (roman), long/short arm (superscript) and source species (subscript). Kreike et al. (1994) studied the resistance to Pa2 and Pa3 found in *S. spegazzinii* BGRC accession 8218 seedling number 15. They found that a QTL (*Qpa*) on chromosome 5 explained 51% and 45% of the variation in resistance to Pa2 and Pa3, respectively, with a high correlation ($r = 0.65$) between resistance to Pa2 and Pa3. There was also a minor QTL on chromosome 7 explaining 6% and 9% of the variation in resistance to Pa2 and Pa3, respectively. Caromel et al. (2003) also studied resistance derived from *S. spegazzinii* BGRC 8218/CGN 2000 and found three QTLs, one on chromosome 5 (= *Gpa*), one on chromosome 6 and one on chromosome 12, explaining 56%, 19% and 15%, respectively, of the variation. Tan et al. (2009) studied the resistance to Pa3 found in *S. tarijense* (BGRC 24717/CGN 18107) and found a QTL on the long arm of chromosome 11 which explained 81.3% of the variation in PCN scores. Another minor QTL on the long arm of chromosome 9 explained 5.3% of the variation but clones with both resistance alleles were no more resistant than those with just the *GpaXI*$_{tar}^{l}$ allele. The presence of *GpaXI*$_{tar}^{l}$ reduced the cyst count from 210 to 10. Finally, Rouppe van der Voort et al. (1998) found a compound locus (*Grp1*) on chromosome 5 that gave resistance to pathotypes Ro5 and Pa2 and partial resistance to Pa3. They concluded that the locus contained more than one gene because the source of the locus was a complex hybrid (AM78-3778) whose progenitor species were not resistant to both *G. rostochiensis* and *G. pallida*. They assumed that the separate introgression of the genes was followed by fortuitous recombination events which resulted in the coupling linkage of the genes. Recently, Park et al. (2019) mapped resistance gene(s) at 26 cM on chromosome 5 that may correspond to *Grp1*. They were looking for resistance to Ro2, a pathotype of *G. rostochiensis* that overcomes *H1* and had been detected on several farms in New York State. They crossed two tetraploid parental clones, NY121 (Ro1, Ro2 and Pa2/3 resistant) and NY115 (Ro1 resistant and Ro2 susceptible), and genotyped 182 F_1 offspring with 8303 SNP markers and assessed them for resistance to Ro2 in a greenhouse pot assay. The log-transformed cyst count data were approximately normally distributed. Analysis with TetraploidSNPMap (Hackett et al. 2017) identified two Ro2 resistance QTLs on chromosome 5: one QTL was located at 26 cM and explained 24.4% of the variation for resistance while the second at 59 cM co-localized with a marker (57R) known to be tightly linked to *H1* and explained 23.8% of the residual variation. Subsequent inoculation with *G. pallida* revealed that the chromosome 5 locus at 26 cM also conferred some resistance against pathotype Pa2/3, explaining 9.2% of the variation (approximately normally distributed after log transformation). A second QTL that increased susceptibility to Pa2/3 was located at 15 cM on chromosome 10 and explained 6.9% of the variation.

There would appear to be very good prospects for combining a few major genes and QTLs to give a very high level of resistance to a broad spectrum of pathotypes of *G. pallida* and *G. rostochiensis*. The observed partial resistance of individual genes/QTLs to *G. pallida* may be due to the genetic heterogeneity of field populations of the nematode or to an ineffective resistance mechanism that allows some 'escapees'. In other words, there may be a classical gene-for-gene interaction between all the resistance genes of the host and the avirulence genes of the pathogen. Furthermore, prospects appear good for a relatively long durability of the broad-spectrum resistance provided that its use is combined with appropriate crop rotations and farm hygiene. Under these circumstances, the multiplication rate and spread of potato cyst nematodes should be limited, but we need experimental evidence. Fournet et al. (2018) reviewed the use of experimental evolution as a tool to anticipate the adaptation of pathogen populations and to identify durable strategies for resistance deployment. They considered adaptation of *G. pallida* to the QTL $GpaV_{vrn}$ (*Gpa5*) which is on the same chromosome as $GpaV_{spl}$ and $GpaV_{spg}$ (*Gpa*). The experimental evolution took place over eight generations (years) using two distinct natural nematode populations (from fields in Saint-Malo and Noirmoutier), in a greenhouse under controlled conditions in which cysts obtained from one potato genotype were used to re-inoculate the same potato genotype. After eight generations, the researchers found that both the Saint-Malo and Noirmoutier lineages adapted to cultivar Iledher ($GpaV_{vrn}$) produced as many females as on the control lineages (i.e. developed on susceptible cultivar Desiree for the same number of generations), showing that the resistance of Iledher was totally overcome. The researchers concluded that the individuals virulent to $GpaV_{vrn}$ were already present in both *G. pallida* populations. However, the rate of resistance breakdown was affected by the genetic background in which $GpaV_{vrn}$ occurred (i.e. clones other than Iledher). Interestingly, the authors also found that adaptation to QTL $GpaV_{vrn}$ resulted in adaptation to the colinear locus $GpaV_{spg}$ but not to $GpaV_{spl}$, with clear implications for the deployment of resistance genes in cultivars.

5.11.2 *Root-Knot Nematodes (*Meloidogyne *spp.)*

Root-knot nematodes (RKN), like PCN, are sedentary endoparasites, but they differ, in that they are parthenogenic, have a wide host range and can produce several generations in a growing season (Phillips 1994). They invade potato roots, establish specialized feeding sites and form galls on roots and within tubers, reducing yields and tuber quality (Davie and Pickup 2018). They survive between crops as eggs which are stimulated to hatch by root diffusates. They have an embryonic stage, four juvenile stages and an adult stage. The second-stage juvenile (J2) is the infective stage of the life cycle. Once a feeding site has been established, the J2 progress through three moults, developing into adult females or males. The mature female deposits her eggs in a gelatinous sac usually on the surface of the galled root and can lay up to 1000 eggs. Unlike cyst nematodes, root-knot nematodes can produce several generations on a crop but their eggs do not survive so long. They can be spread with soil on planting material and farm machinery.

The economic impact of the different species that affect potato production depends on the climatic region (Davie and Pickup 2018). The warm-adapted species of the *M. incognita* (Southern root-knot nematode, SRKN) group are largely confined to the tropics and comprise three closely related species, *M. arenaria*, *M. incognita* and *M. javanica*. Three other economically important species, *M. chitwoodi* Golden (Columbian root-knot nematode, CRKN), *M. fallax* (False Columbian root-knot nematode) and *M. hapla* (Northern root-knot nematode, NRKN) are adapted to cooler climates. They are the most serious root-knot nematodes affecting temperate potato production, such as in the north-western areas of Europe and the USA, for example, *M. chitwoodi* in the Pacific Northwest of the USA (Brown et al. 2009). Finally, *M. luci* has been found to attack potatoes in Portugal.

Current control of root-knot nematodes relies heavily on synthetic nematicides, but as with the cyst nematodes, there is political pressure to reduce their use. Hence, resistant cultivars are an attractive proposition. Janssen et al. (1996) screened over 5000 plants from 64 wild tuber-bearing *Solanum* species, in the Dutch-German potato genebank, for resistance to *M. chitwoodi, M. fallax* and *M. hapla*, and reported many sources of resistance. Two early examples of studies of resistance genes are as follows: Draaistra (2006) mapped a gene from *S. tarijense* ($R_{Mh\text{-}tar}$) to the short arm of chromosome 7 that gave resistance to *M. hapla* population Hb, which is a race A of *M. hapla* collected from a soil sample near the Dutch town of Zwaanshoek. Initially, a single QTL was found that did not explain all of the variation in the amount of egg masses. However, by justifying an 'arbitrary threshold' of 1% reproduction of the total number of inoculated juveniles to distinguish between resistance and susceptibility, it was possible to treat the resistance as a qualitative trait and map the resistance gene more precisely. Draaistra (2006) thought that the practical value of the resistance gene would depend on the allele frequency of a matching virulence within the *M. hapla* populations and the distribution of such nematode populations. Kouassi et al. (2006) mapped a major gene from *S. sparsipilum* genotype 88S.329.15 ($MfaXII_{spl}$) to chromosome 12 that gave resistance to *M. fallax*. When potatoes with this gene are infected, they develop a necrotic reaction at the feeding site of the juveniles which can totally prevent their development to the female stage, as seen in the resistant parent 88S.329.15. The offspring in the mapping population displayed a 1:1 segregation for necrotic to non-necrotic reaction to infection (*NR* locus). The number of females recorded in offspring with a necrotic reaction varied from 0 to 49 with a mean of 14, and in offspring with a non-necrotic reaction from 92 to 157 with a mean of 127, similar to the susceptible parent with a count of 109. When treated as a quantitative trait for number of females, a QTL explaining 94.5% of the variation was mapped to chromosome 12 of the resistant parent, and corresponds to the *NR* locus. The locus is 11.7 cM from *Gpa2* (see above).

Most progress has been reported in breeding for resistance to *M. chitwoodi* in the USDA/ARS programme in Washington, DC, USA, but in 2018, resistant cultivars were still in the pipeline (Graebner et al. 2018). The USDA/ARS researchers have worked on resistance genes in three Mexican wild potato species. The best-characterized resistance gene was the one backcrossed into Tuberosum from *S. bulbocastanum* accession SB22 ($R_{Mc1(blb)}$) that is effective against race 1 of *M. chitwoodi*, the predominant race in the Pacific Northwest of the USA (Brown et al. 2009, 2014; Davies et al. 2015). Brown et al. (2009) found two linked genes on chromosome 11 (38 cM apart) in an advanced backcross clone: one ($R_{Mc1(blb)}$) was a single dominant gene for resistance to race 1 reproduction on the root system, and the other ($R_{Mctuber(blb)}$) was a single dominant gene for tuber resistance to invasion by juveniles, both from weeds (CRKN has wide host range) and from root resistance-breaking pathotypes of races 1 and 2. Under field conditions, in the presence of a root-resistance-breaking pathotype, a tuber-resistant clone produced commercially acceptable tubers. The combination of the two genes, and hence the two phenotypes, should result in a sufficient level of resistance to prevent tuber damage. Brown et al. (2014) found that the three R_{Mc1} genes from diploid *S. bulbocastanum* (*blb*), tetraploid *S. fendleri* (*fen*) and hexaploid *S. hougasii* (*hou*) all mapped to the same position on chromosome 11 (informative markers across species) and conferred resistance to *M. chitwoodi* race 1 in the root system. They hypothesize that the gene in *S. bulbocastanum* was the original one, the diploid species contributing its genome to the polyploids. Davies et al. (2015) did a functional characterization of $R_{Mc1(blb)}$ using avirulent and virulent nematodes and revealed that its resistance is dependent on a hypersensitive response. This involves calcium at the start of signal transduction that results in release of reactive oxygen species that cause cell death and hence prevention of the formation of a nematode feeding site. In summary, the advanced backcross clone of Brown et al. (2009) is proving of value in their conventional breeding programme, but the lack of a commercial potato cultivar with broad-spectrum resistance has prompted Dinh et al. (2015) to explore a transgenic approach. They showed that plant-mediated RNAi silencing of the *Meloidogyne 16D10* effector gene resulted in significant resistance against all of the root-knot nematode species tested (*M. arenaria*, *M. chitwoodi*, *M. hapla*, *M. incognita* and *M. javanica*). We shall look at these results in more detail in the chapter on genetic transformation.

In the meantime, Graebner et al. (2018) have screened 40 plant accessions in the NRSP-6 Potato Genebank, representing nine wild potato species, for their resistance to three isolates of *M. chitwoodi*: two representative of race 1 and race 2, respectively, and the third an isolate of race 1 distinguished by its ability to reproduce on roots of potato plants that have resistance to race 1 conferred by $R_{Mc1(blb)}$. They identified strong resistance to all three isolates of *M. chitwoodi*. Fifteen of the 17 resistant clones were from six accessions of *S. hougasii*, while the other two clones were from *S. bulbocastanum* and *S. stenophyllidium*. The resistance was geographically clustered in and around the Mexican states of Jalisco and Michoacán. Four of the *S. hougasii* clones were non-hosts (reproduction factor < 0.1) for each of the three nematode isolates tested. These are being introgressed into elite potato germplasm.

5.11.3 *Other Nematodes*

Other nematodes mentioned by Davie and Pickup (2018) are as follows, but resistance does not appear to feature as an objective of breeding programmes. The false root nematode (*Nacobbus aberrans*) is an important and widespread pest of potatoes in South America, but does not occur in Europe where it is classified as an A1 quarantine organism. Resistant cultivars are not commercially available. There are over 60 species of root lesion nematode (*Pratylenchus* spp.) of which at least 10 are pests of potato. No highly resistant cultivars are available. The potato rot nematode (*Ditylenchus destructor*) and the stem nematode (*Ditylenchus dipsaci*) are rarely encountered in Western Europe as a result of improvements in seed health, weed control and tuber storage. The virus vector nematodes are the needle nematode (*Longidorus* spp.) and the stubby root nematodes *Trichodorus* and *Paratrichodorus* spp. At least 13 species of *Trichodorus* are able to transmit *Tobacco rattle virus* (TRV) resulting in spraing symptoms in the tubers. Nematicides can be used to control the nematode vectors and weed hosts of TRV should be removed to reduce virus inoculum.

5.12 Conclusions

We have considered 27 diseases and 6 pests for which increased resistance may be required in potatoes. This is a large number, which could be even larger if more potential pest problems were included. So, breeders need to decide which to focus on as the most important for the particular target environments and end uses of their programmes, either because of severe loss of yield or quality in most years with currently available cultivars. Breeders may also want to check that their new cultivars are not particularly susceptible to currently less important pests and diseases. The pathogen (or pest) may be differentiated into species, subspecies, biovars or physiological races. A cultivar or potential cultivar may be resistant to some isolates of the pathogen but susceptible to other isolates (isolate specific) or it may be resistant to most or all known isolates of the pathogen (broad spectrum). A cultivar may be more or less susceptible to different isolates of the pathogen, the isolates thus varying in their aggressiveness. The prerequisites for successful breeding are a source of resistance, a reliable screen for resistance using the most appropriate isolate(s) of the pathogen or population(s) of the pest, and an understanding of the inheritance of resistance. The source of resistance may be a modern cultivar, a cultivated landrace or a wild relative. A reliable test for qualitative resistance is one where there are no escapes so that susceptible control plants are all susceptible. In contrast, a reliable test for quantitative resistance is one in which there is little environmental variation or errors of measurement. Both types of resistance can be assessed on plant parts (e.g. detached leaves), seedlings or whole plants in controlled environment facilities such as growth chambers or specialized glasshouses.

They can also be assessed on plants in the field, often in special nurseries, where the progress of the disease can be followed under more natural but less controlled conditions. For any specific disease test, the best stage of plant growth for inoculation and scoring, inoculum concentration, method of inoculation, and environmental conditions (temperature, day-length, humidity) are all factors to be determined by experiment, if not already known in the literature. Breeders will also need to decide how many seed tubers they require for yield trials and seed-tuber plots in the next clonal generation and hence how many they can make available for assessing disease and pest resistance. The number will be a limiting factor in what can be done. The terms horizontal and vertical resistance are still used, the former usually equated with broad-spectrum quantitative resistance, and the latter with race-specific qualitative oligogenic resistance where, in today's terminology, we often find that a host *R*-gene encodes a receptor for the pathogen's effector molecule specified by an *Avr* gene. Finally, an understanding of the inheritance of resistance comes from appropriate hybridization experiments. Conclusions for specific diseases and pests are as follows.

1. *Late Blight*: Late blight, caused by the oomycete *Phytophthora infestans* (Mont.) de Bary, is still potentially the most serious disease of potatoes worldwide. Genetically diverse global populations of *P. infestans*, comprising both the A1 and A2 mating types, can be monitored using molecular markers as they continue to evolve and form clonal lineages which can become locally dominant. Breeders need to know which avirulence (*Avr*) genes they possess as these specify the effector molecules which are recognized by the host's *R*-genes that provide resistance. The current emphasis in breeding for resistance is on finding and deploying *R*-genes from potato wild relatives that will either singly or in combination provide broad-spectrum resistance that proves durable, unlike the first *R*-genes that were introgressed from *Solanum demissum.* Over 30 *R*-genes from more than 15 wild species are under consideration for introgression breeding or for cloning and deployment through genetic transformation. *R*-genes can be detected in detached leaflet tests or more reliably in whole plant tests. A cultivar will show complete resistance to late blight in the field if it possesses all of the necessary *R*-genes to recognize and effectively stop infection by the *P. infestans* population present in the field. If not, the disease will progress over time (say t in days) and can be monitored as the proportion of the leaf area of plants covered by lesions (y). The graph of $\ln(y/(1\text{-}y))$ against t is a straight line of slope r and intercept k, whereas the graph of y against t is a curve, the disease progress curve. The slope r measures the rate of infection, whereas k is the amount of disease at the start of the epidemic (i.e. the number of spores infecting the cultivar, which will be zero if *R*-genes are effective). Together, k and r determine the amount of disease at time t. Cultivars differ in their r values with ones having low values being said to have quantitative field resistance. A pathologist would analyse quantitative field resistance in terms of the components of infection rate: i.e. the proportion of spores that initiate lesions, rate of spread of lesions, latent period until sporulation occurs, infectious period and

abundance of spore production). A geneticist would do the analysis in terms of Quantitative Trait Loci (QTLs). As QTL × *P. infestans* isolate interactions can occur, there is no guarantee that quantitative field resistance will be durable resistance, the latter remaining either a historical fact or a hope for the future. Finally, breeders need to appreciate that resistance in the foliage does not guarantee resistance in the tubers, and hence, selection for tuber resistance is also required in a breeding programme.

2. *Verticillium Wilt and Early Blight*: Soil-borne Verticillium wilt (*Verticillium dahliae* Kelb.) and air-borne early blight (*Alternaria solani* (Ellis and Martin) Jones and Grout) are warm-climate fungal diseases for which improved resistance is feasible as screening methods have been developed and differences in cultivar resistance/susceptibility have been found, although true resistance needs to be distinguished from late maturity.
3. *Wart*: Wart (*Synchytrium endobioticum* (Schilb.) Perc) is a persistent soil-borne, disfiguring, blemish-forming fungal disease of tubers, with a number of pathotypes in Europe, which can be controlled by quarantine measures and resistant cultivars. As assessment of resistance is laborious and time-consuming, marker-assisted selection for gene *Sen2* (broad-spectrum to eight pathotypes, on chromosome 11), possibly combined with *Sen1* (narrow-spectrum to one pathotype, also on chromosome 11), is an attractive proposition, despite uncertainty over durability of resistance.
4. *Powdery Scab*: Powdery scab (*Spongospora subterranean* (Wallr.) Lagerh) is another persistent soil-borne, disfiguring, blemish-forming fungal disease of tubers in temperate regions which is often a criterion for rejection in seed certification. Improved resistance is feasible as screening methods have been developed and differences in cultivar resistance/susceptibility have been found.
5. *Dry Rot and Gangrene*: Dry rot [(*Fusarium coeruleum* (Lib.) ex Sacc.), *F. sambucinum* (Fuckel) (synonymous with *F. sulphureum* and *Gibberella cyanogena*) and *F. avenaceum* Sacc. (Corda ex. Fr.)] and gangrene (*Phoma foveata* Foister) are tuber-borne fungal diseases which cause spoilage of tubers in storage. Whilst dry rot species occur worldwide, gangrene is associated with northern Europe. Improved resistance to gangrene is relatively easy to achieve, if required, as heritable variation and screening methods are available. Resistance to *F. coeruleum* is different and separate from resistance to *F. sulphureum* and the indications are that breeding for resistance to the former is easier than to the latter.
6. *Black Scurf, Skinspot, Silver Scurf and Blackdot*: Black scurf is a soil- and tuber-borne skin blemish disease caused by the fungus, *Rhizoctonia solani* (Kühn.), which also causes stem canker. Skinspot (*Polyscytalum pustulans* (Owen and Wakef.) M.B. Ellis), silver scurf (*Helminthosporium solani* Dur. and Mont.) and black dot (*Colletotrichum coccodes* (Wallr.) Hughes) are three tuber-blemishing fungal diseases. All three pathogens cause a deterioration of tubers in store and all are mainly or partly tuber-borne as well as soil-borne. Although new cultivars are often tested for resistance/susceptibility, resistance is not usually a high priority in breeding programmes.

7. *Common Scab*: Common scab (*Streptomyces scabies* (Thaxt.) Waksman and Henrici) is a persistent soil-borne, disfiguring, blemish-forming actinomycete disease of tubers in which infection is aided by the production of a phytotoxin, thaxtomin A. Breeding for resistance has been a high priority in North America and resistant cultivars have been produced. Major genes for resistance have not been identified so most of the literature treats it as a quantitative trait for which QTLs have been identified. In field assessments, the potatoes are grown to maturity in a naturally infested or inoculated field (sandy soil) and the harvested tubers scored for scab symptoms. As this is time consuming, genomic selection is an attractive proposition.
8. *Bacterial Wilt and Brown Rot*: Bacterial wilt of plants and brown rot of tubers are caused by *Ralstonia solanacearum* Smith, a soil-borne bacterium that enters via the roots. Five races can be recognized by host range, six biovars by biochemical properties, and four phylotypes can be distinguished in relation to the phylogenetic and geographic origin of strains. Assessment of resistance can be done on plants grown in naturally contaminated fields with susceptible and more resistant cultivars planted as controls. Bacterial wilt and brown rot are the most serious diseases of potato after late blight in the developing world, where breeding for resistance has been a high priority, but with disappointing results to date. Hence, researchers are exploring transgenic approaches to resistance.
9. *Blackleg and Soft Rot*: Blackleg of potato plants is a tuber-borne disease caused by a number of pectolytic bacteria species, most commonly *Pectobacterium atrosepticum* (synonym *Erwinia carotovora* subsp. *atroseptica*). Other bacterial species such as *Pectobacterium carotovorum* (synonym *E. carotovora* subsp. *carotovora*) and *Dickeya solani* (previously classified as *E. chrysanthemi*) can also cause field symptoms, but are more commonly associated with soft rots of tubers in storage. Which *Pectobacterium* and *Dickeya* pathogens dominate depends on the location and prevailing environmental conditions. As resistance to tuber soft rot and that to blackleg are not that closely related, assessment of both is required in breeding for resistance, with laboratory methods available for the former and glasshouse and field methods for the latter. Although quantitative resistance and QTLs have been identified in primitive cultivated potatoes and wild species, and used in breeding with varying degrees of success, it is not obvious from a literature search that we are seeing an increased number of new cultivars with higher levels of resistance to blackleg and soft rot.
10. *Bacterial Ringrot and Zebra Chip Disease*: *Clavibater michiganensis* subsp. *sepedonicus* (Spieck and Kotth.) Davis et al. is the cause of potato ring rot, a recurring seed-tuber-transmitted disease problem in temperate regions, despite many countries treating it as a quarantine disease and having a zero-tolerance policy for the import of infected seed potatoes. Little breeding work appears to have been done although *S. acaule* has been reported as a source of extreme resistance (immunity to infection). Zebra chip disease was first observed in Mexico in 1994 and then spread to the southern parts of the USA, other countries in Central America, and New Zealand. The causal agent of the disease is

believed to be the unculturable bacterium *Candidatus* Liberibacter solanacearum (Lso) which resides in the plant phloem and is primarily vectored by the potato psyllid (lice) *Bactericera cockerelli.* I am not aware of cultivars with resistance, but there is interest in breeding for resistance, although current methods of assessment involve too much work for the early stages of a breeding programme.

11. *Potato Viruses Y, A and V*: PVY is a *Potyvirus* which is transmitted by about 65 aphid species in a non-persistent manner and usually results in severe mosaic symptoms. It is considered the most economically damaging virus of cultivated potatoes. The incidence of PVY strains has been changing with PVY^{O} replaced by PVY^{N} and their recombinants PVY^{NTN} or $PVY^{N\text{-}Wi}$ in some areas, PVY^{NTN} being the strain that also causes potato tuber necrotic ringspot disease (PTNRD) on tubers. Today, breeders' preferred choice of genes for resistance is the major dominant *R*-genes for extreme resistance (*Ry* genes), which are currently effective against all strains of the virus (comprehensive resistance): Ry_{adg} (chromosome 11), Ry_{sto} and $Ry\text{-}f_{sto}$ (chromosome 12) and Ry_{chc} (chromosome 9) which were derived from *S. tuberosum* Group Andigena, *S. stoloniferum* and *S. chacoense*, respectively. Ry_{sto} also provides extreme resistance to viruses PVA and PVV (closely related to PVY) as well as to PVY^{O}, PVY^{N} and PVY^{NTN}.
12. *Potato Viruses X, M and S*: PVX is a *Potexvirus* with a worldwide distribution whose symptoms tend to be mild mosaic. It is one of the few viruses that can be transmitted by contact under natural conditions in the field. Again, in breeding for resistance, major dominant *Rx* genes for extreme resistance (immunity) to virtually all current strains are preferred with the two most widely used being Rx_{adg} (= *Rx1*, chromosome 12) from *S. tuberosum* Group Andigena and Rx_{acl} (= *Rx2*, chromosome 5) from *S. acaule*. PVM and PVS are *Carlaviruses* which are found worldwide, which can also be transmitted by contact, and which can result in symptomless infections. Major dominant genes for resistance have been found.
13. *Potato Leafroll Virus*: PLRV is a *Polerovirus* that causes economic damage in cultivated potatoes throughout the world with leaf rolling a typical symptom. It is aphid-transmitted in a persistent manner by colonizing aphids of which the most important is *Myzus persicae*. Breeding for resistance involves selection for resistance to infection in field exposure trials (secondary infection in tuber progeny plants is also assessed) and/or resistance to accumulation (low virus titre in glasshouse- or field-grown plants) which may in part be due to restricted virus movement. Major genes and QTL alleles of large effect exist for resistance to PLRV infection and accumulation (or movement). Their incorporation into new potato cultivars would greatly reduce the spread of the virus in potato crops and hence be an effective control measure.
14. *Tobacco Rattle Virus and Potato Mop-Top Virus*: TRV is a *Tobravirus* which is distributed in cooler climates and transmitted by free-living nematodes in the genera *Paratrichodorus* and *Trichodorus*, with a preference for light sandy or peaty soils. It results in necrotic symptoms in the tuber flesh, known as spraing or corky ringspot, which may lead to rejection of consignments of potatoes by

processors and supermarkets. Resistance can be assessed in field trials on infective land by scoring tubers for frequency and severity of spraing symptoms, but more reliable results come from a glasshouse pot test using field soil containing viruliferous nematodes and scoring daughter tubers for the severity of spraing symptoms. Extreme resistance, which is known in some processing cultivars, needs to be distinguished from infected but symptomless tubers (tolerance). PMTV infection also results in spraing symptoms in tubers. It is a widely distributed *Pomovirus* which is transmitted in a persistent manner in soil by the fungus *Spongospora subterranea*, the cause of powdery scab. Resistance to PMTV is highly desirable but has proved elusive. Therefore, a wider screening of germplasm would seem worthwhile as well as exploring transgenic approaches.

15. *Potato Tuber-Spindle Viroid*: The viroid PSTVd is the most important pathogen transmitted by True Potato Seed (TPS). It is treated as a quarantine disease in countries where it is not endemic such as the UK. There is a lack of sources of resistance making transgenic approaches worth exploring.
16. *Potato Tuber Moth*: Potato tuber moth (PTM, *Phthorimaea operculella* Zeller) is the most serious insect pest of potatoes worldwide, particularly in warm tropical and sub-tropical climates. Its larvae mine the foliage, stems and tubers in the field, and in storage, with most economic losses occurring in storage, particularly non-refrigerated storage. PTM resistance has been found in diploid breeding material that involves both antibiosis (resistance to feeding by pest when no choice) and antixenosis (non-preference, i.e. tuber unattractive to pest). Antixenosis can be assessed from natural infestation of tubers in open paper bags in storage and antibiosis from larvae on a tuber in a closed plastic container. However, most effort has gone into transgenic resistance which has been provided by the *Bt* proteins encoded by the *cry5* (equivalent to *cry1Ia1*) and *cry1Ac9* genes using the constitutive CaMV 35S promoter. Effective resistance has been demonstrated in field trials under natural infestations and under non-refrigerated storage and laboratory conditions. Despite all of this potential, to the best of my knowledge, transgenic resistance is yet to be deployed by farmers in their fields.
17. *Colorado Potato Beetle*: Colorado potato beetle (CPB, *Leptinotarsa decemlineata* Say) is the best known and most destructive of the leaf-feeding insects attacking potatoes, with both adults and larvae feeding on the leaves. It is a major pest in northern latitudes, particularly in North America, Europe, Russia and Central Asia. Resistance can be assessed under natural infestation in which defoliation is followed over time. A number of resistance factors have been identified in wild species and used in breeding, the best known being glandular trichomes (types A and B), from *S. berthaultii*, and leptine glycoalkaloids from *S. chacoense*. However, cultivars with high levels of resistance do not appear to have been widely grown. Transgenic resistance has been provided by gene *cry3A*, which encodes the Cry3A protein from the bacterium *Bacillus thuringiensis* var. *tenebrionis*, using the constitutive CaMV 35S promoter. Cultivars with this *Bt* gene were commercialized in North America in the late 1990s until

marketing ceased because of consumer concerns over GM potato products. Nevertheless, integrated conventional breeding and transgenic strategies may provide the best opportunity to build durable host resistance to insects such as CPB.

18. *Cyst Nematodes*: Potato cyst nematodes (PCN) (*Globodera rostochiensis* Woll. and *G. pallida* Stone) are the most damaging nematodes of potato worldwide. Their second-stage juveniles hatch from viable eggs in response to chemicals exuded from the roots of growing potato plants. The juveniles penetrate and feed on the roots and thus reduce tuber yield, size and quality. Both species are diploid outbreeders that comprise variable populations that make division into pathogenic variants difficult, although five pathotypes, Ro1 to Ro5, of *G. rostochiensis* and three pathotypes, Pa1, Pa2 and Pa3 of *G. pallida* are recognized. Resistant potatoes inhibit or reduce nematode multiplication rates but suffer root damage and hence, tolerance of damage is also required. However, breeding tends to focus on the former for which rapid and reliable tests allow a large number of potato genotypes to be assessed, for example, a tuber in a canister in sterilized soil inoculated with a known number of eggs. Major genes and QTL alleles of large effect have been found for resistance in wild and cultivated species that, singly or in combination, can provide a very high level of resistance to a broad spectrum of pathotypes of *G. pallida* and *G. rostochiensis.* Currently, more cultivars are available with resistance to *G. rostochiensis* than to *G. pallida.* Furthermore, prospects appear good for a relatively long durability of the broad-spectrum resistance provided that its use is combined with appropriate crop rotations and farm hygiene; for example, the *H1* gene from Group Andigena CPC 1673 has been effective for over 50 years in the UK where Ro1 is the only pathotype of *G. rostochiensis*. However, *G. pallida* is more of a problem and genetically variable populations may harbour a few individuals virulent to the resistant genes available.
19. *Root-Knot Nematodes*: Root-knot nematodes (RKN, *Meloidogyne* spp.), like PCN, are sedentary endoparasites in which the second-stage juveniles are the infective stage of the life cycle. However, they differ, in that they are parthenogenic, have a wide host range and can produce several generations in a growing season. The warm-adapted species of the *M. incognita* (Southern root-knot nematode, SRKN) group are largely confined to the tropics and comprise three closely related species, *M. arenaria*, *M. incognita* and *M. javanica*. Three other economically important species, *M. chitwoodi* Golden (Columbian root-knot nematode, CRKN), *M. fallax* (False Columbian root-knot nematode) and *M. hapla* (Northern root-knot nematode, NRKN) are adapted to cooler climates. Progress has been reported in breeding for resistance to *M. chitwoodi* in the USA with resistant cultivars in the pipeline. Two linked dominant genes (chromosome 11) from *S. bulbocastanum* in an advanced backcross clone provide resistance to race 1 reproduction on the root system ($R_{Mc1(blb)}$) and tuber resistance to invasion by juveniles from root resistance-breaking pathotypes of races 1 and 2 ($R_{Mctuber(blb)}$), respectively. The combination of the two genes, and hence the two phenotypes, should result in a sufficient level of resistance to

prevent tuber damage. A gene for resistance to *M. hapla* race A has been found in *S. tarijense* ($R_{Mh\text{-}tar}$ on chromosome 7) and one for resistance to *M. fallax* in *S. sparsipilum* ($MfaXII_{spl}$ on chromosome 12). Finally, plant-mediated RNAi silencing of the *Meloidogyne 16D10* effector gene has been shown to result in significant resistance against all of the root-knot nematode species tested (*M. arenaria, M. chitwoodi, M. hapla, M. incognita* and *M. javanica*), making this an attractive approach for the future.

Resistance to diseases and pests completes our assessment of breeding objectives. It is now time to turn to breeding methods which will achieve those objectives.

References

Alor N, Tierno R, Cooke DEL, de Galarreta JIR (2019) Characterization of *Phytophthora infestans* isolates of potato crops from Spain. Potato Res 62(4):453–463

Alyokhin A, Vincent C, Giordanengo P (2012) Insect pests of potato: global perspectives on biology and management. Elsevier, San Diego, 616p

Andrivon D, Pellé R, Ellissèche D (2006) Assessing resistance types and levels to epidemic diseases from the analysis of disease progress curves: principles and application to potato late blight. Am J Potato Res 83:455–461

Armstrong MR, Vossen J, Lim TY, Hutten RC, Xu J, Strachan SM, Harrower B, Champouret N, Gilroy EM, Hein I (2019) Tracking disease resistance deployment in potato breeding by enrichment sequencing. Plant Biotechnol J 17(2):540–549. https://doi.org/10.1111/pbi.12997

Bae J, Tai G, Jansky SH (2009) Selection for *Verticillium* wilt resistance in potato breeding populations derived from four cross types. Plant Breed 128(3):290–294. https://doi.org/10.1111/j.1439-0523.2008.01551.x

Ballvora A, Hesselbach J, Niewöhner J, Leister D, Salamini F, Gebhardt C (1995) Marker enrichment and high-resolution map of the segment of potato chromosome VII harbouring the nematode resistance gene *Gro1*. Mol Gen Genet 249:82–90

Ballvora A, Ercolano MR, Weiss J, Meksem K, Bormann CA, Oberhagemann P, Salamini F, Gebhardt C (2002) The *R1* gene for potato resistance to late blight (*Phytophthora infestans*) belongs to the leucine zipper/NBS/LRR class of plant resistance genes. Plant J 30:361–371

Ballvora A, Flath K, Lübeck J, Strahwald J, Tacke E, Hofferbert H-R, Gebhardt C (2011) Multiple alleles for resistance and susceptiblity modulate the defence response in the interaction of tetraploid potato (*Solanum tuberosum*) with *Synchytrium endobioticum* pathotypes 1, 2, 6 and 18. Theor Appl Genet 123:1281–1292

Barker H (1987) Multiple components of the resistance of potatoes to potato leaf roll virus. Ann Appl Biol 111:641–648

Barker H (1996) Inheritance of resistance to potato viruses Y and A in progeny obtained from potato cultivars containing gene *Ry*: evidence for a new gene for extreme resistance to PVA. Theor Appl Genet 93:710–716

Barker H (1997) Extreme resistance to potato virus V in clones of *Solanum tuberosum* that are also resistant to potato viruses Y and A: evidence for a locus conferring broad-spectrum resistance. Theor Appl Genet 95:1258–1262

Barker H, Harrison BD (1985) Restricted multiplication of potato leafroll virus in resistant potato genotypes. Ann Appl Biol 107:205–212

Barker H, Solomon RM (1990) Evidence of simple genetic control in potato of ability to restrict potato leaf roll virus concentration in leaves. Theor Appl Genet 80:188–192

Barker H, Solomon-Blackburn RM, McNicol JW, Bradshaw JE (1994) Resistance to potato leaf roll virus multiplication in potato is under major gene control. Theor Appl Genet 88:754–758

Barone A, Ritter E, Schachtschabel U, Debener T, Salamini F, Gebhardt C (1990) Localization by restriction fragment length polymorphism mapping in potato of a major dominant gene conferring resistance to the potato cyst nematode *Globodera rostochiensis*. Mol Gen Genet 224:177–182

Bendahmane A, Kanyuka K, Baulcombe DC (1997) High-resolution genetical and physical mapping of the *Rx* gene for extreme resistance to *Potato virus X* in tetraploid potato. Theor Appl Genet 95:153–162

Bendahmane A, Kanyuka K, Baulcombe DC (1999) The *Rx* gene from potato controls separate virus resistance and cell death responses. Plant Cell 11:781–791

Bendahmane A, Querci M, Kanyuka K, Baulcombe DC (2000) *Agrobacterium* transient expression system as a tool for the isolation of disease resistance genes: application to the *Rx2* locus in potato. Plant J 21(1):73–81

Bidondo LF, Almasia N, Bazzini A, Colombo R, Hopp E, Vazquez-Rovere C, Godeas A (2019) The overexpression of antifungal genes enhances resistance to *Rhizoctonia solani* in transgenic potato plants without affecting arbuscular mycorrhizal symbiosis. Crop Prot 124. https://doi.org/10.1016/j.cropro.2019.05.031

Black W (1970) The nature of inheritance of field resistance to late blight (*Phytophthora infestans*) in potatoes. Am Potato J 47:279–288

Black W, Mastenbroek C, Mills WR, Peterson LC (1953) A proposal for an international nomenclature of races of *Phytophthora infestans* and of genes controlling immunity in *Solanum demissum* derivatives. Euphytica 2:173–179

Bonierbale MW, Amoros WR, Salas E, de Jong W (2020) Potato breeding. In: Campos H, Ortiz O (eds) The potato crop: its agricultural, nutritional and social contribution to humankind. Springer, ebook, pp 163–218

Bradeen JM (2011) Cloning of late blight resistance genes: strategies and progress. In: Bradeen JM, Kole C (eds) Genetics, genomics and breeding of potato. Science Publishers, Enfield, pp 153–183

Bradshaw JE (2009a) Potato breeding at the Scottish Plant Breeding Station and the Scottish Crop Research Institute: 1920-2008. Potato Res 52:141–172

Bradshaw JE (2009b) Breeding for field resistance to late blight of potato at SCRI. Acta Hortic 834:87–100

Bradshaw JE (2016) Plant breeding: past, present and future. Springer, New York, 693p

Bradshaw JE, Ramsay G (2005) Utilisation of the Commonwealth Potato Collection in potato breeding. Euphytica 146:9–19

Bradshaw JE, Wastie RL, Stewart HE (1996) Assessing general combining ability for gangrene resistance by means of a glasshouse seedling test. Potato Res 39:179–183

Bradshaw JE, Pande B, Bryan GJ, Hackett CA, McLaren K, Stewart HE, Waugh R (2004) Interval mapping of quantitative trait loci for resistance to late blight [*Phytophthora infestans* (Mont.) de Bary], height and maturity in a tetraploid population of potato (*Solanum tuberosum* subsp. *tuberosum*). Genetics 168:983–995

Bradshaw JE, Hackett CA, Lowe R, McLean K, Stewart HE, Tierny I, Vilaro MDR, Bryan GJ (2006) Detection of a quantitative trait locus for both foliage and tuber resistance to late blight [*Phytophthora infestans* (Mont.) de Bary] on chromosome 4 of a dihaploid potato clone (*Solanum tuberosum* subsp. *tuberosum*). Theor Appl Genet 113:943–951

Bradshaw JE, Dale MFB, Mackay GR (2009) Improving the yield, processing quality and disease and pest resistance of potatoes by genotypic recurrent selection. Euphytica 170:215–227

Braun S, Gevens A, Charkowski A, Allen C, Jansky S (2017) Potato common scab: a review of the causal pathogens, management practices, varietal resistance screening methods, and host resistance. Am J Potato Res 94(4):283–296. https://doi.org/10.1007/s12230-017-9575-3

Brierley JL, Hilton AJ, Wale SJ, Peters JC, Gladders P, Bradshaw NJ, Ritchie F, MacKenzie K, Lees AK (2015) Factors affecting the development and control of black dot on potato tubers. Plant Pathol 64:167–177

Brown CR, Thomas EE (1994) Resistance to potato leafroll virus derived from *Solanum chacoense*: characterization and inheritance. Euphytica 74:51–57

Brown CR, Corsini D, Pavek J, Thomas PE (1997) Heritability of field resistance to potato leafroll virus in cultivated potato. Plant Breed 116:585–588

Brown CR, Mojtahedi H, Zhang LH, Riga E (2009) Independent resistant reactions expressed in root and tuber of potato breeding lines with introgressed resistance to *Meloidogyne chitwoodi*. Phytopathology 99(9):1085–1089

Brown CR, Zhang L, Mojtahedi H (2014) Tracking the R_{Mc1} gene for resistance to race 1 of Columbia root-knot nematode (*Meloidogyne chitwoodi*) in three Mexican wild potato species with different ploidies. Am J Potato Res 91:180–185

Brugmans B, Hutten RGB, Rookmaker N, Visser RGF, van Eck HJ (2006) Exploitation of a marker dense linkage map of potato for positional cloning of a wart disease resistance gene. Theor Appl Genet 112:269–277

Bryan GJ, McLean K, Bradshaw JE, De Jong WS, Phillips M, Castelli L, Waugh R (2002) Mapping QTLs for resistance to the cyst nematode *Globodera pallida* derived from the wild potato species *Solanum vernei*. Theor Appl Genet 105:68–77

Bryan GJ, McLean K, Pande B, Purvis A, Hackett CA, Bradshaw JE, Waugh R (2004) Genetical dissection of H3-mediated polygenic PCN resistance in a heterozygous autotetraploid potato population. Mol Breed 14:105–116

Burkhart CR, Christ BJ, Haynes KG (2007) Non-additive genetic variance governs resistance to fusarium dry rot in a diploid hybrid potato population. Am J Potato Res 84:199. https://doi.org/10.1007/BF02986269

Byarugaba AA, Prossy N, Kashaija IN (2013) Identification of potato clones of population B3C2 with durable field resistance to late blight (*Phytophthora infestans*) and high yields in Uganda. Afr J Agric Res 8:3055–3059. https://doi.org/10.5897/AJAR2013.6917

Caligari PDS, Nachmias A (1988) Screening for field resistance to early blight (*Alternaria solani*) in potatoes. Potato Res 31:451–460

Caligari PDS, Wastie R (1985) Assessment of a glasshouse test for measuring the resistance of potato cultivars to common scab. Potato Res 28:379–387

Campos H, Ortiz O (eds) (2020) The potato crop: its agricultural, nutritional and social contribution to humankind. Springer, ebook

Caromel B, Mugniery D, Lefebvre V, Andrzejewski S, Ellisseche D, Kerlan MC, Rousselle P, Rousselle-Bourgeois F (2003) Mapping QTLs for resistance against *Globodera pallida* (Stone) Pa2/3 in a diploid potato progeny originating from *Solanum spegazzinii*. Theor Appl Genet 106(8):1517–1523. https://doi.org/10.1007/s00122-003-1211-6

Caromel B, Mugniéry D, Kerlan M-C, Andrzejewski S, Palloix A, Ellissèche D, Rousselle-Bourgeois F, Lefebvre V (2005) Resistance quantitative trait loci originating from *Solanum sparsipilum* act independently on the sex ratio of *Globodera pallida* and together for developing a necrotic reaction. MPMI 18:1186–1194

Carputo D, Basile B, Cardi T, Frusciante L (2000) *Erwinia* resistance in backcross progenies of *Solanum tuberosum* × *S. tarijense* and *S. tuberosum* (+) *S. commersonii* hybrids. Potato Res 43:135–142

Carroll CP (1982) A mass-selection method for the acclimatization and improvement of edible diploid potatoes in the United Kingdom. J Agric Sci Camb 99:631–640

Castelli L, Ramsay G, Bryan G, Neilson SJ, Phillips MS (2003) New sources of resistance to the potato cyst nematodes *Globodera pallida* and *G. rostochiensis* in the Commonwealth Potato Collection. Euphytica 129:377–386

Chandel RS, Vashisth S, Soni S, Kumar R, Kumar V (2020) The potato tuber moth, *Phthorimaea operculella* (Zeller), in India: biology, ecology, and control. Potato Res 63(1):15–39. https://doi.org/10.1007/s11540-019-09426-z

Cho K-S, Kwon M, Cho J-H, Im J-S, Park Y-E, Hong S-Y, Hwang I-T, Kang J-H (2017) Characterization of trichome morphology and aphid resistance in cultivated and wild species of potato. Hortic Environ Biotechnol 58(5):450–457

Cingel A, Savić J, Lazarević J, Ćosić T, Raspor M, Smigocki A (2016) Extraordinary adaptive plasticity of Colorado potato beetle: "ten-striped spearman" in the era of biotechnological warfare. Int J Mol Sci 17(9):1538. https://doi.org/10.3390/ijms17091538

Cockerham G (1955) Strains of potato virus X. In: Streutgers E, Beemster ABR, van der Want JPH (eds) Proceedings of the 2nd conference on potato virus diseases, Lisse-Wageningen, 1954, H. Veenman & Zonen, Wageningen, pp 89–92

Cooke DEL, Lees AK, Shaw DS, Taylor MC, Prentice MWC, Bradshaw NJ, Bain RA (2008) The status of GB blight populations and the threat of oospores. Proc Crop Prot North Br 2008:217–222

Coombs JJ, Douches DS, Li W, Grafius EJ, Pett WL (2003) Field evaluation of natural, engineered, and combined resistance mechanisms in potato for control of Colorado potato beetle. J Am Soc Hortic Sci 128(2):219–224

Corsini DL, Pavek JJ, Davis JR (1990) Verticillium wilt resistant potato germplasm: A66107-51 and A68113-4. Am Potato J 67(8):517–525

Cruickshank G, Stewart HE, Wastie RL (1982) An illustrated assessment key for foliage blight of potatoes. Potato Res 25:213–214

Czajkowski R, Pérombelon MCM, van Veen JA, van der Wolf JM (2011) Control of blackleg and tuber soft rot of potato caused by *Pectobacterium* and *Dickeya* species: a review. Plant Pathol 60(6):999–1013

Dale MFB, Solomon RM (1988) A glasshouse test to assess the sensitivity of cultivars to tobacco rattle virus. Ann Appl Biol 112:225–229

Dale MFB, Phillips MS, Ayres RM, Hancock M, Holliday M, Mackay GR, Tones SJ (1988) The assessment of the tolerance of partially resistant potato clones to damage by the potato cyst nematode *Globodera pallida* at different sites and in different years. Ann Appl Biol 113:79–88

Davidson TMW (1980) Breeding for resistance to virus disease of the potato (*Solanum tuberosum*) at the Scottish Plant Breeding Station. In: Scottish Plant Breeding Station fifty-ninth annual report 1979–80. Scottish Plant Breeding Station, Pentlandfield, pp 100–108

Davidson MM, Butler RC, Wratten SD, Conner AJ (2004) Resistance of potato transgenic for a *cry*1Ac9 gene, to *Phthorimaea operculella* (Lepidoptera: Gelechiidae) over field seasons and between plant organs. Ann Appl Biol 145:271–277

Davie K, Pickup J (2018) Nematode pests of potatoes. In: Wale S (ed) Achieving sustainable cultivation of potatoes volume 2: production, storage and crop protection. Burleigh Dodds Science, Cambridge, pp 263–283

Davies LJ, Brown CR, Elling AA (2015) Calcium is involved in the R_{Mc1blb}-mediated hypersensitive response against *Meloidogyne chitwoodi* in potato. Plant Cell Rep 34(1):167–177

Davis JA, Radcliffe EB, Ragsdale DW (2007) Resistance to green peach aphid, *Myzus persicae* (Sulzer), and potato aphid, *Macrosiphum euphorbiae* (Thomas), in potato cultivars. Am J Potato Res 84:259–269

Day PR (1974) Genetics of host-parasite interaction. Freeman, San Francisco, 238p

Dianqiu LV, Xiaobao Y (2018) CRISPR, a potential powerful weapon to fight potato spindle tuber viroid. J Pathogen Res 1(1):1

Dinh PTY, Zhang L, Mojtahedi H, Brown CR, Elling AA (2015) Broad *Meloidogyne* resistance in potato based on RNA interference of effector gene *16D10*. J Nematol 47(1):71–78

Douches DS, Grafius EJ (2005) Transformation for insect resistance. In: Razdan MK, Mattoo AK (eds) Genetic improvement of Solanaceous crops volume I: potato. Science Publishers, Enfield, pp 235–266

Douches DS, Pett W, Santos F, Coombs J, Grafius E, Li W, Metry EA, Nasr El-Din T, Madkour M (2004) Field and storage testing *Bt* potatoes for resistance to potato tuberworm (Lepidoptera: Gelichiidae). J Econ Entomol 97(4):1425–1431

Douches DS, Coombs JJ, Lacey LA, Felcher KJ, Pett WL (2011) Evaluations of transgenic potatoes for resistance to potato tuberworm in the laboratory and field. Am J Potato Res 88:91–95. https://doi.org/10.1007/s12230-010-9167-y

Draaistra J (2006) Genetic analysis of root-knot nematode resistance in potato. Thesis Wageningen University, The Netherlands

Du J, Verzaux E, Chaparro-Garcia A, Bijsterbosch G, Keizer LCP, Zhou J, Liebrand TWH, Xie C, Govers F, Robatzek S, van der Vossen EAG, Jacobsen E, Visser RGF, Kamoun S, Vleeshouwers VGAA (2015) Elicitin recognition confers enhanced resistance to *Phytophthora infestans* in potato. Nat Plants 1:15034. https://doi.org/10.1038/nplants.2015.34

Duncan DR, Hammond D, Zalewski J, Cudnohufsky J, Kaniewski W, Thornton M, Bookout JT, Lavrik P, Rogan GJ, Feldman-Riebe J (2002) Field performance of transgenic potato, with resistance to Colorado potato beetle and viruses. HortScience 37(2):275–276. https://doi.org/10.21273/HORTSCI.37.2.275

Dunnett JM (1960) Inheritance of resistance to potato root eelworm in a breeding line stemming from *Solanum multidissectum* Hawkes. Report of the Scottish Plant Breeding Station 1960, pp 39–46

Dupuis B, Bragard C, Schumpp O (2019) Resistance of potato cultivars as a determinant factor of *Potato virus Y* (PVY) epidemiology. Potato Res 62(3):123–138

Dziewońska MA, Waś M (1994) Diploid genotype DW.84-1457, highly resistant to potato leafroll virus (PLRV). Potato Res 37:217–224

Ellenby C (1948) Resistance to the potato root eelworm. Nature 162:704

Ellenby C (1952) Resistance to the potato root eelworm, *Heterodera rostochiensis* Wollenweber. Nature 170:1016

Enciso-Rodriguez F, Douches D, Lopez-Cruz M, Coombs J, de los Campos G (2018) Genomic selection for late blight and common scab resistance in tetraploid potato (*Solanum tuberosum*). Genes Genomes Genetics 8(7):2471–2481. https://doi.org/10.1534/g3.118.200273

Errampalli D, Saunders JM, Holley JD (2001) Emergence of silver scurf (*Helminthosporium solani*) as an economically important disease of potato. Plant Pathol 50(2):141–153. https://doi.org/10.1046/j.1365-3059.2001.00555.x

Falloon RE (2008) Control of powdery scab of potato: towards integrated disease management. Am J Pot Res 85:253–260. https://doi.org/10.1007/s12230-008-9022-6

Fernow KH, Peterson LC, Plaisted RL (1970) Spindle tuber virus in seeds and pollen of infected plants. Am Potato J 47:75–80

Flis B, Hennig J, Strzelczyk-Zyta D, Gebhardt C, Marczewski W (2005) The *Ry-fsto* gene from *Solanum stoloniferum* for extreme resistance to *Potato virus Y* maps to potato chromosome XII and is diagnosed by PCR marker GP122718 in PVY resistant potato cultivars. Mol Breed 15:95–101

Flor HH (1942) Inheritance of pathogenicity in *Melampsora lini*. Phytopathology 32:653–659

Forbes GA (2012) Using host resistance to manage potato late blight with particular reference to developing countries. Potato Res 55:205–216

Foster SJ, Park T-H, Pel MA, Brigneti G, Sliwka J, Jagger L, van der Vossen E, Jones JDG (2009) *Rpi-vnt1.1*, a *Tm-2²* homolog from *Solanum venturii*, confers resistance to potato late blight. Mol Plant Microbe Interact 22:589–600

Fournet S, Eoche-Bosy D, Kerlan M-C, Grenier E, Montarry J (2018) Phenotypic and genomic modifications associated with *Globodera pallida* adaptation to potato resistances. Potato Res 61:65–71

Franco-Lara L, Barker H (1999) Characterisation of resistance to potato leafroll virus accumulation in *Solanum phureja*. Euphytica 108:137–144

Fry W (2008) *Phytophthora infestans* – the plant (and *R* gene) destroyer. Mol Plant Pathol 9:385–402

Fuglie KO (2007) Priorities for potato research in developing countries: results of a survey. Am J Potato Res 84:353–365

Fulladolsa AC, Charkowski A, Cai X, Whitworth J, Gray S, Jansky S (2019) Germplasm with resistance to *Potato virus Y* derived from *Solanum chacoense*: clones M19 (39–7) and M20 (XD3). Am J Potato Res 96(4):390–395

Fuller JM, Howard HW (1974) Breeding for resistance to the white potato cyst nematode, *Heterodera pallida*. Ann Appl Biol 77:121–128

Gallego JR, Gamez M, Cabello T (2020) Potential of the *Blattisocius mali* mite (Acari: Blattisociidae) as biological control agent of potato tubermoth (Lepidoptera: Gelechiidae) in stored potatoes. Potato Res 63(2):241–251

Gao Y (2018) Potato tuberworm: impact and methods for control – mini review. CAB Reviews 2018 13, No. 022

Gebhardt C, Mugniery D, Ritter E, Salamini F, Bonnel E (1993) Identification of RFLP markers closely linked to the H_1 gene conferring resistance to *Globodera rostochiensis* in potato. Theor Appl Genet 85:541–544

GILB (1999) Late blight: a threat to global food security. In: Proceedings of the global initiative on late blight conference, March 16–19, 1999, Quito

Goodwin SB, Drenth A (1997) Origin of the A2 mating type of *Phytophthora infestans* outside Mexico. Phytopathology 87:992–999

Goss EM, Tabima JF, Cooke DEL, Restrepo S, Fry WE, Forbes GA, Fieland VJ, Cardenas M, Grünwald NJ (2014) The Irish potato famine pathogen *Phytophthora infestans* originated in Central Mexico rather than the Andes. PNAS 111:8791–8796

Goverse A, Struik PC (2009) Debate on the exploitation of natural plant diversity to create late blight resistance in potato. Potato Res 52:265–271

Graebner RC, Brown CR, Ingham RE, Hagerty CH, Mojtahedi H, Quick RA, Hamlin LL, Wade N, Bamberg JB, Sathuvalli V (2018) Resistance to *Meloidogyne chitwoodi* identified in wild potato species. Am J Potato Res 95(6):679–686

Groth J, Song Y, Kellermann A, Schwarzfischer A (2013) Molecular characterisation of resistance against potato wart races 1, 2, 6 and 18 in a tetraploid population of potato (*Solanum tuberosum* subsp. *tuberosum*). J Appl Genet 54:169–178

Guchi E (2015) Disease management practice on potato (*Solanum tuberosum* L.) in Ethiopia. World J Agric Res 3(1):34–42

Hackett CA, Bradshaw JE, Bryan GJ (2014) QTL mapping in autotetraploids using SNP dosage information. Theor Appl Genet 127:1885–1904. https://doi.org/10.1007/s00122-014-2347-2. Epub 2014 Jul 1

Hackett CA, Boskamp B, Vogogias A, Preedy K, Milne I (2017) TetraploidSNPMap: software for linkage analysis and QTL mapping in autotetraploid populations using SNP dosage data. J Hered 108:438–442

Hämäläinen JH, Watanabe KN, Valkonen JPT, Arihara A, Plaisted RL, Pehu E, Miller L, Slack SA (1997) Mapping and marker-assisted selection for a gene for extreme resistance to potato virus Y. Theor Appl Genet 94:192–197

Hämäläinen JH, Sorri VA, Watanabe KN, Gebhardt C, Valkonen JPT (1998) Molecular examination of a chromosome region that controls resistance to potato Y and A potyviruses in potato. Theor Appl Genet 96:1036–1043

Haverkort AJ, Boonekamp PM, Hutten R, Jacobsen E, Lotz LAP, Kessel GJT, Vossen JH, Visser RGF (2016) Durable late blight resistance in potato through dynamic varieties obtained by cisgenesis: scientific and societal advances in the DuRPh project. Potato Res 59:35–66

Hehl R, Faurie E, Hesselbach J, Salamani F, Whitham S, Baker B, Gebhardt C (1999) TMV resistance gene *N* homologues are linked to *Synchytrium endobioticum* resistance in potato. Theor Appl Genet 98:379–386

Hein I, Birch PRJ, Danan S, Lefebvre V, Odeny DA, Gebhardt C, Trognitz F, Bryan GJ (2009) Progress in mapping and cloning qualitative and quantitative resistance against *Phytophthora infestans* in potato and its wild relatives. Potato Res 52:215–227

Hohl HR, Iselin K (1984) Strains of *Phytophthora infestans* with A2 mating type behaviour. Trans Br Mycol Soc 83:529–530

Howard HW, Cole CS, Fuller JM (1970) Further sources of resistance to *Heterodera rostochiensis* Woll. in the Andigena potato. Euphytica 19:210–216

Huang S, Vleeshouwers VG, Werij JS, Hutten RC, van Eck HJ, Visser RG, Jacobsen E (2004) The *R3* resistance to *Phytophthora infestans* in potato is conferred by two closely linked *R* genes with distinct specificities. Mol Plant Microbe Interact 17:428–435

Huang S, van der Vossen EA, Kuang H, Vleeshouwers VG, Zhang N, Borm TJ, van Eck HJ, Baker B, Jacobsen E, Visser RG (2005) Comparative genomics enabled the isolation of the *R3a* late blight resistance gene in potato. Plant J 42:251–261

Huet G (2014) Breeding for resistances to *Ralstonia solanacearum*. Front Plant Sci 5:715. https://doi.org/10.3389/fpls.2014.00715

Huijsman CA (1960) Some data on the resistance against the potato root-eelworm (*Heterodera rostochiensis* W.) in *Solanum kurtzianum*. Euphytica 9:185–190

Jacobs JME, van Eck HJ, Horsman K, Arens PFP, Verkerk-Bakker B, Jacobsen E, Pereira A, Stiekema WJ (1996) Mapping of resistance to the potato cyst nematode *Globodera rostochiensis* from the wild potato species *Solanum vernei*. Mol Breed 2:51–60

Janse J, Beld D, van Den H (2004) Introduction to Europe of *Ralstonia solanacearum* biovar 2, race 3 in *Pelargonium zonale* cuttings. J Plant Pathol 86:147–155

Jansky S (2000) Breeding for disease resistance in potato. Plant Breed Rev 19:69–155

Jansky SH (2009) Identification of Verticillium wilt resistance in U.S. potato breeding programs. Am J Potato Res 86:504–512. https://doi.org/10.1007/s12230-009-9107-x

Janssen R, Bakker J, Gommers FJ (1991) Mendelian proof for a gene-for-gene relationship between virulence of *Globodera rostochiensis* and the H_1 resistance gene in *Solanum tuberosum* ssp. *andigena* CPC 1673. Revue Nematol 14:213–219

Janssen GJW, von Norel A, Verkeerk-Bakker B, Janssen R (1996) Resistance to *Meloidogyne chitwoodi, M fallax* and *M hapla* in wild tuber-bearing *Solanum* spp. Euphytica 92:287–294

Jeffries C, Lacomme C (2018) Viruses affecting potatoes. In: Wale S (ed) Achieving sustainable cultivation of potatoes volume 2: production, storage and crop protection. Burleigh Dodds Science Publishing, Cambridge, pp 209–241

Jo KR (2013) Unveiling and deploying durability of late blight resistance in potato from natural stacking to cisgenic stacking. PhD thesis. Wageningen University, Wageningen, The Netherlands

Johnson DA, Geary B, Tsror L (2018) Potato black dot – the elusive pathogen, disease development and management. Am J Potato Res 95(4):340–350

Jones RAC (1979) Resistance to potato leaf roll virus in *Solanum brevidens*. Potato Res 22:149–152

Jones RAC (1982) Breakdown of *Potato virus X* resistance gene NX: selection of a group four strain from strain group three. Plant Pathol 31:325–331

Jones RAC (1985) Further studies on resistance-breaking strains of potato virus X. Plant Pathol 34(2):182–189. https://doi.org/10.1111/j.1365-3059.1985.tb01348.x

Jones RW, Perez FG (2019) Assessing possible mechanisms of resistance to early blight caused by *Alternaria solani*. Potato Res. https://doi.org/10.1007/s11540-019-9420-9

Jones JDG, Witek K, Verweij W, Jupe F, Cooke D, Dorling S, Tomlinson L, Smoker M, Perkins S, Foster S (2014) Elevating crop disease resistance with cloned genes. Phil Trans R Soc B 369:20130087. https://doi.org/10.1098/rstb.2013.0087

Judelson HS, Spilman LJ, Shattock RC (1995) Genetic mapping and non-Mendelian segregation of mating type loci in the Oomycete *Phytophthora infestans*. Genetics 141:503–512

Khaled W, Fekih IB, Nahdi S, Souissi R, Boukhris-Bouhachem S (2018) Transmission efficiency of *Potato leafroll virus* by four potato colonizing aphid species in Tunisian potato fields. Potato Res 61:89–96

Khorrami F, Mehrkhou F, Mahmoudian M, Ghosta Y (2018) Pathogenicity of three different entomopathogenic fungi, *Metarhizium anisopliae* IRAN 2252, *Nomuraea rileyi* IRAN 1020C and *Paecilomyces tenuipes* IRAN 1026C against the potato tuber moth, *Phthorimaea operculella* Zeller (Lepidoptera: Gelechiidae). Potato Res 61:297–308. https://doi.org/10.1007/s11540-018-9378-z

Kim HJ, Lee HR, Jo KR, Mahdi Mortazavian SM, Jan Huigen D, Evenhuis B, Kessel G, Visser RGF, Jacobsen E, Vossen JH (2012) Broad spectrum late blight resistance in potato differential set plants MaR8 and MaR9 is conferred by multiple stacked R genes. Theor Appl Genet 124:923–935. https://doi.org/10.1007/s00122-011-1757-7

Kort J, Ross H, Rumpenhorst HJ, Stone AR (1977) An international scheme to identifying and classifying pathotypes of potato cyst-nematodes *Globodera rostochiensis* and *G. pallida*. Nematologica 23:333–339

Kouassi AB, Kerlan M-C, Caromel B, Dantec J-P, Fouville D, Manzanares-Dauleux M, Ellisseche D, Mugniery D (2006) A major gene mapped on chromosome XII is the main factor of a quantitatively inherited resistance to *Meloidogyne fallax* in *Solanum sparsipilum*. Theor Appl Genet 112:699–707. https://doi.org/10.1007/s00122-005-0173-2

Kreike CM, De Koning JRA, Vinke JH, Van Ooijen JW, Stiekema WJ (1994) Quantitatively inherited resistance to *Globodera pallida* is dominated by one major locus in *Solanum spegazzinii*. Theor Appl Genet 88:764–769

Kriel CJ, Jansky SH, Gudmestad NC, Ronis DH (1995) Immunity species to *Clavibacter michiganensis* subsp. *sepedonicus*: inheritance of immunity in *Solanum acaule*. Euphytica 82(2):133–139

Kroschel J, Schaub B (2013) Biology and ecology of potato tuber moths as major pests of potato. In: Giordanengo P, Vincent C, Alyokhin A (eds) Insect pests of potato: global perspectives on biology and management. Elsevier, Oxford, pp 165–192

Kroschel J, Mujica N, Okonya J, Alyokhin A (2020) Insect pests affecting potatoes in tropical, subtropical, and temperate regions. In: Campos H, Ortiz O (eds) The potato crop: its agricultural, nutritional and social contribution to humankind. Springer, ebook, pp 251–306

Kuhl JC, Hanneman RE Jr, Havey MJ (2001) Characterization and mapping of *Rpi1*, a late-blight resistance locus from diploid (1EBN) Mexican *Solanum pinnatisectum*. Mol Gen Genet 265:977–985

Kumar M, Singh JK, Kumar S, Kumar A (2017) A comprehensive overview on black scurf of potato. Int J Curr Microbiol App Sci 6(10):4981–4994

Lebecka R (2017) Screening for potato resistance to blackleg and soft rot. Plant Breed Seed Sci 75:97–104. https://doi.org/10.1515/plass-2017-0013

Lebecka R, Kistowski M, Dębski J, Szajko K, Murawska Z, Marczewski W (2019) Quantitative proteomic analysis of differentially expressed proteins in tubers of potato plants differing in resistance to *Dickeya solani*. Plant Soil 441:317–329. https://doi.org/10.1007/s11104-019-04125-7

Lees AK (2018) Fungal diseases affecting potato storage. In: Wale S (ed) Achieving sustainable cultivation of potatoes volume 2: production, storage and crop protection. Burleigh Dodds Science Publishing, Cambridge, pp 179–188

Lees AK, Bradshaw JE (2001) Inheritance of resistance to *Fusarium sulphureum* in crosses between *S. tuberosum* potato cultivars measured on field and glasshouse grown tubers. Potato Res 44:147–152

Lees AK, De Maine MJ, Nicolson MJ, Bradshaw JE (2000) Long-day-adapted *Solanum phureja* as a source of resistance to blackleg caused by *Erwinia carotovora* subsp. *atroseptica*. Potato Res 43:279–285

Lees AK, Brierley JL, Stewart JA, Hilton AJ, Wale SJ, Gladders P, Bradshaw NJ, Peters JC (2010) Relative importance of seed-tuber and soilborne inoculum in causing black dot disease of potato. Plant Pathol 59:693–702

Lees AK, Stewart JA, Lynott JS, Carnegie SF, Campbell H, AMI R (2012) The effect of a dominant *Phytophthora infestans* genotype (*13_A2*) in Great Britain on host resistance to foliar late blight in commercial potato cultivars. Potato Res 55:125–134

Leonards-Schippers C, Gieffers W, Salamini F, Gebhardt C (1992) The *R1* gene conferring race-specific resistance to *Phytophthora infestans* in potato is located on potato chromosome V. Mol Gen Genet 233:278–283

Li X, van Eck HJ, Rouppe van der Voort JNAM, Huigen DJ, Stam P, Jacobsen E (1998) Autotetraploids and genetic mapping using common AFLP markers: the *R2* allele conferring resistance to *Phytophthora infestans* mapped on potato chromosome 4. Theor Appl Genet 96:1121–1128

Li C, Wang J, Chien DH, Chujoy E, Song B, VanderZaag P (2010) Cooperation-88: a high yielding, multi-purpose, late blight resistant cultivar growing in Southwest China. Am J Potato Res 88:190–194

Li G, Huang S, GuoX LY, YangY GZ, Kuang H, Rietman H, Bergervoet M, Vleeshouwers VGAA, van der Vossen E, Qu D, Visser RGA, Jacobsen E, Vossen JH (2011) Cloning and characterization of R3b; members of the R3 superfamily of late blight resistance genes show sequence and functional divergence. Mol Plant Microbe Interact 24:1132–1142

Li H, Wang Z, Hu X, Shang W, Shen R, Guo C, Guo Q, Subbarao KV (2019) Assessment of resistance in potato cultivars to Verticillium wilt caused by *Verticillium dahliae* and *Verticillium nonalfalfae*. Plant Dis 103(6):1357–1362. https://doi.org/10.1094/PDIS-10-18-1815-RE

Lima MF, Lopes CA, de Melo PE (1996) Selection of potato genotypes at the seedling stage for resistance to bacterial wilt. Pesq Agrop Brasileira 31(4):249–257

Liu Z, Halterman D (2006) Identification and characterization of RB-orthologous genes from the late blight resistant wild potato species *Solanum verrucosum*. Phys Mol Plant Pathol 69:230–239

Lokossou AA, Park T-H, van Arkel G, Arens M, Ruyter-Spira C, Morales J, Whisson SC, Birch PRJ, Visser RGF, Jacobsen E, van der Vossen EAG (2009) Exploiting knowledge of *R/Avr* genes to rapidly clone a new LZ-NBS-LRR family of late blight resistance genes from potato linkage group IV. Mol Plant-Microbe Interact 22:630–641

Lopes CA, Buso JA, Accatino P (1993) Screening CIP potato germplasm for resistance to bacterial wilt in Brazil: methods and preliminary results. Bacterial Wilt Newsl 9:3–5

Luthra SK, Gupta VK, Tiwari JK, Kumar V, Bhardwaj V, Sood S, Dalamu, Kaur RP, Kumar R, Vanishree G, Kumar D, Mhatre P, Chakrabarti SK (2020) Potato breeding in India. CPRI technical bulletin no 74 (revised), ICAR-Central Potato Research Institute, Shimla, Himachal Pradesh, India

Maharijaya A, Vosman B (2015) Managing the Colorado potato beetle; the need for resistance breeding. Euphytica 204(3):487–501

Malcolmson JF (1969) Races of *Phytophthora infestans* occurring in Great Britain. Trans Br Mycol Soc 53:417–423

Malcolmson JF (1976) Assessment of field resistance to blight (*Phytophthora infestans*) in potatoes. Trans Br Mycol Soc 67:321–325

Malcolmson JF, Black W (1966) New R genes in *Solanum demissum* Lindl. and their complementary races of *Phytophthora infestans* (Mont.) de Bary. Euphytica 15:199–203

Marczewski W, Hennig J, Gebhardt C (2002) The *Potato virus S* resistance gene *Ns* maps to potato chromosome VIII. Theor Appl Genet 105:564–657

Marczewski W, Strzelczyk-Zyta D, Hennig J, Witek K, Gebhardt C (2006) Potato chromosomes IX and XI carry genes for resistance to potato virus M. Theor Appl Genet 112:1232–1238

Martin FN, Zhang Y, Cooke DEL, Coffey MD, Grünwald NJ, Fry WE (2019) Insights into evolving global populations of *Phytophthora infestans* via new complementary mtDNA haplotype markers and nuclear SSRs. PLoS One 14(1):1–24. https://doi.org/10.1371/journal.pone.0208606

McGrath JM, Williams CE, Haberlach GT, Wielgus SM, Uchytil TF, Helgeson JP (2002) Introgression and stabilization of *Erwinia* tuber soft rot resistance into potato after somatic hybridization of *Solanum tuberosum* and *S. brevidens*. Am J Potato Res 79(1):19–24

Meade F, Hutten R, Wagener S, Prigge V, Dalton E, Kirk HG, Griffin D, Milbourne D (2020) Detection of novel QTLs for late blight resistance derived from the wild potato species *Solanum microdontum* and *Solanum pampasense*. Genes 11:732. https://doi.org/10.3390/genes11070732

Merz U, Falloon RE (2017) Proceedings of the 3rd international powdery scab workshop, Einsiedeln, Switzerland, July 18–21, 2016. Potato Res 60:195–215

Merz U, Lees AK, Sullivan L, Schwärzel R, Hebeisen T, Kirk HG, Bouchek-Mechiche K, Hofferbert HR (2012) Powdery scab resistance in *Solanum tuberosum*: an assessment of cultivar × environment effect. Plant Pathol 61(1):29–36. https://doi.org/10.1111/j.13653059.2011.02489.x

Mihovilovich E, Alarcón L, Pérez AL, Alvarado J, Arellano C, Bonierbale M (2007) High levels of heritable resistance to Potato leafroll virus (PLRV) in *Solanum tuberosum* subsp. *andigena*. Crop Sci 47:1091–1103. https://doi.org/10.2135/cropsci2006.06.0369

Minnis ST, Haydock PPJ, Ibrahim SK, Grove IG (2002) Potato cyst nematodes in England and Wales-occurrence and distribution. Ann Appl Biol 140:187–195

Mohammed A, Douches DS, Pett W, Grafius E, Coombs J, Liswidowati W, Li W, Madkour MA (2000) Evaluation of potato tuber moth (Lepidoptera: Gelechiidae) resistance in tubers of *Bt-cry5* transgenic potato lines. J Econ Entomol 93:472–476

Mohan SK, Davis JR, Corsini DL, Sorensen LH, Pavek JJ (1990) Reaction of potato clones and accessions of *Solanum* spp. to *Verticillium dahliae* Kleb. and its toxin. Potato Res 33(4):449–458

Moreira A, Jones RAC, Fribourg CE (1980) Properties of a resistant-breaking strain of potato virus X. Ann Appl Biol 95:93–103

Mori K, Mukojima N, Nakao T, Tamiya S, Sakamoto Y, Sohbaru N, Hayashi K, Watanuki H, Nara K, Yamazaki K, Ishii T, Hosaka K (2012) Germplasm release: Saikai 35, a male and female fertile breeding line carrying *Solanum phureja*-derived cytoplasm and potato cyst nematode resistance (*H1*) and *Potato virus Y* resistance (*Rychc*) genes. Am J Potato Res 89:63–72

Mujica N, Kroschel J (2013) Pest intensity-crop loss relationships for the leafminer fly *Liriomyza huidobrensis* (Blanchard) in different potato (*Solanum tuberosum* L.) varieties. Crop Prot 47:6–16

Muller KO, Black W (1951) Potato breeding for resistance to blight and virus diseases during the last hundred years. Z Pflanzenzuchtung 31:305–318

Mundt CC (2014) Durable resistance: a key to sustainable management of pathogens and pests. Infect Genet Evol 27:446–455. https://doi.org/10.1016/j.meegid.2014.01.011

Nachmias A, Caligari PDS, Mackay GR, Livescu L (1988) The effects of *Alternaria solani* and *Verticillium dahliae* on potatoes growing in Israel. Potato Res 31:443–450

Naess SK, Bradeen JM, Wielgus SM, Haberlach GT, McGrath JM, Helgeson JP (2000) Resistance to late blight in *Solanum bulbocastanum* is mapped to chromosome 8. Theor Appl Genet 101:697–704

Nally DP (2011) Human encumbrances. University of Notre Dame Press, Notre Dame, 348p

Navarro FM, Rak KT, Banks E, Bowen BD, Higgins C, Palta JP (2015) Strategies for selecting stable common scab resistant clones in a potato breeding program. Am J Potato Res 92:326–338

NIAB TAG (2019) Potatoes 2019. NIAB, Cambridge

Obidiegwu JE, Flath K, Gebhardt C (2014) Managing potato wart: a review of present research status and future perspective. Theor Appl Genet 127(4):763–780. https://doi.org/10.1007/s00122-014-2268-0

Ohbayashi K (2019) The *Rx* gene derived USDA 41956 and *Rx1* gene derived CPC 1673 confer equal resistance to the migration of *Potato virus X* from potato leaves to tubers. Euphytica 215:90

Orłowska E, Basile A, Kandzia I, Llorente B, Kirk HG, Cvitanich C (2012) Revealing the importance of meristems and roots for the development of hypersensitive responses and full foliar resistance to *Phytophthora infestans* in the resistant potato cultivar Sarpo Mira. J Exp Bot 63:4765–4779

Ortiz R (2001) The state of the use of potato genetic diversity. In: Cooper HD, Spillane C, Hodgkin T (eds) Broadening the genetic base of crop production. CAB International, Wallingford, pp 181–200

Ortiz R, Iwanaga M, Raman KV, Palacios M (1990) Breeding for resistance to potato tuber moth, *Phthorimaea operculella* (Zeller), in diploid potatoes. Euphytica 50(2):119–125

Paal J, Henselewski H, Muth J, Meksem K, Menéndez CM, Salamini F, Ballvora A, Gebhardt C (2004) Molecular cloning of the potato Gro1-4 gene conferring resistance to pathotype Ro1 of the root cyst nematode Globodera rostochiensis, based on a candidate gene approach. Plant J 38(2):285–297

Park TH, Gros J, Sikkema A, Vleeshouwers VG, Muskens M, Allefs S, Jacobsen E, Visser RG, van der Vossen EAG (2005a) The late blight resistance locus *Rpi-blb3* from *Solanum bulbocastanum* belongs to a major late blight *R* gene cluster on chromosome 4 of potato. Mol Plant Microb Interact 18:722–729

Park TH, Vleeshouwers VG, Huigen DJ, van der Vossen EAG, van Eck HJ, Visser RG (2005b) Characterization and high-resolution mapping of a late blight resistance locus similar to *R2* in potato. Theor Appl Genet 111:591–597

Park J, Hackett CA, Dandurand L-M, Wang X, Jong WS (2019) QTL for resistance to *Globodera rostochiensis* pathotype Ro2 and *G. pallida* pathotype Pa2/3 in autotetraploid potato. Am J Potato Res. https://doi.org/10.1007/s12230-019-09745-4

Parrott DM (1981) Evidence for a gene-for-gene relationship between resistance gene H1 from *Solanum tuberosum* ssp. *andigena* and a gene in *Globodera rostochiensis*, and between H2 from *S. multidissectum* and a gene in *G. pallida*. Nematologica 27:372–382

Parsa S, Ccanto R, Olivera E, Scurrah M, Alcázar J, Rosenheim JA (2012) Explaining Andean potato weevils in relation to local and landscape features: a facilitated ecoinformatics approach. PLoS One 7(5):e36533

Patil VU, Gopal J, Singh BP (2012) Improvement for bacterial wilt resistance in potato by conventional and biotechnological approaches. Agric Res 1(4):299–316. https://doi.org/10.1007/s40003-012-0034-6

Pavek JJ, Corsini DL (1994) Inheritance of resistance to warm-growing-season fungal diseases. In: Bradshaw JE, Mackay GR (eds) Potato genetics. CAB International, Wallingford, pp 403–409

Pel MA, Foster SJ, Park TH, Rietman H, van Arkel G, Jones JDG, Van Eck HJ, Jacobsen E, Visser RGF, Van der Vossen EAG (2009) Mapping and cloning of late blight resistance genes from *Solanum venturii* using an interspecific candidate gene approach. Mol Plant Microbe Interact 22:601–615

Perry RN, Moens M, Jones JT (2018) Cyst nematodes. CABI, Wallingford. 488p

Phillips MS (1994) Inheritance of resistance to nematodes. In: Bradshaw JE, Mackay GR (eds) Potato genetics. CAB International, Wallingford, pp 319–337

Phillips MS, Forrest JMS, Wilson LA (1980) Screening for resistance to potato cyst nematodes using closed containers. Ann Appl Biol 96:317–322. https://doi.org/10.1111/j.1744-7348.1980.tb04782.x

Plaisted RL, Tingey WM, Steffens JC (1992) The germplasm release of NYL 235-4, a clone with resistance to the Colorado potato beetle. Am Potato J 69(12):843–846

Plich J, Przetakiewicz J, Śliwka J, Flis B, Wasilewicz-Flis I, Zimnoch-Guzowska E (2018) Novel gene *Sen2* conferring broad-spectrum resistance to *Synchytrium endobioticum* mapped to potato chromosome XI. Theor Appl Genet 131(11):2321–2331

Priegnitz U, Lommen WJM, van der Vlugt RAA, Struik PC (2019) Impact of positive selection on incidence of different viruses during multiple generations of potato seed tubers in Uganda. Potato Res 62:1–30

Priou S, Gutarra L, Aley P (1999) Highly sensitive detection of *Ralstonia solanacearum* in latently infected potato tubers and soil by post-enrichment ELISA on nitrocellulose membrane. EPPO/OEPP Bull 29:117–125

Przetakiewicz J (2008) Assessment of the resistance of potato cultivars to *Synchytrium endobioticum* (Schilb.) Perc. In: Poland – EPPO Bull, vol 38, pp 211–215

Radcliffe EB, Lagnaoui A (2007) Insect pests of potato. In: Vreugdenhil D (ed) Potato biology and biotechnology advances and perspectives. Elsevier, Oxford, pp 543–567

Rashidi M, Novy RG, Wallis CM, Rashed A (2017) Characterization of host plant resistance to zebra chip disease from species-derived potato genotypes and the identification of new sources of zebra chip resistance. PLoS One 12(8):e0183283. https://doi.org/10.1371/journal.pone.0183283

Rauscher GM, Smart CD, Simko I, Bonierbale M, Mayton H, Greenland A, Fry WE (2006) Characterization and mapping of *Rpi-ber*, a novel potato late blight resistance gene from *Solanum berthaultii*. Theor Appl Genet 112:674–687

Rietman H, Bijsterbosch G, Cano LM, Lee HR, Vossen JH, Jacobsen E, Visser RGF, Kamoun S, Vleeshouwers VGAA (2012) Qualitative and quantitative late blight resistance in the potato cultivar Sarpo Mira is determined by the perception of five distinct RXLR effectors. Mol Plant Microbe Interact 25:910–919

Ristić D, Vučurović I, Kuzmanović S, Pfaf-Dolovac E, Aleksić G, Vučurović A, Starović M (2019) The incidence and genetic diversity of *Potato virus S* in Serbian seed potato crops. Potato Res 62:31–46

Ross H (1962) Über die Vererbung der Resistenz gegen den Kartoffelnematoden (*Heterodera rostochiensis* Woll.) in Kreuzungen von *Solanum famatinae* Bitt. Et Wittm. mit *Solanum tuberosum* L. und mit *S. chacoense* Bitt. Der Züchter 32:74–80

Ross H (1986) Potato breeding – problems and perspectives. Advs in Plant Breed 13, Paul Parey, Berlin/Hamburg

Rouppe van der Voort J, Wolters P, Folkertsma R, Hutten R, van Zandvoort P, Vinke H, Kanyuka K, Bendahmane A, Jacobsen E, Janssen R, Bakker J (1997) Mapping of the cyst nematode resistance locus *Gpa2* in potato using a strategy based on comigrating AFLP markers. Theor Appl Genet 95:874–880

Rouppe van der Voort J, Lindeman W, Folkertsma R, Hutten R, Overmars H, van der Vossen E, Jacobsen E, Bakker J (1998) A QTL for broad-spectrum resistance to cyst nematode species (Globodera spp.) maps to a resistance gene cluster in potato. Theor Appl Genet 96:654–661

Rouppe van der Voort J, van der Vossen E, Bakker E, Overmars H, van Zandroort P, Hutten R, Lankhorst RK, Bakker J (2000) Two additive QTLs conferring broad-spectrum resistance in potato to *Globodera pallida* are localized on resistance gene clusters. Theor Appl Genet 101:1122–1130

Roush RT (1998) Two-toxin strategies for management of insecticidal transgenic crops: can pyramiding succeed where pesticide mixtures have not? Phil Trans R Soc Lond B Biol Sci 353(1376):1777–1786. https://doi.org/10.1098/rstb.1998.0330

Salaman RN (1921) Degeneration of potatoes. Report on the International potato conference. Royal Horticultural Society, London, pp 79–91

Sandbrink JM, Colon LT, Wolters PJCC, Stiekema WJ (2000) Two related genotypes of *Solanum microdontum* carry different segregating alleles for field resistance to *Phytophthora infestans*. Mol Breed 6:215–225

Sato M, Nishikawa K, Komura K, Hosaka K (2006) Potato virus Y resistance gene, *Rychc*, mapped to the distal end of potato chromosome 9. Euphytica 149:367–372

Seenivasan N (2020) Management of potato cyst nematodes using liquid bio-formulations of *Pseudomonas fluorescens, Purpureocillium lilacinum* and *Trichoderma viride*. Potato Res 63(4):479–496

Shaner G, Finney RE (1977) The effect of nitrogen fertilization on the expression of slow-mildewing resistance in Knox wheat. Phytopathology 67:1051–1056

Simko I, Haynes KG (2017) Maturity-adjusted resistance of potato (*Solanum tuberosum* L.) cultivars to Verticillium wilt caused by *Verticillium dahliae*. Am J Potato Res 94:173–177

Simko I, Jansky S, Stephenson S, Spooner D (2007) Genetics of resistance to pests and disease. In: Vreugdenhil D (ed) Potato biology and biotechnology advances and perspectives. Elsevier, Oxford, pp 117–155

Simmonds NW (1969) Prospects of potato improvement. Scottish Plant Breeding Station Forty-Eighth Annual Report 1968-69, pp 18–38

Simmonds NW (1991) Genetics of horizontal resistance to diseases of crops. Biol Rev 66:189–241

Simmonds NW, Wastie RL (1987) Assessment of horizontal resistance to late blight of potatoes. Ann Appl Biol 111:213–221

Sjolund MJ, Kelly R, Saddler GS, Kenyon DM (2018) Bacterial diseases affecting potatoes. In: Wale S (ed) Achieving sustainable cultivation of potatoes volume 2: production, storage and crop protection. Burleigh Dodds Science Publishing, Cambridge, pp 189–208

Śliwka J, Jakuczun H, Lebecka R, Marczewski W, Gebhardt C, Zimnoch-Guzowska E (2006) The novel, major locus *Rpi-phu1* for late blight resistance maps to potato chromosome IX and is not correlated with long vegetation period. Theor Appl Genet 113:685–695

Smilde WD, Brigneti G, Jagger L, Perkins S, Jones JD (2005) *Solanum mochiquense* chromosome IX carries a novel late blight resistance gene *Rpi-moc1*. Theor Appl Genet 110:252–258

Soloman-Blackburn RM, Stewart HE, Bradshaw JE (2007) Distinguishing major-gene from field resistance to late blight (*Phytophthora infestans*) of potato (*Solanum tuberosum*) and selecting for high levels of field resistance. Theor Appl Genet 115:141–149

Solomon-Blackburn RM, Barker H (1993) Resistance to potato leafroll luteovirus can be greatly improved by combining two independent types of heritable resistance. Ann Appl Biol 122:329–336

Solomon-Blackburn RM, Barker H (2001a) A review of host major-gene resistance to potato viruses X, Y, A and V in potato: genes, genetics and mapped locations. Heredity 86:8–16. https://doi.org/10.1046/j.1365-2540.2001.00798.x

Solomon-Blackburn RM, Barker H (2001b) Breeding virus resistant potatoes (*Solanum tuberosum*): a review of traditional and molecular approaches. Heredity 86:17–35

Solomon-Blackburn RM, Barker H, Bradshaw JE, De Jong W (2003/4) Evidence that resistance to potato leafroll virus accumulation in tetraploid *Solanum tuberosum* L. is controlled by one or a few major genes that are not complementary. Potato Res 46:137–145

Solomon-Blackburn RM, Nikan J, Barker H (2008) Mechanism of strong resistance to Potato *leafroll virus* infection in a clone of potato (*Solanum tuberosum*). Ann Appl Biol 152:339–347

Song J, Bradeen JM, Naess SK, Raasch JA, Wielgus SM, Haberlach GT, Liu J, Kuang H, Austin-Phillips S, Buell CR, Helgeson JP, Jiang J (2003) Gene *RB* cloned from *Solanum bulbocastanum* confers broad spectrum resistance to potato late blight. Proc Natl Acad Sci USA 100:9128–9133

Song Y-S, Hepting L, Schweizer G, Hartl L, Wenzel G, Schwarzfischer A (2005) Mapping of extreme resistance to PVY (*Rysto*) on chromosome XII using anther-culture-derived primary dihaploid potato lines. Theor Appl Genet 111:879–887

Stewart HE, Bradshaw JE (1993) A glasshouse test for assessing resistance to early blight (*Alternaria solani*). Potato Res 36:35–42

Stewart HE, Bradshaw JE (2001) Assessment of the field resistance of potato genotypes with major gene resistance to late blight (*Phytophthora infestans* (Mont.) de Bary) using inoculum comprised of two complementary races of the fungus. Potato Res 44:41–52

Stewart HE, Wastie RL (1989) A rapid scoring technique for potato tuber disease assessment. Potato Res 32:353–357

Stewart HE, Flavelle PH, McCalmont DC, Wastie RL (1983a) Correlation between glasshouse and field tests for resistance to foliage blight caused by *Phytophthora infestans*. Potato Res 26:41–48

Stewart HE, McCalmont DC, Wastie RL (1983b) The effect of harvest date and the interval between harvest and inoculation on the assessment of the resistance of potato tubers to late blight. Potato Res 26:101–107

Stewart HE, Taylor K, Wastie RL (1983c) Resistance to late blight in foliage (*Phytophthora infestans*) of potatoes assessed as true seedlings and as adult plants in the glasshouse. Potato Res 26:363–366

Stewart HE, Bradshaw JE, Wastie RL, Mackay GR, Erlich O, Livescu L, Nachmias A (1994a) Assessing progenies of potato for resistance to early blight. Potato Res 37:257–269

Stewart HE, Bradshaw JE, Wastie RL (1994b) Correlation between resistance to late blight in foliage and tubers in potato clones from parents of contrasting resistance. Potato Res 37:429–434

Stewart HE, Bradshaw JE, Pande B (2003) The effect of the presence of R-genes for resistance to late blight (*Phytophthora infestans*) of potato (*Solanum tuberosum*) on the underlying level of field resistance. Plant Pathol 52:193–198

Strachan SM, Armstrong MR, Kaur A, Wright KM, Lim TY, Baker K, Jones J, Bryan G, Blok V, Hein I (2019) Mapping the *H2* resistance effective against *Globodera pallida* pathotype Pa1 in tetraploid potato. Theor Appl Genet 132:1283–1294. https://doi.org/10.1007/s00122-019-03278-4

Świeżynski KM, Domański L, Zarzycka H, Zimnoch-Guzowska E (2000) The reaction of potato differentials to *Phytophthora infestans* isolates collected in nature. Plant Breed 119:119–126

Szajko K, Yin Z, Marczewski W (2019) Accumulation of miRNA and mRNA targets in potato leaves displaying temperature-dependent responses to *Potato virus Y*. Potato Res 62(4):379–392

Tan MYA, Hutten RCB, Celis C, Park TH, Niks RE, Visser RGF, van Eck HJ (2008) The *Rpi-mcd1* locus from *Solanum microdontum* involved in resistance to *Phytophthora infestans*, causing a delay in infection, maps on potato chromosome 4 in a cluster of NBS-LRR genes. Mol Plant Microbe Interact 21:909–918

Tan MYA, Park T-H, Alles R, Hutten RCB, Visser RGF, van Eck HJ (2009) $GpaXI_{tar}^{l}$ originating from *Solanum tarijense* is a major resistance locus to *Globodera pallida* and is localised on chromosome 11 of potato. Theor Appl Genet 119(8):1477–1487. https://doi.org/10.1007/s00122-009-1149-4

Tegg RS, Wilson CR (2010) Relationship of resistance to common scab disease and tolerance to thaxtomin A toxicity within potato cultivars. Eur J Plant Pathol 128(2):143–148

Thresh JM (1998) In memory of James Edward Vanderplank 1909-1997. Plant Pathol 47:114–115

Tomczynska I, Stefanczyk E, Chmielarz M, Karasiewicz B, Kaminski P, Jones JD, Lees AK, Sliwka J (2014) A locus conferring effective late blight resistance in potato cultivar Sarpo Mira maps to chromosome XI. Theor Appl Genet 127:647–657

Toxopeus HJ (1964) Treasure-digging for blight resistance in potatoes. Euphytica 13:206–222

Toxopeus HJ, Huijsman CA (1953) Breeding for resistance to potato root eelworm. Euphytica 2:180–186

Trognitz BR, Bonierbale M, Landeo JA, Forbes G, Bradshaw JE, Mackay GR, Waugh R, Huarte MA, Colon L (2001) Improving potato resistance to disease under the global initiative on late blight. In: Cooper HD, Spillane C, Hodgkin T (eds) Broadening the genetic base of crop production. CAB International, Wallingford, pp 385–398

Tung PX, Zaag PV, Li C, Tang W (2018) Combining ability for foliar resistance to late blight [*Phytophthora infestans* (Mont.) de Bary] of potato cultivars with different levels of resistance. Am J Potato Res 95(6):670–678

Turner SJ (1989) New sources of resistance to potato cyst nematodes in the Commonwealth Potato Collection. Euphytica 42:145–153

Valkonen JPT (2015) Elucidation of virus-host interactions to enhance resistance breeding for control of virus diseases in potato. Breed Sci 65:69–76. https://doi.org/10.1270/jsbbs.65.69

Valluru R, Christ BJ, Haynes KG, Vinyard BT (2006) Inheritance and stability of resistance to fusarium tuber rot in tetraploid potatoes. Am J Potato Res 83(4):335–341

Van der Plank JE (1968) Disease resistance in plants. Academic Press, New York, 206p

Van der Vossen EAG, Sikkema A, Hekkert BTL, Gros J, Stevens P, Muskens M, Wouters D, Pereira A, Stiekema W, Allefs S (2003) An ancient R gene from the wild potato species *Solanum bulbocastanum* confers broad-spectrum resistance to *Phytophthora infestans* in cultivated potato and tomato. Plant J 36:867–882

Van der Vossen EAG, Gros J, Sikkema A, Muskens M, Wouters D, Wolters P, Pereira A, Allefs S (2005) The *Rpi-blb2* gene from *Solanum bulbocastanum* is an *Mi-1* gene homolog conferring broad-spectrum late blight resistance in potato. Plant J 44:208–222

van Eck HJ, Vos PG, Valkonen JPT, Uitdewilligen JGAML, Lensing H, de Vetten N, Visser RGF (2017) Graphical genotyping as a method to map *Ny(o,n)sto* and *Gpa5* using a reference panel of tetraploid potato cultivars. Theor Appl Genet 130(3):515–528. https://doi.org/10.1007/s00122-016-2831-y

Van Soest LJM, Rumpenhorst HJ, Huijsman CA (1983) Resistance to potato cyst-nematodes in tuber-bearing Solanum species and its geographical distribution. Euphytica 32:65–74

Verzaux E, Budding D, de Vetten N, Niks RE, Vleeshouwers VGAA, van der Vossen EAG, Jacobsen E, Visser RGF (2011) High resolution mapping of a novel late blight resistance gene *Rpi-avl1* from the wild Bolivian species *Solanum avilesii*. Am J Potato Res 88:511–519

Villamon FG, Spooner DM, Orrillo M, Mihovilovich E, Perez W, Bonierbale M (2005) Late blight resistance linkages in a novel cross of the wild potato species *Solanum paucissectum* (series Piurana). Theor Appl Genet 111:1201–1214

Vleeshouwers VG, Rietman H, Krenek P, Champouret N, Young C, Oh SK, Wang M, Bouwmeester K, Vosman B, Visser RG, Jacobsen E, Govers F, Kamoun S, van der Vossen EAG (2008) Effector genomics accelerates discovery and functional profiling of potato disease resistance and *Phytophthora infestans* avirulence genes. PLoS One 3:e2875

Vleeshouwers VGAA, Raffaele S, Vossen JH, Champouret N, Oliva R, Segretin ME, Rietman H, Cano LM, Lokossou AA, Kessel GJT, Pel M, Kamoun S (2011) Understanding and exploiting late blight resistance in the age of effectors. Annu Rev Plant Physiol Plant Mol Biol 49:507–531

Vossen JH, Nijenhuis M, Arens-de Reuver MJB, van der Vossen EAG, Jacobsen E, Visser RGF (2011) Cloning and exploitation of a functional *R* gene from *Solanum chacoense.* US patent WO/2011/034433 A1, published by the world intellectual property organization 18 September 2009

Vreugdenhil D, Bradshaw J, Gebhardt C, Govers F, MacKerron DKL, Taylor MA, Ross HA (eds) (2007) Potato biology and biotechnology advances and perspectives. Elsevier, Oxford, 823p

Wang M, Allefs S, van den Berg RG, Vleeshouwers VG, van der Vossen EAG, Vosman B (2008) Allele mining in *Solanum*: conserved homologues of *Rpi-blb1* are identified in *Solanum stoloniferum.* Theor Appl Genet 116:933–943

Wastie RL (1991) Breeding for resistance. Adv Plant Path 7:193–224

Wastie RL (1994) Inheritance of resistance to fungal diseases of tubers. In: Bradshaw JE, Mackay GR (eds) Potato genetics. CAB International, Wallingford, pp 411–427

Wastie RL, Bradshaw JE (1993) Inheritance of resistance to *Fusarium* spp. in tuber progenies of potato. Potato Res 36:189–193

Wastie RL, Bradshaw JE (1995) Comparison of resistance to *Fusarium* spp. of glasshouse- and field-grown tuber progenies of potato. Potato Res 38:345–351

Wastie RL, Caligari PDS, Stewart HE, Mackay GR (1987) A glasshouse progeny test for resistance to tuber blight (*Phytophthora infestans*). Potato Res 30:533–538

Wastie RL, Caligari PDS, Stewart HE, Mackay GR (1988) Assessing the resistance to gangrene of progenies of potato (*Solanum tuberosum* L.) from parents differing in susceptibility. Potato Res 31:355–365

Wastie RL, Stewart HE, Brown J (1989) Comparative susceptibility of some potato cultivars to dry rot caused by *Fusarium sulphureum* and *F. solani* var. *coeruleum.* Potato Res 32:49–55

Wastie RL, Mackay GR, Caligari PDS, Stewart HE (1990) A glasshouse progeny test for resistance to gangrene (*Phoma foveata*). Potato Res 33:131–133

Wastie RL, Stewart HE, Bradshaw JE, Lees AK (1997) Assessing progenies of potato for resistance to skin spot (*Polyscytalum pustulans*) in the glasshouse. Potato Res 40:383–389. https://doi.org/10.1007/BF02357997

Whisson SC, Avrova AO, Boevink PC, Armstrong MR, Seman ZA, Hein I, Birch PRJ (2011) Exploiting knowledge of pathogen effectors to enhance late blight resistance in potato. Potato Res 54:325–340. https://doi.org/10.1007/s11540-011-9197-y

White S, Shaw D (2009) The usefulness of late-blight resistant Sárpo cultivars-a case study. Acta Hortic (834):161–166

Wilson CR, Jones RAC (1992) Resistance to phloem transport of potato leafroll virus in potato plants. J Gen Virol 73:3219–3224

Wolters PJ, de Vos L, Bijsterbosch G, Woudenberg JHC, Visser RGF, van der Linden G, Vleeshouwers VGAA (2019) A rapid method to screen wild *Solanum* for resistance to early blight. Eur J Plant Pathol 154(1):109–114. https://doi.org/10.1007/s10658-019-01741-y

Xue W, Haynes KG, Qu X (2019) Characterization of early blight resistance in potato cultivars. Plant Dis 103(4):629–637. https://doi.org/10.1094/PDIS-05-18-0794-RE

Yencho GC, Kowalski SP, Kennedy GG, Sanford LL (2000) Segregation of leptine glycoalkaloids and resistance to Colorado potato beetle (*Leptinotarsa decemlineata* (say)) in F2 *Solanum tuberosum* (4x) × *S. chacoense* (4x) potato progenies. Am J Potato Res 77:167. https://doi.org/10.1007/BF02853941

Yoshida K, Schuenemann VJ, Cano LM, Pais M, Mishra B, Sharma R, Lanz C, Martin FN, Kamoun S, Krause J, Thines M, Weigel D, Burbano HA (2013) The rise and fall of the *Phytophthora infestans* lineage that triggered the Irish potato famine. eLife 2013:2. https://doi.org/10.7554/eLife.00731

Yuan J, Benoît Bizimungu B, De Koeyer D, Rosyara U, Wen Z, Lagüe M (2020) Genome-wide association study of resistance to potato common scab. Potato Res 63(2):253–266

Zhang X-Y, Yu X-X, Yu Z, Xue Y-F, Qi L-P (2014) A simple method based on laboratory inoculum and field inoculum for evaluating potato resistance to black scurf caused by *Rhizoctonia solani*. Breed Sci 64(2):156–163. https://doi.org/10.1270/jsbbs.64.156

Zhu S, Vossen JH, Bergervoet M, Nijenhuis M, Kodde L, Kessel GJT, Vleeshouwers V, Visser RGF, Jacobsen E (2015) An updated conventional- and a novel GM potato late blight *R* gene differential set for virulence monitoring of *Phytophthora infestans*. Euphytica 202:219–234

Zimnoch-Guzowska E, Lebecka R, Pietrak J (1999) Soft rot and blackleg reactions in diploid potato hybrids inoculated with *Erwinia* spp. Am J Potato Res 76:199. https://doi.org/10.1007/BF02854222

Zimnoch-Guzowska E, Lojkowska E, Perombelon M (2005) Resistance to bacterial pathogens. In: Razdan MK, Mattoo AK (eds) Genetic improvement of Solanaceous crops volume I: potato. Science Publishers, Enfield, pp 339–395

Part III
Breeding Methods

Chapter 6
Introgression, Base Broadening and Potato Population Improvements

6.1 Introduction

Potato breeding since the early twentieth century has comprised the introgression of desirable genes from the landraces and wild relatives of cultivated potatoes, the use of landraces in a more general broadening of the genetic base of breeding and the inter-crossing of the resulting clones and cultivars for further progress in the adaptation of the potato to particular target environments and end uses. The history of this potato breeding was reviewed in Chap. 1. We are now going to look at the theory and practice of introgression, base broadening and population improvement, which can be viewed as long-term crop improvement through increasing the frequencies of desirable combinations of alleles (genes) over sexual generations. This will include the use of molecular markers for marker-assisted introgression and selection of specific alleles and genomic selection of unspecified alleles in population improvement. Then in subsequent chapters, we will look at short-term cultivar production aimed at combining desirable alleles in a single genotype, finishing with genetic transformation to produce genetically modified potatoes. However, it is useful to start with a brief review of the germplasm available, its maintenance (conservation) in situ and ex situ in genebanks and the crossability of this germplasm with cultivated potatoes.

6.2 Germplasm Available

The collection and conservation of germplasm of cultivated species and their wild relatives in genebanks was a direct outcome of the work done from 1920 to 1940 by Nickolay Ivanovich Vavilov (Vavilov 1940), and can be viewed as his legacy, which has continued to the present day. There have been numerous collecting expeditions for tuber-bearing *Solanum* species in Central and South America, and the Southwest

J. E. Bradshaw, *Potato Breeding: Theory and Practice*,
https://doi.org/10.1007/978-3-030-64414-7_6

USA, from those pioneered by the Russians in the 1920s (Hawkes 1990) to the more recent ones of the 1990s (Spooner and Hijmans 2001). A summary can be found in the book chapter by Ellis et al. (2020). They led to the establishment of a number of potato germplasm collections worldwide. The world collection has been held at The International Potato Centre (CIP) (http://www.cipotato.org) since its creation in Lima, Peru in 1971. The other main collections are the Commonwealth Potato Collection (CPC, Dundee, Scotland), the Dutch-German Potato Collection (CGN, Wageningen, The Netherlands), the collection of the French National Institute for Agricultural Research (INRA, France), the Groß Lusewitz Potato Collection (GLKS, IPK, Groß Lusewitz, Germany), the collection of the National Institute of Agrobiological Sciences (NIAS, Japan), the Potato Collection of the Vavilov Institute (VIR, St Petersburg, Russia), the U.S. Potato Genebank (NRSP-6, Sturgeon Bay, USA) and Potato Collections in Argentina, Bolivia, Chile, Colombia and Peru. Ellis et al. (2020) estimated that there are at most 24,500–29,500 unique potato accessions conserved worldwide in genebanks, but the accuracy of such an estimate is difficult to access until a thorough rationalization and comparison is made amongst the ex situ collections. Bamberg et al. (2018) reviewed the role of these genebanks under the headings of acquisition, classification, preservation, evaluation and enhancement, and distribution of potato genetic material. They also explained and discussed the implications of the Convention on Biological Diversity (CBD), which came into force in December 1993 and was founded on three principles: the conservation of biological diversity, the sustainable use of its components and the fair and equitable sharing of the benefits arising out of the utilization of genetic resources. The consequences for potatoes have also been reviewed by Ellis et al. (2020). Here I am briefly going to mention the CIP genebank as the world collection and the CPC as the genebank with which I am most familiar (Bradshaw and Ramsay 2005), as well as sources of information on modern cultivars. Readers can find out about the other genebanks from their websites and from Genesys, the global Gateway to Genetic Resources (https://www.genesys-pgr.org/welcome).

6.2.1 CIP Genebank

The following is a brief summary of information from the CIP genebank website (https://cipotato.org/genebankcip/), the chapter by Salas et al. (2014) in the *Sixth Edition of the World Catalogue of Potato Varieties* (Pieterse and Judd 2014), and the book chapter by Ellis et al. (2020). The genebank maintains clonal and seed collections of potato as a global public good under the International Treaty on Plant Genetic Resources for Food and Agriculture (ITPGRFA). Germplasm is available for research, education and breeding purposes and has been used in breeding programmes in over 100 countries. CIP is the custodian of the world's largest in vitro genebank and houses one of the world's leading herbarium collections and a cryopreservation programme. The herbarium collection has more than 90% of the wild and cultivated species represented from southern Colorado Canyon in the United States to southern Chile.

Today, the active cultivated potato collection totals 4727 accessions including 4354 traditional landrace cultivars from 17 countries, mainly from the Andean region, and improved cultivars. The entire clonal collection is conserved in vitro and distributed internationally as tissue-cultured materials [Ellis et al. (2020) quote a total of 8354 in vitro accessions]. The in vitro conservation methods use treatments for reducing plantlet growth, such as osmotic stress and low temperature for storage, to keep viable plantlets in medium-term storage where transferring to fresh media is not needed as often. Thus, by using low temperature (6–8 °C) and culture medium containing sorbitol as an osmotic agent, the need for fresh transferring is prolonged from once every 6–8 weeks to 2 years. The in vitro accessions are the primary source of germplasm for all purposes. A back-up collection is held at a CIP research station in Huancayo, Peru, a 6- to 8-h drive from Lima, and also at the Brazilian Agricultural Research Corporation (EMBRAPA) labs in Brasilia, Brazil. At both sites, the material is kept in slow growth conditions at 7 °C where the cultures can survive approximately 2 years without sub-culturing. Plants are continually re-evaluated to ensure accurate morphological, taxonomic and cytogenetic characterization to identify duplicates and verify identity. If necessary, viruses can be eliminated by thermotherapy (incubation of in vitro plants at 32–34 °C for about 1 month followed by isolation of 0.2–0.35 mm meristems). The improved cultivars comprise ones released mainly by Latin-American countries and genetic stocks used to identify resistance to diseases (e.g. late blight and bacterial wilt). The global collection is maintained 'In Trust' and is distributed with the Standard Material Transfer Agreement (SMTA) under the terms of the International Treaty for Plant Genetic Resources for Food and Agriculture (ITPGRFA). Cryopreservation of the potato collection is routinely done with over 450 new accessions entered into long-term cryopreservation per year. Shoot tips are stored at ultra-low temperatures in the vapour or liquid phase of nitrogen (−196 °C) with a long-term (50+ year) viability-monitoring programme. More information on cryopreservation can be found in the review by Niino and Arizaga (2015). Also, for long-term conservation, Andean farmers in collaboration with CIP have produced true (botanical) potato seed from their traditional cultivars, which is now safely backed up in the Svalbard Global Seed Vault, which opened in February 2008 (Hawtin and Fowler 2011).

The number of accessions of cultivated species are as follows (2, 3, 4 and 5*x* = diploid, triploid, tetraploid and pentaploid):

S. tuberosum Andigenum Gp [4*x* Andigena (3233), 3*x* Chaucha (121), 2*x* Phureja (206), 2*x* Stenotomum (stenotomum 289, goniocalyx 102)]
S. tuberosum Chilotanum Gp [4*x* Chilean Tuberosum (179)]
2*x* *S. ajanhuiri* (14), 3*x* *S. juzepczukii* (36) and 5*x* *S. curtilobum* (6)
2*x* and 4*x* hybrids (235)

It can be seen that the collection is dominated by Group Andigena. Back in the year 2000, a core set of 306 Group Andigena accessions from the 2379 then held at CIP was established to aid utilization (Huaman et al. 2000a). They were chosen to represent the widest morphological diversity and to maximize geographical representation, following a cluster analysis of morphological data. Evaluation data were taken into account in choosing the representative accession of each cluster. Isozyme analysis of the entire collection confirmed that the core collection had captured a representative sample of the alleles at nine allozyme loci, with only the loss of alleles whose frequency was less than 0.05% (Huaman et al. 2000b). A simulation study revealed, however, that ideally a core collection size of 600 would be required to adequately represent allele frequencies and locus heterozygosity (Chandra et al. 2002).

CIP has developed and characterized a suite of 24 highly informative microsatellite markers (SSRs) with size standards for inter-laboratory standardization of fragment lengths. Ghislain et al. (2009a) fingerprinted 742 potato landraces of all four cultivated potato species, *S. tuberosum* Group Andigenum and Group Chilotanum, *S. ajanhuiri*, *S. curtilobum* and *S. juzepczukii*, with 51 single locus SSR markers. They went on to choose a new potato genetic identity (PGI) kit based on 24 of the SSR markers with two per chromosome separated by at least 10 cM, except for chromosome 7 where the two SSRs were separated by only 3 cM. The number of alleles for each of the 24 SSRs ranged from three to nine with an average of 5.5. The 24 SSRs were able to discriminate 93.5% of the 742 landraces compared to 98.8% with the 51 SSRs. The PGI kit can be used for potato germplasm characterization for a variety of purposes such as identity verification (fingerprinting), studying genetic diversity, anchoring genetic linkage maps, establishment of core collections and studying gene flow. When the PGI kit was used to genotype breeding clones and advanced cultivars of potato, most of the breeding material grouped into a well-defined cluster with landraces of the Chilotanum Group, thus confirming that this group has been used extensively in potato breeding worldwide.

Since 1997, 89 Andean communities have received over 6000 samples of cultivated potato from the CIP genebank, comprising more than 1250 accessions or almost 30% of the CIP collection of native landraces. This process is termed repatriation as the ancestors of the Andean communities had developed and conserved these landraces for several thousand years. CIP has also established a 15,000-ha reserve (Cusco Potato Park) high in the Andes of Peru where the indigenous Quechua people grow 1345 landraces of potato, of which 779 were collected locally, 410 were repatriated from CIP and 157 were received through seed exchanges (Fig. 6.1).

The CIP genebank also maintains and conserves an important collection of potato breeding lines and genetic stocks. Currently about 3400 breeding lines are conserved and distributed as in vitro clones to requestors globally: 82% of the potato clones were developed in Peru with the rest derived from other countries. These clones have characteristics of interest for breeding programmes as potential sources of traits for adaptation to climate change, new consumer demands, increased food security, improved nutritional value and enhancement of the income of poor farmers. Breeding lines and genetic stocks are designated under the International Treaty for Plant Genetic Resources for Food and Agriculture (ITPGRFA) and therefore are

Fig. 6.1 Cusco Potato Park, Peru. (Kindly supplied by John Marshall)

subject to the terms and conditions of the Standard Material Transfer Agreement (SMTA) for distribution and use.

Finally, the genebank (https://cipotato.org/crops/potato/wild-potato-species/) maintains a collection of accessions of wild potato species which are native to North (Mexico and United States), Central (Costa Rica, Guatemala, Honduras and Panama) and South America (Argentina, Bolivia, Brazil, Chile, Colombia, Ecuador, Paraguay, Peru, Uruguay and Venezuela). Unfortunately, the taxonomy of wild potato species is complex and has undergone revision. Hawkes (1990) recognized 219 species, whereas Spooner et al. (2014) now recognize only 107 species. Furthermore, different genebanks use different classifications. The CIP genebank uses the taxonomy *Solanum* L. sec Wild *Petota* Dumort (Solanaceae) grouped in 20 series (alphabetical order) containing 172 species representing 132 diploids, 6 triploids, 24 tetraploids, 10 hexaploids and 17 hybrids. The genebank maintains a total of 2595 accessions of 148 out of the 172 species. The long-term conservation of these species is done with botanical seed, which involves the collection, processing, regeneration, conservation and distribution of seed, as we shall see in the next section with the Commonwealth Potato Collection (CPC). Accessions maintained as true botanical seed are backed up in the Svalbard Global Seed Vault as a sealed package of seed equivalent to one to three times that needed for a regeneration.

6.2.2 Commonwealth Potato Collection

The Commonwealth Potato Collection (CPC) resulted from expeditions to Mexico, USA, Central and South America over the period 1938–1981, with a major involvement of J.G. Hawkes of Birmingham University, UK (Hawkes 1951, 1990). Today it comprises 1500 accessions, two thirds of which are wild species (80 of the ones recognized by Hawkes (1990)) and one third cultivated. It was fully converted to true seed populations in the 1960s. The loss of identity of the native cultivars in the collection was compensated by conserving the alleles present in each accession in a form which is secure against the casual losses suffered by annually planted clones. The collection is now held at the James Hutton Institute in Dundee, Scotland. The annual rejuvenation cycle involves 60 accessions so that the entire collection is done every 25 years (Fig. 6.2).

The size of seed sample to be used in regeneration needs to reflect the genetic composition of the accession, with the effective population size (N_e) chosen to minimize loss of alleles, as discussed by Bradshaw (2016). Sufficient seed to obtain 20 (considered adequate) seedlings of each accession is sown in spring in an insect-proof Quarantine Glasshouse, and the plants tested for quarantine organisms (e.g. viruses/viroids) that could be present in stored seed. The 20 plants are inter-crossed at flowering by applying bulked pollen to all open flowers, repeating the process as necessary. Seed is extracted from ripe berries, dried (6% moisture), packeted and kept in a dedicated store in conditions (low temperature and humidity) designed to

Fig. 6.2 Rejuvenation of part of Commonwealth Potato Collection. (From Bradshaw and Bonierbale 2010 with permission; 4 July 2008)

retain viability for many years. A working collection store is maintained at 4 °C and 10% relative humidity. The CPC database is regularly updated and more information about the collection can be found on its website (https://ics.hutton.ac.uk/germinate-cpc/#home). The FAO WIEWS (**W**orld **I**nformation and **E**arly **W**arning **S**ystem on Plant Genetic Resources for Food and Agriculture) Code for The James Hutton Institute is now **GBR251**. Utilization of the collection in potato breeding was reviewed by Bradshaw and Ramsay (2005). Accessions are available to bona fide researchers, breeders and educators under terms consistent with the international treaties that govern the use of such germplasm. The collection is backed up in the Svalbard Global Seed Vault.

6.2.3 Cultivars

The 2014 World Catalogue of Potato Varieties (Pieterse and Judd 2014) listed more than 4800 cultivars from some 100 countries covering all potato growing regions in the world, although a much smaller number have been widely grown. Furthermore, a number of databases on modern cultivars are available, such as The European Cultivated Potato Database (http://www.europotato.org), which currently (November 2019) has information on 5947 cultivated varieties. The site includes all the botanical and disease data provided by institutes throughout Europe. Google searches will find databases for many countries and organizations around the world. These modern cultivars provide breeders with parents adapted to their target environments, growing seasons and end uses, but which require local evaluation and improvement.

6.3 Crossability of Wild and Cultivated Potatoes

The introgression from wild species of genes for resistances to diseases and pests started early in the twentieth century and to a large extent was a matter of trial and error. However, in Chap. 1, we saw that by the beginning of the 1990s, wild potato relatives were still a largely untapped source of potentially desirable genes, which had proved difficult to successfully utilize in potato breeding, and this book considers only 32 wild species. Nevertheless, by the 1990s knowledge was increasing about how to achieve sexual hybridizations between cultivated potatoes and their wild relatives (Ortiz 1998, 2001; Jansky 2006). Furthermore, molecular marker-assisted introgression offered the possibility of faster progress than can be achieved by traditional backcrossing (Hermsen 1994; Barone et al. 2001; Barone 2004; Iovene et al. 2004). So, let us start by looking at the crossability of species as the start to introgression programmes.

The crossability of species has been determined through artificial pollinations done over many years (Jansky 2006). As explained in Chap. 2, the results can be interpreted primarily but not exclusively in terms of Endosperm Balance Number (EBN), which can be regarded as the effective rather than the actual ploidy of a species (Johnston et al. 1980). Despite attempts to understand the genetic and biological basis of the EBN concept, EBNs are determined experimentally relative to species assigned an arbitrary value of 1 (Hermsen 1994). Five groups of species can be recognized based on ploidy and EBN: diploid EBN = 1, diploid EBN = 2, tetraploid EBN = 2, tetraploid EBN = 4 and hexaploid EBN = 4 (Hawkes and Jackson 1992). The main groups of wild species using the taxonomy of Hawkes (1990) are as follows (Hawkes and Jackson 1992).

22 species:	diploid	EBN = 1	e.g. *S. bulbocastanum* and *S. commersonii*
178 species:	diploid	EBN = 2	e.g. *S. chacoense*, *S. microdontum*, *S. spegazzinii* and *S. vernei*
S. tuberosum:	diploid	EBN = 2	
11 species:	tetraploid	EBN = 2	e.g. *S. acaule* and *S. stoloniferum*
S. tuberosum:	tetraploid	EBN = 4	
8 species:	hexaploid	EBN = 4	e.g. *S. albicans*, *S. demissum* and *S. hougasii*

Taxonomic species within the crossability groups have evolved by means of geographical and ecological isolation rather than by genetic incompatibility, and hence hybridizations are normally successful between species with the same EBN number. Today, potato breeders can usually achieve desired sexual hybridization between *S. tuberosum* and its wild relatives by manipulation of ploidy with due regard to EBN (Ortiz 1998, 2001; Carputo and Barone 2005; Jansky 2006). However, from time to time, potato breeders have unexpected successes and failures when attempting to overcome barriers to hybridization.

EBN can be doubled meiotically through the unreduced 2*n* gametes produced from several naturally occurring recessive meiotic mutations that are common in *Solanum* species (Tai 1994). It can also be doubled mitotically through the use of colchicine for chromosome doubling, something which can also occur naturally during callus culture. EBN can be halved by haploidization using in vitro androgenesis (anther culture) or, more commonly in *S. tuberosum*, by parthenogenesis using pollinations with particular clones of diploid *S. phureja* (Wenzel 1994; Veilleux 2005; Ortiz and Mihovilovich 2020). The first such haploids of tetraploid *S. tuberosum* were produced in 1958 (Hougas et al. 1958) and are often called dihaploids. Embryo rescue can also be used to secure a hybrid where embryo abortion is due to a defective endosperm (Jansky 2006).

As the largest compatibility group is the diploid 2EBN one, it is now common for potato breeders to secure tetraploid hybrids from 4*x* (*S. tuberosum*) x 2*x* (2*x* *S. tuberosum* x 2*x* wild species) crosses in which an unbalanced endosperm prevents the development of triploid embryos. There are, however, other barriers to hybridization, such as interspecific pollen-pistil incompatibility and nuclear-cytoplasmic male sterility, although fertility restorer genes have been found for the latter (Jansky 2009). We dealt with different cytoplasms back in Chap. 1. Unilateral incompatibility is known to occur when a self-incompatible (SI) species is pollinated by a self-compatible (SC) one so that *S. verrucosum* (SC female) x *S. phureja* (SI male) is successful, but the reciprocal cross fails (Hermsen 1994; Jansky 2006). Sometimes, incompatible pollen can be helped to achieve fertilization through a second pollination with compatible pollen, a technique known as mentor pollination (Hermsen 1994; Jansky 2006). These phenomena have been reviewed by Camadro et al. (2004) in the context of how sympatric species maintain their integrity.

Potato breeders can also use somatic (protoplast) fusion to achieve difficult or impossible sexual hybridizations, and in ploidy manipulation to achieve maximum heterozygosity (Wenzel 1994; Veilleux 2005). Protoplast fusion can be induced chemically by the polycation polyethyleneglycol (PEG) or via electrofusion, which is now the preferred method. Somatic hybrids derived from the same fusion combination do show genotypic and phenotypic variation and hence require screening for the desired product.

Let us now look at some of the successful schemes that have been used to achieve crosses between wild and cultivated species followed by introgression.

6.4 Introgression in Practice

6.4.1 Use of Solanum demissum *(6x, EBN = 4)*

Introgression (backcross) breeding started in 1908 in Germany and 1909 in Great Britain with the recognition of *S. demissum* as a source of genes for resistance to late blight, caused by the oomycete pathogen *Phytophthora infestans* (Muller and Black 1951). Although *S. demissum*, like the principal cultivated potato, has an EBN of 4, the first hybridizations and introgressions were by trial and error, as can be seen in the early work at the Scottish Plant Breeding Station (SPBS). Craigs Snow White was registered (released) in 1948 and came from a cross made in 1939 between Craigs Defiance as female parent and W800(2) as male parent, the latter having been produced by J. H. Wilson of St Andrews, Scotland. Craigs Defiance had come from a cross made in 1933 between cultivars Epicure and Pepo. Clone W800(2) traces back to Myatt's Ashleaf × New Zealand Red, *S. commersonii* ($2n = 2x = 24$, EBN = 1) × *S. demissum* ($2n = 6x = 72$, EBN = 4) (and hence a cross that should not work) and *S. maglia* × *S. edinense*, the latter being a hybrid of *S. tuberosum* and *S. demissum*, so named because it was first described from material in the Royal Botanic Garden Edinburgh. Craigs Snow White contains what became known as the *R1* gene for blight resistance from *S. demissum* and was designated the R1 differential once appropriate genetic studies had been done.

The cultivar Pentland Ace (the R3 differential, with gene *R3a* and *R3b* variant $R3b^{G1696/G3111}$, Armstrong et al. 2019) was released from SPBS in 1951 after just three backcrosses to *S. tuberosum*, starting with clone 735 [*S. rybinii* = *S. phureja* (2*x*) × *S. demissum* (6*x*)] in 1937 (14 years). Clone 735 was one of a few artificial tetraploid seedlings secured by hybridization despite the species having different endosperm balance numbers (Bradshaw 2009). Genetic analysis (see later) tells us that three backcrosses is as fast as one can expect to do an introgression with modest but realistic population sizes. Marker-assisted selection is faster only when genotypic selection (for desirable genes and against undesirable genes) is more effective than phenotypic selection. This is because the number of generations and population sizes required for a successful introgression depend upon the frequencies of the desirable products of meiosis during sexual hybridization, which in turn depends upon the number of chromosomes and the number and distribution of chiasmata.

1937 *S. rybinii* (2*x*) × *S. demissum* (6*x*)
↓
1939 735 × Gladstone
↓
1940 885(2) × Pepo
↓
1942 Craigs Defiance × 997a44
↓
1951 Pentland Ace

The commercially more successful cultivar Pentland Dell (with genes *R1*, *R3a*, $R3b^{G1696/G3111}$ and *R2* family member $abpt^{T86}$, Armstrong et al. 2019) traces its pedigree back to pentaploid clone 429a8 [*S. demissum* (6*x*) × cv. The Alness (4*x*)] and tetraploid clone 735 and required five generations over the period 1932–1960 (28 years) (Bradshaw 2009). It entered ware production in Great Britain in 1963 when race 4 was the prevalent race of *P. infestans* but succumbed to late blight in 1967. Nevertheless, it went on to become one of SPBS's most successful cultivars and is still grown in Great Britain for making French fries, albeit under the protection of fungicides. A remarkable 58 out of 72 cultivars bred at SPBS (and its successor SCRI, the Scottish Crop Research Institute) from 1920 to 2008 have *S. demissum* in their pedigrees. *S. demissum* was in fact used extensively throughout Europe and North America in breeding for resistance to late blight (Ross 1986; Plaisted and Hoopes 1989).

Recently, Hosaka and Sanetomo (2018) summarized our knowledge of crossability of *S. demissum* and *S. tuberosum* as follows. Although *S. demissum* has an allohexaploid genome structure ($AADDD^dD^d$), we can easily obtain pentaploid hybrids from *S. demissum* × *S. tuberosum* (AAA^tA^t) when *S. demissum* is used as the female parent. The resultant pentaploid F_1 hybrids (AAA^tDD^d) produce abundant normal-looking pollen grains but are non-functional as males and usually produce seeds only if backcrossed (BC_1) with the pollen of *S. tuberosum*.

Hosaka and Sanetomo (2018) generated 87 BC_1 plants from a pentaploid hybrid ($2n = 5x = 60$; *S. demissum* 5H109–5 × *S. tuberosum* breeding clone Saikai 35) crossed with the pollen of Saikai 35, a highly male and female fertile clone possessing a *S. phureja*-derived P-type cytoplasm. The BC_1 plants were all aneuploids

($2n = 49$–59, with average of 53.5) with different portions of the *S. demissum* chromosomes. The researchers evaluated the crossability (seed set) of the BC_1 plants with the *S. demissum* (D) and *S. tuberosum* (T) parents ($BC_1 \times D$ and $D \times BC_1$, $BC_1 \times T$ and $T \times BC_1$, female first) and performed a quantitative trait locus (QTL) analysis based on the association between *S. demissum*-specific simplex markers and the crossability measurements. The results were complicated, but the researchers found that several major genes and many minor genes controlled the crossability of BC_1 plants. One major gene was localized on *S. demissum* chromosome 9A and had positive effects on the berry-setting rates. They concluded that the EBN is represented by the sum of various genetic effects controlled by a large number of genes, and that *S. demissum* has a slightly lower EBN than *S. tuberosum*, although the two species are grouped in the same 4EBN category. The seed setting results are of interest in their own right (Table 6.1): the backcross $F_1 \times T$ produced large seed, the backcross $T \times F_1$ failed as expected, but the backcross $D \times F_1$ produced some normal sized seed. Interestingly, the two selfs of the parent species had the highest berry-setting rates and the most seed per berry. Perhaps it is not surprising that earlier introgressions from *S. demissum* came from trial and error.

6.4.2 *Use of* S. acaule *and* S. stoloniferum *(both 4x, EBN = 2)*

Two tetraploid, EBN = 2, species well-known to potato breeders are *S. acaule* and *S. Stoloniferum*, the sources, respectively, of the Rx_{acl} gene for immunity to all strains of *Potato virus X* (PVX) and the Ry_{sto} gene for comprehensive resistance to *Potato virus Y* (PVY). For example, these genes entered the breeding programme at the Scottish Plant Breeding Station as tetraploid clones in 1955 and 1957, respectively (Bradshaw 2009), kindly provided by R. W. Ross in Cologne (Rx_{acl} in clone 44/1016/10 and Ry_{sto} segregating in the cross coded R49/272). A more recent

Table 6.1 Seed setting results where T is *S. tuberosum*, D is *S. demissum*, F_1 is $D \times T$ and BC_1 is $F_1 \times T$

Cross	Berry-setting rate %	Seed number/berry	Seed weight mg
T self	68.1	219.5	0.49
D self	92.8	166.8	0.41
$F_1 \times D$	32.7	73.3	0.58
$F_1 \times T$	17.3	32.0	0.94
$D \times F_1$	12.0	20.2	0.49
$T \times F_1$	0.0	–	–
$BC_1 \times D$	0 to 93.8, mean 33.2	83.6	Mean 0.44
$BC_1 \times T$	0 to 100, mean 40.7	55.9	Mean 0.77
$D \times BC_1$	49/82 = 0; 5.0 to 81.8, mean 27.4	22.7	Mean 0.63
$T \times BC_1$	67/74 = 0; 4.8 to 20.0, mean 11.9	34.4	Mean 0.59

Data from Hosaka and Sanetomo (2018) with permission

example from Camadro and Espinillo (1990) is germplasm transfer from *S. acaule* to the cultivated potato *S. tuberosum* using 2*n* eggs. They summarized crossability knowledge in 1990 as follows. The tetraploid subspecies of *S. acaule*, *acaule* and *aemulans* are self-fertile disomic polyploids. Direct crosses between 4*x S. acaule* and the common potato 4*x S. tuberosum* have been largely unsuccessful due to the collapse of the endosperm in the hybrid seed. However, 4*x S. acaule* can be easily crossed with diploid species and 2*x S. tuberosum* (dihaploids) yielding triploid progeny. After chromosome doubling using colchicine, some of these triploid hybrids (now hexaploid) have been satisfactorily crossed with 4*x S. tuberosum*.

4*x* (2EBN) wild species × 2*x* (2EBN) Tuberosum
↓
3*x* (2EBN) hybrids
↓ (colchicine doubling)
6*x* (4EBN) hybrids × 4*x* (4EBN) Tuberosum
↓
5*x* (4EBN) hybrids

Following a consideration of EBN, the approach taken by Camadro and Espinillo (1990) was to screen and select as progenitors, plants of 4*x S. acaule* that produced functional unreduced (2*n*) gametes, rather than doubling the somatic chromosome complement of 4*x S. acaule* prior to performing the cross with 4*x S. tuberosum*.

4*x* (4EBN) Tuberosum × 4*x* (2EBN) wild species × 4*x* (4EBN) Tuberosum
↓ (4*x* pollen) (4*x* eggs) ↓
6*x* (4EBN) hybrids × 4*x* (4EBN) Tuberosum
↓
5*x* (4EBN) hybrids × 4*x* (4EBN) Tuberosum

When they screened 197 plants of 4*x S. acaule* (22 introductions) for pollen of heterogeneous size (presumably *n* and 2*n*), only two plants produced this type of pollen, with 3% and 5% of pollen in the large class, respectively, and hence potential 2*n* pollen producers. However, they could not be used in crosses with 4*x S. tuberosum.* In contrast, of 427 plants (47 introductions) screened for 2*n* eggs through controlled crosses with 4*x S. tuberosum*, 41 (13 introductions) produced 219 berries of which 46 contained at least one plump seed. Seed from berries with low seed set (1 or 2 seed) gave rise to hexaploid hybrids, presumably through the functioning of 2*n* eggs from 4*x S. acaule*, whereas those with a higher seed set were unwanted selfs. A sample of six of these hybrids was successfully backcrossed as female parents to 4*x S. tuberosum*, yielding 0.5 berries per pollination and an average of 83.5 seeds per berry, and giving rise to 5*x* BC_1 progenies. Although the hybrids had high pollen stainability values of about 80%, all 105 pollinations with them as male parent failed (other researchers have succeeded). The BC_1 progenies could be easily backcrossed to 4*x* *S. tuberosum*, yielding tetraploid or near tetraploid BC_2 plants. We shall encounter further hybridizations involving tetraploid 2EBN species in the next section.

6.4.3 Use of S. commersonii (2x, EBN = 1)

Hybridization of diploid 1EBN species and 2*x* *S. tuberosum* can be achieved in two ways as shown below (Carputo and Barone 2005). The 3*x* (2EBN) hybrids produce unreduced pollen (36 chromosomes) from the parallel spindle mutation, but egg production involving omission of the second meiotic division can produce eggs with 24–48 chromosomes and hence hybrids with 48–72 chromosomes. The 4EBN hybrids will cross with tetraploid Tuberosum cultivars.

An example is the breeding research by Barone et al. (2001) and Iovene et al. (2004) using *Solanum commersonii* as a source of several valuable traits including resistance to tuber soft rot caused by *Pectobacterium carotovorum* (formerly *Erwinia carotovora*), resistance to bacterial wilt caused by *Ralstonia solanacearum* Smith, high dry matter content of tubers and tolerance of freezing (and cold-acclimating capacity). They produced triploid hybrids between an accession of *S. commersonii* ($2n = 2x = 24$, 1EBN) doubled ($2n = 4x = 48$, 2EBN) by in vitro tissue culture of leaf discs, and diploid Phureja-Tuberosum hybrids ($2n = 2x = 24$, 2EBN). They found that chromosome numbers in the offspring of their 3*x* (2EBN) hybrids × 4*x* (4EBN) Tuberosum crosses ranged from 58 to 67. They chose a 60-chromosome hybrid for the next backcross to 4*x* (4EBN) Tuberosum cv. Blondy, which resulted in 51 offspring with 48–57 chromosomes, with a mean of 52.4 chromosomes compared with an expected mean of 54 assuming a random transmission of extra chromosomes (Iovene et al. (2004). Of the 51 offspring plants, 46 flowered and were analysed for fertility: in crosses with 4*x* (4EBN) Tuberosum as male parent, 80% of the hybrids set fruits with seed, and in crosses with 4*x* (4EBN) Tuberosum as female parent, 59% of the hybrids set fruits with seed. The percentage of *S. commersonii*-specific AFLP molecular markers in the offspring plants ranged from 59% to 91%, with an average value of 75%. The researchers then chose a 48-chromosome hybrid (one out of two available) for their third backcross to 4*x* (4EBN) Tuberosum cv. Blondy and secured 48-chromosome offspring. Barone et al. (2001) provided evidence from *commersonii*-specific RFLPs for recombination between homoeologous chromosomes for 5 out of the 12 chromosomes; for the others, no more than 1 *commersonii*-specific marker was found. They also determined the extent of the wild genome carried in each backcross generation using *commersonii*-specific RAPD and AFLP markers, the results being 92%, 75% and 28.5% in BC_1, BC_2 and BC_3, respectively, with the 3*x* (2EBN) hybrid (F_1) being 100%. Furthermore, in the third backcross, five out of 20 genotypes were highly resistant to tuber-soft rot and two of them combined this resistance with other desirable characteristics and hence were chosen as the parents of the next generation. In summary, the progressive loss of the wild genome combined with recombination between the *S. commersonii* and *S. tuberosum* genomes made it possible to achieve the practical purpose of the breeding scheme, namely the transfer of useful genes from the wild species *S. commersonii* into the cultivated gene pool through backcrosses.

2*x* (1EBN) wild species

↓ (colchicine or *in vitro* doubling)

4*x* (2EBN) wild species × 2*x* (2EBN) Tuberosum × 2*x* (1EBN) wild species

(2*n* = 2*x* pollen)

4*x* (4EBN) Tuberosum × 3*x* (2EBN) hybrids × 4*x* (4EBN) Tuberosum

↓ (3*x* pollen) (2*x*+ eggs) ↓

5*x* (4EBN) hybrids etc. | 4*x*+ (4EBN) hybrids (BC$_1$)

5*x* (4EBN) hybrid × 4*x* (4EBN) Tuberosum

↓

4*x*+ (4EBN) hybrids (BC$_2$)

4*x* (4EBN) hybrid × 4*x* (4EBN) Tuberosum

↓

4*x* (4EBN) hybrids (BC$_3$)

6.4.4 *Use of* **S. bulbocastanum** *(2x, EBN = 1) Through Bridging Crosses and Somatic Fusion*

Hermsen (1994) describes in detail the long but successful introgression of late blight resistance from diploid *S. bulbocastanum* (BB) into tetraploid *S. tuberosum* (TTTT). The work started in 1959 and involved bridging crosses with tetraploid *S. acaule* (AACC) and diploid *S. phureja* (PP) (strictly Group Phureja) before three successive backcrosses to *S. tuberosum* finally yielded cultivars Toluca and Bionica in 2006 and 2008, respectively, with the *Rpi-blb2* resistance gene. Bionica also has *Rpi-abpt*^{T86}, *Rpi-R3a* and *Rpi-R3b*$^{G1696/G3111}$ (Armstrong et al. 2019). The possibly unexpected successful steps were as follows with Hermsen's explanation.

4*x* (EBN = 2) AACC × 2*x* (EBN = 1) BB
↓ (assume A pairs with A and C with C)
3*x* ACB
↓ Doubling
6*x* AACCBB × 2*x* (EBN = 2) PP
↓
4*x* ACBP × 4*x* TTTT
(assume A pairs with C ↓ but B and P must tend to pair)

½(ACBP)TT just a few viable seeds
32 clones coded ABPT of which 3/24 extremely resistant 4*x*
and 4/4 extremely resistant 6*x* (ACBPTT)

Hermsen concluded that *S. bulbocastanum* and *S. phureja* chromosomes show considerable pairing and that the *S. phureja* was therefore a crucial link in the introgression of late blight resistance. Further backcrosses of resistant ABPT clones with tetraploid *S. tuberosum* cultivars were successful, and late blight resistance was transferred to later generations.

Another introgression route starts with the production of fertile hexaploid hybrids of tetraploid *S. tuberosum* and diploid *S. bulbocastanum* through somatic fusion, as done in Wisconsin by Helgeson et al. (1998). They used PEG-mediated (polyethylene glycol) fusion of protoplasts isolated from leaf cells of in vitro shoots of *S. bulbocastanum* PI 245310 and the tetraploid potato line *S. tuberosum* PI 203900 to produce hexaploid somatic hybrids that retained the late blight resistance of the *S. bulbocastanum* parent. The reader is referred to their paper for the details of the protoplast fusion technique and the regeneration of whole plants. RFLP and RAPD analyses confirmed the hybridity of the materials that morphologically looked like hybrids; for example, combining the purple colouration of the *S. bulbocastanum* stems with the compound leaves of *S. tuberosum.* Four of the somatic hybrids were crossed with potato cultivars Katahdin or Atlantic, of which the most widely used were J101, J103 and J138. The backcross (BC_1) progeny segregated for resistance to the US8 genotype (A2 mating type) of *P. infestans.* Resistant BC_1 lines crossed with susceptible cultivars again yielded populations that segregated for resistance. The second backcross [(*S. bulbocastanum* + *S. tuberosum*) × Katahdin] × Atlantic, for example, was designated J101K6A22. Fourteen BC_1 and BC_2 lines were tested for blight resistance in Toluca, Mexico, in the summer of 1995. Good broad-spectrum resistance was obtained in the Toluca field test with all lines that were resistant in Wisconsin in 1994. Subsequent mapping studies in the backcross populations located a major gene for resistance on chromosome 8 between RFLP markers CT64 and CP53 (Naess et al. 2000). When treated as a quantitative trait, 62% of the observed variability in disease resistance was explained by this chromosomal region; nevertheless, the region should be regarded as having a major gene for resistance (*RB*). Colton et al. (2006) demonstrated that marker-based selection can be used to effectively transfer the *RB* gene into potato using traditional breeding methods, an alternative to deploying the *RB* gene through genetic transformation. They found that one out of six of the PCR primers sets that they tested was specific for the resistance allele at the RB locus.

Similar work was done by Rakosy-Tican et al. (2020). They used a resistant *S. bulbocastanum* accession (GLKS 31741 (*blb*41)) carrying the broad-spectrum resistance genes *Rpi-blb1* and *Rpi-blb3* together with the race-specific genes *R3a* and *R3b*. They produced a large number of somatic hybrids (SHs) with five commercial potato cultivars using protoplast electrofusion. They selected three fertile SHs with cv. 'Delikat', one hexaploid and two between pentaploid and hexaploid, and used cvs. 'Baltica', 'Quarta', 'Romanze' and 'Sarpo Mira' to produce backcross generations BC_1 and BC_2. Using gene-specific markers, they identified plants carrying all four genes that were resistant to foliage blight in greenhouse and field trials. Functionality of the genes was shown by using agro-infiltration with the effectors of corresponding *Avr* genes. The *P. infestans* in the field was highly aggressive and overcame the Black differentials R1 to R8 and R10 and R11, with R9 less infected probably due to its later maturation. For a number of hybrids and BC clones in field trials, yield and tuber number were not significantly different from that of the parent cultivar 'Delikat'. Evaluation of agronomic traits and processing qualities of selected BC_2 clones revealed valuable material for further breeding work.

Where introgression is performed at the tetraploid level, the result may not be a genotype with 48 Tuberosum chromosomes of which one or more has the introgressed gene(s). This need not affect the performance of the genotype and its vegetative propagation but could affect its fertility and use as a parent for further breeding. Tek et al. (2004) analysed the introgression of resistances to both early blight and tuber soft rot from a hexaploid somatic hybrid (A206, $2n = 72$) between a cultivated tetraploid *S. tuberosum* ($2n = 48$) and a diploid non-tuber-bearing *S. brevidens* (= *S. palustre*) ($2n = 24$). The three backcrosses to tetraploid cultivars Katahdin, Katahdin and Atlantic, respectively, resulted each generation in a clone for further crossing: C75 (2n =61, of which 12 are *S. brevidens*, five copies of chromosome 8), C75–5 ($2n = 50$, five copies of chromosome 8 of which one from *S. brevidens*) and C75–5 + 297 ($2n = 47$, four copies of chromosome 8 of which one from *S. brevidens*). Clone C75–5 + 297 was high yielding and resistant to both early blight and tuber soft rot. It had 47 chromosomes, including four copies of chromosome 8, three from potato and one from *S. brevidens* that was the only part of the wild species genome present. Hence, there must be another *S. tuberosum* chromosome that has only three copies. This research used the sequential GISH and FISH approach of Dong et al. (2001) to characterize potato germplasm containing chromosomes derived from wild species. Genomic in situ hybridization (GISH) uses whole genomic DNA from one species to detect its presence in a hybrid with another species, whose unlabelled genomic DNA is added in excess concentrations to block any unwanted hybridization with the probe. In contrast, fluorescence in situ hybridization (FISH) is used to map defined DNA sequences (including genes) physically and precisely to a specific region of a chromosome. Applications in potatoes can be found in the book chapter by Gavrilenko (2011). In the approach of Dong et al. (2001), GISH is first used to reveal the number of chromosomes from the wild species. Then the same cytological preparations are re-probed using BAC (Bacterial

Artificial Chromosome) markers specific to individual potato chromosomes. These markers are called chromosome-specific cytogenetic DNA markers (CSCDM). This sequential approach can reveal the number and identity of the alien chromosomes and their potato homologues in the breeding clones.

6.4.5 *Use of* S. pinnatisectum *(2x, EBN = 1)*

Sanetomo et al. (2019) successfully introduced a broad-spectrum late blight resistance gene, *Rpi-blb3*, which they found to be present in *Solanum pinnatisectum*, to potato by sexually breaking a strict reproductive barrier between *S. pinnatisectum* and cultivated potatoes. They used the *Sli* gene for inhibition of self-incompatibility combined with an EBN adjustment via ploidy manipulation. The introgression route was as follows. Chromosome doubling of diploid 1EBN *S. pinnatisectum* by colchicine treatment of seedlings produced tetraploid 2EBN *S. pinnatisectum*. This was crossed with a South American (Stenotomum/Phureja) diploid 2EBN clone carrying the *Sli* gene from *S. chacoense* to produce a triploid 2EBN hybrid. Chromosome doubling in vitro of the triploid hybrid produced a hexaploid 4EBN hybrid, which was used to successfully pollinate tetraploid potato cultivars to produce BC_1 hybrids, which were assumed to be pentaploid. Resistant BC_1 plants were used as male and female parents in further crosses to tetraploid potato cultivars to produce BC_2 progenies, and the process repeated to produce BC_3 progenies. In a BC_2 population tested for resistance to *P. infestans*, 93 genotypes were resistant and 104 were susceptible, as expected with the segregation of a major gene. The presence of the *Rpi-blb3*–specific marker band perfectly matched the resistance phenotype, thus confirming that the gene was *Rpi-blb3*. The reader is referred to the paper for more details.

In India, eight advanced hybrids (fourth backcross clones) have been developed from a tetraploid somatic hybrid between late blight-resistant *S. pinnatisectum* (2*x*, EBN = 1) and *S. tuberosum* dihaploid 'C – 13' (2*x*, EBN = 2). Four backcrosses to Indian potato cultivars resulted in the advanced hybrids being trialled just 5 years later (Luthra et al. 2020).

6.4.6 *Use of* S. chacoense, S. microdontum, S. spegazzinii *and* S. vernei *(all 2x, EBN = 2)*

Whilst the diploid 1EBN wild species just considered proved difficult to use in introgression breeding, the much more common 2EBN wild species proved relatively straightforward, with examples provided by *S. chacoense* and *S. microdontum* for virus resistance and *S. spegazzinii* and *S. vernei* for potato cyst nematode resistance. Two routes are available for introgression. The first uses colchicine to double the chromosome complement, thus producing a tetraploid 4EBN species, which will cross with tetraploid potato cultivars. The second initially uses a cross between a dihaploid of tetraploid *S. tuberosum* and other diploid germplasm. As a first example, quantitative resistance to both potato cyst nematodes (*Globodera. pallida* and *G. rostochiensis*) was found in accessions of *S. vernei* in the Commonwealth Potato Collection (CPC 2487 and CPC 2488). Colchicine treatment of seed from these and another *S. vernei* produced tetraploid plants of *S. vernei,* which were crossed at the Scottish Plant Breeding Station with Tuberosum potatoes in 1957 and 1958. The resulting hybrids were inter-crossed and also outcrossed to other cultivars and clones for four generations with selection for resistance and other desirable traits, before cultivars Morag and Glenna were released in 1985 and 1987, respectively (Bradshaw 2009). Whilst these were commercially acceptable cultivars, it took longer to produce a commercially successful cultivar with partial resistance to potato cyst nematodes. Clone 10300(13), a parent of Glenna and her sister clone 12288af23, was crossed with cultivar Cara to produce clone 15119 ac5. Then the cross between this clone and one with field resistance to late blight produced the cultivar Lady Balfour, which was National Listed in 2001. Genetic analysis of clone 12288af23 revealed a QTL of large effect for resistance on linkage group V (chromosome 5) (Bryan et al. 2002). Mapping was also done on related germplasm at the diploid level in the cross PDH538 × IVP48, where PDH538 is a primary dihaploid clone derived from a tetraploid clone with the *S. vernei* resistance and IVP48 is a diploid Group Phureja clone. Two QTLs for resistance were found on chromosome 5 and one on chromosome 9. The three QTLs acted in a simple additive fashion and explained 42% of the phenotypic variation.

A second example is similar work in Japan on resistance to *Potato virus Y* (PVY) from *S. chacoense*. Hosaka et al. (2001) found extreme resistance to PVY in the Japanese cultivar Konafubuki. The pedigree of Konafubuki shows that it resulted from a cross between susceptible cultivar Irish Cobbler and a chromosome doubled *S. chacoense*, followed by three backcrosses to susceptible tetraploid Tuberosum cultivars/clones. The 1: 1 segregation ratio of resistant to susceptible plants in the tetraploid population from crossing cultivar Kita-akari (susceptible) with Konafubuki (resistant) indicated a major dominant resistance gene (Ry_{chc}). A RAPD marker 38–530 was linked to Ry_{chc} with a recombination frequency of 16.3%. Sato et al. (2006) mapped the resistance gene to the distal end of chromosome 9. However, to reduce the complexity of the genetic analysis, they did the mapping at the diploid level. They produced dihaploid (haploid) plants of cultivar Konafubuki by pollination with haploid inducing (parthenogenetically) clones of *S. phureja.* A resistant haploid plant 98H20–5 (with RAPD marker 38–530) was crossed as the female parent with a susceptible diploid genotype, which had been selected for the presence of the *Sli* gene, which inhibits self-incompatibility, and absence of the RAPD marker band 38–530. The diploid genotype was descended from a cross between a *S. stenotomum*/*S.phureja* clone and a hybrid between *S. chacoense* 525–3 and *S. phureja* 1.22. The mapping population comprised 114 plants of which 50 were resistant and 64 were susceptible (a 1: 1 ratio). Details of the PVY assay using a Japanese isolate of PVY^N and of the molecular marker assays using RAPD and

RFLP markers can be found in the paper. It was known from other mapping work that RAPD 38–530 was located on chromosome 9. Only one recombinant out of 114 plants was found between Ry_{chc} and RAPD 38–530. However, as the recombination frequency was 16.3% in the tetraploid population, it can be inferred that a large chromosome segment harbouring Ry_{chc} and the RAPD marker 38–530 was introgressed from *S. chacoense* and is still retained in advanced cultivars.

6.4.7 *Introgression from Tetraploid* S. tuberosum *Group Andigena*

We have just seen that introgression of virus and cyst nematode resistance was done at the tetraploid level from *S. chacoense* and *S. vernei*. The landraces of tetraploid *S. tuberosum* Group Andigena have also proved a useful source of resistances to viruses and cyst nematodes, and introgression has been straightforward. A good example is the *H1* gene for resistance to the golden potato cyst nematode, *Globodera rostochiensis*, which started to become a serious problem in Europe in the early 1950s. The resistance to what became known as pathotypes Ro1 and Ro4 of *G. rostochiensis* was found by Ellenby in 1952 (Ellenby 1952) in an Andigena accession of the Commonwealth Potato Collection (CPC 1673) and shown to be a simply inherited dominant major gene (Toxopeus and Huijsman 1953). The *H1* gene was quickly backcrossed into the European cultivated potato, with cultivar Maris Piper released (National Listed) from the Plant Breeding Institute, Cambridge in 1966, and Pentland Javelin from the Scottish Plant Breeding Station, Edinburgh in 1968. Fifteen out of the 50 cultivars released from the Scottish Plant Breeding Station and the Scottish Crop Research Institute from 1968 to 2008 had the *H1* gene (Bradshaw 2009). The gene has been used extensively in potato breeding worldwide. The pedigrees of Maris Piper and Pentland Javelin are as follows.

1952 CPC 1673 × CPC 1673
↓
H4/31 × Ulster Knight Arran Cairn × Herald
↓ ↓
1956 Y22/6 × Clarke G50
↓
1966 Maris Piper

Maris Piper was still the most widely grown potato cultivar in Great Britain in 2019.

1952 Kerr's Pink × CPC 1673
↓
1953 Craigs Defiance × 2296(2.1)
↓
1955 Craigs Royal × 2356(53)
↓
1959 2693ac2 × 11-79
↓
1968 Pentland Javelin

6.4.8 Theory of Introgression at the Tetraploid Level

Where introgression is done at the tetraploid level from an outbreeding diploid species that has been made tetraploid through chromosome doubling, the initial cross will be between the doubled wild species with a desirable resistance gene, *R* say, that is not present in the tetraploid cultivar, and undesirable genes, *A* for example, that likewise are not present in the tetraploid cultivar. Hence, the initial cross will be as follows.

$$RRrr \times rrrr \text{ or } RRRR \times rrrr \text{ and } AAaa \times aaaa \text{ or } AAAA \times aaaa$$

and the initial hybrids will be: *Rrrr* or *RRrr* and *Aaaa* or *AAaa*

During the backcrossing, selection will be for *R*--- which will rapidly become *Rrrr* so that the backcrosses will be *Rrrr* × *rrrr* giving a 1: 1 segregation of *Rrrr* and *rrrr* in the offspring. In other words, the process will be the same as in diploids: *Rr* × *rr* giving a 1: 1 segregation of *Rr* and *rr* in the offspring. But what will happen at all other loci (or at least those unlinked to the one under selection) where no selection is practised (i.e. in absence of use of molecular markers)? Here, there is a difference between tetraploids and diploids if we start with *AAaa* and *Aa*, respectively. With the first backcross we start with *AAaa* × *aaaa* giving progeny *AAaa*, *Aaaa* and *aaaa* in the ratio of 1: 4: 1. It can be seen in Table 6.2 that the change in allele frequency is the same, but that the increase in the frequency of the homozygote is slightly slower in tetraploids than diploids.

Bradshaw and Ramsay (2005) also considered the increase in the frequency of whole chromosomes (four from recurrent parent in tetraploid and two from recurrent parent in diploid) and worked out that the elimination of unwanted whole

Table 6.2 Increase in the frequency (%) of recurrent parent allele (*a*) and homozygosity (*aaaa* in tetraploid 4*x* assuming chromosomal inheritance and *aa* in diploid) on backcrossing, starting with *AAaa* and *Aa* in tetraploid and diploid, respectively

Backcross generation	Frequency of *a* in 4*x*	Frequency of *a* in 2*x*	Frequency of *aaaa*	Frequency of *aa*
1	75.0	75.0	16.7	50.0
2	87.5	87.5	52.8	75.0
3	93.8	93.8	75.5	87.5
4	96.9	96.9	87.6	93.8
5	98.4	98.4	93.8	96.9
6	99.2	99.2	96.9	98.4

From Bradshaw and Ramsay (2005) with permission

chromosomes is faster at the diploid than at the tetraploid level. However, the difference between introgression at the diploid and tetraploid levels is not that big.

6.4.9 *Use of Dihaploids for Introgression*

The production of haploids (also called dihaploids) of *S. tuberosum* from 1958 onwards (Hougas et al. 1958) allowed Peloquin and his co-workers (Hermundstad and Peloquin 1987; Jansky et al. 1990) to develop a novel breeding strategy to introgress specific characteristics and to broaden the genetic base of potato in a way similar to that envisaged by Chase (1963) in his analytic breeding scheme. The hybrids between dihaploids of *S. tuberosum* and diploid wild species with an EBN of 2 will often form tubers in long-day growing conditions (Jansky et al. 2004; Jansky 2009). When they also produce 2*n* gametes by FDR (first division restitution; see Chap. 2), much of the genetic diversity of the wild species can be efficiently transferred to the tetraploid offspring from 4*x* x 2*x* crosses and results in about 25% of the wild species genes in the final product (Tai 1994; Jansky 2009). Tai (1994) concluded, however, that haploid-wild species hybrids need to be improved before they are used in 4*x* x 2*x* crosses, for example, through population improvement by recurrent selection (Rousselle-Bourgeois and Rousselle 1992). In other words, we are moving away from introgression of a few genes towards base broadening or incorporation as defined later in this chapter. In contrast, diploid hybrids with *S. raphanifolium* with resistance to cold-sweetening after storage for 3 months at 2 °C have been used by Hamernik et al. (2009) to produce commercially acceptable tetraploid offspring without extensive backcrossing to cultivated germplasm. These should be valuable parents for producing new cultivars suitable for processing.

As potatoes are heterozygous outbreeders, use of the same recurrent parent during introgression results in a self of the recurrent parent and hence inbreeding depression. This can be avoided by using different Tuberosum parents for each backcross, but result in an entirely new cultivar, which may or may not be the desired outcome. Currently, the only way to introduce a gene into a known cultivar is by genetic transformation, which is discussed in the final chapter. This should be cleaner and faster than backcrossing as there is no linkage drag requiring additional backcross generations to break any tight linkages between the desired allele and undesirable ones from the wild species. However, a new possibility is introgression into diploid inbred lines and the production of diploid F_1 hybrid cultivars for propagation through true botanic seed.

6.4.10 Introgression into Diploid Inbred Lines

The advent and current interest in diploid F_1 hybrid breeding for true potato seed propagation means that there is a place for introgression of major genes from diploid 2EBN species into diploid inbred lines of potato. It was Lindhout et al. (2011, 2018) who demonstrated the feasibility of such breeding through the use of the *Sli* gene, a dominant self-incompatibility inhibitor which was found in *S. chacoense* by Hanneman (1985) and mapped to a distal end of chromosome 12, thus showing that it is independent of the *S* locus on chromosome 1 (Hosaka and Hanneman 1998). Lindhout et al. (2011, 2018) produced inbred lines that combined a high level of homozygosity with good plant performance and good tuber quality and yield. These were then crossed to produce diploid F_1 hybrids, which were uniform and showed good tuber quality and which can be marketed as diploid cultivars. Their first example of introgression was for genes for resistance against *Phytophthora infestans* (Su et al. 2019) and provides an example of molecular marker-assisted introgression. This offers the possibility of faster progress than can be achieved by traditional backcrossing because one can select genotypically against the unadapted species genome, as well as genotypically for the desired gene(s). With adequate molecular marker coverage of all chromosomes, it is possible to estimate the optimal combination of population sizes and number of backcross generations and to select in a very precise way for the desired products of meiosis in each backcross generation (Hospital 2003). Let us look at some simple theory before returning to the actual example.

Let us consider a diploid wild species with a dominant resistance gene *R,* which we want to introgress into a diploid inbred cultivated potato. As the diploid wild species is likely to be a heterozygous outbreeder, let us start with the F_1 between the wild and cultivated potato where we select a resistant individual for the first backcross to the cultivated potato. Let us consider a typical chromosome with four markers *A*, *B*, *C* and *D,* which are unique to the wild species. In addition, let us consider one chromosome out of the 12 where marker *B* is also the resistant allele. There is always just one chiasma (crossover) per chromosome pair. Furthermore, this chiasma is equally likely to occur at any location along the chromosome pair (Fig. 6.3).

If we start with the chromosome with the resistance gene, the gene segregates (*R*: *r*) one to one in the first backcross generation. If phenotypic selection for resistance is 100% effective (or a molecular marker such as *B* occurs within the gene), one half

F_1 × cultivated potato

AB(R)CD × *abcd*
a b (r) c d *abcd*
↓
BC_1
3 *AB(R)CD/abcd*
3 ***a b (r) c d/abcd*** select for abcd on other 11 chromosomes
1 *a B(R)CD/abcd*
1 *A b(r) c d/abcd*
1 ***AB(R)c d/abcd*** select for *B*(*R*) on chromosome with *R*
1 *a b(r) CD/abcd*
1 *AB(R)Cd/abcd*
1 *a b(r) c D/abcd*

abcd × *abcd*	*AB(R)cd* × *abcd*
abcd *abcd*	*ab(r) cd* *abcd*
↓	↓
BC_2	
11 chromosomes	Chromosome with *R*
all *abcd/abcd*	5 *AB(R)cd/abcd*
	5 *a b(r) cd/abcd*
	1 *Ab(r)cd/abcd*
	1 ***aB(R)cd/abcd*** select to self
	aB(R)cd
	a b(r)cd
Self ↓	↓
all ***abcd/abcd***	1 ***aB(R)cd/aB(R)cd*** true breeding on selfing
	2 *aB(R)cd/ab(r)cd* segregates on selfing
	1 *ab(r)cd/ab(r)cd*

Fig. 6.3 Marker-assisted introgression of dominant resistance gene (*R*) from wild into cultivated potato (diploid inbred line), where *A*, *B*, *C* and *D* are molecular markers for a typical wild species chromosome and for chromosome with resistance gene, *R* and *B* are the same locus, and where there is just one chiasma which is equally likely between any two markers

of the selected offspring will have the whole chromosome from the wild species. The other half of the progeny will have varying lengths of this chromosome, depending on where the chiasma occurred, namely *ABC*, *BCD* and *AB*. Hence, selection for absence of *C* and *D* would select for the shortest length of chromosome (*AB*) from the wild species, but with the resistant allele, and would be the obvious parent of the next backcross (frequency 1/12). In the second backcross generation (BC_2), selection of genotype *aB(R)cd/abcd* (absence of *A*, *C* and *D*, but presence of *B*(*R*) (frequency 1/12) would further reduce the length of the contribution of the chromosome from the wild species.

The more markers available, the shorter in theory is the length linked to the *R*-gene that can be sought, but what can be achieved in practice will be depend on the frequency and position of the chiasmata and the size of the backcross population. So, in subsequent backcrosses the length linked to the *R*-gene will reduce as a result of more crossovers, and some of these will be really close to the *R*-gene. Eventually, the selected offspring will have very little of the wild species chromosome attached to the resistance gene. Nevertheless, some undesirable wild species alleles may be retained through their linkage to the selected gene, a phenomenon known as linkage drag. If crossovers do not occur in the region of the *R*-gene, then linkage drag could be more of a problem. Hospital (2003) looked at linkage drag in detail and concluded that a typical marker-assisted backcross scheme should involve three to four backcross generations, rather than just one or two. This permits a more drastic reduction of linkage drag while reducing the genotyping effort and increases the probability of success in advanced backcross generations. Finally, two generations of selfing will ensure that we have an inbred line that is true breeding for the resistance gene (*Rr* selfed gives *RR*, 2*Rr* and *rr*; and *RR* selfed will be true breeding), but otherwise resembles the cultivated potato parent apart from any effects of linkage drag.

Now let us consider a gene (*A/a* say) on one of the other chromosomes. In the absence of selection, half the offspring in the first backcross generation will be *Aa* and half *aa* (*Aa* × *aa* gives *Aa*: *aa* offspring in 1:1 ratio). Then in the absence of selection, one quarter of the offspring in the second backcross will be *Aa* and three quarters *aa* (*Aa* × *aa* and *aa* × *aa* combined give *Aa*: *aa* offspring in 1:3 ratio). Hence, the frequency of the cultivated allele *a* increases with each backcross generation: 0.750, 0.875, 0.938, 0.969, 0.984, 0.992 etc. so that in the sixth backcross generation the cultivated parent's alleles will be at a frequency greater than 99% on the 11 chromosomes not carrying the resistance gene.

In practice in the backcross generations, some offspring will display more cultivated species like phenotypes than others. Hence, some phenotypic selection can be practised in favour of the cultivated parent phenotype and against the wild species phenotype. As a consequence, the cultivated allele frequencies will increase faster than by chance and fewer than six backcrosses may suffice. However, let us now consider the situation where we have whole genome coverage with equally spaced molecular markers: four (*A*, *B*, *C* and *D*) per chromosome in our simple example. The presence of the four alleles *A–D* in an individual tells us that it contains the whole chromosome from the wild species, whereas absence of all four alleles tells us that it contains two chromosomes from the cultivated potato parent (frequency ¼). In order to achieve the latter for all 11 chromosomes not carrying the resistance gene, we would need an impossibly large backcross population as $(¼)^{11}$ multiplied by the 1/12 for the other chromosome is approximately 2×10^{-8}. However, if we select against individuals with chromosomes having all four markers *A*, *B*, *C* and *D* (i.e. whole chromosomes from wild species), for all 11 chromosomes, we find that $(¾)^{11}$ (= 0.0422) of the individuals lack whole chromosomes from the wild species for all 11 chromosomes. When this is multiplied by 1/12, we now have 3.5 individuals in 1000. If we pursue the more modest ambition of eliminating the *ABCD* chromosomes along with those lacking the resistance gene (*r*), we have $¼(¾)^{11}$ (= 0.011). This means that in a modest sized first backcross generation of 200 individuals, on average 2 can be found that harbour the *R*-gene but no whole chromosomes from the wild species. Hence, we have fixed parts of all of the cultivated parent's chromosomes and the average frequency of the cultivated allele can be shown to be around 0.833 compared with 0.75 by chance. In my book on plant breeding (Bradshaw 2016), I showed how the process of eliminating as many as possible of the wild species alleles can be repeated in the second and third backcrosses and gave the results. The conclusion was that in a modest sized backcrossing programme of 200 individuals per generation, in just three generations it is possible to achieve a genotype identical to the cultivated parent, apart from an introgressed *R*-gene and some linked genes from the wild species. I did go on to consider the situation where there are always two chiasmata (crossovers) per chromosome pair, but this is more complicated. The general theory can be found in the account by Hospital (2003) and the process explored by computer simulations with realistic chiasmata frequencies and distributions for any given crop species. He considered both reduction of linkage-drag on the carrier (*R*-gene) chromosome and selection on non-carrier chromosomes. He concluded that dense coverage of the non-carrier chromosomes by molecular markers is not necessary to increase the overall recipient (recurrent parent) genome content; in fact, four markers are sufficient for a chromosome of 100 cM. Furthermore, three or four backcross generations are sufficient to increase the recipient genome content on non-carrier chromosomes above 99%, compared with six backcross generations without selection. The results of simulations showed that three backcross generations of size 117, 171 and 370 are required to provide 98.5% recipient genome content. The total of 658 individuals is not that different from the three backcrosses of 200 individuals (total 600) in my simple example. One final point from Hospital (2003) is worth making here. It is not possible to simultaneously manipulate more than five or six known target genes in a marker-assisted backcross programme because of the population sizes required.

Let us return to the example provided by Su et al. (2019). They separately introgressed four different genes for resistance to *Phytophthora infestans* from diploid wild species into diploid inbred lines of potato. The four wild species and their resistance genes were ***Solanum avilesii*** 478–2 (***Rpi-avl1***, chromosome 11, Verzaux et al. 2011), ***Solanum tarijense*** 852–5 (***Rpi-tar1***, chromosome 10, Haverkort et al. 2016), ***Solanum chacoense*** 543–5 (***Rpi-chc1***, chromosome 10, Vossen et al. 2011, allelic to ***Rpi-tar1***) and ***Solanum venturii*** 283–1 (***Rpi-vnt1***, chromosome 9, Foster et al. 2009; Pel et al. 2009). Linked SNP markers were used to select backcross plants with the resistance genes. The three diploid inbred lines used as recurrent parents, P1, P2 and P3, respectively, had homozygosity estimates of 88%, 88% and 79% based on a set of 67 informative molecular markers (SNPs) equally distributed over the 12 potato chromosomes. They came from Solynta's diploid breeding programme, which used the dominant *Sli* gene from *S. chacoense* to inhibit self-incompatibility. All combinations of the four wild species and three inbred lines gave 12 F_1 populations. The best growing and flowering F_1 plants with the resistance genes were backcrossed to the parental lines and 2000 plants of the six best BC_1 populations were screened with four markers that allowed selection of 45–70 plants per backcross (exact numbers in paper) with a recombination (crossover) close to the resistance gene. The 45–70 BC_1 offspring plants were all genotyped for their overall percentage of recurrent parent genome using the 67 informative markers. Out of approximately 10–20 plants with a high percentage of recurrent parent genome, the best five or so for flowering and vigour were used for backcrossing to the recurrent parent (exact numbers in paper). The resulting BC_2 populations were screened with markers on the other side of the resistance gene (except for *Rpi-avl1* which is near the end of chromosome 11) and those chosen were again all genotyped for their overall percentage of recurrent parent genome. The entire background selection was very successful, because most of the genome of the donor wild species parent was no longer present after two backcross generations (paper gives figures of 75%–97% recurrent parent genome in one backcross-two population). Normally, selfing would be done to fix the resistance alleles in the inbred lines. However, for research purposes, they crossed BC_2 plants with different resistance genes to generate populations with individuals with none, one or two different resistance genes. Late blight testing was done at three sites. The resistance levels of the combined gene combinations *vnt*1 + *chc*1; *tar*1 + *vnt*1; *avl*1 + *tar*1; *avl*1 + *chc*1; and *avl*1 + *vnt*1 were generally higher than the resistance levels in plants with a single resistance gene (again details can be found in the paper).

6.5 Combining Major Genes and QTL Alleles of Large Effect at the Tetraploid Level

We have seen, and will continue to see, the use of molecular markers in the introgression of major gene alleles and QTL alleles of large effect from wild and cultivated species into tetraploid potato clones and then cultivars. Some markers will be in the alleles of interest, or tightly linked to them, so that they can be used in place of expensive, laborious or time-consuming phenotypic screens for the trait of interest. Other markers will be used to track what is happening during meiosis and to eliminate the unwanted wild species genome as quickly as possible. Once the introgression is complete, the breeder will want to combine desirable alleles for different traits into single clones (genotypes) for assessment as new cultivars. It will be important to have easy to use (breeder friendly) molecular markers that lie within the desired allele or are diagnostic of it (i.e. tightly linked only to the desired allele in all germplasm). In earlier chapters, we encountered desired alleles and the markers used to map them, some but not all of which could be considered breeder friendly diagnostic markers (e.g. see Moloney et al. 2010; Gebhardt et al. 2011; Gebhardt 2013; Sharma et al. 2014; Ramakrishnan et al. 2015). Selection of these desirable alleles will be part of a more general potato improvement programme involving cycles of hybridization and selection in which new cultivars are produced each cycle to transfer progress to farmers' fields. In the next chapter, we shall see that each cycle comprises multistage, multitrait selection over as many as eight clonal generations, commonly starting with 100,000 seedlings in a glasshouse from 200 to

300 crosses. The number of potential cultivars is typically reduced to 4000 clones in unreplicated plots at a seed-site and then to 1000 clones in the first year of replicated yield trials. It is in this context that I want to consider the theory and practice of molecular-marker-assisted selection in potato breeding.

6.5.1 Theory for Marker-Assisted Selection

Let us start by considering one locus and all initial crosses in a breeding programme of type *Aaaa* × *aaaa* so that in the absence of double reduction, the offspring are $\frac{1}{2}$ *Aaaa* and $\frac{1}{2}$ *aaaa*. Let us discard all offspring lacking the desirable *A* allele and randomly inter-cross the remaining genotypes (*Aaaa* × *Aaaa*) so that in the absence of double reduction, the offspring are $\frac{1}{4}$ *AAaa*, $\frac{1}{2}$ *Aaaa* and $\frac{1}{4}$ *aaaa*. Once again, let us discard all offspring lacking an *A* allele and randomly inter-cross the remaining genotypes ($\frac{1}{3}$ *AAaa* and $\frac{2}{3}$ *Aaaa*) and keep repeating the process. Each generation gametes *AA*, *Aa* and *aa*, in the ratios r: s: t, are randomly mated to produce genotypes *AAAA*, *AAAa*, *AAaa*, *Aaaa* and *aaaa* in the ratios r^2, 2rs, $2rt + s^2$, 2st and t^2, and genotype *aaaa* is discarded. The results for seven generations are shown in column two of Table 6.3 (Bradshaw 2017). The frequency of *A*— increases from 0.5000 to 0.9524, but even at this high frequency, that of genotype *AAAA* is only 0.0527. This is because genotype *AAAA* approaches fixation slowly as the frequency of allele *A* increases (frequency of *A* is 0.506 in generation seven). The frequencies for two unlinked loci are simply the squares of those for one locus, and likewise as more unlinked loci are added. If just two loci were segregating, one could screen 4000 seedlings from the initial crosses, in the glasshouse with molecular markers, and expect to find about 1000 (4000 × 0.25) with both markers to take to the field for phenotypic evaluation of other traits. If four loci were segregating, one would need to screen 16,000 seedlings in the glasshouse to find about 1000 (16,000 × 0.0625) with all four markers. Alternatively, one could screen 4000 seedlings in the

Table 6.3 Marker-assisted selection for presence of at least one copy of desirable allele (*A*) at from 1 to 16 unlinked loci: frequencies of desired genotypes

Generation	Loci						
	1	2	3	4	8	12	16
1	0.5000	0.2500	0.1250	0.0625	0.003906	0.000244	0.000015
2	0.7500	0.5625	0.4219	0.3164	0.1001	0.03168	0.01002
3	0.8488	0.7205	0.6115	0.5191	0.2694	0.1399	0.07259
4	0.8953	0.8016	0.7176	0.6426	0.4128	0.2652	0.1704
5	0.9226	0.8512	0.7853	0.7245	0.5249	0.3803	0.2756
6	0.9403	0.8842	0.8314	0.7818	0.6112	0.4778	0.3736
7	0.9524	0.9070	0.8638	0.8227	0.6769	0.5569	0.4581

From Bradshaw (2017) with permission

glasshouse and random mate the 250 (4000 × 0.0625) or so with all four markers, and then raise another seedling generation in the glasshouse. This time, one would expect over 1250 (4000 × 0.3164 = 1266) out of 4000 seedlings to have all four markers.

If eight loci were segregating, one would need to screen some 256,000 seedlings in the glasshouse to find about 1000 (256,000 × 0.003906) with all eight markers. However, one could screen 4000 seedlings in the glasshouse and random mate the 16 (4000 × 0.003906) or so with all eight markers, and then raise another seedling generation in the glasshouse. This time, one would expect about 400 (4000 × 0.1001) out of 4000 seedlings to have all eight markers. Another cycle of crossing and selecting would result in over 1000 (4000 × 0.2694 = 1078) out of 4000 seedlings having all eight markers. If there were 16 loci segregating, only 1 (65,536 × 0.000015) in 65,536 seedlings would have all 16 markers. Thus, above about 8 loci segregating, marker-assisted selection soon becomes impractical to get started for all the desired alleles. However, once one is making crosses between clones that all have every marker, a steady increase in the frequencies of such clones occurs over generations simply through elimination of offspring clones not possessing all of them (Table 6.3), but even after seven generations, only about 11 in 4000 seedlings will have reached fixation at two loci. Hence, it is important to think in terms of a long-term population improvement programme to increase the frequencies of desirable combinations of alleles at different loci, and not just in terms of making crosses to breed a new cultivar.

These theoretical results provide an explanation for the lack of potato cultivars in which more than a few major genes and QTL alleles of large effect have been stacked. In the SPBS/SCRI potato breeding programme (Bradshaw 2009), introgression of resistance genes from the wild and cultivated potato species of Latin America started for late blight in 1932, for viruses in 1941 and for potato cyst nematodes in 1952. After over 75 years of breeding, no cultivars had been produced with the number of major genes (8) we have been discussing in this section. Deliberate parental line breeding in 5-year cycles was done at the Plant Breeding and Acclimatization Institute (IHAR)-National Research Institute in Poland from the early 1950s, with the aim of producing parents with multiple resistances to six viruses, late blight and soft rot (Zimnoch-Guzowska et al. 2013). From 1968 to 1993 (25 years), the number of major genes for virus resistance in parental lines was increased from two to five, and recently, a molecular marker (CAPS GP122718) has been made available to commercial breeders for selecting for the *Ry-fsto* gene for complete resistance to all known strains of PVY. In India, Sharma et al. (2014) used diagnostic molecular markers to screen a diverse collection of 126 potato genotypes in the National Germplasm Repository for the presence of three genes for resistance to late blight, two for resistance to PVY and three for resistance to potato cyst nematodes. They found 8 out of the 126 genotypes possessed a maximum of four or five genes. More generally, Vos et al. (2015) found that since 1945, introgression breeding had introduced 24% of the 14,530 SNPs they examined, but by 2005 the highest frequencies of introduced alleles were 16% and 18% for two SNPs introduced before 1962.

6.5.2 Stacking More Than Eight Genes

One way of stacking more than eight genes would be to fix the desired allele at each locus, but this is more easily said than done. Let us explore marker-assisted selection with the ability to recognize and select for allele dosage. Breeding to produce multiplex parents for disease and pest resistance genes was first discussed by Cadman (1942) with respect to the *Nx* gene for race-specific resistance to PVX and explained in detail by Toxopeus (1953) and Bradshaw and Mackay (1994). The scheme is straightforward for a single locus, but takes three generations:

$$Aaaa \times Aaaa \rightarrow \frac{1}{4} AAaa$$

$$AAaa \times AAaa \rightarrow \frac{1}{4}(AAAA, AAAa \text{ in ratio} 1:8)$$

$$AAAa \times AAAa \rightarrow \frac{1}{4} AAAA$$

A more recent example is provided by Kaushik et al. (2013) who developed a triplex (*YYYy*) parental potato line with extreme resistance to *Potato virus Y* (PVY) using marker-assisted selection with a sequence characterized amplified region (SCAR) marker (RYSC3) for the Ry_{adg} gene. The marker was detected in 11 PVY-resistant genotypes, including putative duplex clones, and was absent in eight PVY-susceptible genotypes. Furthermore, the SCAR marker data for resistance agreed with that from a double-antibody sandwich enzyme-linked immunosorbent assay (DAS-ELISA) of plants inoculated mechanically with the PVY° strain using a spray gun method. Based on marker band intensities, three parental clones (YY-3, YY-6 and YY – 13) were selected from the 11 and crossed to ascertain their duplex status and to produce triplex/quadruplex lines. The three crosses gave resistant to susceptible segregation ratios that were not significantly different from that expected for duplex parents and random chromatid segregation (20.8: 1). Five putative triplex clones were selected from the progeny of two duplex × duplex crosses (YY-6 × YY-3 and YY-13 × YY-3) based on their higher marker band intensities. Their triplex status was checked in testcrosses to nulliplex parents, and just one clone (YY-6/3 C11) was confirmed as triplex, with a resistant to susceptible segregation ratio not significantly different from the 27: 1 expected with random chromatid segregation. The paper has a colour photograph and description of the triplex clone.

If we move on to consider fixation at more than one genetic locus, the frequencies for two unlinked loci are simply the squares of those for one locus, and likewise as more unlinked loci are added. If four loci were segregating, 1 in 256 seedlings each (glasshouse) generation would have the desired genotype, so a population of 4096 seedlings would on average provide 16 for the next round of crossing and selection. However, with six unlinked loci, just 1 in 4096 seedlings would have the desired genotype. In order to fix eight loci with similar population sizes, one would need additional generations with crosses of type *AAaa* × *Aaaa*, *AAAa* × *AAaa* and *AAAA* × *AAAa*. Hence, fixing four loci followed by another four, or fixing all eight simultaneously, would take six generations in the glasshouse without selection for any other traits. This is clearly a lot of work so breeders may be content to practise marker-assisted selection with elimination of clones not possessing the desired markers, as described in the previous section as part of a long-term population improvement programme.

6.5.3 *Practice of Marker-Assisted Selection*

Although we have been considering using marker-assisted selection (MAS) in the seedling generation in the glasshouse, the few published practical examples have applied MAS to the second field generation, as recommended by Slater et al. (2013) as cost effective in selecting a gene for resistance to potato cyst nematodes or to PVY. Furthermore, pyramiding of genes to date has been restricted to two or three genes. Ortega and Lopez-Vizcon (2012), for example, gave an account of the work done by the small breeding company Appacale in Spain from 1998. They identified the Ry_{adg} and Ry_{sto} genes for resistance to PVY with SCAR marker RYSC3 and SSR marker STM0003, respectively; the *Gro1–4* and *H1* genes for resistance to the cyst nematode *Globodera rostochiensis* with specific primers and marker TG689, respectively; and the resistant allele at QTL *Gpa5* with both marker SPUD1636 and HC. The original references and details can be found in their paper. The markers were applied to breeding clones in their second year of field trials in unreplicated seven-tuber plots at the seed site of the breeding programme, on a modest scale over the period 2000–2010 with 1940 analyses done with 778 positive results. However, because of the modest scale, by 2010 the company had just four breeding clones with more than one resistance gene for use as parents. The breeders thought that applying the markers in the first year, at the single plant stage when most of the material will be discarded, would be too expensive and not efficient. Moreover, breeding clones analysed in the second year with molecular markers are selected based on their genotype plus agronomic performance. Thus only 199 out of 684 clones with an *Ry* gene advanced beyond the second year.

A second example is provided by the breeding work in Japan described by Mori et al. (2015). In 2015, they were using a multiplex PCR method to simultaneously detect markers for resistance genes *H1* (*Globodera rostochiensis*), *R1* and *R2* (*Phytophthora infestans*), Ry_{chc} (PVY) and *Rx1* (PVX). They point out that the multiplex PCR method saves labour and decreases the number of tests with DNA markers; for example, at the Nagasaki Agricultural and Forestry Technical Development Centre, genotypes with multiple resistances are selected from 300 to 400 clones at the second clonal generation of unreplicated 8-plant plots. The clones selected based on molecular markers are then evaluated for other agronomic characteristics, and disease resistance is confirmed by field evaluation at later stages. Finally, they are actively seeking clones with more than one copy of each resistance gene (*RRrr*, *RRRr* and *RRRR*).

A third and most recent example is the use of kompetitive allele-specific PCR (KASP) markers for disease and pest resistance genes in the Teagasc breeding programme in Ireland, where molecular markers are routinely deployed at the second field generation, which generally comprises 2000–4000 unreplicated clones (Meade et al. 2019). KASP is a fluorescence-based genotyping technology based on allele-specific oligo-extension that is increasingly being used to assay a large number of samples for a limited number of specific loci. Meade et al. (2019) presented an approach to consolidate various types of molecular markers into a single SNP-based platform (KASP) using whole-genome sequencing of carefully constructed pools of DNA (i.e. to convert diagnostic polymorphisms from one format into the KASP platform). The diagnostic marker is first used to genotype tetraploid cultivars and breeding clones in order to carefully construct DNA pools of entries in which the marker is either present or absent. These pools are then subjected to whole-genome resequencing. After alignment of the sequencing reads to the potato genome, the region surrounding the marker is examined for polymorphisms unique to the marker positive pool. Candidate regions are assessed for their suitability for KASP assay design, taking into account the ‘off-target’ variation revealed in each pool. Meade et al. (2019) used as examples molecular markers for the *H1* (*Globodera rostochiensis*) and *R2* (*Phytophthora infestans*) resistance genes. They further showed how the sequence data generated could be used to aid design of KASP assays for diagnostic SNPs for other traits mined from the literature. They used a set of 149 different commercial cultivars and advanced breeding clones for which genotypic information generated with both SCAR and KASP assays was available for comparison, and hence for KASP assay validations. The *H1* KASP assay had a concordance rate of 96% with the 57R marker, and the *R2* KASP assay had a concordance rate of 98% with *R2* SCAR marker. They also presented data from screening DNA from 1505 breeding clones and controls in the second field generation with the KASP assays. A total of 718 samples were positive for the *H1* assay (48%) and134 for the *R2* assay (9%). The two-dimensional clustering of data points showed that the KASP approach has the potential to allow dosage determination. In each assay, the cluster at the bottom of the Y-axis identified nulliplex (susceptible) breeding clones, while the next cluster contained the appropriate simplex positive control for each disease resistance gene. For both the *H1* and *R2* assays, a third cluster suggested some breeding clones were duplex for the marker. Cost, labour requirements and time required were all conducive to use in practical potato breeding programmes such as the one at Teagasc.

A fourth and final example is concerned with markers for improving tuber quality. Li et al. (2013) used a population of 243 tetraploid cultivars and German breeding clones for association mapping of new candidate genes functional in carbohydrate metabolism and the development of allele-specific markers. They found new DNA variants at loci encoding ADP-glucose pyrophosphorylase and the invertase *Pain-1* gene, which were associated with positive or negative effects on chip colour, tuber starch content and starch yield. To facilitate marker-assisted selection (MAS), user-friendly PCR assays were developed for specific candidate gene alleles. MAS and marker validation were performed in two tetraploid breeding populations, using various combinations of seven allele-specific markers associated with the traits. The first multi-parental population comprised 500 advanced tetraploid clones from the fifth to eighth year of phenotypic selection of a commercial breeding programme from which 76 clones were selected for having different combinations of five positive and corresponding negative marker alleles. Genotypes combining five positive marker alleles subsequently performed on average better than genotypes with four negative alleles and one positive allele. When tested individually, seven of eight markers (five plus three additional ones) showed an effect on at least one quality trait. The direction of effect was as expected. Combinations of two to three marker alleles were identified that significantly improved average chip quality after cold storage and tuber starch content. The second population was 576 F_1 clones from a cross between cultivars Diana and Candella in which 18 groups were identified for having in common various combinations of six markers. MAS of 121 clones did not subsequently give the expected phenotypic results when evaluated for 2 years for chip quality and tuber starch content. The authors concluded that MAS for tuber quality should not rely on single-cross combinations but should be applied to multiple parents and their progeny and should rely on the choice of suitable marker combinations. They suggested that MAS can first be used to purify parental populations from negative alleles and to increase frequency and dosage of positive alleles (some alleles with positive trait associations were at a low frequency in the population of 243 cultivars and breeding clones). Furthermore, pre-selection for general plant performance traits such as vigour would reduce the number of progeny clones to be subjected to MAS and phenotypic evaluation of tuber quality could then be performed on the remaining clones at a later stage in the breeding cycle.

Although three of the examples just considered applied MAS to the second field generation, a visit in January 2020 to two large Dutch potato breeding companies convinced me that MAS should be applied to the seedling generation. They can now do MAS on 100,000 seedlings at a very early growth stage. This means that the seedlings can be kept growing slowly in a controlled environment room until the analyses have been done. Then those seedlings with the desired markers can be transplanted to pots and grown in a glasshouse. Hence, combining genes at between 8 and 16 loci is now realistic for potato breeders so that only clones with these genes will be taken to the field. However, the breeder will need to decide the relative resources for marker-assisted introgression and then combining introgressed genes for cultivar production.

6.6 Broadening the Genetic Base of Potatoes

6.6.1 Base Broadening of Potatoes in Europe and North America

During the second half of the twentieth century, recognition was given to the value of the cultivated species of South America for broadening the genetic base of European and North American breeding programmes. One of the earliest examples of base-broadening (also described by some as incorporation) was the production of Neotuberosum (or long-day Andigena) potatoes, but there are different views on its success. As a potato breeder and geneticist, I have tried to give fair assessments of the programmes in my reviews of potato breeding, the most recent of which was in my book on plant breeding (Bradshaw 2016). The first Neotuberosum programme was started by Simmonds from the Commonwealth Potato Collection in 1959 (Simmonds 1969). A gene pool of short-day Andigena potatoes with origins approximately 45% Bolivian accessions from the CPC, 35% south Peruvian, 10% north Peruvian and 10% Colombian, was subjected to recurrent mass selection on a 2-year cycle in outdoor plots in Great Britain under northern latitude, long-day summer conditions. In the first year, large populations of seedlings were grown in the field and selected for high yields of tubers of acceptable sizes, shapes and colours. In the second year, the selected tubers were planted in an isolation site and open-pollinated berries harvested to provide the next generation of seedlings. After five cycles of selection, Glendinning (1975) achieved mean yields per plant of 1.15 kg for Neotuberosum compared with 0.33 kg for unadapted Andigena and 0.76 kg for Tuberosum, albeit from a late harvest. In other words, adaptation to long days occurred quickly. The better Andigena clones were comparable in maturity to long-day Tuberosum cultivars and better on average in terms of late blight resistance. These clones were more variable in tuber shape than modern cultivars and inferior in regularity of tuber shape, but of similar cooking quality. As a consequence of their rather 'rough' appearance, it subsequently proved difficult to breed successful cultivars from crosses of this Neotuberosum material with intensively selected Tuberosum clones, despite yield heterosis in crosses to modern cultivars. Just one cultivar, Shelagh (released 1986), out of 50 released from 1968 to 2008 by the Scottish Plant Breeding Station (SPBS) and Scottish Crop Research Institute (SCRI) (Bradshaw 2009) had a Neotuberosum parent despite such parents being used over the period from 1969 to 1998, with intensive evaluation of hybrids from 1979 to 1987. Neotuberosum needed improving for traits in addition to tubering in long days to have real impact because it was not as good as intensively selected Tuberosum. However, recurrent mass selection was discontinued in 1978, when the emphasis changed to evaluating the potential of this improved material in breeding cultivars, a task that was essentially completed by 1987. In contrast, 15 out of the 50 SPBS/SCRI cultivars starting with Pentland Javelin in 1968 have the *H1* gene for resistance to *Globodera rostochiensis* introgressed from a short-day Andigena accession, as does the most widely grown cultivar in Great Britain, Maris Piper released in 1966. As seen earlier in this chapter, at most three backcross generations were required for the original introgressions that produced cultivars Maris Piper and Pentland Javelin.

Experiences were similar elsewhere in Europe and in North America, as reviewed by Bradshaw and Bonierbale (2010). In the USA, Munoz and Plaisted (1981) achieved an increase in yield over six cycles of selection from 0.32 to 1.37 kg per plant (hill). Plaisted (1987) introduced deliberate pollination of selected clones with bulk pollen to avoid inbreeding depression from selfing in his Neotuberosum programme. He also achieved more vigorous plants for selection by raising his seedlings in pots to produce seed tubers for planting field trials. This enabled him to screen the seedlings in the glasshouse for resistance to late blight and viruses, but it also increased the length of each cycle to 3 years. By 1987, his programme had released just one cultivar, Rosa, which was a Tuberosum × Neotuberosum hybrid, although Neotuberosum germplasm did provide adapted sources of resistance to important pathogens, including *Globodera rostochiensis* (golden potato cyst nematode) and *Potato virus Y* (PVY) (Plaisted 1987). In Canada, Tarn and Tai (1983) concluded that first-generation hybrids between Tuberosum and Neotuberosum were not the best use of their Neotuberosum. They also advocated further improvement of their Neotuberosum and/or backcrosses of the hybrids to Tuberosum. No doubt the debates will continue about the relative merits of introgression and base-broadening (incorporation) in tetraploid potato. However, Ghislain et al. (2009b) provided a final twist to the Neotuberosum story. They were studying six Neotuberosum clones and derived germplasm from Plaisted's programme, which in turn was derived from the original populations developed by Simmonds. Using microsatellite (simple sequence repeat, SSR) and plastid DNA marker data, they found that the Neotuberosum germplasm was closely related to Chilotanum Group landraces from lowland south-central Chile rather than to Andigenum Group germplasm. They interpreted this quite unexpected finding to be the result of strong rapid selection against the original Andigenum clones after unintended hybridization with Chilotanum Group germplasm. In addition, they showed that Neotuberosum and Andigenum Group germplasm did not serve to broaden the overall genetic diversity of advanced potato cultivars, but rather that Neotuberosum clones and those not using this germplasm were statistically identical with regard to genetic diversity as assessed by SSRs.

Similar work was done with diploid potatoes of Groups Phureja/Stenotomum. Carroll (1982) (Fig. 6.4) and Haynes (Haynes and Lu 2005) produced populations adapted to long-day conditions in Europe and North America, respectively. The starting material for Carroll's programme was diploid accessions from the Commonwealth Potato Collection (CPC) comprising approximately 75% Phureja and 25% Stenotomum. The programme was conducted over the period 1962–1979 and relied on natural insect pollination and seed set and included both seedling and tuber populations with a minimum generation cycle of 2 years. In practice, the generation time was 3 years once seedlings were raised in a polythene tunnel rather than the field. The self-incompatibility of the diploids was presumed to ensure cross-pollination. The population adapted to tubering in long days after only one to two generations of selection, possibly because Group Phureja had already been selected in the Andes for faster tuber development. Yield improved across several generations mainly as a result of increase in tuber size without a reduction in tuber numbers, and the proportions of oval and long oval, regular-shaped tubers increased. Direct hybridization of Carroll's improved diploid population with tetraploid potato cultivars via unreduced pollen grains ($4x$ x $2x$ crosses) resulted in tetraploid hybrids, some of which were superior to standard tetraploid cultivars in both total and marketable yield (Carroll and De Maine 1989). Furthermore, diploid cultivars, such as Mayan Gold, have now been produced from this population, but they are targeted at

Fig. 6.4 Long-day-adapted Phureja Group potatoes from the population of Carroll (1982). (From Bradshaw and Bonierbale 2010 with permission; Source: SCRI)

niche markets because their yield is only two thirds that of Tuberosum potatoes, and further selection for longer tuber dormancy is desirable.

The diploid programme of Haynes (Haynes and Lu 2005) started in 1966 from approximately equal proportions of Phureja and Stenotomum potatoes. It involved selection within 72 half-sib families and operated on a 2-year cycle for 10 generations, the first five to six for general adaptation followed by four to five for high specific gravity. The programme was then continued at a new location on a 4-year cycle in which a seedling generation in the glasshouse was followed by 3 years of field evaluations in which the number of clones per family was reduced from 100 through 4 to 1. Open-pollinated seed was collected from the 288 clones evaluated in the second year in the field, but only used from the 72 clones selected after the third year in the field when 2 years of data on yield and specific gravity were available. The 72 clones were in fact selected for the highest specific gravity in each of the 72 families. Either no correlation or a positive correlation was obtained between specific gravity and yield.

Subsequently, the population was selected for three cycles for resistance to late blight (Haynes et al. 2014). Each cycle, four clones from each of the 72 families were evaluated for foliar late blight resistance for 2 years in replicated field trials in Pennsylvania (1996–7, 2003–4 and 2009–10 for cycles one, two and three, respectively). For each family, the clone with the lowest mean area under the disease progress curve (AUDPC) was chosen for the seed nursery. These clones were either open-pollinated in the field (cycles one and two) or bulk-pollinated (cycle 3) in the greenhouse to give rise to true potato seed for the next cycle of selection. The mean AUDPC fell from 652 (cycle 1), through 556 (cycle 2) to 276 (cycle 3) compared with control cultivar Atlantic with values of 863, 854 and 775, respectively. Narrow-sense heritability for resistance was estimated as 0.78, 0.77 and 0.80 in cycles one, two and three, respectively. Recurrent maternal half-sib selection continues to improve the levels of resistance to late blight in this diploid population with no adverse effect on the amount of additive genetic variance. Interestingly, in the fourth cycle of the programme, Haynes and Guedes (2018) found that out of 42 clones evaluated, 32 clones flowered and of these, 20 were successfully self-pollinated. Hence, their population may prove useful as a source of diploid inbred lines for F_1 hybrid breeding of potato for propagation through true potato seed (TPS).

6.6.2 CIP Populations B1, B2, B3 and LTVR

On its establishment in 1971, the International Potato Centre (CIP) in Lima, Peru, recognized the need to make broad-based germplasm and candidate cultivars available to National Programmes in developing countries, particularly germplasm with durable resistance to late blight. The first sets of broad-based germplasm from CIP's breeders were the *R*-gene (race-specific) free B populations with quantitative resistance to late blight. B1 was a population of short-day-adapted Andigena potatoes, B2 was derived from crosses of the Andigena clones with long-day-adapted Tuberosum clones lacking *R*-genes, and B3 was a population which in addition contained resistances from wild species but from which *R*-genes had been removed by progeny testing. The aim was to help with the development of improved cultivars, for a wide range of environments, which possessed high and stable levels of resistance to late blight in combination with resistances to viruses and suitable tuber type, and culinary and processing quality (Trognitz et al. 2001; Landeo 2002). The population improvement involved recurrent selection for quantitative resistance to late blight and other desirable traits under high endemic disease pressure in the Andean highlands together with selection in those geographical areas where the new cultivars were to be grown, for example in short days in East Africa (Mulema et al. 2004, 2008). Evaluation under long days in Argentina (Trognitz et al. 2001) and in South Korea was included to broaden the range of adaptation of the largely short-day adapted population. Clones with good general combining ability were identified for use as parents in the local breeding programmes of the National Agricultural Research Systems in developing countries. Landeo (2002; Landeo et al. 2007) reported good progress in selecting for resistance and other traits in the B populations, large additive genetic variances and high heritability estimates for resistance, and stability of resistance across diverse environments and pathogen populations in tropical environments.

CIP's breeders also developed a lowland tropic virus-resistant population denoted LTVR (Bonierbale et al. 2003; Bradshaw and Bonierbale 2010; Khan et al. 2015). This population combines the heat tolerance and early bulking ability of *S. tuberosum* germplasm bred under the summer conditions of the northern Hemisphere on the one hand, with virus resistance from native *S. tuberosum* Group Andigena and Neotuberosum on the other. Recurrent selection with evaluations in a range of environments on the Peruvian coast led to the development of multiplex progenitors of extreme resistance to PVY and PVX, and the selection in each cycle of virus-resistant clones better adapted to the warm conditions of the lowland tropics. More recently, advanced germplasm from European and other long day breeding programmes was incorporated to improve resistance to *Potato leafroll virus* (PLRV) while enhancing earliness and market traits. Interestingly, the *S. tuberosum* × Neotuberosum crosses made a strong contribution to the LTVR population and may account in part for its adaptation to warm conditions.

Thus, CIP's major breeding populations, aimed to develop table and processing cultivars for sustainable potato production in either tropical highland or lowland areas, faced two basic sets of biotic and abiotic constraints: (1) those encountered in the tropical and subtropical highlands, where late blight, cyst nematodes and frost are limiting, and farmers are increasingly faced with drought and (2) those encountered in warmer, drier production areas, that are challenged by viruses, bacterial wilt, drought, root knot nematodes and other pests (Bonierbale et al. 2020). However, increasing requests to support production in the mid-elevation tropics of Sub Saharan Africa and Asia, as well as the temperate regions of Asia, led to the inclusion of heat tolerance and long-day adaptation into the breeding objectives of a new generation and subgroups of the LTVR population, thus requiring the introduction of new germplasm, screening methods and trial sites. Hence, the present emphasis is on earliness to enable the potato's full potential to contribute to incomes and productive cropping systems, and heat and drought tolerance to enable expansion into mid-elevation regions nearer to centres of population, at the same time helping farmers adapt to climate change. In addition, nutritional value has gained in importance with specific attention to iron and zinc biofortification. Bonierbale et al. (2020) have provided a detailed case study of CIP's 'Agile Potato' project, which seeks to develop potato cultivars for fresh and processing use in the short crop production windows of Asia, currently left fallow in the cereal-based systems of the subtropical lowland and temperate regions. The product profile includes 70- to 90-day maturity, resistance to viruses PVY and PLRV, with PVX resistance desirable, stress tolerance (heat, drought and salinity), resistance to late blight and market-oriented quality (high dry matter and chip processing). One particular target is to select 70- to 80-day heat-tolerant potatoes with competitive yields, and more than 18% dry matter and resistance to potato viruses, for the subtropical lowlands of Southwest Asia. Crossing with long-day adapted parents and evaluation under simulated long day conditions in Peru permitted the identification of LTVR families and advanced clones with adaptation to higher latitudes; for example, in Vietnam where day lengths are intermediate and Uzbekistan and Tajikistan where they are long and combined with high summer temperatures. The reader is referred to the book chapter by Bonierbale et al. (2020) for the details of these programmes.

6.6.3 Indian Populations

In the Central Potato Research Institute of India, there has been interest in Tuberosum (female) × short-day Andigena hybrids in breeding for the sub-tropical plains where the potato crop is grown under short days (Gopal et al. 2000; Kumar and Kang 2006). A number of Indian potato cultivars have already been developed from Tuberosum × Andigena crosses, including Kufri Pukhraj, Kufri Giriraj, Kufri Chipsona II and Kufri Shailja (Kumar et al. 2008). More recent information can be found in the review by Kawar et al. (2018) on breeding early-maturing, short-duration cultivars suited for cultivation in the plains.

6.6.4 Diploid Breeding at CIP for Iron and Zinc Content

In 2004, as explained in Chap. 4, CIP chose iron (Fe) and zinc (Zn) as potato targets in the HarvestPlus Biofortification Challenge Program (Paget et al. 2014). The research of Burgos et al. (2007) and Paget et al. (2014) showed that the cultivated diploid potatoes of the Andes are a good source of high concentrations of Fe and Zn, which can be increased further by conventional breeding. Mineral content was determined by inductively coupled plasma–optical emission spectrophotometry (ICPOES). Concentrations in raw, peeled tubers of native Andean potato accessions ranged from 9 to 37 mg Fe kg^{-1} DW and from 8 to 20 mg Zn kg^{-1} DW, with diploid Phureja accession 703274 and diploid Stenotomum accession 701165 showing the highest levels of Fe and Zn, respectively (Burgos et al. 2007). Subsequently, Paget et al. (2014) made a genetic evaluation of Fe and Zn in diploid potatoes produced from a base population of Stenotomum, Goniocalyx and Phureja accessions from the study of Burgos et al. (2007). All analyses were conducted on peeled tubers, and mineral content was determined by ICPOES. The first generation (G1) comprised 17 full-sib families (16 different female parents) in 4 half-sib groups (4 male parents), with family sizes of 23–36 genotypes. It was grown at Huanuco, Peru, at an altitude of 3800 m. In total, there were 487 analyses for mineral and DM content. Forty individuals (genotypes) with high Fe and Zn contents and desirable agronomic characteristics were selected as potential parents of the next generation (G2). In practice, 32 full-sib families were produced from 2 sets of 4 female and 4 male parents crossed in a factorial design. Seedlings were transplanted into the field in Huancayo to produce tuber families for assessment of clones in three-plant plots at Ayacucho, Peru, at an altitude of 2761 m. In total, 1329 clones were analysed for mineral and DM content, with family size ranging from 19 to 74 genotypes. The maximum mineral contents in G2 were 42.5 mg Fe kg^{-1} DW and 38.9 mg Zn kg^{-1} DW. Additive, but not non-additive, genetic variation was detected for Fe and Zn content of tubers and narrow-sense heritability estimates for individual plants on a DW basis were moderate (0.41 and 0.43 for Fe in G1 and G2 and 0.38 and 0.36 for Zn in G1 and G2). Genetic correlations between minerals were all positive when analysed on a DW basis (0.04–0.72) and also small and positive (0.05–0.18) between tuber DM and minerals on a FW basis but negative (−0.14 to −0.38) on a DW basis. The researchers concluded that recurrent individual plant selection should be effective in raising Fe and Zn contents. Furthermore, the level of within-family variation was such that superior individuals should be identified from within a number of families. The researchers also concluded that the improved material should provide a useful source of dietary Fe and Zn. They estimated that in Peru, the best material would provide 56% of the estimated average requirement (EAR) of Fe for children aged 4–8 years and 28% for female adults aged 19–30. Similarly, consumption would provide 37% of the EAR of Zn for children aged 4–8 years and 22% for female adults aged 19–30.

More recently, Sosa et al. (2018) have shown that X-ray fluorescence spectrometry can be used to estimate the iron and zinc contents of freeze-dried and milled potato tuber samples with high precision. Furthermore, the technique is rapid, low cost and suitable for evaluating a large number of potato clones in a breeding programme in a short time frame, unlike the more accurate ICPOES which is time consuming and requires expensive equipment. The researchers used the technique to estimate the iron and zinc concentrations of hundreds of biofortified potato clones, from CIP's breeding programme, grown in three distinct locations in Peru. Twenty clones with high iron concentrations (above 32 mg/kg DW) and 13 clones with high zinc concentrations (above 25 mg/kg DW) were identified. The most recent update on progress in the CIP programme is the paper by Amoros et al. (2020) on heritability and reported genetic gains over generations (cycles) 1 and 2 of more than 29% for iron and 26% for zinc.

6.7 Potato Breeding as Population Improvement

All of the base-broadening schemes just considered involved the genetic improvement of populations of germplasm with a view to providing parents for use in the breeding of new cultivars. In the long-term, the breeding programmes producing new cultivars can also be viewed as population improvement schemes aimed at increasing the frequencies of desirable combinations of alleles (genes) over sexual generations. In other words, the driving force for potato improvement is the cycles of hybridization and selection, with the new cultivars produced each cycle providing the means to transfer progress to farmers' fields. Hence, when one looks at the pedigrees of recently produced cultivars, one can see the footprints of past introgressions and base broadening as part of the continuing cycles, albeit uneven in terms of cycle times both in years and number of intervening clonal generations (as many as eight).

In a breeding experiment at the Scottish Crop Research Institute from 1991 to 2009, the emphasis was put on making the cycles of crossing and selection as fast and efficient as possible, with each cycle producing both potential cultivars and parents for feeding into commercially funded programmes with specific objectives.

Four cycles of crossing and selection were completed and analysed by Bradshaw et al. (2009). Resistance to late blight (complex races of *P. infestans*) in both foliage and tubers and resistance to potato cyst nematodes (*Globodera rostochiensis* Ro1 and *G. pallida* Pa2/3) were combined with a modest increase in yield through selection for breeders' visual preference and an acceptable fry colour for processing. In the first two generations, full-sib family selection was practised on a 3-year cycle with limited selection within the best families in order to combine as quickly as possible genes from sufficient parents to achieve the programme's objectives. In the third generation, more extensive selection was practised within the best families so that the combined between and within full-sib family selection operated on a 6-year cycle. In the fourth generation, improvement of processing quality was introduced through tuber progeny testing and clonal selection for fry colour after storage at 10 °C for 3 months. New germplasm for this trait and for late blight resistance was introduced and full-sib family selection with limited selection within the best families was again practised on a 3-year cycle. The results will be considered in more detail in the next chapter, but it was clear that progress per trait per annum is slow in potato breeding when selection is practised for a number of quantitative traits and that it is difficult to combine improvements in quality and disease and pest resistance with higher yields.

Slater et al. (2014) proposed combining the use of marker-assisted selection (MAS) for major genes and QTL alleles of large effect with estimated breeding values (EBVs) for quantitative (polygenic) traits (see Chap. 2), calculated from pedigree information, to reduce the breeding cycle to as few as 4 years. Further progress can be anticipated with the introduction of genomic selection (GS) to estimate breeding values of clones for unspecified QTL alleles of small effect, where one wants to combine as many as possible in a new cultivar (Slater et al. 2016). Endelman et al. (2018) explored the use of such genomic selection (GS) to improve the rate of progress in the 12 public potato breeding programmes in the USA with a mandate to release cultivars for commercial production. Currently such cultivars are typically released 10 to 12 years after crossing, by which time dozens of traits have been evaluated. The duration of the potato breeding cycle from sexual hybridization to use of new clones as parents is shorter than the time to cultivar release, but still 5–7 years for most programmes. These breeding programmes have primarily used phenotypic selection in combination with genetic markers for a handful of major resistance genes. Endelman et al. (2018) concluded that their approach was applicable to both predicting total genotypic values of clones for commercialization and their breeding values for use as new parents. Thus, genomic selection (GS) has the potential to improve the efficiency of breeding programmes in two ways. Firstly, to reduce the cycle time between one set of crosses and selecting parents from the progeny for the next set of crosses. Secondly, to reduce the amount of phenotypic evaluation in costly field trials conducted over seasons and locations, aimed at identifying potential cultivars. It is still too early to assess how widely GS will be used in practical potato breeding programmes. Nevertheless, it is worth considering some theory and early experimental results.

6.8 Genomic Selection

Conceptually, genomic selection (GS) is simple. The population of clones of interest is assessed for the economically important quantitative traits being considered for improvement (phenotypic values), which usually including yield. The traits may be considered individually or combined into an index of overall merit for selection purposes, as we shall see in the next chapter. The clones are also genotyped for their molecular marker allele composition at sufficient loci to adequately cover the whole genome for the purpose of detecting tight associations with all of the quantitative trait loci affecting the traits. The markers are commonly single nucleotide polymorphism (SNP) markers with two alleles. The phenotypic values are regressed on to all of the markers to establish relationships (associations) that can be used to predict the breeding values and genetic values of other (related) clones, as first proposed by Meuwissen et al. (2001). In other words, we do a standard multiple regression analysis. As tens-of-thousands of markers are scored there is a computational problem, which needs to be solved. In multiple regression analysis, the number (n) of genotypes (clones) is expected to be larger than the number of markers (p), whereas we are now in a situation where the latter is going to be considerably larger than the former. Fitting this large-p with small-n regression requires using some type of variable selection or shrinkage estimation procedure, details of which can be found in the review by de los Campos et al. (2013). Shrinkage estimation avoids having to select markers for inclusion in model. A number of GS methods have been suggested and their accuracy tested, including the use of ridge regression (Hoerl and Kennard 1970) (RR-BLUP), the Least Absolute Shrinkage and Selection Operator (LASSO) (Tibshirani 1996), their Bayesian counterparts and others mentioned by Crossa et al. (2010). In ridge regression, for example, $\lambda\mathbf{Ip}$ is added to $\mathbf{X'X}$ so that we have a matrix for estimating $\mathbf{b}$ that can be inverted [$\mathbf{b} = (\mathbf{X'X} + \lambda\mathbf{Ip})^{-1}\mathbf{X'y}$ where $\lambda = \sigma^2_e/\sigma^2_b$ (σ^2_e is residual variance and σ^2_b is the variance of marker effects, now considered random effects) and $\mathbf{Ip}$ is the identity matrix of size p]. The methods differ in the extent to which they take account of linkage disequilibrium, population structure and kinship and hence can give different results. Habyarimana et al. (2017) found that genomic best linear unbiased prediction GBLUP), Bayesian LASSO (BL) and reproducing kernel Hilbert spaces (RKHS) showed comparable prediction ability in their statutory VCU trials (Value for Cultivation and Use). Likewise, in a breeding context, Sverrisdóttir et al. (2017) found that GBLUP, BayesA and BayesC had similar prediction accuracies for the two traits they investigated. Unlike fixed linear regression models, linear mixed models have no well-established R^2 statistic for assessing goodness-of-fit and prediction power. Sun et al. (2010) showed that the likelihood-ratio-based R^2 (R^2_{LR}) satisfies several desirable requirements for a test statistic. As it reduces to the regular R^2 for fixed models without random effects other than residual, it provides a general measure for the effect of QTL in mixed-model association mapping. In order to develop, validate and estimate the accuracy of the genomic selection equation, three stages are required. First, a subset of the population (reference or training population) is accurately phenotyped and genotyped with molecular markers to see how much of the phenotypic variation can be explained by the markers. Second, the data are used to estimate the breeding values (or genetic values) of the individuals. As there is a danger of over-fitting the data, cross-validation is required to choose the best method of estimating the breeding values. Hence, the training population is divided into two subgroups, one for generating the equation to predict the breeding values (estimation set ES), the other to validate the predictions and assess their accuracy (test set, TS). Würschum and Kraft (2014), in sugar beet, concluded that the optimum proportion of plants to be used as TS is dependent on the population size, but for commonly used population sizes (a few hundred individuals), five-fold cross-validation (TS = 20%) is appropriate. However, Sverrisdóttir et al. (2017) in potato actually used eight-fold cross-validation. Finally, in the third stage, the rest of the individuals in the population or those in a newly derived one (breeding population) are genotyped for the molecular markers and the best prediction equation used to estimate their breeding values or genetic values. Individuals are then selected as parents of the next generation on the basis of their estimated breeding values or are clonally advanced towards commercialization based on their estimated genetic values.

Genomic selection has become well established in commercial animal breeding (Hayes et al. 2013) since the pioneering paper by Meuwissen et al. (2001). Its promise for crop improvement has been demonstrated by recent research in small grain improvement (Rutkoski et al. 2017), hybrid breeding (Schulthess et al. 2017) and forest tree breeding (Grattapaglia 2017). Research is less advanced for clonally propagated crops (Gemenet and Khan 2017) and in its infancy for potatoes (Slater et al. 2016; Habyarimana et al. 2017; Sverrisdóttir et al. 2017; Endelman et al. 2018). Genomic selection in potato breeding has been made possible by SNP (single nucleotide polymorphism) discovery using next-generation sequencing (Hamilton et al. 2011; Uitdewilligen et al. 2013) and the development of SNP arrays such as the 20 K one used by Vos et al. (2015, 2017). It seems likely that fixed content SNP arrays will become the standard for GS in complex, highly heterozygous genomes such as potato because of the problem of missing data in genotyping-by-sequencing (Grattapaglia 2017). There are two ways of estimating the number of markers required, one based on the decay of linkage equilibrium and the other on the effective size of the breeding population. The Slater et al. (2016) estimates of the number of SNP markers required were 8000 based on decay of linkage disequilibrium (LD) and genome map length, and 14,789 based on the effective population size of their population. It will be recalled from Chap. 2 that Vos et al. (2017) evaluated LD decay and various LD-decay estimators in simulated and SNP-array data for tetraploid potatoes. They assumed 10 founder haplotypes in the potato gene pool they were considering and came up with 40,000 SNPs as an upper bound for QTL discovery. They concluded that although their 20 K SolSTW array, with 14,530 SNPs, did not reach the upper bound, it should still be able to detect sufficient QTLs for Genome Wide Association Studies (GWAS) and genomic selection.

Slater et al. (2016) thought that the diversity of the target breeding population would largely dictate the size of the training population required to achieve adequate genomic prediction accuracy with more diverse populations requiring a larger population, as well as denser markers, given the lower extent of LD. Furthermore, traits with low heritability would require a much larger training population to obtain meaningful prediction accuracies, although population sizes of around 1000 individuals were large enough to improve the accuracy when using the estimated breeding values derived from pedigrees. Their estimates of selection accuracy (based on effective population size of 79 for their breeding programme) ranged from 0.19, for a narrow-sense heritability of 0.23 and a training population size of 500, to 0.77 for a narrow-sense heritability of 0.86 and a training population size of 5000. Even with the low selection accuracy, GS was worthwhile compared with phenotypic selection due to the reduction in cycle time from one generation to the next. Their selection intensity was 10% for parents for crossing, but the effectiveness of much higher intensities (20 out of 4000 clones is 0.5%) would be worth exploring.

Endelman et al. (2018) explored different genomic covariance matrices, estimated from markers, in a mixed model analysis (rather than estimating marker effects directly), to predict the total genotypic value of unobserved clones. The reader is referred to the paper for details; but a brief summary is as follows. Phenotype data for yield, specific gravity and fry colour after storage for 6 months were collected between 2012 and 2017 at the University of Wisconsin for a population of 571 round white clones. This was divided into a set of 168 candidates for selection and a set of 403 clones ancestral to this group as the training set. The selection candidates were further narrowed by excluding clones with insufficiently reliable data for validation, leaving 54 clones for yield, 132 clones for specific gravity and 49 clones for fry colour. Reducing the training population size from 403 reduced the prediction accuracy. Genomic covariance matrices for additive (G), digenic dominant (D), and additive × additive epistatic (G#G Hadamard product of the G matrix) effects were calculated using 3895 markers from a potato SNP array of 8303 markers, and the numerator relationship matrix (A) was calculated from a 13-generation pedigree. After removing uninformative ancestors, there were 185 founders (clones with no parent) and 1138 total clones in the pedigree. Based on model fit and prediction accuracy, mixed model analysis with G was superior to A for yield and fry colour but not specific gravity. The amount of additive genetic variance captured by markers was 20% of the total genetic variance for specific gravity, compared to 45% for yield and for fry colour. Within the training population, including non-additive effects improved accuracy and/or bias for all three traits when predicting total genotypic value, but not by that much over the additive and residual genetic covariance matrix (values for latter were 0.53 for yield, 0.46 for fry colour and 0.33 for specific gravity). When six F_1 populations were used for validation, prediction accuracy ranged from 0.06 to 0.63 and was consistently lower (0.13 on average) without allele dosage information. The authors concluded that genome-wide prediction is feasible in potato and that it will improve selection for breeding value given the substantial amount of non-additive genetic variance in elite germplasm. The theory and results presented by Endelman et al. (2018) require acquaintance with the general mixed model (Chap. 2) and matrix algebra and appropriate computer software. I am going to finish this section with some simple theory and a simple example which I find useful in understanding this complex topic.

6.8.1 Estimated Breeding Values

Let us consider a single locus that affects a quantitative trait in a population in equilibrium with p the frequency of allele 'A' and q the frequency of 'a', where there are main effects of alleles and diallelic interactions but no triallelic and tetra-allelic interactions, and where there is chromosomal segregation. The phenotypic values are the genotypic ones plus or minus environmental contributions (e), which on average will cancel each other out. In addition, let us assume that we have a SNP polymorphism in locus A such that our two marker alleles M and m identify A and a, respectively. The breeding values were derived in Chap. 2, including $\alpha_t = a + 3(q\text{-}p)d$.

Genotypes of individuals	$AAAA$	$AAAa$	$AAaa$	$Aaaa$	$aaaa$
Genotype frequency	p^4	$4p^3q$	$6p^2q^2$	$4pq^3$	q^4
Breeding value	$4q\alpha_t$	$(3q\text{-}p)\alpha_t$	$(2q\text{-}2p)\alpha_t$	$(q\text{-}3p)\alpha_t$	$\text{-}4p\alpha_t$
Genotypic value	$2a$	$a{+}3d$	$4d$	$\text{-}a{+}3d$	$\text{-}2a$
Phenotypic value	$2a \pm e$	$a{+}3d \pm e$	$4d \pm e$	$\text{-}a{+}3d \pm e$	$\text{-}2a \pm e$
Marker genotypes	$MMMM$	$MMMm$	$MMmm$	$Mmmm$	$mmmm$
Frequency of M	1	¾	½	¼	0

We can show that in this simple situation, the regression of phenotypic values (**y**, deviations from population mean $\mu_{(Y)}$) on marker genotypes (**X**, coded for number of copies of M, minus the population mean p) does estimate the breeding values of the clones from their molecular marker composition.

$$\mathbf{y} = \mathbf{Xb} + \varepsilon$$

The first column below indicates the frequency of each row in the column vectors.

$$
\begin{array}{ll}
 & \mathbf{y} \\
\tfrac{1}{2}p^4 & \left| +2a \quad + e - \mu_{(Y)} \right| \\
\tfrac{1}{2}p^4 & \left| +2a \quad - e - \mu_{(Y)} \right| \\
2p^3q & \left| + a + 3d + e - \mu_{(Y)} \right| \\
2p^3q & \left| + a + 3d - e - \mu_{(Y)} \right| \\
3p^2q^2 & \left| + 4d \quad + e - \mu_{(Y)} \right| \\
3p^2q^2 & \left| + 4d \quad - e - \mu_{(Y)} \right| \\
2pq^3 & \left| - a + 3d + e - \mu_{(Y)} \right| \\
2pq^3 & \left| - a + 3d - e - \mu_{(Y)} \right| \\
\tfrac{1}{2}q^4 & \left| -2a \quad + e - \mu_{(Y)} \right| \\
\tfrac{1}{2}q^4 & \left| -2a \quad - e - \mu_{(Y)} \right|
\end{array}
=
\begin{array}{c}
\mathbf{X} \\
\left| 1 - p \right| \\
\left| 1 - p \right| \\
\left| \tfrac{3}{4} - p \right| \\
\left| \tfrac{3}{4} - p \right| \\
\left| \tfrac{1}{2} - p \right| \\
\left| \tfrac{1}{2} - p \right| \\
\left| \tfrac{1}{4} - p \right| \\
\left| \tfrac{1}{4} - p \right| \\
\left| 0 - p \right| \\
\left| 0 - p \right|
\end{array}
\begin{array}{c}
\mathbf{b} \\
\left| b_M \right|
\end{array}
+
\begin{array}{c}
\boldsymbol{\varepsilon} \\
\left| \varepsilon_1 \right| \\
\left| \; \right| \\
\left| \; \right| \\
\left| \; \right| \\
\left| \; \right| \\
\left| \; \right| \\
\left| \; \right| \\
\left| \; \right| \\
\left| \; \right| \\
\left| \varepsilon_n \right|
\end{array}
$$

$\mathbf{X'X} = \tfrac{1}{4}pq$; $(\mathbf{X'X})^{-1} = 4/pq$;

$$
\begin{aligned}
\mathbf{X'y} &= p^4(1-p)(2a-\mu_{(Y)}) + 4p^3q(\tfrac{3}{4}-p)(a+3d-\mu_{(Y)}) + 6p^2q^2(\tfrac{1}{2}-p)(4d-\mu_{(Y)}) + 4pq^3(\tfrac{1}{4}-p)(-a+3d-\mu_{(Y)}) \\
&\quad + q^4(-p)(-2a-\mu_{(Y)}) \\
&= p^4q2a + 4p^3q(\tfrac{3}{4}-p)(a+3d) + 6p^2q^2(\tfrac{1}{2}-p)4d + 4pq^3(\tfrac{1}{4}-p)(-a+3d) + q^4p2a \\
&= pq[a + 3(q-p)d] = pq\alpha_t
\end{aligned}
$$

$\mathbf{b} = (\mathbf{X'X})^{-1}\mathbf{X'y} = 4\alpha_t$

$\hat{\mathbf{y}} = \mathbf{Xb} =$

$$
\begin{array}{c}
\left| 1 - p \right| \\
\left| 1 - p \right| \\
\left| \tfrac{3}{4} - p \right| \\
\left| \tfrac{3}{4} - p \right| \\
\left| \tfrac{1}{2} - p \right| \\
\left| \tfrac{1}{2} - p \right| \\
\left| \tfrac{1}{4} - p \right| \\
\left| \tfrac{1}{4} - p \right| \\
\left| 0 - p \right| \\
\left| 0 - p \right|
\end{array}
\begin{array}{c}
\left| 4\alpha_t \right|
\end{array}
=
\begin{array}{|l|}
4q\alpha_t \\
4q\alpha_t \\
(3q-p)\alpha_t \\
(3q-p)\alpha_t \\
(2q-2p)\alpha_t \\
(2q-2p)\alpha_t \\
(q-3p)\alpha_t \\
(q-3p)\alpha_t \\
-4p\alpha_t \\
-4p\alpha_t
\end{array}
$$

Hence, we have estimated the breeding values of the individuals. Furthermore, we have:

$\mathbf{y} - \hat{\mathbf{y}} =$

$$
\begin{array}{|l|}
e - 12q^2d \\
-e - 12q^2d \\
e + 6q(p-q)d \\
-e + 6q(p-q)d \\
e - 3(\tfrac{2}{3}-4pq)d \\
-e - 3(\tfrac{2}{3}-4pq)d \\
e + 6p(q-p)d \\
-e + 6p(q-p)d \\
e - 12p^2d \\
-e - 12p^2d
\end{array}
$$

In other words, the deviations of the phenotypic values from the breeding values are due to digenic interactions and the environment, as we saw in Table 2.5 in Chap. 2.

6.8.2 *Accuracy of Genomic Selection*

In the simple theory of the previous section, we did estimate the breeding values of the individuals from their marker composition. In reality, the prediction of breeding values of selection candidates from a high density of molecular markers depends on the assumption that all Quantitative Trait Loci (QTLs) are in strong linkage disequilibrium (LD) with at least one marker. The prediction can be viewed as whole-genome regression analysis in which all of the marker effects are estimated simultaneously and in which the correlations among markers (SNPs) are taken into account. Hence, we face the issue of accuracy of genomic selection because this affects the response to genomic selection.

The response to genomic selection based on estimated breeding values from markers (M) depends on the correlation (r_{MA}) between these estimated values and the true values (A), termed the accuracy of genomic selection.

$$R = i r_{\mathrm{MA}} \sigma_A$$

This result can be obtained by considering the response to genomic selection as a correlated response to selection based on markers.

The correlated response in trait *y* (CR_{y}) is the response in trait *x* ($i_{\mathrm{x}} h_{\mathrm{x}} \sigma_{\mathrm{Ax}}$) multiplied by the regression of breeding values in *y* ($\mathrm{A_y}$) on those in *x* ($\mathrm{A_x}$), the latter being the genetic correlation (r_{AyAx}) multiplied by $\sigma_{\mathrm{Ay}}/\sigma_{\mathrm{Ax}}$ (Falconer and Mackay 1996). Thus:

$$CR_{\mathrm{y}} = i_{\mathrm{x}} h_{\mathrm{x}} \sigma_{\mathrm{Ax}} \times r_{\mathrm{AyAx}} \sigma_{\mathrm{Ay}} / \sigma_{\mathrm{Ax}} = i_{\mathrm{x}} h_{\mathrm{x}} r_{\mathrm{AyAx}} \sigma_{\mathrm{Ay}} = i r_{\mathrm{MA}} \sigma_A$$

In the present context, *i* is intensity of genomic selection (depends only on proportion of population selected based on markers); the square root of the heritability of markers (h_{x}) is assumed to be 1; the correlation (*r*) of breeding values ($\mathrm{A_y}$ and $\mathrm{A_x}$) is the correlation of true values and those estimated from markers (A and M); and σ_{Ay} ($= \sigma_A$) is the square root of the additive genetic variance in the population.

The theory of the accuracy of genomic selection has developed from pioneering papers such as those of Daetwyler et al. (2008, 2010) and Goddard (2009). The interested reader is referred to the paper by Rabier et al. (2016) for more recent theory and discussion of the accuracy of genomic selection, and for an example of genomic selection (GS) for plant height on 367 genotypes of perennial ryegrass using 24,957 SNPs from genotyping by sequencing (GBS). Although the theoretical results are interesting, I assume that practical potato breeders will probably be more interested in experimental results on accuracy.

We cannot determine r_{MA} (prediction accuracy) directly because we do not know the true breeding values. However, we can determine the correlation (r_{MP}) between the estimated breeding values and the phenotypic values (prediction ability) in the subgroup of the training population used for cross-validation. If we can assume that the co-variances between the estimated (M) breeding values and the true (A) breeding values and the estimated (M) breeding values and the phenotypic (P) values are the additive genetic variance, then we can derive a simple relationship between r_{MA} and r_{MP}.

$$r_{MA} = \sigma_A^{\ 2} / (\sigma_M \sigma_A) \text{and} r_{MP} = \sigma_A^{\ 2} / (\sigma_M \sigma_P)$$

Therefore

$$\begin{aligned} r_{MA}\sigma_M\sigma_A &= r_{MP}\sigma_M\sigma_P \\ r_{MA}\sigma_A &= r_{MP}\sigma_P \\ r_{MA} &= r_{MP}\sigma_P / \sigma_A \\ &= r_{MP} / h_n \end{aligned}$$

as shown, for example, in the theoretical paper of Rabier et al. (2016).

6.8.3 Comparison of Phenotypic and Genomic Selection

In Chap. 2, we derived the classic formula for the response to phenotypic selection per generation for a set of clones from an equilibrium population:

$R = h_n^2S$, where h_n^2 is the narrow-sense heritability (σ_A^2/σ_P^2), or $R = h_n^2 i\sigma_P = ih_n\sigma_A = ir_{AP}\sigma_A$ (or $ir_{AP}\sigma_A/T$ where T is the generation cycle time in years).

In the previous section, we derived the formula for the response to genomic selection in similar form as $R = ir_{MA}\sigma_A$ (or $= ir_{MA}\sigma_A/T$ where T is the generation cycle time in years).

Let us reconsider from Chap. 2 the set of 16 tetraploid clones shown in Table 6.4. They are from a much larger population that has been scored for a quantitative trait, such as yield, and also for the dosage of many SNP markers of which three (*A*, *B* and *C*) are shown. For simplicity, we assume equal frequencies (0.5) of the two SNP alleles at each locus. The three loci are in equilibrium both within each locus and between loci. The example is artificial in the sense that I have ensured that the frequencies of the five genotypes at each locus are in the exact ratio of 1:4:6:4:1 and that there are no correlations between the numbers of alleles at the three loci (i.e. I have eliminated sampling variation). Furthermore, this time we assume that at each locus, the SNP occurs in a causal gene that affects yield. Again, for simplicity, we assume that each SNP has the same effect on yield and that the 16 clones are assigned at random to 16 plots, eight of which increase yield by one unit and eight

Table 6.4 Phenotypic (S) and genomic selection (B) of best four clones: yields of 16 clones (departures from mean = 3) and SNP marker genotypes at three loci *A*, *B* and *C* in equilibrium (number of *A*, *B* and *C* alleles are each coded 2, 1, 0, −1 and −2, which are also breeding values)

Clone	Yield *Y* G + E = P	Breeding Value	Locus *A*	Locus *B*	Locus *C*
1 B	$1-1 = 0$	2	*AAAa* 1	*BBBB* 2	*Cccc* −1
2 S B	$2\frac{1}{3}+1 = 3\frac{1}{3}$	2	*AAAa* 1	*BBBb* 1	*CCcc* 0
3	$1-1 = 0$	1	*Aaaa* −1	*BBBb* 1	*CCCc* 1
4	$-1-1 = -2$	0	*aaaa* −2	*BBBb* 1	*CCCc* 1
5 S	$1\frac{2}{3}+1 = 2\frac{2}{3}$	1	*AAaa* 0	*BBBb* 1	*CCcc* 0
6	$-\frac{1}{3}+1 = \frac{2}{3}$	−1	*AAaa* 0	*BBbb* 0	*Cccc* −1
7 B	$1\frac{2}{3}-1 = \frac{2}{3}$	2	*AAAA* 2	*BBbb* 0	*CCcc* 0
8 S B	$2\frac{1}{3}+1 = 3\frac{1}{3}$	3	*AAAa* 1	*BBbb* 0	CCCC 2
9	$-2\frac{1}{3}-1 = -3\frac{1}{3}$	−2	*AAaa* 0	*BBbb* 0	cccc −2
10	$\frac{1}{3}+1 = 1\frac{1}{3}$	0	*Aaaa* −1	*BBbb* 0	*CCCc* 1
11	$-1\frac{2}{3}-1 = -2\frac{2}{3}$	−2	*Aaaa* −1	*BBbb* 0	*Cccc* −1
12 S	$1+1 = 2$	1	*AAAa* 1	*Bbbb* −1	*CCCc* 1
13	$-\frac{1}{3}-1 = -1\frac{1}{3}$	−1	*AAaa* 0	*Bbbb* −1	*CCcc* 0
14	$-3+1 = -2$	−3	*Aaaa* −1	*Bbbb* −1	*Cccc* −1
15	$-\frac{1}{3}-1 = -1\frac{1}{3}$	−1	*AAaa* 0	*Bbbb* −1	*CCcc* 0
16	$-2\frac{1}{3}+1 = -1\frac{1}{3}$	−2	*AAaa* 0	*bbbb* −2	*CCcc* 0

of which decrease yield by one unit (this is adequate for my purpose: I do not need to assume a set of environmental deviations that are normally distributed). There are additive effects (with $a = 1$) and digenic interactions (with $d = \frac{1}{3}a = \frac{1}{3}$) so that $\alpha_t = a + 3(q-p)d = a = 1$, at each locus; $V_G = V_A + V_D = 3(1 + \frac{1}{6}) = 3\frac{1}{2}$, when summed over 3 loci; $V_P = V_G + V_E = 3\frac{1}{2} + 1 = 4\frac{1}{2}$; and $h_n^2 = V_A/V_P = 3/4\frac{1}{2} = \frac{2}{3}$.

The numbers have been simply chosen so that when the number of A, B and C alleles at each locus (4, 3, 2, 1 and 0) are coded 2, 1, 0, −1 and − 2, these are also the breeding values ($\alpha_t = 1$, and $p = q = \frac{1}{2}$ at each locus). The genotypic value (G) of each clone is the sum of the genotypic values at each locus to which an environmental deviation (E) is added to get the phenotypic value (P). The breeding value of each clone is the sum of the breeding values for each locus.

If we select the four best clones on phenotype (2, 5, 8 and 12), we find that their mean value is $2\frac{5}{6}$ (which is also departure from mean of 16 clones, which is 0). If we multiply this by the narrow sense heritability ($2\frac{5}{6} \times \frac{2}{3}$), we get a predicted response to selection in next sexual generation of 1.889. If, in contrast, we select the four best clones on breeding value (1, 2, 7 and 8), we find that the mean breeding value is 2.250, which is also the predicted response to selection. However, in this simple example, this selection results in relatively large changes in allele frequency, so the actual equilibrium responses to selection based on the changes in allele frequency are 1.453 and 1.703, respectively. Either way, in this simple example, genomic selection is superior to phenotypic selection; but we assumed that we had already got completely accurate estimates of the breeding values from our training population, which is very unlikely. Nevertheless, if r_{AP} and r_{MA} are similar in magnitude, the advantage of genomic selection comes from a greater intensity of selection (more clones can be genotyped than phenotyped) and a reduced cycle time (genomic selection can be practised on seedlings in the glasshouse, whereas phenotypic selection requires assessment of clones in field plots and preferably in replicated trials). These are effects which accumulate over generations.

Before leaving Table 6.4, it is worth modifying it to bring home the importance of robust prediction equations based on cross-validation. In Table 6.5, we have uncorrelated loci B and C explaining all of the variation in yield but markers for loci A and B. We have two populations of 16 clones, one for estimating the parameters in our prediction model and the other for testing our predictions. It is an artificial example in which the only difference between the two populations is that the random samples for locus A have been chosen to give a chance correlation with yield in the estimation set but no correlation in the test set, which is what would happen on average with many test samples (the reader may need to refer back to the section in Chap. 2 on GWAS).

The intercorrelation matrix for the pairs of columns ($\mathbf{y}$, $\mathbf{x}_A$ and $\mathbf{x}_B$) in the estimation set is:

$$\mathbf{R} = \begin{array}{c} \\ y \\ x_A \\ x_B \end{array} \begin{array}{c} \begin{array}{ccc} y & x_A & x_B \end{array} \\ \begin{vmatrix} 1.000 & 0.265 & 0.707 \\ 0.265 & 1.000 & 0.000 \\ 0.707 & 0.000 & 1.000 \end{vmatrix} \end{array}$$

Table 6.5 Estimation (ES) and Test (TS) Sets in Training Population for Genomic Selection, based on *B* and *C* in Table 6.4, but where yields of 16 clones (deviations from mean) are determined by loci *B* and *C* and SNP markers are available for loci *A* and *B* (number of *A* and *B* alleles are each coded 2, 1, 0, −1 and −2)

Clone	Yield *y* ES and TS	Locus *A* ES	Predicted yield ES	Predicted yield ES	Locus *A* TS	Predicted yield TS	Predicted yield TS
	B + *C*		*A* + *B*	*B*		*A* + *B*	*B*
1	1	*Aaaa* −1	1.625	2	*AAAa* 1	2.375	2
2	1	*AAAa* 1	1.375	1	*AAAa* 1	1.375	1
3	2	*Aaaa* −1	0.625	1	*Aaaa* −1	0.625	1
4	2	*AAaa* 0	1.000	1	*aaaa* −2	0.250	1
5	1	*AAAa* 1	1.375	1	*AAaa* 0	1.000	1
6	−1	*AAaa* 0	0.000	0	*AAaa* 0	0.000	0
7	0	*AAaa* 0	0.000	0	*AAAA* 2	0.750	0
8	2	*AAAA* 2	0.750	0	*AAAa* 1	0.375	0
9	−2	*Aaaa* −1	−0.375	0	*AAaa* 0	0.000	0
10	1	*Aaaa* −1	−0.375	0	*Aaaa* −1	−0.375	0
11	−1	*AAAa* 1	0.375	0	*Aaaa* −1	−0.375	0
12	0	*AAaa* 0	−1.000	−1	*AAAa* 1	−0.625	−1
13	−1	*AAaa* 0	−1.000	−1	*AAaa* 0	−1.000	−1
14	−2	*aaaa* −2	−1.750	−1	*Aaaa* −1	−1.375	−1
15	−1	*AAAa* 1	−0.625	−1	*AAaa* 0	−1.000	−1
16	−2	*AAaa* 0	−2.000	−2	*AAaa* 0	−2.000	−2

The correlation of 0.707 between *y* and x_B is statistically significant ($P < 0.01$), whereas the correlation between *y* and x_A is not statistically significant ($P > 0.1$). Nevertheless, when we fit the model $\hat{y} = b_A x_A + b_B x_B$, as explained in Chap. 2, where $\hat{y}$ is the estimated yield, we get:

$$\hat{y} = 0.375 x_A + 1.000 x_B$$

and we find that the correlation between the actual (*y*) and estimated ($\hat{y}$) yields in the estimation set is 0.755 and that we have explained 57% of the variation in actual yields. In comparison, fitting the model $\hat{y} = 1.000 x_B$ explains 50% of the variation (correlation is 0.707), and it is tempting to conclude that $\hat{y} = 0.375 x_A + 1.000 x_B$ is the better model. However, when we use these two equations to predict the yields of the 16 clones in the test set, we find that the correlations between the actual and estimated yields are 0.662 and 0.707 for the A + *B* and *B* models, respectively. The predictions using just the marker *B* locus are better than using *A* + *B* in the test set and hence the model is more robust. Explaining 50% of the variation is the best we can do because the other 50% is explained by locus *C* for which we do not have a marker. In other words, we over-fitted the data in the estimation set because we included a marker *A* that had no true association with yield, just a chance association in our estimation set of small size (the non-significant correlation coefficient of 0.265). The true situation was revealed in test set:

$$\mathbf{R} = \begin{array}{c|ccc|} & y & x_A & x_B \\ y & 1.000 & 0.000 & 0.707 \\ x_A & 0.000 & 1.000 & 0.000 \\ x_B & 0.707 & 0.000 & 1.000 \end{array}$$

Although this is an artificially contrived example, it does demonstrate the danger of over-fitting data. In this simple example, we were warned by the non-significant correlation, but with real data, small non-significant correlations will include ones that really do affect yield, hence the need for robust prediction equations based on cross-validation.

The paper of Sverrisdóttir et al. (2017) provides data of relevance to the way genomic selection is likely to be done in a practical breeding programme. They describe the results of genomic prediction in tetraploid potato for two important quality traits, starch content assessed over 2 years and chipping (crisping) quality of tubers assessed over a single season as fry colour after storage at 4 °C for 2 months. The potatoes came from field trials at Vandel, Denmark in 2013 and 2014, of a population of 762 individuals from a diallel set of crosses of 18 potato cultivars. Parent-offspring regressions gave narrow-sense heritabilities of 0.90 for starch content and 0.78 for fry colour. Genotyping-by-sequencing (GBS) of the population provided 171,859 SNPs. Genomic estimated breeding values (GEBVs) were calculated with eightfold cross-validation by three methods, GBLUP, BayesA and BayesC, which gave very similar results. The cross-validation was used to determine the robustness of the GEBVs, with their accuracy determined as their correlation with the observed phenotypes. The correlations were 0.56 for fry colour and 0.73 for starch content. In addition, a test panel of 74 historical breeding clones, which were unrelated to the population, was selected and genotyped with the same markers to determine their GEBVs, whose accuracy was also determined by their correlation with phenotypic data acquired over the period 1997 to 2014. The correlations were between 0.30 and 0.31 for fry colour and 0.42 and 0.43 for starch content. The test panel comprised cultivars, which had been selected for several traits including starch content. Finally, GS performance with all the SNPs was compared with that of SNP markers selected for significant associations with the traits from a genome-wide association study (GWAS). This found 372 SNPs significantly associated with fry colour and 612 with starch content. The SNPs were mainly found on chromosome 10 for both traits. Using GWAS selected SNPs did not improve prediction accuracy when predicting the performance of individuals in the test panel and has the disadvantage of requiring a separate marker set for each trait. The authors concluded that a GS breeding programme should be established by assembling a training population of potential parents and cultivars that are relevant to the breeding programme, rather than using unrelated material. Furthermore, due to the high allelic diversity of potato, it will be necessary to construct large training populations that include entries covering the range of a trait's variance.

In a study with smaller population sizes (45–219), Byrne et al. (2020) accumulated fry-colour data on a population of potato lines from the Teagasc tetraploid breeding programme in Ireland and combined them with genotypic data for a GWAS and to evaluate accuracy of genomic prediction. They identified a major QTL on chromosome 10 for fry colour and could predict fry colour with moderate accuracy using genome-wide markers. The mean predictive ability ranged between 0.11 and 0.77 for fry colour 'off-the-field' and between 0.24 and 0.66 for fry colour after long-term storage at 4.5 °C. Furthermore, their results provided evidence that it is possible to identify a small subset of SNPs for processing characteristics that can give moderate predictive ability, albeit lower than that achieved with genome-wide markers. The reader is referred to the paper for more details.

We shall see in the next chapter that potato breeders commonly handle 4000 clones in unreplicated plots and 1000 clones in replicated yield trials. Hence, with a determined effort, a training population of 1000–4000 clones is feasible at the start of a new breeding programme, but the phenotyping would be greatly helped by the development of appropriate high-throughput phenotyping for use in the field and on harvested material, as reviewed by Araus and Cairns (2014). The phenotypic data would need to be incorporated into an index of overall merit for selection purposes, unless a prediction equation was developed for each trait and independent culling levels used for selection, as discussed in the next chapter. Having determined the equation(s) to predict the genomic estimated breeding values (GEBVs) in the first cycle of crossing and selection, in subsequent cycles it would be used on seedlings in the glasshouse for a very short generation time in years (2 or even 1 year) as there would be no intervening clonal generations. However, new cultivars would be sought each cycle by a thorough assessment of clones over clonal generations, and this process could accommodate checking and modifying the genomic selection equation(s). Furthermore, a second genomic selection equation(s) for genetic values could be developed for use in clonal selection. Ideally the seedling population size would need to be large enough to combine the genomic selection (GS) with the marker-assisted selection (MAS) of major genes and QTL alleles of large effect, which we discussed earlier. If the loci used in MAS could be matched to markers generated in GS genotyping, then a single genotyping platform could be used for MAS and GS.

6.9 Conclusions

After reviewing the germplasm available for potato breeding and the crossability of wild species to cultivated potatoes, we looked at its utilization in introgression breeding and base broadening (incorporation), the aim being to provide better parents for use in the breeding of new cultivars. In the long-term, the breeding programmes producing the new cultivars can be viewed as population improvement schemes aimed at increasing the frequencies of desirable combinations of alleles (genes) over sexual generations. In other words, the driving force for potato improvement is the cycles of hybridization and selection, with the new cultivars produced each cycle providing the means to transfer progress to farmers' fields. Furthermore, when one looks at the pedigrees of recently produced cultivars, one can see the footprints of past introgressions and base broadening as part of the continuing cycles, albeit uneven in terms of cycle times both in years and number of intervening clonal generations. In modern potato improvement, the breeding work can be done with greater speed, efficiency and effectiveness with the aid of molecular markers. These can be used for marker-assisted introgression and selection of specific alleles (major genes and quantitative trait loci alleles of large effect) and for genomic selection of unspecified alleles in population improvement. Within this overall framework, there are a number of specific conclusions.

1. *Modern Cultivars*: Modern cultivars provide breeders with parents that are reasonably well adapted to their target environments, growing seasons and end uses, but which require further improvement by breeding. The 2014 World Catalogue of Potato Varieties (Pieterse and Judd 2014) listed more than 4800 cultivars from 100 countries covering all potato growing regions in the world; but there are more, for example The European Cultivated Potato Database had information on 5947 cultivars in November 2019. Modern cultivars are descended from the Chilotanum Group of landraces, which have been used extensively in potato breeding worldwide. The genetic base of modern cultivars has been widened through the use of the Andigenum Group landraces and wild relatives with scope for greater use in the future.
2. *Landraces and Wild Relatives*: The genebank at the International Potato Centre, Lima, Peru (CIP), maintains clonal collections (4354 landraces in vitro) and botanical seed collections (2595 accessions of 148 out of the 172 wild species they recognize; see below) of potato as a global public good under the International Treaty on Plant Genetic Resources for Food and Agriculture (ITPGRFA). Germplasm is available for research, education and breeding purposes and has been used in breeding programmes in over 100 countries. It is estimated that there may be 24,500–29,500 unique potato accessions conserved worldwide in genebanks (ex situ conservation). CIP also has a cryopreservation programme and a herbarium collection of 90% of the wild and cultivated species of potato, represented from southern Colorado Canyon in the United States to southern Chile. Finally, CIP has established a 15,000-ha reserve (Cusco Potato Park) high in the Andes of Peru where the indigenous Quechua people grow 1345 landraces of potato (in situ conservation).
3. *Potato Genetic Identity (PGI) Kit*: CIP has developed and characterized a suite of 24 highly informative microsatellite markers (SSRs), with two per chromosome and three to nine alleles per locus, that was able to discriminate 93.5% of the 742 landraces examined. The PGI kit can be used for potato germplasm characterization, for identity verification (fingerprinting), studying genetic diversity, anchoring genetic linkage maps, establishment of core collections and studying gene flow.

4. *Potato Taxonomy*: Unfortunately, the taxonomy of wild potato species is complex and has undergone revision. Hawkes (1990) recognized 219 species, whereas Spooner et al. (2014) now recognize only 107 species. Breeders need to be aware of the changes.
5. *Crossability*: Potato breeders can usually achieve desired sexual hybridizations between *S. tuberosum* and its wild relatives by manipulation of ploidy with due regard to EBN (Endosperm Balance Number), although unexpected successes and failures do occur, and other barriers do exist. Somatic hybridization (protoplast fusion) can be used to achieve difficult or impossible sexual hybridizations.
6. *Introgression Breeding*: Successful introgression of genes, primarily for disease and pest resistance, into tetraploid Tuberosum Group has been achieved from *Solanum demissum* (6*x*, EBN = 4), *S. acaule* and *S. stoloniferum* (both 4*x*, EBN = 2), *S. bulbocastanum*, *S. commersonii* and *S. pinnatisectum* (all 2*x*, EBN = 1), *S. chacoense*, *S. microdontum*, *S. spegazzinii* and *S. vernei* (all 2*x*, EBN = 2; or chromosome doubled to 4*x*, EBN = 4) and *S. tuberosum* Andigena Group (4*x*, EBN = 4). However, the diploid 1EBN species proved difficult and involved ploidy manipulation and sometimes bridging crosses or somatic hybridization. Today gene cloning and genetic transformation offer a much faster and precise route.
7. *Inbreeding and Linkage Drag*: As the potato is an outbreeding species, different recurrent parents need to be used during introgression to avoid inbreeding, which would also occur on selfing to fix the desired allele. Hence, the outcome is a new cultivar, not a particular cultivar with just the addition of the desired allele. The latter can be achieved through genetic transformation (cisgenesis), which also avoids the problem of linkage drag and additional backcross generations to break any tight linkages between the desired allele and undesirable ones from the wild species. Another possibility is introgression into diploid inbred lines.
8. *Marker-Assisted Introgression*: Molecular markers can be used during introgression to select the desirable gene and manipulate the background genotype. Application of the theory to introgression into diploid inbred lines shows that with a modest sized backcrossing programme of 200 individuals per generation, it is possible in just three generations to achieve a genotype identical to the cultivated parent, apart from the introgressed gene and some linked genes from the wild species. Dense marker coverage of the non-carrier chromosomes is not necessary to increase the overall recipient (recurrent parent) genome content; in fact, four markers are sufficient for a chromosome of length 100 cM.
9. *Combining Major Genes and QTL Alleles of Large Effect at the Tetraploid Level*: Once introgressed into parental material, major genes and QTL alleles of large effect can be stacked in new cultivars using marker-assisted selection (MAS). Theory shows that the population sizes involved start to become large above eight loci segregating. However, once one is making crosses between clones that all have every marker, a steady increase in the frequencies of such clones occurs over generations simply through elimination of offspring clones

not possessing all of them. Hence, it is important to think in terms of a long-term population improvement programme to increase the frequencies of desirable combinations of alleles at different loci, and not just in terms of making crosses to breed a new cultivar. The theoretical results provide an explanation for the lack of potato cultivars in which more than a few major genes and QTL alleles of large effect have been stacked. The few published practical examples have applied MAS in the second field generation, but MAS on 100,000 seedlings is feasible and desirable, before transplanting those selected to pots in a glasshouse.

10. *Base Broadening*: Since the 1960s, breeders globally have been making an effort to incorporate more Andigenum Group germplasm (landraces) into their breeding programmes, with selection where necessary for the ability to form tubers in longer days. Most of these efforts have involved improvement of broad-based populations with the aim of feeding parental material into the breeding of finished cultivars, both locally and internationally. Examples of this approach are the production of Neotuberosum (tetraploid long-day Andigena) in Europe and North America; the production of diploid potatoes from Groups Phureja/Stenotomum adapted to long-day conditions in Europe and North America; CIP's broad-based *R*-gene-free B populations with quantitative resistance to late blight and their lowland tropic virus-resistant population (LTVR), which combines heat tolerance and virus resistance with other desirable traits; CPRI's (India) use of hybrids of Tuberosum (female) and short-day Andigena in breeding for the sub-tropical plains; and CIP's use of Group Phureja/Stenotomum germplasm for high concentrations of iron and zinc in their biofortification programme. The value and status of Neotuberosum remains controversial, but the more general incorporation of landraces into potato breeding programmes has become established.
11. *Genomic Selection*: Thinking in terms of population improvement provides the framework for the use of genomic selection for quantitative traits in potato breeding, where one wants to use genomic estimated breeding values (GEBVs) to combine as many desirable, but unspecified, QTL alleles of small effect as possible in a new cultivar. The idea is simple and involves three stages: first, a subset of the population of clones (training population) is accurately phenotyped and genotyped with molecular markers to see how much of the phenotypic variation can be explained by the markers; second, the training population is divided into two subgroups (for cross-validation), one for generating the equation to predict the breeding values (estimation set), the other to validate the predictions and assess their accuracy (test set, often 20% of training population); and third, the rest of the clones in the population or those in a newly derived one (breeding population) are genotyped for the molecular markers and the best prediction equation used to estimate their breeding values in order to select parents for crossing. The advantage of genomic selection over phenotypic selection comes from a greater intensity of selection (more clones can be genotyped than phenotyped) and a reduced cycle time (genomic selection can be practised on seedlings in the glasshouse, whereas phenotypic selection

requires assessment of clones in field plots and preferably in replicated trials). Furthermore, these are effects which accumulate over generations and will result in the superiority of genomic selection over phenotypic selection, providing the true breeding values are estimated by molecular markers with similar accuracy (prediction accuracy) to estimation by phenotype. My view is that with a determined effort, phenotyping a training population of between 1000 and 4000 clones is feasible. Genomic selection can also be used to select clones as potential cultivars by estimating genetic rather than breeding values.

In the next chapter, it is time to move onto the breeding of finished cultivars.

References

Amoros W, Salas E, Hualla V, Burgos G, De Boeck B, Eyzaguirre R, zum Felde T, Bonierbale M (2020) Heritability and genetic gains for iron and zinc concentration in diploid potato. Crop Sci. https://doi.org/10.1002/csc2.20170

Araus JL, Cairns JE (2014) Field high-throughput phenotyping: the new crop breeding frontier. Trends Plant Sci 19:52–61. https://doi.org/10.1016/j.tplants.2013.09.008

Armstrong MR, Vossen J, Lim TY, Hutten RC, Xu J, Strachan SM, Harrower B, Champouret N, Gilroy EM, Hein I (2019) Tracking disease resistance deployment in potato breeding by enrichment sequencing. Plant Biotechnol J 17(2):540–549. https://doi.org/10.1111/pbi.12997

Bamberg J, Jansky S, del Rio A, Ellis D (2018) Ensuring the genetic diversity of potatoes. In: Wang-Pruski G (ed) Achieving sustainable cultivation of potatoes Volume 1: Breeding improved varieties. Burleigh Dodds, Cambridge, pp 57–79

Barone A (2004) Molecular marker-assisted selection for potato breeding. Am J Potato Res 81:111–117

Barone A, Sebastiano A, Carputo D, della Rocca F, Frusciante L (2001) Molecular marker-assisted introgression of the wild *Solanum commersonii* genome into the cultivated *S. tuberosum* gene pool. Theor Appl Genet 102:900–907

Bonierbale MW, Amoros W, Landeo J (2003) Improved resistance and quality in potatoes for the tropics. Acta Hortic 619:15–22

Bonierbale MW, Amoros WR, Salas E, de Jong W (2020) Potato breeding. In: Campos H, Ortiz O (eds) The potato crop: its agricultural, nutritional and social contribution to humankind. Springer, ebook, pp 163–218

Bradshaw JE (2009) Potato breeding at the Scottish Plant Breeding Station and the Scottish Crop Research Institute: 1920-2008. Potato Res 52:141–172

Bradshaw JE (2016) Plant breeding: past, present and future. Springer, Cham

Bradshaw JE (2017) Review and analysis of limitations in ways to improve conventional potato breeding. Potato Res 60:171–193. https://doi.org/10.1007/s11540-017-9346-z

Bradshaw JE, Bonierbale M (2010) Potatoes. In: Bradshaw JE (ed) Root and tuber crops, handbook of plant breeding, vol 7. Springer, New York, pp 1–52

Bradshaw JE, Mackay GR (1994) Breeding strategies for clonally propagated potatoes. In: Bradshaw JE, Mackay GR (eds) Potato genetics. CAB International, Wallingford, pp 467–497

Bradshaw JE, Ramsay G (2005) Utilisation of the Commonwealth Potato Collection in potato breeding. Euphytica 146:9–19

Bradshaw JE, Dale MFB, Mackay GR (2009) Improving the yield, processing quality and disease and pest resistance of potatoes by genotypic recurrent selection. Euphytica 170:215–227

Bryan GJ, McLean K, Bradshaw JE, De Jong WS, Phillips M, Castelli L, Waugh R (2002) Mapping QTLs for resistance to the cyst nematode *Globodera pallida* derived from the wild potato species *Solanum vernei*. Theor Appl Genet 105:68–77

Burgos G, Amoros W, Morote M, Stangoulis J, Bonierbale M (2007) Iron and zinc concentration of native Andean potato cultivars from a human nutrition perspective. J Sci Food Agric 87:668–675. https://doi.org/10.1002/jsfa.2765

Byrne S, Meade F, Mesiti F, Griffin D, Kennedy C, Milbourne D (2020) Genome-wide association and genomic prediction for fry color in potato. Agronomy 10:90. https://doi.org/10.3390/agronomy10010090

Cadman CH (1942) Autotetraploid inheritance in the potato: some new evidence. J Genet 44:33–52

Camadro EL, Espinillo JC (1990) Germplasm transfer from the wild tetraploid species *Solanum acaule* bitt. to the cultivated potato, *S. tuberosum* L. using 2N eggs. Am Potato J 67(11):737–749

Camadro EL, Carputo D, Peloquin SJ (2004) Substitutes for genome differentiation in tuber-bearing *Solanum*: interspecific pollen-pistal incompatibility, nuclear-cytoplasmic male sterility, and endosperm. Theor Appl Genet 109:1369–1376

Carputo D, Barone A (2005) Ploidy level manipulations in potato through sexual hybridisation. Ann Appl Biol 146:71–79

Carroll CP (1982) A mass-selection method for the acclimatization and improvement of edible diploid potatoes in the United Kingdom. J Agric Sci Camb 99:631–640

Carroll CP, De Maine MJ (1989) The agronomic value of tetraploid F1 hybrids between potatoes of group Tuberosum and group Phureja/Stenotomum. Potato Res 32:447–456

Chandra S, Huaman Z, Krishna H, Ortiz R (2002) Optimal sampling stategy and core collection size of the Andean tetreploid potato based on isozyme analysis data-a simulation study. Theor Appl Genet 104:1325–1334

Chase SS (1963) Analytic breeding in *Solanum tuberosum* L. - a scheme utilizing parthenotes and other diploid stocks. Can J Genet Cytol 5:359–363

Colton LM, Groza HI, Wielgus SM, Jiang J (2006) Marker-assisted selection for the broad-spectrum potato late blight resistance conferred by gene *RB* derived from a wild potato species. Crop Sci 46:589–594

Crossa J, de los Campos G, Perez P, Gianola D, Burgueno J, Araus JL, Makumbi D, Singh RP, Dreisigacker S, Yan J, Arief V, Banziger M, Braun H-J (2010) Prediction of genetic values of quantitative traits in plant breeding using pedigree and molecular markers. Genetics 186:713–724

Daetwyler HD, Villanueva B, Woolliams JA (2008) Accuracy of predicting the genetic risk of disease using a genome-wide approach. PLoS One 3(10):e3395

Daetwyler HD, Pong-Wong R, Villanueva B, Woolliams JA (2010) The impact of genetic architecture on genome-wide evaluation methods. Genetics 185(3):1021–1031

de los Campos G, Hickey JM, Pong-Wong R, Daetwyler HD, Calus MPL (2013) Whole-genome regression and prediction methods applied to plant and animal breeding. Genetics 193:327–345

Dong F, McGrath JM, Helgeson JP, Jiang J (2001) The genetic identity of alien chromosomes in potato breeding lines revealed by sequential GISH and FISH analyses using chromosome specific cytogenetic DNA markers. Genome 44:729–734

Ellenby C (1952) Resistance to the potato root eelworm, *Heterodera rostochiensis* Wollenweber. Nature 170:1016

Ellis D, Salas A, Chavez O, Gomez R, Anglin N (2020) Ex situ conservation of potato [*Solanum* Section *Petota* (Solanaceae)] genetic resources in genebanks. In: Campos H, Ortiz O (eds) The potato crop: its agricultural, nutritional and social contribution to humankind. Springer, ebook, pp 109–138

Endelman JB, Schmitz Carley CA, Bethke PC, Coombs JJ, Clough ME, da Silva WL, De Jong WS, Douches DS, Frederick CM, Haynes KG, Holm DG, Miller JC, Muñoz PR, Navarro FM, Novy RG, Palta JP, Porter GA, Rak KT, Sathuvalli VR, Thompson AL, Craig Yencho G (2018) Genetic variance partitioning and genome-wide prediction with allele dosage information in autotetraploid potato. Genetics 209(1):77–87. https://doi.org/10.1534/genetics.118.300685

Falconer DS, Mackay TFC (1996) Introduction to quantitative genetics, 4th edn. Longman, Harlow, 464p

Foster SJ, Park TH, Pel M, Brigneti G, Sliwka J, Jagger L, van der Vossen E, Jones JD (2009) *Rpi-vnt1.1*, a Tm-2^2 homolog from *Solanum venturii*, confers resistance to potato lateblight. Mol Plant Microbe Interact 22(5):589–600. https://doi.org/10.1094/MPMI-22-5-0589

Gavrilenko T (2011) Application of molecular cytogenetics in fundamental and applied research of potato. In: Bradeen JM, Kole C (eds) Genetics, genomics and breeding of potato. Science Publishers, Enfield, pp 184–206

Gebhardt C (2013) Bridging the gap between genome analysis and precision breeding in potato. Trends Genet 29:248–256

Gebhardt C, Urbany C, Li L, Stich B, Paulo J, Draffehn A, Ballvora A (2011) Molecular diagnostics for complex pest and disease resistance and tuber quality traits: concept, achievements and perspectives. Potato Res 54:313–318

Gemenet DC, Khan A (2017) Opportunities and challenges to implementing genomic selection in clonally propagated crops. In: Varshney RK, Roorkiwal M, Sorrells ME (eds) Genomic selection for crop improvement. Springer, Cham, pp 185–198

Ghislain M, Núñez J, del Rosario Herrera M, Pignataro J, Guzman F, Bonierbale M, Spooner DM (2009a) Robust and highly informative microsatellite-based genetic identity kit for potato. Mol Breed 23:377–388

Ghislain M, Núñez J, del Rosario Herrera M, Spooner DM (2009b) The single Andigenum origin of Neo-Tuberosum potato materials is not supported by microsatellite and plastid marker analyses. Theor Appl Genet 118:963–969. https://doi.org/10.1007/s00122-008-0953-6

Glendinning DR (1975) Neo-Tuberosum: new potato breeding material. 2. A comparison of Neo-Tuberosum with unselected Andigena and with Tuberosum. Potato Res 18:343–350

Goddard M (2009) Genomic selection: prediction of accuracy and maximisation of long term response. Genetica 136:245–257. https://doi.org/10.1007/s10709-008-9308-0

Gopal J, Chahal GS, Minocha JL (2000) Progeny mean, heterosis and heterobeltiosis in *Solanum tuberosum* × *tuberosum* and *S. tuberosum* × *andigena* families under a short day sub-tropic environment. Potato Res 43:61–70

Grattapaglia D (2017) Status and perspectives of genomic selection in forest tree breeding. In: Varshney RK, Roorkiwal M, Sorrells ME (eds) Genomic selection for crop improvement. Springer, Cham, pp 199–249

Habyarimana E, Parisi B, Mandolino G (2017) Genomic prediction for yields, processing and nutritional quality traits in cultivated potato (*Solanum tuberosum* L.). Plant Breed 136:245–252. https://doi.org/10.1111/pbr.12461

Hamernik AJ, Hanneman RE, Jansky SH (2009) Introgression of wild species germplasm with extreme resistance to cold sweetening into the cultivated potato. Crop Sci 49:529–542

Hamilton JP, Hansey CN, Whitty BR, Buell CR (2011) Single nucleotide polymorphism discovery in elite North American potato germplasm. BMC Genomics 12(1):302. https://doi.org/10.1186/1471-2164-12-302

Hanneman RE Jr (1985) Self fertility in *Solanum chacoense*. Am Potato J 62:428–429

Haverkort AJ, Boonekamp PM, Hutten R, Jacobsen E, Lotz LAP, Kessel GJT, Vossen JH, Visser RGF (2016) Durable late blight resistance in potato through dynamic varieties obtained by cisgenesis: scientific and societal advances in the DuRPh project. Potato Res 59:35–66

Hawkes JG (1951) The Commonwealth Potato Collection. Am Potato J 28:465–471

Hawkes JG (1990) The potato: evolution, biodiversity & genetic resources. Belhaven Press, London. 259p

Hawkes JG, Jackson MT (1992) Taxonomic and evolutionary implications of the endosperm balance number hypothesis in potatoes. Theor Appl Genet 84:180–185

Hawtin G, Fowler L (2011) The global crop diversity trust. In: Frison C, Lόpez F, Esquinas-Alcázar JT (eds) Plant genetic resources and food security. FAO, Bioversity International and Earthscan, Abingdon, pp 209–221

Hayes BJ, Lewin HA, Goddard ME (2013) The future of livestock breeding: genomic selection for efficiency, reduced emissions intensity, and adaptation. Trends Genet 29:206–214

Haynes KG, Lu W (2005) Improvement at the diploid species level. In: Razdan MK, Mattoo AK (eds) Genetic improvement of Solanaceous crops, volume I: potato. Science Publishers Inc, Enfield, pp 101–114

Haynes KG, Guedes ML (2018) Self-compatibility in a diploid hybrid population of *Solanum phureja – S. stenotomum*. Am J Potato Res 95:729–734

Haynes KG, Qu X, Christ BJ (2014) Two cycles of recurrent maternal half-sib selection reduce foliar late blight in a diploid hybrid *Solanum phureja – S. stenotomum* population by two-thirds. Am J Potato Res 91:254–259

Helgeson JP, Pohlman JD, Austin S, Haberlach GT, Wielgus SM, Ronis D, Zambolim L, Tooley P, McGrath JM, James RV, Stevenson WR (1998) Somatic hybrids between *Solanum bulbocastanum* and potato: a new source of resistance to late blight. Theor Appl Genet 96:738–742

Hermsen JGT (1994) Introgression of genes from wild species, including molecular and cellular approaches. In: Bradshaw JE, Mackay GR (eds) Potato genetics. CAB International, Wallingford, pp 515–538

Hermundstad SA, Peloquin SJ (1987) Breeding at the 2*x* level and sexual polyploidization. In: Jellis GJ, Richardson DE (eds) The production of new potato varieties. Cambridge University Press, Cambridge, pp 197–210

Hoerl AE, Kennard RW (1970) Ridge regression: biased estimation for nonorthogonal problems. Technometrics 12:55–67

Hosaka K, Hanneman RE Jr (1998) Genetics of self-compatibility in a self-incompatible wild diploid potato species *Solanum chacoense*. 2. Localization of an S locus inhibitor (*Sli*) gene on the potato genome using DNA markers. Euphytica 103:265–271

Hosaka K, Sanetomo R (2018) Effects of *Solanum demissum* chromosomes on crossability in the backcross progeny to *Solanum tuberosum*. Euphytica 214:113. https://doi.org/10.1007/s10681-018-2186-3

Hosaka K, Hosaka Y, Mori M, Maida T, Hiroshi Matsunaga H (2001) Detection of a simplex RAPD marker linked to resistance to *Potato virus Y* in a tetraploid potato. Am J Potato Res 78:191–196

Hospital F (2003) Marker-assisted breeding. In: Newbury HJ (ed) Plant molecular breeding. Blackwell, Oxford, pp 30–59

Hougas RW, Peloquin SJ, Ross RW (1958) Haploids of the common potato. J Hered 49:103–107

Huaman Z, Ortiz R, Gomez R (2000a) Selecting a *Solanum tuberosum* subsp. *andigena* core collection using morphological, geographical, disease and pest descriptors. Am J Potato Res 77:183–190

Huaman Z, Ortiz R, Zhang D, Rodriguez F (2000b) Isozyme analysis of entire and core collections of *Solanum tuberosum* subsp. *andigena* potato cultivars. Crop Sci 40:273–276

Iovene M, Barone A, Frusciante L, Monti L, Carputo D (2004) Selection for aneuploid potato hybrids combining a low wild genome content and resistance traits from *Solanum commersonii*. Theor Appl Genet 109:1139–1146

Jansky S (2006) Overcoming hybridization barriers in potato. Plant Breed 125:1–12

Jansky S (2009) Breeding, genetics and cultivar development. In: Singh J, Kaur L (eds) Advances in potato chemistry and technology. Academic Press, Burlington, pp 27–62

Jansky SH, Yerk GL, Peloquin SJ (1990) The use of potato haploids to put 2*x* wild species germplasm into a usable form. Plant Breed 104:290–294

Jansky SH, Davis GL, Peloquin SJ (2004) A genetic model for tuberization in potato haploid-wild species hybrids grown under long-day conditions. Am J Potato Res 81:335–339

Johnston SA, der Nijs TPM, Peloquin SJ, Hanneman RE Jr (1980) The significance of genic balance to endosperm development in interspecific crosses. Theor Appl Genet 57:5–9

Kaushik SK, Sharma R, Garg ID, Singh BP, Chakrabarti SK, Bhardwaj V, Pandey SK (2013) Development of a triplex (YYYy) parental potato line with extreme resistance to potato virus Y using marker assisted selection. J Hortic Sci Biotechnol 88:580–584

Kawar PG, Kardile HB, Raja S, Dutt S, Kumar R, Manivel P, Bhardwaj V, Singh BP, Govindakrishnan PM, Chakrabarti SK (2018) Developing early-maturing and stress-resistant potato varieties. In: Wang-Pruski G (ed) Achieving sustainable cultivation of potatoes. Burleigh Dodds, Cambridge, pp 143–167

Khan MA, Munive S, Bonierbale M (2015) Early generation in vitro assay to identify potato populations and clones tolerant to heat. Plant Cell Tissue Organ Cult 121:45–52. https://doi.org/10.1007/s11240-014-0677-z

Kumar R, Kang GS (2006) Usefulness of Andigena (*Solanum tuberosum* ssp. *andigena*) genotypes as parents in breeding early bulking potato cultivars. Euphytica 150:107–115

Kumar R, Kumar V, Gopal J et al (2008) Inventory of potato germplasm (Group Andigena) collection. Technical Bulletin No. 86, CPRI, Shimla

Landeo JA (2002) Durable resistance: quantitative/qualitative resistance. In: Lizarraga, C. (ed) Proceedings of the global initiative on late blight conference, 11–13 July 2002, Hamburg, Germany, pp 29–36

Landeo JA, Gastelo M, Díaz L et al (2007) Phenotypic stability for horizontal resistance to late blight in potato. In: African potato association conference proceedings, vol 7. Alex, Egypt, pp 58–62

Li L, Tacke E, Hofferbert HR, Lübeck J, Strahwald J, Draffeh AM, Walkemeier B, Gebhardt C (2013) Validation of candidate gene markers for marker-assisted selection of potato cultivars with improved tuber quality. Theor Appl Genet 126(4):1039–1052. https://doi.org/10.1007/s00122-012-2035-z

Lindhout P, Meijer D, Schotte T, Hutten RCB, Visser RGF, van Eck HJ (2011) Towards F1 hybrid seed potato breeding. Potato Res 54:301–312

Lindhout P, de Vries M, ter Maat M, Ying S, Viquez-Zamora M, van Heusden S (2018) Hybrid potato breeding for improved varieties. In: Wang-Pruski G (ed) Achieving sustainable cultivation of potatoes Volume 1: breeding improved varieties. Burleigh Dodds, Cambridge, pp 99–122

Luthra SK, Gupta VK, Tiwari JK, Kumar V, Bhardwaj V, Sood S, Dalamu, Kaur RP, Kumar R, Vanishree G, Kumar D, Mhatre P, Chakrabarti SK (2020) Potato breeding in India. CPRI Technical Bulletin No 74 (revised), ICAR-Central Potato Research Institute, Shimla, Himachal Pradesh

Meade F, Byrne S, Griffin D, Kennedy C, Mesiti F, Milbourne D (2019) Rapid development of KASP markers for disease resistance genes using pooled whole genome resequencing. Potato Res. https://doi.org/10.1007/s11540-019-09428-x

Meuwissen THE, Hayes BJ, Goddard ME (2001) Prediction of total genetic value using genome-wide dense marker maps. Genetics 157(4):1819–1829

Moloney C, Griffin D, Jones PW, Bryan GJ, McLean K, Bradshaw JE, Milbourne D (2010) Development of diagnostic markers for use in breeding potatoes resistant to *Globodera pallida* pathotype Pa2/3 using germplasm derived from *Solanum tuberosum* ssp. *andigena* CPC 2802. Theor Appl Genet 120:679–689. https://doi.org/10.1007/s00122-009-1185-0

Mori K, Asano K, Tamiya S, Nakao T, Mori M (2015) Challenges of breeding potato cultivars to grow in various environments and to meet different demands. Breed Sci 65(1):3–16. https://doi.org/10.1270/jsbbs.65.3

Mulema JMK, Olanya OM, Adipala E, Wagoire W (2004) Stability of late blight resistance in population B potato clones. Potato Res 47:11–24

Mulema JMK, Adipala E, Olanya OM, Wagoires W (2008) Yield stability analysis of late blight resistant potato selections. Expl Agric 44:145–155

Muller KO, Black W (1951) Potato breeding for resistance to blight and virus diseases during the last hundred years. Z Pflanzenzuchtung 31:305–318

Munoz FJ, Plaisted RL (1981) Yield and combining abilities in Andigena potatoes after six cycles of recurrent phenotypic selection for adaptation to long day conditions. Am Potato J 58:469–479

Naess SK, Bradeen JM, Wielgus SM, Haberlach GT, McGrath JM, Helgeson JP (2000) Resistance to late blight in *Solanum bulbocastanum* is mapped to chromosome 8. Theor Appl Genet 101:697–704

Niino T, Arizaga MV (2015) Cryopreservation for preservation of potato genetic resources. Breed Sci 65(1):41–52. https://doi.org/10.1270/jsbbs.65.41

Ortega F, Lopez-Vizcon C (2012) Application of molecular marker-assisted selection (MAS) for disease resistance in a practical potato breeding programme. Potato Res 55:1–13

Ortiz R (1998) Potato breeding via ploidy manipulation. Plant Breed Rev 16:15–86

Ortiz R (2001) The state of the use of potato genetic diversity. In: Cooper HD, Spillane C, Hodgkin T (eds) Broadening the genetic base of crop production. CAB International, Wallingford, pp 181–200

Ortiz R, Mihovilovich E (2020) Genetics and cytogenetics of the potato. In: Campos H, Ortiz O (eds) The potato crop: its agricultural, nutritional and social contribution to humankind. Springer, ebook, pp 219–247

Paget M, Amoros W, Salas E, Eyzaguirre R, Alspach P, Apiolaza L, Noble A, Bonierbale M (2014) Genetic evaluation of micronutrient traits in diploid potato from a base population of Andean Landrace Cultivars. Crop Sci 54:1949–1959

Pel MA, Foster SJ, Park TH, Rietman H, van Arkel G, Jones JDG, Van Eck HJ, Jacobsen E, Visser RGF, Van der Vossen EAG (2009) Mapping and cloning of late blight resistance genes from *Solanum venturii*using an interspecific candidate gene approach. Mol Plant Microbe Interact 22:601–615

Pieterse L, Judd J (2014) World catalogue of potato varieties, 6th edn. Agrimedia GmbH, Clenze

Plaisted RL (1987) Advances and limitations in the utilization of Neotuberosum in potato breeding. In: Jellis GJ, Richardson DE (eds) The production of new potato varieties. Cambridge University Press, Cambridge, pp 186–196

Plaisted RL, Hoopes RW (1989) The past record and future prospects for the use of exotic potato germplasm. Am Potato J 66:603–627

Rabier C-E, Barre P, Asp T, Charmet G, Mangin B (2016) On the accuracy of genomic selection. PLoS One 11(6):e0156086. https://doi.org/10.1371/journal.pone.0156086

Rakosy-Tican E, Thieme R, König J, Nachtigall M, Hammann T, Denes T-E, Kruppa K, Molnár-Láng M (2020) Introgression of two broad-spectrum late blight resistance genes, *Rpi-Blb1* and *Rpi-Blb3*, from *Solanum bulbocastanum* Dun plus race-specific *R* genes into potato pre-breeding lines. Front Plant Sci. https://doi.org/10.3389/fpls.2020.00699

Ramakrishnan AP, Ritland CE, Sevillano RHB, Riseman A (2015) Review of potato molecular markers to enhance trait selection. Am J Potato Res 92:455–472

Ross H (1986) Potato breeding – problems and perspectives. Advs in Plant Breed 13, Paul Parey, Berlin and Hamburg

Rousselle-Bourgeois F, Rousselle P (1992) Creation et selection de populations diploide de pomme de terre (*Solanum tuberosum* L.). Agronomie 12:59–67

Rutkoski JE, Crain J, Poland J, Sorrells ME (2017) Genomic selection for small grain improvement. In: Varshney RK, Roorkiwal M, Sorrells ME (eds) Genomic selection for crop improvement. Springer, Cham, pp 99–130

Salas A, Quispe V, Poma E, Vargas F, Chavez O, Ellis D (2014) The importance of the biodiversity of the wild *Solanum* sec *Petota* (*Solanaceae*) conserved at the International Potato Center (CIP). In: Pieterse L, Judd J (eds) World catalogue of potato varieties, 6th edn. Agrimedia GmbH, Clenze, pp 305–407

Sanetomo R, Habe I, Hosaka K (2019) Sexual introgression of the late blight resistance gene *Rpi-blb3* from a Mexican wild diploid species *Solanum pinnatisectum* Dunal into potato varieties. Mol Breed 39:13

Sato M, Nishikawa K, Komura K, Hosaka K (2006) Potato virus Y resistance gene, *Rychc*, mapped to the distal end of potato chromosome 9. Euphytica 149:367–372

Schulthess AW, Zhao Y, Reif JC (2017) Genomic selection in hybrid breeding. In: Varshney RK, Roorkiwal M, Sorrells ME (eds) Genomic selection for crop improvement. Springer, Cham, pp 149–183

Sharma R, Bhardwaj V, Dalamu D, Kaushik SK, Singh BP, Sharma S, Umamaheshwari R, Baswaraj R, Kumar V, Gebhardt C (2014) Identification of elite potato genotypes possessing multiple disease resistance genes through molecular approaches. Sci Hortic 179:204–211

Simmonds NW (1969) Prospects of potato improvement. Scottish Plant Breeding Station Forty-Eighth Annual Report 1968-69, pp. 18–38

Slater AT, Cogan NOI, Forster JW (2013) Cost analysis of the application of marker-assisted selection in potato breeding. Mol Breed 32:299–310

Slater AT, Cogan NOI, Hayes BJ, Schultz L, Dale MFB, Bryan GJ, Forster JW (2014) Improving breeding efficiency in potato using molecular and quantitative genetics. Theor Appl Genet 127:2279–2292

Slater AT, Cogan NOI, Forster JW, Hayes BJ, Daetwyler HD (2016) Improving genetic gain with genomic selection in autotetraploid potato. Plant Genome 9. https://doi.org/10.3835/plantgenome2016.02.0021

Sosa P, Guild G, Burgos G, Bonierbale M, zum Felde T (2018) Potential and application of X-ray fluorescence spectrometry to estimate iron and zinc concentration in potato tubers. J Food Compos Anal 70:22–27

Spooner DM, Hijmans RJ (2001) Potato systematics and germplasm collecting, 1989-2000. Am J Potato Res 78:237–268

Spooner DM, Ghislain M, Simon R, Jansky SH, Gavrilenko T (2014) Systematics, diversity, genetics, and evolution of wild and cultivated potatoes. Bot Rev 80:283–383

Su Y, Viquez-Zamora M, den Uil D, Sinnige J, Kruyt H, Vossen J, Lindhout P, van Heusden S (2019) Introgression of genes for resistance against *Phytophthora infestans* in diploid potato. Am J Potato Res https://doi.org/10.1007/s12230-019-09741-8

Sun G, Zhu C, Kramer MH, Yang S-S, Song W, Piepho H-P, Yu J (2010) Variation explained in mixed-model association mapping. Heredity 105:333–340

Sverrisdóttir E, Byrne S, Sundmark HER, Johnsen HØ, Kirk HG, Asp T, Janss L, · Kåre L. Nielsen KL (2017) Genomic prediction of starch content and chipping quality in tetraploid potato using genotyping-by-sequencing. Theor Appl Genet 130:2091–2108. https://doi.org/10.1007/s00122-017-2944-y

Tai GCC (1994) Use of 2*n* gametes. In: Bradshaw JE, Mackay GR (eds) Potato genetics. CAB International, Wallingford, pp 109–132

Tarn TR, Tai TCC (1983) Tuberosum × Tuberosum and Tuberosum × Andigena potato hybrids: comparisons of families and parents, and breeding strategies for Andigena potatoes in long-day temperate environments. Theor Appl Genet 66:87–91

Tek AL, Stevensen WR, Helgeson JP, Jiang J (2004) Transfer of tuber soft rot and early blight resistances from *Solanum brevidens* into cultivated potato. Theor Appl Genet 109:249–254

Tibshirani R (1996) Regression shrinkage and selection via the LASSO. J R Stat Soc B 58:267–288

Toxopeus HJ (1953) On the significance of multiplex parental material in breeding for resistance to some diseases in the potato. Euphytica 2:139–146

Toxopeus HJ, Huijsman CA (1953) Breeding for resistance to potato root eelworm. Euphytica 2:180–186

Trognitz BR, Bonierbale M, Landeo JA, Forbes G, Bradshaw JE, Mackay GR, Waugh R, Huarte MA, Colon L (2001) Improving potato resistance to disease under the global initiative on late blight. In: Cooper HD, Spillane C, Hodgkin T (eds) Broadening the genetic base of crop production. CAB International, Wallingford, pp 385–398

Uitdewilligen JGAML, Wolters A-MA, D'hoop BB, Borm TJA, Visser RGF, van Eck HJ (2013) A next generation sequencing method for genotyping-by-sequencing of highly heterozygous autotetraploid potato. PLoS One 8(5):e62355. https://doi.org/10.1371/journal.pone.0062355

Vavilov NI (1940) The theory of the origin of cultivated plants after Darwin. Nauka [Science] 2:55–75

Veilleux RE (2005) Cell and tissue culture of potato (Solanaceae). In: Razdan MK, Mattoo AK (eds) Genetic improvement of Solanaceous crops, volume I: potato. Science Publishers Inc, Enfield, pp 185–208

Verzaux E, Budding D, de Vetten N, Niks RE, Vleeshouwers VGAA, van der Vossen EAG, Jacobsen E, Visser RGF (2011) High resolution mapping of a novel late blight resistance gene *Rpi-avl1* from the wild Bolivian species *Solanum avilesii*. Am J Potato Res 88:511–519

Vos PG, Uitdewilligen JGAML, Voorrips RE, Visser RGF, van Eck HJ (2015) Development and analysis of a 20K SNP array for potato (*Solanum tuberosum*): an insight into the breeding history. Theor Appl Genet 128:2387–2401

Vos PG, Paulo MJ, Voorrips RE, Visser RGF, van Eck HJ, van Eeuwijk FA (2017) Evaluation of LD decay and various LD-decay estimators in simulated and SNP-array data of tetraploid potato. Theor Appl Genet 130:123–135. https://doi.org/10.1007/s00122-016-2798-8

Vossen JH, Nijenhuis M, Arens-de Reuver MJB, van der Vossen EAG, Jacobsen E, Visser RGF (2011) Cloning and exploitation of a functional *R* gene from *Solanum chacoense*. US patent WO/2011/034433 A1, published by the world intellectual property organization 18 September 2009

Wenzel G (1994) Tissue culture. In: Bradshaw JE, Mackay GR (eds) Potato genetics. CAB International, Wallingford, pp 173–195

Würschum T, Kraft T (2014) Cross-validation in association mapping and its relevance for the estimation of QTL parameters of complex traits. Heredity 112:463–468

Zimnoch-Guzowska E, Yin Z, Chrzanowska M, Flis B (2013) Sources and effectiveness of potato PVY resistance in IHAR's breeding research. Am J Potato Res 90:21–27. https://doi.org/10.1007/s12230-012-9289-5

Chapter 7
Breeding and Multiplication of Clonally Propagated Potato Cultivars

7.1 Introduction

Since the early 1900s, potato breeding programmes have comprised cycles of artificial hybridization followed by multistage, multitrait selection over as many as eight clonal generations, primarily among developing elite germplasm. Desirable genes for disease and pest resistance have been introgressed from landraces and wild relatives, and a few of the parents used in the hybridizations have come from more general base broadening (Bradshaw and Bonierbale 2010). The driving force for potato improvement has therefore been the cycles of hybridization and selection, with the new cultivars produced each cycle providing the means to transfer progress to farmers' fields. A typical example (Fig. 7.1) is the potato breeding done at the Scottish Plant Breeding Station (SPBS) and the Scottish Crop Research Institute (SCRI) from 1920 to 2008, where the first cultivar The Alness is separated by 11 generations spanning 78 years from the last cultivar to be reviewed, Vales Everest (Bradshaw 2009b). In between are SPBS cultivars Craigs Defiance, Roslin Chania and Roslin Sasumua, both for Kenya, Pentland Crown (most widely grown cultivar in Great Britain from 1970 to 1979), Pentland Dell (widely used for making French fries) and Pentland Ivory (an early maincrop cultivar).

Cultivars Majestic and Shamrock were, so far as is known, derived from earlier British cultivars, Pepo was of European origin, and Abundance and Epicure each have Magnum Bonum as their female parent. Its female line traces back through Early Rose and Garnet Chili to Rough Purple Chili and the year 1851 and is one of the sources of Chilean Tuberosum cytoplasm. *Solanum demissum* was used for resistance to late blight (*Phytophthora infestans*), 11–79 for resistance to *Potato virus Y* and H1H3 (from Group Andigena) for resistance to cyst nematodes (*Globodera rostochiensis* and *G. pallida*, respectively). In total, 72 cultivars were bred with their pedigrees and more details of the breeding programmes given by Bradshaw 2009b. Likewise, for India, Luthra et al. (2020) have given a detailed account of the 64 cultivars bred from hybridizations at the ICAR-Central Potato

J. E. Bradshaw, *Potato Breeding: Theory and Practice*,
https://doi.org/10.1007/978-3-030-64414-7_7

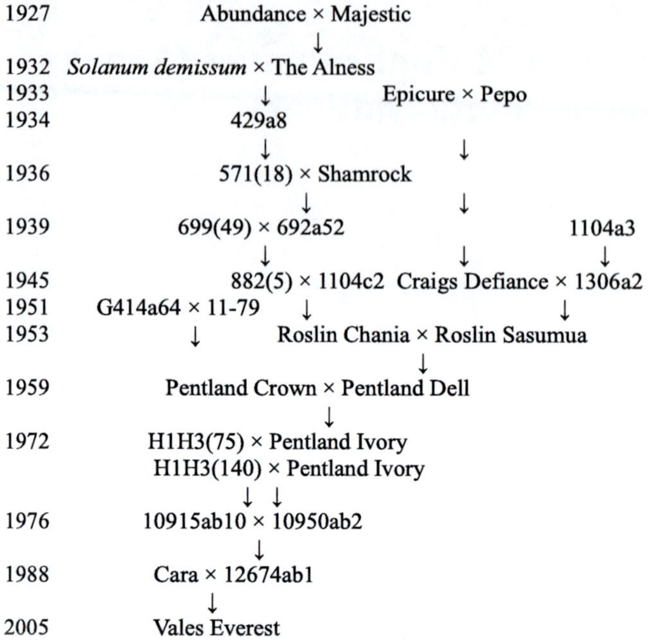

Fig. 7.1 Potato breeding at the Scottish Plant Breeding Station and the Scottish Crop Research Institute from cultivar The Alness to cultivar Vales Everest. (From pedigree information in Bradshaw 2009b)

Research Institute over the 70-year period 1949–2019, including a connection with Dr. Black's programme at SPBS (e.g. Lady Balfour is related to Kufri Jyoti). The parentage of the 64 cultivars shows that 25 have both parents of Indian origin, 28 have one parent of Indian origin and one of foreign origin, and 11 have both parents of foreign origin, with of total of 98 parents used.

Globally, there has certainly been no shortage of new cultivars, with the 2014 World Catalogue of Potato Varieties (Pieterse and Judd 2014) listing more than 4800 cultivars from some 100 countries covering all the potato growing regions of the world, although a much smaller number have been widely grown. The success of an individual breeding programme can be judged by the success of the cultivars that it produces. For example, in 1977, the Pentland series of 16 cultivars from SPBS occupied approximately 40% of the British potato acreage (Mackay 2003). However, of more significance globally is the number of cultivars available for a particular agro-ecological zone and end use, and the extent to which they can be considered an improvement on the previous generation of cultivars. Four examples will suffice. Jin (2018) reported four rounds of cultivar substitution in China, from the 1960s to 2010. The fourth round saw a particular breakthrough with new processing cultivars being widely grown. Love et al. (1998) revealed progress in breeding for chip (crisp) quality in North America by comparing 44 cultivars released over the period from 1876 to 1997. There was evidence for a slight improvement over time in tuber solids, large improvements in reducing sugar content (lower) and chip colour, and a recent improvement in ability to produce defect-free chips. In contrast, Bradshaw (2008) found a complicated picture when he examined the effect of new cultivars on the disease and pest resistance of the British potato crop over the period 1966–2006. New cultivars gained favour with growers and end users primarily because of their higher yields and either their greater suitability for processing or their greater attractiveness on supermarket shelves. Sometimes, improved disease and pest resistance helped to establish a new cultivar: Pentland Crown was bred to have good resistance to viruses PLRV and PVY and Maris Piper was the first cultivar to have the *H1* gene for resistance to the golden potato cyst nematode (*Globodera rostochiensis*) from Group Andigena. However, other changes in cultivars increased the susceptibility of the British potato crop: for example, Maris Piper is particularly susceptible to common scab, Estima to blackleg and Pentland Dell to TRV (*Tobacco rattle virus*). Finally, Mori et al. (2015) reviewed progress in breeding cultivars in Japan where potatoes are grown in summer in the northern part of the country (e.g. Hokkaido), in winter in the southernmost part of the country (e.g. Okinawa), and double cropping occurs in spring and autumn in other regions (e.g. Kyushu). Cross-breeding began in Hokkaido in 1918, following introductions of foreign cultivars from 1902, and the first starch cultivar was released in 1938. Breeding for double cropping started in 1947. Leading cultivars for chip (crisp) processing, starch production and French fries were released in 1976, 1981 and 1981, respectively; and likewise, for double cropping in 1971 and 1978. After the golden potato cyst nematode (*Globodera rostochiensis* pathotype Ro1) was found in Hokkaido in 1972, a major breeding effort resulted in the release of many cultivars from 1986 onwards with the *H1* gene for resistance, including double cropping cultivars from 1997.

Given the wide range of agro-ecological zones and regions within zones in which potatoes are grown, and the different potato characteristics needed for the various end uses of potato crops, globally a large number of breeding programmes is required to produce the necessary range of adapted cultivars. In other words, modern breeding programmes need to produce locally adapted cultivars in the same way that farmers selected locally adapted landraces, even though breeders would like their cultivars to be grown and used as widely as possible, a point that can be seen in the context of breeding strategies for table potato production in Sweden and the Fennoscandian Region (Eriksson et al. 2016).

Up until the 1960s, breeding programmes were relatively small in size by modern standards and the time to breed a new cultivar was relatively short, often just 7–10 years from crossing to registration or National Listing (Bradshaw 2009b). Then, from the 1960s, breeders felt the need to increase the sizes of their programmes, both to assess more traits (around 20 at SPBS/SCRI) and to assess potential cultivars in more environments, and so raising 100,000 seedlings each year from 200 to 300 crosses became the norm (Mackay 2005; Bradshaw 2009b; Jansky 2009; Bradshaw and Bonierbale 2010; Lindhout et al. 2011; Eriksson et al. 2016) and is still common today. For example, during the 19th Joint Meeting of EAPR Breeding and Varietal Assessment Section and EUCARPIA Section Potatoes in

Rostock-Warnemuende, Germany, in December 2018, we visited the breeding programme of Norika in Gross Luesewitz. The programme was very similar to the SCRI one described by Bradshaw and Mackay (1994) to which I will refer to as typical in this chapter. However, there are two main weaknesses of such breeding programmes: the number of clonal generations required to select a new cultivar and the inability to affect most economically important traits, which are quantitative in nature, during intense early-generation selection. There is just no way that 100,000 clones, all potential new cultivars, can be assessed in replicated yield trials for all of the desired traits, so visual selection is practised by the breeders.

In this chapter, we are going to look at the theory and practice of this conventional breeding and ways in which it can be made faster and more efficient and effective. The emphasis will be on breeding tetraploid cultivars for clonal propagation, although the same considerations apply to breeding diploid cultivars for clonal propagation. Then, in the next chapter, we will consider breeding cultivars for propagation through true potato seed. For ease of access, I am going to use examples from the SPBS/SCRI breeding programme when I require actual data, but the results and conclusions will be generally applicable.

7.2 Choice of Parents and Crosses

At the start of a new programme, the breeder needs to define their breeding objectives for new cultivars in terms of the target environment (agroecological zone) and target end use (staple food, vegetable for cash, crop for processing). These broad objectives then need to be translated into selection criteria (yield, quality, resistance to pests and diseases, tolerance of abiotic stresses) and into the environments in which potential cultivars will be assessed. Finally, the breeder needs to assess potential parents, decide which crosses to make, and decide how many offspring to raise from each cross. In other words, the breeder requires a hybridization strategy. Here, we are going to consider a crossing strategy for quantitative traits where the number of desired alleles is unknown and individual alleles are not recognized, whilst recognizing that in future we are likely to see strategies based on an increasing use of marker-assisted selection and genomic selection, as outlined in the previous chapter. Crosses will be between parents which complement each other for desirable traits or possess the desired traits in varying degrees. It would be advantageous to predict the amount of variation that will be generated within crosses as those combining high variances (transgressive segregation giving offspring superior to the better parent) with high means would be the best ones from which to seek new cultivars (Bradshaw 2009a). However, experiments have shown that only the use of mid-parental values to predict offspring (cross) means can be recommended as a guide to the usefulness of crosses (Bradshaw et al. 1998a). In the simple example in Tables 7.1 and 7.2, it can be seen that the mid-parental values and offspring means are the same when there are just additive genetic differences (*a*), but differences occur with the introduction of diallelic interactions (*d*).

Table 7.1 Parental genotypic frequencies, values and general combining abilities (GCA) for random mating population in equilibrium with two alleles *A* and *a* at frequencies *p* and *q* under chromosomal segregation so that gamete frequencies are p^2 (*AA*), 2*pq* (*Aa*) and q^2 (*aa*), population mean (Mean) is 2*a*(*p* – *q*) + 12*pqd* and GCAs are means of offspring from mating genotypes with gametes in equilibrium frequencies, minus the population mean

Genotype	*AAAA*	*AAAa*	*AAaa*	*Aaaa*	*aaaa*
Frequency	p^4	$4p^3q$	$6p^2q^2$	$4pq^3$	q^4
Value	$2a$	$a + 3d$	$4d$	$-a + 3d$	$-2a$
Value-mean	$4aq - 12pqd$	$a(-p + 3q) - 3(4pq - 1)d$	$2a(-p + q) - 4(3pq - 1)d$	$-a(3p - q) - 3(4pq - 1)d$	$-4ap - 12pqd$
GCA	$2aq - 6pqd + 4q^2d$	$\frac{1}{2}a(-p + 3q) - 1\frac{1}{2}(4pq - 1)d - 2q(p - q)d$	$a(-p + q) - 2(3pq - 1)d + (2/3 - 4pq)d$	$-\frac{1}{2}a(3p - q) - 1\frac{1}{2}(4pq - 1)d - 2p(q - p)d$	$-2ap - 6pqd + 4p^2d$
Selfed value	$2a$	$a + 2\frac{1}{2}d$	$3\frac{1}{3}d$	$-a + 2\frac{1}{2}d$	$-2a$

Table 7.2 Mid-parent values, mid-self values and offspring means for random mating population in equilibrium with two alleles *A* and *a* at frequencies *p* and *q* under chromosomal segregation (cross frequencies can be obtained from parental frequencies in Table 7.1)

Female parent	Male parent	Mid-parent	Mid-self	Offspring
AAAA	*AAAA*	$2a$	$2a$	$2a$
AAAA	*AAAa*	$1\frac{1}{2}a + 1\frac{1}{2}d$	$1\frac{1}{2}a + 1\frac{1}{4}d$	$1\frac{1}{2}a + 1\frac{1}{2}d$
AAAA	*AAaa*	$a + 2d$	$a + 1\frac{2}{3}d$	$a + 2\frac{2}{3}d$
AAAA	*Aaaa*	$\frac{1}{2}a + 1\frac{1}{2}d$	$\frac{1}{2}a + 1\frac{1}{4}d$	$\frac{1}{2}a + 3\frac{1}{2}d$
AAAA	*aaaa*	0	0	$0 + 4d$
AAAa	*AAAa*	$a + 3d$	$a + 1\frac{1}{4}d$	$a + 2\frac{1}{2}d$
AAAa	*AAaa*	$\frac{1}{2}a + 3\frac{1}{2}d$	$\frac{1}{2}a + (35/12)d$	$\frac{1}{2}a + 3\frac{1}{6}d$
AAAa	*Aaaa*	$0 + 3d$	$0 + 2\frac{1}{2}d$	$0 + 3\frac{1}{2}d$
AAAa	*aaaa*	$-\frac{1}{2}a + 1\frac{1}{2}d$	$-\frac{1}{2}a + 1\frac{1}{4}d$	$-\frac{1}{2}a + 3\frac{1}{2}d$
AAaa	*AAaa*	$0 + 4d$	$0 + 3\frac{1}{3}d$	$0 + 3\frac{1}{3}d$
AAaa	*Aaaa*	$-\frac{1}{2}a + 1\frac{1}{2}d$	$-\frac{1}{2}a + (35/12)d$	$-\frac{1}{2}a + 3\frac{1}{6}d$
AAaa	*aaaa*	$-a + 2d$	$-a + 1\frac{2}{3}d$	$-a + 2\frac{2}{3}d$
Aaaa	*Aaaa*	$-a + 3d$	$-a + 2\frac{1}{2}d$	$-a + 2\frac{1}{2}d$
Aaaa	*aaaa*	$-1\frac{1}{2}a + 1\frac{1}{2}d$	$-1\frac{1}{2}a + 1\frac{1}{4}d$	$-1\frac{1}{2}a + 1\frac{1}{2}d$
aaaa	*aaaa*	$-2a$	$-2a$	$-2a$

It can also be seen in Tables 7.1 and 7.2 that the same applies to using general combining abilities (GCAs) to predict offspring means. When there are just additive genetic differences (*a*), the parental values are simply twice their GCAs, but this is no longer true with the introduction of diallelic interactions (*d*). Furthermore, GCAs need to be determined from hybridizations that generate appropriate half-sib families and hence are not available in advance of such hybridizations. Likewise, there is no advantage over mid-parental values of mid-selfed values, and hence no reason to go to the trouble of selfing the parents, despite some authors suggesting the use of mid-self values (Gopal 2018). When there are just additive genetic differences (*a*), the mid-self and mid-parent values are the same, whereas in the presence of diallelic interactions (*d*) just 4 out of 15 predictions are better with mid-self values (Table 7.2).

In combining ability analysis, the GCA of a parent is the mean value of its offspring when mated with a random sample of gametes from all of the parents, minus the mean value of all of the parents. The offspring (cross) mean of a cross can be predicted from the sums of the parental GCAs, with departures from the predictions attributed to the specific combining ability (SCA) of the two parents, the result of allelic and non-allelic interactions. Thus:

$$\text{cross}(\text{offspring})\text{mean} = \mu(\text{overall mean}) + GCA_{\text{female parent}} + GCA_{\text{male parent}} + SCA_{\text{two parents}}$$

Back at the beginning of Chap. 3, we examined the results of a 6 × 6 half diallel set of crosses with selfs which was assessed for yield at Mylnefield and Gourdie Farms, Dundee (56.46°N, 2.97°W) in 1993 and 1994, respectively (Bradshaw 2009a). There were 218 clones from the 15 hybridizations of the six parents. SCA (component of variance 0.457) accounted for slightly more variation than GCA (0.354), and their interactions with years were smaller. There was a high correlation between parental yields and their GCAs ($r = 0.875$) and a moderate one between offspring means and mid-parent values ($r = 0.69$). The highest yielding offspring clone came from the second highest yielding cross (4th out of 15 for mid-parent value) and the second highest yielding offspring clone came from the highest yielding cross (3rd for mid-parent value). The components of the clonal yields were as follows:

$$\begin{aligned}14.30(\text{best clone}) = {} & 8.57(\text{mean of all clones}) + 0.69(GCA_{\text{Cara}}) \\ & + 0.34(GCA_{\text{P.Squire}}) + 0.56(SCA) + 4.14(\text{clone in cross})\end{aligned}$$

$$\begin{aligned}13.73(\text{second best clone}) = {} & 8.57(\text{mean of all clones}) + 0.22(GCA_{\text{Desiree}}) \\ & + 0.66(GCA_{\text{M.Piper}}) + 1.54(SCA) + 2.74(\text{clone in cross})\end{aligned}$$

In both crosses, there was transgressive segregation and both parents had positive GCAs and combined with a positive SCA, but the only information available before the crosses were made was the mid-parental values.

So, the initial choice of pair crosses needs to be based on differences in mid-parent (phenotypic) values for the set of desired traits, as this is the only readily-available predictor of progeny means. How good a predictor of progeny means depends on the narrow-sense heritabilities of the traits. For a random mating population in equilibrium the regression of offspring mean on mid-parental value is an estimate of the narrow-sense heritability. In other words, we imagine that we have taken pairs of clones at random from an equilibrium population and used them in pair crosses to create a new population of offspring progenies. Parents and offspring progenies are then assessed in the same trials to avoid unwanted complications from genotype × environment interactions. The slope of the regression line (*b*) is estimated as the covariance of offspring (cross) means and mid-parent values divided by the variance of the mid-parent values. For autotetraploid potato, with chromosomal segregation, the covariance is $\frac{1}{2}\sigma^2_A + (1/6)\sigma^2_D$ and the variance is $\frac{1}{2}\sigma^2_A + \frac{1}{2}\sigma^2_D + \frac{1}{2}\sigma^2_T + \frac{1}{2}\sigma^2_Q + \frac{1}{2}\sigma^2_e/r$, where σ^2_A, σ^2_D, σ^2_T and σ^2_Q are the variances due to the main effects and diallelic, triallelic and tetra-allelic interactions of alleles A_i, A_j, A_k and A_l at a locus '*A*', summed over loci, and σ^2_e/r is the plot to plot environmental variation divided by the number of replicates *r* of each parent in the assessment trial (Kempthorne 1957; Bradshaw, 1994). Hence:

$$b = \left(\sigma^2_A + \frac{1}{3}\sigma^2_D\right) / \left(\sigma^2_A + \sigma^2_D + \sigma^2_T + \sigma^2_Q + \sigma^2_e / r\right)$$

where the numerator contains a contribution from σ^2_D because diallelic interactions can be transmitted through diploid gametes, and the denominator is the phenotypic variation that exists between the parents and includes the environmental variation, which can be reduced by increasing the number of replicates in the assessment trials. The regression line can also be considered the predicted response to selection seen in the next generation. However, as one generation of random mating does not establish single locus genetic equilibrium with tetrasomic inheritance, the immediate response to selection is not the equilibrium one we derived in Chap. 2, where the narrow sense heritability (h_n^2) was as follows:

$$h_n^2 = \sigma^2_A / \left(\sigma^2_A + \sigma^2_D + \sigma^2_T + \sigma^2_Q + \sigma^2_e / r\right)$$

Desired traits can be combined into an index of overall merit but have usually been considered individually by potato breeders. Either way, it is clearly important to have as much information as possible about potential parents and to assess them as accurately as possible, both phenotypically and genotypically with molecular markers with a view to incorporating marker-assisted and genomic selection into the breeding programme.

Let us take an example from the breeding and research programme at the Scottish Crop Research Institute. The genetic variation available in a new breeding programme was assessed through a 15 × 15 half diallel set of 105 crosses made in 1992. The programme was primarily designed to combine quantitative resistances to late blight (*Phytophthora infestans*) and the white potato cyst nematode (*Globodera pallida*) with commercially acceptable tuber yield and quality, but also included parents with resistance to *Potato leafroll virus* (PLRV) and *Potato virus Y* (PVY). Resistance to late blight, the white potato cyst nematode and gangrene (*Phoma foveata*), as well as breeders' visual assessment of tubers (preference), were assessed in seedling progeny tests (Bradshaw et al. 1995, 1996). The assessment of yield and other economically important traits was done on tuber progenies in replicated ware trials in 1995 and 1996, the seedling progenies having been grown in a glasshouse in 1993 and maintained at a high-grade seed site in 1994 and 1995 (Bradshaw et al. 2000). Each cross (progeny) was maintained as two rows (separate replicates) of 18 spaced plants (36 unique genotypes). Each yield trial had an RCB (randomized complete block) design with two replicates of each parent (15 parents) and progeny (105 progenies) in two-drill plots of 18 tubers (2 rows of 9) spaced 45 cm apart within drills with 75 cm between drills. Each progeny plot should have contained tubers from 18 different offspring but there were departures from this ideal, as explained in the paper by Bradshaw et al. (2000) along with all of the cultural, scoring, harvesting and assessment details.

The regressions of offspring mean (averaged over replicates and also over 2 years for yield) on mid-parental mean are shown for fry colour and total yield in Fig. 7.2, the former having been assessed in December 1996, but not in 1995, on 18 tubers from each plot which had been kept in a potato store at ambient temperature. Each tuber was cut in half and a 2-mm-thick slice taken and fried in vegetable oil at 175 °C in a thermostatically-controlled chip fryer until all of the water had boiled off (bubbling ceased). Colour was assessed on a 1–9 scale by comparison with a standardized colour chart (1 = extremely dark to 9 = extremely pale), where 5 and above is an acceptably light colour.

The mid-parent value was a good predictor of the offspring mean for fry colour ($b = 0.944 \pm 0.051$) but not so good for total yield where there was a larger scatter of points ($b = 0.518 \pm 0.060$). Mid-parent value was also quite a good predictor for dry-matter content (measured as specific gravity) ($b = 0.762 \pm 0.058$). However, the regressions were in the range $b = 0.359$ to $b = 0.683$ for all of the other traits (crop emergence and maturity; tuber size, regularity of shape, resistance to common scab (*Streptomyces scabies*) and uniformity of size and appearance; and tuber sprouting and keeping quality in storage). Interestingly, the offspring-midself correlations were slightly less than the offspring-midparent correlations for all traits, as predicted above, so there would be no advantage in selfing the parents and assessing the selfed progenies in order to improve the prediction of offspring means. The results, together with the combining ability analyses, provided useful information about likely progress in the breeding programme in selecting for the various traits both

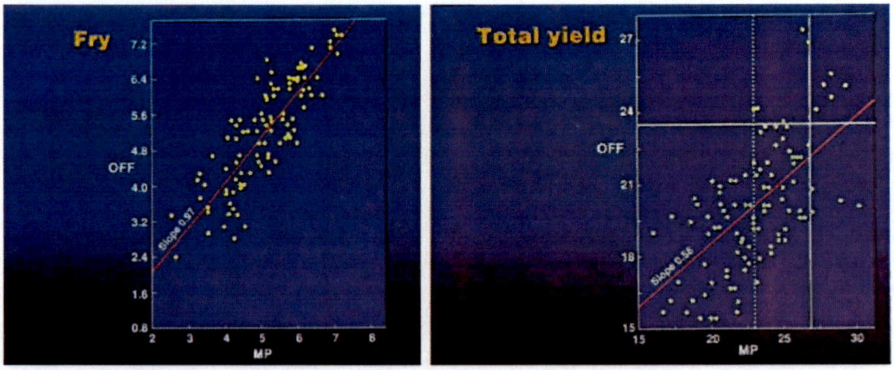

Fig. 7.2 The regressions of offspring mean (averaged over replicates and also over 2 years for yield) on mid-parental mean for fry colour and total yield in a 15 × 15 half diallel set of 105 crosses made at the Scottish Crop Research Institute (SCRI) in 1992. (Source: SCRI)

within generations and over generations. However, they did not really provide much useful information about a general crossing strategy for quantitative traits. This is because a breeder does not make one small set of crosses to improve fry colour with its high heritability and a different larger set of crosses to improve yield with its moderate heritability, and so on for each trait; although one wouldn't make crosses if there was no evidence of genetical variation for a trait! The breeder wants to make the best set of crosses for improving all of the traits desired in a new cultivar and for continued potato improvement over many generations. Furthermore, heritabilities are germplasm dependent and hence will vary from one breeding programme to another. Therefore, in practice, breeders make a relatively large number of crosses to ensure that they include the best ones possible from the available parents. As for the number offspring to raise from each cross, we can get some guidance from our imaginary random mating population in equilibrium. The average amount of genetical variation within each cross (full-sib family) is the same ($\sigma^2_D = \sigma^2_T = \sigma^2_Q = 0$) or slightly larger than the genetical variation between crosses (full-sib families), so the progeny size should be the same or slightly larger than the number of crosses.

$$\text{Genetic variation between crosses} = (1/2)\sigma^2_A + (2/9)\sigma^2_D + (1/12)\sigma^2_T + (1/36)\sigma^2_Q$$

$$\text{Genetic variation within crosses} = (1/2)\sigma^2_A + (7/9)\sigma^2_D + (11/12)\sigma^2_T + (35/36)\sigma^2_Q$$

Wricke and Weber (1986) used a different approach to hybridization strategy in their book on selection in plant breeding which may still be of interest to readers. They considered maximization of the response to one-step selection for individual clones, using a linear model for the phenotypic value (P_{jk}) of genotype (clone) k from cross j, in a crossing strategy involving m crosses each producing progeny of size n for a total of N (= mn) offspring.

$$P_{jk} = \mu + C_j + S_{jk} + \bar{e}_{jk}$$

where μ is the overall mean, C_j is the effect of cross j, S_{jk} is the effect of genotype k within cross j and $\bar{e}_{jk}$ is the environmental variation appropriate for the trial design.

There is an intraclass correlation within a cross (related clones which give information about the cross) which influences the selection response. Given the values of the genetic variation between and within crosses, and the environmental variation, Wricke and Weber (1986) found numerically the optimum values of m and n for a given N. The number of crosses increased with the genetic variance between crosses and decreased with an increase in the environmental variance. Fifty crosses each of size 24 for 1200 offspring was optimum when the three variances were equal and the selected proportion was from 10 to 20 per cent. In other words, the number of crosses was twice the progeny size of each cross, but the authors acknowledge that their examples are only a rough guide since the exact variances were not known, indeed, are not known until the crosses have been made and assessed. However, I am going to explore the case for selection between crosses before selection within the chosen crosses.

7.3 Methods of Artificial Hybridization

Before considering the size of a breeding programme, let us briefly consider making artificial hybridizations.

The floral characteristics of potatoes and methods of artificial hybridization and self-pollination have been described by Plaisted (1980). An illustrated account of the practice of making crosses can be found in the book *Potato Breeding* edited by Tiemens-Hulscher et al. (2013). Breeders will establish their own routine for making crosses and extracting and storing seeds.

The following is a brief account of some key points and not a practical manual. A temperature of 19 °C (certainly below 25 °C) and 16 h of daylength are recommended for crosses involving *S. tuberosum* Group Tuberosum, and hence, an extension of natural daylength is required for crossing in the tropics and subtropics. In contrast, Group Andigena is in effect day-neutral for flowering, but not for tuberization. In Scotland, crossing can be done in summer in a glasshouse (Fig. 7.3) whereas in a warmer climate (e.g. south of England), it may be necessary to do the crossing in the cooler months of spring. Breeders usually encourage flowering in a glasshouse by the periodic removal of daughter tubers, a task made easier by placing the parental tuber on a brick and covering it with compost (Fig. 7.3a, b). Sometimes, breeders encourage flowering of parents known to produce few flowers by grafting young potato shoots onto tomato or other compatible *Solanaceous* plants. Stems with flowers attached can also be cut in the field and brought to the glasshouse and maintained in jars of water, with an anti-bacterial agent to reduce contamination, for use as female parents (Peloquin and Hougas 1959). As most potatoes can self-pollinate, emasculation is necessary before artificial pollination. Unripe anthers are best removed with a pair of tweezers 1 or 2 days before the flower opens. Pollen can be shaken (e.g. using a modified electric toothbrush as a vibrator) from the anthers of the male parent and collected on a flat surface (Fig. 7.3c) or in a small glass vessel. When flowering of the male and female parents does not coincide, pollen can be stored in Eppendorf tubes at −20 °C in a freezer until required, if necessary, for between 1 and 3 years. A cluster of open flowers can be pollinated 1 or 2 days after emasculation by bringing the pollen into contact with the stigmas (Fig. 7.3d); for example, by dipping the stigmas into a small vessel containing the collected pollen. After pollination a string tag with the crossing code is looped around the stem of the flowers (Fig. 7.3e). As a cluster of berries starts to form, they are placed in a bag (cellophane or net) to prevent them from dropping on the ground (Fig. 7.3f). The clusters of berries are harvested before they begin to rot and then held in cold storage. Ripe (soft) berries are squeezed in a beaker of water, or a blender with blunt blades, so that the seeds will sink to the bottom allowing the pulp to be decanted away. The seeds are dried on filter paper and then stored in clearly labelled paper or plastic bags.

In India, the parents selected for hybridization are planted in a crossing block to obtain flowering under natural day-length conditions in the hills around Shimla, and in artificially illuminated conditions in the sub-tropical plains (Fig. 7.4) (Luthra

Fig. 7.3 Artificial hybridization of potatoes at SCRI. (a) Three tubers of same parent on bricks to be covered with compost (20 May 2008). (b) Lower leaves removed from plants and stems tied to canes (26 June 2008). (c) Anther (pollen) being taken from male parent (28 July 2008). (d) Emasculated female parent being pollinated. (e) Pollinated flowers of hybridization have been labelled. (f) Developing berries in a cellophane bag (28 July 2008)

et al. 2020). Normal agronomical practices are followed for raising the plants but more space is provided in between two beds to facilitate movement of technicians doing the hybridizations. Flowering in the hills generally starts in the third week of June with July the peak flowering time for doing the hybridizations. In the plains,

Fig. 7.4 Crossing block at Modipuram centre of the ICAR-Central Potato Research Institute, India (13 December 2008)

the crossing block is planted in the first week of October and flowering starts in the last week of November with the peak in December. Luthra et al. (2020) give details of pollination method, seed extraction and seed storage in polythene lined aluminium foil covers or double polythene bags which are sealed and kept in desiccators containing calcium chloride in a refrigerator at 6–10 °C.

Not all desired crosses are successful. Problems can arise from clones failing to flower, buds and flowers dropping either before or after fertilization, low pollen production, and failure to produce viable pollen (male sterility which is not uncommon). In cycles 1–3 of the SCRI breeding programme mentioned earlier, the success rates achieved were 120/428, 125/428 and 130/351 desired combinations of crosses, a typical success rate of just over 30% (Bradshaw et al. 2009). When a cross is successful, however, over 600 botanical seeds can be obtained from three to five berries, the outcome of pollination of just three to five flowers. The complete set of 105 crosses from the 15 × 15 half diallel set mentioned earlier was achieved through a careful choice of parents for male and female fertility.

7.4 Size of Breeding Programme

Let us start by seeing if there is a theoretical justification for producing 100,000 seedlings from 200 to 300 crosses. We can use the binomial distribution as an approximation of the normal distribution to construct a smaller but conveniently sized population of 65,536 seedlings/clones from 256 crosses each with an offspring size of 256. If the genetic variation between crosses is the same as the genetic variation between the offspring in each cross, we can achieve the distribution shown in Fig. 7.5 with deviations of −4, −3, −2, −1, 0, 1, 2, 3 and 4 in the ratios 1:8:28:56:70:56:28:8:1. The variances between crosses and within each cross both have a value of 2 so that the total variance between the 65,536 seedlings/clones has a value of 4. The population mean can be set at an arbitrary value, say 10, for our quantitative trait which could, for example, be breeders' visual assessment of merit (preference) or yield as shown in Fig. 7.5. When we later introduce environmental variation, we will not worry about the field layout of the crosses and clones other than to assume that the experimental design is such that there is no environmental variation between crosses additional to that between clones.

Two assumptions were made in constructing the population of seedlings/clones. The first assumption was that the amount of genetic variation between and within crosses is the same with each normally distributed. It is based on the additive genetic variances between and within full-sib families being equal in a random mating population in equilibrium. It is therefore unlikely to be true in practice but has a theoretical basis in the absence of real data. The second assumption was that the amount of genetic variation within each cross is the same. Although unrealistic, it will do for our purpose. It is the simplest assumption that can be made, given that one cannot accurately predict or estimate within cross variances. It implies that there is no correlation between the cross mean and within cross variance.

If we select those clones with a value greater than 13 (i.e. ≥14), we select 2517 clones, which is 3.8%. There is, however, no way in practice that a population of 65,536 clones can be accurately assessed for yield and other economically

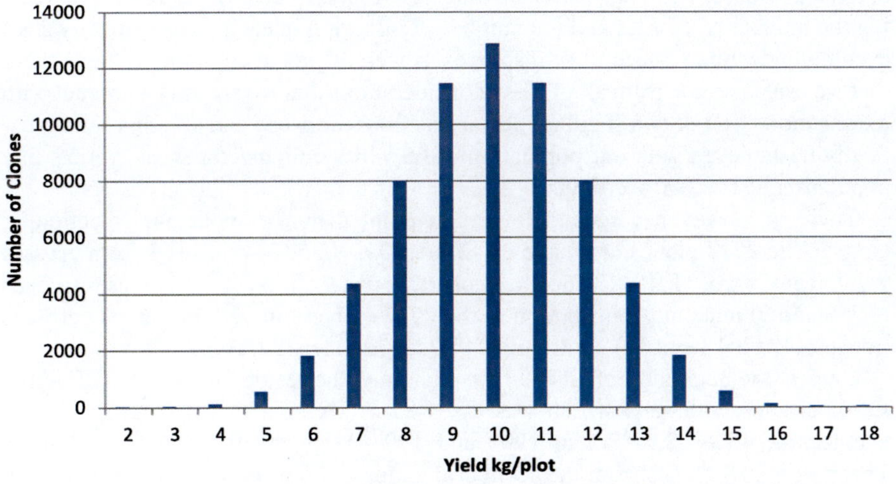

Fig. 7.5 Yields of 65,536 clones (binomial distribution with mean = 10, variance = 4) comprising 256 clones from each of 256 crosses where variation between crosses equals variation within each cross (both are binomial distributions with variance = 2). (From Bradshaw 2017 with permission)

important traits. Hence, in the programme at SPBS/SCRI before 1982, intense visual selection of the seedlings in a glasshouse and subsequent spaced plants at a seed site, was used to reduce 100,000 seedlings to 4000 clones (4%) for further evaluation (Bradshaw and Mackay 1994). Although crosses were made for specific purposes, in practice, they were used to generate a population of clones (4000) from which to seek new cultivars by selection for economically important traits. This meant that crosses made for a specific objective could fortuitously contribute offspring to a different objective, an option that still appeals to breeders even if it is not the most efficient way to go about breeding. Haynes et al. (2012) quote proportions of 2–3% for similar selection in breeding programmes in the USA. Therefore, the 3.8% selection in our theoretical example is realistic.

The first point to make about selecting clones with a value of 14 or more, is that all selected clones will have a value of at least two standard deviations from the mean of all clones (the standard deviation of the distribution in Fig. 7.5 is 2). The mean of the selected clones is in fact 14.34, which is 4.34 from the mean of 10, or 2.17 standard deviations. In selection theory, 14.00 is called the truncation point (T), which is 2 (x) standard deviations from the mean, the difference of 4.34 is called the selection differential (S) and the value of 2.17 in standard deviation units is called the intensity of selection (i). If the proportion of clones selected is expressed as a percentage ($p\% = 3.8\%$ in our example), the relationships between $p\%$, x and i can be determined from the properties of the normal distribution and have been tabulated (Falconer and Mackay 1996). Thus, for $p\% = 3.8\%$ we find that i is 2.175, which is very close to our actual value of 2.17. In other words, our binomial distribution is a good approximation to a normal distribution. The intensity of selection is proportional to the logarithm of population size required for the percentage selected and hence is a declining function of numbers ($i = 1.755$, 2.665, 3.367 and 3.960 for 1 in 10, 1 in 100, 1 in 1000 and 1 in 10,000 selected). Therefore, increasing the number of crosses and the number of offspring in each cross (progeny size) eventually becomes wasteful of resources as well as practically impossible. I think that we can conclude from all of these considerations that we should be content with a truncation point of two standard deviations from the mean and that this can realistically be achieved with our population of 65,536 seedlings/clones, as well as in a population of 100,000 seedlings.

There is, however, a second important point to make about our selection of clones. The 2517 clones are made up of 70, 504, 1036, 744 and 163 from crosses with means of 10, 11, 12, 13 and 14. In other words, 93/256 (36.3%) crosses (means 6, 7, 8 and 9) make no contribution to the 2517 clones and 93/256 (36.3%) crosses (means 11, 12, 13 and 14) contribute 2447/2517 clones (97.2%).

Back at the beginning of Chap. 3 we examined the results of a 6 × 6 half diallel set of crosses with selfs which was assessed at Mylnefield and Gourdie Farms, Dundee (56.46°N, 2.97°W) in 1993 and 1994, respectively (Bradshaw 2009a). There were 218 clones from the 15 hybridizations of the six parents. Furthermore, 17 out-yielded the highest yielding parent when averaged over the 2 years and the two replicates in each year. There was more genetic variation for yield within (component of variance 3.168) than between crosses (0.954) and also smaller amounts of variation for the years × crosses interaction (0.232) and the years × clones within crosses interaction (0.410). Despite the greater amount of variation within than between crosses, the top five crosses out of the 15 (33%) contributed 12 out of the 17 superior clones and the bottom five crosses (33%) just one.

Simmonds (1996) demonstrated statistically that the proportion of segregates within crosses superior to standards (controls) falls away quickly with rank (best to worse) of crosses, when realistic standards (best cultivars currently available) have means close to the means of the best crosses. In the 6 × 6 half diallel, the mean of the four best cultivars (Maris Piper, Pentland Squire, Cara and Desiree) was 11.08 kg/plot and the means of the two best crosses were 10.99 and 10.17 kg/plot. These two crosses contributed 5 and 3 clones, respectively, out of the 17 selected. Hence, both theory and practice confirm that the best few crosses contribute most of the superior segregates. The conclusion is clear, once inferior crosses have been recognized, all of their offspring can be discarded as unlikely to contain superior segregates. So, let us look at selection between crosses for quantitative traits in more detail and compare it with selection between clones to determine if one should deliberately eliminate the worst crosses rather than allow this to happen as a consequence of clonal selection.

7.5 Selection Between Crosses Versus Selection Between Seedlings/Clones for Quantitative Traits

Let us start by considering the mean yields of 256 crosses, each comprising 256 offspring, shown in Fig. 7.6 (binomial distribution with mean = 10, variance = 2). Now let us select the best 9 crosses (cross means ≥13) out of the 256 (i.e. 3.5%). As the standard deviation of the distribution has a value of 1.414, a cross mean of 13 is over two standard deviations (SDs) from the mean of all crosses (12.828 is mean plus 2 SDs). The mean of the 9 crosses is 13.11 (2.2 standard deviations from the mean of all crosses), and they provide 2304 clones which also have a mean of 13.11 as no selection was practised among the 256 clones within each cross.

We can compare the distribution of the 2304 clones with a random sample of 2517 from the population of 65,536 clones to see the effect of selecting the 9 best crosses (Fig. 7.7). We can also compare this distribution with that of the best 2517 clones out of the 65,536 (Fig. 7.7). The mean of the best 2517 clones selected from the population of 65,536 clones is 14.34 compared with 13.11 for the 2304 clones from the selection of the best 9 out of the 256 crosses. The ratio of the responses to between cross selection (3.11) and between clone selection (4.34) is 0.717, which is close to that (0.707) expected from selection theory (Falconer and Mackay 1996). In practice, there will be environmental variation as well as genetical variation between seedlings/clones. As the environmental variation increases, the selection between seedlings/clones becomes less effective, and in the extreme one would end up with a random sample of clones and no progress. In contrast, the environmental

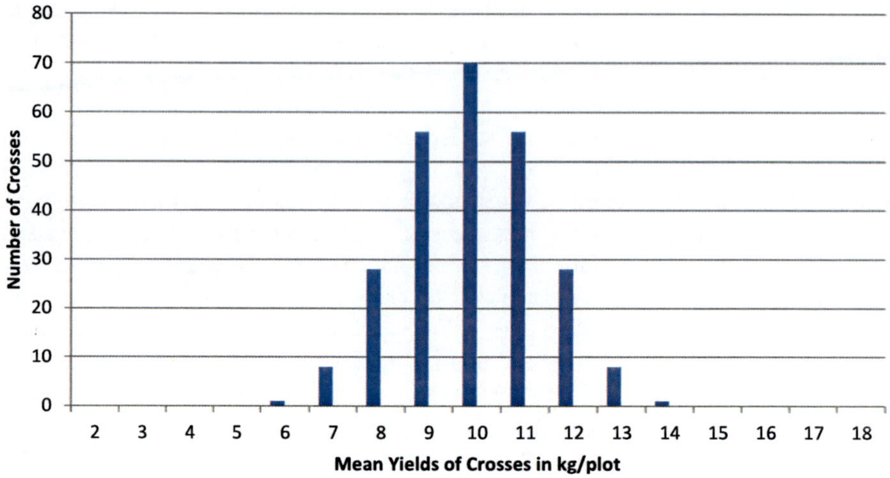

Fig. 7.6 Mean yields of 256 crosses each comprising 256 offspring (binomial distribution with mean = 10, variance = 2). (From Bradshaw 2017 with permission)

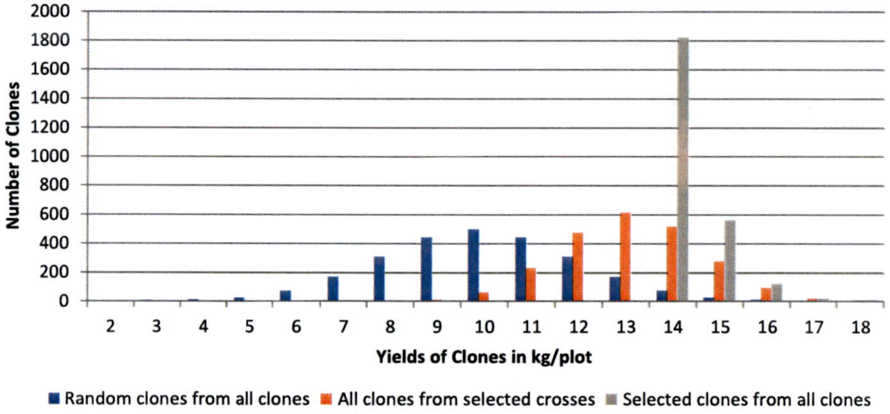

Fig. 7.7 Comparison of random sample of 2517 clones from all 65,536 clones with all 2304 clones from best 9 crosses and best 2517 clones out of all 65,536. (Data from Bradshaw 2017)

deviations for clones within crosses would tend to cancel out, so that most of the variation between crosses would still be genetic, and in the extreme of a very large number of clones within crosses, all of the variation between crosses would be genetic. Hence, we can envisage a situation in which selection between crosses would be effective whereas selection between clones would be ineffective. So how much environmental variation needs to occur before selection between crosses is superior to selection between clones.

7.5.1 Theory for Comparing Selection Between Crosses with Selection Between Clones

The theory is based on comparing family selection with individual selection given by Falconer and Mackay (1996).

The phenotypic variation between the means of crosses is:

$$\sigma_x^2 + \left(\sigma_o^2 + \sigma_e^2\right)/n$$

where σ_x^2 is the genetic variation between crosses, σ_o^2 is the genetic variation between offspring clones within crosses, σ_e^2 is the environmental variation between offspring clones within crosses and n is the number of offspring clones in each cross.

The phenotypic variation between all clones is:

$$\sigma_c^2 + \sigma_e^2$$

where σ_c^2 is the genetic variation and σ_e^2 is the environmental variation between clones.

We are going to assume that all of the crosses have the same within cross variance and that the experimental design is such that there is no environmental variation between crosses additional to that between clones. Furthermore, we are going to assume that the genetic variation between crosses is equal to the genetic variation within crosses:

$$\sigma_x^2 = \sigma_o^2 = \frac{1}{2}\sigma_g^2 \text{ so that the total genetic variation between all clones is } \sigma_c^2 = \sigma_g^2 = \left(\sigma_x^2 + \sigma_o^2\right)$$

If n is sufficiently large, the phenotypic variation between the means of crosses will be $\approx \frac{1}{2}\sigma_g^2$.

In these circumstances, the responses to selection between crosses (R_x) and between all clones (R_c) will be:

$R_x = 1 \times i_x \times 0.707\sigma_g$, where the heritability is 1, i_x is the intensity of cross selection and $0.707\sigma_g$ is the phenotypic standard deviation, and
$R_c = \sigma_g^2/(\sigma_g^2 + \sigma_e^2) \times i_c \times (\sigma_g^2 + \sigma_e^2)^{1/2}$, where the heritability is $\sigma_g^2/(\sigma_g^2 + \sigma_e^2)$, i_c is the intensity of clonal selection and $(\sigma_g^2 + \sigma_e^2)^{1/2}$ is the phenotypic standard deviation.

If the two intensities of selection are equal (as they are in the example we have been considering), the responses to selection will be equal when:

$\sigma_g^2/(\sigma_g^2 + \sigma_e^2)^{1/2} = 0.707\sigma_g$, which on rearranging gives:
$\sigma_g/(\sigma_g^2 + \sigma_e^2)^{1/2} = 0.707$, which occurs when $\sigma_g^2 = \sigma_e^2$.

In other words, they are equal when the broad-sense heritability of the differences between seedlings/clones is one-half, and between clonal selection is superior to between cross selection when the heritability is >½. So, let us look at selection between seedlings in the glasshouse and between spaced plants and unreplicated plots at a seed-site.

7.5.2 *Early-Generation Selection*

There is no way that a breeder can assess 100,000 clones for yield and other economically important traits. Hence, potato breeders typically practised visual selection on their seedlings in the glasshouse (Fig. 7.8), spaced plants at a seed-site and small unreplicated plots at the seed-site in what they called the glasshouse/seedling (GH/SG), first clonal year/generation (FCY/FCG) and second clonal year/generation (SCY/SCG), respectively. The number of potential cultivars (clones) entering replicated yield trials was thereby greatly reduced, for example, at SCRI (Scottish Crop Research Institute) to 1000 (1%).

During the 1980s, a number of breeders did experiments to study the nature and effectiveness of visual selection, of which the following are a representative sample: Plaisted et al. (1984); Swiezynski (1984); Tai and Young (1984); Brown et al. (1988); Neele et al. (1989); and Gopal et al. (1992). This type of selection was attractive for two main reasons. Firstly, large numbers of genotypes could be quickly assessed and a decision, reject or select, taken immediately. Secondly, the shortcomings of every visual character could be integrated into the final, albeit subjective, selection decision. However, the experimental studies confirmed that visual selection, as traditionally practised in the GH, FCY and SCY, was not very effective; but there was no general consensus on how to address the problem (Bradshaw and

Fig. 7.8 Glasshouse/seedling generation (GH/SG) being grown in glasshouse at Scottish Crop Research Institute (SCRI). (Source: SCRI)

Mackay 1994). Hence, such visual selection is still common practice and readers may therefore like to refer to some of the original papers for the ideas that were put forward or consult the review of these papers by Bradshaw and Mackay (1994), or a more recent review by Gopal (2018).

The ideas were as follows: (1) screening seedling and first clonal generations for resistance to diseases so long as clean seedstocks of promising clones could be maintained (Plaisted et al. 1984); (2) screening seedlings for resistance to viruses PVX, PVY, PVA and PVS in parental line breeding and also individual tubers for high starch content, smooth shape, tuber size and early tuber bulking (Swiezynski 1984); (3) practising a moderate selection intensity for visual preference over several years of testing rather than intense selection over fewer clonal generations, but not family (cross) selection in preference to individual selection because of a lower predicted response to selection (Tai and Young 1984); (4) cross selection based simply on the mean of breeders' preference scores, averaged over scorers and clones within progenies, estimated on seedlings or first clonal year plants, in preference to individual clone selection in the first clonal year (Brown et al. 1988); (5) practising moderate (32%) rather than high (<10%) selection intensity for breeders' visual preference in the first clonal year if an early harvest is required to produce virus-free seed tubers which could result in selection for early maturity and low underwater weight, combined with selection at ware potato harvest time in second clonal generation, possibly including underwater weight (specific gravity) and foliage maturity (Neele et al. 1989); and (6) discarding seedlings of poor vigour before transplantation to the field (in programmes where this is done), discarding seedlings when harvested on basis of undesirable tuber colour, shape, eye depth and cracking, and then first generation clones (unreplicated four-plant plots) for cracking, poor yield and poor average tuber weight, and finally second generation clones (unreplicated 12-plant plots) for undesirable number of tubers (Gopal et al. 1992). In summary, the suggestions fell into three categories: selection for disease resistance in the seedling generation, less intense selection over more generations to capture more of the desirable clones, and selection for components of breeders' visual preference.

After the Scottish Crop Research Institute (SCRI) breeding programme moved from Edinburgh to Dundee in the 1980s, the opportunity was taken to re-examine early-generation selection through the following experiment (Bradshaw et al. 1998a). In the summer of 1992, 72 seedlings from each of 198 pair crosses (progenies) were grown in a glasshouse as a randomized complete block (RCB) trial with four replicates, in which the experimental unit was 18 seedlings grown in two adjacent rows of nine 10-cm square pots (SG). The seedlings were grown to maturity and the senesced foliage was removed. The tubers in each pot were independently assessed by two breeders using a 1–9 scale of increasing visual preference. Three groups of four progenies with high, medium and low mean scores were chosen to progress the experiment through the first and second clonal generations (FCG and SCG) to the third clonal generation (TCG) without selection. The experiment thus comprised 864 clones (18 clones × 4 replicates × 12 progenies), with the FCG and SCG grown at a high-grade seed site (Blythbank Farm, West Linton, Peeblesshire) and the TCG grown at a ware site (Gourdie Farm, Dundee). The FCG had an RCB design with the same four replicates as in the glasshouse but with new randomisations for progenies within replicates and clones within plots. Each plot comprised a single drill of 18 plants (clones) at 50 cm spacing with 75 cm between drills. The produce of all plants was independently assessed by two breeders on a 1–9 scale of increasing visual preference. The SCG clones were grown in four-plant plots (37.5 cm spacing between plants) in the same four replicates as the previous year but with complete individual randomization of clones within replicates. The produce of plots was placed in wooden boxes and scored independently in the field by the same two breeders who made visual assessments on a 1 (low) to 9 (high) scale for yield, tuber size, regularity of tuber shape (appearance) and resistance to growth cracking as an aid to determining overall preference on the same scale. Tuber shape was also recorded using a 1–4 scale for round, oval, long oval and extra-long oval. In order to keep the experiment to a manageable size, two of the four replicates were each progressed the next year to a replicated yield trial at the ware site. Each trial had an RCB design with two replicates and single-drill plots of five tubers spaced 45 cm apart with 75 cm between drills. The trials were scored for emergence on a 1 (none) to 9 (all five plants well established) scale and also for maturity on a 1 (all five plants dead-early maturity) to 9 (all five plants still green-late maturity) scale. At harvest, the produce of each plot was placed in a net bag and taken into a potato store for weighing and assessment at the beginning of December. Visual assessments were made by a breeder on a 1 (low) to 9 (high) scale for tuber size, regularity of shape (appearance), resistance to growth cracking, resistance to sprouting and resistance to common scab (*Streptomyces scabies*). Three large tubers from each plot were cut open to check for internal defects and were given a score on a 1 (all three tubers with severe defects) to 9 (no defects) scale. The plots were then independently given an overall preference score by two breeders on a 1–9 scale of increasing visual preference, which took account of yield but not internal condition and hence, it was comparable to earlier generation scores. Two tubers from each plot were steamed for 45 min and scored for sloughing (1 severe disintegration to 9 none) before being cut open. After-cooking blackening was scored 4–5 h later (1 completely black to 9 none). Lastly, at the beginning of the last week of February, all of the plots were given a final score for sprouting (1 extensive to 9 none) and keeping quality (1 tubers very soft to 9 tubers still hard). The reader is referred to the paper (Bradshaw et al. 1998a) for the cultural details and the statistical analyses, and for a discussion of how the results compare with those of the previous research referred to above.

This kind of experiment allows one to calculate the response (*R*) in clonal generation Y to selection in clonal generation X (*S*) as shown in a simple theoretical example in Fig. 7.9. The data could be individual clones or cross means.

The response *R* in clonal generation Y to selection in clonal generation X can be determined from the intensity of selection ($i = S/\sigma_X$) and either the regression of Y on X (b_{YX}) and the square root of the phenotypic variance in generation X (σ_X), or the correlation between X and Y (r_{XY}) and the square root of the phenotypic variance in generation Y (σ_Y):

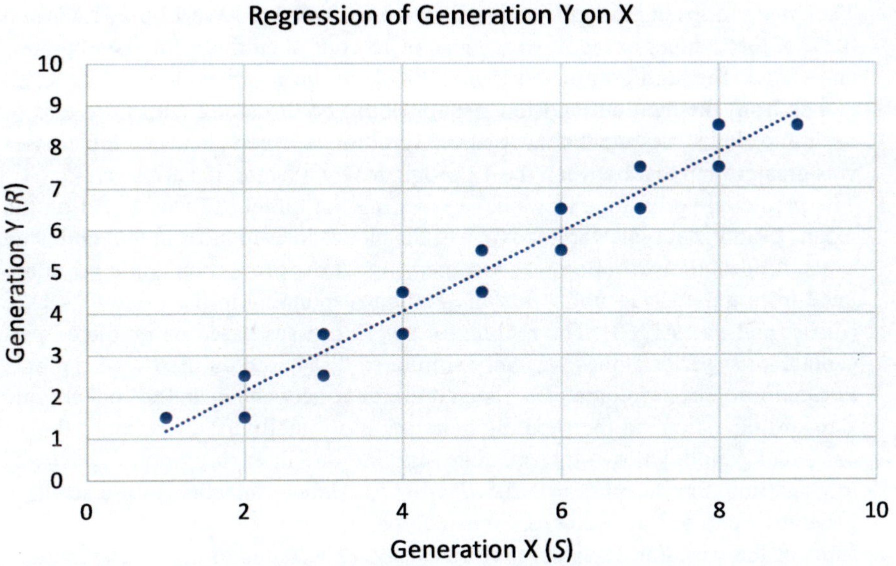

Fig. 7.9 Regression of visual preference scores on 1 to 9 scales of increasing preference for 16 clones in clonal generation Y (mean of many replicates) on scores in clonal generation X (unreplicated)

$$R = b_{YX}S = ib_{YX}\sigma_X = ir_{XY}\sigma_Y, \text{where} b_{YX} = \text{cov}_{XY} / \sigma_X{}^2 \text{and} r_{XY} = \text{cov}_{XY} / \sigma_X\sigma_Y.$$

In the theoretical example in Fig. 7.9, $\sigma_X{}^2 = 5.5$, $\sigma_Y{}^2 = 5.25$, $Cov_{XY} = 5.25$, $b_{YX} = 0.955$, $r_{XY} = 0.977$ and $R = 2.239i$.

In the SCRI experiment, Generation Y was the TCG which had an RCB design with two replicates and Generation X was SG, FCG or SCG. As the SCG clones were grown unreplicated in a block with complete individual randomization that was different from the two replicates in the TCG, the covariance between these generations is an estimate of the genetic variance between clones because the environmental covariance should be zero (in my simple example the true variance was 2.50 and the estimated variance 2.78). Hence, the slope of the regression line (b_{YX}) is an estimate of broad-sense heritability. The situation is slightly more complicated for the SG and FCG where 18 clones from a cross were grown together in a plot, albeit with a different randomization each generation. Furthermore, I have ignored genotype × generation (year and site) interactions.

So, let us look at some of the conclusions from the SCRI experiment:

1. The differences for visual preference between the high, medium and low groups in the seedling generation (SG: 5.19, 4.52 and 3.76, respectively) were maintained over the first, second and third clonal generations, and the correlations between the means of the 12 progenies in the seedling generation and in the first, second and third clonal generations were 0.875, 0.889 and 0.865, respectively. Hence, between cross selection in the seedling generation was effective.

2. The three groups in the replicated trials were predicted to have 13.6%, 1.8% and 0.2% of their clones exceeding the mean of 13 control cultivars for visual preference. Thus, the high group contributed 90.7% of the selected clones. Hence, all clones from the medium and low groups could be discarded with little loss of desirable clones, despite the fact that the variation between clones within crosses was greater than that between the 12 progenies by a factor of 1.87.
3. The predicted percentage of clones exceeding the target (T) was based on the mean ($\bar{x}$) and genotypic variance (σ_c^2) of the clones in each cross in the replicated trials. A normal distribution was assumed so that the proportion could be calculated from the normal probability integral corresponding to the value $(T - \bar{x})/\sigma_c$ (Fisher and Yates 1963). The ranking of the 12 crosses based on predicted percentage of superior clones was very similar to that based on their means alone. Furthermore, the correlation for visual preference between σ_c and the phenotypic standard deviations in the seedling generation originally used to choose the 12 progenies (one high, two intermediate and one low in each group of four) was low and non-significant ($r = 0.205$, $P > 0.05$). Hence, selection in the seedling generation can be based on cross mean alone.
4. Most of the variation in visual preference scores was due to visual assessments of yield with varying contributions from tuber size and regularity of tuber shape (appearance). Not surprisingly, then, differences between the three groups were seen in the second and third clonal generations for yield, visual size and visual appearance, with the high group having the greater yield, larger tuber size and better appearance. Hence, there is scope for the breeder to make changes to their subjective components of visual preference. Other correlated responses to visual selection appeared in the third clonal generation: the high group was faster to emerge, later to mature, had more scab, was earlier to sprout in store and kept less well in store. If selection of the high group of crosses in the seedling generation was combined with further selection of the best 34 out of 120 clones (in high group) for visual preference in the second clonal generation, the total response in the third clonal generation was 1.00 compared with 1.74 for the best 34 clones actually present. The largest correlated responses in the third clonal generation to this within cross selection in the second clonal generation were for faster emergence, higher yield and greater regularity of shape (appearance), but also for earlier maturity (presumably a reflection of the shorter growing season at the seed-site).
5. Finally, the responses to cross selection were compared with individual clonal selection for visual preference using the correlations (r) between the third clonal generation and the seedling (0.865 and 0.432 for crosses and clones, respectively), first (0.743, 0.351) and second clonal generations (0.782, 0.549), and the phenotypic standard deviations (σ_Y) in the third clonal generation (0.528 and 0.912 for crosses and clones, respectively) ($R = ir_{XY}\sigma_Y$, from above). Assuming the same intensity of selection (i), the ratios of the responses in the third clonal generation (Y) to cross and clonal selection in the seedling, first and second clonal generations (X) were 1.16, 1.23 and 0.82. In other words, cross selection was superior in the seedling and first clonal generations but not in the second clonal generation.

7.6 Selection Between Crosses for Quantitative Traits

Once the decision is made to deliberately eliminate the worst crosses rather than allow this to happen as a consequence of clonal selection, a breeder is not restricted to visual selection. Progenies (crosses) can be evaluated as pots of seedlings for disease and pest resistance and as plots of tuber progenies at normal spacing in ware trials, without attempting to identify individual clones within crosses. The latter can still be done in parallel on seedling progenies in the glasshouse and tuber progenies of spaced plants at the seed site, or at a later stage by sowing more seed of the selected crosses. When crosses (progenies) are assessed in plots, the relevant environmental variation is between plots, but the effect of this variation on progeny means can be reduced by replication with a consequent increase in broad-sense heritability (h^2_b).

The breeding programme operated at the Scottish Crop Research Institute from 1991 to 2009 was introduced earlier in this chapter. Four cycles of selection were practised between 120 and 145 crosses, with 20 per cent on average selected (Bradshaw et al. 2009). The crosses were assessed in seedling progeny tests with RCB designs and two replicates for breeders' visual preference and resistance to the white potato cyst nematode (PCN), foliage blight and tuber blight as described by Bradshaw et al. (1995, 2003). In the tests for blight and PCN resistance, the selfed progenies of a resistant and susceptible clone were used as controls. The results of the 1998 progeny tests, from the 1997 cycle 3 crossing, are shown in Fig. 7.10 and summarized in Table 7.3.

A total of 122 out of 145 progenies were assessed in all of the progeny tests and 12 of these were selected using an index based on Smith's (1936) discriminant function for plant selection. The relative economic weights in phenotypic standard deviation units were 1: 1: ½: ½ for breeders' preference, PCN, foliage blight and tuber blight, respectively, so that the two blight scores together were given a similar weight to each of the other two traits. The index also took into account the small correlation between foliage and tuber resistance ($r = -0.35$, with negative sign as the resistance scales were in opposite directions). Later in this chapter, we will take a more detailed look at Smith's (1936) discriminant function and discuss the merits of index selection compared with independent culling levels.

More plants of each selected progeny (140–213) were raised in blocks in a glasshouse in 1999, in 12.5-cm square pots arranged in a honeycomb grid of pots and spaces for ease of checking for stolons trying to grow from one pot to another, as well as for securing larger tubers. Four tubers from each pot were grown as a four-plant plot at the seed site in 2000 (total 2178 clones). By 'missing out' the spaced-plants generation at the seed site (FCG), the time to four-plant plots was the same as in the conventional breeding programme. This was possible because by selecting 10% of the crosses, it was unnecessary to grow singles in the field and select 10% of them to achieve the required manageable number of four-plant plots.

It can be seen in Fig. 7.10 and Table 7.3 that selecting four traits reduces the progress that can be made with each trait. Nevertheless, progenies with low

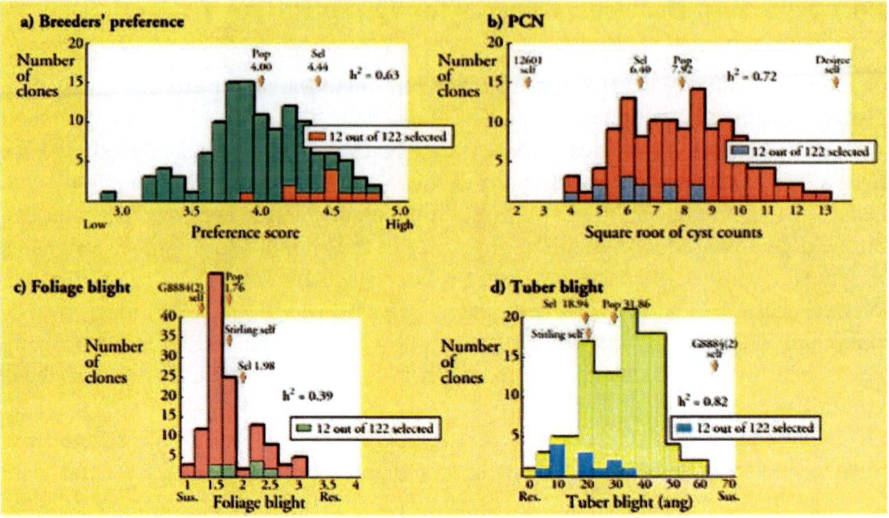

Fig. 7.10 Seedling progeny tests for 122 crosses in third cycle of a breeding programme at the Scottish Crop Research Institute (the 'Y' axes should read 'Number of crosses' not 'Number of clones'). (Bradshaw et al. 1998b; Source: SCRI)

Table 7.3 Summary of results of progeny tests shown in Fig. 7.10 (a high score for foliage blight and low scores for PCN and tuber blight represent resistance)

				Heritability
1998 Progenies	122 Assessed	12 Selected	Best for trait	Broad-sense
Preference	4.00	4.44	4.85	0.63
PCN	7.92	6.40	3.91	0.72
Foliage blight	1.76	1.98	3.00	0.39
Tuber blight	31.9	18.9	0.00	0.82

From Bradshaw et al. (2009) with permission

preference scores and poor resistance to PCN and blight were eliminated. The actual shift in the means of the population of clones resulting from the 12 selected crosses, compared with that from all 122 crosses, can be predicted from the theory for index selection, as we shall see once I've added fry colour to the index. It will be seen that the changes for PCN and tuber blight are quite large, that for breeders' preference and fry colour are moderate and that for foliage blight quite small.

The selection intensity could be increased by assessing more and selecting fewer crosses (progenies). However, the latter will be limited by the minimum numbers of progenies that need to be selected to avoid loss of favourable alleles and inbreeding depression through genetic drift, say around 20. The maximum number of families that could be assessed in progeny tests was limited by resources to around 200. Hence, 20 out of 200 was considered the best selection intensity ($i = 1.742$; Falconer and Mackay 1996) that could realistically be achieved.

In our simple theoretical example, a higher selection intensity of 9 out of 256 crosses was used to achieve a percentage of clones (3.5%) similar to that from selection of individual seedlings/clones (4%), and also to that used in the SCRI breeding programme (4%). If half the crosses were eliminated for each of eight traits, one would be left with one cross out of the 256 whereas to achieve 20 out of 200 crosses from equal selection of eight traits would mean eliminating only one quarter for each trait ($i = 0.424$ for each trait). This means that the breeder needs to give priority to the most important traits to limit the number undergoing selection. The breeder will also need to practise recurrent selection for a number of sexual generations, with the length of each cycle as short as possible, so that small improvements accumulate into worthwhile ones (Bradshaw et al. 2003).

Yield, agronomic and quality traits can be assessed in tuber progeny tests, as done by Bradshaw et al. (2000) for their 15 × 15 diallel set of crosses. The FCG, grown in progeny rows of spaced-plants at the seed-site, provided the bulks of tubers required for a yield trial at a ware site the next year, as well as tubers for maintaining the progenies at the seed site. There is an argument for doing the ware trials in the target environment(s) for more than 1 year to reduce the effects of genotype × environment interactions. Potential difficulties in scoring a large number of variable progenies did not arise in practice, with one exception; it was not possible to steam sufficient tubers to determine after-cooking blackening, whereas it was possible to cope with fry colour. Hence, virtually all of the same traits can be scored on progenies as are scored on individual clones. However, the 2 years of yield trials plus 1 year for the resowings of the best progenies in a glasshouse to provide tubers for four-plant plots the next year, makes a total of 3 years. During this time in the conventional breeding programme, there would have been 3 years of clonal trials at a ware site to reduce the number of potential cultivars from 1000 to 40 clones. In a later section, we will compare the relative progress from these two routes and find that progress is faster with clonal selection than with tuber progeny tests. Furthermore, we saw in the last chapter that genomic selection of clones opens up the possibility of even faster progress.

7.6.1 Addition of Fry Colour

Fry colour is an important quality trait that cannot be reliably assessed on tubers from seedlings. Nevertheless, Bradshaw et al. (2003) showed how the use of mid-parent values and a tuber-progeny test could be integrated with the seedling progeny tests. The 1998 progenies in the SCRI breeding programme came from crosses made in 1997 from a pool of 108 parental clones which had been selected at the single-plant stage from the previous cycle of crosses. These parents advanced to unreplicated four-plant plots in 1997 and to a replicated yield trial in 1998. Hence, mid-parental values for fry-colour were available in 1998 at the same time as the results of the seedling progeny tests (Table 7.3). Hence, selection of the progenies to advance to the seed-site as tuber progenies and to be resown in the glasshouse could be based on the results of the seedling progeny tests plus the mid-parent values for fry colour. Then in 1999, the tuber progenies at the seed-site were assessed for fry-colour and any with disappointing results discarded. The results presented by Bradshaw et al. (2003) were based on 96 progenies which were included in all of the 1998 seedling progeny tests and whose 62 parents had been assessed in the replicated yield trial in 1998. The mid-parental vales for fry colour, after storage of tubers at 10 °C for 2 months, were added to Smith's (1936) discriminant function for selecting the best crosses (progenies) with selection index I:

$$I = 1.782(\text{bp} - 3.978) + 0.356(\text{fb} - 1.766) - 0.0428(\text{tb} - 31.75)$$
$$-0.350(\text{pcn} - 7.917) + 1.424(\text{fry} - 4.017)$$

where the trait scores are deviations from their means and high values are desirable for bp (breeders' preference), fb (foliage blight) and fry (fry colour) and low ones for tb (tuber blight) and pcn (white potato cyst nematode). As the tests used different samples of seed of the progenies and were independent of each other, all of the environmental covariances were zero (including those with fry colour).

The analyses and results can be found in the paper by Bradshaw et al. (2003). The predicted responses to selecting 12 out the 96 progenies on the basis of the index are shown in Table 7.4.

Furthermore, 36 out of the 145 progenies in the 1998 seedling progeny tests, were grown in a tuber progeny trial with two replicates of 15 plants of each progeny at the high-grade seed site in 1999. Two replicate bulks of 15 tubers (one tuber per plant) were assessed for fry colour after storage at 10 °C for 4 months. The parents of 30 out of the 36 progenies had been assessed for fry colour in the 1998 yield trial and the mid-parent values and progeny means of these progenies were compared. The results can be summarized as follows. Inclusion of fry colour in the selection index identified two progenies with desirable scores for fry colour which had not been included in the 12 shown in Table 7.3 on the basis of their preference, PCN and blight scores, and hence had been lost from the programme. There was a moderate correlation of $r = 0.534$ ($P < 0.01$) between mid-parent and offspring mean for the 30 progenies, which rose to $r = 0.708$ ($P < 0.001$) when an outlier was omitted. The

Table 7.4 Heritability and predicted response to selection of 12 out of 96 progenies for five traits included in selection index (breeders' preference (bp), foliage blight (fb), tuber blight (tb) white potato cyst nematode (pcn) and fry colour (fry)): SCRI breeding programme 1998

	bp	fb	tb	pcn	fry
Scale	1–9	1–4	Degrees	Square root cysts	1–9
Heritability	0.641	0.408	0.835	0.736	0.878
Mean of all progenies	3.978	1.766	31.75	7.917	4.017
Predicted response	0.294	0.103	−6.91	−0.998	0.603

Data from Bradshaw et al. (2003)

outlier progeny, which had the lowest progeny mean, could be eliminated from the programme by not advancing it from the glasshouse resowings to the field. The two progenies selected on the basis of having the highest mid-parent values, were ranked 3rd and 8th out of the 30 in the progeny test. Six of the top eight progenies for fry colour in 1999 were included in the 96 progenies assessed the previous year where they ranked 24, 1, 5, 30, 35 and 2 on their index scores. Thus, advancing between a quarter (24) and a third (32) of the 96 seedling progenies to the seed site on the basis of the selection index would have included four or five of the top six progenies for fry colour.

Finally, Mackay (2005) gave an example from SCRI of discarding whole progenies (crosses) in a successfully targeted and accelerated breeding programme for processing potatoes which started with just 3100 seedlings from 43 crosses. Seventeen progenies (crosses) were selected out of the 43 which were grown at a seed site and assessed for fry colour after storage of their tubers at 4 and 10 °C. Superior clones were then identified within these 17 progenies on the basis of tuber traits, fry tests, dry matter and trialling as ware at two sites proximal to processing plants in the third and subsequent field generations. Three cultivars (Golden Millennium, Harborough Harvest and Montrose) with enhanced low-temperature storage characteristics were placed on the UK National List in 1999 just 8 years after being raised as seedlings in a glasshouse.

7.6.2 Summary of Selection Between Seedlings/Clones Compared with Selection Between Crosses

Let us return to our simple theoretical example and make a summary of selection between seedlings/clones compared with selection between crosses (Fig. 7.11). The numbers are those used in the theoretical example apart from putting in a number for spaced plants at seed site in FCG which the example did not specify. The traits are those actually used in the breeding and research programme at the Scottish Crop Research Institute. Both schemes lead to the same number of clones in unreplicated four-plant plots in the same year. Other breeders may have different priorities and operate programmes of a different size. Nevertheless, Fig. 7.11 shows the essence of the difference between selection between seedlings/clones and selection between crosses. As to which is the superior scheme, that is a matter to be determined by experiment, but I suspect the answer will be neither! The problem of ineffective early generation selection can be solved by the use of molecular markers, for marker-assisted selection of major genes and quantitative trait loci alleles of large effect, and for genomic selection of many unspecified alleles of small but cumulative effect. Shown in bold are 2517 plants available as parents for the next round of crossing and for assessment as potential new cultivars (see previous chapter).

	Selection between seedlings/clones	Selection between crosses
Year 1	256 crosses	
Year 2	65,536 seedlings in glasshouse (SG)	256 crosses in progeny tests:
	Breeders' visual preference	Breeders' visual preference PCN Foliage blight and tuber blight
1 ↕← 4	**[65,536 seedlings assessed with molecular markers pre-glasshouse: 2517 plants in large pots in glasshouse**]	Plus mid-parent values for fry colour
Year 3	25,170 spaced plants at seed site FCG)	9 crosses × 256 seedlings of each = 2304 seedlings in glasshouse
	Breeders' visual preference	Plus 9 crosses in tuber progeny test for fry colour at seed site
Year 4	2517 clones in unreplicated four-plant plots at seed site (SCG)	2304 clones in unreplicated four-plant plots at seed site
	Visual selection by breeder	Visual selection by breeder
Year 5	655 clones in unreplicated plots at seed site and replicated yield trials at a ware site which provide information on yield, agronomic and quality traits including fry colour (TCG)	

Fig. 7.11 Selection between seedlings/clones compared with selection between crosses for 65,536 seedlings from 256 crosses in which 655 (1%) clones enter replicated yield trials

7.7 Selection Between Clones in the Intermediate Generations

In the previous sections, we considered three routes, including the use of molecular markers, by which we could reduce 100,000 seedlings in a glasshouse to 4000 clones in unreplicated plots at a seed-site, which was the maximum number that had been handled at that stage in the conventional breeding programme at SCRI prior to 1982 (Bradshaw and Mackay 1994). It is now time to consider the next stage in the search for new cultivars among these 4000 clones. In the potato programme at the Scottish Crop Research Institute prior to 1982 (Bradshaw and Mackay 1994), 1000 clones entered replicated yield trials (Year 5 or TCG), having been visually selected from the 4000 in unreplicated plots at a seed site (Table 7.5). Decreasing numbers (1000, 360, 120) of potential cultivars were then assessed for 3 years (3rd, 4th and 5th CGs) in replicated yield trials (with plots of 5–10 tubers) at a local ware site before the most promising ones (40, 20) underwent 2 years (6th and 7th CGs) of more extensive testing at a number of sites, including overseas ones. Sometimes there was a third year of such trials (Mackay 2005). One or a few clones (4) would then be entered into 2 years of official statutory trials (National List Trials) and

Table 7.5 Assessment and multiplication of clones from second clonal generation

Generation	Seed site	Clones	Ware site
(Clonal)	Number of plants	Number	Number of plots and plants per plot
2nd	4	4000	–
3rd	6	1000	2 × 5
4th	20	360	2 × 10
5th	100	120	2 harvests of 2 × 10
6th	300	40	Ware trials at a number of sites
7th	700	20	Ware trials at a number of sites
8th	2000	4	Official statutory trials

Based on SCRI potato breeding programme: Bradshaw and Mackay (1994)

registered for Plant Breeders' Rights. Multiplication from disease-free stock would start with a view to commercialization. During these intermediate and final stages of selection, the production of seed tubers was separated from the trials that were grown under ware conditions, designed as far as possible to resemble those of good commercial practice. Clones undergoing selection were assessed for their yield and agronomic performance, external and internal defects, cooking and processing characteristics, and for their disease and pest resistance in special tests, as discussed in Chaps. 3, 4 and 5. Selection was thus both multistage and multitrait over as many as eight clonal generations. Before looking at the theories of multistage and multitrait selection, we need to consider some practical matters.

7.7.1 Practical Considerations

The main practical considerations arise out of the limited amount of planting material of each clone in each generation and the overall resources that can be allocated to a breeding programme. Typical numbers in the SCRI breeding programme are shown in Table 7.5. Other breeders will have their own schemes with different numbers, but overall a similar strategy. A relatively slow rate of natural vegetative reproduction of ×8 on average was assumed, with variation between clones. Therefore, on average, the four plants of each clone at the seed site in the second clonal generation produced 32 tubers. Six of these were used for multiplication of the clone at the seed site the next year and 10 were used for a replicated yield trial at a ware site, leaving a few tubers of each clone for assessment of traits like fry colour and resistance to PCN. Seed tubers were produced at a site and under a management regime conducive to the production of disease-free planting material. Assessments were made under ware conditions designed as far as possible to approximate to those of good commercial practice. The ten-plant plots produced enough tubers to make grading feasible and hence to partition total yield into marketable fractions. The tubers from the ware trials were used to assess quality traits including cooking and processing quality, as described earlier in this chapter for experiments into breeding methods. Testing for resistance to major pests and diseases was done in special tests that required differing numbers of tubers to be supplied from the plots at the seed site. Hence, the availability of tubers determined at which clonal generation the testing could be done: for example, resistance to PCN and foliage blight as early as the third clonal generation, but other diseases at later stages, as described in Chap. 5. Two harvests are shown in Table 7.5 for the 5th clonal generation as a way to divide clones into their suitability as first early, second early, early maincrop, maincrop and late maincrop cultivars. Another possibility would be to grow clones at more than one site or in near optimal conditions compared with a combination of stresses. By the 6th clonal generation, enough planting material was available for the final stages of selection to be done at a number of contrasting sites in collaboration with potential end users, before deciding which potential cultivars were entered into Official Statutory Trials with a view to registration and commercialization. Once we get to the 6th clonal generation and ware trials at a number of sites, we are into the considerations of genotype × environment interactions presented in Chap. 3, namely identifying the best new cultivars for particular target environments and end uses and making recommendations to growers and end users. Here, however, we are more concerned with multistage, multitrait selection from the 2nd to the 6th clonal generation. Clearly, it is not possible to assess all desirable traits in unreplicated plots and the first year of replicated trials, however desirable this may be from a theoretical point of view (once material has been discarded on one trait, it cannot be assessed for other traits, and each clonal generation is a loss of 1 year in producing a new cultivar). Some traits may have to wait until the number of clones being assessed is small enough and the number of tubers available of each is large enough for accurate testing; for example: black spot bruising, taste using a trained panel and some disease tests. Again, the use of molecular markers offers a solution to this problem in multitrait selection, but phenotypic data will be required for commercialization. For now, let us look at how many traits it is possible to assess in the first year of replicated trials and their heritabilities.

7.7.2 Broad-Sense Heritabilities

Broad-sense heritabilities are unique to a particular breeding programme because they depend on both the genotypes and assessment environments being used by the breeders. Hence, I am not going to present a survey of the literature, but just report one set of results from the SCRI programme which demonstrates what can be done in practice and also provides us with some data for our theoretical considerations.

As part of the potato genetics research programme at SCRI from 1992 to 2008, an extensive assessment was made of offspring clones from the cross between processing clone 12601ab1 and table cultivar Stirling (Bradshaw et al. 2008; Hackett et al. 2014). Foliage resistance to late blight of 226 clones was assessed in a field nursery (two replicates of two-plant plots), tuber resistance to late blight of 200 clones was assessed on tubers from glasshouse-grown plants (two replicates of eight

Fig. 7.12 The cross between processing clone 12601ab1 and table cultivar Stirling: population of 225 clones assessed in yield trials with randomized complete block designs with two replicates and single-drill plots of five tubers in 1999 (below), 2000 and 2001 in Dundee, Scotland. (Source: SCRI)

plants) and resistance to the white potato cyst nematode of 227 clones was assessed in containers (four containers). Broad-sense heritabilities of clone means were all high, 0.878, 0.870 and 0.937, respectively.

Yield, agronomic and quality traits were assessed on 217–225 clones in ware trials in Dundee, Scotland over the 3 years 1999 (Fig. 7.12), 2000 and 2001. The trials had randomized complete block designs with two replicates and single-drill plots of five tubers spaced 45 cm apart. The distance between adjacent drills was 75 cm in 1999 and 90 cm in 2000 and 2001. The trials were therefore equivalent to the third clonal generation in the breeding programme and also to the third, fourth and fifth generations without any selection. Details can be found in the paper by Bradshaw et al. (2008). The data were analysed by the method of residual maximum likelihood (REML) with years, clones and blocks as random effects. Hence, variance components were estimated for clones (σ_c^2), clones × years interaction (σ_{cy}^2) and residual plot to plot variation (σ^2). These can be used to estimate two broad-sense heritabilities (h^2_b) for clonal means:

1. in one generation/year (mean of two replicates) $h^2_b = \sigma_c^2/(\sigma_c^2 + \sigma_{cy}^2 + \frac{1}{2}\sigma^2)$,
2. over 3 years (mean of 3 years × two replicates) $h^2_b = \sigma_c^2/(\sigma_c^2 + \frac{1}{3}\sigma_{cy}^2 + \frac{1}{6}\sigma^2)$.

The heritability results are shown in Table 7.6.

Table 7.6 The cross between processing clone 12601ab1 and table cultivar Stirling: population of 225 clones assessed in yield trials with randomized complete block designs with two replicates in 1999, 2000 and 2001 in Dundee, Scotland; components of variance for clones (σ_c^2), clones × years interaction (σ_{cy}^2) and residual plot to plot variation (σ^2) and the broad-sense heritability (h^2_b) for clonal means 1) $\sigma_c^2/(\sigma_c^2 + \sigma_{cy}^2 + ½\sigma^2)$ and 2) $\sigma_c^2/(\sigma_c^2 + ⅓\sigma_{cy}^2 + ⅙\sigma^2)$, or $\sigma_c^2/(\sigma_c^2 + ½\sigma_{cy}^2 + ¼\sigma^2)$ for 2 years and $\sigma_c^2/(\sigma_c^2 + ½\sigma^2)$ for 1 year's data

Trait	σ_c^2	σ_{cy}^2	σ^2	h^2_b (1)	h^2_b (2)
Emergence	2.119	0.498	1.126	0.666	0.849
Height	68.65	15.32	24.20	0.715	0.876
Maturity (2 years)	2.476	0.160	0.507	0.857	0.916
Yield	4.440	0.627	1.640	0.754	0.896
DM[1]	1.677	0.560	1.068	0.605	0.807
Size	0.6862	0.0962	0.3648	0.711	0.873
Shape (regularity)	0.4417	0.0530	0.3454	0.662	0.845
Growth cracks	0.1450	0.0872	0.0776	0.535	0.766
Common scab (1 year)	0.957	–	0.982	–	0.661
Internal condition	0.502	0.534	1.333	0.295	0.544
OverY[2]	0.7700	0.1298	0.4048	0.699	0.867
Over[3]	0.6604	0.0863	0.3881	0.702	0.868
Sprouting	1.796	0.683	0.926	0.610	0.816
Keeping quality (2 years)	0.8026	0.4660	0.6833	0.498	0.653
ACB[4]	0.4563	0.4108	0.9154	0.344	0.599
Fry4 °C storage	0.5558	0.1585	0.5149	0.572	0.790
Fry10 °C storage	0.4443	0.1541	0.4001	0.556	0.778

Data from Bradshaw et al. (2008)
1. Dry matter concentration, 2. overall preference including yield but not internal condition, 3. overall preference including internal condition but not yield, and 4. after-cooking blackening

Heritability (1) is the most relevant broad-sense heritability as it refers to a single clonal generation but allows for the effects of clones × years interaction in the denominator. The traits had heritabilities in the range 0.295 to 0.857 with most (13/16) above 0.5. There was little in the way of growth cracks and internal defects so a breeder would simply discard the few clones showing these defects. For keeping quality and after-cooking blackening, relatively high clones × years interactions as well as plot-to-plot variation contributed to the lower heritabilities, and the breeder would need to take this into consideration when making selections. The remaining traits had heritabilities greater than 0.55.

Heritability (2) shows the more accurate assessments that can be obtained from 3 years of trials, but we shall see that the advantage of multistage selection is lost with moderate heritabilities because the breeder loses time in years (for two traits there were only 2 years of data and for one trait only 1 year of data, with heritability (2) calculated for means of 2 and 1 year, respectively).

Clone means over years were based on the best linear unbiased prediction (BLUP) of clone effects from the REML analysis and were used to examine correlations between traits. There were just 12 out of 136 correlations with $|r|$ greater than

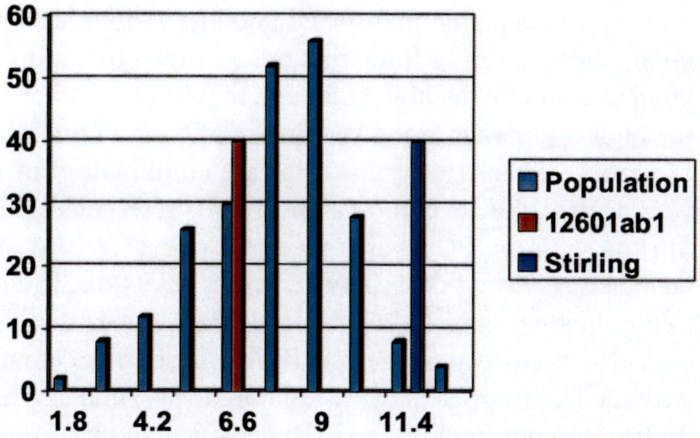

Fig. 7.13 Yields (kg/plot) of clones in cross 12601ab1 × Stirling (12601ab1 7.25 kg/plot, Stirling 12.01 kg/plot and population mean 7.747 kg/plot). (Data from Bradshaw et al. 2008)

0.4, and no real surprises. The two visual preference scores (OverY and Over) were highly correlated ($r = 0.941$) and both were correlated with Yield (0.857 and 0.718, respectively), despite the fact that scorers were not meant to include yield in Over. Hence, the three traits are all assessments of yield. Tuber size is a component of yield and not surprisingly Size was correlated with Yield (0.609) and the preference scores (0.619 and 0.606, respectively). Fry colour after tuber storage at 4 °C (Fry4) was correlated with fry colour after storage at 10 °C (Fry10) (0.701). Keeping quality was correlated with sprouting resistance (0.768), fast emergence was correlated with high Yield (0.579) and OverY (0.451) and early sprouting in store (i.e. lack of dormancy) (0.460). Finally, height and late maturity were correlated (0.674).

This information on genetic variation, heritabilities and genetic correlations between traits can be used to predict the response to clonal selection. Let us take yield as an example. The yield data are summarized as a histogram in Fig. 7.13 and are the means of 3 years. Yields per plot ranged from 1.425 to 12.818 with a mean of 7.747. The genetic variation between clones (σ_c^2) was 4.440 and the broad-sense heritability (h^2_b (2)) was 0.896 (Table 7.6). Hence, the predicted response to selection ($R = ih_b\sigma_c$) was 1.99i. If we select 8 out of 200 clones for higher yield (equivalent to 40 out of 1000 in breeding scheme of Table 7.5), the intensity of selection is 2.127 and the expected response to selection is a 4.233 kg/plot increase on 7.747 kg/plot (i.e. 55% increase). The expected yield of selected clones in the next clonal generation is therefore 11.980 kg/plot, which is virtually the yield of cultivar Stirling (12.01). However, if one wanted to retain 8 out of 200 clones from equal selection of eight traits, that would mean eliminating only one third for each trait and hence a low intensity of selection for each trait ($i = 0.549$). For yield, that would mean an increase of 1.093 kg/plot (i.e. 14% increase) and an expected yield of selected clones in the next clonal generation of 8.840 kg/plot, which is above clone 12601ab1 (7.25) but lower than the mid-parental value of 9.63 kg/plot.

This kind of information should help a breeder focus on traits where most progress can be made and decide on priorities based on the predicted responses to selection. For example, high yield, high dry matter concentration, regularity of tuber shape, good keeping quality and good fry colour after storage at 4 °C would seem a good combination of traits in a cultivar for processing. In other words, it would be good to combine the high yield and regularity of tuber shape of cultivar Stirling with the high dry matter, good keeping quality and good fry colour after storage at 4 °C of clone 12601ab1; but that is a rare combination of traits, particularly when one would also like to combine the blight resistance (foliage and tuber) of cultivar Stirling with the PCN resistance of clone 12601ab1. We therefore reach the same conclusion as we did for between cross selection. The breeder needs to give priority to the most important traits to limit the number undergoing selection. The breeder will also need to practise recurrent selection for a number of sexual generations, with the length of each cycle as short as possible, so that small improvements accumulate into worthwhile ones (Bradshaw et al. 2003).

7.7.3 Assessment of Traits

It is worth briefly describing the assessments of traits that were done on the five-plant plots and their produce as this shows what can be done in a practical plant breeding programme (Bradshaw et al. 2008). Others breeders may use larger plots, take larger samples and measure traits in a different way to achieve more accurate assessments. My purpose is simply to make the point that acceptable heritabilities can be obtained with small plots and small sample sizes which allow many clones to be assessed for many traits with limited planting material. Readers may also like to refer to the protocols in the book *Potato breeding a practical manual for the potato chain* edited by Tiemens-Hulscher et al. (2013) as they are illustrated with excellent colour photographs together with tables of scores for standard cultivars. I have added some comments from their book. Assessments should be recorded using pocket (portable) computers so that the data can be transferred directly into databases for analysis.

Emergence was scored on a 1 (none: very slow emergence) to 9 (all plants in a plot well established: very rapid emergence) scale at an appropriate time during crop emergence (for example, emergence in 90% of plots).

Canopy height from the top of the drill was measured about 1 month before the crop was burnt down prior to harvest. Some breeders may prefer to assess extent of development of foliage on a 1–9 scale (Tiemens-Hulscher et al. 2013).

Foliage maturity was scored on a 1 (all plants in a plot dead: early maturity) to 9 (all plants still green: late maturity) scale at an appropriate time before the crop was burnt down prior to harvest. Sometimes, breeders use the reverse scale with 9 for early maturity (Tiemens-Hulscher et al. 2013).

Yield: At harvest, the tubers from each plot were placed in a net bag, taken into a potato store at ambient temperature and weighed within 3 months. Fresh weights in kg/plot were recorded automatically by a balance connected to a portable computer. With larger plot sizes, tubers can be graded into size fractions before weighing each fraction. For an accurate assessment of yield some breeders consider it necessary to have a plot with 24 plants (four rows of six plants of which the two middle rows are harvested) (Tiemens-Hulscher et al. 2013). However, with limited planting material, single-row plots of five plants have a surprisingly high heritability, despite some inter-plot competition (see Chap. 3).

Dry matter content (DM) was determined either by measuring the specific gravity of a 3.63 kg sample of tubers with a hydrometer at the same time as the plots were weighed, or by weighing five tubers in air (x) and then underwater (y) [SG = $x/(x-y)$]. Dry matter content was then estimated from the relationship with specific gravity found by von Scheele et al. (1937). Some breeders use sample sizes as large as 5 kg to determine under water weight (Tiemens-Hulscher et al. 2013).

A number of visual assessments of tubers were made when the plots were weighed.

Tuber size, regularity of tuber shape and resistance to growth cracks were scored on a 1 (low) to 9 (high) scale. Tubers with deep eyes would get a low score for Shape as well as ones that showed the more common occurrence of departures from a regular round, oval or long oval shape. When common scab (*Streptomyces scabies*) was present on tubers, resistance was scored on a 1 (susceptible) to 9 (resistant) scale. Some breeders record depth of eye as a separate trait and also record uniformity of tuber size, both on a 1–9 scale (Tiemens-Hulscher et al. 2013).

Internal defects (incipient hollow heart, hollow heart, internal necrosis and flecking) were assessed by cutting open three large tubers from each plot and internal condition was scored on a 1 (all three tubers with a severe defect such as hollow heart) to 9 (no defects) scale.

Overall preference scores were given by two breeders on a 1–9 scale of increasing preference. One score (OverY) included yield but not internal condition while the other (Over) included internal condition but not yield.

Also, when the plots were weighed, samples of three and five tubers were taken from each plot and stored at 4 and 10 °C, respectively, for a minimum of 3 months.

Sprouting, keeping quality and after-cooking blackening (ACB) were determined on the samples stored at 10 °C. Sprouting was scored on a 1 (extensive: short dormancy) to 9 (none: long dormancy) scale and keeping quality was also scored on a 1 (tubers very soft) to 9 (tubers still hard) scale. Two tubers were steamed for 45 min and usually scored for sloughing on a 1 (severe disintegration) to 9 (none) scale before being cut open. ACB was scored 4–5 h later on a 1 (completely black) to 9 (none) scale. In cooking quality evaluation, four (A, B, C and D) broad categories are recognized with A (quite firm salad type) and D (very mealy, mushy type) the extremes (Tiemens-Hulscher et al. 2013).

Fry colour assessment after storage of tubers at 4 °C (Fry4) and 10 °C (Fry10) was done on three tubers. Each tuber was cut in half and a 1–1.5 mm-thick slice was taken and fried in vegetable oil at 175 °C in a thermostatically controlled chip fryer until all of the water had boiled off (bubbling ceased). Colour was assessed by comparison with a standardized chart on a 1 (extremely dark) to 9 (extremely pale) scale, where five and above is an acceptably light colour (Mackay and Dale 1990).

A start can be made at this stage to recording descriptions of potential cultivars for characteristics such as foliar type, flower colour and extent of flowering, extent of berry production, tuber shape (round, round-oval, oval, long-oval, long), skin colour (white, yellow, brown, red and purple) and flesh colour (white, cream, light yellow, yellow, dark yellow and orange, as well as red and purple). Sometimes, these traits may be selection criteria such as tuber shape for end use (long-oval for French fries and round to round-oval for chips/crisps) or skin and flesh colour for consumer preference. Together, the traits provide a clone description for use when submitting a potential cultivar for registration. They also allow the breeder in subsequent clonal generations to check for planting mistakes and admixture of clones.

As high-throughput phenotyping develops, it may be possible to assess and select clones grown in small plots in yield trials for disease and physiological traits, including stress tolerance (e.g. through nutritional and water status), as well as their produce for quality traits. The different methodological approaches used to evaluate these traits fall into three categories: (1) remote sensing and imaging proximal to the crop, (2) laboratory analyses of samples and (3) near-infrared reflectance spectroscopy (NIRS) analysis in the harvestable part of the crop (Araus and Cairns 2014). Besides the choice of the most appropriate traits, it is also crucial to determine the key time(s) for their evaluation. The remote sensing and imaging can be classified as spectral, thermal or digital (camera) and requires empirical calibration against the traits of interest. Simultaneous measurements of all plots within a trial can only be done from the air.

Prashar et al. (2013) used infra-red thermography (thermal imaging) to assess differences in stomatal behaviour of a diploid mapping population of potato comprising 188 clones grown in a field trial at Balruddery Farm, Dundee, Scotland (56.48°N, 3.13°W). In addition to the main yield trial with two replicates and five-plant plots, there were five adjacent trials each with two replicates and two-plant plots, the purpose of which is explained in the paper. Thermal images were obtained using a ThermaCAM P25 infrared camera. They were taken on three dates from stolon initiation to tuber initiation, from a fork-lift at about 8 m height and at an angle to maximise the number of individual plots in any image. Plot canopy temperatures were estimated from the thermal images. Variation in canopy temperature explained 29.2% of the variation in yield at maturity. The results showed a clear negative association between canopy temperature and final tuber yield. Although the population was grown under an ample supply of moisture, the approach should also be suitable for evaluating plant response to stress under field conditions.

NIRS is being developed for rapid analyses of potato tuber quality. Bonierbale et al. (2009), for example, developed, validated and applied NIRS to the characterization of 152 Phureja accessions (colour range: cream, light yellow and deep yellow) for total and individual carotenoid contents of raw tubers. NIRS estimated total carotenoids (103–2135 μg 100 g^{-1} FW) and zeaxanthin concentrations (trace to 1290 μg 100 g^{-1} FW) with high accuracy and could differentiate accessions with low, medium and high concentrations of violaxanthin (trace to 278 μg 100 g^{-1} FW), antheraxanthin (3–354 μg 100 g^{-1} FW), lutein (55–189 μg 100 g^{-1} FW) and β-carotene (trace to 18 μg 100 g^{-1} FW). NIRS can therefore be used to assess large numbers of genotypes (clones) in a breeding programme, in preference to costlier and more time consuming HPLC (high performance liquid chromatography). Likewise, Bernhard et al. (2016) used 117 samples of starch potatoes, food potatoes and landraces for the development of NIRS calibrations for crude protein content (CPC) and dry matter content (DMC) of tubers prepared in different ways (grating, mashing, and drying and grounding). DMC varied between 14.1% and 35.2% and CPC between 4.93% and 12.31% of dry matter and between 1.43% and 2.87% of fresh matter. Accurate calibrations were obtained for CPC in dried potato flour (coefficient of determination of cross validation 0.936) and for DMC (0.939) and CPC (0.684) of fresh potatoes. The latter was thought to be useful for the selection of potato cultivars with comparatively high or low tuber protein content.

Now that we have seen what can be done in practice, let us look at the theories of multistage and multitrait selection and consider if we can make any improvements to current practice.

7.8 Theory of Clonal Selection: Multistage

The classic paper by Finney (1958) on plant selection covers the theory and practice of multistage clonal selection. He recognized three stages of selection. In the first stage, a breeder (visually) discards seedlings and clones with obvious defects whilst building up material for more detailed assessment. This is the stage when the environmental variation is large compared with the genetic variation. Finney's sobering conclusion was that two-stage selection is little better than one-stage selection under conditions of very high experimental error and that only a small gain can be achieved in one-stage selection. Hence, it is worth discarding a large proportion of clones at random before assessment starts to reduce the experimental error (larger plots or more replicates) for those clones that do go into the yield trial, even though intense selection is not now possible. The experimental evidence presented earlier suggested that breeders were discarding a lot of clones at random when practising intense visual selection on seedlings in the glasshouse and spaced plants at the seed site. In the second stage, trials are done to find those clones with the highest merit among those that survive the preliminary screening. Yield assessment is a major component of merit and Finney concentrated on yield. In the third phase, potential cultivars undergo extensive evaluation in official replicated trials over sites and seasons before the best are commercialized if they are better than current cultivars. Finney's paper was concerned with the second stage, which is also our current interest. He concluded that most of the possible gain can be achieved in two rather than more stages with moderate experimental error and plant selection. So, let us look at his hypothesis about the optimum selection strategy.

7.8.1 *Finney's Optimum Strategy*

The goal is to reduce N clones to n over k clonal generations where π is the proportion selected so that $n = \pi N$. Finney's simple tentative conclusion to a complex problem was as follows. The optimum strategy for maximizing the gain from selection is never far from the following one. Select an equal proportion (P) of clones each generation so that $P_1 = P_2 = \ldots = P_k = \pi^{(1/k)}$. Furthermore, with finite resources (e.g. land area = A), divide them equally between generations so that $A_1 = A_2 = \ldots = A_k = A/k$. The number of generations of selection is a balance between diminishing gains from each generation being offset by delays in the commercialization of new cultivars. Finney thought that it would be difficult to justify more than five generations of selection and that two or three would usually be adequate. A simple example with three generations would be as follows: 1080 clones are assessed in small plots in an unreplicated trial and the best 360 (1/3) are selected; these are assessed in a trial with 3 replicates and the same size plots (i.e. same area) and the best 120 (1/3) are selected; finally, these are assessed in a trial with 3 replicates and plots 3 times as large (i.e. same area) and the best 40 (1/3) are selected.

The strategy is based on maximizing the response to selection rather than maximizing the chance of selecting the best clone. Once environmental as well as genetical variation is present, we can no longer be certain of selecting the best clones, and this is what happens in practice. The risk of phenotypically selecting a genetically inferior clone increases as the environmental variation increases, as can be shown with simple examples (Bradshaw 2016). Finney thought that a breeder should accept this situation as inevitable and concentrate on maximizing the gain from selection, rather than worry about maximizing the probability that a good clone will be selected and not rejected. The latter, however, is a consideration when it comes to the credibility of an official trials system (stage three of selection). Breeders want to see a fair trials system in which a genetically superior cultivar is selected and an inferior one is rejected, as discussed in Chap. 3 (Talbot 1997). As the precision of the trial system is reduced, the probability of rejecting a good genotype increases as does the risk of wrongly selecting poor genotypes. However, let us return to maximizing the response to selection and compare single-stage with two-stage selection in a simple example.

7.8.2 *Single-Stage and Two-Stage Selection*

Let us select 15 clones as potential new cultivars out of the 256 shown in Fig. 7.14. The figure shows the phenotypic distribution of the yields (kg/plot) of the 256 clones. As it is a theoretical example, I know the genotypic values and environmental deviations of all the clones.

V_G (the genetic variation) = 1, V_E (the environmental variation) = 4 (unreplicated plots) and V_P (the phenotypic variation) = 5; hence the broad-sense heritability = 0.2.

The phenotypic mean of the best 15 clones = 12.4; hence the selection differential = 4.4 and the predicted response to selection in the next clonal generation = 0.88 (based on the normal distribution).

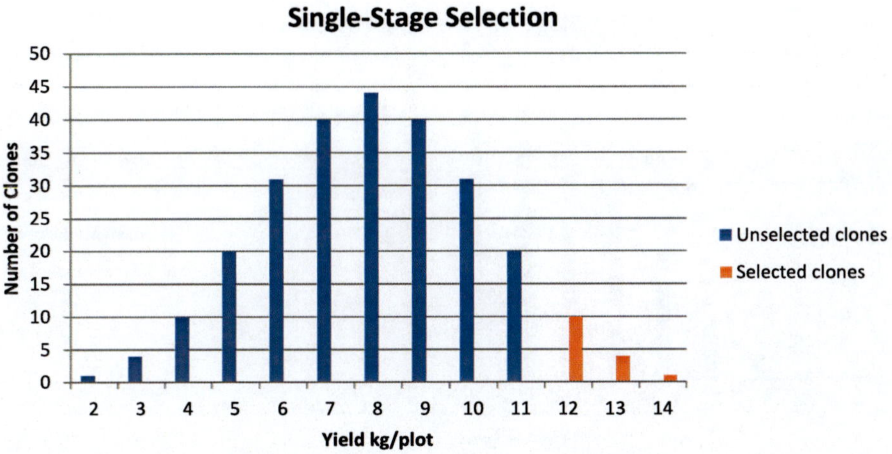

Fig. 7.14 Single-stage selection: progeny of size 256 clones: genetic values 6, 7, 8, 9 and 10 in a binomial distribution with ratios 1:4:6:4:1 and environmental deviations of −4, −2, 0, 2, 4 in a binomial distribution with ratios 1:4:6:4:1; population mean is 8

The observed response in the next clonal generation can be calculated from the table used to create the example (not shown): the mean genotypic value of the 15 selected clones = 8.9333, giving a response to selection = 0.9333 (based on the binomial distribution).

Out of the 15 clones selected, 6, 4 and 5 had genotypic values of 8, 9 and 10, whereas there were 16 clones among the 256 with a genotypic value of 10.

Can we do better through two-stage selection? This time let us select the best 66 clones in the first stage, followed by the best 15 out of the 66 in the second stage (Fig. 7.15). The 66 clones selected in stage one comprises 1, 4, 30, 20 and 11 clones with genetic values of 6, 7, 8, 9 and 10, giving them a genetic mean of 8.5454, which is therefore their mean in the second stage of selection ($R = 0.5454$). This time, we have retained 11 out of the 16 clones with a genotypic value of 10. The clones retain their genotypic values in the second stage, but there is now less environmental variation as they can be assessed in a trial with four replicates because we are allocating the same resources (area of land) for the 66 clones as we did for the 256 (unreplicated plots). The environmental component of the phenotypic variation in clonal means is therefore one quarter of the plot to plot variation. In other words, the clonal means are assessed with greater precision. The genetic variance is also less as a result of selecting 66 out of 256 clones.

The outcome is $V_G = 0.793$ (reduced from 1), $V_E = 1$ (reduced from 4), $V_P = 1.793$ and a broad-sense heritability of 0.442, which is more than double the 0.200 with single-stage selection.

If we now select the best 15 clones (10 out of the 11 with score 10 are selected) we find that they have a phenotypic mean of 10.358 and a genotypic mean of 9.370.

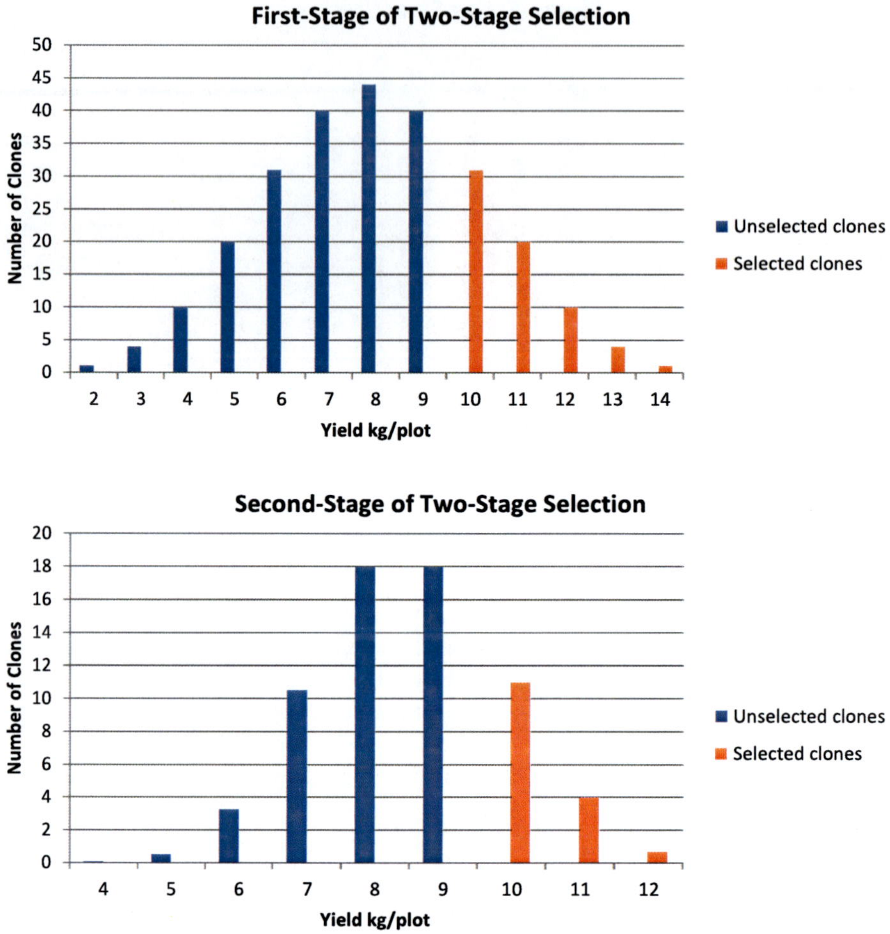

Fig. 7.15 Two-stage selection: 66 clones from stage one of which 1, 4, 30, 20 and 11 have genetic values 6, 7, 8, 9 and 10 and environmental variation (values −2, −1, 0, 1, 2 in a binomial distribution: ratios 1:4:6:4:1)

Hence, $S = 1.8126$ (10.358–8.5454), $R = 0.8246$ (9.370–8.5454) and $R/S = 0.455$ (the realized heritability), which is very close to the estimated heritability for the second stage of selection. It can be seen that the response in the second stage is greater than in the first because of the higher heritability, the proportion selected being similar.

Out of the 15 clones selected, 2, 6 and 7 (rounded to whole numbers) had genotypic values of 8, 9 and 10, compared with 6, 4 and 5 in one-stage selection.

Furthermore, the genotypic mean of 9.370 is higher than the 8.933 achieved for 15 clones from one-stage selection. However, this advantage is achieved at the expense of taking 2 years instead of one to do the selection. If this increases the length of the breeding cycle from 4 (SCG) to 5 years (TCG), the relative responses per year would be 0.933/4 = 0.233, compared with 1.370/5 = 0.274. There is an 18% advantage in two-stage selection.

If the whole exercise is repeated with a different heritability in the first year of assessment, one gets a different answer. For example, if you start with single-stage selection with $V_G = 1$, $V_E = 1$ (unreplicated plots say) and $V_P = 2$, and hence a heritability of 0.5, and select 16 (for convenience) out of 256 clones, you find that the response to selection is 1.3125. In two-stage selection with 65 (for convenience) selected out of 256, followed by 16 out of 65, the response is 1.65 (0.8615 + 0.7885), which is greater than that for one-stage selection. However, if we once again assume that two-stage selection increases the length of the breeding cycle from 4 to 5 years, we find that the responses per year are identical (1.3125/4 = 0.33, compared with 1.65/5 = 0.33). In other words, as the heritability increases the advantage of two-stage selection is lost and the response in the first stage becomes larger than in the second stage.

So, what does this mean for practical potato breeding? When selection is started in unreplicated plots, two-stage selection over the unreplicated plots and a yield trial in the following generation is desirable (say 4000 to 1000 to 360 clones). However, let us return to Table 7.6 and the cross between processing clone 12601ab1 and table cultivar Stirling and use the components of variance to compare one-stage selection (1080 to 40 clones) with three-stage selection (1080 to 360 to 120 to 40 clones) in replicated trials. Let us consider the components of variance for yield as an example.

Let us assume that the best we could do would be to assess all 1080 clones for 3 years and then select the 40 highest yielding ones. The intensity of selection i would be 2.186 (1 in 27 = 3.704%), the broad-sense heritability h^2_b would be 0.896 $(\sigma_c^2/(\sigma_c^2 + \frac{1}{3}\sigma_{cy}^2 + \frac{1}{6}\sigma^2)$ and the phenotypic standard deviation σ_p would be 2.219 $(\sigma_c^2 + \frac{1}{3}\sigma_{cy}^2 + \frac{1}{6}\sigma^2)^{\frac{1}{2}}$, where σ_c^2, σ_{cy}^2 and σ^2 are the components of variance for clones, clones × years interaction and residual plot to plot variation.

The response to selection (R) is $R = ih^2_b\sigma_p = 2.186 \times 0.896 \times 2.219 = 4.346$

The yield of the selected clones in the next clonal generation would be 12.093 (7.747 + 4.346) kg/plot compared with 12.01 for the highest yielding parent, cultivar Stirling.

The appropriate heritability (0.754) for one-stage selection is $\sigma_c^2/(\sigma_c^2 + \sigma_{cy}^2 + \frac{1}{2}\sigma^2)$, where we acknowledge the presence of the clones × years interaction in the denominator and we divide the plot to plot variation by two as we have two replicates. Likewise, the phenotypic standard deviation σ_p is 2.426 $(\sigma_c^2 + \sigma_{cy}^2 + \frac{1}{2}\sigma^2)^{\frac{1}{2}}$ and the intensity of selection is again 2.186.

The response to one-stage selection is $R = ih^2_b\sigma_p = 2.186 \times 0.754 \times 2.426 = 3.999$

This means that 92% of the response from selecting 40 out of 1080 clones assessed over 3 years is achieved in one-stage selection, in year 5 as opposed to year 7, so the responses per year are 0.800 for single-stage selection and 0.621 for assessment over 3 years.

For three-stage selection, we have three generations to consider. If we select a third of the clones each generation, we have the same intensity of selection each generation, $i = 1.097$ (1 in 3 = 33.3%). The amount of genetic variation between clones reduces at each stage so that $\sigma_c^2(1) > \sigma_c^2(2) > \sigma_c^2(3)$, and for simplicity we will assume that the clones by years interaction reduces by a similar proportion. We have more tubers to plant at each stage so the plot to plot variation (σ^2) can be reduced by more replication (r). In Table 7.6, we showed an increase in the number of tubers from 10 to 20 to 40. If these are grown in plots of five plants, σ^2/r will be $½\sigma^2$, $1/4\sigma^2$ and $1/8\sigma^2$ for the three generations. The new genetic variance (σ_c^{2*}) as a result of selection can be obtained from the formula in Falconer and Mackay (1996):

$\sigma_c^{2*} = (1 - h^2_b k)\sigma_c^2$ where $k = i(i - x)$, x being the deviation of the point of truncation from the population mean. Using the table in Falconer and Mackay (1996), $k = 0.721$ in our example.

The response to the first stage of selection is $R = ih^2_b\sigma_p = 1.097 \times 0.754 \times 2.426 = 2.007$

The response to the second stage of selection is $R = ih^2_b\sigma_p = 1.097 \times 0.744 \times 1.650 = 1.347$

($\sigma_c^{2*} = (1 - (0.754 \times 0.721)) \times 4.440 = 2.026$ and $\sigma_p^2 = (\sigma_c^{2*} + \sigma_{cy}^{2*} + ¼\sigma^2) = 2.026 + 0.286 + 0.410 = 2.722$ so that $h^2_b = 0.744$ and $\sigma_p = 1.650$)

The response to the third stage of selection is $R = ih^2_b\sigma_p = 1.097 \times 0.735 \times 1.130 = 0.911$.

($\sigma_c^{2*} = (1 - (0.744 \times 0.721) \times 2.026 = 0.939$ and $\sigma_p^2 = (\sigma_c^{2*} + \sigma_{cy}^{2*} + 1/8\sigma^2) = 0.939 + 0.133 + 0.205 = 1.277$ so that $h^2_b = 0.735$ and $\sigma_p = 1.130$)
The total response to three-stage selection is 2.007 + 1.347 + 0.911 = 4.265

This means that one-stage selection achieves 94% of the total response from three-stage selection, which in turn achieves 98% of response for 1080 clones assessed over 3 years. Hence, in this particular example with relatively high heritabilities, there is no advantage in three-stage selection in which the number of replicates increases each clonal generation over assessing all 1080 clones for 3 years with no increase in replication. Furthermore, we have already concluded that the loss in time with 3 years of assessment means that the response per year is greater with single-stage selection.

With ACB, a trait with a broad-sense heritability (1) of <0.5 (0.344), the total response from three-stage selection (0.434 + 0.401 + 0.354 = 1.189) is slightly more (105%) than for 1080 clones assessed over 3 years (1.131). Although one-stage selection achieves only 73% (0.866) of the response from three-stage selection, the response per year is still slightly better (0.866 ÷ 5 = 0.173 compared with 1.189 ÷ 7 = 0.170).

Readers may like to do similar calculations with data from their own breeding programmes, but it looks as though one generation of selection in replicated yield trials may be sufficient to pick the parents for the next cycle of hybridization and selection as well as potential cultivars for assessment in multisite trials. If so, the breeder may need to alter the balance between the number of tubers used for multiplication at the seed site and the number used for ware trials or use more rapid methods of multiplication of potential cultivars. Furthermore, we saw earlier that the use of molecular markers for marker-assisted and genomic selection could result in even earlier selection of parents and potential cultivars than the first year of replicated yield trials.

7.8.3 Other Theoretical Approaches to Multistage Selection

Wricke and Weber (1986) compared Finney's approach to that of Cochran (1951) who regarded selection in two stages as selection for two traits using independent culling levels (see next section). They concluded that the selection response using Finney's values is only slightly less than the response using the optimum values for the selection fractions and number of replicates calculated by Cochran's method. Furthermore, little improvement comes from using information from the first stage in the second stage. Hence, plant breeders can still benefit from reading the whole of Finney's paper, despite the fact that it was written over 60 years ago.

A more recent paper by Ishii and Yonezawa (2006) has shown how the work of Finney and other researchers can be extended to provide more general guidelines for multistage yield selection, particularly in the presence of genotype × environment interactions. It is more relevant to Finney's third stage of plant breeding where potential cultivars undergo extensive evaluation in replicated trials over sites and seasons before the best are commercialized if they are better than current cultivars. The authors used extensive Monte Carlo calculations (simulations), based on the principle of maximizing the probability of success (S) that an entry with the highest yielding ability is selected for a target region from a population of N candidates. They considered N in the range 20–100, with examples given for 50, and three-stage yield selection. The entries were assessed in replicated trials at a number of locations in the target region. The results confirmed that the total number and allocation of locations determined S. With the same total number of locations ($s_1 + s_2 + s_3$), a uniform allocation over stages ($s_1 = s_2 = s_3$) or decreasing number ($s_1 > s_2 > s_3$) in the presence of a high magnitude of genotype × location interaction, was nearly if not exactly the best. If circumstances permit, new locations from the target region should be chosen in each stage. The common allocation practice of 1 ($=s_1$) $< s_2 < s_3$ proved undesirable, and hence should only be used when the amount of planting material is the limiting factor. The optimum selected fractions ($p = 1/N = p_1p_2p_3$) and proportions of total plots ($\alpha_1 + \alpha_2 + \alpha_3 = 1$; where total was predetermined as $N\beta$ with β in the range 2–10) used at each stage were in the order $p_1 < p_2 < p_3$ and $\alpha_1 > \alpha_2 > \alpha_3$. However, $p_1 > p_2 > p_3$ instead of $p_1 < p_2 < p_3$ is appropriate when genotype × year interaction is the major error component in the yield scores of entries. Replications should be allocated once all of the other decisions have been made. These considerations are probably more relevant to the people running Official Trials than to breeders. Nevertheless, as explained in Chap. 3, the selection environments for the second (intermediate) stage of a breeding programme are important for success but are restricted by the amount of planting material available. Again, this problem would be solved with genomic selection if the training population of clones could be phenotyped in more than environment; for example, absence of stress (maximum yield environment) and most likely combination of stresses in target region for breeding programme.

7.9 Theory of Clonal Selection: Multitrait

During the intermediate generations of their programmes, potato breeders will assess their clones (potential cultivars) for yield, quality traits and resistance or tolerance to the most serious abiotic and biotic stresses. There are three broad strategies for such multitrait selection: tandem selection, independent culling levels and index selection, and the same applies to genomic selection (i.e. use of separate equations for each trait or a single equation for merit as an index of all traits). In tandem selection, only one trait is selected in each clonal generation, say trait one in the first generation, trait two in the second, and so on. With independent culling levels, each clonal generation, all clones are discarded which have a phenotypic value below the culling level for at least one trait, for example, low yield, poor quality or poor disease resistance. With index selection, the aim is to improve traits simultaneously in such a way as to make the biggest possible improvement in overall merit each clonal generation. The differences between the three types of selection are shown graphically in Figs. 7.16, 7.17 and 7.18 for the selection of 64 out of 512 clones, using the binomial rather than the normal distribution for simplicity. It is assumed that the

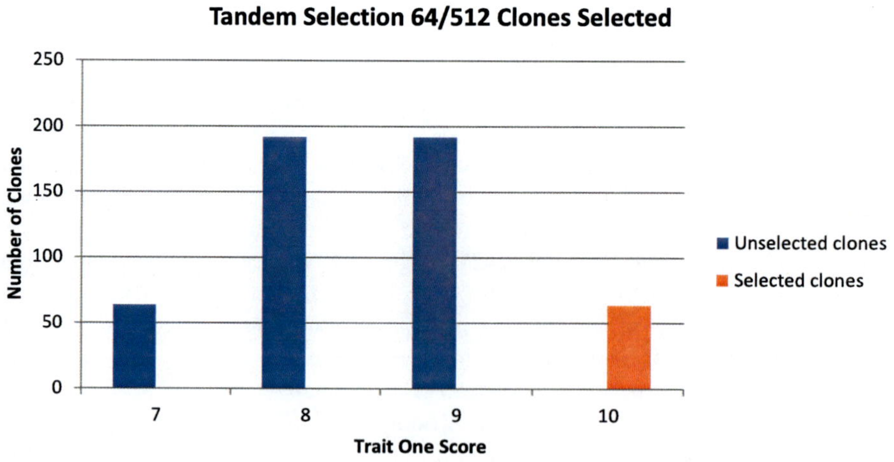

Fig. 7.16 Tandem selection: 64 out of 512 clones (genotypes) selected for one trait in one clonal generation (mean of selected clones minus population mean = 10–8.5 = 1.5). (From Bradshaw 2016 with permission)

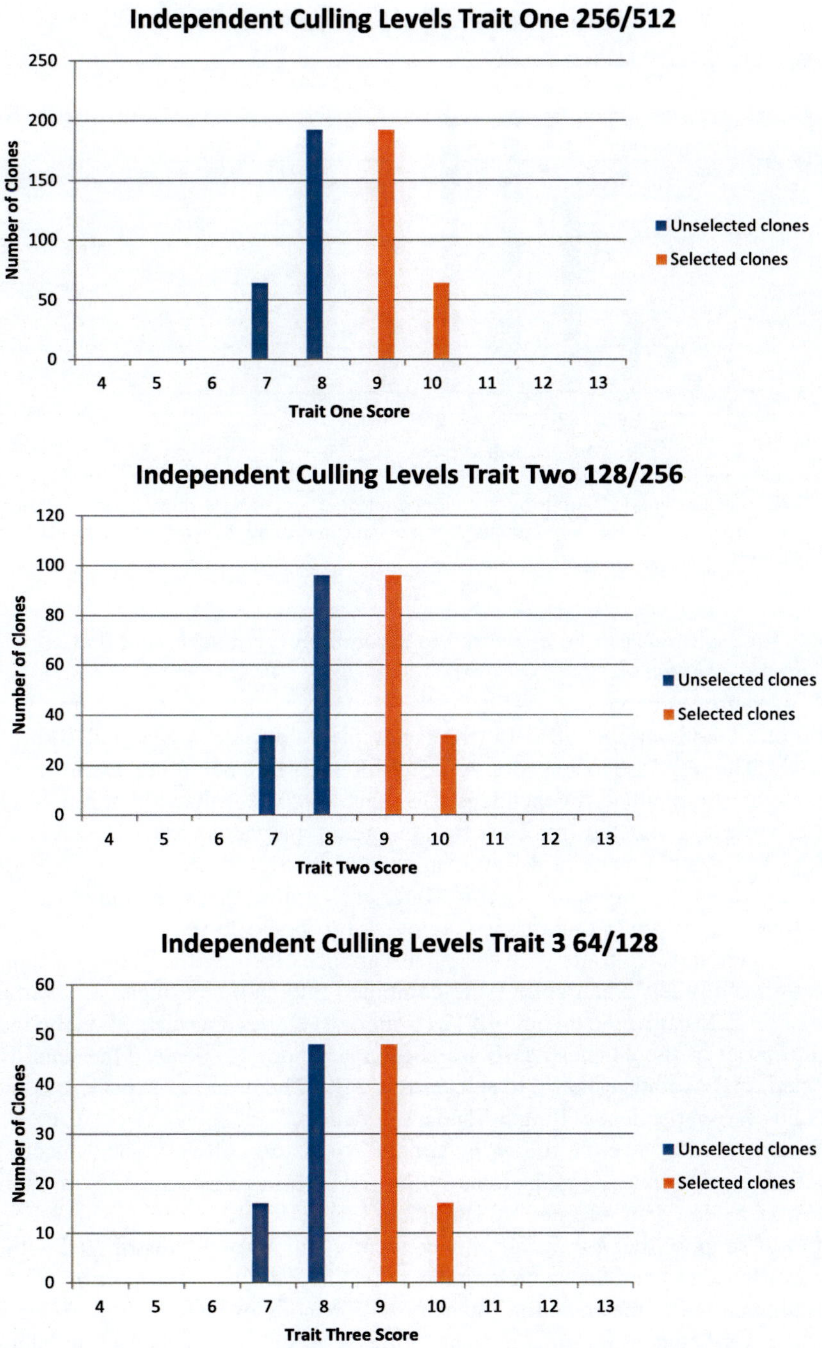

Fig. 7.17 Independent culling levels: half of clones culled for trait one (256/512), then half for trait two (128/256) and finally half for trait three (64/128), all in one clonal generation (for each trait, mean of selected clones minus population mean = 9.25–8.50 = 0.75, so total for all three traits is 3 × 0.75 = 2.25)

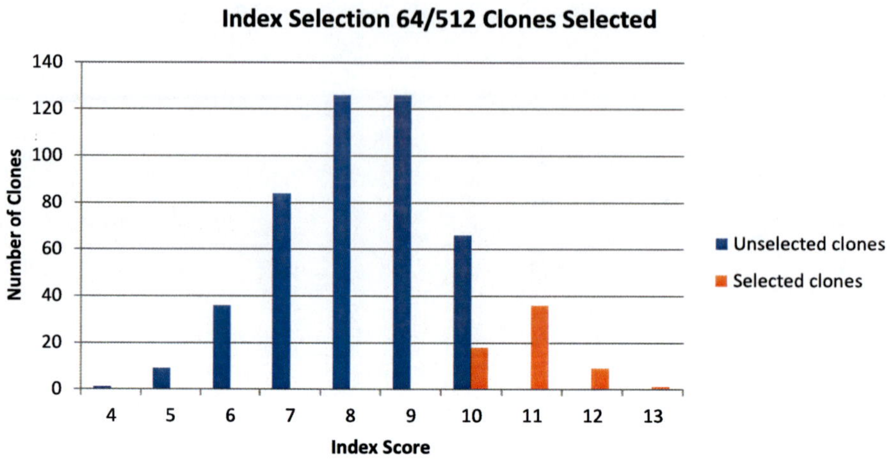

Fig. 7.18 Index selection: 64 out of 512 clones selected on index of three traits in one clonal generation (mean of selected clones minus population mean = 10.89–8.50 = 2.39). (From Bradshaw 2016 with permission)

scores for each trait can be converted to a common merit scale and that all of the variation is genetic.

In tandem selection, one eighth of the clones are selected for one trait with a mean of 8.5 and variance of 0.75 (the genotypic values are 7, 8, 9 and 10 in ratio 1:3:3:1). The response to selection is 1.5. With independent culling levels, one half of the clones are culled independently for each of three traits so that one in eight clones are again selected (for each trait the genotypic values are 7, 8, 9 and 10 in ratio 1:3:3:1 with a mean of 8.5 and variance of 0.75). Of these, on average, three quarters score 9 and one quarter score 10 for each trait so that selection of each trait contributes 0.75 to merit and hence the overall response to selection is 2.25. With index selection, three traits with the same variances for merit as those for tandem selection and independent culling are combined into an index (mean 8.5, variance $3 \times 0.75 = 2.25$) and then 64 out of 512 (1 out of 8) clones are selected on the index. Eighteen out of the 84 scoring 10 are chosen at random to achieve the total of 64 selected. The overall response to selection is 2.39. The order, as expected, is Index selection > Independent culling > Tandem selection.

Bulmer (1980) gave the following simple explanation of why index selection is superior to selection by independent culling, which in turn is superior to tandem selection, as first demonstrated by Hazel and Lush (1942).

Let us suppose that k traits are independently and normally distributed with the same variance (V) and the same heritability (h^2). If we select a proportion (p) from a large population, the selection intensity (i) for index or tandem selection is Z/p, where Z is the standard normal ordinate corresponding to p. The selection intensity can therefore be determined from tables of the properties of the normal distribution. Under independent culling, we set the threshold for each trait in such a way that the probability of its being exceeded is $p^* = p^{1/k}$. This ensures that the proportion selected is p. The selection intensity (i) for any trait considered by itself is Z^*/p^*, where Z^* is the standard normal ordinate corresponding to p^*. The responses to selection in this simple example can be calculated from the formula for a single trait: $R = h^2S = h^2i\sigma_P = h^2(Z^*/p^*)\sigma_P$, where S is the selection differential and $\sigma_P = V^{1/2}$.

With tandem selection, only one trait is selected each generation, so the genetic gain per generation is:

$$R = h^2\left(Z/p\right)V^{\text{‰}}.$$

In our simple example 1/8 ($p = 0.125$, $Z/p = i = 1.647$) of clones were selected from population with $V = 0.75$ ($V^{1/2} = 0.866$) and $h^2 = 1$, so that $R = 1.43$, compared with our value of 1.5 based on the binomial distribution.

With independent culling levels, k traits are selected. For each trait $R = h^2(Z^*/p^*)V^{1/2}$, and assuming each trait makes the same contribution to overall merit, on the scale of overall merit, the total response is simply the response for each trait multiplied by the number of traits:

$$R = kh^2\left(Z^*/p^*\right)V^{\frac{1}{2}}.$$

In our simple example with three traits ($k = 3$), ½ ($p = 0.5$, $Z^*/p^* = i = 0.798$) of clones were selected for each trait with $V = 0.75$ ($V^{1/2} = 0.866$) and $h^2 = 1$, so that $R = 2.07$, compared with our value of 2.25 based on the binomial distribution.

With index selection, the contributions to overall merit from each trait for each clone are first determined, and then selection practised on the merit score. As the k traits are independently and normally distributed with the same variance (V), the variance of the merit scores is kV. Hence, the response to selection for merit is:

$$R = h^2\left(Z/p\right)\left(kV\right)^{\text{‰}}.$$

In our simple example, with three traits ($k = 3$), 1/8 ($p = 0.125$, $Z/p = i = 1.647$) of clones were selected from population with $V = 0.75$ for each trait ($kV = 2.25$, $(kV)^{1/2} = 1.5$) and $h^2 = 1$, so that $R = 2.47$, compared with our value of 2.39 based on the binomial distribution.

The relative efficiency of tandem selection to index selection is $h^2(Z/p)V^{1/2}$ divided by $h^2(Z/p)(kV)^{1/2}$, which equals $1/k^{1/2}$. With $k = 4$, for example, the relative efficiency is 0.5.

The relative efficiency of independent culling levels to index selection is likewise $k^{1/2}(Z^*/p^*)(p/Z)$. With $k = 4$ and $p = 0.1$, the relative efficiency is 0.8 ($2 \times 0.697 \times 0.570$).

For all k and p, the genetic gain is in the rank order Index selection > Independent culling > Tandem selection, as found in our example: 2.47 > 2.07 > 1.43 (normal distribution); 2.39 > 2.25 > 1.5 (binomial distribution).

This conclusion can be extended to the more general case where the traits are correlated and of varying economic value and variability. Young (1961) found that the superiority of index selection over the other methods increases with an increasing number of traits under selection, but decreases with increasing differences in their relative importance, its superiority being at a maximum when the traits are of equal importance. The superiority of index selection over independent culling levels decreases with increasing selection intensity, but its superiority over tandem selection is independent of the intensity. The relative efficiency of index selection over the other methods is much affected by the phenotypic correlations between traits when the traits are of equal importance, the relative efficiency being higher when the phenotypic correlation is low or negative.

The theoretical superiority of both independent culling levels and index selection over tandem selection confirms the undesirability of intense early generation visual selection for breeders' preference because it does not assess most economically important traits. However, before drawing more conclusions, let us look at independent culling levels and index selection in a little more detail.

7.9.1 *Independent Culling Levels*

The general theory of the prediction of the response to selection with independent culling levels is difficult (Wricke and Weber 1986), except in the situation when traits are uncorrelated so that $R = h^2(Z^*/p^*)V^{1/2}$ for each trait. So, let us consider the simple example in Table 7.7 which shows the bivariate distribution of phenotypes for 256 clones assessed for two uncorrelated traits. All of the variation for trait 1 is genetic (genetic values 6, 7, 8, 9 and 10 in ratio 1:4:6:4:1) whereas only half of the variation for trait 2 is genetic (genetic values 7, 8 and 9 in ratio 1:2:1, each with environmental deviations of −1, 0 and 1 in ratio 1:2:1). Now let us cull all those clones with a phenotypic score of 8 or less for trait 1 and also all of those clones with a phenotypic score of 8 or less for trait 2, leaving 25 clones (in bold and underlined). The mean of the 25 clones is 9.2 ($S = 1.2$) for trait 1 and also 9.2 ($S = 1.2$) for trait 2. However, their genotypic means are 9.2 and 8.6, respectively, and these

Table 7.7 Independent culling levels for two traits: phenotypic values of 256 clones for two uncorrelated traits with same means (8) and phenotypic variances (1), with all of variation genetic for trait 1 but only half of variation (½) genetic for trait 2, and phenotypic culling levels ≤8 for both traits

				Trait 2		
	Phenotypes	6	7	8	9	10
	Genotypes	7	½ 7, ½ 8	⅙ 7, ⅔ 8, ⅙ 9	½ 8, ½ 9	9
	6	1	4	6	4	1
	7	4	16	24	16	4
Trait 1	8	6	24	36	24	6
	9	4	16	24	**<u>16</u>**	**<u>4</u>**
	10	1	4	6	**<u>4</u>**	**<u>1</u>**

would be the means of the selected clones in the next clonal generation. These responses to selection (*R*) are as expected for traits with broad-sense heritabilities of 1 and 0.5, respectively ($R = h^2S$). Furthermore, if we take a closer look at Table 7.7, we see that for trait 1, all 25 clones are above the culling point in the next clonal generation, whereas for trait 2, this is true of only 15 out of the 25 clones. Had there been a strong and positive genetic correlation between the two traits, more clones would have made the culling points, whereas a strong and negative genetic correlation would have resulted in fewer clones.

The practical message is that care must be taken in using independent culling levels. Firstly, the standard control cultivars in the assessment trials must be adequately replicated in order to set accurate culling points. Secondly, broad-sense heritabilities of traits should be determined and clonal values (as departures from the mean of all clones) multiplied by these heritabilities before doing the culling. The assessment of clones should be done in a trial with a minimum of two replicates in order to estimate the heritabilities, but more than two replicates should only be used if resources allow all clones to be assessed, or for fewer clones, if the balance between the intensity of selection and heritability of clonal means results in a greater response to selection. Thirdly, genetical correlations between traits should be determined and ideally used to predict the different responses to selection with different culling levels. However, a potentially simpler and better method is to use index selection, which automatically takes these correlations into account and also forces the breeder to decide the relative importance of different traits.

7.9.2 Index Selection

The aim of index selection is to improve several traits simultaneously in such a way as to make the biggest possible improvement in overall merit, which could be the cost of producing 40 tonnes of high-quality potatoes or the cost of producing a bag of chips (crisps). Ideally one would weight the genetic improvement of each trait according to its effect on this economic value, but in the absence of such information, the weighting will be subjective, although based on the latest information available from farmers and end users. The basic theory and use of various indices can be found in the review by Lin (1978) and in books such as those by Wricke and Weber (1986) and Falconer and Mackay (1996). Here, we will just consider Smith's (1936) discriminant function which is also known as the Optimum Index of Smith. Wricke and Weber (1986) provide a simple worked example showing the superiority of the optimum index over the base index of Williams (1962), the restricted index of Kempthorne and Nordskog (1959) and the desired gains index of Pesek and Baker (1969), as well as over independent culling levels.

The genetic value for merit is symbolized by *H* (genotypic index) and for each clone this aggregate genetic value is the sum of the true clonal means of all the traits (y's) to be improved, as deviations from their population means, weighted (a's) by their relative economic values. Hence, for five traits (1–5):

$$H = a_1y_1 + a_2y_2 + a_3y_3 + a_4y_4 + a_5y_5$$

$$I = b_1x_1 + b_2x_2 + b_3x_3 + b_4x_4 + b_5x_5$$

The phenotypic index (*I*) for merit is the sum of the phenotypic clonal means (*x*'s), as deviations from their population means, weighted (*b*'s) to maximize the correlation between *I* and *H*, so that selection using *I* maximizes the change in *H*. This is equivalent to minimizing the sum of squared deviations in the regression of *H* on *I* (i.e. $\Sigma(I - H)^2$). This is done by a standard statistical procedure in which the partial derivatives with respect to the *b*'s are set equal to zero. The result is a set of simultaneous equations, as many as there are traits, whose solution gives the values of the *b*'s to be used in the index. The *b*'s are the partial regression coefficients of the individual clones' genetic values on each measurement. In matrix algebra:

$$\mathbf{b} = \mathbf{P}^{-1}\mathbf{Ga}$$

where **P** and **G** are the phenotypic and genetic variance-covariance matrices.

If the clones are grown in a replicated trial, the phenotypic and genetic variance-covariance matrices can be obtained from appropriate analyses of variance and covariance of the traits. The genetic (G) variance-covariance matrix is estimated by subtracting the environmental (E) variance-covariance from the phenotypic (P) one.

With no phenotypic or genotypic covariances (correlations) between traits we have:

$$\begin{vmatrix} b_1 \\ b_2 \\ b_3 \\ b_4 \\ b_5 \end{vmatrix} = \begin{vmatrix} 1/P_1 & 0 & 0 & 0 & 0 \\ 0 & 1/P_2 & 0 & 0 & 0 \\ 0 & 0 & 1/P_3 & 0 & 0 \\ 0 & 0 & 0 & 1/P_4 & 0 \\ 0 & 0 & 0 & 0 & 1/P_5 \end{vmatrix} \begin{vmatrix} G_1 & 0 & 0 & 0 & 0 \\ 0 & G_2 & 0 & 0 & 0 \\ 0 & 0 & G_3 & 0 & 0 \\ 0 & 0 & 0 & G_4 & 0 \\ 0 & 0 & 0 & 0 & G_5 \end{vmatrix} \begin{vmatrix} a_1 \\ a_2 \\ a_3 \\ a_4 \\ a_5 \end{vmatrix} = \begin{vmatrix} a_1G_1/P_1 \\ a_2G_2/P_2 \\ a_3G_3/P_3 \\ a_4G_4/P_4 \\ a_5G_5/P_5 \end{vmatrix}$$

In other words, the weight for each trait is its economic weight multiplied by its broad-sense heritability, which makes sense. When both desirable and undesirable correlations occur between traits, these are accommodated by the index. Furthermore, a new index can be calculated every clonal generation.

The predicted response to index selection (*R*) is estimated from the regression of *H* on *I* multiplied by the selection differential ΔI, i.e. the mean of selected clones minus the mean of all clones. The regression coefficient is the covariance of *H* and *I* divided by the variance of *I* (σ_I^2). The covariance and variance can be shown to be $\mathbf{a'Gb}$ and $\mathbf{b'Pb}$, respectively. The intensity of selection *i* is $\Delta I/\sigma_I$. Hence:

$$R = \mathbf{a'Gb}\Delta I / \sigma_I^2 = \mathbf{a'Gb}i / \sigma_I = \mathbf{a'Gb}i / (\mathbf{b'Pb})^{1/2}$$

The responses to selection for individual traits (R_1, R_2, R_3, R_4 and R_5) are:

$\mathbf{R} = \mathbf{Gb}\Delta I/\sigma_I^2$ where **R** is a column vector with elements R_1, R_2, R_3, R_4 and R_5.

$$R = a_1R_1 + a_2R_2 + a_3R_3 + a_4R_4 + a_5R_5$$

As a simple example, I am going back to the cross between processing clone 12601ab1 and table cultivar Stirling and the components of variance in Table 7.6. Let us consider high yield (b_1), high dry matter concentration (b_2), regularity of tuber shape (b_3), resistance to sprouting (b_4) and good fry colour after storage at 4 °C (b_5), as a good combination of traits in a processing cultivar. Furthermore, as all of the genetic correlations between these traits were less than $|r| = 0.4$, let us set them to zero for simplicity. Then from Table 7.6:

$|b_1| = |a_1G_1/P_1| = |a_1 0.754|$
$|b_2| \quad |a_2G_2/P_2| \quad |a_2 0.605|$
$|b_3| \quad |a_3G_3/P_3| \quad |a_3 0.662|$
$|b_4| \quad |a_4G_4/P_4| \quad |a_4 0.610|$
$|b_5| \quad |a_5G_5/P_5| \quad |a_5 0.572|$

and $I = b_1x_1 + b_2x_2 + b_3x_3 + b_4x_4 + b_5x_5$.

In the absence of actual economic weights, traits can be given equal weights in standard deviations by making the a's $1/\sigma_p$ where σ_p is $(\sigma_c^2 + \sigma_{cy}^2 + ½\sigma^2)^{½}$ in our example. Hence:

$$I = 0.754x_1/\sigma_{p1} + 0.605x_2/\sigma_{p2} + 0.662x_3/\sigma_{p3} + 0.610x_4/\sigma_{p4} + 0.572x_5/\sigma_{p5}$$

As the variates (x's) in the index are independent, the variance of the index (σ_I^2) is:

$$\sigma_I^2 = (0.754)^2\sigma_{x1}^2/\sigma_{p1}^2 + (0.605)^2\sigma_{x2}^2/\sigma_{p2}^2 + (0.662)^2\sigma_{x3}^2/\sigma_{p3}^2 + (0.610)^2\sigma_{x4}^2/\sigma_{p4}^2 + (0.572)^2\sigma_{x5}^2/\sigma_{p5}^2, \text{and } as\ \sigma_{x1}^2 = \sigma_{p1}^2, etc.$$

$$\sigma_I^2 = (0.754)^2 + (0.605)^2 + (0.662)^2 + (0.610)^2 + (0.572)^2 = 2.072 \text{and} \sigma_I = 1.439$$

If we select 40 out of 1080 clones, the intensity of selection (i) is 2.186, and **R** = **Gb**i/σ_I is as follows:

Yield (kg/plot) $R_1 = (\sigma_{c1}^2 a_1 0.754)\ i/\sigma_I = 2.096$
DM (%) $R_2 = (\sigma_{c2}^2 a_2 0.605)\ i/\sigma_I = 0.925$
Shape (1–9) $R_3 = (\sigma_{c3}^2 a_3 0.662)\ i/\sigma_I = 0.544$
Sprouting (1–9) $R_4 = (\sigma_{c4}^2 a_4 0.610)\ i/\sigma_I = 0.971$
Fry4 °C (1–9) $R_5 = (\sigma_{c5}^2 a_5 0.572)\ i/\sigma_I = 0.490$

These are respectable responses to selection, but clearly, the response for each trait is less than the response for selection for that trait alone; for example, 2.096 kg/plot for yield compared with our earlier estimate of 3.999 kg/plot for yield alone.

As it is easy with modern computers and computer software to construct an index, index selection in which all of the traits in the first year of replicated trials are included is preferable to independent culling levels.

7.10 Intermediate Generations: The Selection Environment

Our tentative conclusion from the theory and practice of multistage, multitrait selection is that as many traits as possible should be measured on the clones (about 1000) in the first year of replicated trials and that index selection should be used to select clones (about 40) for use as parents in the next round of crossing and for assessment as potential cultivars in multisite trials. There will be traits than can only be measured in further clonal generations when enough tubers become available for testing, including yield in multiple-row plots for increased accuracy. However, there is no point in delaying a new cycle of crossing. Should a clone chosen to be a parent subsequently prove to have additional quality characteristics or resistance to additional pests and diseases, its offspring can be tested for these traits. During the multisite trials, the remaining clones can be assessed for genotype × environment interactions with a view to making recommendations to farmers and end users about the general or specific adaptability of those chosen to be released as new cultivars. However, as explained in Chap. 3, genotype × environment interactions have implications for the choice of selection environments in the intermediate generations and for the eventual success of the breeding programme. There will only be enough tubers for assessment of clones in one or two environments. At SCRI (Table 7.5), for example, unreplicated four-plant plots at a seed site provided enough tubers for a larger seed plot (six tubers) the next year, a ware trial with two replicates of five-tuber plots (10 tubers) and two lots of four tubers for testing for resistance to late blight and PCN. If the size of the unreplicated plot was increased to six tubers, then there would be enough tubers for ware trials at two contrasting sites, for example, one in the presence and the other in the absence of a combination of stresses such as drought, heat and salt. However, the breeder might prefer a single ware trial complemented with in vitro tests for stress tolerance. The breeder might also prefer to seek alleles of large effect for stress tolerance at Quantitative Trait Loci (QTLs) with a view to Marker Assisted Selection (MAS), or genes to target for Genetic Modification. It strikes me that the practical restrictions imposed by the intermediate generations mean that separate breeding programmes will be required for different agroecological zones and even areas within zones, as well as for different end uses, as seen, for example, in Japan where there are four public breeding stations and two private breeding companies (Mori et al. 2015). But let us finish this section with three examples of breeding for general adaptability within these constraints.

Haynes et al. (2012) concluded from their research on early generation selection in the eastern USA, that evaluation in multiple locations should be done as early as possible to identify clones with the broadest adaptation and hence the greatest likelihood of going on to become new cultivars. They could achieve this in their programme by reducing numbers assessed in the first field generation (FCG) (16,500–11,000) but increasing them in the second generation (SCG) (350–1650). Mild selection in the second generation would then allow a substantial number of clones (>350) to be assessed in unreplicated plots of six to eight plants at five or six locations in the third field generation (TCG).

In the SCRI example considered earlier of a successfully targeted and accelerated breeding programme for processing potatoes, Mackay (2005) was able to progress 136 clones from four-plant plots in the second clonal generation (SCG) to replicated ware trials at two contrasting sites (one in Scotland and one in England) in the third clonal generation (TCG), as well as maintaining them at a seed-site.

Where one is breeding for double cropping, as in Japan (Mori et al. 2015), two clonal generations can be grown in 1 year in spring and autumn in contrasting environments. In spring cropping, the potato is planted when temperatures are relatively low and the day length is relatively short. Thus, potatoes grow under conditions of increasing temperatures and day lengths. In contrast, in autumn cropping, the potato is planted when temperatures are high and day lengths are relatively long, and both decrease during the growing season. As the tubers harvested from the spring crop are used as seed tubers for the autumn crop, tuber dormancy of less than 60–80 days is desirable. By taking all clones forward from spring to autumn, selection can be practised for high adaptability to these contrasting environments.

7.11 Next Cycle of Crossing and Selecting

In one cycle of crossing and selection, we have identified a number of opportunities to influence the commercial worth of the clones identified as potential cultivars after completion of the replicated trial(s) at a ware site(s) in the third clonal generation. These opportunities occur in the choice of parents and crosses, in between cross selection in the seedling and first clonal generations and in within cross selection in both the second and third clonal generations. We have also seen that visual selection for breeders' preference can result in a correlated response for higher yield before selection for yield per se starts in the third clonal generation. However, when the clones selected as potential cultivars are used as parents in the next round of crossing, it is only the superiority of their breeding values that is transmitted to the next generation. In other words, we need to consider narrow-sense rather than broad-sense heritabilities. As well as crossing the clones from the third clonal generation, we could consider saving time by practising between full-sib family selection after the first clonal generation; but which is best?

Let us use the total yield data from the 15 × 15 half diallel set of 105 crosses (full-sib families) made at the Scottish Crop Research Institute (SCRI) in 1992 and illustrated in Fig. 7.2 (Bradshaw et al. 2000). Furthermore, let us assume that the full-sib families could have been assessed in a yield trial with two replicates at a ware site in the first clonal generation (i.e. as a tuber progeny trial), and that the clones from the selected families could have been assessed 3 years later in a yield trial with two replicates at a ware site in the third clonal generation. In other words, full-sib family selection for yield could be practised on a three-year cycle whereas full-sib family selection combined with clonal selection within the best families would require a six-year cycle. Although one of the 105 crosses in the diallel was 12601ab1 × Stirling, I am going to use the components of variance from the diallel to generate the relevant within cross variation rather than use the actual components from just one cross that was investigated in detail (Table 7.6). The relevant components of variance are as follows:

σ^2_{GCA} = 1.529 (general combining ability)
σ^2_{SCA} = 1.146 (specific combining ability)
σ^2_{YGCA} = 0.557 (year × general combining ability interaction)
σ^2_{YSCA} = 0 (year × specific combining ability interaction)
σ^2_{e} = 9.763 (plot to plot variation)
r = 2 (number of replicates in yield trials)

The (equilibrium) response to between full-sib family selection in which the families are assessed in just 1 year is (see Chap. 2):

$$R = i\left(2\sigma^2_{GCA} - \frac{1}{3}\sigma^2_{SCA}\right) \Big/ \left(2\sigma^2_{GCA} + \sigma^2_{SCA} + 2\sigma^2_{YGCA} + \sigma^2_{YSCA} + \sigma^2_{e}/r\right)^{\frac{1}{2}} = 0.838i$$

on the assumption that triallelic and tetra-allelic interactions are negligible.

The (equilibrium) response to within full-sib family selection in which the clones within families are assessed in just 1 year is (see Chap. 2):

$$R = i\left(2\sigma^2_{GCA} - \frac{1}{3}\sigma^2_{SCA}\right) \Big/ \left(2\sigma^2_{GCA} + 4\frac{1}{3}\sigma^2_{SCA} + 2\sigma^2_{YGCA} + 4\frac{1}{3}\sigma^2_{YSCA} + \sigma^2_{e}\Big/r\right)^{\frac{1}{2}} = 0.715i$$

on the assumption that triallelic and tetra-allelic interactions are negligible.

The narrow-sense heritabilities for between and within family selection are 0.262 and 0.191, respectively, compared with broad-sense heritabilities of 0.412 and 0.572, respectively.

Earlier, we concluded that selecting 20 out of 200 crosses (full-sib families) was realistic, giving an intensity of selection (i) of 1.742. Likewise, we concluded that selecting 40 out of 1000 clones could be done in the third clonal generation, giving an intensity of selection of 2.154.

Hence, the response to between family selection would be 0.838 × 1.742 = 1.460 kg/plot on a three-year cycle and the response to combined between and within family selection would be (0.838 × 1.742) + (0.715 × 2.154) = 1.460 + 1.540 = 3.000 kg/plot on a six-year cycle, giving similar responses per year of 0.487 and 0.500 kg/plot, respectively. As the average yield of the 105 families was 20.16 kg/plot, we are talking of a 2.5% increase in yield per annum. However, if yield was just one of eight traits being selected equally, the intensities of selection per trait would be 0.424 and 0.549, respectively, giving responses per year of 0.118 and 0.125 kg/plot, respectively, or approximately 0.625%. In practice, the combined selection would allow more traits to be assessed and selected and hence is preferable, but progress in improving each trait would be slow. Although results will vary with trait and breeding programme, it is almost certain that breeders will need to practice recurrent selection for a number of sexual generations so that small improvements accumulate into worthwhile ones. In the previous chapter, we saw that one way to both increase i and reduce T (length of cycle in years) is through genomic selection, provided there is adequate selection accuracy (r) as the response to selection per unit of time is proportional to ir/T (Bassi et al. 2016; Slater et al. 2016).

7.11.1 SCRI Breeding Programme

In the breeding programme which operated at the Scottish Crop Research Institute from 1991 to 2009, four cycles of crossing and selection were completed and analysed by Bradshaw et al. (2009). Resistance to late blight (complex races of *P. infestans*) in both foliage and tubers and resistance to potato cyst nematodes (*Globodera rostochiensis* Ro1 and *G. pallida* Pa2/3) were combined with a modest increase in yield through selection for breeders' visual preference and an acceptable fry colour for processing. In the first two generations, full-sib family selection was practised on a three-year cycle with limited selection within the best families in order to combine as quickly as possible genes from sufficient parents to achieve the programme's objectives. In the third generation, more extensive selection was practised within the best families so that the combined between and within full-sib family selection operated on a six-year cycle (a five-year cycle is possible with clonal selection for foliage blight but not tuber blight). In the fourth generation, improvement of processing quality was introduced through tuber progeny testing and clonal selection for fry colour after storage at 10 °C for 3 months. New germplasm for this trait and for late blight resistance was introduced and full-sib family selection with limited selection within the best families was again practised on a three-year cycle. The reader is referred to the paper for all of the results. For our purpose, the brief summary in Table 7.8 will suffice, ending in clones selected for commercially funded breeding.

The 108 clones selected from generation (cycle) 4 were grown at a high-grade seed site in 2006 as four-plant plots and again in 2007 as six-plant plots to provide seed-tubers for further evaluations of traits that had undergone selection, including a ware trial at Gourdie Farm, Dundee in 2007. Included in all of the evaluations were the original 36 parents from 1991, as well as the 15 clones selected from generation (cycle) 3 and nine other controls. The 15 selected clones were on average 14% higher yielding than the original 36 parents (16.68 kg/plot compared with 14.66 kg/plot) but slightly inferior for regularity of tuber shape and a breeders' preference score which did not include yield, thus confirming that the preference scores throughout the programme were dominated by yield. Thus, the apparent increase in yield was 14% over 12 years, or just over 1% per annum. However, when

Table 7.8 Summary of four cycles of SCRI potato breeding programme from 1991 to 2009 (Bradshaw et al. 2009)

Cycle	Year	Germplasm	Assessed	Selected	Criteria
1	1991	Crosses			
	1992	Progenies	120	36	Preference, PCN (Pa2/3), blight
	1993	Clones	1620	108	Preference
2	1994	Crosses			
	1995	Progenies	137	29	Preference, PCN (Pa2/3), blight
	1996	Clones	870	108	Preference, PCN (Pa2/3, Ro1)[d]
3	1997	Crosses			
	1998	Progenies	145	31	Preference, PCN (Pa2/3), blight
	1999	Clones	930	144	Preference
	1999	12 Resows[c]			
	2000	4-plant plots	2178 + 144		Preference
	2001/2			15	PCN (Pa2/3, Ro1), blight
4	2003	Crosses			
	2004	Progenies	122	36	Preference, PCN (Pa2/3), blight
	2005	Progenies	36	25	Fry colour
	2005	Clones	900	108	Preference
	2006/7[b]	Clones	108	8	Yield trial, PCN (Pa2/3), blight
Com[a]	2008	Crosses			

[a]Commercially funded breeding programmes
[b]Unreplicated four-plant plots at seed site in 2006
[c]More seed of best 12 progenies sown in glasshouse
[d]PCN testing done 30 January to 16 April 1997

15 of the 36 parents were used to produce 105 progenies in a half diallel set of crosses, the mean of the offspring means was 11% less than the mean of the mid-parent values (20.16 kg/plot compared with 22.61 kg/plot) (Bradshaw et al. 2000). In other words, selection was operating against a decrease which would occur in the absence of selection, presumably because desirable combinations of genes were being broken up at the same time as new ones assembled. The 15 selected clones were superior for resistance to foliage blight (5.73 compared with 4.41 on 1 to 9 scale of increasing resistance) and tuber blight (18.63% compared with 33.94% infected tubers), when tested with an aggressive isolate of *P. infestans* (genotype 'blue' 13_A2 race 1,2,3,4,5,6,7,10,11) which had dramatically increased in Britain over the 3 years 2005 to 2007 to account for more than 70% of the *P. infestans* population (Cooke et al. 2008). The 15 clones as a group were also more resistant to PCN than the original 36 parents (4.03 compared with 5.07 for square root of number of cysts), but surprisingly only eight were significantly more resistant than the susceptible control. Nevertheless, two of the 15 clones combined high levels of resistance to *G. pallida* with high levels of foliar and tuber resistance to late blight, a combination not present in any of the 36 original parents. Furthermore, 23 out of the 108 fourth generation (cycle) clones also had this combination of traits, and of these, 16

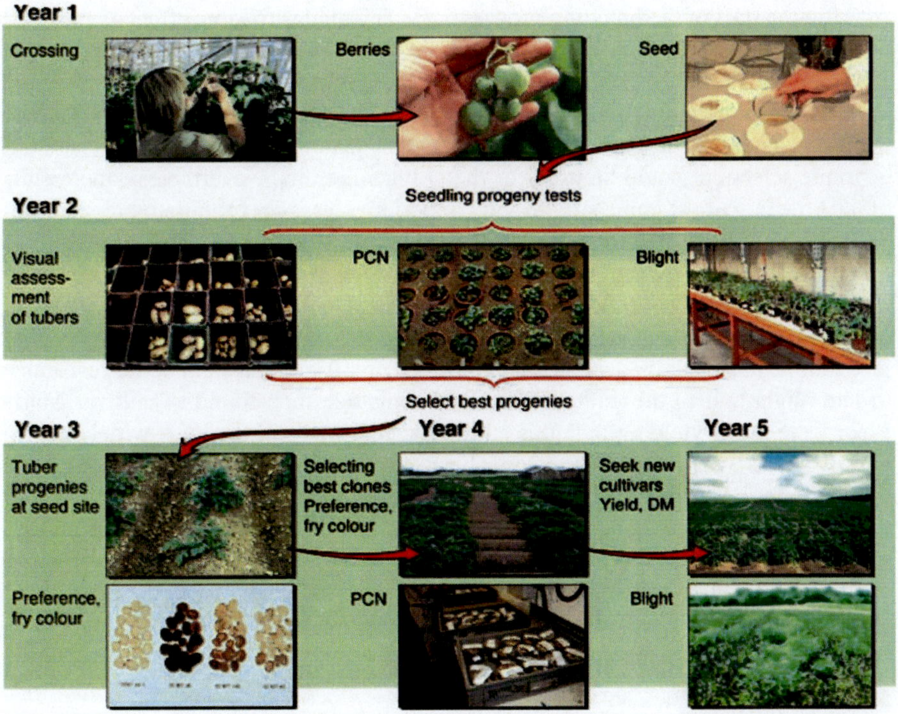

Fig. 7.19 Potato breeding at SCRI on five-year cycle: Year 1 is Crossing; Year 2 is Seedling Generation in Glasshouse (SG); Year 3 is First Clonal Generation of Spaced Plants at Seed Site (FCG); Year 4 is Second Clonal Generation in Unreplicated Plots at Seed Site (SCG); and Year 5 is Third Clonal Generation in Replicated Yield Trial at Ware Site (TCG). Clonal tests for PCN and late blight are shown in years 4 and 5, respectively. (From Bradshaw and Bonierbale 2010 with permission; Source: SCRI)

also had an acceptable fry colour after storage at 10 °C. However, only five of the 16 were not significantly lower yielding than cultivar Maris Piper, the most widely grown one in Great Britain. Eight clones were used as parents in commercially funded breeding programmes in 2008.

An important conclusion from the breeding programme was that clonal selection for disease and pest resistance was worthwhile in addition to between progeny selection. One round of clonal selection for PCN resistance appeared to have produced as much change as three rounds of between progeny selection, with limited or no within progeny selection, while clonal selection for foliage and tuber blight produced about 30% and 40%, respectively, of the change from three rounds of between progeny selection. Hence, combined between and within progeny selection in potato breeding can be recommended on a five-year cycle, as shown in Fig. 7.19. Potential new parents can be included in the clonal assessments and used in crossing if they prove superior to the clones undergoing selection. Their progenies will survive if superior to those produced from the selected clones.

After the start of the breeding programme, QTL alleles of large effect were found for the resistances to late blight and the white potato cyst nematode (Bryan GJ et al. 2002; Hackett et al. 2014), but not for yield, agronomic and quality traits, with the exception of maturity (Bradshaw et al. 2008). Today, marker-assisted selection of the QTL alleles of large effect would be used instead of the progeny tests and genomic selection would be used for the polygenic traits. Nevertheless, the results of the breeding programme still provide important messages for potato breeders and set a baseline for faster progress from genomic selection.

In summary, progress per trait per annum is slow in potato breeding when selection is practised for a number of traits and it is difficult to combine improvements in quality and disease and pest resistance with higher yields. In the SCRI breeding programme just described, clones combining better fry colour with better resistance to late blight and to the white potato cyst nematode than found in cultivar Maris Piper were lower yielding than this cultivar, which was still the most widely grown one in Great Britain in 2019, having been released in 1966 and having held the number one spot for over 40 years. In contrast, the reader will recall from Chap. 3 that the high-yielding clones identified at SCRI in a 6 × 6 half diallel set of crosses were not good enough for other traits to be selected as a new cultivar. In the previous chapter, we saw that the use of molecular markers can help us solve the problem of slow progress, not just in Scotland but globally! However, we also saw that marker-assisted selection and genomic selection will also have their limitations.

7.12 Cultivar Registration and Multiplication

Once the number of potential cultivars has been reduced to around 40, they can be assessed in multisite trials covering all of the target environments of the breeding programme. As explained in Chap. 3, the resulting knowledge about genotype × environment interactions allows recommendations to be made to farmers and end users about the general or specific adaptability of those clones chosen to be released as new cultivars. Furthermore, as we also saw in Chap. 3, estimates of the components of variance from a series of trials can be used to evaluate the efficiency and effectiveness of the trials system in terms of the number of locations, years and replicates used, and to make improvements where necessary (Talbot 1984, 1997).

Once a few promising clones have been identified for potential commercialization, they need to be registered as new cultivars and multiplied to meet farmer demand for seed tubers to plant. The process will vary according to the regulatory system of the country or area in which the breeder wishes to register a new cultivar; for example, in the European Union, it has to be entered into the European Common Catalogue of Varieties to be eligible for seed certification and marketing in all member states. It must be shown to be Distinct, Uniform and Stable (DUS) compared with existing cultivars and to have Value for Cultivation and Use (VCU). The DUS examination only describes morphological characteristics of the potato plant. The criteria for VCU can be found by consulting the website of the testing authority. The breeder needs to produce an Approved Seed Stock for submission to these statutory trials which usually take 2 years, but sometimes a third year is required. The stock submitted for DUS becomes the definitive stock of the new cultivar for use during seed certification and for Plant Breeders' Rights. The breeder is likely to apply for Plant Breeders' Rights on the new cultivar in order to receive royalty income on the sale of seed tubers for a period of 25 years with the possibility of extending to 30 years. In addition to passing DUS, the cultivar's proposed name needs to be accepted for the granting of PBR. A brief account of registration of cultivars in The Netherlands can be found in the book *Potato breeding* by Tiemens-Hulscher et al. (2013).

7.12.1 Plant Breeders' Rights

Plant Breeders' Rights were established in 1961 to encourage commercial breeding with the adoption in Paris of the International Convention for the Protection of New Varieties of Plants (UPOV, from the French name). Unauthorized commercial propagation of cultivars was forbidden, with two exceptions. Farmers retained their right to sow seed saved from their own crops and breeders could use protected cultivars as parents for further breeding. However, the 1991 UPOV Act introduced some significant changes. It restricted the right of farmers to freely use (botanical) seeds or other propagation material (tubers) for further cultivation by making it an optional exception that is limited to countries which make it a special provision. Plants, and cultivars in some countries, could now be patented in addition to having Plant Breeders' Rights. Finally, and ambiguously, anyone using a protected variety in breeding has to make major changes or else the 'new' variety will be considered an 'essentially derived' variety, falling to the ownership of the first breeder. The intention is to discourage small and insignificant changes in the variety's characteristic from being passed off as true innovation. UPOV headquarters are in Geneva (Switzerland) and by December 2017, 75 countries were members (www.upov.int).

7.12.2 Seed Certification and Multiplication

Potato yields and quality are certainly best when crops are planted with disease-free seed tubers in the correct physiological state. Experience around the world during the twentieth century showed that this is most likely to be achieved through statutory seed certification schemes operating in areas where potatoes are grown only for seed. Such areas will usually be geographically and climatically less favourable to the aphid vectors of viruses which cause systemic infection (Jeffries et al. 2006). However, low-income countries are characterized by informal seed systems with very low use of certified seed, often below 10% (Forbes et al. 2020). Furthermore, there is a conflict of interest between those promoting a commercial seed industry, cultivar protection and harmonization of seed standards, and those promoting farmers rights to save and trade their own seed (tubers). Nevertheless, here we are going to consider a typical seed tuber production scheme with certification in which a genetically unique seedling is maintained, multiplied and grown as a new cultivar. A useful account of the various multiplication procedures can be found in a special issue of *Potato Research, The Canon of Potato Science* (Struik et al. 2007).

Seed tuber production starts from pathogen-free microplants which are produced by the Certifying Authority (or under licence) in sterile laboratory conditions. In Scotland, for example, SASA (http://www.sasa.gov.uk) is the Certifying Authority for the Seed Potato Classification Scheme (SPCS). Single-node cuttings from tuber sprouts, from tubers supplied by the breeder, can be taken to provide rooted plantlets for pathogen testing. The breeder's stock should have been grown to a high health status and officially inspected and approved. If pathogen (mainly virus) elimination is required, larger plants can be grown from the plantlets and exposed to thermotherapy (e.g. heat treatment at 36–40 °C for 3 weeks; Bamberg et al. 2018) or to chemotherapy (e.g. Virozole) prior to meristem-tip culture. Dissected portions of the meristematic region of a shoot tip are placed on a liquid nutrient medium for plant regeneration. Once the resulting, or original, plantlets have been confirmed to be pathogen-free, they can be used for rapid multiplication by in vitro nodal cuttings under artificial, aseptic conditions in the laboratory. The healthy plant material is cut into individual stem portions, each with one axial bud and an attached leaf (single-node cutting), which are placed on agar media in jars. The axial bud grows into stems with numerous buds from which more single-node cuttings can be taken. Once sufficient have been produced, they are allowed (with the help of growth regulators) to develop into fully rooted in vitro plantlets (transplants). These are then used by licenced commercial growers to produce minitubers (Pre-basic TC (Tissue Culture)) in a greenhouse or screenhouse with high health status. The transplants can be grown in soil where they are likely to produce from two to five minitubers per plant. However, many more (up to 40) uniform tubers can be achieved from frequent harvests when the transplants are grown in aeroponic, hydroponic or nutrient film cultures. In aeroponic culture (Fig. 7.20), the 'below ground' plant parts are

Fig. 7.20 Aeroponic culture of minitubers in Chengdu, China (8 April 2008)

Fig. 7.21 (**a**) Minitubers being produced on inter-nodal cuttings from microplants in screenhouse in North East of Scotland. (**b**) pre-basic seed production in field in North East Scotland (farmer grows three field generations) (14 July 2011)

suspended in air and intermittently misted with nutrient solution. In hydroponic culture the transplants are grown in static nutrient solution whereas the nutrient flows along the lower roots in nutrient film culture. The minitubers are then grown by officially approved commercial growers in the field to produce pre-basic (field grown) seed tubers for further field multiplication. Subsequent generations result in grades of basic seed and finally certified seed in the amount required by ware growers who produce potatoes for consumption (Fig. 7.21).

Most European countries plant whole seed tubers whereas cut tuber pieces are commonly used in African, American and Asian countries. There are usually from four to six field generations. In Scotland, seed crops are officially inspected twice during the growing season and tubers must also be inspected for diseases and disorders and meet the required standards before they can receive the official label required for marketing. The land used for seed crops must be free from wart disease and also tested and confirmed free from cyst nematodes. The interval between potato crops in the rotation must be 7 years for pre-basic seed crops and 5 years for basic seed crops. Certified seed can only be sold for ware, it cannot be replanted in Scotland. It is common practice for the certifying authority to hold in vitro pathogen-free nuclear stocks of cultivars under multiplication, and also to fingerprint cultivars with molecular markers for unique identification (http://www.sasa.gov.uk).

After harvesting, seed tubers are usually stored for several months before planting. Storage can be done either in the dark at low temperature (2–4 °C) to prolong dormancy, suppress sprouting and prevent losses of water and storage carbohydrates and proteins, or in diffused-light storage facilities where some level of cooling is combined with the sprouting-inhibiting and ageing-delaying effect of light. (Gottschalk and Ezekiel 2006; Struik 2018). Diffuse light storage (DLS) of seed tubers at ambient temperatures has become established in developing countries where cold-store facilities are not available (Johansen and Molmann 2018). Seed tubers have been successfully stored for up to 8–9 months in small rustic ventilated stores in tropical regions with mean temperatures well above 20 °C. Interestingly, in Norway (69.7°N, 18.9°E), Johansen and Molmann (2018) have demonstrated the possibility of successful long-term storage (7 months) of potato seed tubers in light at elevated temperatures (15 °C) and a potential for earlier harvests and higher early yields from such treatments.

7.12.3 Seed Production in China

Seed potato production in China takes place mainly in the northern single-cropping zone at high latitudes and the south-western mixed-cropping zone at high altitudes. Virus-free seed potato production started in the 1970s, but by 2018, only accounted for about 35% of potatoes planted by farmers, with most farmers still using seed potatoes they had produced themselves for the next season (Jin 2018). However, virus-free seed potato production is increasing as larger enterprises become involved with their own complete systems for seed potato propagation. Jin (2018) has described the seed propagation system in China which is similar to one described above for Scotland.

7.12.4 Seed Production in India

The following account is based on Singh et al. (2020) unless otherwise stated. About 90% of the total potato area in India is located in the sub-tropical plains, with 6% in the hills and 4% in the plateau region. Up until 1970, the hills were the only source of quality seed potatoes for the plains and plateau regions. The October produce of the hills was partially dormant and hence could not be used immediately, but could be used for raising the spring crop in the plains, which was planted during December–January. However, the crop was exposed to a high aphid population which resulted in an accumulation of aphid transmitted viruses. Nevertheless, when the spring crop was harvested during March–April, some of the infected produce was kept in cold stores for planting the autumn crop during September–October. Thus, seed from the northern hills had higher productivity than the seed produced in spring in the plains. This led to a high demand for hill seed for growing potatoes in the sub-tropical plains. The availability of good quality seed tubers at affordable prices became a major issue for potato cultivation because the hill area was insufficient to meet the large requirements from the plains.

Systematic work on the development of seed production was started after the establishment of the ICAR-Central Potato Research Institute (ICAR-CPRI) in 1949. Surveys of the important potato growing regions of India from 1952 onwards revealed a remarkable consistency in the build-up of the aphid population in the North-Western and Central Indo-Gangetic plains. The aphid population remained very low during October to December. This finding led to the development from 1959 of what became known as 'the seed plot technique' (separate plots for seed) for the production of quality seed potatoes in the Indo-Gangetic plains from 1970 and the current region-wide seed supply chains. Today, however, there is concern about white fly (*Bremisia tabaci*) transmitted viruses (e.g. tomato leaf curl New Delhi virus, ToLCNVD) becoming a major problem and threatening the disease-free seed production in areas where a sufficient aphid-free period is available for quality seed production (Subramanian 2020). The current annual demand for certified seed in India is about 5.25 to 6.30 million tonnes, but if a seed replacement rate of 25% rather than 100% can be accepted, this reduces to around 1.33 million tonnes. ICAR-CPRI produces nucleus and breeder seed of notified cultivars every year by both a conventional seed system and a hi-tech one, but the latter is in the process of replacing the former. Breeder seed is supplied to different state governmental agencies (state seed corporations, National Seed Corporation and other agencies) for further multiplication cycles as Foundation seed-1, Foundation seed-2 and finally Certified Seed for supplying to farmers. However, shortages of quality planting materials can result in large areas being planted with local, degenerated seed. Hence, to cater for the growing demand for quality seed, private enterprises are becoming involved and they prefer rapid multiplication methods instead of the conventional practice of slow field multiplication where the multiplication rate is only one to five or six.

The conventional seed system of ICAR-CPRI involved clonal selection, tuber indexing and multiplication of healthy indexed tubers over four field generations on their farms to produce breeder seed. The process started with the indexing of four tubers per plant from the field for contagious and insect transmitted viruses by testing plants raised from a single eye by ELISA (PVX, PVS, PVM, PVA, PVY and PLRV). Healthy indexed tubers were then multiplied over four field generations; in the first year as spaced plants, then as clonal rows followed by two bulk multiplications. Crop inspection, rouging of diseased plants and immune diagnosis were key features of the programme, as was proper management and strict health standards in subsequent multiplications. Today, however, hi-tech seed production systems can reduce the open field exposure in the production of breeder seed from 4 to 2 years. The process starts with micropropagation of mother plants which have been raised from healthy looking tubers, under controlled conditions, and shown to be free of viruses by ELISA and PCR tests. If no virus-free plants are found, a virus-free stock of the cultivar can be raised from meristem tip culture coupled with chemotherapy (Ribavarin) and thermotherapy. Mericlones are indexed/tested by ELISA, PCR and ISEM (Immuno sorbent electron microscopy) and healthy plants sub-cultured 8–10 times to get the desired number of healthy plants of each cultivar. These microplants are used to produce minitubers via microtubers (induced in the axils of leaves of microplants), minitubers from in-vitro plantlets or aeroponic minitubers, all under protected net house conditions. Aeroponic systems are gaining in popularity in India and microtubers can be of value for germplasm conservation and for storage and exchange of germplasm, as well as in potato research. The minitubers (Generation-0, nucleus seed) are planted in the field (or nethouse) the next year (Generation-1, pre-basic) and the produce planted in the field the following year (Generation-2) to provide breeder (basic) seed (tubers) at harvest. Further multiplication can be over three generations (Foundation seed-1, Foundation seed-2 and Certified Seed) or reduced to two (Generation-3 and Generation-4) by rapid multiplication methods. Hi-tech seed production can result in epigenetic/somaclonal changes (e.g. in tuber shape) that cause problems during seed certification and possibly later with consumers, so monitoring is required. Further information on hi-tech seed potato production systems can be found in the article by Buckseth et al. (2019), and an account of potato seed certification in India can be found in the article by Singh et al. (2019). It covers both hill seed, grown in the high hills generally 2500 m above mean sea level, and plain seed grown in areas where aphid infestation is low during the crop growing season, as well as potato-tissue culture raised minitubers.

7.12.5 Seed Production in Brazil

Brazil has three potato growing seasons per year in a climate where a high population density of virus vectors occurs, hence the importance of virus-free seed tubers for planting (Kreuze et al. 2020). Brazil does have a seed certification scheme and currently some of the planted seed originates from tissue culture and greenhouse-produced minitubers. Kreuze et al. (2020) describe an effective low-cost alternative system that involves potato sprouts. These have to be detached from virus-free seed potatoes when they reach at least 3–5 cm in length, by hand-removing at the stolon and with primordial root formation on the base. After removal, they can easily be packaged into sealed polyethylene bags for transport. They are stored under favourable conditions for sprout growth, such as a dark room at 20 °C and 70–80% relative humidity. They can be directly planted in soil-substrate under greenhouse conditions, or in a hydroponics or aeroponics system, to produce minitubers. The interested reader is referred to the book chapter for details and references (Kreuze et al. 2020).

7.12.6 *Apical Cuttings*

It is easy to take and root stem cuttings from mature potato plants (i.e. plants with compound leaves) but they generally only yield 3–5 tubers per cutting. In contrast, apical cuttings provide a means of rapid multiplication that can be used as an alternative to minitubers as starting material for seed production, as currently being done in Kenya (Parker and Reddy 2020). The process begins with tissue culture plants that are not allowed to mature (produce compound leaves) and produce minitubers. Cuttings are produced from shoots from the tissue culture plantlets by continuously taking single or double node cuts, rooting the shoot, then transplanting to the field (Fig. 7.22). Hence, there are two stages in the system, production of cuttings in a screenhouse which are then sold to seed producers or farmers to plant in the field to produce tubers.

Fig. 7.22 Healthy tissue culture microplants and apical cuttings (CIP display in AgriExpo at Global Potato Conclave 2020, Mahatma Mandir, Gandhinagar, Gujarat, India, 28–31 January 2020) (30 January 2020)

The starting tissue culture microplants can be used to establish more 'mother plants' over a two-month period (5–20 mother plants per tissue culture plant at 5 × 5 cm spacing in compost in a suitable crate), before going into production of cuttings for the field by continuously taking cuttings from the mother plants as they grow over a further two-month period (i.e. leaving bottom leaves and buds for further growth). These cuttings are rooted in plugs and hardened off, which takes 2–3 weeks, before being planted in the field with immediate irrigation for establishment. The last cuttings to be taken often produce both large tubers and many tubers in the field. Typically, apical cuttings produce 10–25 or more tubers per cutting compared with 5–10 tubers per minituber, although the latter are more versatile and easier to handle. In Kenya, productivity ranges from 50 to 100 cuttings per tissue culture plant. A second field generation of multiplication can provide enough seed tubers for planting crops. At harvest farmers can grade out small tubers for their next season's seed and sell the larger tubers to markets, but should renew their seed by purchasing new cuttings when yields start declining. There is some evidence that apical cuttings work better for Andigena than Tuberosum potatoes (Monica Parker, personal communication). Propagation by apical cuttings is quite popular in Vietnam and could have a place in the north eastern states of India which lack an aphid-free period for seed potato production and hence rely on seed from elsewhere in India, although the same reasoning would apply to minitubers produced using aeroponics. In southwestern Uganda, where smallholder farmers commonly use seed potato tubers from the informal sector (seed recycled over several generations), Priegnitz et al. (2019, 2020) found some benefits from positive seed selection for three generations over the farmers' traditional method of selection. In positive selection, the healthiest looking plants in ware potato crops are pegged and selected just before full flowering to identify plants whose tubers will serve as seed for the next generation, provided they are still without symptoms 2 weeks later and their tubers look healthy after storage. In contrast, traditional farmer selection just involves visual selection of tubers from the bulk of the potato harvest. The researchers found lower levels of *Potato leafroll virus* (PLRV) and *Potato virus X* (PVX) as a result of positive seed selection, but not of *Potato virus S* (PVS) and *Potato virus M* (PVM). The average yield increase resulting from positive selection was 12%, but yield increases were variable, ranging from −5.7% to +36.9%, and in the individual experiments were often not significant. These yield increases were due to higher yields per plant, mostly through higher weights per tuber rather than higher numbers of tubers per plant. There were, however, a varying number of plants that did not emerge or had to be rogued because of bacterial wilt, as well as some unaccountable missing plants.

7.12.7 Global Exchanges of Germplasm

Finally, it is important to point out that strict quarantine procedures are required when potatoes are transferred from one country to another to prevent the introduction of diseases, particularly non-indigenous ones. Again, advances with in vitro techniques are proving useful. Lang (2001) has described how CIP supplies its new cultivars to farmers in East Africa through a seed multiplication scheme that starts in Kenya with nodal cuttings being taken from virus-free sprouts. These are supplied in vitro from CIP headquarters in Lima, Peru to the Quarantine Station in Kenya. The shipping of disease-free in vitro material has become the standard for international shipment of clonal material from genebanks (Bamberg et al. 2018).

7.13 Conclusions

In this chapter, we have looked at the theory and practice of conventional breeding of tetraploid potato cultivars for clonal propagation. Globally, a large number of breeding programmes is required to produce the necessary range of adapted cultivars for the wide range of agro-ecological zones and regions within zones in which potatoes are grown, and the different potato characteristics needed for the various end uses of potato crops, namely staple food, vegetable for cash and crop for processing. For any given programme, breeding should be viewed as cycles of artificial hybridization followed by multistage, multitrait selection over as few as possible clonal generations. Consideration of a crossing and selection strategy for quantitative traits led us to conclude that progress per trait per annum is slow in potato breeding when selection is practised for a number of traits, and that it is difficult to combine improvements in quality and disease and pest resistance with higher yields. We recognized that the use of molecular markers for marker-assisted and genomic selection can help us solve the problem of slow progress. Finally, we recognized the importance of providing disease-free planting material of new cultivars and considered ways in which this can be done. More specific conclusions are as follows.

1. *Hybridization Strategy*: It is important to have as much information as possible about potential parents and to assess them as accurately as possible. The choice of pair crosses needs to be based on differences in mid-parent (phenotypic) values for the set of desired traits, as this is the only readily available true predictor of progeny means (i.e. prediction before crosses are made). How good a predictor of progeny means depends on the narrow-sense heritabilities of the traits. Hence, in practice, crosses will be between parents which complement each other for desirable traits or possess the desired traits in varying degrees. Furthermore, a relatively large number of crosses will be made to ensure that they include the best ones possible from the available parents. In future, potential parents will also be genotyped with molecular markers and form the training population for genomic selection.
2. *Size of Breeding Programme*: A theoretical justification for the common practice of producing 100,000 seedlings from 200 to 300 crosses can be based on the relationship between the intensity of selection and size of breeding programme. Increasing the number of crosses and the number of offspring in each cross (progeny size) eventually becomes wasteful of resources as well as practically

impossible. Hence, in practising selection, breeders should be content with a truncation point of two standard deviations from the mean. The number of crosses and size of each progeny should be roughly equal on the assumptions that the amount of genetic variation between and within crosses is the same, with each normally distributed, and that the amount of genetic variation within each cross is the same. These assumptions are simple but unrealistic although they have a theoretical basis, yet nevertheless are useful in the absence of real data.

3. *Selection Between Crosses* Versus *Selection Between Seedlings/Clones*: The conclusion from theoretical considerations and experimental results is that about one-third of the crosses (superior) will contribute most (over 90%) of the genetically superior offspring clones and that a further one third of crosses (inferior) will make no contribution. Further considerations indicate that the worst crosses should be identified and deliberately eliminated rather than allow this to happen as a consequence of clonal selection. This can be done through seedling progeny tests for breeders' visual preference scores and important pests and diseases. These tests could therefore replace the common but ineffective practice of reducing 100,000 seedlings to 4000 clones in unreplicated plots through visual selection of seedlings in a glasshouse and spaced plants at a seed site: there is simply no way that a breeder can assess 100,000 clones for yield and other economically important traits. By sowing more seeds (200 per cross) of the best 20 out of 200 crosses and progressing the tubers from the seedlings directly to 4000 four-plant plots at the seed site, there is no loss of time compared with current practice. In contrast, incorporation of tuber progeny tests for yield, agronomic and quality traits would potentially add 3 years to between cross selection. In future, however, assessment of seedlings with molecular markers will solve the problem of ineffective early generation selection.
4. *Selection Between Clones in the Intermediate Generations*: During the intermediate generations of clonal selection, some 1000 out of 4000 clones in unreplicated plots are selected (primarily visually) to enter replicated yield trials in which they are reduced to 40 potential cultivars for assessment in multisite trials. Practical considerations arise out of the limited amount of planting material of each clone in each generation and the overall resources that can be allocated to a breeding programme. Nevertheless, five-plant plots and their produce can be used to assess many traits and with two-replicates, broad-sense heritabilities can be acceptably high. As high-throughput phenotyping develops, it may be possible to assess and select clones grown in these small plots for physiological traits, including stress tolerance, as well as their produce for quality traits. Multistage selection theory indicates that one-stage selection of 40 out of 1000 clones is superior to three-stage selection with increasing replication when judged by the response per year. Furthermore, index selection in which all of the traits in the first year of replicated trials are included, is preferable to independent culling levels, particularly as it is easy with modern computers to construct an index. However, if a trial is conducted at only one site in favourable conditions (high yields), selection for tolerance of relevant abiotic stresses will require it to be complemented with in vitro tests for stress tolerance, unless the breeder opts to

seek alleles of large effect for stress tolerance at Quantitative Trait Loci (QTLs) with a view to Marker Assisted Selection (MAS), or genes to target for Genetic Modification. In future, the phenotypic data from the replicated trial(s) will be an important part of validating genomic selection.

5. *Next Cycle of Crossing and Selecting*: The 40 clones identified from the intermediate clonal generations as potential cultivars will be assessed in multisite trials before one or two of them are entered into 2 years of official statutory trials (National List Trials) and registered for Plant Breeders' Rights, as well as multiplication starting from disease-free stock with a view to commercialization. The 40 clones should also be used as parents in the next cycle of crossing and selecting. There is no point in delaying a new cycle of crossing to wait for information on traits that can only be measured in further clonal generations when enough tubers become available for testing. Should a clone chosen to be a parent prove to have additional quality characteristics or resistance to additional pests and diseases, its offspring can be tested for these traits. It is only the superiority of parental breeding values that is transmitted to the next generation so the response to selection depends on narrow-sense rather than broad-sense heritabilities. Both theory and experimental results indicate that selection between families (progenies) combined with selection between clones in the best families on a five- or six-year cycle is superior to between family selection on a three-year cycle, when the responses per year are compared. Furthermore, combined selection allows more traits to be assessed and selected, but progress in improving each trait is slow. Breeders will need to practise recurrent selection for a number of sexual generations so that small improvements accumulate into worthwhile ones. Genomic selection will have its biggest impact by reducing the cycle time to a single year, although the genomic estimated breeding values will need checking as the cycles progress.
6. *Seed Certification and Multiplication*: Potato yields and quality are highest when crops are planted with disease-free seed tubers in the correct physiological state. This is best achieved through statutory seed certification and multiplication schemes operating in areas where potatoes are grown only for seed and which are geographically and climatically less favourable to the aphid vectors of viruses which cause systemic infection. Today, the starting materials for multiplication are disease-free microplants from which minitubers or apical cuttings are produced under protected screenhouse conditions.

In the next chapter, we will look at an alternative means of propagating new cultivars through true potato seed (TPS).

References

Araus JL, Cairns JE (2014) Field high-throughput phenotyping: the new crop breeding frontier. Trends Plant Sci 19:52–61. https://doi.org/10.1016/j.tplants.2013.09.008

Bamberg J, Jansky S, Ellis D (2018) Ensuring the genetic diversity of potatoes. In: Wang-Pruski G (ed) Achieving sustainable cultivation of potatoes. Burleigh Dodds, Cambridge, pp 57–79

Bassi FM, Bentley AR, Charmet G, Ortiz R, Crossa J (2016) Breeding schemes for the implementation of genomic selection in wheat (Triticum spp). Plant Sci 242:23–36. https://doi.org/10.1016/j.plantsci.2015.08.021

Bernhard T, Truberg B, Friedt W, Snowden R, Wittkop B (2016) Development of near-infrared reflection spectroscopy calibrations for crude protein and dry matter content in fresh and dried potato tuber samples. Potato Res 59:149–165

Bonierbale M, Grüneberg W, Amoros W, Burgos G, Salas E, Porras E, zum Felde T (2009) Total and individual carotenoid profiles in *Solanum phureja* cultivated potatoes: II. Development and application of near-infrared reflectance spectroscopy (NIRS) calibrations for germplasm characterization. J Food Compost Anal 22:509–516

Bradshaw JE (1994) Quantitative genetics theory for tetrasomic inheritance. In: Bradshaw JE, Mackay GR (eds) Potato genetics. CAB International, Wallingford, pp 71–99

Bradshaw JE (2008) Effect of new cultivars on the disease and pest resistance of the British potato crop. Proc Crop Prot North Br 2008:223–229

Bradshaw JE (2009a) A genetic perspective on yield plateau in potato. Potato J 36:79–94

Bradshaw JE (2009b) Potato breeding at the Scottish Plant Breeding Station and the Scottish Crop Research Institute: 1920–2008. Potato Res 52:141–172

Bradshaw JE (2016) Plant breeding: past, present and future. Springer, New York, 693p

Bradshaw JE, Bonierbale M (2010) Potatoes. In: Bradshaw JE (ed) Root and tuber crops, handbook of plant breeding 7. Springer, New York, pp 1–52

Bradshaw JE (2017) Review and analysis of limitations in ways to improve conventional potato breeding. Potato Res 60:171–193

Bradshaw JE, Mackay GR (1994) Breeding strategies for clonally propagated potatoes. In: Bradshaw JE, Mackay GR (eds) Potato genetics. CAB International, Wallingford, pp 467–497

Bradshaw JE, Stewart HE, Wastie RL, Dale MFB, Phillips MS (1995) Use of seedling progeny tests for genetical studies as part of a potato (*Solanum tuberosum* subsp. *tuberosum*) breeding programme. Theor Appl Genet 90:899–905

Bradshaw JE, Wastie RL, Stewart HE (1996) Assessing general combining ability for gangrene resistance by means of a glasshouse seedling test. Potato Res 39:179–183

Bradshaw JE, Dale MFB, Swan GEL, Todd D, Wilson RN (1998a) Early-generation selection between and within pair crosses in a potato (*Solanum tuberosum* subsp. *tuberosum*) breeding programme. Theor Appl Genet 97:1331–1339

Bradshaw JE, Chapman IM, Dale MFB, Mackay GR, Solomon-Blackburn RM, Phillips MS, Stewart HE, Swan GEL, Todd D, Wilson RN (1998b) Applied potato genetics and breeding: potato improvement by multitrait genotypic recurrent selection. Scottish Crop Research Institute Annual Report 1998:92–96

Bradshaw JE, Todd D, Wilson RN (2000) Use of tuber progeny tests for genetical studies as part of a potato (*Solanum tuberosum* subsp. *tuberosum*) breeding programme. Theor Appl Genet 100:772–781

Bradshaw JE, Dale MFB, Mackay GR (2003) Use of mid-parent values and progeny tests to increase the efficiency of potato breeding for combined processing quality and disease and pest resistance. Theor Appl Genet 107:36–42

Bradshaw JE, Hackett CA, Pande B, Waugh R, Bryan GJ (2008) QTL mapping of yield, agronomic and quality traits in tetraploid potato (*Solanum tuberosum* subsp. *tuberosum*). Theor Appl Genet 116:193–211

Bradshaw JE, Dale MFB, Mackay GR (2009) Improving the yield, processing quality and disease and pest resistance of potatoes by genotypic recurrent selection. Euphytica 170:215–227

Brown J, Caligari PDS, Dale MFB, Swan GEL, Mackay GR (1988) The use of cross prediction methods in a practical potato breeding programme. Theor Appl Genet 76:33–38

Bryan GJ, McLean K, Bradshaw JE, De Jong WS, Phillips M, Castelli L, Waugh R (2002) Mapping QTLs for resistance to the cyst nematode *Globodera pallida* derived from the wild potato species *Solanum vernei*. Theor Appl Genet 105:68–77

Buckseth T, Singh RK, Sharma AK, Tiwari JK, Chakrabarti SK (2019) Hi-tech seed potato production system. Indian Horticulture, November–December 2019:18–21

Bulmer MG (1980) The mathematical theory of quantitative genetics. Clarendon Press, Oxford, 255p

Cochran WG (1951) Improvement by means of selection. Proceedings of the second Berkeley symposium on mathematical statistics and probability, pp 449–470

Cooke DEL, Lees AK, Shaw DS, Taylor MC, Prentice MWC, Bradshaw NJ, Bain RA (2008) The status of GB blight populations and the threat of oospores. Proc Crop Prot North Br 2008:217–222

Eriksson D, Carlson-Nilsson U, Ortíz R, Andreasson E (2016) Overview and breeding strategies of table potato production in Sweden and the Fennoscandian region. Potato Res 59:279–294

Falconer DS, Mackay TFC (1996) Introduction to quantitative genetics, 4th edn. Longman, Harlow, 464p

Finney DJ (1958) Plant selection for yield improvement. Euphytica 7:83–106

Fisher RA, Yates F (1963) Statistical tables, sixth edition. Oliver and Boyd, Edinburgh, 146p

Forbes GA, Charkowski A, Andrade-Piedra J, Parker ML, Schulte-Geldermann E (2020) Potato seed systems. In: Campos H, Ortiz O (eds) The potato crop: its agricultural, nutritional and social contribution to humankind. Springer, ebook, pp 431-447

Gopal J (2018) Advances in conventional potato-breeding techniques. In: Wang-Pruski G (ed) Achieving sustainable cultivation of potatoes. Burleigh Dodds, Cambridge, pp 81–98

Gopal J, Gaur PC, Rana MS (1992) Early generation selection for agronomic characters in a potato breeding programme. Theor Appl Genet 84:709–713

Gottschalk K, Ezekiel R (2006) Storage. In: Gopal J, Khurana SMP (eds) Handbook of potato production, improvement and postharvest management. The Haworth Press, Binghampton/New York, pp 489–522

Hackett CA, Bradshaw JE, Bryan GJ (2014) QTL mapping in autotetraploids using SNP dosage information. Theor Appl Genet 127:1885–1904. https://doi.org/10.1007/s00122-014-2347-2. Epub 2014 Jul 1

Haynes KG, Gergela DM, Hutchinson CM, Yencho GC, Clough ME, Henninger MR, Halseth DE, Sandsted E, Porter GA, Ocaya PC (2012) Early generation selection at multiple locations may identify potato parents that produce more widely adapted progeny. Euphytica 186:573–583. https://doi.org/10.1007/s10681-012-0685-1

Hazel LN, Lush JL (1942) The efficiency of three methods of selection. J Hered 33:393–399

Ishii T, Yonezawa K (2006) Optimizing the procedure of multi-stage yield selection for selecting the best from a population of candidate entries. Breed Sci 56:7–15

Jansky S (2009) Breeding, genetics and cultivar development. In: Singh J, Kaur L (eds) Advances in potato chemistry and technology. Academic, Burlington, pp 27–62

Jeffries C, Barker H, Khurana SMP (2006) Viruses and viroids. In: Gopal J, Khurana SMP (eds) Handbook of potato production, improvement, and postharvest management. Food Products Press, New York, pp 387–448

Jin L (2018) Potato production and breeding in China. In: Wang-Pruski G (ed) Achieving sustainable cultivation of potatoes. Burleigh Dodds, Cambridge, pp 259–270

Johansen TJ, Molmann JAB (2018) Seed potato performance after storage in light at elevated temperatures. Potato Res 61:133–145

Kempthorne O (1957) An introduction to genetic statistics. Wiley, New York, 545p

Kempthorne O, Nordskog AW (1959) Restricted selection indices. Biometrics 15:10–19

Kreuze JF, Souza-Dias JAC, Jeevalatha A, Figueira AR, Valkonen JPT, Jones RAC (2020) Viral diseases in potato. In: Campos H, Ortiz O (eds) The potato crop: its agricultural, nutritional and social contribution to humankind. Springer, ebook, pp 389–430

Lang J (2001) Notes of a potato watcher. Texas A&M University Press, College Station, 420p

Lin CY (1978) Index selection for genetic improvement of quantitative characters. Theor Appl Genet 52:49–56

Lindhout P, Meijer D, Schotte T, Hutten RCB, Visser RGF, van Eck HJ (2011) Towards F1 hybrid seed potato breeding. Potato Res 54:301–312

Love SL, Pavek JJ, Thompson-Johns A, Bohl W (1998) Breeding progress for potato chip quality in North American cultivars. Am J Potato Res 75:27–36

Luthra SK, Gupta VK, Tiwari JK, Kumar V, Bhardwaj V, Sood S, Dalamu, Kaur RP, Kumar R, Vanishree G, Kumar D, Mhatre P, Chakrabarti SK (2020) Potato breeding in India. CPRI technical bulletin no 74 (revised), ICAR-Central Potato Research Institute, Shimla, Himachal Pradesh, India

Mackay GR (2003) Potato breeding at SCRI during the last quarter of the 20th century. Scott Crop Res Inst Annu Rep 2001/2002, 83–92

Mackay GR (2005) Propagation by traditional breeding methods. In: Razdan MK, Mattoo AK (eds) Genetic improvement of Solanaceous crops volume I: potato. Science Publishers, Enfield, pp 65–81

Mackay GR, Dale MFB (1990) Quality in potatoes. SCRI Annual Report 1990:9–12

Mori K, Asano K, Tamiya S, Nakao T, Mori M (2015) Challenges of breeding potato cultivars to grow in various environments and to meet different demands. Breed Sci 65(1):3–16. https://doi.org/10.1270/jsbbs.65.3

Neele AEF, Nab HJ, de Jongh de Leeuw MJ, Vroegop AP, Louwes KM (1989) Optimising visual selection in early clonal generations of potato based on genetic and economic considerations. Theor Appl Genet 78:665–671

Parker ML, Reddy R (2020) Apical cuttings to boost potato seed systems. In: souvenir, global potato conclave 2020, Mahatma Mandir, Gandhinagar, Gujarat, India 28–31 January 2020, pp 276–278

Peloquin SJ, Hougas RW (1959) Decapitation and genetic markers as related to haploidy in *Solanum tuberosum*. Eur Potato J 2:176–183

Pesek J, Baker RJ (1969) Desired improvement in relation to selection indices. Can J Plant Sci 49:803–804

Pieterse L, Judd J (2014) World catalogue of potato varieties, 6th edn. Agrimedia GmbH, Clenze

Plaisted RL (1980) Potato. In: Fehr WR, Hadley HH (eds) Hybridization of crop plants. The American Society of Agronomy, Inc., Madison, pp 483–494

Plaisted RL, Thurston HD, Brodie BB, Hoopes RW (1984) Selecting for resistance to diseases in early generations. Am Potato J 61:395–403

Prashar A, Yildiz J, McNicol JW, Bryan GJ, Jones HG (2013) Infra-red thermography for high throughput field phenotyping in *Solanum tuberosum*. PLoS One 8(6):e65816. https://doi.org/10.1371/journal.pone.0065816

Priegnitz U, Lommen WJM, van der Vlugt RAA, Struik PC (2019) Impact of positive selection on incidence of different viruses during multiple generations of potato seed tubers in Uganda. Potato Res 62:1–30. https://doi.org/10.1007/s11540-018-9394-z

Priegnitz U, Lommen WJM, van der Vlugt RAA, Struik PC (2020) Potato yield and yield components as affected by positive selection during several generations of seed multiplication in southwestern Uganda. Potato Res 63(4):507–543

Simmonds NW (1996) Family selection in plant breeding. Euphytica 90:201–208

Singh RK, Buckseth T, Sharma AK, Tiwari JK, Chakrabarti SK (2019) Potato seed certification in India. Indian Horticulture, November–December 2019:54–58

Singh BP, Venkatasalam EP, Singh RK (2020) Seed potato production in India: then and now. In: Souvenir, global potato conclave 2020, Mahatma Mandir, Gandhinagar, Gujarat, India 28–31 January 2020, pp 270–275

Slater AT, Cogan NOI, Forster JW, Hayes BJ, Daetwyler HD (2016) Improving genetic gain with genomic selection in autotetraploid potato. Plant Genome 9. https://doi.org/10.3835/plantgenome2016.02.0021

Smith HF (1936) A discriminant function for plant selection. Ann Eugenics 7:240–250

Struik PC (2018) Understanding ageing processes in seed potatoes. In: Wang-Pruski G (ed) Achieving sustainable cultivation of potatoes. Burleigh Dodds, Cambridge, pp 33–55

Struik PC, Lommen WJM, Haverkort AJ, Storey RMJ (2007) The canon of potato science. Potato Res 50

Subramanian S (2020) Virus-vectors relationship with special reference to vectors of potato. In: Souvenir, global potato conclave 2020, Mahatma Mandir, Gandhinagar, Gujarat, India 28–31 January 2020, pp 232–235

Swiezynski KM (1984) Early generation selection methods used in Polish potato breeding. Am Potato J 61:385–394

Tai GC, Young DA (1984) Early generation selection for important agronomic characteristics in a potato breeding population. Am Potato J 61:419–434

Talbot M (1984) Yield variability of crop varieties in the U.K. J Agric Sci (Camb) 102:315–321

Talbot M (1997) Resource allocation for selection systems. In: Kempton RA, Fox PN (eds) Statistical methods for plant variety evaluation. Chapman and Hall, London, pp 162–174

Tiemens-Hulscher M, Delleman J, Eising J, van Bueren ETL (2013) Potato breeding. Aardappelwereld BV, The Hague, 171p

Von Scheele C, Svensson G, Rasmussen J (1937) Die bestimmung des starkegehalts und der trockensubstanz der kartoffel mit hilfe des spezifischen gewichts. Landw Vers Sta 127:67–96

Williams JS (1962) The evaluation of a selection index. Biometrics 18:375–393

Wricke G, Weber WE (1986) Quantitative genetics and selection in plant breeding. Walter de Gruyter, Berlin/New York, 406p

Young SSY (1961) A further examination of the relative efficiency of three methods of selection under less-restricted conditions. Genet Res 2:106–121

Chapter 8
Breeding and Multiplication of TPS-Propagated Potato Cultivars

8.1 Introduction

Potatoes reproduce sexually through botanical seed and asexually (vegetatively) by producing tubers. Breeders make hybridizations and then grow genetically unique seedlings from the resulting true botanical seeds. The seedlings grow into plants and each produce a number of genetically identical tubers (a clone). The breeder can then decide which clones to replant as seed tubers the next season. The breeder can thus establish and evaluate distinct clones as potential cultivars which will also be maintained by asexual (vegetative) reproduction and multiplied and sold to farmers as seed tubers for planting. We saw in the last chapter that the starting materials for multiplication are disease-free microplants from which minitubers or apical cuttings are produced under protected screenhouse conditions. This is currently the normal method of potato propagation. However, there are circumstances where propagation by true potato seed (TPS) is an attractive proposition. Indeed, in 1994, Golmirzaie et al. (1994) thought that TPS technology was at the starting-gate of the race to develop a new green revolution, designed specifically to improve the production and consumption of potatoes in the developing countries of the torrid zone. More recently, Jansky et al. (2016) have argued that it is worth considering the possibility of converting the potato from an asexually propagated tetraploid crop into an inbred seed-propagated diploid crop to take full advantage of modern genetics and genomics tools for breeding work, as well as for the production of F_1 hybrid cultivars. They recognize that this would be an extremely ambitious long-term programme of research and development, but with many potential benefits. TPS has certainly been attracting enthusiasts for 50 years! In this chapter, we are going to look at the circumstances that make TPS an attractive proposition, the ways in which cultivars can be bred for TPS propagation and the ways in which farmers can grow such crops from TPS.

J. E. Bradshaw, *Potato Breeding: Theory and Practice*,
https://doi.org/10.1007/978-3-030-64414-7_8

8.2 Advantages and Disadvantages of TPS

When CIP (International Potato Centre) was founded in 1971, TPS propagation appeared most attractive in the torrid zones of the lowland tropics and subtropics. Here, the difficulty of establishing a TPS crop, later maturation (15–20 days) and less uniformity, could be outweighed by three advantages (Golmirzaie et al. 1994). First, seed costs are reduced due to the much smaller amounts of planting material required, which can be easily stored and transported. Depending on the method used, from 50 to 250 g of TPS can be used to replace the 2 tonnes or so of seed tubers required to grow 1 hectare of potatoes. Second, planting time is flexible because the farmer does not have to consider the physiological age and condition of seed tubers. For example, TPS can be sown successively at intervals and in amounts to provide a continuous supply of potatoes to the market-place. Finally, tubers are free from tuber-borne diseases with the possible exception of the few caused by true seed-borne viruses. In practice, TPS potatoes have been established in Bangladesh, China, Egypt, India, Indonesia, Nicaragua, Peru, Philippines, southern Italy and Vietnam (Almekinders et al. 1996; Ortiz 1997; Simmonds 1997). They have proved extremely useful for food security, particularly for small scale farmers (Chujoy and Cabello 2009). Chilver et al. (2005) reviewed on-farm profitability and prospects for TPS, based on the results of on-farm research in Egypt, India, Indonesia and Peru conducted in the mid-1990s. The emphasis was on those costs that change and the attributes of output that make one technology more or less profitable than another. Assumptions were made about the variables conditioning comparative profitability. Yield was considered the same or positive for TPS because the crop would be largely free of disease. Per-unit harvest value was seen as strongly negative for TPS because botanical seed generates a relatively large number of small-size tubers which lack uniformity of size and shape and are harvested late. In contrast, seed costs were seen as strongly positive for TPS. Fungicide costs were also seen as positive for TPS because it was thought that less fungicide would be required to control late blight in a genetically heterogeneous crop. Other variable costs were considered the same. The actual results were interesting. The one that I am going to highlight is on yield. Seedling tubers of good hybrids were high yielding at levels exceeding 20 t/ha, but the same was true of competing clones where seed quality was not as big a problem as might have been expected. TPS transplants in the north-eastern plains and hills of India produced fairly good yields of 15 to 20 t/ha, but farmers in the same regions were producing 25–30 t/ha with clonally propagated material. The authors concluded that widespread geographical adoption of TPS was unlikely in the immediate future, but that investment in a small but sustained TPS breeding effort could be justified in both China and India. It would be interesting to see the results of a similar analysis done in 2020. Nevertheless, farmers always need to be persuaded of the benefits of TPS for them to switch from a clonally propagated crop.

8.3 Breeding Cultivars for TPS

Breeding cultivars for TPS started at CIP in 1972 with the aim of high yields, acceptable uniformity for agronomic and reproductive traits, and resistance to important diseases. Breeding strategies were based on the theory of inbreeding and crossbreeding under tetrasomic inheritance, which is more complicated than that for disomic inheritance (see Chap. 2), and in particular on research reporting higher heterotic levels with the most heterozygous material (Golmirzaie et al. 1994). Nevertheless, Golmirzaie et al. (1994) thought that simply crossing unrelated genetic material, as a means of exploiting heterosis in a TPS programme, offered promise for rapid progress. None of the breeding methods available at the time could deliver genetic uniformity and hence all of them involved selection for acceptable uniformity. Methods and progress were reviewed by Golmirzaie et al. (1994), Ortiz (1997), Simmonds (1997) and Gopal and Ortiz (2006). The following is based on the review by Simmonds (1997) which contains an extensive list of references, but the interested reader is referred to all four for more information. Simmonds considered five types of tetraploid population for TPS production, two of which he dismissed because of inbreeding depression, namely deliberately selfed tetraploids and open-pollinated materials, which even in mixtures in the presence of bumble bees are very largely selfs. When he compared the yields of hand-pollinated (cross-pollination) and open-pollinated progenies with the same female parents from eight samples in the literature, he found that the latter were 54, 60, 65, 66, 70, 71, 72 and 84% of the former (average 67.4%), and attributed this to the inferiority of selfs. The three types of population that he considered were therefore hand-pollinated crosses between 4*x* parents, with or without emasculation of the female parent, tetraploid crosses in which crossing is assisted by male sterility in one parent, and crosses on 4*x* females, hand-pollinated by diploid/dihaploid stocks which produce unreduced diploid pollen grains by first division restitution (FDR). Interestingly, Simmonds did not rule out the use of pure diploid crosses ($2x \times 2x$) in which the self-incompatibility mechanism could be used to maximize crossing and minimize selfing.

The breeding of cultivars for TPS production therefore involved the selection from breeding programmes of parents that also had appropriate female and male fertility, and which produced progenies with high seed germination and fast seedling growth, as well as high yields, acceptable uniformity for agronomic and reproductive traits, and resistance to important diseases. Golmirzaie et al. (1994) reported on a CIP project that evaluated and selected parental lines for agronomic and reproductive traits under the normal short-day length and high temperatures at San Ramon, Peru (800 m altitude, 11°08'S). The result of selection was TPS parental lines that flower well and set berries under warm tropical conditions. Golmirzaie et al. (1994) thought that it was important to identify parental clones with a high general combining ability (GCA) for yield and tuber uniformity and earliness in their progenies. Simmonds (1997) concluded from his review that seed production per plant (seed fertility) was greater in diploids than in tetraploids; and that in tetraploids, Andigena potatoes were superior to Tuberosum potatoes. He reminded readers that seed and tuber dormancy were also issues and that they were polygenically controlled and correlated. He also gave references to the reported hybrid vigour that could be exploited in crosses between Andigena and Tuberosum potatoes and to the potential value of $4x \times 2x$ crosses. However, Simmonds also expressed three views, which with hindsight seem important. The first was that the use of improved diploids per se in $2x \times 2x$ crosses to produce diploid cultivars was an attractive proposition; the second was that it is not certain that diploids will always be inferior to tetraploids; and third that it is exceedingly doubtful whether the overall uniformity (i.e. for many traits) of a clone can ever be approached by seedling families which will also be inferior in yields to standard cultivars, because the genetically variable family will contain superior genotypes that can be clonally propagated. More recently, Lindhout et al. (2011, 2018), Lindhout (2020) and Stockem et al. (2020) have demonstrated the feasibility of producing diploid inbred lines, through the introduction of a dominant gene (*Sli*) from *S. chacoense* that inhibits gametophytic self-incompatibility, and true diploid F_1 hybrids for maximum heterosis and uniformity. I am therefore going to concentrate on breeding diploid potato cultivars and leave the reader interested in breeding tetraploid cultivars to consult the earlier reviews mentioned above.

8.4 Breeding Diploid Potato Cultivars

The first potatoes to be domesticated, in the highlands of southern Peru, were diploids (Spooner et al. 2005). The result was a diploid cultigen, *S. tuberosum* L. Stenotomum Group ($2n = 2x = 24$), from which all other cultivated potatoes were derived. Although tetraploid potatoes came to be selected and preferred by Andean farmers, they continued to cultivate and improve diploid landraces/cultivars. They selected Phureja from Stenotomum potatoes for lack of tuber dormancy and faster tuber development so that they could grow up to three crops a year in the lower, warmer, eastern valleys of the Andes. Phureja potatoes were therefore able to spread into northern Ecuador, Colombia and Venezuela and are the second most widely cultivated type in South America after Andigena (Hawkes 1990). Interestingly, Ghislain et al. (2006) found that 32 out of 102 tested accessions of Phureja in the CIP (International Potato Centre) collection of landraces were triploid or tetraploid, not diploid, in agreement with Hawkes (1990) that not all Phureja are diploid. Goniocalyx (diploid) has been recognized as a northern subgroup of Stenotomum, noted for tubers with a bright yellow flesh (Hawkes 1990). Today, the CIP collection of cultivated potatoes contains 206 Phureja, 289 Stenotomum and 102 Goniocalyx accessions (https://cipotato.org/genebankcip/). It was, however, tetraploid potatoes that were introduced into Europe and then the rest of the world from the end of the sixteenth century, so that today, the principal cultivated potato is tetraploid *Solanum tuberosum*, derived from Andigena and Chilean Tuberosum but also with introgressions from a few wild species.

In the 1960s, modern breeders in North America and Europe successfully widened the genetic base of their breeding programmes through the production of populations of long-day adapted diploid populations of Groups Phureja/Stenotomum, with the Andean landraces held in genebanks as their starting materials (Carroll 1982; Haynes and Lu 2005). The population improvement schemes were relatively simple and done on generation (cycle) times of two to four years. Diploid cultivars have been produced from such schemes; for example, Mayan Gold, Inca Sun, Inca Dawn, Mayan Queen, Mayan Star and Mayan Twilight from the SPBS/SCRI programme in Scotland (Bradshaw 2009). However, these cultivars are targeted at niche markets because their yield is only 60% that of tetraploid Tuberosum potatoes. White et al. (2018) compared the fresh-weight yields of 36 diploid Phureja cultivars/clones with those of 27 tetraploid Tuberosum cultivars/clones in three trials at Gourdie Farm, Dundee (56°28′N, 03°07′W) over the three years 2006, 2007 and 2008 and found that the average Phureja yields were 65% of the average Tuberosum yields. Furthermore, since 2004 CIP breeders have been working on biofortification of potatoes for iron (Fe) and Zinc (Zn) content, and reported good initial progress (Paget et al. 2014; Amoros et al. 2020) from a base population of Stenotomum, Goniocalyx and Phureja accessions from the study of Burgos et al. (2007).

Since the 1960s, it has also been possible to reduce tetraploid *S. tuberosum* to the diploid level through crosses with 2*x* Group Phureja ‘pollinators’ that induce dihaploid production by a form of parthenogenesis (Hougas and Peloquin 1958; Hougas et al. 1958). Subsequently, the frequency of dihaploids (also referred to as haploids) was increased by the use of elite pollinators (Hougas et al. 1964). It was soon found that dihaploids (EBN = 2) could be crossed with many wild tuber-bearing species which were also diploid (EBN = 2), as well as with cultivated diploid *S. tuberosum* (EBN = 2). Hence, much breeding and genetical research could be done at the diploid level, although it was assumed that the production of unreduced gametes in diploids would be used to return to the tetraploid level for further breeding and cultivar production. However, most dihaploids, diploid species and their hybrids displayed gametophytic self-incompatibility (*S* locus on chromosome 1), which prevented the production of the true breeding lines by selfing that could have been even more useful for breeding and genetical research.

There was, however, another method of producing both dihaploids from tetraploid potatoes and monoploids from diploid potatoes, namely anther/microspore culture. The review of the topic by Veilleux (2005) is still useful. It also covers the possibility of producing somatic embryos from a standard cultivar, and hence synthetic seed that produces a crop as uniform as the one from seed tubers of the cultivar, but with all of the advantages of TPS. Veilleux (2005) pointed out that anther culture had been possible from 1972, but by 2005, it was still common to find that agronomically desirable cultivars had been only marginally responsive or totally recalcitrant, whereas primitive cultivars or selected wild species had responded with an abundance of embryos and subsequent plantlets. Of particular interest in the current context are the papers by Paz and Veilleux (1997, 1999). They doubled the chromosome number, by in vitro shoot regeneration, of anther-derived monoploids of diploid Group Phureja potatoes to generate homozygous lines, which were all male sterile but varied in female fertility. Although seed set was low, it was adequate for using the doubled-monoploids as parental genotypes in practical breeding. They evaluated 14 families from crossing five doubled-monoploids with three heterozygous diploids with a broad genetic base. The evaluation was done in North America in the field from transplanted seedlings, so the comparison with tetraploid cultivar Kennebec, grown from tubers, was not entirely fair. Nevertheless, the results are of interest. The average yields of the 14 families ranged from 174 to 404 g/plant compared with 1320 for Kennebec. Average tuber numbers varied from 13 to 49 tubers/plant compared with 9 for Kennebec and average tuber weight ranged from 8 to 17 g/tuber compared with 147 for Kennebec. It would seem timely to see if a larger breeding effort could result in male fertile as well as female fertile homozygous lines for improvement in yield and other traits. Assuming that the self-incompatibility system operated in these lines, it could be exploited to produce diploid F_1 hybrids for TPS propagation without the need for hand-pollinations. In the meantime, a new method of producing such inbred lines has opened up.

A dominant self-incompatibility inhibitor gene (*Sli*) was found in *S. chacoense* by Hanneman (1985) and mapped to a distal end of chromosome 12, thus showing that it is independent of the *S* locus on chromosome 1 (Hosaka and Hanneman 1998a, b). Lindhout et al. (2011, 2018) subsequently used the *Sli* gene for the production of diploid inbred lines which could be used to produce true F_1 hybrid cultivars for maximum heterosis and genetic uniformity. They started work in 2008 by producing inter-species hybrids between diploid potato germplasm and a homozygous accession of the wild species *S. chacoense* carrying the *Sli* gene. The hybrids were extremely vigorous and about half of them produced many berries upon self-pollination. Many of the selfed offspring showed weak growth, but intense selection over the following selfed generations, and new crosses between F_3 and F_5 inbred lines followed by more intense selection over the subsequent selfed generations, resulted in inbreds with improved performance. The researchers made crosses between the best inbred lines and tested 45 hybrids in a yield trial with two replicates and two-plant plots in the Netherlands in 2015 together with 20 inbred lines and four control cultivars. The results were encouraging with 13 hybrids having higher fresh-weight yields than any of the inbreds and also out-performing the tetraploid control cultivars Lady Anna and Mozart, but not Lady Rosetta and Felsina. Further encouraging results were obtained from a new series of 216 experimental hybrids produced mainly from F_6 inbred lines in the winter season of 2014/2015.

The researchers' first scientific paper on the contribution and stability of yield components of diploid hybrid potatoes (Stockem et al. 2020) came from crosses made in the winter of 2015/2016 between inbred parents that had been selfed for five to seven generations. The botanical seeds were sown into plug trays in a greenhouse in the first week of April 2016. Six weeks after sowing, the seedlings were transplanted into an open field in the Netherlands at 30 cm spacing in ridges of 75 cm width. Irrigation was applied at 1 and 3 days after planting to prevent the seedlings from drying out. The haulm was killed 90 days after planting and the tubers lifted 3 weeks later and stored at 4 °C until required the next season. Out of

572 hybrids grown, a panel of 65 was selected for uniformity, but not for yield. These 65 diploid hybrids were compared with a range of 19 tetraploid cultivars in trials at five sites in 2017, three in the Netherlands, one in Belgium and one in France. The diploid hybrids yielded on average between 16 and 52 t/ha, with an overall mean of 31.4 t/ha, and the tetraploid cultivars between 52 and 101 t/ha, with an overall mean of 72.1 t/ha. Hence, on average, the diploid yield was 43.6% that of the tetraploids. The lower fresh-weight yield was a result of lower weight of individual tubers (71 vs. 177 g on average). The diploids had a higher number of stems (25.6 vs. 14.9 stems/m^2) but fewer tubers per stem (1.8 vs. 2.9 tubers) resulting in approximately equal numbers of tubers per unit area (45.2 vs. 43.5 tubers/m^2). Interestingly, it was the number of tubers per stem which explained 80% of the variation in yield in the tetraploids and 90% in the diploid hybrids. In contrast, in the SPBS/SCRI diploid breeding programme in Scotland, yield improved across several generations mainly as a result of increases in tuber size without a reduction in tuber numbers.

One other early result from the Solynta team is worth mentioning. Vries et al. (2016) reported on the potential of hybrid potato for the tropical highlands of East Africa where clean planting material is a problem and hence TPS is an attractive proposition. They trialled 10 representatives (out of 203) of their first experimental hybrids (crosses made in winter 2014/15) in East Africa (Nioka, Albertine Rift in Ituri Province, Democratic Republic of Congo at 1680 masl). The hybrids were crosses between F_4 lines tracing back to the three heterozygous diploid founders (D1, D2 and D16) and the *Sli*-donor DS. In Nioka, seedlings were raised from botanical seeds sown on 25 January 2016 and transplanted to the field 6 weeks later into irrigated ridges at 30 cm spacing with 75 cm between ridges, with three replicates of each hybrid. Tubers were harvested 4 months later when all of the plants had died. The diploid hybrids yielded between 8 and 29 t/ha compared with average yields of potatoes in East Africa of 5 to 14 t/ha. The best hybrid had yielded 26 t/ha in the Netherland's trial the previous year. Cultivars in the Netherlands yield from 35 to 40 t/ha. These are just preliminary results, but indicate the potential of diploid hybrids in East Africa and the yield to be made up on tetraploid cultivars in the Netherlands.

More recently, Ye et al. (2018) created self-compatible diploid potatoes by knocking out the self-incompatibility gene *S-RNase*, using the CRISPR–Cas9 system, in the diploid self-incompatible *S. tuberosum* Group Phureja S15–65 clone. The single guide RNA was designed to target the relatively conserved domain of the S-RNase protein and could target S_{p3} and S_{p4} alleles simultaneously. They ultimately obtained five mutant diploid clones, each containing mutations in at least one *S-RNase* allele. As *S-RNase* is a co-dominant gene, the mutation of either allele would enable pollen containing the same *S*-haplotype to extend into the style, thus resulting in self-compatibility. Although the rate of fruit setting varied among the lines, the number of seeds per fruit was sufficient for further propagation, ranging from 67 to 288 seeds per fruit. These results demonstrated that inducing loss-of-function mutations in *S-RNase* genes resulted in self-compatibility in potato. The growth vigour and plant morphology of the mutant lines did not differ from those of the wild type, indicating that they could be used directly for breeding. The percentage of (T_1) plants in the selfed families without the Cas9 cassette varied from 3.6% to 24.5%, thus demonstrating that self-compatibility can be achieved without introducing any exogenous DNA. Using the same method, the researchers also obtained *S-RNase* mutants in two more *S. tuberosum* Group Phureja clones (S15-47 and S15-76) and two *S. tuberosum* Group Stenotomum clones (S15-48 and S15-107). Hence, improved clones from a population improvement scheme can be made self-fertile for the purpose of inbreeding and crossbreeding to produce F_1 hybrid cultivars.

Rather than dwell on the lower yield of the diploids, I want to consider inbreeding and crossbreeding as a method of both diploid population improvement and cultivar production for comparison with other methods. We shall see that the advantage of such a breeding method is that it gives us the option of producing genetically uniform cultivars for propagation either through true potato seed (TPS) or seed tubers.

8.4.1 Population Improvement

Let us start our consideration of population improvement over a number of cycles of crossing and selection with the simple example of one cycle shown in Fig. 8.1. It was generated using the binomial distribution (Table 8.1) and shows the genetic variation (i.e. no environmental variation) among the clones (genotypes) in a diploid breeding population and the inbred lines that can be derived from the population. There are eight ($n = 8$) unlinked loci of equal effect, each with two alleles at equal frequencies and complete dominance for increasing effect (e.g. yield). Thus, $a = d = 1$ at all eight loci with $AA = Aa = 2$ and $aa = 0$ and their frequencies in the ratio $\frac{1}{4}:\frac{1}{2}:\frac{1}{4}$. The mean of the population of clones is 12 and the variance is 6, and we know from the genetical model that the additive genetic variance $V_A = 4$ ($V_A = \Sigma \frac{1}{2} a^2 = \frac{1}{2} na^2$) and the dominance variance $V_D = 2$ ($V_D = \Sigma \frac{1}{4} d^2 = \frac{1}{4} nd^2$). If we produce a set of true breeding lines from this population of clones by continuous selfing it will have a mean of 8 and a variance of 8 ($2V_A = \Sigma a^2 = na^2$). As expected, we observe inbreeding depression on selfing with the population mean falling from 12 to 8.

Now if we select all of the clones with a score (yield) of 16 (10%) and inter-cross them in all combinations, we create a new population of clones with a mean of 14.22. In other words, the response to selection is 2.22 and not 4 (16–12) because genes and not genotypes are transmitted to the next generation and the selected clones are a mixture of homozygotes and heterozygotes. Put another way, while the broad-sense heritability is 1, the critical narrow-sense heritability is $\frac{2}{3}$ [V_A/($V_A + V_D$)]. At each of the eight loci we are selecting AA and $2Aa$ and rejecting aa and thus increasing the desired allele (A) frequency from $\frac{1}{2}$ to $\frac{2}{3}$ and the genotype frequencies of AA, Aa and aa from $\frac{1}{4}:\frac{1}{2}:\frac{1}{4}$ to $(\frac{2}{3})^2$: $2(\frac{2}{3})(\frac{1}{3})$: $(\frac{1}{3})^2$ in the new

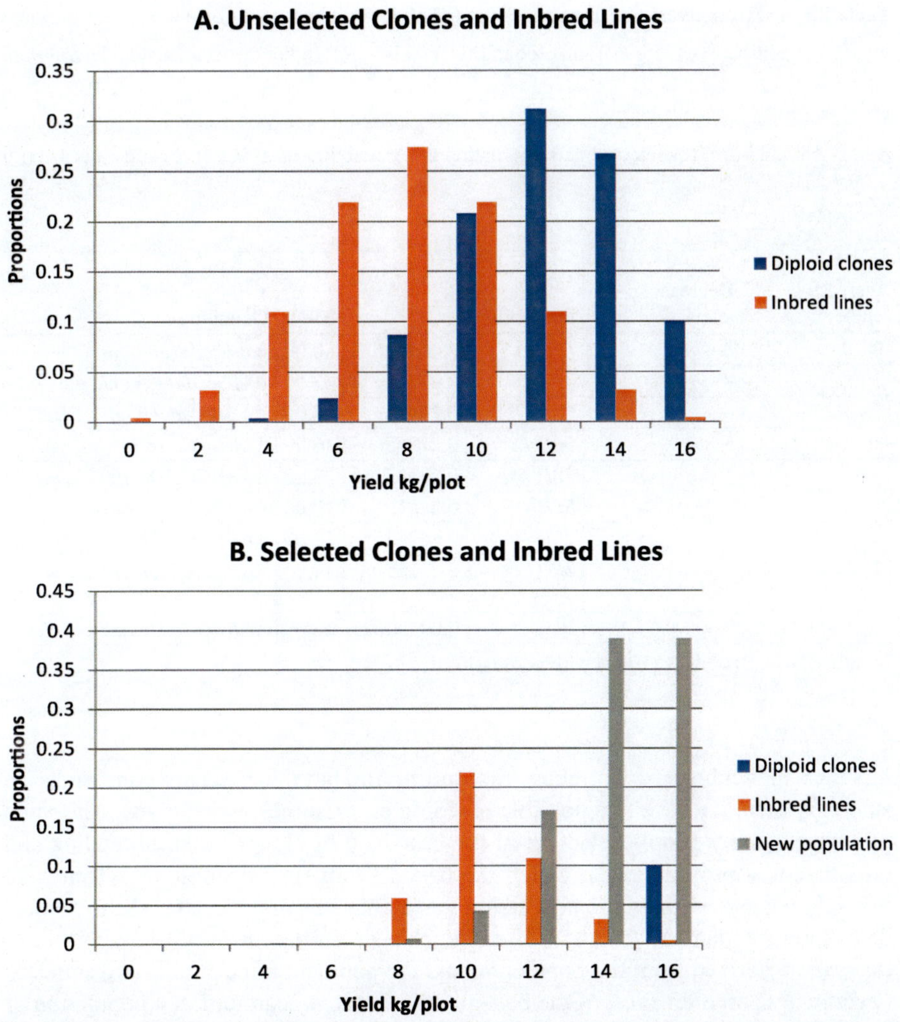

Fig. 8.1 Distribution (proportions) of genotypes in a diploid population of clones and the inbred lines derived from them for eight unlinked loci each of equal effect, each with two alleles at equal frequencies and complete dominance for increasing effect (e.g. yield): (**a**) Unselected, (**b**) Selected clones and inbred lines and new population formed from either crossing clones or inbred lines

population so that each locus contributes 0.278 (1.778–1.5) to the increase in population mean, making a total of 2.22 (8×0.278).

However, if we form the next generation by selecting the best 42% (the proportion 27/64 to be exact) of the inbred lines, we get exactly the same new population because we have increased the allele frequencies by exactly the same amount (the frequency of *A* in each of the nine groups of inbred lines in Table 8.1 increases from 0 to 1 in increments of $\frac{1}{8}$). The variance of the new population will in practice be

Table 8.1 Proportions of true breeding lines for eight unlinked genes of equal effect on yield where probability of *AA* is $\frac{1}{2}$ (*P*) and probability of *aa* is $\frac{1}{2}$ (*Q*) (other distributions in Fig. 8.1 were generated as follows: for unselected clones the probability of *A*- is $\frac{3}{4}$ (*P*) and *aa* is $\frac{1}{4}$ (*Q*) where *A*- is $\frac{1}{3}$ *AA* and $\frac{2}{3}$ *Aa*; for selected new population the probability of *A*- is 8/9 (*P*) and *aa* is 1/9 (*Q*) where *A*- is $\frac{1}{3}$ *AA* and $\frac{2}{3}$ *Aa*

Increasing allele	Decreasing allele	Probability	$P = Q = \frac{1}{2}$	Yield	Genotype
0	8	Q^8	0.00390625	0.00	*aabbccddeeffgghh*
1	7	$8PQ^7$	0.03125	2.00	*AAbbccddeeffgghh*, etc.
2	6	$28P^2Q^6$	0.109375	4.00	*AABBccddeeffgghh*, etc.
3	5	$56P^3Q^5$	0.21875	6.00	*AABBCCddeeffgghh*, etc.
4	4	$70P^4Q^4$	0.2734375	8.00	*AABBCCDDeeffgghh*, etc.
5	3	$56P^5Q^3$	0.21875	10.00	*AABBCCDDEEffgghh*, etc.
6	2	$28P^6Q^2$	0.109375	12.00	*AABBCCDDEEFFgghh*, etc.
7	1	$8P^7Q$	0.03125	14.00	*AABBCCDDEEFFGGhh*, etc.
8	0	P^8	0.00390625	16.00	*AABBCCDDEEFFGGHH*

Modified from Bradshaw (2016) with permission

less than shown because of linkage disequilibrium, but this need not concern us. In summary then, we have two possible breeding programmes which give us the same progress, namely simple selection of the best 10% of clones or an inbreeding and crossbreeding programme in which the best 42% of inbred lines are selected. In Fig. 8.2, we can see how the population mean changes with the increase in favourable allele frequency. We can also see the changes in the mean of the inbred lines that can be derived from the population and the population mean plus two standard deviations as an indication of the best genotype that we can find in a population of modest size. The improvements in the population mean exploit the additive genetic variance in the population [$\Sigma 2pq\alpha^2$ where $\alpha = a + d(q-p)$] whereas the improvements in the best genotype exploit the additive and dominance [$\Sigma(2pqd)^2$] genetic variance, where summation is over loci (8 unlinked loci in our example).

Clearly, it can be argued that if we select the best 10% of inbred lines, we will make faster progress with this scheme than with the one selecting the best clones, but the cycle times in years may be different and this affects the progress per annum. Lindhout (2020) reported a 5-year cycle time for F_1 hybrid breeding (3 years for inbred line selection and 2 years for selection on hybrid performance) whereas Carroll (1982) reported a 3-year cycle time for his mass selection programme (crossing to produce seed, raising seedlings to produce tubers, assessing clones in field). However, the important point is that when designing diploid population improvement schemes, cycles of inbreeding with selection and crossbreeding, can

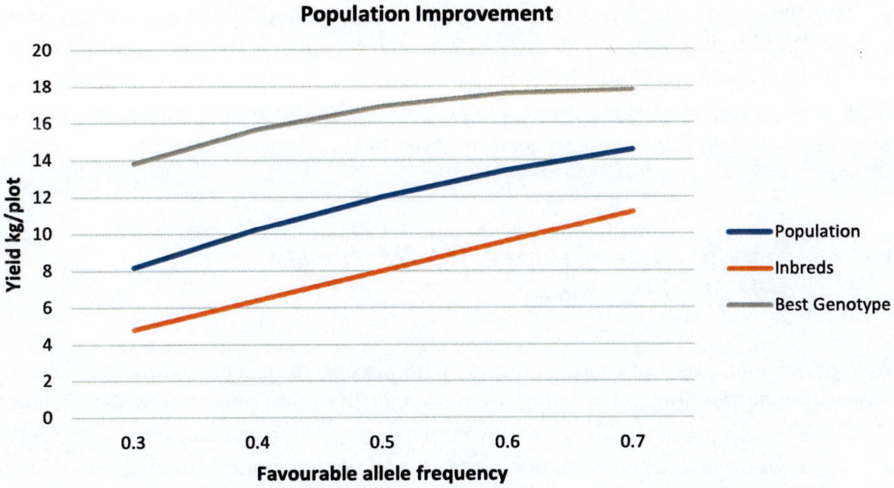

Fig. 8.2 Population improvement through change in favourable allele frequency (p): mean of inbreds = $\Sigma(p-q)a$, population mean = $\Sigma(p-q)a + \Sigma 2pqd$ and best genotype is population mean plus two standard deviations [$a = d = 1$ at all 8 loci and mean of inbred lines at $p = 0.5$ is set at 8; $a = ½(AA-aa)$ and $d = Aa - ½(AA + aa)$]

be included for consideration along with cycles of crossbreeding with selection of clones from between and within families. The theory, including various complications, can be found in books on plant breeding, including mine (Bradshaw 2016). Estimates from experiments of the genetical (additive and non-additive) and environmental components of variation allow the responses to selection to be predicted and compared and the best scheme to be chosen. When doing this it is important to take into account practical considerations, such as the time to complete one cycle and the ease of production of botanical seed and seed tubers which could affect the level of replication of test material and the intensity of selection that can be practised. As predicting the response to selection over a number of generations is difficult from theory, the latter can be complemented with computer simulations to guide the design of the breeding programme. Although the aim is to choose the best scheme for making progress, all schemes can be adjusted to result in the same increases in desirable alleles at many loci and hence the same rate of progress, which will depend on the additive genetic variance. A breeder can increase the rate of progress by increasing the intensity of selection (i) and to a certain extent the narrow-sense heritability (h_n^2) (replication to reduce environmental variation) for the trait(s) of interest, but without additive genetical variation (σ_A^2) there will be no progress at all. The breeders' equation for the response to mass selection (R) per cycle takes the form:

$R = ih_n\sigma_A = ir_{AP}\sigma_A$ where r_{AP} is the correlation between phenotypic and true breeding values

or in the twenty-first century,

$R = ir_{MA}\sigma_A$ where r_{MA} is the correlation between estimated genomic and true breeding values

The amount of additive genetic variation is determined by the starting material used in the breeding programme, so it is important to choose the right germplasm as well as the right breeding objectives. The breeder wants the starting population to have high means for the desired traits and high additive genetic variances as it is genes and not genotypes that are passed to the next generation.

8.4.2 *Genetic Basis of Inbreeding Depression in Diploid Potatoes*

Modern molecular genotyping combined with phenotyping is providing insight into inbreeding depression and opening up the possibility of targeted removal of deleterious alleles. Zhang et al. (2019), for example, explored the genetic basis of inbreeding depression in diploid potatoes using three potato clones (PG6226 = E and PG6235 = RH from Wageningen University, and PG6359 = C151 from CIP) from which selfed progenies were produced of sizes 104, 223 and 201, respectively; PG6235 and PG6359 being natural self-compatible mutants, while the selfing population of PG6226 was generated by forced self-pollination at the young bud stage. The researchers first evaluated the mutation burden in the three parental clones and 148 diploid landraces by genome sequencing and SNP calling followed by amino acid conservation modelling. They found that deleterious mutations in potato are enriched in the pericentromeric regions and are line specific, which means that there will be a high degree of heterosis when two inbred lines from different lineages are crossed. Then using the three F_2 populations, they identified 15 genomic regions with severe segregation distortions due to gametic (4) and zygotic (11) selection. Five regions contained recessive lethal mutations and four regions contained mutations affecting growth vigour. Most of the deleterious recessive alleles affecting survival and growth vigour were located in regions with high recombination rates, and hence can be removed by selection following appropriate hybridizations. In addition to the alleles with substantial deleterious effects, there were numerous mutations that had minor effects on fitness.

8.4.3 *The Importance of the Additive Genetic Variance*

It is worth pausing for a moment and going back to tetraploid breeding and the arguments during the second half of the twentieth century about the apparent lack of progress in potato breeding, particularly with regard to increasing yield. The problem was seen as a lack of additive genetic variance, albeit from different viewpoints with different solutions.

Simmonds (1969) attributed the lack of progress to the narrow genetic base of modern potato breeding and started the first Neotuberosum programme in 1959 with the aim of recreating long-day adapted potatoes from a wider gene pool of short-day adapted Andigena landraces held in the Commonwealth Potato Collection. Peloquin thought that there was a lack of additive compared with non-additive genetic variation and advocated breeding for maximum heterozygosity (Werner and Peloquin 1991; Bonierbale et al. 1993). However, conventional potato breeding was in effect based on the crop physiologists' view of actual and potential potato yields (Haverkort and Struik 2015). If cultivars are available which match their growth cycle from planting until crop maturity with the length of the available growing season, then there may not be much scope to increase potential yields. These high potential yields will be due to a high harvest index, adaptation to tubering in different day-lengths and in the temperatures of the growing season, and tuber initiation and rate of bulking matched to the available growing season. Under these circumstances, the need is to reduce the gap between potential and actual yields by providing farmers with disease-free (clean) planting material, adequate irrigation if rain is insufficient, adequate fertilizers and a means of controlling pests and diseases, including those that affect the planting material. Hence, breeding was concentrated on adaptation to the growing season and inbuilt resistance to pests and diseases, and more recently to the requirements of processors.

There is still a way to go in closing the yield gap. In 2016, actual yields typically ranged over climatic regions from 24% to 72% (globally about 50%) of potential yields, which varied according to the length of growing season, day-lengths and temperatures (Beukema and van der Waag 1990; Haverkort and Struik 2015). Nevertheless, in 2016, the average yield of potatoes on a dry-weight basis (3.92 t/ha) compared favourably with those of maize (4.79), wheat (2.90) and rice (3.94), for which so much has been made about genetic increases in yield. Indeed, averaged over 2017 and 2018 and corrected to a dry-weight basis, maize yields in the USA were 10.09 t/ha (11.87 × 0.85) and potato yields were 9.82 t/ha (49.08 × 0.20), an almost identical figure (http://faostat.fao.org). Perhaps maize breeders needed F_1 hybrids to get their yields up to those of potatoes! The interested reader can refer to Mueller et al. (2019) for an analysis of the simultaneous gains in grain yield and nitrogen efficiency over 70 years of maize genetic improvement. They compared in multiple N fertilizer treatments, seven maize hybrids from a single seed company which were commercialized between 1946 and 2015. The rate of yield gain was much greater when 220 kg N ha^{-1} was applied (0.12 Mg ha^{-1} $year^{-1}$) compared to when there was no fertilizer added (0.05 Mg ha^{-1} $year^{-1}$). There was an increase in kernel number primarily between 1946 and 1976 and in kernel weight from 1976 to 2015, accompanied by increases in harvest index. Modern hybrids produced more grain per unit of accumulated N by more efficiently remobilizing N stored in stems than in leaves to support kernel growth.

Let us finish this section with a point about getting diploid potato yields up to those of tetraploid potatoes. I am going to use a simple example where on average a population of diploids (clones or hybrids with same genotypes) has half the yield of a population of tetraploids (clones), with the best diploids having the same yields as the worst tetraploids, and where the two populations have the same variation (Fig. 8.3). Although the two populations are shown in the figure as binomial

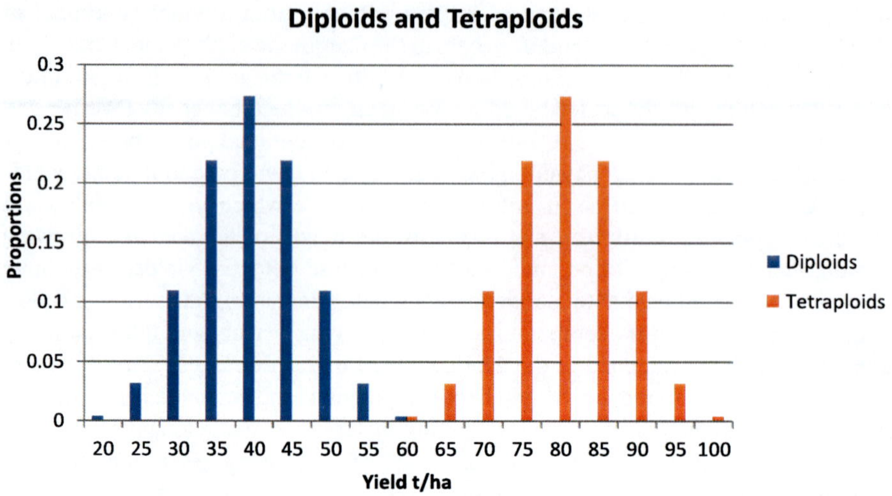

Fig. 8.3 Theoretical comparison of diploids and tetraploids where on average diploids have half the yield of the tetraploids, the same variation and the best diploids have the same yields as the worst tetraploids

distributions, I am going to assume that a large number of loci are segregating so that we have normal distributions. Let us start with the diploid population, and for simplicity assume mass selection of clones to avoid the complications of selfing. The population has a mean of 40 t/ha and a variance of 50. If we assume a narrow-sense heritability of $\frac{2}{3}$ (non-additive genetic variation but no environmental variation), the additive genetic variance is $33\frac{1}{3}$, and if we select the best 4.3% of diploids, the selection intensity is 2.121, and the response to selection each generation is:

$$R = ih_n\sigma_A = 2.121 \times \left(\frac{2}{3}\right)^{\frac{1}{2}} \times \left(33\frac{1}{3}\right)^{\frac{1}{2}}$$
$$= 2.121 \times 0.8165 \times 5.7735 = 10.00 \ \ \text{t/ha}$$

In other words, it will take four cycles (generations) of crossing and selection to increase the population mean of the diploids to that of the current tetraploids.

In practice, it may take fewer or more generations. We may be able to practise a higher intensity of selection, the heritability may be higher and there may be more additive genetic variance which we can better exploit through inbreeding, or we may be set back by genotype × environment interactions. Nevertheless, it should be obvious after four generations of crossing and selection what progress can be made, remembering that the more traits you try to improve, the less progress will be made with each trait.

In the meantime, cycles of crossing and selection may also have made progress at the tetraploid level. The response to clonal mass selection at the tetraploid level is also:

$R = ih_n\sigma_A$, but σ_A^2 in tetraploids is:
$\Sigma 4pq\alpha_t^2$ where $\alpha_t = [a + 3(q-p)d]$ with a equal to ¼($AAAA-aaaa$) compared with $\Sigma 2pq\alpha^2$ and $\alpha = [a + (q-p)d]$ with a equal to ½($AA-aa$) in diploids (d is appropriate digenic interaction).

The relative sizes of the additive genetic variances in diploid and tetraploid populations need to be determined by experiment, although it will be recalled from the real tetraploid example in Chap. 3 that the predicted increase per cycle was 11% (i.e. there appeared to be adequate additive variance for yield). Later in this chapter, we will see that Maris (1990) reported mean coefficients of variation for tuber yield, tuber numbers and mean tuber weight that were 18%, 20% and 14% higher, respectively, at the diploid level than at the tetraploid level.

Another way of looking at the difference between selection in diploids and tetraploids is through the change in allele frequency. For a single locus with two alleles *A* and *a* in equilibrium, for tetraploids we have:

Genotype	*AAAA*	*AAAa*	*AAaa*	*Aaaa*	*aaaa*
Frequency	p^4	$4p^3q$	$6p^2q^2$	$4pq^3$	q^4

And for diploids we have:

Genotype	*AA*	*Aa*	*aa*
Frequency	p^2	$2pq$	q^2

If we select half of the populations on superior phenotype under an additive model where all genotypes can be distinguished and a dominance model where we can only distinguish *A*--- and *aaaa* and *A*- and *aa*, respectively, we get the results shown in Table 8.2.

Although we are dealing with very simple models, it can be seen that the difference between diploids and tetraploids is not that great with an additive model at the chosen allele frequencies, but can be large with a dominance model where the superiority of the diploids in increasing the allele frequency is clear.

Table 8.2 Changes in frequency of allele *A* from different starting frequencies in diploids (2*x*) and tetraploids (4*x*) when 50% of population selected with additive (Add) and dominance (Dom) models

	4*x* Add	2*x* Add	4*x* Dom	2*x* Dom
Initial *A*	New *A*	New *A*	New *A*	New *A*
0.25	0.41	0.50	0.37	0.50
0.50	0.79	0.75	0.53	0.67
0.75	0.91	1.00	0.68	0.80

8.4.4 Diploid Inbreeding and Crossbreeding in Other Crops

There is no shortage of information on diploid inbreeding and crossbreeding from other crops. In my book on plant breeding (Bradshaw 2016), I covered maize, kale and the vegetables of *Brassica oleracea*, onions and carrots, rye and sugar beet, as well as the relevant theory which I will not repeat here. Two important points of relevance for diploid F_1 hybrid potatoes are as follows.

Firstly, too many inbred lines are produced to cross in all combinations so that selection is practised during inbreeding involving both line per se evaluation and selection and testcross evaluation and selection for general combining ability; the latter starting in the earliest generation where it is practically possible. The inbred line per se selection is required as the inbreds must have sufficient vigour and fertility to be grown as parents for the F_1 hybrid seed production. Inbred line per se selection is most effective on additively inherited traits with high heritability whereas testcross selection is more effective for traits which display directional dominance and have low heritability (e.g. yield).

Secondly, the breeder may be fortunate and find two populations which prove on crossing to be heterotic groups. In other words, group one by group two hybrids are superior to both group one by group one and group two by group two hybrids. Hence, on average, there is an advantage in making such hybrids. Simple theoretical considerations show that the additional heterosis gets larger as the allele frequencies diverge between the two populations and is at its maximum when different alleles are fixed in the two populations. It also increases with the extent of directional dominance and epistasis for increased trait value (e.g. yield). In these circumstances, breeding plans emphasize crossing inbred lines within heterotic groups in order to maintain favourable genetic combinations that enhance heterosis. Selection during inbreeding within each heterotic group is for combining ability with the other heterotic group, using a sample of bulked pollen or the best inbreds from that group. The establishment of heterotic groups saw the development of selection schemes designed to improve the cross between two populations from different groups, rather than the populations per se. The schemes are known as reciprocal recurrent selection (RRS). Selection of parents (usually their selfed progeny) for the next cycle of crossing within one population is based on their combining ability with the other population, and the response to selection of interest is in the cross of the two populations. The advantage of the method over those that select independently within each population occurs for those loci that display overdominance ($d > a$) or pseudo-overdominance from tight repulsion linkage (*Ab*/*aB*) of alleles of increasing effect. With RRS, the two eventual outcomes for two alleles at each of two loci, for example, are fixation of '*A*' in one population and '*a*' in the other, and likewise and independently (if unlinked) for '*B*' and '*b*'. In contrast, independent selection within each population would result in a stable equilibrium of the two alleles and a lower level of heterosis. At those loci which do not display overdominance (or pseudo-overdominance), the alleles of increasing effect would eventually be fixed in both populations, and this is the explanation for the continuing improvement of the inbred lines.

Interestingly, albeit at the tetraploid level, Bonierbale et al. (2020) reported that CIP have started a reciprocal recurrent selection scheme to benefit from intercrossing the best parents from two complementary populations, B3 (quantitative resistance to late blight) and LTVR (lowland tropic virus resistant), which were covered in Chap. 6.

8.5 Diploid Cultivar Production Plus Relevant Tetraploid Results

It is now time to turn our attention to finding and producing cultivars from our improved diploid population. The improved population can be viewed as a set of genotypes (clones or hybrids) which need to be assessed to find the best one(s) for multiplication (propagation) as a new cultivar(s). This can be done through the standard set of multilocational trials that breeders undertake for a few years (two or three), as described in the previous chapter. All of the genetic variation within the improved population, both additive and non-additive, is available in the selection process. Once the best genotype(s) is found it can be multiplied in either of two ways: directly as a clone or indirectly by crossing its two parental inbred lines to produce hybrid TPS.

The production of disease-free clonally propagated cultivars was dealt with in the last chapter. The starting material is disease-free microplants. These can be used in two ways. In the first they are used to produce minitubers either directly or via microtubers (induced in the axils of leaves of microplants), under protected net house conditions. In the second, they are not allowed to mature (produce compound leaves) and produce minitubers. Rather, they are the source of apical cuttings that can be rooted and subsequently transplanted to the field. The material for propagation by TPS is indeed true potato seed.

8.5.1 Production of True Potato Seed

Gopal and Ortiz (2006) reviewed the commercial production of TPS in tetraploid potatoes, emphasizing the importance of the parental genotypes flowering for sufficient time and producing berries with enough seed, with genotype, daylength (photoperiod 14–18 hours) and temperature (15–20 °C at night) the main determinants. In the tropics and subtropics, high altitudes (above 1500 m) during summer are conducive to flowering and fruiting. On average a berry has 100–200 seeds. The interested reader can also find information on factors affecting TPS quality, vigour, storability (dry seed) and dormancy (TPS is dormant at harvest) in the review by

Gopal and Ortiz (2006). Golmirzaie et al. (1994) mentioned the possibility of using protoplast fusion to produce a cybrid that can transfer cytoplasmic male sterility of the tetrad type from a donor clone into a recipient TPS parental line. Tetrad sterile pollen attracts bees and therefore natural cross-pollination is encouraged. Female and male rows of potatoes can be inter-planted in various combinations and berries harvested from the female rows. If required, hand pollinations can be done without emasculation in this type of material.

Lindhout et al. (2018) are producing diploid hybrid potato seeds mainly by hand pollinations. They report that the seed yield per plant varies from hundreds to many thousands of seeds. Each successful hand pollination produces a berry with 50–150 seeds and each plant produce 5–50 berries. The emasculation of the flowers is a time-consuming part of the hand pollinations. Hence, the use of male sterility, or self-incompatible sister inbred lines to the self-compatible ones, would reduce the costs of seed production. Lindhout et al. (2018) also point out that issues need to be resolved over the transport of potato seeds to other countries and the registration process for plant breeders' rights in the European Union.

8.5.2 *Production of Planting Material for Farmers*

Simmonds (1997) pointed out that methods of propagation fall naturally into two classes, those in which there is no pre-propagation before commercial production and those in which a multiplication phase intervenes. The first includes direct drilling of seed and also transplanting seedlings. Although direct drilling is common in other crops, transplanting proved the superior method and became the preferred one by about 1980, despite potato seedlings being small and rather delicate, and thus requiring a high level of horticultural skills. In other words, the preferred method was the one that potato breeders use, although pelleted and primed seeds are possibilities for the future. However, by about 1980 a multiplication phase looked even more attractive. Potato plants grown in dense nurseries at about 100 plants per square metre produced small but usable tuberlets (minitubers with average size of 10–15 g) which could be chitted before being planted. Only 750 kg of seed tubers of average size 15 g are required to plant 1 ha of potatoes (Gopal and Ortiz 2006). Later, during the 1980s, the idea of rapid multiplication of minitubers became established. Finally, Simmonds (1997) thought that socio-economics would determine the balance between what he called cleanish seedlings and infected tubers (clones). Chilver et al. (2005) pointed out that substituting seedling tubers for tuber seed is a relatively small step for most farmers who can cultivate seedling tubers without making significant changes in their production practices. They replace a clone with a family of individuals, but the mode of propagation is still vegetative.

Moving on to 2018, Lindhout et al. (2018) considered a number of ways to circumvent the exposure of weak seedlings to harsh conditions. Seedling tubers can be produced under greenhouse conditions by transplanting seedlings into appropriately sized pots. Alternatively, seedling tubers can be produced by aeroponics as described for clonal multiplication. These seedling tubers are equivalent to minitubers that are produced from microplants. However, they are cheaper to produce and give the flexibility of starting production where and when required. Furthermore, fewer rounds of multiplication are required to satisfy the commercial demand for planting material, ideally just one round and then release to commercial farmers. Under tropical conditions seeds can be sown in a simple greenhouse, or in the field under plastic cover with plant densities of 100 per square metre, and the resulting seedlings at 5–10 cm length can be transplanted directly at a suitable density into ridges in the field to produce seed tubers. The transplanting needs to be done in a frost-free season and with irrigation. Lindhout et al. (2018) did also consider the possibility of a new cropping system that would allow commercial ware production from transplants, citing results from Kenya as an early success.

8.6 The Return of Old Questions

In this chapter we have concentrated on diploid F_1 hybrid breeding for the production of potato cultivars that can be propagated from TPS because it is new and solves the problem of uniformity; but it does raise some old questions about diploids versus tetraploids and TPS versus vegetative propagation. We have shown that theoretically there is no difference between the diploid genotypes that can be produced by F_1 hybrid breeding and those produced in other types of diploid breeding. Although F_1 hybrid breeding allows a superior genotype to be multiplied from botanical seed, the farmer may be supplied with disease-free minitubers to plant in the same way as for a vegetatively propagated cultivar. Under these circumstances, the superiority of the two breeding methods and types of cultivar will depend upon the economics of the seed system for the supply of disease-free planting material. If we do reach the point where (pelleted) botanical seed can be directly drilled, then we are into the economics of two different farming methods, and diploid F_1 hybrid potatoes will find their favoured places around the world. We showed from genetical considerations in Chap. 2 that it is not possible to produce genetically uniform tetraploid cultivars for TPS propagation. Hence, if uniformity is essential, or preferred, the choice will be diploid F_1 hybrids for TPS propagation.

That leaves the question of how to compare diploid F_1 hybrids for TPS propagation with tetraploid cultivars for vegetative propagation from tubers. The answer to this question, I think, is indirectly. As just done, we compare TPS and vegetative propagation at the diploid level and then diploid and tetraploid cultivars produced by vegetative propagation. If tetraploid cultivars are higher yielding, then we have to consider the advantage of higher yield potential against the advantages of TPS per se. So, let us finish with the issue of the yield of diploids and tetraploids.

8.6.1 *Relative Yields of Diploids and Tetraploids*

As the first dihaploids of Tuberosum tended to be male sterile, many breeders and researchers worked with diploid hybrids between diploid Tuberosum (T) and diploid Phureja/Stenotomum (P). Hence, the few early papers on the relative yields of diploids and tetraploids took the following form.

P/T × P/T → P/P: 2 P/T: TT
↓ ↓
PP/TT PP/TT → PPPP: 8 PPPT: 18 PPTT (PPTT: 16 PTPT: TTPP): 8 PTTT: TTTT

In addition to the cross being made at the diploid level, the two parents would be treated with colchicine to double their chromosome number and the cross also made at the tetraploid level. The comparison of the parents (P/T versus PP/TT) would be used to judge the effect of ploidy per se, and the comparison of the offspring means would be used to judge the effect of allelic diversity (maximum heterozygosity). If the two diploid parents have four different alleles at a locus *A*, the diploid cross ($A_1A_2 \times A_3A_4$) would have equal frequencies of the following genotypes which are all heterozygotes with two alleles:

A_1A_3, A_1A_4, A_2A_3 and A_2A_4.

However, the progeny of the tetraploid cross ($A_1A_1A_2A_2 \times A_3A_3A_4A_4$) would have one each of $A_1A_1A_3A_3$. $A_1A_1A_4A_4$. $A_2A_2A_3A_3$. $A_2A_2A_4A_4$, four each of $A_1A_1A_3A_4$. $A_2A_2A_3A_4$. $A_1A_2A_3A_3$. $A_1A_2A_4A_4$ and 16 $A_1A_2A_3A_4$; in other words, just 4/36 genotypes with only two different alleles.

One of the most quoted and typical papers was that of Maris (1990) which was done at the University of Wisconsin, Sturgeon Bay, USA. It involved two primary crosses which were made at both the diploid and tetraploid level:

(Phureja PI 195198.16 × US-W 1) × (Phureja PI 225710.2 × US-W I) and
(Phureja PI 225695.1 × US-W 42) × (Phureja PI 195198.13 × US-W 1).

US-W 1 was a dihaploid of cultivar Katahdin and US-W42 was a dihaploid of cultivar Chippewa, and unusually both were male fertile. The two crosses at both ploidy levels provided 30 F_1 clones for evaluation and the production of further sib and backcross populations, a total of 12 families at each ploidy level. The second-generation clonal material was evaluated in 1969 in a trial with three replicates and two-plant plots of each clone (six plants for parents). Individual plants were assessed for 12 traits including yield per plant and specific gravity. The coefficient of variation (CV) of the diploid families was higher than that for the tetraploid families for all traits except eye depth. The mean CVs for tuber yield, tuber numbers and mean tuber weight were 18%, 20% and 14% higher, respectively, at the diploid level than at the tetraploid level. Hence, the response to selection will be greater at the diploid than at the tetraploid level. However, the tetraploid families generally had taller plants, later maturity (7 days later), fewer tubers (23.3 vs. 26.7 per plant), and higher tuber yield (855 vs. 659 g per plant, or 77% of former), mean tuber weight (40.9 vs. 28.1 g) and dry matter yield than their diploid counterparts with almost identical specific gravities. There was a positive correlation between the corresponding diploid and tetraploid families for nearly all traits, indicating that the best performing combinations at the diploid level would tend to be the best at the tetraploid level. In contrast to the families, however, the tetraploid parents were not superior to their diploid counterparts. In fact, for the two high-yielding diploids, the tetraploids were much lower yielding as a result of having less than half the number of tubers. Hence, these results support the hypothesis that the higher yield of tetraploids is due to a larger number of different alleles at each locus (maximum heterozygosity) rather than ploidy level per se. Maris (1990) also concluded from comparing his results with those in the literature that the best results can be expected of tetraploid hybrids from crosses of tetraploid cultivars with long-day adapted diploid species (Phureja/Stenotomum) that form $2n$ gametes by first division restitution (FDR).

The advent of molecular markers allowed further testing of the maximum heterozygosity hypothesis. Bonierbale et al. (1993), for example, using Tuberosum, Neotuberosum and Phureja germplasm from the Cornell University potato breeding programme, found that specific combinations of individual molecular-marker fragments were more important than maximum heterozygosity for tuber yield heterosis. Fragments whose segregation patterns showed significant associations with the measured yield characteristics corresponded to 26 mapped loci from seven of the 12 potato chromosomes (linkage groups 1, 2, 4, 5, 7, 9 and 12).

Stupar et al. (2007) looked at changes associated with autopolyploidization. They developed a synthetic autopolyploid series in potato (primarily Group Phureja) that included one monoploid ($1x$) clone, two diploid ($2x$) clones and one tetraploid ($4x$) clone. Interestingly, the diploid plants were the most vigorous and generated the greatest biomass with the monoploid inferior to both the diploids and the tetraploid. However, the diploid and tetraploid plants had similar gene expression patterns (about 9000 genes).

Taken together, these results suggest that the superiority of tetraploid potatoes comes from their genetic make-up, rather than polyploidy per se, and is not simply a matter of what has been termed maximum heterozygosity. Further research and breeding are required to determine if diploid cultivars can eventually be as good as tetraploid ones. Interestingly, in 2017 the Dutch breeding company Bejo (http://www.bejo.com/magazine/bejo-introduces-its-first-true-potato-seed-variety) introduced a tetraploid cultivar (Oliver F1) for TPS propagation in some African countries, clearly thinking that there is still a place for such tetraploids.

8.7 Conclusions

The principal cultivated potato is a tetraploid that is vegetatively propagated through tubers. Diploid potatoes are still cultivated in their place of origin in the Andes, as well as in other parts of the world such as Europe and North America, but are lower yielding than tetraploid potatoes. Likewise, some potatoes are seed (TPS) propagated using the tetraploid progenies of $4x \times 4x$ and $4x \times 2x$ crosses, mainly in the torrid zones of the lowland tropics and subtropics. The choice between the two methods of propagation was seen as balancing the difficulty of establishing a TPS crop, later maturation (15–20 days) and less uniformity against the benefits of cheap, easily stored and transported seed, flexibility of planting time and freedom from tuber-borne diseases. The advent of diploid F_1 hybrid cultivars for seed propagation has solved the problem of lack of genetic uniformity, and the production of minitubers for farmers to plant overcomes the problem of crop establishment, albeit once removed from actual seed or seedlings. Likewise, the production of minitubers and apical cuttings from disease free microplants has solved the problem of clean planting material for vegetative propagation, at least in the absence of extensive field multiplication. Hence, the choice of method of propagation is now a question of the relative ease and economics of producing clean planting material from seed or microplants and the relative yields of diploid and tetraploid potatoes. Four cycles (generations) of crossing and selecting diploid potatoes should reveal how quickly and to what extent their yield (and other traits) can be improved relative to tetraploid potatoes. A positive outcome could alter the extent to which diploid cultivars are grown and the extent to which they are seed-propagated rather than vegetatively propagated. Other conclusions are as follows:

1. *Diploid Inbred Lines*: These can be produced by self-pollination after the introduction of the dominant gene (*Sli*) from *S. chacoense* that inhibits gametophytic self-incompatibility, or after knocking out the self-incompatibility gene, *S-RNase*, using the CRISPR–Cas9 system. Monoploids have been produced by anther culture of diploid potatoes and then doubled-monoploids selected during *vitro* shoot regeneration, but they were male sterile although of varied female fertility. Thus, self-pollinations with selection is the current route to fertile inbred lines for use in hybrid breeding.
2. *Diploid Population Improvement*: Any of the population improvement schemes available for diploid outbreeding crops can be used for potatoes with good progress dependent on the presence of adequate additive genetic variation (variance). The choice of starting material is therefore of key importance, as well as choosing the right breeding objectives. The advantage of cycles of inbreeding and crossbreeding is that this method allows cultivars to be bred for both seed and vegetative propagation and is likely to be as good as, if not better, than other schemes for population improvement. It is the changes in favourable allele frequencies that drive the improvements in the inbred lines, the population mean and the best genotype that can realistically be found in an improved population.
3. *Diploid Inbreeding and Crossbreeding*: Large inbreeding programmes necessitate both line per se evaluation and selection and testcross evaluation and selection for general combining ability, the latter starting in the earliest generation that is practically possible, as done in maize. Where two populations exist and display additional heterosis in crosses between them, selection within each population should be for combining ability with the other population and the

response to selection of interest will be in the cross of the two populations (reciprocal recurrent selection).

4. *Yield Difference Between Diploids and Tetraploids*: The average fresh-weight yield of 36 diploid Phureja cultivars/clones was 65% of the average of 27 tetraploid Tuberosum cultivars/clones in 3 years of trials in Dundee, Scotland. The first published data on 65 diploid hybrids from the Solynta programme showed that on average their yield was 44% of that of 19 tetraploid cultivars in trials in one year at five sites, three in the Netherlands, one in Belgium and one in France. These figures give an indication of the yield improvement required in diploid potatoes. In tetraploids the aim of breeding is likely to continue to focus on factors that reduce yield potential, provided that currently grown cultivars are adapted to the available growing season. The superiority of tetraploid potatoes appears to come from their genetic make-up, rather than polyploidy per se, and is not simply a matter of what has been termed maximum heterozygosity. Whether or not diploid potatoes can eventually be as good as tetraploid ones is an open question.
5. *Production of TPS*: Expensive hand pollination is commonly used, but the use of either male sterility or self-incompatibility (sister lines or doubled-monoploids) should be considered.
6. *Production of Planting Material for Farmers*: Commercial growing of TPS cultivars could be through direct drilling of seed or transplanting of seedlings, but the production of seedling tubers from seedlings in pots or in aeroponics provides the farmer with familiar planting material. These seedling tubers are equivalent to the minitubers produced from microplants, but should be cheaper and more flexible to produce. Furthermore, it should be possible to satisfy commercial demand from one round of field multiplication.

In the next, and last, chapter we shall look at gene editing and genetic transformation as our final ways of genetically improving potatoes.

References

Almekinders CJM, Chilver AS, Renia HM (1996) Current status of the TPS technology in the world. Potato Res 39:289–303

Amoros W, Salas E, Hualla V, Burgos G, De Boeck B, Eyzaguirre R, zum Felde T, Bonierbale M (2020) Heritability and genetic gains for iron and zinc concentration in diploid potato. Crop Sci. https://doi.org/10.1002/csc2.20170

Beukema HP, van der Waag DE (1990) Introduction to potato production. Pudoc, Wageningen, pp 23–24

Bonierbale MW, Plaisted RL, Tanksley SD (1993) A test of the maximum herterozygosity hypothesis using molecular markers in tetraploid potatoes. Theor Appl Genet 86:481–491

Bonierbale MW, Amoros WR, Salas E, de Jong W (2020) Potato breeding. In: Campos H, Ortiz O (eds) The potato crop: its agricultural, nutritional and social contribution to humankind. Springer, pp 163–218

Bradshaw JE (2009) Potato breeding at the Scottish Plant Breeding Station and the Scottish Crop Research Institute: 1920–2008. Potato Res 52:141–172

Bradshaw JE (2016) Plant breeding: past, present and future. Springer, Cham

Burgos G, Amoros W, Morote M, Stangoulis J, Bonierbale M (2007) Iron and zinc concentration of native Andean potato cultivars from a human nutrition perspective. J Sci Food Agric 87:668–675. https://doi.org/10.1002/jsfa.2765

Carroll CP (1982) A mass-selection method for the acclimatization and improvement of edible diploid potatoes in the United Kingdom. J Agric Sci (Camb) 99:631–640

Chilver A, Walker TS, Khatana V, Fano R, Suherman R, Rizk A (2005) On-farm profitability and prospects for true potato seed (TPS). In: Razdan MK, Mattoo AK (eds) Genetic improvement of Solanaceous crops, volume I: potato. Science Publishers Inc, Enfield, pp 39–63

Chujoy E, Cabello R (2009) The canon of potato science: 29. True potato seed (TPS). Potato Res 50:323–325

Ghislain M, Andrade D, Rodriguez F, Hijmans RJ, Spooner DM (2006) Genetic analysis of the cultivated potato *Solanum tuberosum* L. Phureja Group using RAPDs and nuclear SSRs. Theor Appl Genet 113:1515–1527

Golmirzaie AM, Malagamba P, Pallais N (1994) Breeding potatoes based on true seed propagation. In: Bradshaw JE, Mackay GR (eds) Potato genetics. CAB International, Wallingford, pp 499–513

Gopal J, Ortiz R (2006) True potato seed. In: Gopal J, Khurana SMP (eds) Handbook of potato production, improvement and postharvest management. The Haworth Press, Binghampton, pp 557–585

Hanneman RE Jr (1985) Self fertility in *Solanum chacoense*. Am Potato J 62:428–429

Haverkort AJ, Struik PC (2015) Yield levels of potato crops: recent achievements and future prospects. Field Crop Res 182:76–85

Hawkes JG (1990) The potato: evolution, biodiversity & genetic resources. Belhaven Press, London

Haynes KG, Lu W (2005) Improvement at the diploid species level. In: Razdan MK, Mattoo AK (eds) Genetic improvement of Solanaceous crops, volume I: potato. Science Publishers Inc, Enfield, pp 101–114

Hosaka K, Hanneman RE (1998a) Genetics of self-compatibility in a self-incompatible wild diploid potato species *Solanum chacoense*. 1. Detection of an S-locus inhibitor (Sli) gene. Euphytica 99:191–197. https://doi.org/10.1023/A:1018353613431

Hosaka K, Hanneman RE (1998b) Genetics of self-compatibility in a self-incompatible wild diploid potato species *Solanum chacoense*. 2. Localization of an S-locus inhibitor (*Sli*) gene on the potato genome using DNA markers. Euphytica 103:265–271. https://doi.org/10.1023/A:1018380725160

Hougas RW, Peloquin SJ (1958) The potential of potato haploids in breeding and genetic research. Am Potato J 35:701–707

Hougas RW, Peloquin SJ, Ross RW (1958) Haploids of the common potato. J Hered 49:103–107

Hougas RW, Peloquin SJ, Gabert AC (1964) Effect of seed-parent and pollinator on the frequency of haploids in *Solanum tuberosum*. Crop Sci 4:593–595

Jansky SH, Charkowski AO, Douches DS, Gusmini G, Richael C, Bethke PC, Spooner DM, Novy RG, De Jong H, De Jong WS, Bamberg JB, Thompson AL, Bizimungu B, Holm DG, Brown CR, Haynes KG, Sathuvalli VR, Veilleux RE, Miller JC, Bradeen JM, Jiang J (2016) Reinventing potato as a diploid inbred line–based crop. Crop Sci 56:1412–1422. https://doi.org/10.2135/cropsci2015.12.0740

Lindhout P (2020) Diploid breeding: a paradigm shift in potato. In: Souvenir, global potato conclave 2020. Mahatma Mandir, Gandhinagar, Gujarat, India 28–31 January 2020, pp 146–148

Lindhout P, Meijer D, Schotte T, Hutten RCB, Visser RGF, van Eck HJ (2011) Towards F1 hybrid seed potato breeding. Potato Res 54:301–312

Lindhout P, de Vries M, ter Maat M, Ying S, Viquez-Zamora M, van Heusden S (2018) Hybrid potato breeding for improved varieties. In: Wang-Pruski G (ed) Achieving sustainable cultivation of potatoes. Burleigh Dodds, Cambridge, pp 99–122

Maris B (1990) Comparison of diploid and tetraploid potato families derived from *Solanum phureja* x dihaploid *S. tuberosum* hybrids and their vegetatively doubled counterparts. Euphytica 46:15–33

Mueller SM, Messina CD, Vyn TJ (2019) Simultaneous gains in grain yield and nitrogen efficiency over 70 years of maize genetic improvement. Sci Rep 9:9095. https://doi.org/10.1038/s41598-019-45485-5

Ortiz R (1997) Breeding for potato production from true seed. Plant Breed Abstr 67:1355–1360

Paget M, Amoros W, Salas E, Eyzaguirre R, Alspach P, Apiolaza L, Noble A, Bonierbale M (2014) Genetic evaluation of micronutrient traits in diploid potato from a base population of Andean Landrace Cultivars. Crop Sci 54:1949–1959

Paz MM, Veilleux RE (1997) Genetic diversity based on randomly amplified polymorphic DNA (RAPD) and its relationship with the performance of diploid potato hybrids. J Am Soc Hortic Sci 122(6):740–747

Paz MM, Veilleux RE (1999) Influence of culture medium and *in vitro* conditions on shoot regeneration in *Solanum phureja* monoploids and fertility of regenerated doubled monoploids. Plant Breed 118:53–57

Simmonds NW (1969) Prospects of potato improvement. Scottish Plant Breeding Station Forty-Eighth Annual Report 1968–69, pp 18–38

Simmonds NW (1997) A review of potato propagation by means of seed, as distinct from clonal propagation by tubers. Potato Res 40:191–214

Spooner DM, McLean K, Ramsay G, Waugh R, Bryan GJ (2005) A single domestication for potato based on multilocus amplified fragment length polymorphism genotyping. Proc Natl Acad Sci 102:14694–14699

Stockem J, de Vries M, van Nieuwenhuizen E, Lindhout P, Struik PC (2020) Contribution and stability of yield components of diploid hybrid potatoes. Potato Res 63(3):345–366

Stupar RM, Bhaskar PB, Yandell BS, Rensink WA, Hart AL, Ouyang S, Veilleux RE, Busse JS, Erhardt RJ, Buell CR, Jiang J (2007) Phenotypic and transcriptomic changes associated with potato autopolyploidization. Genetics 176:2055–2067

Veilleux RE (2005) Cell and tissue culture of potato (Solanaceae). In: Razdan MK, Mattoo AK (eds) Genetic improvement of Solanaceous crops volume I: potato. Science Publishers, Enfield, pp 185–208

Vries M, ter Maat M, Lindhout P (2016) The potential of hybrid potato for East-Africa. Open Agric 1(1):151–156. https://doi.org/10.1515/opag-2016-0020

Werner JE, Peloquin SJ (1991) Significance of allelic diversity and 2n gametes for approaching maximum heterozygosity in 4x potatoes. Euphytica 58:21–29

White PJ, Bradshaw JE, Brown LK, Dale MFB, Dupuy LX, George TS, Hammond JP, Subramanian NK, Thompson JA, Wishart J, Wright G (2018) Juvenile root vigour improves phosphorus use efficiency of potato. Plant Soil 432:45–63

Ye M, Peng Z, Tang D, Yang Z, Li D, Xu Y, Zhang C, Huang S (2018) Generation of self-compatible diploid potato by knockout of S-RNase. Nat Plant 4:651–654

Zhang C, Wang P, Tang D, Yang Z, Lu F, Qi J, Tawari NR, Shang Y, Li C, Huang S (2019) The genetic basis of inbreeding depression in potato. Nat Genet 51:374–378

Chapter 9
Gene Editing and Genetic Transformation of Potatoes

9.1 Introduction

The driving force for the genetic improvement of potatoes since the early 1900s has been cycles of hybridization and selection, with the new cultivars produced each cycle providing the means to transfer progress to farmers' fields. Desirable genes particularly for disease and pest resistance have been introgressed from landraces and wild relatives, and some of the parents used in the hybridizations have come from more general base broadening. In other words, potato breeders have primarily used sexual reproduction to generate new combinations of genetic variants on which to practise selection, the variants being the products of past mutations. Nevertheless, other genetic routes to crop improvement have been developed, some of which could be of increasing importance for potatoes. In particular, we need to consider gene editing and genetic transformation, but cannot avoid the issue of whether or not all of their products should be considered genetically modified (GM) potatoes and hence governed by the GM regulations, which vary from country to country. First, however, let us start with brief sections on mutation breeding and somaclonal variation.

9.1.1 Mutation Breeding

Mutation breeding dates from the work of Stadler with X-rays on maize and barley in 1928 (Stadler 1928a, b). Since the 1960s gamma rays have been the most widely used physical mutagen and account for half of all mutant cultivars (3088 in 190 species by 2009). Mutation breeding got a boost in 1964 with the establishment of research programmes by the joint FAO/IAEA Division of Nuclear Techniques in Food and Agriculture; IAEA being the International Atomic Energy Agency based in Vienna. FAO/IAEA replaced their *Mutation Breeding Manual* in 2012 with a new

J. E. Bradshaw, *Potato Breeding: Theory and Practice*,
https://doi.org/10.1007/978-3-030-64414-7_9

book *Plant Mutation Breeding and Biotechnology* edited by Shu et al. (2012). Important developments since the 1990s have made mutation breeding a more attractive proposition: (1) heavy-ion beam mutagenesis (since 1990s) for inducing DNA double-strand breaks which result in a limited number of null mutations (Kazama et al. 2011); (2) chemical mutagenesis coupled with a high-throughput screen of the mutagenized population to detect point mutations in target DNA sequences, with many examples of TILLING in the literature since 2003 (Bradshaw 2016), including the use of ethyl methanesulphonate (EMS) to produce an allelic series (missense mutations) with the possibility of finding an allele with a new function; and (3) site-directed DNA sequence modifications known as gene editing, genome editing and genome engineering (Mahfouz et al. 2016), with the CRISPR/Cas9 system first demonstrated in plants in 2013 (Bortesi and Fischer 2015). The last development is a major step forward because until then, for all practical purposes, mutagenesis could be considered to occur at random in the genome.

As a clonally propagated tetraploid crop, the potato is not considered an attractive proposition for deliberate mutation breeding, although limited success has been achieved; for example, selecting gamma-ray mutants of the cultivar Lemhi Russet for improved resistance to blackspot bruise and low-temperature sweetening (Love et al. 1993). However, if more breeding work is done at the diploid level, particularly with inbred lines, then mutation breeding may be used more often, particularly gene editing which can also be done at the tetraploid level. An account of mutation breeding can be found in my book on plant breeding (Chapter 16, Bradshaw 2016). In this chapter, I concentrate on gene editing.

9.1.2 Somaclonal Variation

Cell culture is a key technique in the use of biotechnology for crop improvement, especially for gene transformation. However, plants regenerated from cultured cells often show a high frequency of alterations in their phenotypes, particularly when there is a callus phase, a phenomenon known as somaclonal variation. Hence, there is a risk that regenerated plants will carry background mutations for undesirable characteristics that need to be screened out of the crop improvement programme. In contrast, there have been examples of desirable traits that could be used in plant breeding. After such variation was first reported in protoclones of potato cultivar Russet Burbank (Shepard et al. 1980), it was investigated as a breeding method in its own right. Desirable improvements on parent cultivars were reported and assessed under field conditions, but most somaclonal variation comprised undesirable changes that did not merit further breeding effort (Kumar 1994; Veilleux 2005). Hence, today somaclonal variation tends to be viewed as a source of undesirable variants that need to be screened out of breeding programmes that involve plant regeneration, including the in vitro stage of cultivar multiplication. A summary of the possible causes of somaclonal variation and methods of detection can be found in the review by Bairu et al. (2011).

9.2 Gene Editing

In gene editing, engineered nucleases are designed to catalyse a double-stranded break (DSB) at a specific location in the genome, and this stimulates the desired DNA modifications to occur at or near the break site. Such genome editing can therefore accelerate plant breeding by allowing the introduction of precise and predictable modifications directly into an elite cultivar background. A reference genome sequence, efficient genetic transformation and good functional genomics resources are all desirable and now available in potato with further improvements anticipated. The early reviews by Curtin et al. (2012), Lusser et al. (2012) and Bortesi and Fischer (2015) are still useful, and so is the EFSA scientific opinion addressing the safety assessment of plants developed using zinc finger nuclease 3 and other site-directed nucleases with similar function (EFSA 2012b). The following brief account of gene editing is taken from these sources.

9.2.1 *Engineered Nucleases*

Zinc finger nucleases (ZFNs), transcription activator-like effector nucleases (TALENs) and meganucleases (MNs) can all be regarded as types of site-directed nucleases (SDNs) in terms of their modes of action. Meganucleases are also called LAGLIDADG homing endonucleases (LHEs). In all SDNs, one domain of the protein is responsible for binding to a specific DNA-sequence whilst a second domain is responsible for the DNA cleavage at the binding site. ZFNs and TALENs are proteins which have been engineered to combine two such domains whereas MNs are naturally occurring proteins with both domains, but the latter has proved difficult to engineer to target novel DNA sites. Both ZFNs and TALENs are designed to work in pairs that bind two opposing DNA target sites separated by a space. Hence, they can recognize two DNA sequences of 9–18 bps (ZFNs) and 12–30 bps (TALENS), each separated by 5–7 bps. MNs are characterized by a recognition site of 12–40 bp. Depending on the level of specificity of the SDN used and presence/absence of sequences homologous to the recognition site in the genome, off-target modifications at other sites in the genome can be expected to occur. Until effective methods can be developed to introduce the SDNs themselves into plants, the gene encoding the SDN needs to be stably inserted by transformation or transiently expressed. Furthermore, inducible expression may be required if constitutive expression has undesirable off-target effects. TALENs have proved more successful and hence more popular than ZFNs.

The type II clustered regularly interspaced short palindromic repeat (CRISPR/Cas9) system from *Streptococcus pyogenes* was discovered as part of the adaptive immune system of bacteria and archaea, protecting them against invading nucleic acids such as viruses by cleaving the foreign DNA in a sequence-dependent manner. The system consists of a single monomeric protein (Cas9) and a chimeric guide

RNA (gRNA) which binds with the target DNA sequence. The chimeric RNA combines a CRISPR RNA (crRNA) for target specificity and a trans-activating CRISPR RNA (tracrRNA), which stabilizes the structure and activates Cas9 to cleave homologous double-stranded DNA sequences known as protospacers. Sequence specificity is conferred by a 20-nucleotide (nt) sequence in the gRNA of which the so-called seed sequence of approximately 12 nt is thought to be particularly important for the pairing between RNA and target DNA. The presence of a protospacer adjacent motif (PAM) directly downstream from the target DNA is a prerequisite for DNA cleavage by Cas9. The PAM is either a 5′-NGG-3′ sequence or less frequently a NAG sequence. Activity of the system is dependent on cell type and delivery method. Functional gRNA and Cas9 protein can be delivered directly into plant cells or cells can be transformed with gRNA and Cas9 transgenes on a separate chromosome to the targeted locus so that they can subsequently be removed by segregation. Multiple traits can be modified simultaneously because multiple gRNAs with different sequences can be used to achieve high-efficiency multiplex genome engineering at different loci simultaneously. Off-target effects appear to be rare and the careful selection of specific gRNA sequences should minimize the risk of such unwanted genome modifications. Hence, CRISPR/Cas9 is proving more popular than the other systems.

In a heterozygous genome, whether tetraploid or diploid, removal of the editing tools by crossing, whether a TALEN or CRISPR/Cas9, would result in a new cultivar rather than an edited version of the original cultivar. Hence, the editing tools must access the genome without integration of foreign DNA. Ghislain and Douches (2020) point out in their book chapter that this can be done by transient expression through PEG-mediated protoplast transfection or *Agrobacterium*-mediated leaf infiltration, but the absence of a selectable agent and somatic variation of plants regenerated from protoplasts can make these strategies labour-intensive. In contrast, in a homozygous genome such as those becoming available in diploid potatoes, the editing tools could be removed by crossing, which would be easiest if they had integrated into a different chromosome to the one targeted for editing.

9.2.2 Gene Knockouts Through Double-Strand Break Repair by Non-homologous End Joining

The outcome of gene editing depends on the type of natural DNA repair that occurs and this can be influenced by the presence of appropriate template DNA. The simplest form of targeted modification is non-homologous end joining (NHEJ) mediated gene knockouts.

DNA sequences near the ends of double-strand breaks (DSBs) are rapidly degraded, with the consequence that they expand into gaps that cannot simply be re-ligated to restore the original sequence. They are, however, re-joined end to end in what appears to be a random fashion. The process is sometimes referred to as 'illegitimate recombination', but is more commonly called non-homologous end joining (NHEJ). It is thought to be the main method of DSB repair in the vegetative tissues (including meristem initials) of higher plants. The processing of the DNA to produce ends suitable for ligation (re-joining) can result in short deletions or insertions and is the reason why NHEJ-based repair often results in mutations in the repaired DNA. Insertions or deletions introduced into the nuclease target sites (break sites), located in a gene's open reading frame, commonly cause frame shift mutations with loss of gene function (a gene knockout). More extensive chromosome rearrangements can also occur. Recent research indicates that at least four different end joining pathways operate in plants but only one is known in detail, and these details can be found in the review by Mannuss et al. (2012).

If the double-strand break occurs in just one of the homologous chromosomes, the result will be a heterozygous mutant. This plant(s) will need to be selfed and the segregating offspring genotyped to identify individuals that are homozygous for the mutation and also no longer carry the transgene for the site-directed nucleases if this was its delivery mechanism. However, in both a tetraploid and a diploid outbreeder, this would create a new cultivar as already explained. Whole-genome sequencing is required to confirm the absence of any transgene fragments in the resulting lines.

9.2.3 Gene Replacement or Editing

The aim is to modify a gene rather than eliminate its function by using the homologous recombination (HR) pathway to repair the nuclease-induced double-stranded break (DSB). Several alternative HR pathways may exist in plants, but there is good evidence for repair by synthesis-dependent strand annealing (SDSA) (Waterworth et al. 2011; Mannuss et al. 2012). Furthermore, if a double-strand break occurs between two repeated sequences in the same orientation, then single-strand annealing (SSA) can be used because the repeats provide sufficient homology for annealing. Adjacent to the break, single-stranded DNAs are created by 5′ digestion so that the repeated sequences can be used as complementary strands to anneal the ends of the break, after which non-homologous tails are detached and the nicks ligated. The genetic information between the two repeats is lost.

Gene replacement/editing is therefore achieved by supplying a DNA donor template that contains sequences homologous to those flanking the break site; albeit currently at a low frequency of about 1% as NHEJ still operates. This strategy is more complex than mutagenesis in that the donor template has to be delivered to the cell at the same time as the nuclease mediates cleavage. The HR pathway uses the homology in the donor template to repair the DSB, thereby incorporating the donor sequence into the chromosome. The donor sequence is designed to achieve the desired changes in the gene, either to the promoter or to the coding regions that alter the catalytic activity of an enzyme. In other words, a new allele is produced. The excitement surrounding the technique comes from the possibility of the cumulative addition of improved alleles to an elite cultivar without the need for sexual hybridization.

9.2.4 Gene Insertion

The aim is to introduce a transgene or transgenes at a specific chromosomal location through homologous recombination (HR). In practice, a recombinant DNA molecule is constructed in which the DNA fragment or the gene cassette of interest is sandwiched between stretches of DNA that are homologous with the DNA sequences flanking the DSB site. Sites of gene insertion are chosen that are conducive to high levels of gene expression. Furthermore, the insertion of multiple transgenes at the same chromosomal locus (stacking) makes it easy to move the transgenes into other germplasm by crossing, because the linked transgenes behave genetically as a single locus. EFSA (EFSA 2012b) concluded that the main difference between the SDN-3 (gene insertion) technique and transgenesis is that the insertion of DNA is targeted to a predefined region of the genome. Therefore, the SDN-3 technique can minimize hazards associated with the disruption of genes and/or regulatory elements in the recipient genome. Whilst the SDN-3 technique can induce off-target changes in the genome of the recipient plant, these would be fewer than those occurring with most mutagenesis techniques. Furthermore, where such changes occur, they would be of the same types as those produced by conventional breeding techniques.

9.2.5 Site-Directed Structural Changes

Although perhaps less common, a breeder might want to induce specific structural changes to the genome. A nuclease or a combination of two nucleases could be targeted to cleave neighbouring locations along a chromosome. Some of the breaks would be repaired by joining the respective outer breakpoints, thereby deleting the sequence separating the two cleavage sites. Alternatively, it is possible that the released DNA fragment would invert before DSB repair, thereby resulting in an inversion of sequence between the two cleavage sites. Similar methods can be used to stimulate chromosomal translocations if the two nucleases target sites on different chromosomes.

9.2.6 Potato Examples

At the time of writing, there are relatively few potato examples; for example, the review by Hameed et al. (2018) has 10 examples of knockouts, of which six use TALENS and four use CRISPR/Cas9. Hence, it is reassuring that gene editing of alleles and gene insertion, as well as gene knockouts, have been demonstrated in maize by Svitashev et al. (2015). The use of TALENs involved transgenic expression or protoplast delivery and the use of CRISPR/Cas9 was through *Agrobacterium*-mediated transformation of cells with CRISPR constructs (Hameed et al. 2018). Here, I am going to give just two examples which illustrate the effectiveness of the techniques in producing gene knockouts in all four alleles of tetraploid potatoes as well as the amount of work involved.

Clasen et al. (2016) used TALENs to knockout the vacuolar invertase gene (*VInv*) in the commercial potato cultivar Ranger Russet. The gene encodes a protein that breaks down sucrose to glucose and fructose, and the aim was to solve the problem of cold-induced sweetening in storage in which reducing sugars accumulate in tubers. When processed at high temperatures, reducing sugars form dark-pigmented products that are bitter and unacceptable to consumers, and also react with the free amino acid asparagine to form acrylamide, a potential cancer-causing agent. Due to the high levels of heterozygosity in the potato genome, the task of simultaneously targeting multiple alleles required careful TALEN design and optimization.

The researchers therefore used 454 pyrosequencing of exon 1 of *VInv* from cultivar Ranger Russet to assess allele diversity. TALENs were synthesized based on the consensus target sequences among the three *VInv* alleles found. Three TALEN pairs were tested for activity through three TALEN-encoding plasmids which were individually introduced into Ranger Russet protoplasts by polyethylene glycol (PEG)-mediated transformation. The protoplasts were sacrificed 48 h post-transformation and genomic DNA isolated to determine the frequency of mutations at the TALEN recognition site. The best TALEN pair (9.5% mutations) was selected for generating potato lines with knockout mutations in *VInv*. Ranger Russet protoplasts (100,000–200,000) were transformed with the TALEN-encoding plasmid and regenerated into whole plants. At the time that plantlets were transferred to rooting medium, a small leaf sample was removed and DNA was isolated for genotyping of the *VInv* locus. Following shoot and root formation, potato plantlets were transferred to soil. From a total of approximately 600 regenerated shoots, 18 plants were identified that likely harboured mutations in at least one *VInv* allele. Of the 18 candidate plants, six contained mutations in one *VInv* allele (2 with no TALEN DNA), five contained mutations in two *VInv* alleles (2 with no TALEN DNA), two contained mutations in three *VInv* alleles (1 with no TALEN DNA) and five contained mutations in all four *VInv* alleles (2 with no TALEN DNA). Tubers cold-stored at 4 °C for 14 days from full *VInv*-knockout plants had undetectable levels of reducing sugars (but raised sucrose) whereas a mutation in only one allele resulted in similar levels to wild-type. Lines with mutations in two or three alleles had intermediate values. The processed chips (crisps) from full *VInv*-knockout plants contained reduced levels of acrylamide (1820 μg/kg) compared with wild-type (6820 μg/kg) and were lightly rather than darkly coloured. Furthermore, seven of the 18 modified plant lines appeared to contain no TALEN DNA insertions in the potato genome (i.e. the plasmid DNA had not integrated into the plant genome). This means that they have no foreign DNA and hence are not different from naturally occurring mutations, the ultimate source of genetic variation used by breeders!

Yasumoto et al. (2020) infected potato cultivar Sassy with *Agrobacterium tumefaciens* harbouring a TALEN-expression vector targeting *sterol side chain reductase 2* (*SSR2*) gene (see later in this chapter) and regenerated shoots without selection. They obtained 7 out of 371 regenerated lines with a disrupted-*SSR2* gene but without the transgene of the TALEN gene, revealing that the disruption was caused by transient gene expression. Two of the seven lines had no intact *SSR2* sequences (complete knockouts) and very much lower levels of α-chaconine and α-solanine than the controls. The researchers refer to their technique as Agrobacterial mutagenesis.

Andersson et al. (2017) developed full gene knockouts of tetraploid potato using transient expression of the CRISPR-Cas9 system designed to target a *GBSS* gene. Granule-bound starch synthase (GBSS) is the single enzyme responsible for the synthesis of amylose and in potato, as in most other plants, the GBSS enzyme is encoded by a single locus (*GBSSI*) having four alleles in the cultivated potato. The reader is referred to the original paper for the design of the CRISPR-Cas9 constructs that took account of allelic variants in parts of the *GBSS* gene in cultivar Kuras. The researchers demonstrated that transient expression of the CRISPR-Cas9-system in protoplasts of cultivar Kuras yielded mutations in all four alleles of the gene in up to 2% of regenerated lines from a single PEG-mediated transfection with naked vector DNA. The resulting calluses were incubated in liquid media before being transferred to solid medium for shoot development. Whilst the majority of the viable shoots developed into plantlets indistinguishable from the parental phenotype, some of the regenerated shoots failed in elongation and did not develop into plantlets. High-resolution fragment analysis (HFRA) was used for screening of mutations for indels using PCR and capillary electrophoresis. Out of 129 lines identified as mutated by the HFRA analysis, 29 were subjected to genotyping to determine the exact size and location of the mutations. Most mutations resulted in small deletions of 1–10 bp, but there were six single-bp insertions, and DNA inserts of 34, 125 and 236 bp were found in three lines and shown to be random parts of vector DNA. Full knockout of GBSS enzyme activity was confirmed in four-allele mutated lines by phenotypic studies of starch which was confirmed as being of amylopectin quality by the red-brown staining with iodine. One remaining wild-type allele was shown sufficient to maintain enough GBSS enzyme activity to produce significant amounts of amylose. Subsequently, Johansen et al. (2019) improved CRISPR/Cas9 editing efficacies in the granule-bound starch synthase gene in tetraploid potato protoplasts through the use of endogenous potato *U6* promoters driving expression of the CRISPR component. Hence, the CRISPR-Cas9 system is an efficient way to fix a recessive mutant (achieve four copies) by site-directed mutagenesis (gene knockouts), but more research is required to see if it can be used for gene editing to create and fix novel alleles in potatoes.

9.3 History of Genetically Modified Potatoes

We are now going to put the clock back to before the use of site-directed nucleases and consider genetic transformation to produce what became known as genetically modified (GM) potatoes, which are also referred to as biotech crops or transgenic crops. The transformation process results in the insertion of one or more agriculturally important genes into the potato genome and thus allows targeted improvements of successful and widely grown cultivars. The first GM crops were planted in 1996 and by 2017 occupied 189.8 million hectares in 24 countries, which is over 10% of all cropland in the world (ISAAA 2017). However, four crops (soybean, maize, cotton and canola), two traits (herbicide tolerance and Bt insect resistance) and five countries (USA, Brazil, Argentina, Canada and India) still dominated. The story for potatoes is one of reluctance to accept GM.

Monsanto were the first to commercialize GM potatoes, all processing cultivars under the trade name NewLeaf™ in North America from 1995. NewLeaf™ cultivars had resistance to the potentially devastating Colorado potato beetle (CPB, *Leptinotarsa decemlineata*) whose larvae feed on potato foliage. These were followed in 1997 by NewLeaf™ Plus with combined resistance to CPB and *Potato leafroll virus* (PLRV) and in 1999 by NewLeaf™ Y with combined resistance to CPB and *Potato virus Y* (PVY). The US Department of Agriculture (USDA), the US Environmental Protection Agency (EPA) and the US Food and Drug Administration (FDA) approved and de-regulated the products based on substantial equivalence to the untransformed cultivars. Nevertheless, leading processing and fast food outlet companies in North America were reluctant to purchase GM potatoes, because of consumer concerns over GM potato products, and Monsanto stopped marketing them in 2001.

There was also considerable resistance to GM potatoes in the European Union. The first GM potato for non-food use to be cultivated in Europe was expected to be Amflora. It was a pure amylopectin starch potato developed by BASF Plant Science by switching off the gene for granule-bound starch synthase (GBSS), the key enzyme for the synthesis of amylose. However, in 2009, there were delays in its approval for commercial production because of concerns over the antibiotic resistance marker which it contains. Although the EU Commission did then give authorization, Amflora GM potato was banned by the EU General Court in 2013. However, BASF had already stopped selling Amflora potatoes in the EU because it was struggling to gain a market share amid widespread popular and political resistance. BASF had also developed a late blight (*Phytophthora infestans*) resistant potato, Fortuna, by using genetic transformation to introduce two resistance genes from the wild tuber-bearing Mexican species *Solanum bulbocastanum*. BASF applied on 31 October 2011 for EU approval for Fortuna for both cultivation and for food and feed use, but announced on 16 January 2012 that it was stopping all of its GM crop developments targeted for the European market due to lack of acceptance of GM technology in Europe. Early in 2013, BASF further announced that the company was no longer seeking to introduce Fortuna onto the European market. BASF Plant Science relocated from Germany to the USA.

Innate™ potatoes produced by the J.R. Simplot Company in the USA are the latest attempt to establish GM potatoes for human consumption in North America, but it is not yet clear if they will be acceptable to companies such as McDonald's. Simplot's Generation 1 Innate™ potatoes were deregulated by the USDA on 7 November 2014 and then voluntarily reviewed by the FDA who concluded on 20 March 2015 that they were as safe and nutritious as their conventional counterparts. They are processing cultivars (Russet Burbank, Ranger Russet, Atlantic and Snowden) with resistance to black spot bruise and lower levels of asparagine in their tubers which results in greatly reduced production of acrylamide, a potential carcinogen. Then on 13 January 2016, the FDA completed its food and feed safety assessment of Simplot's Generation 2 Innate™ potatoes (Russet Burbank, Ranger Russet and Atlantic) which have resistance to late blight as well as lower levels of reducing sugars which contribute to a reduction in acrylamide whilst enhancing cold storage capacity. The blight resistance was developed in collaboration with the John Innes Centre in the UK. In March 2017, the EPA gave approval for planting. The ISAA Brief 53 (ISAAA 2017) reported that Generation 1 potatoes were planted on 160, 2500 and 1618 ha in 2015, 2016 and 2017, respectively, and that Generation 2 potatoes were planted on 809 ha for the first time in 2017. The Generation 1 and 2 potatoes have now been approved for growing in Canada by the Canadian Food Inspection Agency and Health Canada. We shall look at the Simplot programme in more detail in a later section, but now it is time to look at the regulations covering GM crops.

9.4 Regulations Governing GM Crops, Food and Feed

Regulations vary according to country and breeders and biotechnologists will need to be familiar with those governing the target environments of their proposed GM crops as well as those governing their research and development in the laboratory, glasshouse and field. For example, in the USA and Canada, no separate or additional regulatory approval is necessary for commercializing hybrid stacks produced from already approved GM plants, whereas in the EU and Japan stacks are considered new events which require an assessment of additional risks that could arise from the combined genes. Here, we will consider two contrasting sets of regulations, the very stringent ones for the European Union (EU) and the less stringent ones for the USA which focus on the product and not the technology used for production. An issue in risk assessment has been whether or not the 'precautionary principle' should apply to GM crops. Some argue that the complexity and novelty associated with GM technology mean that the only concern should be appropriate measures to anticipate and prevent harm whilst others think that the consequences of not growing a GM crop should also be considered; for example, a GM crop with improved disease resistance or improved nutritional value.

9.4.1 EU Legislation

EU Directive 2001/18/EC of the European Parliament and of the Council of Ministers of 12 March 2001 deals with the deliberate release into the environment of genetically modified organisms and replaced Council Directive 90/220/EEC. It sets out procedures for considering applications to release GMOs (genetically modified organisms) into the environment for research or commercial purposes. The current definition of genetic modification in the European Union (EU) includes the techniques of inducing mutations and fusing protoplasts, but in Annex 1b of the EU Directive 2001/18/EC, both techniques have been exempted. Given that inducing mutations has been exempted, it would seem logical to exempt gene editing that does not involve insertion of genes, but this had not happened by the time of writing.

Regulation (EC) No 1829/2003 of the European Parliament and of the Council of Ministers of 22 September 2003 deals with genetically modified food and feed. It creates a specific harmonized procedure for the scientific assessment and authorization of GM food and feed products. It also requires labelling of all GM food and feed which contains or consists of GMOs, is produced from GMOs (like glucose syrup from GM maize), or contains ingredients produced from GMOs (e.g. tomato paste and lecithin from GM soya, which is used as an emulsifier in chocolate bars).

Regulation (EC) No 1830/2003 of the European Parliament and of the Council of Ministers of 22 September 2003 concerns the traceability and labelling of genetically modified organisms and the traceability of food and feed products produced from genetically modified organisms and amends Directive 2001/18/EC. It sets out EU requirements for a document audit trail to account for and identify approved GM products throughout the marketing chain, with the aim of facilitating accurate labelling.

The EU directives and regulations are implemented in member states through their own appropriate legislation. On 13 January 2015, members of the European Parliament passed a new law to allow EU member states from spring 2015 to restrict or ban the cultivation of GM crops in their own territory, even if this is allowed at EU level, thus allowing greater flexibility within the EU. The European Food Safety Authority (EFSA) does scientific evaluations of potential GM crops and products for the European Commission and has produced useful guidance on the EU regulations. The guidance is rigorous and as comprehensive as possible. The European Commission (EC) is the executive body of the European Union responsible for proposing legislation, implementing decisions, upholding the EU treaties and managing the day-to-day business of the EU.

The 2010 EFSA document (EFSA 2010) produced by the GMO panel provides guidance for the environmental risk assessment (ERA) of genetically modified (GM) plants submitted within the framework of Regulation (EC) No. 1829/2003 on GM food and feed or under Directive 2001/18/EC on the deliberate release into the environment of genetically modified organisms (GMOs). In particular, it provides guidance to risk assessors for assessing potential effects of releasing GM plants into the environment and the rationales for data requirements in order to complete a comprehensive ERA, and to draw conclusions for post-market environmental monitoring (PMEM).

The 2011 EFSA document (EFSA 2011) produced by the GMO panel provides updated guidance for the risk assessment of food and feed containing, consisting or produced from genetically modified (GM) plants, submitted within the framework of Regulation (EC) No 1829/2003 on GM food and feed. The proposed risk assessment strategy for GM plants and derived food and feed seeks to deploy appropriate approaches to compare GM plants and derived food and feed with their respective comparators. The underlying assumption of this comparative approach is that traditionally cultivated crops have gained a history of safe use for consumers and/or domesticated animals. More generally, the safety of novel foods and food products has been assessed in terms of whether the novel food product is substantially equivalent to an existing analogous food product, but the application of substantial equivalence to GM foods has been criticized. The substantially equivalent approach was strengthened in the new regulations (1829/2003) by requiring more evidence. The EFSA document provides guidance on how to perform the comparative analysis of the relevant characteristics of the GM plant. In addition, every section of the document addresses specifically the requirements for GM plants containing a combination of transformation events, providing guidance on how to establish that the combination is stable and that no interactions occur between the events that may raise safety concerns.

9.4.2 USA Legislation

In the USA, regulations are implemented by three federal agencies: the US Department of Agriculture (USDA), the US Environmental Protection Agency (EPA) and the US Food and Drug Administration (FDA). The nature of the GM modification and its intended use determine which agency or agencies are involved in its regulation. Some US states require additional regulation of GM crops.

The Animal and Plant Health Inspection Service (APHIS) of USDA, through the Federal Plant Pest Act, determines whether a transgenic crop is likely to be an agricultural or environmental pest. Field tests of GM crops can often be conducted if APHIS is simply notified and certain criteria regarding safety of the GM crop are met, provided the applicant meets certain standards to ensure biological confinement. Otherwise, a more detailed and involved process is required to obtain a permit. On the other hand, GM crops for commercialization can be granted non-regulated status by APHIS after presentation of extensive data on the nature of the genetic modification and its effects on the crop and the wider ecosystem. The GM cultivar and any offspring can then be released in the USA without APHIS review, but APHIS can prevent the continued use of a GM crop if it is considered a risk of becoming a pest.

The EPA, through the Federal Insecticide, Fungicide and Rodenticide Act, is responsible for ensuring the safety of pesticides including those produced biologically such as plants genetically engineered to be pest resistant. The EPA regulates the testing and commercialization of pest-resistant GM crops and sets tolerances for herbicide residues on herbicide-resistant crops.

The FDA, through the Federal Food, Drug and Cosmetic Act, regulates foods and animal feed derived from GM crops. Foods from GM crops have to meet the same safety standards as all other foods, but do not need FDA pre-market approval if their GM products are substantially equivalent to those in other foods.

9.4.3 Cultivation of GM Potatoes in Centre of Origin and Diversity

The cultivation of GM cultivars in the centre of origin and diversity of the potato has long been strongly opposed, partly based on cultural values and partly because of concerns that transgene flow into landraces would alter the natural gene pool and diminish biodiversity. Numerous studies have demonstrated that gene flow can occur in potato between commercial potato cultivars and native landraces, and even wild species (Scurrah et al. 2008), but the fate of foreign genes and their impact on biodiversity is not so clear. Ghislain et al. (2015) have shown that Andean farmers can grow a modern cultivar in the same field as their native landraces and maintain the integrity of both. They screened 1671 DNA samples from more than 400 native cultivars (landraces) from three regions in the Peruvian Andes where the commercial, fertile cultivar Yungay and landraces have coexisted for 15–25 years. Simple sequence repeat (SSR) markers were used to identify putative hybrids based on allele sharing with those of Yungay. None of the landraces were hybrids derived from a successful gene-flow event between 'Yungay' and a native potato. The authors concluded that even when all conditions for gene flow are met, farmers have not unintentionally incorporated genes from the commercial variety 'Yungay' into their native potato germplasm. These results will probably not allay the fears of those opposed to GM crops, but they do provide some evidence for the view that Andean farmers could grow transgenic potatoes among their landraces without destroying the integrity of those landraces.

9.4.4 Summary of Issues

In summary, the regulations governing GM crops are more stringent than those for conventionally bred ones. Commercialization requires the demonstration that GM crops and products are safe to grow and eat or feed to animals. Regulations governing GM crops, food and feed, are also often designed to address any other public concerns and to allow farmer and consumer choice (labelling and traceability) over whether or not to grow GM crops and eat GM products. Nevertheless, politicians, pressure groups and the general public still need to be convinced that GM crops are of real benefit to farmers and consumers alike. There has also been concern that much of the work on GM crops has been done by multinational companies and the processes and products protected with patents, despite the emergence of some public–private partnerships designed to deliver GM crops to developing countries and poor farmers. In his book *Seeds of Science*: *Why We Got It So Wrong on GMOs*, environmentalist Mark Lynas (Lynas 2018) tells the story of how he changed his mind about GMOs. Having been one of the original GM field wreckers in the 1990s, in 2013 he apologized for having destroyed GM crops. Having been convinced of the importance of climate change by rigorous peer-reviewed science, he looked at the science of GMOs and to his surprise found that respected scientists were not anti-GMO, and changed his mind. The (US) National Academies of Sciences, Engineering and Medicine (2016) reviewed the experiences with GM crops over the last 20 years in terms of agronomic and environmental effects, human health effects and social and economic effects. You will find detailed analyses of all the available information, but no sweeping statements and generalizations. I think that it is fair to conclude that progress in crop improvement requires appropriate use of conventional breeding and genetic engineering jointly rather than in isolation. We need to consider when both could be used to enhance a crop trait, when only conventional breeding could be used because of the large number of genes involved, and when only genetic engineering could be used because the trait is not available through sexual hybridization. Future progress will come from the appropriate use of sexual reproduction in further conventional breeding and introgression; from mutation breeding, cisgenesis and gene editing; and from genetically modified crops. So, let us return to potatoes and see what genetic transformation has to offer.

9.5 Genetic Transformation

Agrobacterium-mediated and direct gene transfer methods have been developed for the stable transformation of plants. *Agrobacterium*-mediated transformation uses the natural ability of this bacterial pathogen of dicotyledonous (broad-leafed) plants to transfer genes into plant genomes. Direct gene transfer methods rely on the delivery of large amounts of naked DNA whilst the plant cell is transiently permeabilized. Particle bombardment (biolistics) is the major direct gene transfer method which has commonly been used for the transformation of monocotyledonous plants, particularly cereals. Other direct gene transfer methods are electroporation of plant cells and protoplasts, treatment of protoplasts with polyethylene glycol (PEG) and use of silicon carbide fibres to penetrate the cell wall and plasma membrane. *Agrobacterium*-mediated transformation has been the main method used for potatoes.

9.5.1 Agrobacterium-*Mediated Transformation Procedures*

Agrobacterium tumefaciens (*Rhizobium radiobacter*) is a soil-borne, Gram-negative, rod-shaped, motile bacterium that is the causative agent of crown-gall disease of crops such as apples and grapes. It is found around the roots of plants and normally survives on the nutrients they release. However, if the plant becomes wounded or damaged at the stem base, it can infect the plant at the wound site and cause disease symptoms. Crown-gall formation depends on the presence in *A. tumefaciens* of a Ti plasmid (tumour-inducing, $\approx$ 200 kb in size) whose T-DNA (transfer DNA, $\approx$ 20 kb in size) region is transferred from the bacterium into the plant cell and then into its nucleus, where it becomes integrated into the host's nuclear genome. The T-DNA has genes that encode proteins involved in hormone biosynthesis (auxin and cytokinin), the cause of cellular proliferation and gall formation (hence called oncogenes), and the production of opines (amino acid derivatives) and agropines (sugar derivatives) as a source of nutrients for the bacterium. Analysis showed that one or more copies of the T-DNA could be integrated into the host genome and that the insertions were bordered by small (24 bp) imperfect left- and right-border repeats which also border the T-DNA in the Ti plasmid. The borders in the plasmid are recognized by an enzyme complex and one of the enzymes (VirD2) produces single-stranded nicks in the plasmid DNA, and then becomes covalently attached to the 5′ end of the displaced strand. Repair synthesis replaces the displaced strand. The T-DNA strand and associated proteins pass through the nuclear pore and the T-DNA strand is converted to double-stranded form. Integration into the host genome at naturally occurring chromosome (double-stranded) breaks occurs by non-homologous end joining (NHEJ). As a consequence, there is no preference for integration of the donor genes into similar nucleotide sequences in the recipient genome; but the process may not be entirely at random. Different strains of *A. tumefaciens* exist and provide variations on this theme.

Procedures for *Agrobacterium*-mediated transformation were first published in 1984 by De Block et al. (1984) and Horsch et al. (1984). They were possible because the only features of the T-DNA required for integration into the host plant genome are the short border sequences; removal of the oncogene sequences enables plants to be regenerated from transformed plant tissue by manipulating the plant hormone composition of the medium; and the virulence genes in the plasmid responsible for the transfer of the T-DNA into the host plant cell function in *trans* (i.e. they could be on a separate plasmid). The techniques of molecular biology are used to convert the Ti plasmid into a vector for plant transformation. The details need not concern us here, but two points are worth making. Firstly, the initial isolation and cloning of the desired gene(s) and the genetic engineering of the plasmid are commonly done in *Escherichia coli* before transfer to *A. tumefaciens*. Hence, the plasmid must be able to replicate in both these bacteria. Secondly, the genes that are going to be integrated into the genome of the host plant may come from prokaryotes or non-plant eukaryotes. Hence, they made need to be given appropriate promoters and terminator sequences (to stop transcription) to ensure their expression in the host plant. Today, binary vectors are commonly used in which the transfer apparatus (the virulence genes) and the T-DNA are located on separate plasmids. The former plasmid is maintained in a suitable strain of *A. tumefaciens* whilst the latter plasmid is normally maintained in *E. coli* and has the gene of interest (transgene) cloned into it. This plasmid can then be transferred into the *A. tumefaciens* strain for the plant transformation. Advances are continually being made in the design of plant transformation vectors and the regeneration of transformed plants. Furthermore, numerous factors can influence the level of donor gene expression in the plant, but the three most important ones are the characteristics of the genetic construct introduced, the number of copies inserted and the site of insertion. Hence, it is important for potato breeders and biotechnologists to check the scientific literature for the methods currently being used for potatoes.

9.5.2 Agrobacterium-*Mediated Transformation of Potatoes*

A prerequisite for transformation is the isolation and cloning of the desired gene. This will probably include production of a cDNA library from the mRNA present in the cellular source of the desired gene, cloning the cDNA into plasmids that are maintained in colonies of *Escherichia coli*, screening the colonies for the desired gene, and further genetic engineering of the colony with the desired gene. However, the details need not concern us here.

Potato transformation using *Agrobacterium*-mediated systems was developed during the 1980s, first with *A. rhizogenes* (Ooms et al. 1986) and then more successfully with *A. tumefaciens* which today is the method of choice (Stiekema et al. 1988). The gene of interest would be incorporated into the bacterial plasmid along with a promoter and selectable marker, such as resistance to an antibiotic or herbicide; the bacterium would be co-cultured with freshly cut tuber discs or leaf or internode explants of the potato; and regeneration of shoots with the selectable marker would take place in plant tissue culture in the presence of the selectable agent to ensure selection of transformants. Regeneration could be a problem with some genotypes so the easily transformed cultivar Désirée was a frequent choice for research purposes (Davies 2002). Both Chang et al. (2002) and Morris et al. (2006) were able to simultaneously co-transform potatoes with two genes using a single selectable marker. The transgene constructs were multiplied separately in *A. tumefaciens* clones and cultures of the *Agrobacterium* were mixed and incubated with the potato internode explants. From 300 explants, Morris et al. (2006) generated 38 independent transformants of which 4 contained both transgenes. Clearly, this increases the speed and efficiency of transformation as a breeding method and makes it a more attractive proposition. The choice of promoter is important for gene expression. The CaMV 35S promoter has frequently been used for constitutive gene expression, but others have been developed for higher constitutive expression and for leaf or tuber expression (Douches and Grafius 2005). Selecting the best transgenic potato clones for commercialization required a lot of work; namely, seeking the best constructs and promoters and eliminating any undesirable transformants or somaclonal variants that arose from the transformation per se and in vitro regeneration, respectively. Davies (2002) recommended the production of several hundred independently transformed lines and full and effective field selection as normally done by plant breeders in their programmes. In the Monsanto NewLeaf™ Plus and NewLeaf™ Y programmes, for example, each started with about 3000 original clones from which six were finally selected for commercialization some 7 years later (Davies 2002).

9.5.3 Clean-Gene Technology and Gene Stacking

Initially, it was common practice to use antibiotic- or herbicide-resistant genes as markers for the selection of transformants. However, public concerns arose over the risk to human and animal health of using antibiotic resistance marker genes. Today, marker-free transformants can be produced, most easily by using transformation vectors without selection markers, followed by PCR analysis of regenerated plants for the presence of the transferred gene(s). We will consider such an example in the section on late blight resistant potatoes (Jo et al. 2014). However, if necessary, marker genes can be eliminated from the final plant by several techniques involving recombination during sexual hybridization (EFSA 2012a, b).

Initially, genetic transformation involved single genes, but these are now being combined (stacked) in new GM cultivars as well as the genes required for new biosynthetic capacity such as the production of provitamin A in Golden Potato. There are four basic strategies for doing this: (1) hybrid stacking in which a plant containing one or more transgenes is crossed with another plant containing other transgenes and the offspring selected for the presence of all the transgenes, and multi-stack hybrid production by iterative hybridizations which can be labour-intensive and time-consuming; (2) co-transformation in which a plant is transformed with two or more independent transgenes in separate gene constructs which are delivered to the plant simultaneously; (3) linked genes or multigene cassette transformation in which a plant is transformed with a single construct that contains two or more linked transgenes, with four to six considered the current limit; and (4) re-transformation in which a plant containing a transgene is transformed with other transgenes. The largest combination of genes considered to date in potatoes is eight. A new project based at the Sainsbury Laboratory in Norwich (http://www.tsl.ac.uk/news/new-potato-at-thesainsbury-laboratory/), for example, aims to stack eight genes in cultivar Maris Piper, three which confer resistance to late blight, two which confer resistance to nematodes and a further three which switch off the production of asparagine and reducing sugars and polyphenol oxidase. The result will be a disease and pest-resistant potato with desirable processing characteristics, otherwise almost identical to the most widely grown cultivar in the UK. In other words, major improvements will have been made to a cultivar that farmers and end users know how to handle. In future, transformation at a site-directed double-stranded break

(DSB) will be possible, rather than random insertion, and this will allow gene stacking at a single site to create a supergene that will be inherited as a Mendelian gene.

9.6 Cisgenic, Intragenic and Transgenic Potatoes

A distinction is made between transgenic, intragenic and cisgenic crops in general and potatoes in particular. The distinction lies in the source of the genes as cisgenesis and intragenesis make use of the same transformation techniques as transgenesis. Transgenic crops contain genes (transgenes) derived from organisms (e.g. bacteria and fungi) or other plant species that are not cross-compatible with the cultivated crop, whereas intragenic and cisgenic crops do not contain such genes. The following definitions have been used by the EFSA Panel on Genetically Modified Organisms (EFSA 2012a and online www.efsa.europa.eu/efsajournal). The additional comments have also been taken from this source.

'Cisgenesis is the genetic modification of a recipient organism with a gene from a crossable, sexually compatible, organism (same species or closely related species). This gene includes its introns and is flanked by its native promoter and terminator in the normal sense orientation. Cisgenic plants can harbour one or more cisgenes, but they do not contain any parts of transgenes or inserted foreign sequences. To produce cisgenic plants any suitable technique used for the production of transgenic organisms may be used. Genes must be isolated, cloned and transferred back into a recipient where stably integrated and expressed'. This definition implies the use of marker-free transformation. Sometimes, the term cisgenesis is also used to describe an *Agrobacterium*-mediated transfer of a gene from a crossable, sexually compatible plant where T-DNA borders may remain in the resulting organism after transformation (cisgenesis with T-DNA borders). The T-strand transferred from *Agrobacterium* to the plant normally contains three nucleotides from the right border (RB) repeat and up to 22 nucleotides from the left border (LB) repeat, but their deletion can occur during integration. Database similarity searches indicate that sequences similar to the 22-bp LB repeat can be found in plants. Hence, any hazards related to these sequences would not differ from those in conventional plant breeding. It is also possible for part of the vector backbone to remain attached to the border and to be co-inserted into the plant genome. This needs to be checked in any transformant as by definition cisgenic (and intragenic) plants must not contain vector backbone sequences of bacterial origin.

'Intragenesis is a genetic modification of a recipient organism that leads to a combination of different gene fragments from donor organism(s) of the same or a sexually compatible species as the recipient. These may be arranged in a sense or antisense orientation compared to their orientation in the donor organism. Intragenesis involves the insertion of a reorganized, full or partial coding region of a gene frequently combined with another promoter and/or terminator from a gene of the same species or a crossable species'. Again, the definition implies the use of marker-free transformation. Insertion has been done, in potatoes, using plant-derived transfer DNAs (P-DNAs) which resemble *Agrobacterium* T-DNA borders that support DNA transfer from *Agrobacterium* to plant cells (Rommens et al. 2007). Intragenesis includes the use of gene-silencing approaches such as RNA interference which introduces inverted DNA repeats.

Some researchers have argued that cisgenesis with only one or a few cisgenes should be exempted from the definition of genetic modification (Jacobsen 2007). This is because the transformation process creates a GM plant whose genetic content is within the gene pool used by classical plant breeding, as also happens with induced mutations and protoplast fusion, both of which are exempt. The EFSA Panel on Genetically Modified Organisms did a safety assessment of plants developed through cisgenesis and intragenesis and published their opinion in 2012 (EFSA 2012a and online www.efsa.europa.eu/efsajournal). The Panel concluded that similar hazards can be associated with cisgenic and conventionally bred plants, whilst novel hazards can be associated with intragenic and transgenic plants. For example, the insertion of DNA fragments into plant genomes will create new open reading frames (ORFs) which are new nucleotide sequences uninterrupted by the presence of a stop codon. Potentially new sequence combinations, and hence novel proteins, could arise in transgenic and intragenic plants which would not normally arise in conventionally bred and cisgenic plants. The Panel concluded that their previous *Guidance for risk assessment of food and feed from genetically modified plants* (EFSA 2011) and *Guidance on the environmental risk assessment of genetically modified plants* (EFSA 2010) are applicable for the evaluation of food and feed products derived from cisgenic and intragenic plants and for performing an environmental risk assessment and do not need to be developed further. The Panel envisaged that on a case-by-case basis lesser amounts of (transformation) event-specific data may be needed for the risk assessment. One further point made by The Panel and worth repeating here is as follows. Cisgenesis and intragenesis do not introduce additional genes associated with linkage drag in introgression breeding from wild relatives, and thus avoid introducing any unwanted traits and hazards associated with these other genes. In introgression breeding, any potential cultivars producing such undesirable phenotypes are removed by breeders during testing and selection.

9.7 Gene Silencing

Transgene-mediated gene silencing through RNA interference (RNAi) offers a directed way of achieving some desirable phenotypes for crop improvement by inactivating one or several specific genes. Such gene silencing mimics recessive loss of function mutations, but in a dominant way so that it can easily be used in polyploids such as potato, in F_1 hybrid cultivars, and also for silencing whole families of genes.

RNA interference (RNAi) and post-transcriptional gene silencing (PTGS) were first observed in plants when it was found that interactions between multiple foreign transgenes or between transgenes and endogenous genes led to the degradation of specific mRNAs. Details can be found in Slater et al. (2008). Briefly, the degradation was sequence-specific and resulted in the silencing of transgenes and endogenous genes. It was subsequently shown that RNA silencing and related gene-silencing phenomena depended on the production of small (21–26 nucleotides) double-stranded RNA molecules as part of an RNA degradation pathway. These small interfering RNAs (siRNAs) are produced from long, double-stranded RNA molecules which can arise from several different mechanisms. For example, in transgene-induced gene silencing the product of transcription from antisense transgenes can bind to the complementary mRNA from the sense-orientated endogenous gene. Whatever the mechanism, the long, double-stranded RNA is cleaved by endonucleases to yield multiple siRNA duplexes of two sizes and ultimately results in cleavage of target RNAs. Whilst unintended gene silencing presents problems for genetic modification that need to be overcome in vector design, vectors can also be deliberately designed for gene silencing. This is done by producing vectors that directly or indirectly lead to the production of double-stranded RNA. The simplest RNAi method is to produce antisense RNA by placing the sequence to be expressed in the reverse orientation with respect to the promoter. The promoter directs the production of antisense RNA which is complementary to the endogenous mRNA and hence they form double-stranded RNA. However, a more effective way of directing RNAi is by producing a vector in which the transgene sequence has been cloned as an inverted repeat with the repeat units separated by a spacer (usually an intron) sequence, and expression is driven by a strong promoter. In other words, the order is promoter, then sense sequence, then spacer, and finally antisense sequence. Transcription of the inverted repeat allows the two complementary regions to form a double-stranded hairpin RNA which is extremely efficient in RNAi-induced destruction of any homologous mRNA.

It is worth pointing out that microRNAs (miRNAs) are short (21–24 nucleotides) RNAs which are involved in the post-transcriptional gene silencing of endogenous RNAs. They are encoded by the plant genome and exhibit a high degree of complementarity to the endogenous mRNAs which they silence by cleavage. They are formed as duplexes in the nucleus and then exported to the cytoplasm where they are processed into a single strand that interacts with the complementary target mRNA. When discovered they were a novel class of RNA involved in regulating gene expression. All RNA silencing pathways are capable of directing the methylation of DNA and thereby silencing gene expression at the transcriptional level.

9.8 Potato Examples

It is now time to look at some potato examples. The list of potential targets for genetic transformation is steadily increasing but good examples of certain types of traits are currently missing, namely yield, photosynthetic efficiency and efficiencies of nutrient and water use. Some crop ecologists have even expressed doubts about biotechnology providing major improvements to these traits because any simple (major gene) solutions would already have been found by millions of years of natural selection (Denison 2012). Hence, it seems premature to try to provide a definitive list of potato traits for improvement by genetic transformation; rather, I am going to concentrate on examples of disease and pest resistance, improved processing quality and improved nutritional value. Even here I am not going to try to provide a comprehensive list of examples. Some that I included in a book chapter in 2007 (Bradshaw 2007) have been updated whilst others have been dropped as they do not provide examples of new ideas or approaches, and regrettably have not been commercialized. One topic not covered, but which has been explored in many crop species, is the expression of inducible promoters in response to various plant pathogens (strictly speaking in response to chemicals such as elicitors) that provides broad-spectrum, pathogen-inducible resistance. The interested reader is referred to a recent review by Ali and Kim (2019) on synthetic promoters (constitutive and tissue-specific, as well as inducible) where genetic components found in nature are rearranged.

9.8.1 Late Blight

Today, there is much interest in the *R*-genes found in potato wild relatives and their introduction into commercially successful cultivars by genetic transformation as well as their use in conventional breeding. A good example is the DuRPh project at Wageningen University in the Netherlands which was done from 2006 through to 2015. A detailed account of the project and the results can be found in the paper by Haverkort et al. (2016). The paper deals with cloning of *R*-genes, transformation of potato cultivars with *R*-genes, selection of cisgenic true-to-type events, resistance management and communication to society, and is well worth reading. The project provided the first scientific report on the production and functional evaluation of cisgenic *R*-gene stacking in different potato cultivars (Jo et al. 2014). As an example, from the project, the researchers used marker-free transformation of the cisgenes *Rpi-vnt1.1* (from *S. venturii* from Argentina) and *Rpi-sto1* (from *S. stoloniferum* from Mexico) in combination with PCR selection. Two hundred stem explants from each of three selected cultivars (Atlantic, Bintje and Potae9) were prepared and co-cultivated with an *A. tumefaciens* strain carrying only the two cisgenic *R*-genes between the T-DNA borders of a binary plasmid (the genetic engineering involved is described in the paper). From 1515 shoots collected and screened by PCR with *Rpi-vnt1.1* and *Rpi-sto1* primers, 27 PCR-positive shoots were selected. They came from different explants, indicating that they were independent transformation events. From eight of these (two from Atlantic, five from Bintje and one from Potae9), normal looking plants were produced that were vector backbone-free (vector backbone gene-specific PCR analysis) and in which both genes were functionally expressed (based on *Avrvnt1* and *Avrsto1* infiltration). The eight plants were resistant to all five *P. infestans* isolates tested (detached leaf tests). One of the cisgenic Atlantic plants was re-transformed with a third *R*-gene, *Rpi-chc1* (from

S. chacoense from Bolivia/Argentina/Paraguay/Uruguay). Out of 20 marker-free transformation events, four cisgenic events were selected that functionally expressed all three *R*-genes. In other words, a cisgenic cultivar with three *R*-genes had been produced. Eventually, transformation with three genes in one T-DNA was achieved (Haverkort et al. 2016). All three genes had broad-spectrum resistance in their own right. In field trials in Wageningen in 2014 and 2015, potato clones carrying stacked *R*-genes, including *Rpi-vnt1.1* and *Rpi-sto1*, were not sprayed with fungicide and did not become infected with *P. infestans*, unlike susceptible cultivar Désirée. Allowing 2–3 years of field trials, it takes a total of just 3–4 years to produce cisgenic late blight resistant potato cultivars which can be released for seed tuber multiplication, which is less than conventional introgression breeding.

Another good example of stacking three *R*-genes is provided by Ghislain et al. (2018) in the context of African highland potato production in Uganda, where high humidity and low temperatures of 14–25 °C during the two rainy seasons (October–January and March–June) favour the spread of *Phytophthora infestans*. Late blight (LB) is thought to cause about 15–30% annual yield loss in sub-Saharan Africa SSA, affecting mainly smallholder farmers. The LB population in SSA is remarkably uniform with lineage US-1 predominant before 2007 but subsequently being replaced by 2_A1 (KE-1); both A1 mating type. The researchers assembled three resistance genes [*Rpi-blb1* (*RB*) and *Rpi-blb2* from *Solanum bulbocastanum* and *Rpi-vnt1.1* from *S. venturii*] in the same transcription direction in a T-DNA vector, along with selectable markers [*nptII* (kanamycin resistance) and *aad*], and transferred the vector into *Agrobacterium tumefaciens* by electroporation. Then using *Agro*-infection they produced 331 and 149 transgenic events in cultivars Désirée and Victoria (known as Asante in Kenya), of which 26 (12 used in field trial) and 15 (1 used), respectively, met their criteria for use: absence of backbone vector sequences, presence of three *R*-genes, presence of left and right border sequences, single copy of *nptII* gene, and extreme resistance to LB in whole-plant bioassay. Although the T-DNA was large at 18,585 bp, transformation efficiencies were acceptable at 3.4% and 1.5% for Désirée and Victoria, respectively. Resistance to natural infection by *P. infestans* was assessed in a confined field trial (Kabale, south-western Uganda 1°15′19.0″S 29°56′34.3″E, 2200 masl) in a randomized complete block design with three replicates in each of three seasons. The disease usually started 30–45 days after planting and was observed only on non-transgenic materials in all three trials. Moderately resistant cultivar Cruza 148 reached 40–45% affected foliage, whereas cultivars Désirée and Victoria were dead 20 days after the disease had started. None of the transgenic plants had lesions caused by *P. infestans* at any time. All isolate samples collected during the three seasons (2015–2016) indicated complete displacement of US-1 and confirmed that the *P. infestans* strain present at the confined field trial belonged to the new aggressive strain 2_A1, which expresses all cognate avirulence effector genes and hence can be recognized by the *R*-genes. In the third season, it was possible to do a yield assessment 110 days after planting in unsprayed plots. Transgenic Désirée and transgenic Victoria had total yields of 29 and 45 t/ha, respectively, whereas their unsprayed non-transgenic equivalents had yields of <10 t/ha and did not produce any marketable tubers. The researchers concluded that their late blight resistant potato cultivars, which are the farmers' preferred cultivars, could be rapidly adopted and bring significant income to smallholder farmers in sub-Saharan Africa.

9.8.2 Viruses

Among Monsanto's NewLeaf™ potatoes were NewLeaf™ Plus (available 1997) and NewLeaf™ Y (available 1999) which combined resistance to the Colorado potato beetle (CPB) with *Potato leafroll virus* (PLRV) and *Potato virus Y* (PVY), respectively. The former was achieved using the constitutive Figwort Mosaic Virus (FMV) promoter within a construct designed to prevent virus replication, and the latter by using the FMV promoter to express the PVY coat protein gene. These types of resistance were mentioned in a review by Solomon-Blackburn and Barker (2001) of traditional and molecular approaches to breeding virus resistant potatoes, the latter including coat protein-mediated resistance to PVX, PVY and PLRV, movement protein-mediated resistance to PLRV at the RNA level, polymerase-mediated resistance that inhibits virus replication in PLRV, ribozyme-mediated resistance to PSTVd and post-transcriptional gene silencing of PVY protease gene. Solomon-Blackburn and Barker (2001) thought that the most efficient way to use transgenes in a breeding programme might be to insert a series of genes for different disease resistances/attributes in selected advanced clones nearing cultivar selection, thus allowing time for trialling after transformation. The interested reader is referred to their review for more details and also to the review by Thieme and Thieme (2005). An account of mechanisms of transgenic virus resistance can be found in Slater et al. (2008).

Now let us move on 12 years to a paper that considers transgenic resistance to more than one virus. Chung et al. (2013) generated transgenic potato plants of cultivar Vales Sovereign (susceptible to PVY and PLRV) expressing fused, tandem, 200 bp segments derived from the capsid protein coding sequences of potato virus Y (PVY strain O) and potato leafroll virus (PLRV Canadian isolate) as well as the cylindrical inclusion body coding sequences of potato virus A (PVA strain U) as inverted-repeat double-stranded RNAs, separated by an intron. Internodal stem segments of 1 cm, and in some cases, leaf disks of tissue culture plantlets of Vales Sovereign, were used for plant transformation with *Agrobacterium tumefaciens* harbouring engineered plasmids containing the cassette of viral segments. The orientation of the expressed double-stranded RNAs in the transgenic plants was either sense–intron–antisense or antisense–intron–sense RNAs. Molecular characterization confirmed that the transgenic plants contained transgenes expressing hairpin RNAs that were degraded in planta by the RNA silencing system into small RNAs of 21–24 nt. Four lines of these transgenic potato plants were assessed for resistance to infection by PVY-O, PLRV, or PVA, all transmitted by a natural vector, the green-peach aphid, *Myzus persicae*. Resistance was assessed by the absence of detectable virus accumulation in the foliage. All four transgenic potato lines tested showed

100% resistance to infection by either PVY-O or PVA, but variable resistance to infection by PLRV, ranging from 72% to 96% in different lines (untransformed controls showed 9%, 64% and 24% resistance, respectively). This was regardless of the orientation of the viral inserts in the construct used to generate the transgenic plants and the gene copy number of the transgene. Hence, the researchers had demonstrated the potential for using tandem, fused viral segments and the inverted-repeat expression system to achieve multiple virus resistance in potato to viruses transmitted by aphids.

9.8.3 Colorado Potato Beetle and Potato Tuber Moth

Resistance to the Colorado potato beetle (CPB, *Leptinotarsa decemlineata*) (Davies 2002; Duncan et al. 2002) and to the potato tuber moth (PTM, *Phthorimaea operculella* Zeller) (Mohammed et al. 2000; Davidson et al. 2004; Douches and Grafius 2005) have been achieved with genes that encode proteins from the bacterium *Bacillus thuringiensis*.

The following brief account of the use of *cry* endotoxin genes from *B. thuringiensis* (*Bt*) is taken from Slater et al. (2008) where more detailed information can be found. *B. thuringiensis* was discovered by Shigetane Ishiwatari in 1901 in diseased silkworms. The toxic effect of the bacteria on insect larvae was subsequently shown to be due to insecticidal crystal (Cry) proteins (δ-endotoxins) which form inclusion bodies during sporulation. After ingestion by an insect larva, the crystals dissolve in its midgut and bind to receptors in the epithelium cells, killing them as a consequence by osmotic lysis. The toxicity of the crystals to other animals including humans is extremely low. The various strains of *B. thuringiensis* produce a wide range of different Cry proteins which share a common active core comprising three domains. They are encoded by *cry* genes carried on plasmids. Individual Cry proteins are active against the larvae of particular classes of insect. Preparations of *B. thuringiensis* spores or isolated crystals have been used as 'organic' pesticides for over half a century and this use aided the rapid development of the *Bt*-strategy based on genetic modification. However, the prokaryotic gene sequences of the *cry* genes had to be modified to obtain high levels of stable expression in plants. The specificity of Cry proteins allowed the targeting of specific pests by particular transgenes.

Transgenic resistance to the Colorado potato beetle has been provided by gene *cry3A*, which encodes the Cry3A protein from the bacterium *Bacillus thuringiensis* var. *tenebrionis* using the constitutive CaMV 35S promoter (see review by Douches and Grafius 2005). The first GM cultivars with this *Bt* gene to be commercialized by Monsanto in North America from 1995 onward were Russet Burbank, Atlantic, Snowden and Superior (Duncan et al. 2002). Monsanto advised growers to plant 20% of their field area with a non-Bt-containing potato as a refuge for susceptible beetles and to rotate crops and fields in which transgenic potatoes were grown to extend the durability of resistance (Duncan et al. 2002).

Coombs et al. (2003) and Douches and Grafius (2005) presented results from field studies in Michigan and Long Island, NY in 1998, 1999 and 2000 that compared natural (leptine glycoalkaloids and glandular trichomes), engineered (*Bt-cry3A* and *Bt-cry5* transgenic potato clones) and combined [(*Bt-cry3A* or *Bt-cry5*) + glandular trichomes and *Bt-cry3A* + leptine glycoalkaloids] plant resistance mechanisms of potato for control of CPB. The field trials were a choice situation under natural infection with CPB. The researchers considered defoliation over time to be the best estimate of the host plant's resistance against CPB and used the RAUDC (relative area under the defoliation curve) as the measure of host plant resistance. The results showed that the *Bt-cry3A* transgenic clones, the high leptine clone (USDA8380-1) and the high foliar glycoalkaloid clone (ND5873-15) were most effective for controlling defoliation by CPB adults and larvae. The *Bt-cry5* clone (SPc5-G2) was not as effective as the *Bt-cry3A* transgenic clones ('Russet Burbank Newleaf' RBN15 and YGc3.1). The glandular trichome (NYL235-4) and *Bt-cry5* plus glandular trichome clones proved to be ineffective, which was surprising given the results mentioned earlier. Douches and Grafius (2005) concluded that the ability to integrate conventional breeding and transgenic strategies (including constitutively expressed plant-based proteins) may provide the best opportunity to build durable host resistance to insects such as CPB.

Transgenic resistance to the potato tuber moth has been provided by the *Bt* proteins encoded by the *cry5* (equivalent to *cry1Ia1*) (Mohammed et al. 2000; Douches et al. 2004) and *cry1Ac9* (Davidson et al. 2004) genes using the constitutive CaMV 35S promoter. Douches et al. (2004) evaluated their *Bt-cry1Ia1* potato lines for resistance to potato tuber moth in Egypt under natural infestations in field trials with four replicates of 10-plant, single-row plots (a row of a non-transgenic cultivar was planted between each plot). A total of 27 *Bt* transgenic potato lines from six different *Bt* constructs were evaluated over a 5-year period starting in 1997, but only the trial in 1999 gave clear-cut results, demonstrating the difficulty of doing such trials. The *Bt-cry1Ia1*-Spunta lines (Spunta-G2, Spunta-G3 and Spunta-6a3) were the most resistant with 99–100% of tubers free of damage (Spunta control 73% free). These same three lines were also over 90% free of tuber moth damage (Spunta control 0% free) after 100 undamaged tubers of each (from field trial in 2001) were placed for 3 months in a traditional, above ground, non-refrigerated potato storage building (a Nawhalla). Further laboratory and field evaluations of the three lines confirmed their resistance in the Pacific Northwest of the USA, where potato tuber moth was starting to become a problem (Douches et al. 2011). Davidson et al. (2004), in New Zealand, grew small field plots of 13 transgenic potato lines from five cultivars in each of three seasons to provide young and mature leaves and newly harvested and stored tubers for a larval growth analysis in the laboratory (tubers were punctured to facilitate access by larvae). They demonstrated that their transgenic potato lines exhibited stable resistance to larvae across field seasons, between affected plant organs and between plant organs of different ages. Despite all of this potential, to the best of my knowledge, transgenic resistance is yet to be deployed in farmers' fields.

Finally, let us consider a novel way to produce potatoes that are resistant to CPB. Zhang et al. (2015) found that transplastomic potato plants producing double-stranded RNAs (dsRNAs) targeted against the β-actin gene of CPB were protected from herbivory and were lethal to its larvae. It was known that dsRNAs targeted against essential genes can trigger a lethal RNA interference (RNAi) response in insect pests. However, when expressed from the nuclear genome, an endogenous plant RNAi pathway is able to process these dsRNAs into short interfering RNAs. In contrast, long dsRNAs can be stably produced in chloroplasts, a cellular compartment that appears to lack an RNAi machinery. Hence, when expressed from the chloroplast genome, the dsRNAs accumulated to as much as 0.4% of the total cellular RNA, and it was possible to engineer resistant plants. Another advantage of engineering the chloroplast is that most of the expression occurs in the leaves and not the tubers which are used for human consumption.

9.8.4 Nematodes

Ali et al. (2017) have provided a review of transgenic strategies for enhancement of nematode resistance in plants. Here I am going to consider two examples. Urwin et al. (2003) were able to demonstrate that constructs based on a cysteine proteinase inhibitor (cystatin) from sunflower and a variant of a rice cystatin conferred similar levels of resistance to a natural mixture of the cyst nematodes *Globodera pallida* and *G. rostochiensis*, as chicken egg white cystatin (CEWC) under the control of the CaMV35S promoter. The cystatin works by impairing digestion of dietary protein by the nematodes. The constructs were used to transform susceptible cultivar Désirée. In a field trial at the University of Leeds Field Station, the levels of resistance were similar to that provided by the natural partial resistance of cultivar Sante. Transformation of Sante and the partially resistant South American (Bolivia) cultivar Maria Huanca with CaMV35S/CEWC raised the status of both cultivars from partial to full resistance. The transgenic Maria Huanca plants were tested in containment trials with a *G. pallida* population from Toralapa, Cochabamba province, Bolivia. The plants were grown for 42 days on a 13–11 h light–dark cycle to mimic growing conditions at Toralapa over the growing season. Subsequent research demonstrated that genetically modified, nematode-resistant (GMNR) plants did not harm above ground organisms or soil microarthropods in potato fields (Green et al. 2012). Green et al. (2012) then went on to investigate the potential of a new transgenic trait for control of *G. pallida* that suppresses root invasion. The new transgenic potato plants secreted a non-lethal disulphide-constrained peptide (nAChRbp) from their roots that disrupts the chemoreception that cyst nematodes require to locate host plants by binding to their acetylcholine receptors. Furthermore, expression of the peptide was restricted to root tips by using a tissue-specific promoter from the *Arabidopsis MDK4–20* gene to enhance biosafety. Cultivar Désirée was transformed with the MDK-peptide construct. Three out of 40 MDK-peptide potato lines were selected for further evaluation on the basis of their level of active peptide secretion and normal growth phenotype. The researchers established that the new transgenic trait provided resistance to *G. pallida* in both a containment glasshouse and a field trial, with the most effective line providing 77% resistance. In addition, molecular (DNA barcoding and quantitative PCR) nematode faunal analysis established that the peptide and the cystatin that offer distinct bases for transgenic plant resistance to *G. pallida* do so without impact on the non-target nematode soil community. Hence, the researchers concluded that transgenic potatoes for potato cyst nematode control can replace pesticide use without impact on soil quality and that combining the two types of resistance was expected to give about 95% control. However, their final conclusion was that European Union (EU) regulations (Directive 2001/18/EC) discourage small-scale trials because of the detailed data and risk assessments they demand and hence hinder the development of transgenic approaches that could help conserve the soil quality of productive arable fields in the EU.

My second example is another one involving RNA interference. Dinh et al. (2015) were able to generate broad-spectrum resistance to root-knot nematodes (*Meloidogyne* spp.) based on an RNA interference (RNAi) transgene capable of silencing the *16D10 Meloidogyne* effector gene. Endogenous nematode genes can be silenced by expressing double-stranded RNA (dsRNA) that is complementary to nematode genes in planta. Since root-knot nematodes are able to ingest molecules of up to 140 kDa, there is an opportunity for oral delivery of dsRNA or small interfering RNA (siRNA) into the nematode. During feeding, the nematodes take up plant cytoplasm containing dsRNA or siRNA that may then silence the respective target gene in the parasite. Nematode effector genes represent a very specific and therefore attractive target. Effector genes are expressed in the oesophageal gland cells of plant–parasitic nematodes and form the basis for the molecular interactions between the parasite and its host. The *16D10* effector gene is 95% to 98% identical between *M. arenaria*, *M. hapla*, *M. incognita* and *M. javanica*, but the *M. chitwoodi* orthologue is only 70% identical to *M. incognita*. Importantly, however, all orthologues found so far carry a highly conserved region of about 21 nt that is targeted by the 16D10i-2 silencing construct. The researchers transformed potato stem segments (about 1 cm in length) of cultivar Russet Burbank by coincubation with *Agrobacterium tumefaciens* strain GV3101 carrying silencing vector pART27(16D10i-2) or empty vector pART27 as control. The best performing transgenic line (D21) out of 22 for resistance to *M. chitwoodi* had a Pf/Pi ratio (multiplication ratio) of 0.22. Southern blots indicated that line D21 carried a single copy of the 16D10i-2 transgene. Northern blots confirmed a strong expression of 16D10-specific small RNAs in line D21. It was tested for resistance against *M. arenaria*, *M. hapla*, *M. incognita* and *M. javanica* with empty vector line E34, wild-type PA99N82-4 (with resistance gene $R_{Mc1(blb)}$) and Russet Burbank as controls. At 35 days after inoculation (DAI), the number of egg masses per plant was significantly reduced by 65% to 97% in the RNAi line compared to wild-type and empty vector controls. The largest reduction was observed in *M. hapla* and the smallest in *M. javanica*. Likewise, at 55 DAI the number of eggs per plant was significantly reduced by 66% to 87%, with the largest and smallest reductions in *M. hapla* and

M. arenaria, respectively, compared to wild-type and empty vector controls. So, plant-mediated RNAi silencing of the *16D10* effector gene resulted in significant resistance against all of the root-knot nematode species tested, whereas $R_{Mc1(blb)}$, the best characterized *Meloidogyne* resistance gene in potato, did not have a broad-spectrum resistance effect. Line PA99N82-4 ($R_{Mc1(blb)}$) significantly lowered the number of eggs on a per plant and per gram root basis for *M. hapla*, but not for the other root-knot nematodes tested. The attraction of *M. incognita* second-stage juveniles (J2) to 16D10i-2 potato line D21, PA99N82-4, empty vector E34 and wild-type Russet Burbank controls was also investigated as well as subsequent root invasion. Silencing of *16D10* did not interfere with the attraction of the J2 to roots nor did it reduce root invasion. This result means that 16D10i-2-mediated resistance mechanisms occur after nematodes have invaded host roots and become sedentary.

9.8.5 Acrylamide, Black Spot Bruise and Cold-Induced Sweetening

It is now time to turn to processing quality, starting with the Simplot programme that led to Innate™ potatoes. The programme took 14 years of scientific development, safety assessments, and extensive field tests. It used marker-free transformants and intragenic approaches to silence genes by RNA interference that are key to the expression of black spot bruise, asparagine and reducing sugars in tubers. In RNA interference, as we have seen, double-stranded RNA triggers the degradation of gene transcripts. The method is intragenic because the inserted genes came from cultivated potatoes (*S. tuberosum*) or a sexually compatible wild relative (*S. verrucosum* for *Ppo5*). No foreign genes, no antibiotic resistance markers and no vector backbone sequences were incorporated into the plant genome. A plasmid vector called pSIM1278 was used to separately incorporate two silencing 'cassettes' into cultivars Ranger Russet, Russet Burbank and Atlantic and into two proprietary chipping (crisping) cultivars. Expression of the first cassette lowered the transcript levels for the *Asn1* (asparagine synthetase-1, = *Ast1*) (Rommens et al. 2008) and *Ppo5* (polyphenol oxidase-5) genes and consequently limited the formation of the acrylamide precursor asparagine and the formation of impact-induced black spot bruise that occurs when the enzyme polyphenol oxidase oxidizes phenols to produce dark pigments. This browning also normally occurs when untransformed potatoes are cut and left in the air. The presence of black spot bruise results in lower quality and subsequent production losses during processing into French fries or chips (crisps). Asparagine is converted into acrylamide, a chemical compound that has been linked to cancer, when potatoes are fried at high temperatures. Expression of the second cassette reduced the tuber formation of reducing sugars (glucose and fructose) from starch by lowering transcript levels for genes *PhL* (amyloplast-targeted phosphorylase-L) and *R1* (water dikinase) (Rommens et al. 2006). Benefits included improved quality by contributing to the desired golden-brown colours required by most French fry and chip customers. Lower reducing sugars in stored potatoes also contribute to lower acrylamide production because they react with the amino acid asparagine in the Maillard reaction to produce acrylamide. The research on asparagine synthetase is particularly revealing.

Chawla et al. (2012), of Simplot Plant Sciences, showed that tuber-specific silencing of *asparagine synthetase 1* (by RNA interference) reduces the acrylamide-forming potential of potatoes grown in the field without affecting tuber shape and yield. Initially, Rommens et al. (2008) found that simultaneous silencing of *asparagine synthetase −1* and *−2* (*Ast1* and *Ast2*) limited the formation of asparagine and reduced the acrylamide-forming potential of tubers. However, although the phenotype of silenced lines appeared normal in the greenhouse, unexpectedly, field-grown tubers were small and cracked. This prompted a more thorough analysis of the role of the two asparagine genes in potato. Chawla et al. (2012) showed that tuber-specific silencing of *Ast1* reduced asparagine levels by 60% to 80% in processing cultivars Russet Ranger and Russet Burbank. This was sufficient to reduce acrylamide levels by up to 70% compared with the controls and hence is one possible solution to the acrylamide problem.

Vacuolar invertase (hydrolysis of sucrose to glucose and fructose) has been another target for reducing acrylamide formation. Zhu et al. (2016) examined the impact of silencing vacuolar invertase and asparagine synthetase genes on acrylamide formation of fried potato products; glucose, fructose and asparagine being the precursors of acrylamide. They generated a large number of silencing lines of potato cultivar Russet Burbank by targeting the vacuolar invertase gene *VInv* and the asparagine synthetase genes *StAS1* (= *Ast1*) and *StAS2* (= *Ast2*) with a single RNA interference construct. The construct contained a 311-bp cDNA fragment from the *StAS1* gene and a 311-bp cDNA fragment from the *VInv* gene. The *StAS1* fragment was selected from a region that shares 79% sequence similarity with the corresponding region of *StAS2* and represents the region with the highest sequence similarity between the two genes. The construct was driven by the CaMV 35S promoter. Seven out of 127 RNAi lines that showed a representative range of expression for the three genes in cold-stored tubers (14 days at 4 °C) were selected for field evaluation in Wisconsin in 2013 and in-depth analyses of tubers at harvest and 1, 3 and 6 months after harvest. Harvested tubers were stored for 50 days at 13 °C and cooled over 24 days to a final storage temperature of 9 °C, which is a storage temperature profile typical of that used in commercial practice. The amounts of glucose and fructose were well correlated with the acid invertase activity ($R = 0.81$–0.95) and with the *VInv* expression level ($R = 0.78$–0.89) at the stem end of tubers at harvest and 1, 3 and 6 months of storage. After 3 months of storage, six out of the seven lines had lower levels of reducing sugars and four had lower levels of asparagine and reducing sugars, two of which were particularly low. Differences in asparagine content mirrored the differences in *StAS1* expression at the tuber stem end. A strong correlation ($R = 0.91$) between asparagine content and the level of *StAS1* expression was found in stem end of the tubers 3 months after harvest, but a less strong correlation was found between asparagine content and the level of *StAS2* expression ($R = 0.45$). All seven lines had lower levels of acrylamide in fried strips of stem-end tissues, with two lines particularly low at 84 and 89 μg/kg acrylamide, respectively, compared to 1233 μg/kg of the Russet Burbank control. The fried strips of the best two lines were also free of sugar-end defects at all sampling periods. Although normal phenotypes were associated with all seven RNAi lines in the 2013 field trial when they were planted as transplants, lines with the greatest extent of silencing were stunted and chlorotic in a repeat field trial in 2014 when they were planted as seed tubers. However, tubers from these plants developed normally. Nevertheless, more research is required. Interestingly, the extent of acrylamide reduction of the best triple silenced lines was similar to that of the best *VInv*-single silenced lines (88 μg/kg of acrylamide) developed previously from the same potato cultivar Russet Burbank. These results show that an acrylamide mitigation strategy focused on developing potato cultivars with low reducing sugars is likely to be an effective and sufficient approach for minimizing the acrylamide-forming potential of French fry processing potatoes.

Readers may also be interested in the paper by Wiberley-Bradford et al. (2016) on temperature-dependent regulation of sugar metabolism in wild-type and low-invertase transgenic chipping potatoes during and after cooling for low-temperature storage.

9.8.6 Modified Starch and Inulin

In countries with a starch industry there are considerable opportunities to breed or genetically engineer potatoes with novel starches for use in food and non-food products. Some of the functionalities of starch are unique to the polymer as a result of the structure and organization of its linear amylose (17–25% of starch) and highly branched amylopectin (Li et al. 2006). Genetic engineering has already generated the two extreme types of potato starch. First, high amylopectin starch was produced using antisense constructs to down-regulate the granule-bound starch synthase gene (*gbss1*, *GBSSI*) that controls amylose synthesis (Visser et al. 1991). Second, high-amylose starch was produced by the consecutive transformation of potato with two different gene constructs for the concurrent down-regulation of two starch branching enzymes *Sbe1* (*SBE B*) and *Sbe2* (*SBE A*), respectively (Schwall et al. 2000). Hofvander et al. (2004) achieved the coordinated inhibition of both *Sbe1* and *Sbe2* with one hybrid gene construct using one round of transformation. In the gene construct, the 3′ end of *Sbe2* (1.3 kb) and the 3′ end of *Sbe1* (1.8 kb) were cloned in tandem and in the antisense direction in relation to the *gbss* promoter of potato. The introduction of the gene constructs into potato was performed by conventional *Agrobacterium tumefaciens* transformation. Potato cultivars Prevalent, Producent, Dinamo and Kuras, which are commercially used for starch production, were transformed with the gene construct. Forty-three lines derived from the four different cultivars were selected for greenhouse propagation to produce seed tubers for subsequent field trials and to verify the results with regard to the higher amylose content previously determined for 700 independent transgenic lines using in vitro-generated microtubers. Eight of the transgenic lines had amylose contents of 54–94% compared with 25–27% for the controls. Quantitative polymerase chain reaction (PCR) indicated that the eight lines had an insert copy number of between four and seven. Staining of starch granules with iodine and visual inspection under a microscope indicated that the granule morphology was changed in the transgenic lines, with seemingly smaller granules and irregular shapes. Field trials were conducted in southern Sweden over 3 years. The high-amylose lines had a higher fresh weight tuber yield than their respective parental cultivars by 20–48%. However, starch content decreased with increasing amylose. After harvest and temporary storage, potato tubers were processed in pilot-sized starch production machinery. The smaller granule size of high-amylose starch was confirmed at 32–39 μM compared with 50–62 μM for common potato starch. A strong negative correlation was found between the amylose content and the degree of branching. In general, an increase in amylose content resulted in a strong decrease in dry matter content, attributed to a lower starch content. Fructose levels were increased by 40–150% and glucose levels by 40–133%. The researchers concluded that potatoes containing starch with a very low degree of branching, such as 0.3% compared with controls at 2.23% to 2.36%, were not suitable for commercial cultivation due to severe starch yield reduction and other effects. With slightly more branching, however, the effects were much reduced and the modified starch was considered suitable for biodegradable plastics, expanded products and film-forming operations.

It is worth mentioning that the introduction of inulin to potato is another way to improve its nutritional value by reducing its energy density through increased dietary fibre. Inulin is a mixture of linear fructose polymers with different chain lengths and a glucose molecule at each C2-end. Inulin belongs to the fructan group of polysaccharides and serves as a storage carbohydrate in many plant species. Compounds such as inulin reduce the energy density of food and are used to enrich food with dietary fibre or to replace sugar and fat. Globe artichoke (*Cynara scolymus*) synthesizes inulin molecules with a chain length of up to 200, which is the highest degree of polymerization of inulin molecules known for plants. Hellwege et al. (2000) transferred the ability to synthesize high molecular weight inulin to potato plants (cultivar Désirée) via constitutive expression (cauliflower mosaic virus 35S promoter) of the *1-SST* (encodes sucrose: sucrose 1-fructosyltransferase) and the *1-FFT* (encodes fructan: fructan 1-fructosyltransferase) genes of globe artichoke (*Cynara scolymus*). The researchers found that the transgenic potato tubers were able to synthesize the full range of inulin molecules naturally occurring in globe artichoke. The inulin (fructan) made up about 5% of the combined sugar, fructan and starch content (in hexose equivalents) of the transgenic tubers. Although inulin accumulation did not influence the sucrose concentration in leaves or tubers, a reduction in starch content of about 18% occurred in transgenic tubers, indicating that inulin synthesis did not increase the storage capacity of the tubers. No adverse phenotypic effects were observed in the transgenic plants nor a reduction in tuber yield, albeit with greenhouse grown plants.

9.8.7 Glycoalkaloids

Genetic modification can also be used to control glycoalkaloid concentrations. About 95% of the total content is accounted for by α-solanine and α-chaconine, which are differently glycosylated forms (galactose and glucose, respectively) of the aglycone, solanidine. Although not all of the steps in their biosynthesis are known, several intermediates have been identified (Itkin et al. 2013; Kaminski et al. 2016), and virtual elimination of tuber glycoalkaloids by gene silencing is feasible (Fig. 9.1).

Cholesterol is synthesized from acetyl-CoA, converted to solanidine and then by two separate pathways to α-solanine and α-chaconine. Altered glycoalkaloid content has been associated with a number of candidate genes. Kaminski et al. (2016) used next generation sequencing, bulk segregant analysis to identify a genomic region (chromosome 1) with candidate genes responsible for differential glycoalkaloid content in a diploid mapping population of *S. tuberosum* crossed with

Fig. 9.1 Genes mentioned in text for biosynthesis of α-solanine and α-chaconine. (From Bradshaw 2019 with permission)

Acetyl-CoA
↓
↓
Cycloartenol
SSR2 ↓ ↓ *SMT1*
Cycloartanol 24-Methylenecycloartanol
SMO ↓
SD ↓
Cholesterol
PGA2 ↓
PGA1 ↓
Furostanol-intermediate
GAME4 ↓
Solanidine
SGT1 ↓ ↓ *SGT2*
γ-Solanine γ-Chaconine
β-Solanine β-Chaconine
↓ *SGT3* ↓
α-Solanine α-Chaconine

S. sparsipilum and backcrossed to *S. tuberosum*. The differences in total glycoalkaloid levels came equally from α-solanine and α-chaconine. Genes were found encoding critical enzymes in the synthesis of cholesterol: sterol 24-C-methyltransferase (*SMT1*), sterol desaturase (*SD*) and C-4 sterol methyl oxidase (*SMO*). Sawai et al. (2014) used RNA interference to silence the *SSR2* gene encoding sterol side chain reductase 2 which reduces cycloartenol to cycloartanol. The resulting tubers had significantly lower levels of cholesterol and glycoalkaloids (about 10% of controls) without affecting plant growth. Similar results were achieved by Yasumoto et al. (2020) who infected potato cultivar Sassy with *Agrobacterium tumefaciens* harbouring a TALEN-expression vector targeting *sterol side chain reductase 2* (*SSR2*) gene (see earlier in chapter) and regenerated shoots without selection.

Umemoto et al. (2016) focused on the biosynthetic steps from cholesterol to solanidine. They identified two cytochrome P450 monooxygenase genes, *PGA1* (chromosome 6) and *PGA2* (chromosome 7), that encode enzymes mediating two oxidation steps. Silencing *PGA1* or *PGA2* by RNA interference resulted in a large reduction in glycoalkaloid content and the creation of novel phenotypes, including the suppression of flower development and tuber sprouting. In other respects, the transgenic plants grew normally in the vegetative stages and produced tuber yields in the greenhouse similar to control plants. It was inferred from the metabolites that accumulated in the transgenic plants, and were produced by in vitro enzyme assays, that *PGA1* and *PGA2,* respectively, encoded enzymes that catalysed the 26- and 22-hydroxylation steps in the biosynthetic pathway. Itkin et al. (2013) also investigated a gene (*GAME4*) encoding an enzyme involved in the conversion of cholesterol to glycoalkaloids by oxidation of a furostanol-intermediate to its 26-aldehyde. Again, silencing the gene by RNA interference showed a large reduction to very low levels (<0.5 mg 100 g^{-1}) of α-solanine and α-chaconine without adverse effects.

Finally, Shepherd et al. (2015) focused on the genes that encode the enzymes for the biosynthesis of γ-solanine from UDP-galactose and solanidine (*SGT1*), γ-chaconine from UDP-glucose and solanidine (*SGT2*), and both α-solanine and α-chaconine from UDP-rhamnose and β-solanine and β-chaconine, respectively (*SGT3*). They used antisense fragments of cDNA to down-regulate *SGT1*, *SGT2* or *SGT3* in cultivar Désirée. Down-regulation of *SGT1* reduced the concentration of α-solanine, but had no effect on α-chaconine (previous work had found an increase in α-chaconine); down-regulation of *SGT2* reduced the concentration of α-chaconine, but increased the concentration of α-solanine; and down-regulation of *SGT3* reduced the concentrations of both. By using metabolite profiling, the researchers not only revealed expected changes in specific glycoalkaloid levels but also significant changes in several other metabolites, some of which could be explained in terms of known pathways.

9.8.8 Protein Quantity and Quality

On a dry weight (DW) basis, potatoes contain about 10% protein (2% on FW basis), equal to that of most cereals such as rice or wheat, and of good quality for a plant protein. Its biological value of between 90 and 100 is close to whole egg protein which is given a value of 100 (Kanter and Elkin 2019), biological value being the proportion of absorbed protein from a food which becomes incorporated into the proteins of the human body relative to a standard. Potato protein is a good source of lysine, but low contents of sulphur amino acids (methionine and cysteine) limit its nutritive value (Friedman 1996). Both protein quantity and quality can be improved by genetic modification.

An attractive source of a well-balanced protein in terms of amino acid composition is Amaranth Albumin 1. Chakraborty et al. (2010) developed transgenic potatoes with enhanced nutritive value by tuber-specific expression of a gene (*AmA1*) encoding a seed protein, Amaranth Albumin 1 (AmA1) from *Amaranthus hypochondriacus* in seven Indian cultivars suitable for cultivation in different agroclimatic regions (Fig. 9.2). A construct having the full-length *AmA1* gene under the control of a tuber-specific promoter (*GBSS*: granule-bound starch synthase) was used for large-scale *Agrobacterium tumefaciens*-mediated transformation, albeit using kanamycin as a selectable marker. Real-time PCR using Taqman chemistry

Fig. 9.2 Transgenic Indian potato cultivars with improved nutritive value by tuber-specific expression of a gene (*AmA1*) encoding a seed protein, Amaranth Albumin 1 (AmA1), from *Amaranthus hypochondriacus*: Modipuram centre of the ICAR-Central Potato Research Institute (13 December 2008)

revealed the presence of a single copy of the transgene in most of the transgenic events with very few having two to three copies. RT–PCR analysis revealed that the *AmA1* transcript was most abundant in tubers, followed by the stolon and stem with the leaf showing the least *AmA1* transcript, thus confirming the largely tissue-specific expression of the transgene. Furthermore, although the transcript was found in different organs, the protein products could be detected only in the field-grown tuber. In two field trials of the seven cultivars in Modipuram and New Delhi, India, the transgenic tubers had a 25–60% increase in total protein content and a concomitant increase in several of the essential amino acids that are otherwise limited in potato, including methionine and cysteine. The transgenic potatoes also exhibited enhanced photosynthetic activity with a concomitant increase in total biomass and for some a moderate increase in tuber yield. The data on field performance and safety evaluation indicated that the transgenic potatoes are suitable for commercial cultivation. Furthermore, in vitro and in vivo studies on experimental animals demonstrated that the transgenic tubers are also safe for human consumption. As AmA1 is a non-allergenic protein that originated from an edible crop, transgenic potato crops expressing AmA1 should receive public acceptability.

9.8.9 Genetic Engineering of Carotenoids: β-Carotene and Astaxanthin

Vitamin A deficiency remains a major health problem in some parts of the world where it is the major cause of blindness in children. Beta-carotene (provitamin A), the precursor of vitamin A, is therefore a target for genetic engineering of carotenoids in potatoes as it is either absent or present only in low concentrations in *S. tuberosum* and closely related *Solanum* species.

Ducreux et al. (2005) achieved β-carotene contents of 10.30 μg g^{-1} DW and 6.51 μg g^{-1} DW in tetraploid cultivar Désirée and diploid cultivar Mayan Gold, respectively, by expression of an *Erwinia uredovora CrtB* gene encoding phytoene synthase specifically in the tubers. Diretto et al. (2007a) went further and transformed potato cultivar Désirée with a mini-pathway of key genes (enzymes) for the synthesis of β-carotene from geranylgeranyl diphosphate. Expression of three bacterial genes from *Erwinia herbicola* encoding phytoene synthase (*CrtB*), phytoene desaturase (*CrtI*) and lycopene beta-cyclase (*CrtY*), under tuber-specific (patatin) promoter control, resulted in tubers with deep yellow flesh without any adverse leaf phenotypes (Fig. 9.3). Carotenoid levels in one transformant had increased from 5.8 to 114.4 μg g^{-1} DW and β-carotene from 0.013 to 47.4 μg g^{-1} DW (and α-carotene 0 to 6.2, lutein 1.0 to 23.1, violaxanthin 0.7 to 5.6), compared with 'Golden Rice 2' at 31 μg g^{-1} DW. Assuming a β-carotene to retinol conversion of 6:1 this is sufficient to provide 50% of the recommended daily allowance (RDA) of vitamin A from 250 g FW of 'Golden Potatoes' (GP) (Diretto et al. 2007a). More recently, Chitchumroonchokchai et al. (2017) provided data on bioavailability from transgenic

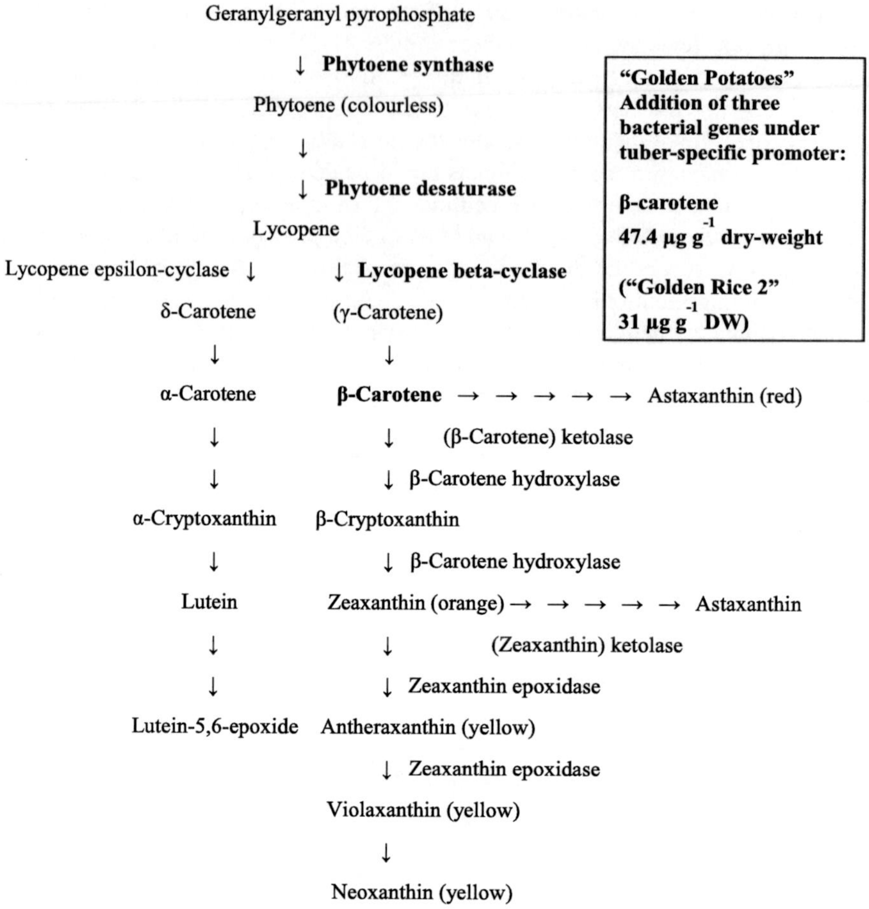

Fig. 9.3 Carotenoid biosynthetic pathway (simplified) in potatoes transformed to produce β-carotene (provitamin A) and astaxanthin. (From Bradshaw 2019 with permission)

lines GP17 and GP30. Beta-carotene (>36 µg g^{-1} DW) and α-carotene (>31 µg g^{-1} DW) were the most abundant carotenoids in GP17 and GP30 with total concentrations of these provitamin A carotenoids (68 and 91 µg g^{-1} DW in GP17 and GP30, respectively) being more than 100-fold greater than that in cultivar Désirée. Golden Potatoes also contained 2–3 times more lutein (>19 µg g^{-1} DW). Furthermore, α-tocopherol, an isomer of vitamin E was unexpectedly detected in high concentrations in GP17 and GP30 (>78 µg g^{-1} DW), which were 15 times greater than in cultivar Désirée. Retention of all of these compounds on boiling was greater than 82%. Boiled potatoes were subjected to in vitro digestion (simulated oral, gastric and small intestinal phases of digestion) to determine the extent of transfer of carotenoids and α-tocopherol from the food matrix to micelles (i.e. their bioaccessibility). The average figures for GP17 and GP30 were 14% for the provitamin A carotenoids, 31% for lutein and 24% for α-tocopherol. The micelles were then added to monolayers of Caco-2 human intestinal epithelial cells to investigate apical uptake (i.e. bioavailability). Apical uptake during incubation for 4 h was around 15% for provitamin A carotenoids, 20% for lutein and 53% for α-tocopherol. The researchers concluded that for children and women of reproductive age, a 150-g serving of boiled Golden Potatoes has the potential to contribute 42% and 23%, respectively, of their daily requirement for provitamin A (measured as retinol activity equivalents) as well as 34% and 17%, respectively, of their daily vitamin E requirement.

Diretto et al. (2007b) also produced more modest increases in total carotenoid and β-carotene levels by the tuber-specific (patatin promoter) silencing of the genes encoding β-carotene hydroxylases CHY1 and CHY2 in cultivar Désirée. Beta-carotene levels increased from 0.00225 up to 0.085 μg g^{-1} DW and total carotenoids from 4.9 up to 14.3 μg g^{-1} DW. These changes were accompanied by a decrease in the immediate product of β-carotene hydroxylation, zeaxanthin, but not of downstream violaxanthin and neoxanthin.

Finally, Diretto et al. (2006) found that tuber-specific (patatin promoter) silencing in cultivar Désirée of the gene encoding the enzyme for the first dedicated step in lutein biosynthesis, *lycopene epsilon cyclase* (*LCY-e*), also resulted in modest increases of β-carotene from 0.00317 up to 0.044 μg g^{-1} DW and of total carotenoids from 4.68 up to 9.98 μg g^{-1} DW. These changes were not accompanied by a decrease in lutein, suggesting that the enzyme lycopene epsilon-cyclase is not rate-limiting for lutein accumulation.

There is also interest in engineering potatoes to accumulate ketocarotenoids. Astaxanthin, for example, has been associated with a range of health benefits and is used extensively as a feed additive in aquaculture (Morris et al. 2006). In nature, the main sources of astaxanthin are marine bacteria and microalgae. The conversion of β-carotene to astaxanthin requires the action of a hydroxylase as well as a ketolase. Morris et al. (2006) described the levels and types of carotenoid that accumulate in both tetraploid Tuberosum cultivar Désirée and diploid Phureja cultivar Mayan Gold on transformation with an algal *β-carotene oxygenase* gene (*bkt1* encoding a β-ketolase from *Haematococcus pluvialis*). A patatin promoter was used for tuber-specific expression. Two major ketocarotenoids, ketolutein and astaxanthin, absent in untransformed controls, were detected and gave tubers a reddish colour. The level of unesterified astaxanthin reached 13.9 μg g^{-1} DW and ketolutein reached 9.8 μg g^{-1} DW in some *bkt1* expressing Phureja lines but were much lower (0.6 and 0.5 μg g^{-1} DW, respectively) in the Tuberosum background. As astaxanthin biosynthesis requires the concerted action of both a carotene 4,4′ oxygenase (ketolase) and a 3,3′ hydroxylase, the researchers concluded that high activity of the 3,3′ hydroxylase occurred naturally in potato tubers. Subsequently, Campbell et al. (2015) investigated approaches to optimizing tuber astaxanthin content. They achieved a nutritionally significant concentration of 77 μg g^{-1} DW, which has the potential to provide the recommended 5 mg daily intake of astaxanthin from 55 g DW of potato (ca. 220 g FW). They selected a clone (01H15.57) from diploid Phureja cross 01H15

[Inca Sun (DB378(1), deep yellow) × Yema de Huevo (deep yellow)] with a high zeaxanthin content of 41.6 μg g^{-1} DW (total carotenoid content 53.9). They then managed to increase its astaxanthin content from 0 to 77.1 μg g^{-1} DW (total carotenoid content 93.5) by co-expressing the cauliflower *Or* transgene (tuber-specific GBSS promoter) and the *Brevundimonas crtZ* (hydroxylase) and *crtW* (ketolase) transgenes in tandem (*crtZW*) (constitutive 35S CaMV promoter). The *crtW* gene came from the marine bacterium *Brevundimonas* sp., strain SD212 and encodes an enzyme (ketolase) which accepts both β-carotene and zeaxanthin as substrates. Co-expression with the CrtZ enzyme from the same species results in efficient conversion of β-carotene to astaxanthin. The cauliflower *Or* (orange) gain of function gene mutation encodes a DnaJ cysteine-rich domain-containing protein whose transgene had been shown to increase total carotenoid levels in cultivar Désirée from about 4–24 μg g^{-1} DW and to produce β-carotene as a major carotenoid (Lu et al. 2006). The *crtZW* transgene alone increased astaxanthin content from 0 to 44.1 μg g^{-1} DW (total carotenoid content 57.4), whereas *Or* alone increased zeaxanthin content to 66.7 μg g^{-1} DW (total carotenoid content 88.3) but did not produce any astaxanthin. Astaxanthin levels were higher in summer than winter grown plants and could be associated with transcriptome and metabolome restructuring. Campbell et al. (2014) also made a genome-wide QTL and bulked transcriptomic analysis of population 01H15 and found a novel major QTL on chromosome 9 for increased carotenoid content.

9.9 Integration of Genetic Transformation into Conventional Breeding Programmes

Given the potential delays in modifying new cultivars just before or after their release, we need to consider if the genetic transformation could be incorporated at an earlier stage into the breeding programme. As widely grown cultivars will be used as parents in the conventional breeding, one can envisage genetically modified ones being used as parents, followed by selection of offspring containing the introduced gene(s) for further assessment. Where selection for quantitative traits is first practised between crosses, followed by selection within the most promising ones, the following can be done. The untransformed cultivars can be used in crosses to determine the best ones, and then the transformed versions used in repeats of the best crosses to provide more seed on which to practise within progeny selection. Field trialling of the former would not be governed by any GM regulations, and selection within the latter could initially be done in a greenhouse. However, the same problem would occur as with marker-assisted selection for major genes and QTL alleles of large effect. Once more than eight genes are segregating, the required population sizes start to become large. However, in the future, the use of site-directed transformation (rather than random insertion) should reduce the number of undesirable transformants and allow gene stacking at a single site to create a supergene that will be inherited as a Mendelian gene (Mahfouz et al. 2016). It should then be possible to stack genes more quickly at relatively few sites. The segregation of such units will not pose any major logistical problems in a potato improvement programme combining genetic transformation with conventional breeding that includes marker-assisted and genomic selection.

9.10 Conclusions

The prerequisites for successfully producing and marketing a GM crop are as follows: understanding the biochemistry of the trait of interest; understanding the control of gene expression for the genes that encode the key enzymes for delivering the required biochemistry; the ability to clone the key gene(s) and to do the necessary genetic engineering in bacterial plasmids to produce vectors containing transgene expression cassettes; the ability of the vector to deliver the expression cassette into plant cells; the ability to select or produce marker-free transformants; the ability to regenerate whole plants from transformed cells and tissues; the ability to stack combinations of transgenes; demonstration that the resulting GM crop and products are safe to grow and eat or feed to animals; meeting all of the regulations governing GM crops; and last but possibly most difficult of all, convincing politicians, pressure groups and the general public that the GM crop is of real benefit to farmers and consumers alike. Finally, with the advent of engineered nucleases designed to catalyse a double-stranded break (DSB) at a specific location in the genome, desired DNA modifications (gene editing and genetic transformation) can be stimulated to occur at or near the break site rather than at random. Specific conclusions are as follows.

1. *Mutation Breeding*: As a clonally propagated tetraploid crop, the potato is not considered an attractive proposition for deliberate mutation breeding. However, if more breeding work is done at the diploid level, particularly with inbred lines, then modern physical and chemical mutation breeding methods may be used more often.
2. *Somaclonal Variation*: Somaclonal variation tends to be viewed as a source of undesirable rather than desirable variants that need to be screened out of breeding programmes that involve plant regeneration, including the in vitro stage of cultivar multiplication.
3. *Gene Editing*: TALENS and CRISPR-Cas9 can be designed to target all four alleles of a gene in tetraploid potato and hence achieve a knock-out mutation. They can be delivered as a plasmid or naked DNA into protoplasts (polyethylene glycol (PEG)-mediated transfection) which are regenerated into whole plants lacking the foreign DNA because it has not been integrated into the potato genome. More research, however, is required to see if gene editing can be used to create and fix novel alleles in potatoes.

4. *Production of Genetically Modified Potatoes*: Potato transformation using *Agrobacterium*-mediated systems was developed in the 1980s, with *A. tumefaciens* becoming the species of choice. Marker-free transformants can be produced by using transformation vectors without selection markers, followed by PCR analysis of regenerated plants for the presence of the transferred gene(s). There are four basic strategies for stacking transgenes: first, by sexual hybridization of plants containing different transgenes; second, by co-transformation of a plant with two or more independent transgenes in separate gene constructs which are delivered to the plant simultaneously; third, by multigene cassette in which a plant is transformed with a single construct that contains two or more linked transgenes, with four to six considered the current limit; and fourth, by re-transformation in which a plant containing a transgene is transformed with other transgenes. Currently up to eight transgenes are being stacked.
5. *Commercialization of Genetically Modified Potatoes*: The period from 1995 to 2015 saw the production of the first genetically modified (GM) potatoes but also the failure to market them in North America and Europe because of consumer concerns over GM potato products and technology. Generation 1 and 2 of Innate™ potatoes (processing cultivars) are now being marketed in North America but it is too early to predict how widely they will be grown.
6. *Regulations Governing GM Potatoes*: Regulations vary according to country; for example, being very stringent in the European Union (EU) and less stringent in the USA where the focus is on the product and not the technology used for production. Currently, the EU does not distinguish between transgenic, intragenic and cisgenic crops in general and potatoes in particular. Transgenic crops contain genes (transgenes) derived from organisms (e.g. bacteria and fungi) or other plant species that are not cross-compatible with the cultivated crop, whereas intragenic and cisgenic crops do not contain such genes and are marker-free. A cisgene includes its introns and is flanked by its native promoter and terminator in the normal sense orientation, whereas an intragene can be a combination of different gene fragments from a donor organism(s) of the same, or a sexually compatible, species as the recipient, arranged in a sense or antisense orientation. Intragenesis includes the use of gene-silencing approaches such as RNA interference. Regulations should make sense when compared with those governing conventionally bred cultivars.
7. *GM Potatoes in Centre of Origin and Diversity*: Although Andean farmers can maintain the integrity of their landraces in the presence of a pollen-fertile commercial cultivar, this will not allay the fears of those opposed to GM potato crops being grown in their centre of origin and diversity.
8. *Gene Silencing*: Transgene-mediated post-transcriptional gene silencing (PTGS) through RNA interference (RNAi) offers a directed way of achieving some desirable phenotypes for potato improvement by inactivating one or several specific genes. Such gene silencing mimics recessive loss of function mutations, but in a dominant way so that it can easily be used in polyploids such as potato. Transformation vectors can be deliberately designed for gene silencing through the production of double-stranded RNA. An extremely efficient RNAi-induced destruction of any homologous mRNA is achieved through an

inverted repeat that allows the two complementary regions to form a double-stranded hairpin RNA which the silencing system degrades into small interfering RNAs (siRNAs).

9. *Examples of GM Potatoes*: The list of potential targets for genetic transformation of potatoes is steadily increasing but good examples of certain types of traits are currently missing, namely yield, photosynthetic efficiency and nutrient and water use efficiencies. Hence, it seems premature to try to provide a definitive list of potato traits for improvement by genetic transformation; but good examples for disease and pest resistance, improved processing quality and improved nutritional value can be found.
10. *Disease and Pest Resistance*: Three *R*-genes for resistance to late blight (*Phytophthora infestans*) from potato wild relatives have been stacked in commercial cultivars both in the Netherlands (e.g. *Rpi-vnt1.1*, *Rpi-sto1* and *Rpi-chc1*) and Uganda (e.g. *Rpi-blb1*, *Rpi-blb2* and *Rpi-vnt1.1*). Transgenic resistance is available to *Potato leafroll virus* (PLRV), *Potato virus Y* (PVY) and *Potato virus X* (PVX), and fused viral segments in the inverted-repeat (silencing) expression system may provide multiple virus resistance to viruses transmitted by aphids. Resistance to the Colorado potato beetle (CPB, *Leptinotarsa decemlineata*) and to the potato tuber moth (PTM, *Phthorimaea operculella* Zeller) has been achieved with genes that encode insecticidal crystal (Cry) proteins from the bacterium *Bacillus thuringiensis* with *Bt-cry3A* best for CPB and *Bt-cry5* (equivalent to *Bt-cry1Ia1*) best for PTM; resistance to CPB has also been achieved in plants whose chloroplasts have been engineered to produce double-stranded RNAs targeted against the β-actin gene of CPB. Resistance to cyst nematodes (*Globodera pallida* and *G. rostochiensis*) can be achieved by combining a transgene that encodes a cysteine proteinase inhibitor (cystatin) with one that encodes a peptide that is secreted from potato roots and disrupts the chemoreception that cyst nematodes require to locate host plants. Broad-spectrum resistance to species of root-knot nematodes (*Meloidogyne* spp.) can be achieved by using RNA interference (RNAi) to silence the *16D10 Meloidogyne* effector gene which is important in the molecular interactions between the parasite and its host.
11. *Improved Processing Quality*: Processing quality for French fries and chips (crisps) can be improved by using RNA interference to silence genes that are key to the formation of black spot bruise [*Ppo5* (polyphenol oxidase-5)], asparagine [*Asn1* (asparagine synthetase-1, = *Ast1*), tuber-specific silencing] and reducing sugars [*PhL* (amyloplast-targeted phosphorylase-L) and *R1* (water dikinase) and *VInv* (vacuolar invertase)] in tubers; reducing sugars and asparagine being the precursors of acrylamide, which is probably carcinogenic to humans, and high reducing sugars being the cause of low-temperature sweetening. For the starch industry, high amylopectin starch can be produced using antisense constructs to down-regulate the granule-bound starch synthase gene (*GBSSI*) that controls amylose synthesis, and high-amylose starch by concurrent down-regulation of two starch branching enzymes, *Sbe1* (*SBE B*) and *Sbe2* (*SBE A*).

12. *Improved Nutritional Value*: The dietary fibre of potato can be increased by the introduction of the ability to synthesize inulin molecules through the expression of the *1-SST* (encodes sucrose: sucrose 1-fructosyltransferase) and the *1-FFT* (encodes fructan: fructan 1-fructosyltransferase) genes from globe artichoke (*Cynara scolymus*). Virtual elimination of toxic tuber glycoalkaloids is feasible by silencing genes encoding key enzymes in the synthesis of cholesterol from acetyl-CoA, its conversion to solanidine, and then by two separate pathways to α-solanine and α-chaconine. Protein quantity and quality can be increased by tuber-specific expression of a gene (*AmA1*) encoding a seed protein, Amaranth Albumin 1 (AmA1) from *Amaranthus hypochondriacus*. Provitamin A (β-carotene) levels in potatoes can be increased from negligible to 47.4 $\mu g\ g^{-1}$ DW (compared with 'Golden Rice 2' at 31 $\mu g\ g^{-1}$ DW) by expression of three bacterial genes from *Erwinia herbicola* encoding phytoene synthase (*CrtB*), phytoene desaturase (*CrtI*) and lycopene beta-cyclase (*CrtY*) under tuber-specific promoter control; these Golden Potatoes have a deep yellow flesh colour and can help relieve vitamin A deficiency. Potatoes can also be genetically engineered to produce astaxanthin (colour red), a feed additive in aquaculture, from *β-carotene* either by transformation with an algal *β-carotene oxygenase* gene (*bkt1* encoding a β-ketolase from *Haematococcus pluvialis*) under the control of a tuber-specific promoter, or by transformation with two marine bacterium (*Brevundimonas*) genes, *crtZ* (hydroxylase) and *crtW* (ketolase) in tandem under a constitutive promoter; with the highest levels produced by transforming Phureja potatoes with yellow/orange flesh.

My overall conclusion on breeding methods is that progress in potato improvement requires appropriate use of conventional breeding and genetic engineering jointly rather than in isolation. We need to consider when both could be used to enhance a crop trait, when only conventional breeding could be used because of the large number of genes involved, and when only genetic engineering could be used because the trait is not available through sexual hybridization. Future progress will come from the appropriate use of sexual reproduction in further conventional breeding with introgression and incorporation of genes from potato landraces and wild relatives; from cisgenesis and gene editing; and from genetically modified potatoes. New cultivars will be produced at both the diploid and tetraploid level; the former primarily as F_1 hybrids for propagation through true potato seed (TPS) and the latter primarily as clones for vegetative propagation starting with disease-free microplants.

References

Ali MA, Azeem F, Abbas A, Ahmad F, Li H, Dababat AA (2017) Transgenic strategies for enhancement of nematode resistance in plants. Front Plant Sci 8. https://doi.org/10.3389/fpls.2017.00750

Ali S, Kim W-C (2019) A fruitful decade using synthetic promoters in the improvement of transgenic plants. Front Plant Sci 10:1433. https://doi.org/10.3389/fpls.2019.01433

Andersson M, Turesson H, Nicolia A, Fält A-S, Samuelsson M, Hofvander P (2017) Efficient targeted multiallelic mutagenesis in tetraploid potato (*Solanum tuberosum*) by transient CRISPR-Cas9 expression in protoplasts. Plant Cell Rep 36:117–128

Bairu MW, Aremu AO, Van Staden J (2011) Somaclonal variation in plants: causes and detection methods. Plant Growth Regul 63:147–173

Bortesi L, Fischer R (2015) The CRISPR/Cas9 system for plant genome editing and beyond. Biotechnol Adv 33:41–52

Bradshaw JE (2007) Breeding potato as a major staple crop. In: Kang MS, Priyadarshan PM (eds) Breeding major food staples. Blackwell Publishing, Oxford, pp 277–332

Bradshaw JE (2016) Plant breeding: past, present and future. Springer, Cham

Bradshaw JE (2019) Improving the nutritional value of potatoes by conventional breeding and genetic modification. In: Qureshi AMI, Dar ZA, Wani SH (eds) Quality breeding in field crops. Springer, Cham, pp 41–84

Campbell R, Pont SD, Morris JA, McKenzie G, Sharma SK, Hedley PE, Ramsay G, Bryan GJ, Taylor MA (2014) Genome-wide QTL and bulked transcriptomic analysis reveals new candidate genes for the control of tuber carotenoid content in potato (*Solanum tuberosum* L.). Theor Appl Genet 127:1917–1933

Campbell R, Morris WL, Mortimer CL, Misawa N, Ducreux LJ, Morris JA, Hedley PE, Fraser PD, Taylor MA (2015) Optimising ketocarotenoid production in potato tubers: effect of genetic background, transgene combinations and environment. Plant Sci 234:27–37

Chakraborty S, Chakraborty N, Agrawal L, Ghosh S, Narula K, Shekhar S, Naik PS, Pande PC, Chakrborti SK, Datta A (2010) Next-generation protein-rich potato expressing the seed protein gene *AmA1* is a result of proteome rebalancing in transgenic tuber. Proc Natl Acad Sci U S A 107:17533–17538. https://doi.org/10.1073/pnas.1006265107

Chang MM, Culley D, Choi JJ, Hadwiger LA (2002) *Agrobacterium*-mediated co-transformation of a pea beta-1,3-glucanase and chitinase genes in potato (*Solanum tuberosum* L. cv. Russet Burbank) using a single selectable marker. Plant Sci 163:83–89

Chawla R, Shakya R, Rommens CM (2012) Tuber-specific silencing of *asparagine synthetase-1* reduces the acrylamide-forming potential of potatoes grown in the field without affecting tuber shape and yield. Plant Biotechnol J 10:913–924

Chitchumroonchokchai C, Diretto G, Parisi B, Giuliano G, Failla ML (2017) Potential of golden potatoes to improve vitamin A and vitamin E status in developing countries. PLoS One 12(11):e0187102. https://doi.org/10.1371/journal.pone.0187102

Chung BN, Yoon J-Y, Palukaitis P (2013) Engineered resistance in potato against potato leafroll virus, potato virus A and potato virus Y. Virus Genes 47:86–92. https://doi.org/10.1007/s11262-013-0904-4

Clasen BM, Stoddard TJ, Luo S, Demorest ZL, Li J, Cedrone F, Tibebu R, Davison S, Ray EE, Daulhac A, Coffman A, Yabandith A, Retterath A, Haun W, Baltes NJ, Mathis L, Voytas DF, Zhang F (2016) Improving cold storage and processing traits in potato through targeted gene knockout. Plant Biotechnol J 14(1):169–176

Coombs JJ, Douches DS, Li W, Grafius EJ, Pett WL (2003) Field evaluation of natural, engineered, and combined resistance mechanisms in potato for control of Colorado potato beetle. J Am Soc Hort Sci 128(2):219–224. https://doi.org/10.21273/JASHS.128.2.0219

Curtin SJ, Voytas DF, Stupar RM (2012) Genome engineering of crops with designer nucleases. Plant Genome 5:42–50

Davidson MM, Butler RC, Wratten SD et al (2004) Resistance of potatoes transgenic for a *cry*1Ac9 gene, to *Phthorimaea operculella* (Lepidoptera: Gelechiidae) over field seasons and between plant organs. Ann Appl Biol 145:271–277

Davies HV (2002) Commercial developments with transgenic potato. In: Valpuesta V (ed) Fruit and vegetable biotechnology. Woodhead Publishing Limited, Cambridge, pp 222–249

De Block M, Herrera-Estrella L, Van Montagu M, Schell J, Zambryski P (1984) Expression of foreign genes in regenerated plants and in their progeny. EMBO J 3:1681–1689

Denison RF (2012) Darwinian agriculture. Princeton University Press, Princeton

Dinh PTY, Zhang L, Mojtahedi H, Brown CR, Elling AA (2015) Broad *Meloidogyne* resistance in potato based on RNA interference of effector gene *16D10*. J Nematol 47(1):71–78
Diretto G, Tavazza R, Welsch R, Pizzichini D, Mourgues F, Papacchioli V, Beyer P, Giuliano G (2006) Metabolic engineering of potato tuber carotenoids through tuber-specific silencing of lycopene epsilon cyclase. BMC Plant Biol 6:13. https://doi.org/10.1186/1471-2229-6-13
Diretto G, Al-Babili S, Tavazza R, Papacchioli V, Beyer P, Giuliano G (2007a) Metabolic engineering of potato carotenoid content through tuber-specific overexpression of a bacterial mini-pathway. PLoS One 2(4):e350
Diretto G, Welsch R, Tavazza R, Mourgues F, Pizzichini D, Beyer P, Giuliano G (2007b) Silencing of beta-carotene hydroxylase increases total carotenoid and beta-carotene levels in potato tubers. BMC Plant Biol 7:11. https://doi.org/10.1186/1471-2229-7-11
Douches DS, Grafius EJ (2005) Transformation for insect resistance. In: Razdan MK, Mattoo AK (eds) Genetic improvement of Solanaceous crops, Volume I: Potato. Science Publishers Inc, Enfield, pp 235–266
Douches DS, Pett W, Santos F, Coombs J, Grafius E, Li W, Metry EA, Nasr El-Din T, Madkour M (2004) Field and storage testing Bt potatoes for resistance to potato tuberworm (Lepidoptera: Gelichiidae). J Econ Entomol 97(4):1425–1431
Douches DS, Coombs JJ, Lacey LA, Felcher KJ, Pett WL (2011) Evaluations of transgenic potatoes for resistance to potato tuberworm in the laboratory and field. Am J Potato Res 88:91–95. https://doi.org/10.1007/s12230-010-9167-y
Ducreux LJ, Morris WL, Hedley PE, Shepherd T, Davies HV, Millam S, Taylor MA (2005) Metabolic engineering of high carotenoid potato tubers containing enhanced levels of β-carotene and lutein. J Exp Bot 56:81–89
Duncan DR, Hammond D, Zalewski DJ, Cudnohufsky J, Kaniewski W, Thornton M, Bookout JT, Lavrik PB, Rogan GJ, Feldman-Riebe J (2002) Field performance of transgenic potato, with resistance to Colorado potato beetle and viruses. HortScience 37:275–276
EFSA (2010) EFSA Panel on Genetically Modified Organisms (GMO). Scientific opinion. Guidance on the environmental risk assessment of genetically modified plants. EFSA J 8(11):1879
EFSA (2011) EFSA Panel on Genetically Modified Organisms (GMO). Scientific opinion. Guidance for risk assessment of food and feed from genetically modified plants. EFSA J 9(5):2150
EFSA (2012a) EFSA Panel on Genetically Modified Organisms (GMO). Scientific opinion. Addressing the safety assessment of plants developed through cisgenesis and intragenesis. EFSA J 10(2):2561
EFSA (2012b) EFSA Panel on Genetically Modified Organisms (GMO). Scientific opinion. Addressing the safety assessment of plants developed using Zinc Finger Nuclease 3 and other Site-Directed Nucleases with similar function. EFSA J 10(10):2943
Friedman M (1996) Nutritional value of proteins from different food sources. A review. J Agric Food Chem 44:6–29
Ghislain M, Douches DS (2020) In: Campos H, Ortiz O (eds) The potato crop: its agricultural, nutritional and social contribution to humankind. Springer, ebook, pp 139–162
Ghislain M, Montenegro JD, Juarez H, del Rosario HM (2015) Ex-post analysis of landraces sympatric to a commercial variety in the center of origin of the potato failed to detect gene flow. Transgenic Res 24(3):519–528. https://doi.org/10.1007/s11248-014-9854-4
Ghislain M, Byarugaba A, Magembe E, Njoroge A, Rivera C, Roman ML, Tovar JC, Gamboa S, Forbes G, Kreuze J, Barekye A, Kiggundu A (2018) Stacking three late blight resistance genes from wild species directly into African highland potato varieties confers complete field resistance to local blight races. Plant Biotech J 17(6):1119–1129. https://doi.org/10.1111/pbi.13042
Green J, Wang D, Lilley CJ, Urwin PE, Atkinson HJ (2012) Transgenic potatoes for potato cyst nematode control can replace pesticide use without impact on soil quality. PLoS One 7(2):e30973. https://doi.org/10.1371/journal.pone.0030973

Hameed A, Zaidi SS, Shakir S, Mansoor S (2018) Applications of new breeding technologies for potato improvement. Front Plant Sci:29 June 2018. https://doi.org/10.3389/fpls.2018.00925

Haverkort AJ, Boonekamp PM, Hutten R, Jacobsen E, Lotz LAP, Kessel GJT, Vossen JH, Visser RGF (2016) Durable late blight resistance in potato through dynamic varieties obtained by cisgenesis: scientific and societal advances in the DuRPh project. Potato Res 59:35–66

Hellwege EM, Czapla S, Jahnke A, Willmitzer L, Heyer AG (2000) Transgenic potato (*Solanum tuberosum*) tubers synthesize the full spectrum of inulin molecules naturally occurring in globe artichoke (*Cynara scolymus*) roots. PNAS 97(15):8699–8704. https://doi.org/10.1073/pnas.150043797

Hofvander P, Andersson M, Larsson C-T, Larsson H (2004) Field performance and starch characteristics of high-amylose potatoes obtained by antisense gene targeting of two branching enzymes. Plant Biotechnol J 2:311–320. https://doi.org/10.1111/j.1467-7652.2004.00073.x

Horsch RB, Fraley RT, Rogers SG, Sanders PR, Lloyd A, Hoffmann N (1984) Inheritance of functional foreign genes in plants. Science 223:496–498

ISAAA (2017) Global status of commercialized biotech/GM crops in 2017. ISAAA Brief:53

Itkin M, Heinig U, Tzfadia O, Bhide AJ, Shinde B, Cardenas PD, Bocobza SE, Unger T, Malitsky S, Finkers R, Tikunov Y (2013) Biosynthesis of antinutritional alkaloids in solanaceous crops is mediated by clustered genes. Science 341:175–179

Jacobsen E (2007) The canon of potato science. 6. Genetic modification and cis- and transgenesis. Potato Res 50:227–230

Jo K-R, Kim C-J, Kim S-J, Kim T-Y, Bergervoet M, Jongsma MA, Visser RGF, Jacobsen E, Vossen JH (2014) Development of late blight resistant potatoes by cisgene stacking. BMC Biotechnol 14:50. https://doi.org/10.1186/1472-6750-14-50

Johansen IE, Liu Y, Jørgensen B, Bennett EP, Andreasson E, Nielsen KL, Blennow A, Petersen BL (2019) High efficacy full allelic CRISPR/Cas9 gene editing in tetraploid potato. Sci Rep 9:17715. https://doi.org/10.1038/s41598-019-54126-w

Kaminski KP, Kørup K, Andersen MN, Sønderkær M, Andersen MS, Kirk HG, Nielsen KL (2016) Next generation sequencing bulk segregant analysis of potato support that differential flux into the cholesterol and stigmasterol metabolite pools is important for steroidal glycoalkaloid content. Potato Res 59:81–97. https://doi.org/10.1007/s11540-015-9314-4

Kanter M, Elkin C (2019) Potato as a source of nutrition for physical performance. Am J Potato Res 96:201–205. https://doi.org/10.1007/s12230-018-09701-8

Kazama Y, Hirano T, Saito H, Liu Y, Ohbu S, Hayashi Y, Abe T (2011) Characterization of highly efficient heavy-ion mutagenesis in *Arabidopsis thaliana*. BMC Plant Biol 11:161. https://doi.org/10.1186/1471-2229-11-161

Kumar A (1994) Somaclonal variation. In: Bradshaw JE, Mackay GR (eds) Potato genetics. CAB International, Wallingford, pp 197–212

Li X-Q, Scanlon MG, Liu Q, Coleman WK (2006) Processing and value addition. In: Gopal J, Khurana SMP (eds) Handbook of potato production, improvement, and postharvest management. Food Products Press, New York, pp 523–555

Love SL, Thompson-Johns A, Baker T (1993) Mutation breeding for resistance to blackspot bruise and low temperature sweetening in the potato cultivar Lemhi Russet. Euphytica 70:69–74

Lu S, Van Eck J, Zhou X, Lopez AB, O'Halloran DM, Cosman KM, Li L (2006) The cauliflower *Or* gene encodes a DnaJ cysteine-rich domain-containing protein that mediates high levels of β-carotene accumulation. Plant Cell 18:3594–3605

Lusser M, Parisi C, Plan D, Rodríguez-Cerezo E (2012) Deployment of new biotechnologies in plant breeding. Nat Biotechnol 30:231–239

Lynas M (2018) Seeds of science: why we got it so wrong on GMOs. Bloomsbury Sigma, London, 304p

Mahfouz MM, Cardi T, Stewart CN (2016) Next-generation precision genome engineering and plant biotechnology. Plant Cell Rep 35:1397–1399. https://doi.org/10.1007/s00299-016-2009-8

Mannuss A, Trapp O, Puchta H (2012) Gene regulation in response to DNA damage. Biochim Biophys Acta 1819:154–165

Mohammed A, Douches DS, Pett W, Grafius E, Coombs J, Liswidowati W, Li W, Madkour MA (2000) Evaluation of potato tuber moth (Lepidoptera: Gelechiidae) resistance in tubers of Bt-cry5 transgenic potato lines. J Econ Entomol 93:472–476

Morris WL, Ducreux JM, Fraser PD, Millam S, Taylor MA (2006) Engineering ketocarotenoid biosynthesis in potato tubers. Metab Eng 8:253–263

National Academies of Sciences, Engineering, and Medicine (2016) Genetically engineered crops: experiences and prospects. The National Academies Press, Washington, DC. https://doi.org/10.17226/23395

Ooms G, Bossen ME, Burrell MM, Carp A (1986) Genetic manipulation in potato with *Agrobacterium rhizogenes*. Potato Res 29:367–379

Rommens CM, Ye J, Richael C, Swords K (2006) Improving potato storage and processing characteristics through all-native DNA transformation. J Agric Food Chem 54:9882–9887

Rommens CM, Haring MA, Swords K, Davies HV, Belknap WR (2007) The intragenic approach as a new extension to traditional plant breeding. Trends Plant Sci 12:397–403

Rommens CM, Yan H, Swords K, Richael C, Ye J (2008) Low-acrylamide French fries and potato chips. Plant Biotechnol J 6:843–853

Sawai S, Ohyama K, Yasumoto S, Seki H, Sakuma T, Yamamoto T, Takebayashi Y, Kojima M, Sakakibara H, Aoki T, Muranaka T, Saito K, Umemoto N (2014) Sterol side chain reductase 2 is a key enzyme in the biosynthesis of cholesterol, the common precursor of toxic steroidal glycoalkaloids in potato. Plant Cell 26:3763–3774

Schwall GP, Safford R, Westcott RJ, Jeffcoat R, Tayal A, Shi YC, Gidley MJ, Jobling SA (2000) Production of very high amylose potato starch by inhibition of SBE A and B. Nat Biotechnol 18:551–554

Scurrah M, Celis-Gamboa C, Chumbiauca S, Salas A, Visser RGF (2008) Hybridization between wild and cultivated potato species in the Peruvian Andes and biosafety implications for deployment of GM potatoes. Euphytica 164:881–892

Shepard JF, Bidney D, Shahin E (1980) Potato protoplasts in crop improvement. Science 208:17–24

Shepherd LVT, Hackett CA, Alexander CJ, McNicol JW, Sungurtas JA, Stewart D, McCue KF, Belknap WR, Davies HV (2015) Modifying glycoalkaloid content in transgenic potato – metabolome impacts. Food Chem 187:437–443

Shu QY, Forster BP, Nakagawa H (2012) Plant mutation breeding and biotechnology. CABI, Wallingford, 608p

Slater A, Scott NW, Fowler MR (2008) Plant biotechnology: the genetic manipulation of plants, 2nd edn. Oxford University Press, Oxford, 376p

Solomon-Blackburn RM, Barker H (2001) Breeding virus resistant potatoes (*Solanum tuberosum*): a review of traditional and molecular approaches. Heredity 86:17–35

Stadler LJ (1928a) Genetic effects of X rays in maize. Proc Natl Acad Sci 14:69–75

Stadler LJ (1928b) Mutations in barley induced by X-rays and radium. Science 68:186–187

Stiekema WJ, Heidekamp F, Louwerse JD, Verhoeven HA, Dijkhuis P (1988) Introduction of foreign genes into potato cultivars Bintje and Desiree using an *Agrobacterium tumefaciens* binary vector. Plant Cell Rep 7:47–50

Svitashev S, Young JK, Schwartz C, Gao H, Falco SC, Cigan AM (2015) Targeted mutagenesis, precise gene editing, and site-specific gene insertion in maize using Cas9 and guide RNA. Plant Physiol 169:931–945

Thieme T, Thieme R (2005) Resistance to viruses. In: Razdan MK, Mattoo AK (eds) Genetic improvement of Solanaceous crops, Volume I: Potato. Science Publishers Inc, Enfield, pp 293–337

Umemoto N, Nakayasu M, Ohyama K, Yotsu-Yamashita M, Mizutani M, Seki H, Saito K, Muranaka T (2016) Two cytochrome P450 monooxygenases catalyze early hydroxylation steps in the potato steroid glycoalkaloid biosynthetic pathway. Plant Physiol 171:2458–2467

Urwin PE, Green J, Atkinson HJ (2003) Expression of a plant cystatin confers partial resistance to *Globodera*, full resistance is achieved by pyramiding a cystatin with natural resistance. Mol Breed 12:263–269

Veilleux RE (2005) Cell and tissue culture of potato (Solanaceae). In: Razdan MK, Mattoo AK (eds) Genetic improvement of Solanaceous crops, Volume I: Potato. Science Publishers Inc, Enfield, pp 185–208

Visser RGF, Somhorst I, Kuipers GJ et al (1991) Inhibition of the expression of the gene for granule-bound starch synthase in potato by antisense constructs. Mol Gen Genet 225:289–296

Waterworth WM, Drury GE, Bray CM, West CE (2011) Repairing breaks in the plant genome: the importance of keeping it together. New Phytol 192:805–822

Wiberley-Bradford AE, Busse JS, Bethke PC (2016) Temperature-dependent regulation of sugar metabolism in wild-type and low-invertase transgenic chipping potatoes during and after cooling for low-temperature storage. Postharvest Biol Technol 115:60–71

Yasumoto S, Sawai S, Lee HJ, Mizutani M, Saito K, Umemoto N, Muranaka T (2020) Targeted genome editing in tetraploid potato through transient TALEN expression by *Agrobacterium* infection. Plant Biotechnol 37(2):205–211. https://doi.org/10.5511/plantbiotechnology.20.0525a

Zhang J, Khan SA, Hasse C, Ruf S, Heckel DG, Bock R (2015) Full crop protection from an insect pest by expression of long double-stranded RNAs in plastids. Science 347(6225):991–994

Zhu X, Gong H, He Q, Zeng Z, Busse JS, Jin W, Bethke PC, Jiang J (2016) Silencing of vacuolar invertase and asparagine synthetase genes and its impact on acrylamide formation of fried potato products. Plant Biotechnol J 14(2):709–718

Index

J. E. Bradshaw, *Potato Breeding: Theory and Practice*,
https://doi.org/10.1007/978-3-030-64414-7